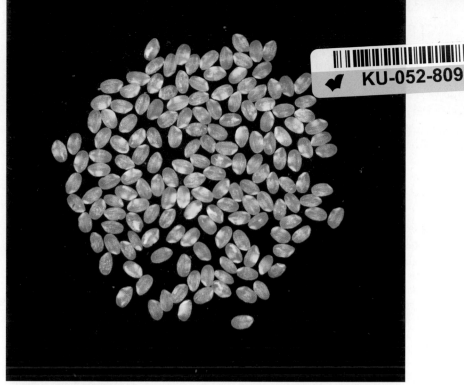

This golden rice has been nutritionally enhanced by genetic engineering. *(Courtesy of Ingo Potrykus and Peter Beyer)*

FOURTH EDITION

Essentials of Genetics

William S. Klug
The College of New Jersey

Michael R. Cummings
University of Illinois at Chicago

With contributions by

Jon Herron, *University of Washington*
Charlotte Spencer, *Cross Cancer Institute*
Sarah M. Ward, *Colorado State University*

Prentice Hall
Upper Saddle River, New Jersey 07458

Library of Congress Cataloging-in-Publication Data
Klug, William S.
    Essentials of genetics / William S. Klug, Michael R. Cummings ; with essays
contributed by Mark Shotwell, Charlotte Spencer.— 4th ed.
      p. cm.
    Includes bibliographical references and index.
    ISBN 0-13-091264-6 (pbk.)
    1. Genetics.   I. Cummings, Michael R.   II. Title.
QH430.K576   2002
576.5—dc21   2001034589
                          CIP

Editor in Chief: Sheri L. Snavely
Executive Editor: Gary Carlson
Editorial Assistant: Lisa Tarabokjia
Production Editor: Donna Young
Production Assistant: Nancy Bauer
Photo Researchers: Alexandra Truitt and Jerry Marshall
Photo Coordinator: Debbie Hewitson
Vice President of Production and Manufacturing: David W. Riccardi
Executive Managing Editor: Kathleen Schiaparelli
Assistant Managing Editor, Science Media: Alison Lorber
Executive Marketing Manager: Jennifer Welchans
Project Manager: Karen Horton
Manufacturing Manager: Trudy Pisciotti
Manufacturing Buyer: Michael Bell
Director of Creative Services: Paul Belfanti
Director of Design: Carole Anson
Art Director: Jonathan Boylan
Interior Designer: Anne Flanagan
Cover Designer: Anne Flanagan
Managing Editor, Audio/Video Assets: Grace Hazeldine
Illustrations: Imagineering Scientific and Technical Artworks, Boston Graphics
Photo Editor: Beth Boyd-Brenzel
Copy Editor: Luana Richards
Proofreader: Jennefer Vecchione
Composition: Vicki L. Croghan, Envision Inc.
About the cover: Stages of mitosis in cells of the Indian muntjak, a deer species with only a few chromosomes. (Fluorescent micrographs: S. Hauf and J.-M. Peters/*Science* Magazine)

 © 2002, 1999, 1996 by William S. Klug and Michael R. Cummings
Published by Prentice-Hall, Inc.
Upper Saddle River, NJ 07458

Previous edition © 1993 by Macmillan Publishing Company, a division of Macmillan, Inc.

Printed in the United States of America

10 9 8 7 6 5 4 3 2 1

ISBN 0-13-091264-6

Pearson Education LTD., *London*
Pearson Education Australia PTY, Limited, *Sydney*
Pearson Education Singapore, Pte. Ltd.
Pearson Education North Asia Ltd, *Hong Kong*
Pearson Education Canada, Ltd, *Toronto*
Pearson Educación de Mexico, S.A. de C.V.
Pearson Education—Japan, *Tokyo*
Pearson Education Malaysia, Pte. Ltd

Dedication

The appearance of this edition marks the tenth version of the textbooks *Essentials of Genetics* and *Concepts of Genetics*. The continued success of these books is due in no small part to the creative thinking and many diverse efforts of our various editors over the past 20 years.

They have provided leadership, good judgment, lots of patience, and many pleasant interactions. Collectively, they had to find us, guide us, and occasionally steer us through difficult times. Even after several of them have, for one reason or another, become separated from the project, they have remained supportive and continued as our colleagues and friends. It is to this group of highly talented individuals that we dedicate this book.

In particular, we thank Bob Lakemacher, Bob Rogers, Sheri Snavely, and Gary Carlson for their guidance and encouragement during the development and production of these texts.

We are indeed fortunate to have encountered each of you when we did, and we are grateful for the role that you have played during the evolution of these textbooks.

Steve Klug and Mike Cummings

About the Authors

William S. Klug is currently Professor of Biology at The College of New Jersey (formerly Trenton State College) in Ewing, New Jersey. He served as Chairman of the Biology Department for 17 years, a position to which he was first elected in 1974. He received his B.A. degree in Biology from Wabash College in Crawfordsville, Indiana and his Ph.D. from Northwestern University in Evanston, Illinois. Prior to coming to Trenton State College, he returned to Wabash College as an Assistant Professor, where he first taught genetics as well as general biology and electron microscopy. His research interests have involved ultrastructural and molecular genetic studies of oogenesis in *Drosophila*. He has taught the genetics course as well as the senior capstone seminar course in human and molecular genetics to undergraduate Biology majors for each of the last 32 years.

Michael R. Cummings is currently Associate Professor in the Department of Biological Sciences and in the Department of Molecular Genetics at the University of Illinois at Chicago. He has also served on the faculty at Northwestern University and Florida State University. He received his B.A. from St. Mary's College in Winona, Minnesota, and his M.S. and Ph.D. from Northwestern University in Evanston, Illinois. He has also written textbooks in human genetics and general biology for non-majors. His research interests center on the molecular organization and physical mapping of human acrocentric chromosomes. At the undergraduate level, he teaches courses in Mendelian genetics, human genetics, and general biology for non-majors. He has received numerous teaching awards given by the university and by student organizations.

Brief Contents

Contents

x   Contents

Genetics, Technology, and Society   300

Chernobyl's Legacy

15

Regulation of Gene Expression   306

16

Recombinant DNA Technology   326

Genetics, Technology, and Society   345

DNA Fingerprints in Forensics: The Case of the
Telltale Palo Verde

17

Chromosome Structure and DNA Sequence Organization   348

18

Genomics and Proteomics   365

Genetics, Technology, and Society   389

Completion of the Human Genome Project:
The Hype and the Hope

——— 24 ———
Conservation Genetics   495

Genetics, Technology, and Society   506

Gene Pools and Endangered Species: The Plight of the Florida Panther

Preface

Essentials of Genetics is written for courses requiring a text that is shorter and more basic than its more comprehensive companion, *Concepts of Genetics*. While coverage is thorough, current, and of high quality, *Essentials* is written to be more accessible to biology majors early in their undergraduate careers, as well as by a mixture of students majoring in agriculture, forestry, wildlife management, chemistry, psychology, and so on. Because the text is shorter than many other books, *Essentials of Genetics* will be more manageable in one-quarter and one-semester courses.

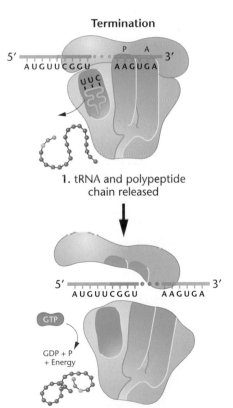

Termination

1. tRNA and polypeptide chain released

2. GTP-dependent termination factors activated; components separate; polypeptide folds into protein

Goals

Although *Essentials of Genetics* is almost 300 pages shorter than its companion volume, our goals are the same for both books. Specifically, we seek to

- Emphasize concepts rather than excessive detail
- Write clearly and directly to students in order to provide understandable explanations of complex, analytical topics
- Establish a careful organization within and between chapters
- Maintain constant emphasis on scientific analysis as the means to illlustrate how we know what we know
- Propagate the rich history of genetics that so beautifully illustrates how information is acquired and extended within the discipline as it develops and grows
- Create inviting, engaging, and pedagogically useful full-color figures enhanced by equally helpful photographs to support concept development

These goals serve as the cornerstones of *Essentials of Genetics*. This pedagogic foundation allows the book to accommodate courses with many different approaches and lecture formats. Chapters are written to be as independent of one another as possible, allowing instructors to utilize them in various sequences. We believe that the varied approaches embodied in these goals work together to provide students with optimal support for their study of genetics.

Features of the Fourth Edition

- Online Media Tutorials—Students are guided in their understanding of important concepts by working through what are simply the best animations, tutorial exercises, and self-assessment tools available.
- Length—Once again we have managed to streamline the text. This new *Essentials* is 508 pages, 7 pages shorter than the previous edition.
- Revised Organization—We provide an improved chapter sequence designed to flow smoothly from start to finish, including an early (the first chapter) introduction to DNA as well a cohesive sequence of chapters centering on the genetic role of DNA, its structure, replication, expression, and regulation.
- New Chapters—Sex Determination and Sex Chromosomes combines new information with parts of chapters from the previous edition; Genomics and Proteomics is a totally new chapter that gives students concepts and tools necessary to understand the information explosion occurring in these fields. The chapter entitled Conservation Genetics represents the first coverage of this emerging discipline in any genetics textbook.

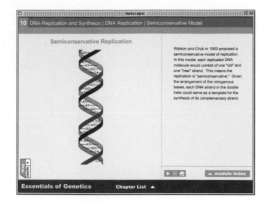

• Redesign of the Art Program—The pedagogic value, to say nothing of the beautiful execution, of the new art program will be readily apparent to users of the previous edition.

• Section Numbers—All sections are numbered making it easier to assign topics and for students to find topics within chapters

• New Photographs—An even greater number of photographs illustrate and enhance this edition.

• Emerging Topics in Genetics—Coverage of cutting edge topics includes comparative genomics, which analyzes the recently sequenced genomes of a number of organisms, including our own species (Chapter 18); proteomics, which attempts to define the potential role of the genes discovered during the Human Genome Project (Chapter 18); and conservation genetics, which assesses and attempts to maintain genetic diversity in the many endangered species on our planet (Chapter 24).

• Modernization of Topics—In addition to the areas considered in the section above on emerging topics, modernization is particularly evident in the discussions of recombinant DNA technology (Chapter 16), the organization of repetitive DNA sequences in the human genome (Chapter 17), the role of genetics in the origin of cancer (Chapter 21), and the analysis of HIV infection and resistance in population genetic studies (Chapter 22).

• New "Genetics, Technology, and Society" Essays—New topics include "Genetically Modified Foods," "DNA at the Millennium," "Completion of the Human Genome Project: The Hype and the Hope," and "Endangered Species: The Florida Panther."

• Emphasis on Problem Solving—"Insights and Solutions" sections at the ends of chapters guide students in how to think analytically about problems. "Problems and Discussion Questions" have been expanded to offer more opportunity for study.

MITOSIS

(g) Plant cell telophase   (a) Interphase   (b) Prophase

Cell plate

Kinetochore fiber
Kinetochore

(f) Telophase

(c) Prometaphase

(e) Anaphase   (d) Metaphase

Emphasis on Concepts

As in its companion volume, *Essentials of Genetics* continues to emphasize the conceptual framework of genetics. Our experience with this approach shows that students more easily comprehend and take with them to succeeding courses the most important ideas in genetics as well as an analytic view of biological problems. To aid students in identifying conceptual aspects of a major topic, each chapter begins with a section called "Chapter Concepts," which in a few sentences captures the essence of the most important ideas about to be presented. Then, each chapter ends with a "Chapter Summary," which enumerates the five to ten key points that have been covered. These two features help to ensure that students focus on concepts and are not distracted by the many, albeit important, details of genetics. Specific examples and carefully designed figures support this approach throughout the book.

Insights and Solutions

Genetics, more than any other discipline within biology, requires problem solving and analytical thinking. At the endof each chapter we include what has become an extremely popular and successful section called "Insights and Solutions." In this section we stress:

Problem solving
Quantitative analysis
Analytical thinking
Experimental rationale

Problems or questions are posed and detailed solutions or answers are provided. This feature primes students for moving on to the "Problems and Discussion Questions" section that concludes each chapter.

Problems and Discussion Questions

In order to optimize the opportunities for student growth in the important areas of problem solving and analytical thinking, each chapter concludes with an extensive collection of problems and discussion questions. These represent various levels of difficulty, with the most challenging problems located at the end of each section. Brief answers to half the problems are in Appendix A. The Student Handbook is available to students for faculty who wish to expose their students to detailed answers to all problems and questions.

Acknowledgments

All comprehensive texts are dependent on the valuable input provided by many reviewers and colleagues. While we take full responsibility for any errors in this book, we gratefully acknowledge the help provided by those individuals who reviewed or otherwise contributed to the content and pedagogy of this and previous editions.

In particular, we thank Sarah Ward at Colorado State University for creating Chapter 24—Conservation Genetics and Jon Herron at the University of Washington for his input into Chapter 22—Population Genetics and Chapter 23—Genetics and Evolution. As in past editions, Charlotte Spencer at the Cross Cancer Institute in Alberta wrote or revised most of the Genetics, Technology, and Society essays. Others essays were previously contributed by Mark Shotwell at Slippery Rock University.

Robert W. Adkinson, Louisiana State University
Janice Bossart, The College of New Jersey
Paul Bottino, University of Maryland
Jim Bricker, The College of New Jersey
Hugh Britten, University of South Dakota
Aaron Cassill, University of Texas, San Antonio
Jimmy D. Clark, University of Kentucky
Stephen J. D'Surney, University of Mississippi
Scott Erdman, Syracuse University
Nancy H. Ferguson, Clemson University
Kim Gaither, Oklahoma Christian Univeristy
David Galbreath, McMaster University
Derek J. Girman, Sonoma State University
Elliot Goldstein, Arizona State University
Mark L. Hammond, Campbell University
Mike Hoopman, The College of New Jersey
David Hoppe, University of Minnesota, Morris
John A. Hunt, University of Hawaii, Honolulu
David Kass, Eastern Michigan University
Arlene Larson, University of Colorado at Denver
Beth A. Krueger, Monroe Community College
Hsiu-Ping Liu, Southwest Missouri State University
Paul F. Lurquin, Washington State University
Sally Mackenzie, University of Nebraska
Terry C. Matthews, Millikan University
Cynthia Moore, Washington University
Janet Morrison, The College of New Jersey
Michelle A. Murphy, University of Notre Dame
Marcia L. O'Connell, The College of New Jersey
Malcolm Schug, University of North Carolina, Greensboro
Ralph Seelke, University of Wisconsin, Superior
Gurel S. Sidhu, California State University
Gerald Schlink, Missouri Southern State College
Randy Scholl, Ohio State University
Mark Sturtevant, Northern Arizona University
Christine Tachibana, University of Washington
R. C. Woodruff, Bowling Green State University
Marie Wooten, Auburn University

For the Student

Online Media Tutorials (New)

The most sophisticated learning and tutorial package available for students of genetics, this online tutorial support system addresses students' most difficult concepts as identified through a survey of instructors. Concepts and processes begin with an overview that usually includes animations, proceeds to one or a series of interactive exercises, followed by self

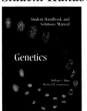

MEDIA TUTORIAL Transcription

quizzes. Each chapter contains a glossary, help function, search function, web links to fascinating and useful web sites, plus additional problem-solving questions. Students who experience difficulty with exercises or quizzes will be directed to specific sections of the text for review. An online demonstration of these amazing tutorials can be found at http://csep10.phys.utk.edu/genetics_demo

Look for this icon to identify media tutorials in the book (placed vertically on the edge of the page):

The following topics are addressed in the interactive media tutorials:

- Mitosis
- Meiosis
- Segregation
- Monohybrid Cross

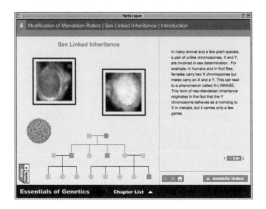

- Independent Assortment
- Meiosis and Mendel
- Extensions of Mendelian Inheritance: codominance, incomplete dominance, multiple alleles, gene interaction
- Probability: simple probability, sum rule, product rule, binomial probability
- Sex-Linked Inheritance: X-linked, Y-linked
- Linkage and Recombinantion: linkage versus independent assortment
- Linked Genes: mapping linked genes, construction of linkage maps
- Mapping a Three-Point Cross: includes calculation of recombination frequency, interference
- Virtual Crossover Laboratory
- Chromosome Aberrations: overview, inversions, translocations, deletions, duplications
- Phage Genetics: life cycle (lytic, lysogenic) phage cross, phage complementation
- Bacterial Genetics: transformation, conjugation
- DNA Structure: helix, components (sugar, base, phosphate), nucleotides
- DNA Replication: components, bidirectional replication, fork formation

- DNA Recombination (steps, structures)
- Transcription: components, initiation, elongation, termination, genetic code, Prokaryotes versus eukaryotes
- Translation: initiation, elongation, termination, prokaryotes versus eukaryotes, summary: DNA to RNA to protein
- Regulation of Gene Expression: prokaryotes (lac operon)
- Regulation of Gene Expression: eukaryotes: transcriptional regulation, alternative splicing, translational, posttranslational control
- Mutation: gene/protein colinearity, effect on protein structure (hemoglobin)
- Mutation at the DNA Level: nucleotide substitution, frameshift mutations
- DNA Repair
- Recombinant DNA Technology: restriction enzymes, cloning in plasmid vector, selection
- PCR: mechanism, uses
- Restriction Fragment Length Polymorphisms: codominant trait, transmission, linkage to disease loci
- DNA Fingerprinting
- Chromatin, Chromosome Structure: nucleosomes, fibers, scaffolding, isochores
- Population Genetics: allele frequencies, Hardy–Weinberg, selection
- Analysis of Human Pedigrees: autosomal dominant, recessive, X-linked dominant, recessive, mitochondrial
- Nondisjunction: normal, meiosis I, meiosis II gamete formation, fertilization outcomes
- Chemical Mutagenesis, Ames Test
- Restriction Mapping

Student Handbook and Solutions Manual

Harry Nickla, Creighton University
(0-13-093338-4)
Completely reviewed and checked for accuracy, this valuable handbook provides a detailed step-by-step solution or extended discussion for every problem in the text in a chapter-by-chapter format. The handbook also contains extra study problems and a thorough review of concepts and vocabulary.

New York Times Themes of the Times: Genetics and Molecular Biology

Coordinated by Harry Nickla, Creighton University
(0-13-060462-3)
This exciting newspaper-format supplement brings together recent genetics and molecular biology articles from the pages of the highly respected New York Times. This free supplement, available

through your local representative, encourages students to make the connections between genetic concepts and the latest research and breakthroughs in the field. This resource is updated regularly.

Science on the Internet: A Student's Guide

Andrew Stull and Harry Nickla
(0-13-028253-7)
The perfect guide to help your students take advantage of explosion of our *Essentials of Genetics* home page on the World Wide Web. This resource gives clear steps to access our regularly updated genetics resource area as well as an overview of general navigation and research strategies.

For the Instructor

Instructor's CD-ROM (0-13-065850-2)

For *Essentials* adopters, this CD-ROM is partitioned into sections that contain

- Content from the website
- All figures from the text
- PowerPoint (TM) format for figures
- The complete Instructor's Manual

Instructors will be able to coordinate lectures presentations with text content knowing students will be studying using the same animations, based upon the text. No more searching for the Instructor's Manual—It's on the CD-ROM.

Instructor's Resource Manual with Testbank

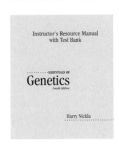

This manual and test bank contains over 800 questions and problems an instructor can use to prepare exams. The manual also provides optional course sequences, a guide to audiovisual supplements, and a section on searching the web. The testbank portion of the manual is also available in electronic format for both Windows and Macintosh users. Prentice Hall Custom Test allows instructors to create and tailor exams to their own needs. With the Online Testing option, exams can also be administered online and data can then be automatically transferred for evaluation. A comprehensive desk reference guide is included, along with online assistance.

Transparencies

200 figures from the text are included in the transparency package: 150 four-color transparencies from the text plus 50 transparency masters. The font size of the labels has been increased for easy viewing from the back of the classroom.

WebCT Course for Essentials of Genetics, 4th edition

The Prentice Hall WebCT course content for *Essentials* helps you meet the challenge of creating robust, interactive and educationally rich online courses. Our WebCT course material provides you with high quality, class-tested material pre-programmed and fully functional in the WebCT environment. Whether used as an online supplement to either a campus-based or distance learning course, our pre-assembled course content gives you a tremendous head start in developing your own online courses.

Blackboard Course for Essentials of Genetics, 4th edition

The *Essentials* Blackboard course contains web-based content and resources such as online study guides, assessment databanks, and lecture resource material. The abundant online content from *Essentials*, combined with Blackboard's popular tools and easy-to-use interface, result in a robust web-based course that is easy to implement, manage, and use—taking your courses to new heights in student interaction and learning. The Blackboard course management solution enable you to quickly add an online component to your campus-based course to provide you with a sophisticated technology base for total customization, scalability, and integration into your distance learning course.

CourseCompass Course for Essentials of Genetics, 4th edition

The *Essentials* CourseCompass™ course is the perfect course management solution that combines quality content with state-of-the-art Blackboard technology! CourseCompass™ is a dynamic, interactive online course management tool powered by Blackboard but hosted by Pearson Education. This exciting product allows you to teach with market-leading *Essentials* content in an easy-to-use customizable format.

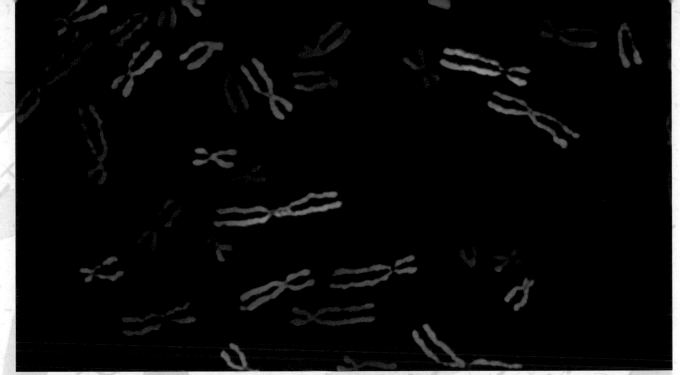

Human metaphase chromosomes, each composed of two sister chromatids joined at a common centromere. *(Omikron/Science Source/Photo Researchers, Inc.)*

1

An Introduction to Genetics

CHAPTER CONCEPTS

Genetics is the science of heredity. The discipline has a rich history and involves investigations of molecules, cells, organisms, and populations, using many different experimental approaches. Not only does genetic information play a significant role during evolution, its expression influences the functioning of individuals at all levels. Genetics thus unifies the study of biology and has had a profound impact on human affairs.

As the new millennium approached, a controversy was brewing among the 270,000 residents of the remote island nation of Iceland. Following months of fierce and sometimes heated debate, Iceland's Parliament passed a law giving a local biotechnology company the rights to develop a large database containing detailed DNA profiles of every resident of the country. This genetic information will be correlated with the genealogical and medical records of each entry and marketed to researchers around the world over the next decade.

This is not a passage drawn from Aldous Huxley's *Brave New World*, but rather an example of the current interface between genetics and society as we enter the new century. There are many interesting reasons why such a possible invasion of genetic privacy is about to occur to the residents of this small outpost country. The people of Iceland represent a nearly unique case of genetic uniformity seldom seen and accessible to scientific investigation. Indeed, for a variety of reasons, most all Icelanders share a close genetic resemblance, not only to one another, but to their Viking ancestors who settled this country over 1000 years ago. Because of this, geneticists believe that the Icelandic population is a tremendous asset in studying the link between genetics and disease. Because of the state-supported health-care system, medical records exist for all residents, as far back as 1900. Genealogical information is available for every living resident as well as for over 500,000 of the estimated 750,000 individuals who have ever lived in Iceland.

On the flip side are issues of privacy, consent, and commercialism—issues that will be central to most future dilemmas and controversies arising from applications of newly acquired genetic technology. The central question currently being asked throughout the global scientific community is what we will actually do with the vast genetic information that we are acquiring and where society-at-large factors into our decision. For example, how will knowledge of the complete nucleotide sequence of the human genome be used? More than at any other time in the history of science, addressing the ethical questions surrounding an emerging technology is clearly as important as the information that arises from that technology.

As you launch your study of the discipline of genetics, try to remain sensitive to issues like those described above. As you proceed through this text, the knowledge you gain will become a rich resource for achieving a thorough understanding of modern-day genetics. On the one hand, never has there been a more exciting time to be immersed in the study of this science. On the other hand, never has the need for caution been so apparent to society. Along the way, enjoy your studies, but take your responsibility as a novice geneticist most seriously.

1.1　The Historical Context of Genetics

Because genetic processes are fundamental to the comprehension of life itself, the discipline of genetics sits at the center of the field of biology. Genetic information directs cellular function, determines an organism's external appearance, and serves as the link between generations in every species. As such, knowledge of genetics is essential to the thorough understanding of other disciplines of biology, including molecular biology, cell biology, physiology, evolution, ecology, systematics, and behavior. Genetics thus unifies biology and serves as its "core."

It is thus not surprising that genetics has a long, rich history. In the chapters that follow, we will discuss the nature of chromosomes, the way in which genetic information is transmitted from one generation to the next, and the way in which this information is stored, altered, expressed, and regulated. The most significant scientific findings, which serve as the foundation for our discussions, were obtained in the nineteenth century. As the twentieth century dawned, certain discoveries were made that began to clarify our understanding of the physical basis of living organisms and their relationship to one another. Several related ideas were gaining acceptance and were particularly significant: (1) Matter is composed of atoms; (2) cells are the fundamental units of living organisms; (3) nuclei somehow serve as the "life force" of cells; and (4) chromosomes housed within nuclei somehow play an important role in heredity. When these ideas were correlated with the newly rediscovered genetic findings of Gregor Mendel and integrated with Darwin's theory of natural selection and the origin of species, a more complete picture of life at the level of the individual and of the population emerged. The era of modern-day biology was initiated on this foundation.

But what of the many important ideas and hypotheses that served as forerunners of nineteenth-century thought? In the following short discussion, we consider some of these, several of which can be traced back well over 1000 years!

Prehistoric Times: Domesticated Animals and Cultivated Plants

While we don't know when people first recognized the existence of heredity, various types of archeological evidence (e.g., primitive art, preserved bones and skulls, and dried seeds) provide insights. Such evidence documents the successful domestication of animals and cultivation of plants thousands of years ago. These ancient endeavors represent artificial selection of genetic variants within populations. For example, between 8000 and 1000 B.C., horses, camels, oxen, and various breeds of dogs (derived from the wolf family) were domesticated and selective breeding soon followed. Cultivation of many plants, including maize, wheat, rice, and the date palm, began around 5000 B.C. Remains of maize dating to this period have been recovered in caves in the Tehucan Valley of Mexico. Assyrian art depicts artificial pollination of the date palm, thought to have originated in Babylonia (Figure 1–1). Such cultivation very likely led to the types of date palms found in that region today, where over 400 varieties exist in just four oases in the Sahara Desert. They differ from one another in various traits such as fruit taste.

Prehistoric evidence of cultivated plants and domesticated animals documents our ancient ancestors' successful attempts

FIGURE 1 1 Relief carving depicting artificial pollination of date palms (800 B.C.). *(The Metropolitan Museum of Art, Gift of John D. Rockefeller, Jr., 1932. (32.143.3) Photograph © 1983 The Metropolitan Museum of Art.)*

to manipulate the genetic composition of useful species. There is little doubt that people soon learned that desirable and undesirable traits are passed to successive generations and more desirable varieties of animals and plants could be selected. Human awareness of heredity was thus apparent during prehistoric times.

The Greek Influence: Hippocrates and Aristotle

Few, if any, significant ideas were put forward to explain heredity during prehistoric times; however, during the Golden Age of Greek culture, philosophers directed much more attention to this subject as it relates to the origin of humans, and of reproduction and heredity in particular. This is evident in the writings of the Hippocratic school of medicine (500–400 B.C.) and subsequently of the philosopher and naturalist Aristotle (384–322 B.C.).

For example, the Hippocratic treatise *On the Seed* argued that active "humors" resided in various parts of the body, serving as the bearers of hereditary traits. Drawn from various parts of the male body to the semen, they could be healthy or diseased, the latter condition accounting for the appearance of newborns with congenital disorders or deformities. It was further believed that these humors could be altered in individuals before they were passed on to offspring, explaining how newborns could "inherit" traits that their parents had "acquired" because of their environment.

Aristotle (Figure 1–2), who studied under Plato for some 20 years, was more critical and more expansive than Hippocrates in his analysis of human origins and heredity. Aristotle proposed that the generative power of male semen resided in a "vital heat" that it contained. This vital heat had the capacity to produce offspring of the same "form" (i.e., basic structure and capacities) as the parent. Aristotle believed that it cooked and shaped the menstrual blood produced by the fe-

FIGURE 1–2 Aristotle describes the animals Alexander has sent him. (A fresco in The Main Hall of The Assemblée Nationale in Paris. Painting by Eugene Delacroix.) *(Eugene Delacroix, frescos from the spandrels of the main hall of the Assemblée Nationale, Paris, France. © Photograph by Erich Lessing/Art Resource, NY.)*

male, which was the "physical substance" giving rise to an offspring. The embryo developed not because it already contained the parts in miniature (as some Hippocratics had thought), but because of the shaping power of the vital heat. These ideas constituted only one part of Aristotelian philosophy of order in the living world.

Although the ideas of Hippocrates and Aristotle sound primitive and naive today, we should recall that, prior to the 1800s, neither sperm nor eggs had yet been observed in mammals. Thus, the Greek philosophers' ideas were worthy ones in their time and for centuries to come.

1600–1850: The Dawn of Modern Biology

During the ensuing 1900 years (from 300 B.C. to 1600 A.D.), the theoretical understanding of genetics was not extended by new and significant ideas. However, in Roman times, plant grafting and animal breeding were common. By the Middle Ages, naturalists, well aware of the impact of heredity on organisms they studied, were faced with reconciling their findings with current religious beliefs. The theories of Hippocrates and Aristotle still prevailed and, when applied to humans, they no doubt conflicted with the prevailing religious doctrines.

Between 1600 and 1850, major strides were made that provided much greater insights into the biological basis of life, setting the scene for the revolutionary work and principles presented by Charles Darwin and Gregor Mendel. In the 1600s, the English anatomist William Harvey (1578–1657) wrote a treatise on reproduction and development, patterned after Aristotle's work. In it he is credited with the earliest statement of the theory of **epigenesis**—that an organism is derived from substances present in the egg that differentiate into adult structures during embryonic development. Epigenesis holds that structures such as body organs are not initially present in the early embryo, but instead are formed *de novo* (anew).

The theory of epigenesis conflicts directly with the theory of **preformation**, first put forward in the seventeenth century. Stating that sex cells contain a complete miniature adult called the **homunculus** (Figure 1–3), perfect in every form, preformation was popular well into the eighteenth century. However, work by the embryologist Casper Wolff (1733–1794) and others clearly disproved this theory, thus favoring epigenesis. Wolff was convinced that several structures, such as the alimentary canal, were not present in the earliest embryos he studied, but were instead formed later during development.

During this same period, other significant chemical and biological findings affected future scientific thinking. In 1808,

FIGURE 1–3 Depiction of the "homunculus," a sperm containing a miniature adult, perfect in proportion and fully formed. *(Hartsoeker, N. Essay de dioptrique, Paris, 1694, p. 230. National Library of Medicine.)*

John Dalton expounded his **atomic theory**, which stated that all matter is composed of small invisible units called atoms. Around 1830, Matthias Schleiden and Theodor Schwann, using improved microscopes, proposed the **cell theory**, stating that all organisms are composed of basic visible units called cells, which are derived from similar preexisting structures. By this time, the idea of **spontaneous generation**, the creation of living organisms from nonliving components, had clearly been disproved, and living organisms were considered to be derived from preexisting organisms and to consist of cells made up of atoms.

Another prevailing notion prevalent in the nineteenth century, the **fixity of species**, was influential. According to this doctrine, animal and plant groups remain unchanged in form from the moment of their appearance on earth. Embraced particularly by those who also adhered to a belief in special creation, this doctrine was popularized by several people, including the Swedish physician and plant taxonomist Carolus Linnaeus (1707–1778), who is better known for devising the binomial system of classification.

The influence of the tenet of fixity of species is illustrated by considering the work of the German plant breeder Joseph Gottlieb Kolreuter (1733–1806), who worked with tobacco. He crossbred two groups and derived a new hybrid form, which he then converted back to one of the parental types by repeated backcrosses. In other breeding experiments using carnations, he clearly observed segregation of traits, which was to become one of Mendel's principles of genetics. These results seemed to contradict the idea of "fixed species" that do not change with time. While Kolreuter was puzzled by these outcomes, because of his belief in both special creation and the fixity of species, he failed to recognize the real significance of his findings.

Charles Darwin and Evolution

With the preceding information as background, we conclude our coverage of the historical context of genetics with a brief discussion of the work of Charles Darwin, who in 1859 published the book-length statement of his evolutionary theory, *The Origin of Species*. Darwin's many geological, geographical, and biological observations convinced him that existing species arose by descent with modification from other ancestral species. Greatly influenced by his now famous voyage on the HMS *Beagle* (1831–1836), Darwin's thinking culminated in his formulation of the theory of **natural selection**, a theory that attempts to explain the causes of evolutionary change. Formulated and proposed at the same time (but independently) by Alfred Russel Wallace, natural selection is based on the observation that populations tend to consist of more offspring than the environment can support, leading to a struggle for survival among them. Those organisms with heritable traits that allow them to adapt to their environment are better able to survive and reproduce than those with less adaptive traits. Over a long period of time, slight but advantageous variations will accumulate. If a population bearing these inherited variations becomes reproductively isolated, a new species may result.

The primary gap in Darwin's theory was a lack of understanding of the genetic basis of variation and inheritance, a gap that left it open to reasonable criticism well into the twentieth century. Aware of this weakness in his theory, in 1868, Darwin published a second book, *Variations in Animals and Plants under Domestication*, in which he attempted to provide a more definitive explanation of how heritable variation arises gradually over time. Even though Darwin never understood the basis for inherited variation, his ideas concerning evolution may be the most influential theory ever put forward in the history of biology. His ability to distill his extensive observations and synthesize his ideas into a cohesive hypothesis describing the origin of diversity of organisms populating the earth constitutes a major achievement in the history of science.

As Darwin's work ensued, Gregor Johann Mendel (Figure 1–4) conducted his experiments between 1856 and 1863, and published his classic paper in 1866. In it, Mendel demonstrated a number of statistical patterns underlying inheritance and developed a theory involving hereditary factors in the germ cells to explain these patterns. His research was virtually ignored until it was partially duplicated and then cited by Carl Correns, Hugo de Vries, and Eric Von Tschermak around 1900 and subsequently championed by William Bateson.

By the early part of the twentieth century, chromosomes were discovered and support for the epigenetic interpretation of development had grown considerably. It gradually became clear that heredity and development were dependent on "information" contained in chromosomes, which were contributed by gametes to each individual. The "gap" in Darwin's theory had narrowed considerably. In Chapter 3, we will return to a thorough analysis of Mendel's findings, which to this day serve as the foundation of genetics.

1.2   The Molecular Basis of Genetics

In a following chapter, we consider in detail the experiments of Gregor Mendel and subsequently the extensive research that established the basic tenets of transmission genetics. Together, this body of work established the **chromosomal theory of inheritance**, which states that inherited traits are controlled by genes that reside in chromosomes, which are faithfully transmitted through gametes to future generations. It is useful to give a brief overview of the molecular basis for genetic function that serves as the underpinnings of the more classical information. This will provide you with a foundation for a more comprehensive understanding of Mendel's work as well as that which followed it, as presented in ensuing chapters. As you will see, it is indeed remarkable that so much was learned without the knowledge of molecular genetics, which today we take for granted.

The Trinity of Molecular Genetics

The way in which genes control inherited variation is best understood in terms of three molecules, sometimes referred to as the trinity of molecular genetics: **DNA**, **RNA**, and **protein**. The nucleic acid DNA (deoxyribonucleic acid) serves as the genetic material in all living organisms as well as in most viruses. DNA is organized into genes and stores genetic information. As part of the chromosomes, the information contained in genes can be transmitted faithfully by parents through gametes to their offspring. For the gene's DNA to subsequently influence an inherited trait, the stored genetic information in the DNA must first be transferred to a closely related nucleic acid, RNA (ribonucleic acid). In eukaryotic organisms, RNA carries the genetic information out of the nucleus, where chromosomes reside, into the cytoplasm of the cell. There, the information in RNA is translated into proteins, which serve as the end products of most all genes. Ultimately, it is the diverse functions of proteins that determine the biochemical identity of cells and that ultimately determine the expression of inherited traits. The process of storage and expression of genetic information, upon which life on earth is based, may be summarized as

DNA makes RNA that makes Proteins

The process of transferring information from DNA to RNA is called **transcription**. The subsequent conversion of the genetic information contained in RNA into a protein is called **translation**.

FIGURE 1–4 Gregor Johann Mendel, who in 1866 put forward the major postulates of transmission genetics as a result of experiments with the garden pea. *(Archiv/Photo Researchers, Inc.)*

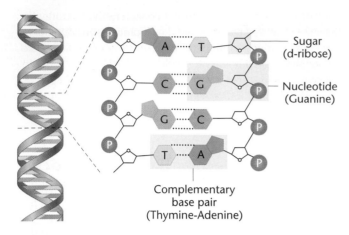

FIGURE 1–5   Summary of the structure of DNA, illustrating on the left the nature of the double helix, and on the right the chemical components making up both strands.

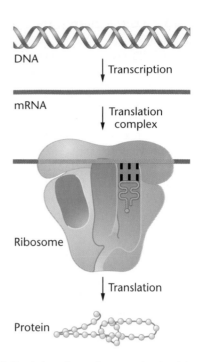

FIGURE 1–6   Depiction of genetic expression involving transcription of DNA into mRNA, and the translation of mRNA on a ribosome into a protein.

The Structure of Nucleic Acids

Several forms of the structure and components of DNA are shown in Figure 1–5. The molecule exists in cells as a long, coiled ladderlike structure described as a double helix. Each strand of the helix consists of a linear polymer made up of genetic building blocks called **nucleotides**, of which there are four types. Nucleotides vary, depending upon which of four nitrogenous bases is part of the molecule— A (adenine), G (guanine), T (thymine), or C (cytosine). These comprise the genetic alphabet, which in various combinations ultimately specify the components of proteins. One of the great discoveries of the twentieth century was made in 1953 by James Watson and Francis Crick, who established that the two strands of their proposed double helix are exact complements of one another, such that the rungs of the ladder always consist of either A $=$ T or G $\equiv$ C base pairs. As we shall see when we pursue molecular genetics in depth later in the text, this **complementarity** between adenine and thymine nitrogenous base pairs and between guanine and cytosine nitrogenous base pairs, attracted to one another by **hydrogen bonds**, is critical to genetic function. Complementarity serves as the basis for both the replication of DNA and for the transcription of DNA into RNA. During both processes, under the direction of the appropriate enzyme, DNA strands serve as templates for the synthesis of the complementary molecule.

RNA is chemically very similar to DNA. However, it demonstrates a small variation in its component sugar (ribose vs. deoxyribose), and it contains the nitrogenous base uracil in place of thymine. Additionally, in contrast to the double helix of DNA, RNA is generally single-stranded. Importantly, it can form complementary structures with a strand of DNA. In such cases, uracil base-pairs with adenine. As noted earlier, this complementarity is the basis for transcription of the chemical information in DNA into RNA. This process is shown in part (a) of Figure 1–6.

The Genetic Code and RNA Triplets

Once an RNA molecule complementary to one strand of a gene's DNA is transcribed, the RNA behaves as a messenger that directs the synthesis of proteins. This is accomplished during the association of this RNA molecule (called messenger RNA, or mRNA, for short) with a complex cellular structure called the **ribosome**. The process by which proteins are synthesized under the direction of mRNA, as mentioned earlier, is called translation. Ribosomes serve as nonspecific workbenches for protein synthesis.

Proteins, as the end product of genes, are linear polymers made up of amino acids, of which there are 20 different types in living organisms. A major question is how information present in mRNA is encoded to direct the insertion of specific amino acids into protein chains as they are synthesized. The answer is now quite clear. The genetic code consists of a linear series of triplet nucleotides present in mRNA molecules. Each triplet reflects, through complementarity, the information stored in DNA and specifies the insertion of a specific amino acid as the mRNA is translated into the growing protein chain [Figure 1–6(b)]. A key discovery in how this is accomplished involved the identification of a series of adapter molecules called **transfer RNA (tRNA)**. Within the ribosome, these adapt the information encoded in the mRNA triplets to the specific amino acid during translation.

As the preceding discussion documents, DNA makes RNA that makes protein, and with great specificity. Using an alphabet of only four letters (A, T, C, and G), a language exists that directs the synthesis of highly specific proteins that collectively serve as the basis for all biological function.

Proteins and Biological Function

As we have mentioned, proteins are the end products of genetic expression. They are the molecules responsible for imparting the properties that we attribute to the living process. The potential for achieving the diverse nature of biological function rests with fact that the alphabet used to construct

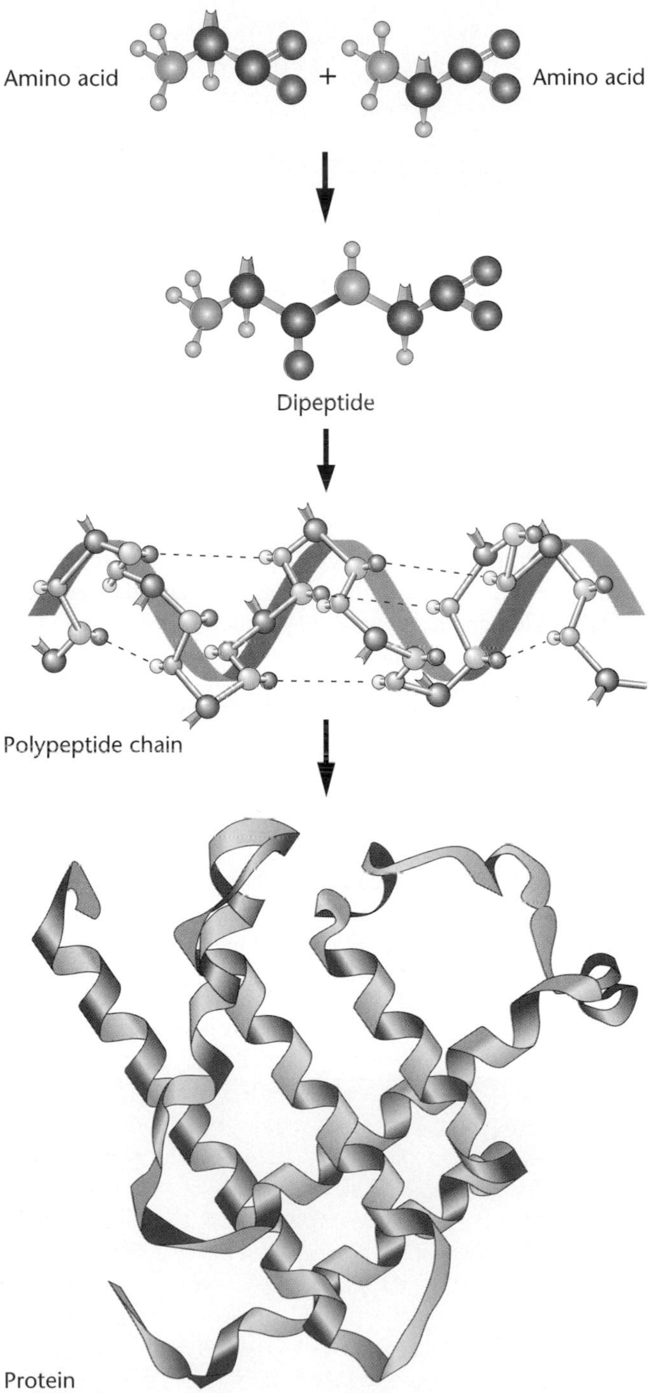

FIGURE 1–7 Depiction of three steps leading to the formation of a protein. Initially, two amino acids are joined to form a dipeptide. As synthesis continues, a longer polypeptide chain is formed which most often coils into a right-handed alpha helical structure. This polypeptide then folds into a 3-dimensional conformation specific to the protein's function.

proteins consists of 20 letters (amino acids), which combine to create words that can be thousands of letters long (Figure 1–7). If we consider a protein chain that is just 100 amino acids in length, and at each position there can be any one of 20 amino acids, then the number of different molecules, each with a unique sequence, is equal to

$$20^{100}$$

Since 20^{10} exceeds 5×10^{12}, or over 5 trillion, imagine how large the above number is! Obviously, evolution has seized on a class of molecules that have the potential for enormous structural diversity as they serve as the mainstay in biological systems.

The main category of proteins is that which includes **enzymes**. These molecules serve as biological catalysts, essentially allowing biochemical reactions to proceed at rates that sustain life under the conditions that exist on earth. For example, by lowering the energy of activation in reactions, metabolism is able to proceed under the direction of enzymes at body temperature (37°C) in humans, where in the absence of enzymes, these chemical reactions would proceed at rates thousands of times more slowly. As a result, enzymes, each under the control of one or more specific genes, are capable of directing both the anabolism (the synthesis) and catabolism (the breakdown) of all organic molecules in the cell, including carbohydrates, lipids, nucleic acids, and proteins themselves.

There are countless proteins other than enzymes that are critical components of cells and organisms. These include such diverse examples as **hemoglobin**, the oxygen-binding pigment in red blood cells; **insulin**, the pancreatic hormone; **collagen**, the connective tissue molecule; **keratin**, the structural molecule in hair; **histones**, the proteins integral to chromosome structure in eukaryotes; **actin** and **myosin**, the contractile muscle proteins; and **immunoglobulins**, the antibody molecules of the immune system. Specific proteins are also critical components of all membranes and serve as molecules that regulate genetic expression. The potential for such diverse functions rests with the enormous variation of three-dimensional conformation that may be achieved by proteins. The final conformation of a protein is the direct result of the unique linear sequence of amino acids that constitute the molecule. To come full circle, this sequence is dictated by the stored information in the DNA of a gene that is transferred to RNA, which then directs the synthesis of a protein. DNA makes RNA that then makes protein.

1.3 Investigative Approaches in Genetics

It will be useful in your study of genetics to know something about the various research approaches that have advanced our knowledge of the field. Investigations have involved viruses, bacteria, and a wide variety of plants and animals and have spanned all levels of biological organization, from molecules to populations. Although some overlap exists, most have used one of four basic approaches.

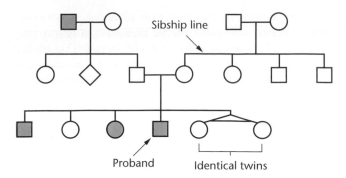

FIGURE 1–8 A representative human pedigree tracing a genetic characteristic through three generations.

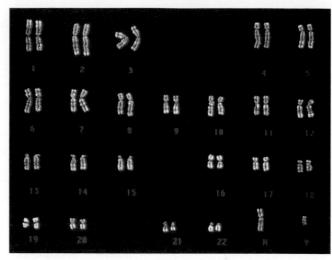

FIGURE 1–9 The human karyotype. *(Sovereign/Phototake)*

The classical investigative approach is the study of **transmission genetics**, in which the patterns of inheritance of traits are examined. Experiments are designed so that the transmission of traits from parents to offspring can be analyzed through several generations. Patterns of inheritance are sought that will provide insights into genetic principles. The first significant experimentation of this kind to have a major impact on understanding heredity was performed by Gregor Mendel in the middle of the nineteenth century. Information derived from his work serves today as the foundation of transmission genetics. In human studies, where designed matings are neither possible nor desirable, **pedigree analysis** is often useful. As shown in (Figure 1–8), patterns of inheritance are traced through several generations, leading to inferences concerning the mode of inheritance of the trait under investigation.

The second approach involves **cytogenetics**—the study of chromosomes. The earliest such studies used light microscopy. The initial discovery of chromosome behavior during mitosis and meiosis, early in the twentieth century, was a critical event in the history of genetics, because of the important role these observations played in the rediscovery and acceptance of Mendelian principles. The light microscope continues to be useful in the investigation of chromosome structure and abnormalities and is instrumental in preparing **karyotypes**, which illustrate the chromosomes characteristic of any species arranged in a standard sequence (Figure 1–9). With the advent of electron microscopy, the repertoire of investigative approaches in genetics has grown. In high-resolution microscopy, genetic molecules and their behavior during gene expression can be visualized directly.

The third general approach involves **molecular genetic analysis**, which has had the greatest impact on the recent growth of genetic knowledge. Molecular studies, initiated in the early 1940s, have consistently expanded our knowledge of the role of genetics in life processes. Although experiments initially relied on bacteria and the viruses that infect them, extensive information is now available concerning the nature, expression, replication, and regulation of the genetic information in eukaryotes as well. The precise nucleotide

sequence has been determined for many genes cloned in the laboratory. Recombinant DNA studies (Figure 1–10), in which genes from another organism are literally spliced into bacterial or viral DNA and then cloned, serve as the basis of a far-reaching research technology used in molecular genetic investigations. Building on this approach, the new field of DNA biotechnology now exists whereby genes are identified, sequenced, cloned, and manipulated. Furthermore, using the most recent technology, it is now possible to probe gene function in extreme detail. Such molecular and biochemical analysis has created the potential for **gene therapy**, and has profound implications for medicine, agriculture, and bioethics.

Perhaps the most striking achievement thus far in the history of biotechnology occurred in 1996 at the Roslin Institute in Scotland, when the world's most famous lamb Dolly was born (Figure 1–11). The first animal ever to be cloned from an adult somatic cell, Dolly was the result of the research of

FIGURE 1–10 Visualization of DNA fragments under ultraviolet light. The bands were produced using recombinant DNA technology. *(Dan McCoy/Rainbow)*

FIGURE 1–11 Dolly, a Finn Dorset sheep cloned from the genetic material of an adult mammary cell, shown next to her first-born lamb, Bonnie. *(Photo courtesy of Roslin Institute)*

Ian Wilmut, who fused the nucleus of a frozen udder cell taken from a 6-year-old sheep with an enucleated oocyte of another sheep. Following implantation into a surrogate mother, complete embryonic and fetal development was accomplished under the direction of the genetic material of the udder cell. While the implications of this event are far reaching and raise numerous ethical concerns, the ultimate goal of Wilmut's research is to use cloned animals as models to study human disease and to produce therapeutic drugs beneficial to humans. Since this work, other organisms, including cows, have been successfully cloned.

The final investigative approach involves the study of **population genetics**. In these investigations, scientists attempt to define how and why certain genetic variation (Figure 1–12) is maintained in populations, while other variation diminishes or is lost with time. Such information is critical to the understanding of evolutionary processes. Population genetics also allows us to predict gene frequencies in future generations.

Together, these varied approaches used to investigate genetics have transformed a subject that was only poorly understood in 1900 into one of the most advanced scientific disciplines today. As a result, the impact of genetics on society has been, and will continue to be, immense. We will discuss applications of genetics in the following section and throughout the text.

1.4 Genetics and Society

In addition to acquiring information for the sake of extending knowledge in any discipline of science—an experimental approach called **basic research**—scientists conduct investigations to solve problems facing society or simply to improve the well-being of members of our society—an approach called **applied research**. Together, both types of genetic research have combined to: (1) enhance the quality of our existence on this planet; and (2) provide a more thorough understanding of life processes. As we shall see throughout this text, there is very little in our lives that genetics fails to touch.

Eugenics and Euphenics

There is always the danger that scientific findings will be used to formulate policies and/or actions that are unjust or even tragic. In this section we review such a case, which began near the end of the nineteenth century. At that time, Darwin's theory of natural selection strongly influenced some people's thinking concerning the human condition. Our story recounts the initial attempt to apply genetic knowledge directly for the improvement of human existence. Championed in England by Francis Galton, the general approach is called **eugenics**, a term Galton coined in 1883.

Galton, a cousin of Charles Darwin, believed that many human characteristics are inherited and subject to artificial selection if human matings could be controlled. **Positive eugenics** encouraged parents displaying favorable characteristics to have large families. Superior intelligence, intellectual achievement, and artistic talent are examples. **Negative eugenics**, on the other hand, attempted to restrict reproduction of parents displaying unfavorable characteristics. Low intelligence, mental retardation, and criminal behavior are examples.

In the United States, the eugenics movement was a significant social force and led to state and federal laws that permitted sterilization of those considered "genetically inferior." Over half of the states passed such laws, commencing in 1907 with Indiana. Sterilization was mandated for "imbeciles, idiots, convicted rapists, and habitual criminals." By 1931, involuntary sterilization also applied to "sexual perverts, drug fiends, drunkards, and epileptics." Immigration to the United States from certain areas of Europe and from Asia was also restricted, to prevent the influx of what were regarded as genetically inferior people.

Besides violating fundamental human rights, such policies were seriously flawed by an inadequate understanding of the genetic basis of various characteristics. Formulation of eugenic policies was premised on the mistaken notions that

FIGURE 1–12 Genetic variation exhibited in the skin of corn snakes. The wild type (normal) variety displays orange and black markings. *(Zig Leszczynski/Animals Animals/Earth Scenes)*

"superior" and "inferior" traits are totally under genetic control and that genes deemed unfavorable could be removed from a population by selecting against (sterilizing) individuals expressing those traits. Environmental influences as well as genetic theory underlying population genetics were largely ignored as eugenic policies were developed.

In Nazi Germany in the 1930s, the concept of achieving a superior, racially pure group was an extension of the eugenics movement. Initially applied to those considered socially and physically defective, the underlying rationale of negative eugenics was soon applied to entire ethnic groups, including Jews and Gypsies. Fueled by various forms of racial prejudice, Adolf Hitler and the Nazi regime took eugenics to a horrific extreme by instituting policies aimed at the extinction of these "impure" human populations. This deplorable disregard for human life was preceded by incremental policies involving forced sterilization and mercy killings. This movement, based on scientifically invalid premises, culminated in the mass murder of the Holocaust.

Even before the Nazi Party rose to power in 1933, English and American geneticists began distancing themselves from the eugenics movement. They were concerned about the validity of the premises underlying the movement and the evidence in support of these premises. Thus, many geneticists chose not to study human genetics for fear of being grouped with those who supported eugenics.

However, since the end of World War II, tremendous strides have been made in human genetics research. Today, a new term, **euphenics**, has replaced eugenics. Euphenics refers to medical and/or genetic intervention designed to reduce the impact of defective genotypes on individuals. The use of insulin by diabetics and the dietary control of newborn phenylketonurics are longstanding examples. Today, "genetic surgery" to replace defective genes looms clearly on the horizon. Furthermore, social policies now have a solid genetic foundation on which they may be based. Nevertheless, caution is still required to ensure that our expanded knowledge of human genetics does not obscure the role played by the environment in determining an individual's phenotype.

Genetic Advances in Agriculture and Medicine

As a result of research in genetics, major benefits have accrued to society in the fields of agriculture and medicine. Although cultivation of plants and domestication of animals had begun long before, the rediscovery of Mendel's work in the early twentieth century spurred scientists to apply genetic principles to these human endeavors. The use of selective breeding and hybridization techniques has had the most significant impact in agriculture.

Plants have been improved in four major ways: (1) enhanced potential for more vigorous growth and increased yields, including the unique genetic phenomenon of hybrid vigor (heterosis); (2) increased resistance to natural predators and pests, including insects and disease-causing micro-

organisms; (3) production of hybrids exhibiting a combination of superior traits derived from two different strains or even two different species (Figure 1–13); and (4) selection of genetic variants with desirable qualities such as increased protein value, increased content of limiting amino acids, which are essential in the human diet, or smaller plant size, reducing vulnerability to adverse weather conditions.

Over the past five decades, these improvements have resulted in a tremendous increase in yield and nutrient value in such crops as barley, beans, corn, oats, rice, rye, and wheat. It is estimated that in the United States the use of improved genetic strains has led to a threefold increase in crop yield per acre. In Mexico, where corn is the staple crop, a significant increase in protein content and yield has occurred. A substantial effort has also been made to improve the growth of Mexican wheat. Led by Norman Borlaug, a team of researchers has developed varieties of wheat that incorporate favorable genes from other strains found in various parts of the world, revolutionizing wheat production in Mexico and other underdeveloped countries. Because of this effort, which led to the well-publicized "Green Revolution," Borlaug received the Nobel Peace Prize in 1970. There is little question that this application of genetics has contributed to the well-being of our own species by improving the quality of nutrition worldwide.

FIGURE 1–13 *Triticale*, a hybrid grain derived from wheat and rye, produced as a result of applied genetic breeding experiments. *(Grant Heilman/Grant Heilman Photography, Inc.)*

FIGURE 1–14 The effects of breeding and selection, as illustrated by the production of this Vietnamese pot-bellied pig. *(Renee Lynn/Photo Researchers, Inc.)*

Applied research in genetics has also resulted in the development of superior breeds of livestock (Figure 1–14). Selective breeding has produced chickens that grow faster, yield more high-quality meat per chicken, and lay greater numbers of larger eggs. In larger animals, including pigs and cows, the use of artificial insemination has been particularly important. Sperm samples derived from a single male with superior genetic traits can now be used to fertilize thousands of females located in all parts of the world.

Equivalent strides have been made in medicine as a result of advances in genetics, particularly since 1950. Numerous disorders in humans have been discovered to result from either a single mutation or a specific chromosomal abnormality. For example, the genetic basis of disorders such as sickle-cell anemia, erythroblastosis fetalis, cystic fibrosis, hemophilia, muscular dystrophy, Tay–Sachs disease, Down syndrome, and many metabolic disorders is now well documented and often understood at the molecular level. The importance of acquiring knowledge of inherited disorders is underscored by the estimate that more than 10 million children or adults in the United States suffer from some form of genetic affliction and that every child-bearing couple stands an approximately 3 percent risk of having a child with some form of genetic anomaly.

Additionally, it is now apparent that all forms of cancer have a genetic basis. Although cancer is not usually an inherited disorder, it is very clear that *cancer is a genetic disorder at the somatic cell level.* Most cancers are derived from somatic cells that have undergone some type of genetic change; malignant tumors are then derived from the genetically altered cell. In some cases, a genetic predisposition to cancer also exists.

The recognition of the molecular basis of human genetic disorders and cancer is providing the impetus for the development of methods for detection and treatment. Genetic counseling gives couples objective information on which they can base informed decisions about child-bearing. In the case of cancer, recent genetic discoveries have already led to more effective early detection and more efficient treatment.

Applied research in genetics is also providing other medical benefits. Advances in immunogenetics have made possible compatible blood transfusions as well as organ transplants. In conjunction with immunosuppressive drugs, the number of successful transplant operations involving human organs, including the heart, liver, pancreas, and kidney, is increasing annually.

The most recent advances in human genetics have been dependent on the application of DNA biotechnology. First developed in the 1970s, recombinant DNA techniques paved the way for manipulating and cloning a variety of genes, including those that encode many medically important molecules, such as insulin, blood clotting factors, growth hormone, and interferon. Human genes are isolated and spliced into vectors and transferred to host cells that serve as "production centers" for the synthesis of these proteins.

Recombinant DNA techniques are being extended considerably. DNA of any organism of interest is routinely manipulated in the laboratory. Human genes responsible for inherited disorders such as cystic fibrosis and Huntington disease have been identified, isolated, cloned, and studied. It is hoped that such research will pave the way for gene therapy, whereby genetic disorders are treated by inserting normal copies of genes into the cells of afflicted individuals.

Perhaps the most far-reaching use of DNA biotechnology involves the Human Genome Project, in which the entire genetic complement (the **genome**) of several species, including our own, has been sequenced. The genomic sequence of several bacterial species, as well as that of yeast, the fruit fly, the mustard plant (*Arabidopsis*), and the nematode (*Caenorhabditis elegans*) is now known.

In later chapters, the applications of DNA biotechnology to agriculture and medicine are discussed in greater detail. Although other scientific disciplines are also expanding in knowledge, none has paralleled the growth of information that is occurring in genetics. As we noted at the outset of this chapter, while there never has been a more exciting time to be immersed in the study of genetics, the potential impact of this discipline on society has never been more profound. By the end of this course, we are confident you will agree that the present truly represents the "Age of Genetics."

Genetics, Technology, and Society

The Frankenfood Debates

Until recently, North Americans paid little attention to the nature or extent of genetic engineering of agricultural products. We have assumed that genetically engineered, or genetically modified (GM), foods are equivalent to nonengineered foods and that they pose no particular threats to human or environmental health. However, this attitude is gradually changing as we learn more about the controversies surrounding GM foods in Europe and as more countries ban the importation of GM seeds, crops, and foods.

Genetic engineering of plants involves recombinant DNA techniques. Scientists clone a gene of interest from a plant or animal, attach the gene to a suitable promoter sequence, and introduce the gene into plant cells that are growing in tissue culture. After the plant cells have successfully integrated the new gene, the cultured cells are grown into whole plants. In this way, every cell of the new GM plant contains the new gene. The GM plants are then extensively tested for growth and expression of the transgene.

GM crops burst on the agricultural scene in the mid-1990s. By 1999, half of U.S. soybeans, 40 percent of U.S. corn, and half of Canada's canola crop were grown from genetically engineered seeds. Other common GM crops include cotton, potatoes, and tomatoes. It is estimated that over 60 percent of all processed foods in North America contain ingredients derived from GM plants. A dozen other GM crops are in the pipeline for approval and marketing. Virtually all GM crops currently used have been developed and are marketed by large international agrochemical companies—companies that hold patents on the seeds and on the technology to generate the seeds.

At present, the two genetic modifications found in the majority of GM crops are *Bt* pest resistance and glyphosate herbicide resistance. The *Bt* gene is derived from a bacterium (*B. thuringiensis*), and encodes a toxin that kills certain insects. This toxin does not appear to affect other animals and plants. Plants that make their own *Bt* toxin do not require spraying for these insects. In theory, insecticide use will decrease, yields will increase, and farmer's profits will rise. Glyphosate (known commercially as RoundUp) is an herbicide that kills all plants, except for those with natural or engineered resistance. Use of glyphosate-resistant crops should allow farmers to kill weeds that compete with the GM crop, thereby increasing yields and profits.

Proponents of GM crops argue that genetic engineering of plants will reduce the use of pesticides and herbicides, improve the quality of foods, and increase yields sufficiently to feed the world's hungry population. Critics are less optimistic, even labeling GM foods as "Frankenfoods." The most frequent criticism of GM foods is a perceived threat to human health. Polls suggest that over half of North Americans worry about the safety of GM foods. In Europe, the figure is closer to 90 percent. The U.K. government recently banned all commercial GM crops until the year 2003. Many British and European supermarket chains are removing GM foods from their stores. The European Parliament is considering legislation to make agrochemical companies financially responsible for any ill effects from GM food consumption. Unfortunately, there has been little animal research, and no human research, on the long-term health effects of any GM food. So far, animal studies appear to show no ill effects of GM foods; however, each GM crop is different and few have been tested.

Environmental criticisms of GM crops are numerous. Many scientists argue that the continuous presence of *Bt* toxin in *Bt*-engineered crops will keep insect pests under constant pressure to evolve resistance. Once insects are resistant, the advantages of growing *Bt*-engineered crops will vanish, requiring farmers to resort to using other, more toxic, pesticides. Recent data suggest that *Bt*-expressing plants leak the *Bt* toxin through their roots into the soil, with unknown effects on soil ecology. And, *Bt*-expressing plants and their pollen may be toxic to beneficial insects, as suggested by recent studies of monarch butterflies and *Bt*-corn pollen. Another criticism is that glyphosate-resistance may be transferred to wild relatives of the glyphosate-resistant plants through cross-pollination, creating "superweeds" that cannot be killed with glyphosate. Similarly, engineered traits such as virus or fungus resistance may be transferred to weeds, with unpredictable effects. These scenarios are not impossible, as genetic exchange between cultivated and wild plants has been documented for decades. Interestingly, herbicide-resistant canola has already become a common weed in Canadian wheat fields. If wild and weedy plants acquire glyphosate resistance, it will be necessary to use higher levels of glyphosate, or other, more toxic, herbicides to control weeds. If these resistance scenarios are correct, the benefits of *Bt*-engineered and glyphosate-engineered crops will be short-lived.

Critics argue that GM crops will do little to feed the world's hungry. They contend that the millions of people who are malnourished are too poor to afford patented seed, the petrochemicals required to produce the crop, and GM foods. They feel that GM technologies favor wealthier farmers, large monocultures, and exports. In a similar vein, critics of GM crops worry that an increasingly large proportion of agricultural biotechnology will be controlled by a small number of global corporations, leading to knowledge that is proprietary and patented. This may lead to a kind of "bio-serfdom" of the world's farmers. Understandably, biotechnology companies want to protect their investments in genetic engineering by patenting the technology and preventing farmers from saving seeds for replanting. To ensure that farmers cannot replant GM seeds, some companies are developing "terminator technologies" that render GM seeds sterile. These restrictions, though, counter the argument that GM crops will feed the world's hungry. In addition, the concept that life can be patented and privatized is a profound moral issue for some people.

The major problem with the Frankenfood Debates is that there is very little solid information about either the benefits or perils of this new technology. GM foods have been developed and marketed so quickly that there has been little time for research on long-term medical or ecological effects of each genetically engineered crops. If this new genetic technology is to deliver benefits as promised, we need time to address the scientific, ethical and political questions that surround it.

References

Chrispeels, M. J. 2000. Biotechnology and the poor. *Plant Physiol.* **124**: 3–6.

Ferber, D. 1999. Risks and benefits: GM crops in the cross hairs. *Science* **286**: 1662–66.

Serageldin, I. 1999. Biotechnology and food security in the 21st century. *Science* **285**: 387–89.

Website

"Living in a GM World", (a compilation of articles on GM crops and biotechnology). *New Scientist* website.
http://www.nsplus.com/gm/gm.jsp

Chapter Summary

1. The history of genetics, which emerged as a fundamental discipline of biology early in the twentieth century, dates back to prehistoric times.
2. Molecular genetics, based on the concept that DNA makes RNA that makes protein, serves as the underpinnings of the more classical work referred to as transmission genetics.
3. The four investigative approaches most often used in the study of genetics are transmission genetic studies, cytogenetic analyses, molecular experimentation, and inquiries into the genetic structure of populations.
4. Genetic research can be either basic or applied. Basic genetic research extends our knowledge of the discipline; the objective of applied genetics research is to solve specific problems affecting the quality of our lives and society in general.
5. Eugenics, the application of knowledge of genetics for the improvement of human existence, has a long and controversial history. Euphenics, genetic intervention designed to ameliorate the impact of genotypes on individuals, represents the modern eugenic approach.
6. Genetic research has had a highly positive impact on many facets of agriculture and medicine.
7. DNA biotechnology is greatly expanding our research capability. It has also had a profound impact on the elucidation of inherited diseases, has made possible the mass production of medically important gene products, and will serve as the foundation on which gene therapy is developed.

Key Terms

actin, 7

applied research, 9

atomic theory, 4

basic research, 9

cell theory, 4

chromosomal theory of inheritance, 5

collagen, 7

complementarity, 6

cytogenetics, 8

DNA (deoxyribonucleic acid), 5

enzyme, 7

epigenesis, 4

eugenics, 9

euphenics, 10

fixity of species, 4

gene therapy, 8

genome, 11

hemoglobin, 7

histone, 7

homunculus, 4

hydrogen bonds, 6

immunoglobulins, 7

insulin, 7

karyotype, 8

keratin, 7

molecular genetic analysis, 8

myosin, 7

natural selection, 4

negative eugenics, 9

nucleotide, 6

pedigree analysis, 8

population genetics, 9

positive eugenics, 9

preformation, 4

protein, 5

ribosome, 6

RNA (ribonucleic acid), 5

spontaneous generation, 4

transcription, 5

transfer RNA (tRNA), 6

translation, 5

transmission genetics, 8

Problems and Discussion Questions

1. Describe and contrast the ideas of Hippocrates and Aristotle relating to the genetic basis of life.
2. Define and contrast epigenesis and preformationism.
3. Which ideas and doctrines that preceded Darwin were central to his thinking?
4. Describe Darwin's and Wallace's theory of natural selection. What information was missing from it? That is, what gap remained in their theory?
5. Describe the "trinity" of molecular genetics and how it serves as the basis of this discipline.
6. Describe the four major investigative approaches used in studying genetics.
7. Contrast basic and applied research.
8. Norman Borlaug received the Nobel Peace Prize for his work in genetics. Why do you think he was awarded this prize?
9. Contrast positive and negative eugenics. Which of these categories includes the approach called euphenics? Define this approach.
10. How has genetic research been applied to agriculture and to medicine?

Selected Readings

Allen, G. E. 1996. Science misapplied: The eugenics age revisited. *Technol. Rev.* 99:23–31.

Anderson, W. F., and Dircumakos, E. G. 1981. Genetic engineering in mammalian cells. *Sci. Am.* (July) 245:106–21.

Borlaug, N. E. 1983. Contributions of conventional plant breeding to food production. *Science* 219:689–93.

Bowler, P. J. 1989. *The Mendelian revolution: The emergence of hereditarian concepts in modern science and society*. London: Athione.

Cocking, E. C., Davey, M. R., Pental, D., and Power, J. B. 1981. Aspects of plant genetic manipulation. *Nature* 293:265–70.

Day, P. R. 1977. Plant genetics: Increasing crop yield. *Science* 197:1334–39.

Dunn, L. C. 1965. *A short history of genetics*. New York: McGraw-Hill.

Gardner, E. J. 1972. *History of biology*, 3rd ed. New York: Macmillan.

Garver, K. L., and Garver, B. 1991. Eugenics: Past, present, and future. *Am. J. Hum. Genet.* 49:1109–18.

Gasser, C. S., and Fraley, R. T. 1989. Genetically engineering plants for crop improvement. *Science* 244:1293–99.

—————. 1992. Transgenic crops. *Sci. Am.* (June) 266:62–69.

Horgan, J. 1993. Eugenics revisited. *Sci. Am.* (June) 268:123–31.

Jones, S. 2000. *"The Origin of Species"—Updated*. New York: Random House.

Keller, E. F. 2000. *The century of the gene*. Cambridge MA: Harvard University Press.

King, R. C., and Stansfield, W. D. 1997. *A dictionary of genetics*, 5th ed. New York: Oxford University Press.

Kolata, G. 1998. *Clone—The road to Dolly, and the path ahead*. New York: William Morrow.

Olby, R. C. 1985. *Origins of Mendelism*, 2nd ed. London: Constable.

Silver, L. 1998. *Remaking Eden: Cloning and beyond in a brave new world*. New York: Avon Books.

Stubbe, H. 1972. *History of genetics: From prehistoric times to the rediscovery of Mendel* (transl. by T. R. W. Waters). Cambridge, MA: MIT Press.

Torrey, J. G. 1985. The development of plant biotechnology. *Am. Sci.* 73:354–63.

Vasil, I. K. 1990. The realities and challenges of plant biotechnology. *Bio/Technology* 8:296–301.

Weinberg, R. A. 1985. The molecules of life. *Sci. Am.* (Oct.) 253:48–57.

Chromosomes in the prometaphase stage of mitosis, derived from a cell in the flower of *Haemanthus*. *(Dr. Andrew S. Bajer, University of Oregon)*

2

Mitosis and Meiosis

CHAPTER CONCEPTS

Genetic continuity between cells and organisms of any sexually reproducing species is maintained by the processes of mitosis and meiosis. The processes are orderly and efficient, serving to produce diploid somatic cells and haploid gametes, respectively. It is during these division stages that the genetic material is condensed into discrete, visible structures called chromosomes.

2.1 **Cell Structure**

2.2 **Homologous Chromosomes, Haploidy, and Diploidy**

2.3 **Mitosis and Cell Division**
Interphase and the Cell Cycle
Prophase
Prometaphase and Metaphase
Anaphase
Telophase

2.4 **Cell Cycle Control**

2.5 **Meiosis and Sexual Reproduction**
An Overview of Meiosis
The First Meiotic Division: Prophase I
Metaphase, Anaphase, and Telophase I
The Second Meiotic Division

2.6 **Spermatogenesis and Oogenesis**

2.7 **The Significance of Meiosis**

2.8 **The Relationship Between Chromatin and Chromosomes**

In every living thing there exists a substance referred to as the genetic material. Except in certain viruses, this material is composed of the nucleic acid DNA. A molecule of DNA is organized into units called genes, the products of which direct the metabolic activities of cells. DNA, with its array of genes, is organized into structures called chromosomes, which serve as vehicles for transmitting genetic information. The manner in which chromosomes are transmitted from one generation of cells to the next, and from organisms to their descendants, must be exceedingly precise. In this chapter we consider exactly how genetic continuity is maintained between cells and organisms.

Two major processes are involved in eukaryotes: **mitosis** and **meiosis**. Although the mechanisms of the two processes are similar in many ways, the outcomes are quite different. Mitosis leads to the production of two cells, each with the same number of chromosomes as the parent cell. Meiosis, on the other hand, reduces the genetic content and the number of chromosomes by precisely half. This reduction is essential if sexual reproduction is to occur without doubling the amount of genetic material at each generation. Strictly speaking, mitosis is that portion of the cell cycle during which the hereditary components are precisely and equally divided into daughter cells. Meiosis is part of a special type of cell division that leads to the production of sex cells: **gametes** or **spores**. This process is an essential step in the transmission of genetic information from an organism to its offspring.

Normally, chromosomes are visible during a cell's life only during mitosis and meiosis. When cells are not undergoing division, the genetic material making up chromosomes unfolds and uncoils into a diffuse network within the nucleus, generally referred to as chromatin. We will briefly review the structure of cells, emphasizing those components that are of particular significance to genetic function. Then we devote the remainder of the chapter to the behavior of chromosomes during cell division.

2.1   Cell Structure

Before describing mitosis and meiosis, a brief review of the structure of cells will be helpful. As we shall see, many components, such as the nucleolus, ribosome, and centriole, are involved directly or indirectly with genetic processes. Other components, the mitochondria and chloroplasts, contain their own unique genetic information. It is also useful to compare the structural differences between the prokaryotic bacterial cell and the eukaryotic cell. Variation in the structure and function of cells is dependent on specific genetic expression by each cell type.

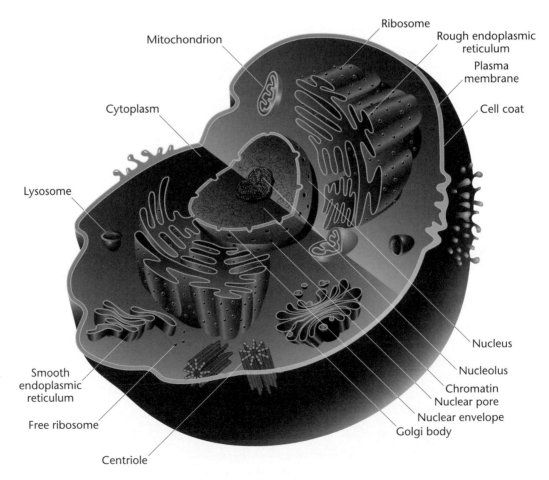

FIGURE 2–1 A generalized animal cell. The cellular components discussed in the text are emphasized here.

Before 1940, our knowledge of cell structure was limited to what we could see with the light microscope. Around 1940, the transmission electron microscope was in its early stages of development, and by 1950, many details of cell ultrastructure had emerged. Under the electron microscope, cells were seen as highly organized, precise structures. A new world of whorling membranes, organelles, microtubules, granules, and filaments was revealed. These discoveries revolutionized thinking in the entire field of biology. We will be concerned with those aspects of cell structure that relate to genetic study. The typical animal cell shown in Figure 2–1 illustrates most of the structures we will discuss.

All cells are surrounded by a **plasma membrane**, an outer covering that defines the cell boundary and delimits the cell from its immediate external environment. This membrane is not passive, but instead actively controls the movement of materials into and out of the cell. In addition to this membrane, plant cells have an outer covering called the **cell wall**. One major component of this rigid structure is a polysaccharide called **cellulose**.

Many, if not most, animal cells have a covering over the plasma membrane, referred to as the **cell coat**. Consisting of glycoproteins and polysaccharides, its chemical composition differs from comparable structures in either plants or bacteria. The cell coat, among other functions, provides biochemical identity at the surface of cells. These forms of cellular identity at the cell surface are under genetic control. For example, various antigenic determinants such as the **AB** and **MN antigens** are found on the surface of red blood cells. In other cells, **histocompatibility antigens**, which elicit an immune response during tissue and organ transplants, are present. A variety of **receptor molecules** are also important components at the surface of cells. These constitute recognition sites that transfer specific chemical signals across the cell membrane into the cell.

The presence of a nucleus and other membranous organelles characterizes eukaryotic cells. The nucleus houses the genetic material, DNA, which is complexed with an array of acidic and basic proteins into thin fibers. During nondivisional phases of the cell cycle, these fibers are uncoiled and dispersed into **chromatin**. As we will soon discuss, during mitosis and meiosis, chromatin fibers coil and condense into structures called **chromosomes**. Also present in the nucleus is the **nucleolus**, an amorphous component where ribosomal RNA is synthesized and where the initial stages of ribosomal assembly occur. The areas of DNA encoding rRNA are collectively referred to as the **nucleolus organizer region** or the **NOR**.

The lack of a nuclear envelope and membraneous organelles is characteristic of **prokaryotes**. In bacteria such as *Escherichia coli*, the genetic material is present as a long, circular DNA molecule that is compacted into an area referred to as the **nucleoid** area. Part of the DNA may be attached to the cell membrane, but in general the nucleoid constitutes a large area throughout the cell. Although the DNA is compacted, it does not undergo the extensive coiling characteristic of the stages of mitosis where, in eukaryotes, chromosomes become visible. Nor is the DNA in these organisms associated as extensively with proteins as is eukaryotic DNA. Figure 2–2, in which two bacteria are forming during cell division, illustrates the nucleoid regions that house the bacterial chromosome.

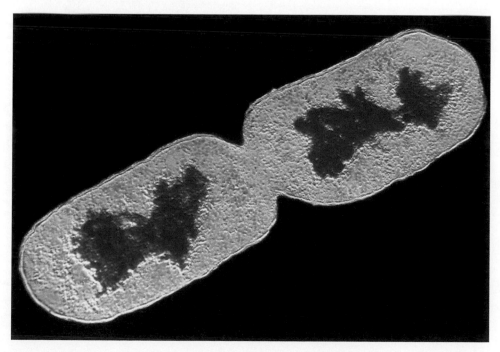

FIGURE 2–2 Color-enhanced electron micrograph of *E. coli* undergoing cell division. Particularly prominent are the two chromosomal areas (shown in red), called nucleoids, that have been partitioned into the daughter cells. (*CNRI/Science Photo Library/Photo Researchers, Inc.*)

Prokaryotic cells do not have a distinct nucleolus, but do contain genes that specify rRNA molecules, the initial step in ribosome assembly.

The remainder of the eukaryotic cell enclosed by the plasma membrane, excluding the nucleus, is composed of **cytoplasm** and all associated **cellular organelles**. Cytoplasm consists of a nonparticulate, colloidal material referred to as the cytosol, which surrounds and encompasses the cellular organelles. Beyond these components, an extensive system of tubules and filaments comprising the cytoskeleton provides a lattice of support structures within the cytoplasm. Consisting primarily of tubulin-derived microtubules and actin-derived microfilaments, this structural framework maintains cell shape, facilitates cell mobility, and anchors the various organelles.

One such organelle, the membranous **endoplasmic reticulum (ER)**, compartmentalizes the cytoplasm, greatly increasing the surface area available for biochemical synthesis. The ER may appear smooth, in which case it serves as the site for synthesizing fatty acids and phospholipids, or it may appear rough because it is studded with ribosomes. Ribosomes serve as sites where genetic information contained in messenger RNA (mRNA) is translated into proteins.

Three other cytoplasmic structures are very important in the eukaryotic cell's activities: mitochondria, chloroplasts, and centrioles. **Mitochondria** are found in both animal and plant cells and are the sites of the oxidative phases of cell respiration. These chemical reactions generate large amounts of adenosine triphosphate (ATP), an energy-rich molecule. **Chloroplasts** are found in plants, algae, and some proto-zoans. This organelle is associated with photosynthesis, the major energy-trapping process on earth. Both mitochondria and chloroplasts contain a type of DNA that is distinct from that found in the nucleus. Furthermore, these organelles can duplicate themselves and transcribe and translate their genetic information. It is interesting to note that the genetic machinery of mitochondria and chloroplasts closely resembles that of prokaryotic cells. This and other observations have led to the proposal that these organelles were once primitive free-living organisms that established a symbiotic relationship with a primitive eukaryotic cell. This theory, which describes the evolutionary origin of these organelles, is called the **endosymbiont hypothesis**.

Animal cells and some plant cells also contain a pair of complex structures called the **centrioles**. These cytoplasmic bodies, located in a specialized region called the centrosome, are associated with the organization of spindle fibers that function in mitosis and meiosis. In some organisms, the centriole is derived from another structure, the basal body, which is associated with the formation of cilia and flagella. Over the years, many reports have suggested that centrioles and basal bodies contain DNA, which could be involved in the replication of these structures. Currently, this is thought not to be the case.

The organization of spindle fibers by the centrioles occurs during the early phases of mitosis and meiosis. Composed of arrays of microtubules, these fibers play an important role in the movement of chromosomes as they separate during cell division. The microtubules consist of polymers of polypeptide subunits of the protein tubulin.

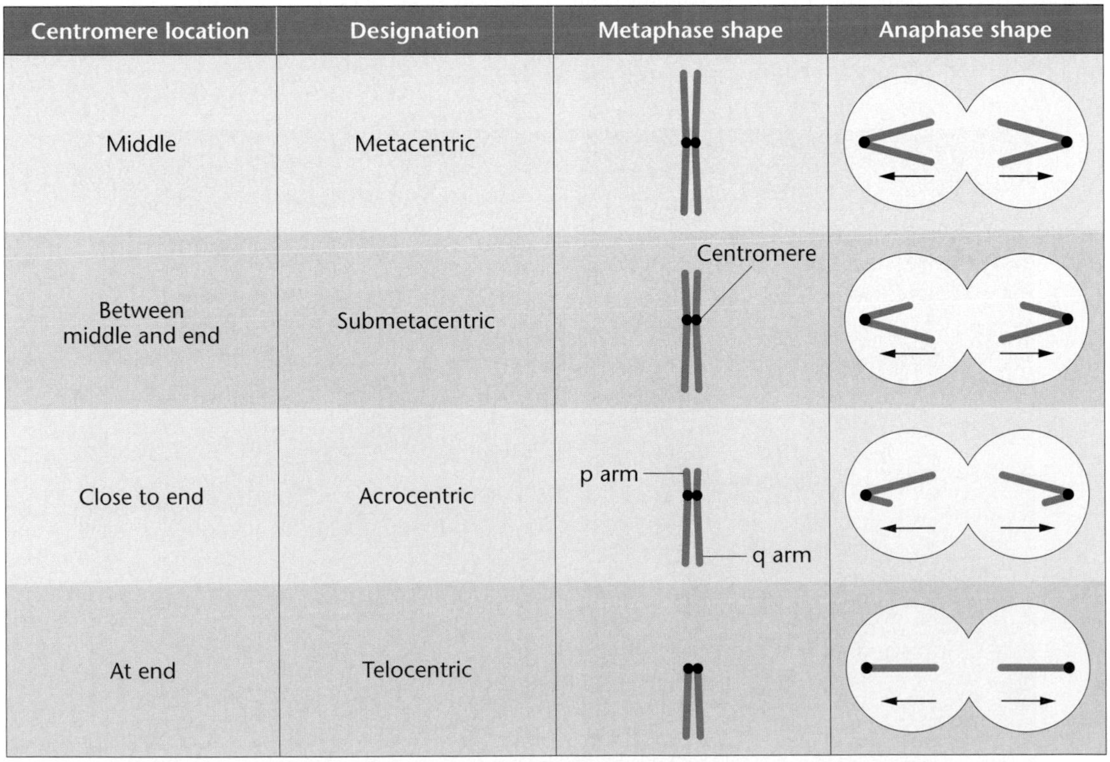

Centromere location	Designation	Metaphase shape	Anaphase shape
Middle	Metacentric		
Between middle and end	Submetacentric	Centromere	
Close to end	Acrocentric	p arm / q arm	
At end	Telocentric		

FIGURE 2–3   Centromere locations and designations of chromosomes based on centromere location. Note that the shape of the chromosome during anaphase is determined by the position of the centromere.

2.2   Homologous Chromosomes, Haploidy, and Diploidy

As we discuss the processes of mitosis and meiosis, it is important that you understand the concept of homologous chromosomes. Such an understanding will also be of critical importance in our future discussions of Mendelian genetics. Chromosomes are most easily visualized during mitosis. When they are examined carefully, they are seen to take on distinctive lengths and shapes. Each contains a condensed or constricted region called the **centromere**, which establishes the general appearance of each chromosome. Figure 2–3 shows chromosomes with centromere placements at different points along their lengths. Extending from either side of the centromere are the arms of the chromosome. Depending on the position of the centromere, different arm ratios are produced. As Figure 2–3 illustrates, chromosomes are classified as **metacentric**, **submetacentric**, **acrocentric**, or **telocentric** on the basis of the centromere location. The shorter arm, by convention, is shown above the centromere and is called the **p arm** (p stands for "petite"). The longer arm is shown below the centromere and is called the **q arm** (q being the next letter in the alphabet).

When studying mitosis, several other observations are of particular relevance. First, all somatic cells derived from members of the same species contain an identical number of chromosomes. In most cases, this represents the **diploid number (2n)**. When the lengths and centromere placements of all such chromosomes are examined, a second general feature is apparent. Nearly all of the chromosomes exist in pairs with regard to these two criteria. The members of each pair are called **homologous chromosomes**. For each chromosome exhibiting a specific length and centromere placement, another exists with identical features. There are exceptions to the rule of chromosomes in pairs. Bacteria and viruses have but one chromosome, and organisms such as yeasts and molds, and certain plants such as bryophytes (mosses) spend the predominant phase of the life cycle in the haploid stage. That is, they contain only one member of each homologous pair of chromosomes during most of their lives.

Figure 2–4 illustrates the physical appearance of different pairs of homologous chromosomes. There, the human mitotic chromosomes have been photographed, cut out of the print, and matched up, creating a **karyotype**. As you can see, humans have a 2n number of 46 and, on close examination, do exhibit a diversity of sizes and centromere placements. Note also that each of the 46 chromosomes is

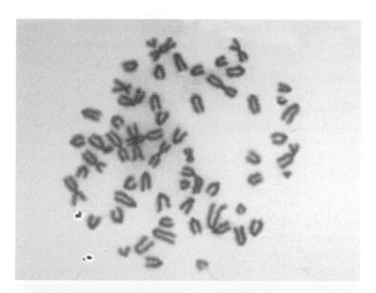

FIGURE 2–4 A metaphase preparation of chromosomes derived from a dividing cell of a human male (top photo), and the karyotype derived from the metaphase preparation. All but the X and Y chromosomes are present in homologous pairs. Each chromosome is clearly a double structure, constituting a pair of sister chromatids joined by a common centromere. *(Top: Cytographics/ Visuals Unlimited; bottom: L. Lisco-D.W. Fawcett/Visuals Unlimited)*

TABLE 2.1 The Haploid Number of Chromosomes for a Variety of Organisms

Common Name	Scientific Name	Haploid Number	Common Name	Scientific Name	Haploid Number
Black bread mold	*Aspergillus nidulans*	8	House mouse	*Mus musculus*	20
Broad bean	*Vicia faba*	6	Human	*Homo sapiens*	23
Cat	*Felis domesticus*	19	Jimson weed	*Datura stramonium*	12
Cattle	*Bos taurus*	30	Mosquito	*Culex pipiens*	3
Chicken	*Gallus domesticus*	39	Mustard plant	*Arabidopsis thaliana*	5
Chimpanzee	*Pan troglodytes*	24	Pink bread mold	*Neurospora crassa*	7
Corn	*Zea mays*	10	Potato	*Solanum tuberosum*	24
Cotton	*Gossypium hirsutum*	26	Rhesus monkey	*Macaca mulatta*	21
Dog	*Canis familiaris*	39	Roundworm	*Caenorhabditis elegans*	6
Evening primrose	*Oenothera biennis*	7	Silkworm	*Bombyx mori*	28
Frog	*Rana pipiens*	13	Slime mold	*Dictyostelium discoidium*	7
Fruit fly	*Drosophila melanogaster*	4	Snapdragon	*Antirrhinum majus*	8
Garden onion	*Allium cepa*	8	Tobacco	*Nicotiana tabacum*	24
Garden pea	*Pisum sativum*	7	Tomato	*Lycopersicon esculentum*	12
Grasshopper	*Melanoplus differentialis*	12	Water fly	*Nymphaea alba*	80
Green alga	*Chlamydomonas reinhardi*	18	Wheat	*Triticum aestivum*	21
Horse	*Equus caballus*	32	Yeast	*Saccharomyces cerevisiae*	16
House fly	*Musca domestica*	6			

clearly a double structure consisting of two parallel **sister chromatids** connected by a common centromere. Had these chromosomes been allowed to continue dividing, the sister chromatids, which are replicas of one another, would have separated into the two new cells as division continued.

The haploid number (*n*) of chromosomes is equal to one-half the diploid number. Collectively, the total set of genes contained in a haploid set of chromosomes constitutes the **genome** of the species. The examples listed in Table 2.1 demonstrate the wide range of *n* values found in plants and animals.

Homologous pairs of chromosomes have important genetic similarities. They contain identical gene sites along their lengths, each called a **locus** (pl. **loci**). Thus, they are identical in their genetic potential. In sexually reproducing organisms, one member of each pair is derived from the maternal parent (through the ovum) and one is derived from the paternal parent (through the sperm). Therefore, each diploid organism contains two copies of each gene as a consequence of **biparental inheritance**. As we shall see in the following chapters on transmission genetics, the members of each pair of genes, while influencing the same characteristic or trait, need not be identical. In a population of members of the same species, many different alternative forms of the same gene, called **alleles**, can exist.

The conceptual issues of haploid number, diploid number, and homologous chromosomes are important in understanding the process of meiosis. During the formation of gametes or spores, meiosis converts the diploid number of chromosomes to the haploid number. As a result, haploid gametes or spores contain precisely one member of each homologous pair of chromosomes—that is, one complete haploid set. Following fusion of two gametes in fertilization, the diploid

number is reestablished; that is, the zygote contains two complete sets of haploid chromosomes. The constancy of genetic material is thus maintained from generation to generation.

There is one important exception to the concept of homologous pairs of chromosomes. In many species, one pair, the **sex-determining chromosomes**, is often not homologous in size, centromere placement, arm ratio, or genetic content. For example, in humans, females carry two homologous X chromosomes, while males carry one Y chromosome in addition to one X chromosome (Figure 2–4). The X and Y chromosomes are not strictly homologous. The Y is considerably smaller and lacks most of the gene sites contained on the X. Nevertheless, in meiosis they behave as homologs so that gametes produced by males receive either one X or one Y chromosome.

2.3 Mitosis and Cell Division

The process of mitosis is critical to all eukaryotic organisms. In some single-celled organisms, such as protozoans and some fungi and algae, mitosis (as a part of cell division) provides the basis for asexual reproduction. Multicellular diploid organisms begin life as single-celled fertilized eggs called **zygotes**. The mitotic activity of the zygote and the subsequent daughter cells is the foundation for the development and growth of the organism. In adult organisms, mitotic activity is prominent in wound healing and other forms of cell replacement in certain tissues. For example, the epidermal skin cells of humans are continuously sloughed off and replaced. Cell division also results in the continuous production of reticulocytes that eventually shed their nuclei and replenish the supply of red blood cells in vertebrates. In abnormal situations, somatic cells may lose control of cell division, forming a tumor.

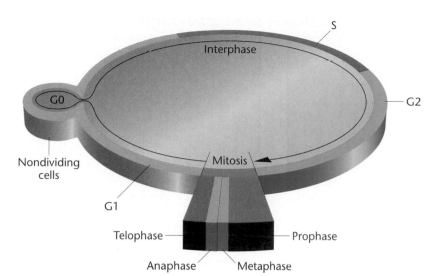

FIGURE 2–5 The intervals comprising an arbitrary cell cycle. Following mitosis (M), cells enter the G1 stage of interphase, initiating a new cycle. Cells may become nondividing (G0) or continue through G1, where they become committed to begin DNA synthesis (S) and complete the cycle (G2 and M). Following mitosis, two daughter cells are produced and the cycle begins anew for both cells.

The genetic material is partitioned into daughter cells during nuclear division or **karyokinesis**. This process is quite complex and requires great precision. The chromosomes must first be exactly replicated and then accurately partitioned. The end result is the production of two daughter nuclei, each with a chromosome composition identical to that of the parent cell.

Karyokinesis is followed by cytoplasmic division, or **cytokinesis**. The less complex division of the cytoplasm requires a mechanism that partitions the volume into two parts, then encloses both new cells in a distinct plasma membrane. Cytoplasmic organelles either replicate themselves, arise from existing membrane structures, or are synthesized *de novo* (anew) in each cell. The subsequent proliferation of these structures is a reasonable and adequate mechanism for reconstituting the cytoplasm in daughter cells.

Usually following cell division, the initial size of each new daughter cell is approximately one-half the size of the parent cell. The nucleus of each new cell is not appreciably smaller than the nucleus of the original cell, however. Quantitative measurements of DNA confirm that there is an amount of genetic material in the daughter nuclei equivalent to that in the parent cell.

Interphase and the Cell Cycle

Many cells undergo a continuous alternation between division and nondivision. The events that occur from the completion of one division until the beginning of the next division constitute the **cell cycle** (Figure 2–5). We will consider the initial stage of the cycle, called **interphase**, as the interval between divisions. It was once thought that the biochemical activity during interphase was devoted solely to the cell's growth and its normal function. However, we now know that another biochemical step critical to the ensuing mitosis occurs during interphase: *the replication of the DNA of each chromosome*. Occurring before the cell enters mitosis, this period during which DNA is synthesized is called the **S phase**. The initiation and completion of synthesis can be detected by monitoring the incorporation of radioactive precursors into DNA.

Investigations of this nature show two periods during interphase when no DNA synthesis occurs, one before and one after S phase. These are designated **G1 (gap I)** and **G2 (gap II)**, respectively. During both of these intervals, as well as during S, intensive metabolic activity, cell growth, and cell differentiation occur. By the end of G2, the volume of the cell has roughly doubled, DNA has been replicated, and mitosis (M) is initiated. Following mitosis, continuously dividing cells then repeat this cycle (G1, S, G2, M) over and over, as shown in Figure 2–5.

Much is known about the cell cycle based on *in vitro* (in glass) studies. When grown in culture, many cell types in different organisms traverse the complete cycle in about 16 hours. The actual process of mitosis occupies only a small part of the overall cycle, often less than an hour. The length of the S and G2 phase of interphase are fairly consistent among different cell types. Most variation is seen in the length of time spent in the G1 stage. Figure 2–6 shows the relative length of these intervals in a typical cell.

G1 is of great interest in the study of cell proliferation and its control. At a point late in G1, all cells follow one of two paths. They either withdraw from the cycle, become quiescent, and enter the **G0 stage** (see Figure 2–5), or they become committed to initiating DNA synthesis and completing

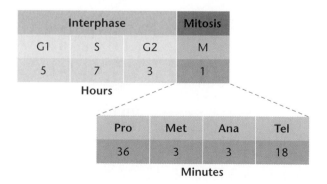

Interphase			Mitosis
G1	S	G2	M
5	7	3	1

Hours

Pro	Met	Ana	Tel
36	3	3	18

Minutes

FIGURE 2–6 The time spent in each phase of one complete cell cycle of a human cell in culture. Times vary according to cell types and conditions.

the cycle. Cells that enter G0 remain viable and metabolically active but are nonproliferative. Cancer cells apparently avoid entering G0 or pass through it very quickly. Other cells enter G0 and never reenter the cell cycle. Still others remain in G0, but they can be stimulated to return to G1, and thereby reenter the cycle.

Cytologically, interphase is characterized by the absence of visible chromosomes. Instead, the nucleus is filled with chromatin fibers that have formed as the chromosomes have uncoiled and dispersed following the previous mitosis. This is diagrammed in Figure 2–7(a). Once G1, S, and G2 are completed, mitosis is initiated. Mitosis is a dynamic period of vigorous and continual activity. For discussion purposes, the entire process is subdivided into discrete stages, and specific events are assigned to each one. These stages, in order of occurrence, are **prophase**, **prometaphase**, **metaphase**, **anaphase**, and **telophase**. Like interphase, these stages are also diagrammed in Figure 2–7. A photograph of each stage is shown in Figure 2–8.

Prophase

Often, over half of mitosis is spent in prophase, a stage characterized by several significant activities. One of the early events in prophase of all animal cells involves the migration of two pairs of centrioles to opposite ends of the cell. These

structures are found just outside the nuclear envelope in an area of differentiated cytoplasm called the **centrosome**. It is thought that each pair of centrioles consists of one mature unit and a smaller, newly formed centriole.

The direction of migration of the centrioles is such that two poles are established at opposite ends of the cell. Following their migration, the centrioles are responsible for organizing cytoplasmic microtubules into a series of **spindle fibers** that are formed and run between these poles. This creates an axis along which chromosomal separation occurs.

Interestingly, cells of most plants (there are a few exceptions), fungi, and certain algae seem to lack centrioles. Spindle fibers are nevertheless apparent during mitosis. Therefore, centrioles are not universally responsible for the organization of spindle fibers.

As the centrioles migrate, the nuclear envelope begins to break down and gradually disappears. In a similar fashion, the nucleolus disintegrates within the nucleus. While these events are taking place, the diffuse chromatin fibers begin to condense, continuing until distinct threadlike structures, or chromosomes, become visible. It becomes apparent near the end of prophase that each chromosome is actually a double structure split longitudinally except at a single point of constriction, called the **centromere** (see Figure 2–4). The two parts of each chromosome are called **chromatids**. Because the DNA contained in each pair of chromatids represents the du-

FIGURE 2–7 Mitosis in an animal cell with a diploid number of 4. The events occurring in each stage are described in the text. Of the two homologous pairs of chromosomes, one contains longer, metacentric members and the other shorter, submetacentric members. The maternal chromosome and the paternal chromosome of each pair are shown in different colors. In (g), the telophase stage in a plant cell illustrates the formation of the cell plate and lack of centrioles.

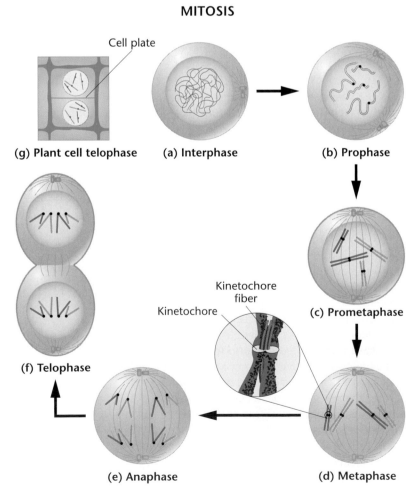

MITOSIS

Cell plate

(g) Plant cell telophase   (a) Interphase   (b) Prophase

Kinetochore fiber

Kinetochore

(c) Prometaphase

(f) Telophase

(e) Anaphase   (d) Metaphase

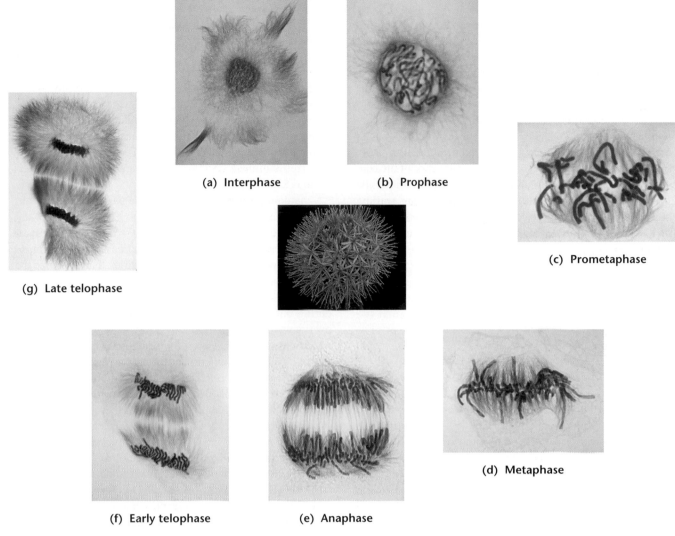

(a) Interphase **(b)** Prophase **(c)** Prometaphase

(g) Late telophase

(d) Metaphase

(f) Early telophase **(e)** Anaphase

FIGURE 2–8 Light micrographs illustrating the stages of mitosis depicted in Figure 2–7. These stages are derived from the flower of *Haemanthus*, shown in the center. *(Dr. Andrew S. Bajer, University of Oregon)*

plication of a single chromosome, these chromatids are genetically identical. Therefore, they are called **sister chromatids**. In humans, with a diploid number of 46, a cytological preparation of late prophase will reveal 46 chromosomes randomly distributed in the area formerly occupied by the nucleus.

Prometaphase and Metaphase

The distinguishing event of the ensuing stages is the migration of each chromosome, led by the centromeric region, to the equatorial plane. In some descriptions, the term **prometaphase** refers to the period of chromosome movement, and **metaphase** is applied strictly to the chromosome configuration following migration, as depicted in Figures 2–7(c) and 2–8(c). The equatorial plane, also referred to as the **metaphase plate**, is the midline region of the cell, a plane that lies perpendicular to the axis established by the spindle fibers.

Migration is made possible by the binding of spindle fibers to a structure associated with the centromere of each chromosome called the **kinetechore**. This structure, consisting of multilayered plates of proteins, forms on opposite sides of each centromere, intimately associating with the two sister chromatids of each chromosome. Once attached to microtubules making up the spindle fibers, the sister chromatids are now ready to be pulled to opposite poles during the ensuing anaphase stage.

At the completion of metaphase, each centromere is aligned at the plate with the chromosome arms extending outward in a random array. This configuration is shown in Figures 2–7(d) and 2–8(d).

Anaphase

Events critical to chromosome distribution during mitosis occur during shortest stage of mitosis, **anaphase**. During this phase, sister chromatids of each chromosome *disjoin* (separate) from each other and migrate to opposite ends of the cell. For complete disjunction to occur, each centromeric region must be split in two. This event signals the initiation of anaphase. Once it occurs, each chromatid is referred to as a **daughter chromosome**.

Movement of daughter chromosomes to the opposite poles of the cell is dependent upon the centromere–spindle fiber attachment. Recent investigations reveal that chromosome migration results from the activity of a series of specific proteins, generally called motor proteins. These proteins use the energy generated by the hydrolysis of ATP, and their activity is said to constitute **molecular motors** in the cell. These motors act at several positions within the dividing cell, but all are involved in the activity of microtubules and ultimately serve to propel the chromosomes to opposite ends of the cell. The centromeres of each chromosome *appear* to lead the way during migration, with the chromosome arms trailing behind. The location of the centromere determines the shape of the chromosome during separation, as you can see in Figure 2–3.

The steps occurring during anaphase are critical in providing each subsequent daughter cell with an identical set of chromosomes. In human cells, there would now be 46 chromosomes at each pole, one from each original sister pair. Figures 2–7(e) and 2–8(e) show anaphase prior to its completion.

Telophase

Telophase is the final stage of mitosis and is depicted in Figures 2–7 (f) and (g) and 2–8(f) and (g). At its beginning, there are two complete sets of chromosomes, one at each pole. The most significant event is **cytokinesis**, the division or partitioning of the cytoplasm. Cytokinesis is essential if two new cells are to be produced from one. The mechanism differs greatly in plant and animal cells. In plant cells, a **cell plate** is synthesized and laid down across the region of the metaphase plate. Animal cells, however, undergo a constriction of the cytoplasm in much the same way a loop of string might be tightened around the middle of a balloon. The end result is the same: Two distinct cells are formed.

It is not surprising that the process of cytokinesis varies among cells of different organisms. Plant cells, which are more regularly shaped and structurally rigid, require a mechanism for depositing new cell wall material around the plasma membrane. The cell plate, laid down during telophase, becomes the **middle lamella**. Subsequently, the primary and secondary layers of the cell wall are deposited between the cell membrane and middle lamella on both sides of the boundary between the two daughter cells. In animals, complete constriction of the cell membrane produces the **cell furrow** characteristic of newly divided cells.

Other events necessary for the transition from mitosis to interphase are initiated during late telophase. They represent a general reversal of those that occurred during prophase. In each new cell, the chromosomes begin to uncoil and become diffuse chromatin once again, while the nuclear envelope reforms around them. The nucleolus gradually re-forms and becomes visible in the nucleus during early interphase. The spindle fibers also disappear. At the completion of telophase, the cell enters interphase.

2.4 Cell Cycle Control

The cell cycle, including mitosis, is fundamentally the same in all eukaryotic organisms. The similarity of the events leading to cell duplication in many diverse organisms suggests that the cell cycle is governed by a genetic program that has been conserved throughout evolution and is therefore genetically regulated. Elucidation of this genetic program provides information basic to our understanding of the nature of living organisms. Furthermore, because disruption of this regulation can lead to the uncontrolled cell division that characterizes malignancy, interest in how genes regulate the cell cycle is keen.

A mammoth research effort over the past decade has paid high dividends, and we now have knowledge of many of the genes that control the cell cycle. As with other studies of genetic input into essential biological processes, investigation is focusing on the discovery of mutations that interrupt the cell cycle and the subsequent study of the effects of these mutations. As we will return to this subject in Chapter 21 during our consideration of cancer, what follows is a brief overview.

Many mutations that exert their effect at various stages of the cell cycle are now known. First discovered in yeast but now evident in all organisms, including humans, such mutations were originally designated as *cdc* **mutations** (**cell division cycle mutations**). The study of these mutations has established that during the cell cycle, at least three major **checkpoints** exist, where the cell is monitored or "checked" before it can proceed to the next stage of the cycle.

The products of many of these genes are enzymes called *cdc* **kinases** that can add phosphates to other proteins. They serve as "master control" molecules that work in conjunction with proteins called **cyclins**. These kinases phosphorylate cyclins and influence their activity at the cell cycle checkpoints. These activities thus regulate the cell cycle. When a *cdc* kinase works in conjunction with a cyclin, it is called a **Cdk protein**, for **Cyclin-dependent kinase protein**.

Figure 2.9 identifies three "checkpoints" within the cell cycle. The first is the **G1/S checkpoint**, which monitors the cell size achieved following the previous mitosis, and whether the DNA has been damaged. If the cell has not achieved an adequate size, or if the DNA has been damaged, further progress through the cycle is arrested until these conditions are "corrected," so to speak. If both conditions are initially "normal," then the checkpoint is traversed and the cell proceeds to the S phase of the cycle.

The second important checkpoint is the **G2/M checkpoint**, where physiological conditions in the cell are monitored prior to entering mitosis. If DNA replication or repair to any DNA damage has not been completed, the cell cycle is arrested until these processes are completed. The final checkpoint occurs during mitosis and is called the **M checkpoint**. Here, both the successful formation of the spindle fiber system and the attachment of spindle fibers to the kinetechores associated with the centromeres are monitored. If spindle fibers are not properly formed or attachment is inadequate, mitosis is arrested.

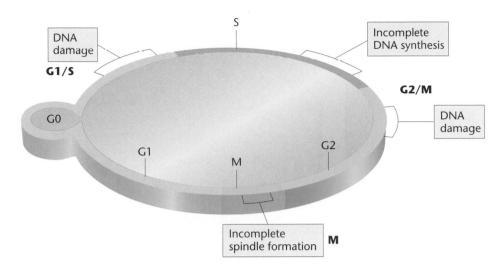

The importance of cell cycle control and these checkpoints can be demonstrated by considering what happens when this regulatory system is impaired. If, for example, a cell that has incurred damage to its DNA is allowed to proceed through the cell cycle, the damage may lead to uncontrolled cell division—precisely the definition of a cancerous cell. As we saw above, such a damaged cell would normally be arrested at either the G1/S or the G2/M checkpoint.

An interesting related finding involves the protein product of the *p53* gene in humans and its involvement during scrutiny at the G1/S checkpoint. This protein functions during the regulation of **apoptosis**, the genetic process whereby programmed cell death occurs. When the normal *p53* gene product is present, a proliferative cell that has incurred severe damage to its DNA will be targeted for programmed cell death at the G1/S checkpoint and thus effectively removed from the cell population. This surveillance is dependent on the product of the *p53* gene. However, if the gene has mutated, resulting in abnormal function of its gene product, the damaged cell may proceed through the checkpoint and continue to proliferate in an uncontrolled manner.

In fact, a high percentage of human cancers has been found to contain mutations in the *p53* gene. These include a wide variety of cancers, including colon, breast, lung, and bladder malignancies. In the language of cancer genetics, *p53* is referred to as a **tumor-suppressor gene**. The role of this and other genes involved in the cell cycle has sparked great interest in cancer research, as discussed in Chapter 21.

2.5 Meiosis and Sexual Reproduction

The process of meiosis, unlike mitosis, reduces the amount of genetic material by one-half. Whereas in diploids, mitosis produces daughter cells with a full diploid complement, meiosis produces gametes or spores with only one haploid set of chromosomes. During sexual reproduction, gametes then combine in fertilization to reconstitute the diploid complement found in parental cells. Figure 2–10 compares the two processes by following two pairs of homologous chromosomes.

Meiosis must be highly specific since, by definition, haploid gametes or spores contain precisely one member of each homologous pair of chromosomes. Successfully completed, meiosis ensures genetic continuity from generation to generation.

The process of sexual reproduction also ensures genetic variety among members of a species. As you study meiosis, you will see that this process results in gametes with many unique combinations of maternally and paternally derived chromosomes among the haploid complement. With such a tremendous genetic variation among the gametes, a large number of chromosome combinations are possible at fertilization. Furthermore, we shall see that the meiotic event referred to as **crossing over** results in genetic exchange between members of each homologous pair of chromosomes. This creates intact chromosomes that are mosaics of the maternal and paternal homologs from which they are derived, further enhancing the potential genetic variation in gametes and the offspring derived from them. Sexual reproduction therefore reshuffles the genetic material, producing offspring that often differ greatly from either parent. This process constitutes the major form of genetic recombination within species.

An Overview of Meiosis

In the preceding discussion, we established what might be considered the goals of meiosis. Before we consider the phases of this process systematically, we will briefly examine how diploid cells give rise to haploid gametes or spores. You should refer to the meiotic portion of Figure 2–10 during the following discussion.

You have seen that in mitosis each paternally and maternally derived member of any given homologous pair of chromosomes behaves autonomously during division. By contrast, early in meiosis, homologous chromosomes form pairs; that is, they **synapse**. Each synapsed structure is

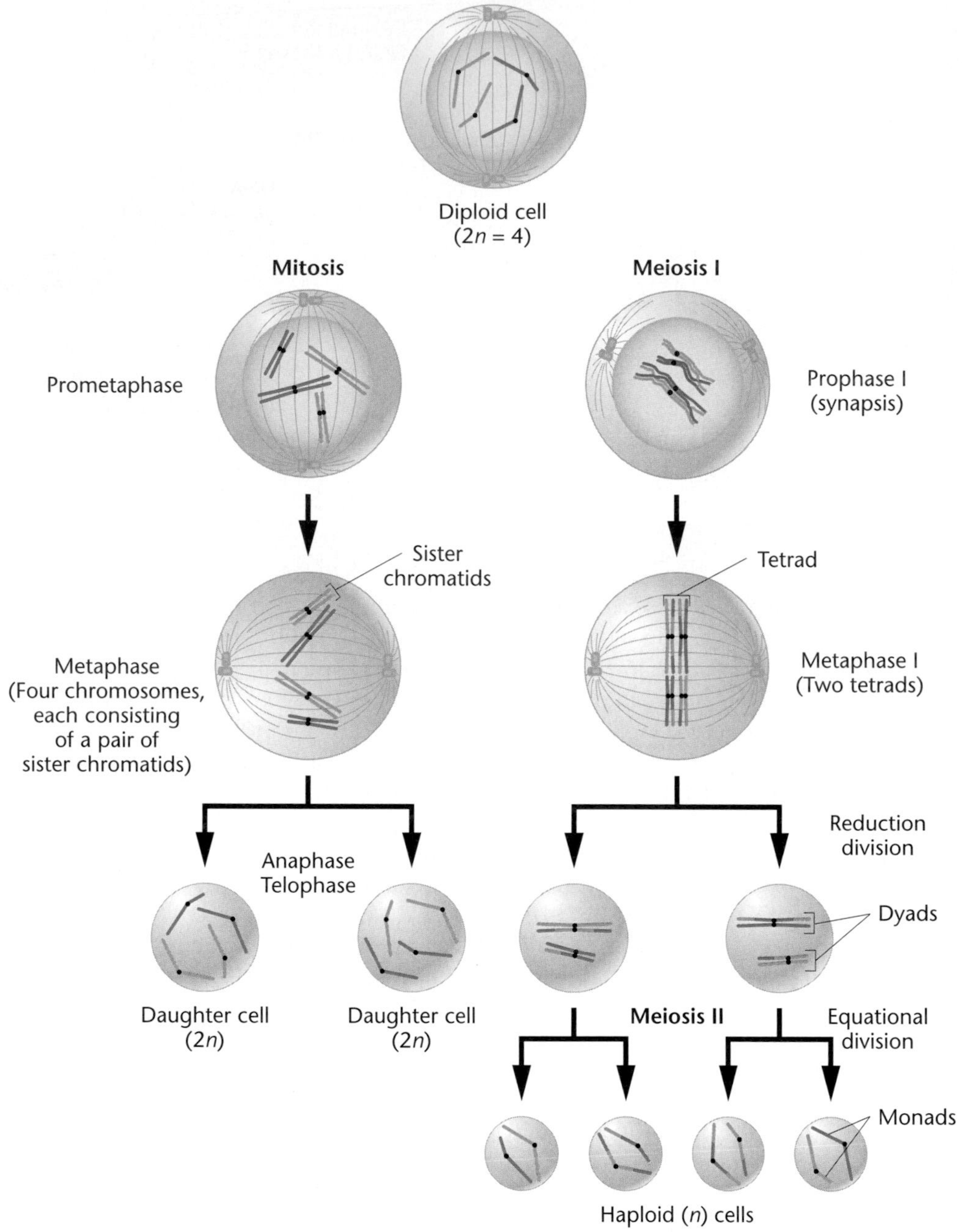

FIGURE 2–10 Overview of the major events and outcomes of mitosis and meiosis. As in Figure 2–7, two pairs of homologous chromosomes are followed.

initially called a **bivalent**, which eventually gives rise to a unit, the **tetrad**, consisting of four chromatids. The presence of four chromatids demonstrates that both homologs (making up the bivalent) have, in fact, duplicated. Therefore, in order to achieve haploidy, two divisions are necessary. In meiosis I, described as a **reductional division** (because the number of centromeres, each representing one chromosome, is *reduced* by one-half following this division), components of each tetrad—representing the two homologs—separate, yielding two **dyads**. Each dyad is composed of two sister chromatids joined at a common centromere. During meiosis II, described as **equational** (because the number of centromeres remains *equal* following this division), each dyad

splits into two **monads** of one chromosome each. Thus, the two divisions potentially produce four haploid cells.

The First Meiotic Division: Prophase I

We turn now to a detailed account of meiosis. As in mitosis, meiosis is a continuous process. We name the parts of each stage of division only to facilitate discussion. From a genetic standpoint, three events characterize the initial stage, prophase I (Figure 2–11). First, as in mitosis, chromatin present in interphase thickens and coils into visible chromosomes. Second, unlike mitosis, members of each homologous pair of chromosomes undergo synapsis. Third, crossing over, an ex-

Meiotic prophase I

Leptonema

Zygonema

Pachynema

Diplonema

Diakinesis

FIGURE 2–11 The substages of meiotic prophase I for the chromosomes depicted in Figure 2–7.

change process, occurs between synapsed homologs. Because of the complexity of these genetic events, this stage of meiosis has been further divided into five substages: leptonema,* zygonema,* pachynema,* diplonema,* and diakinesis. As we discuss these substages, be aware that, even though it is not immediately apparent in the earliest phases

---

*These are the noun forms of these substages. The adjective forms (leptotene, zygotene, pachytene, and diplotene) are also used in the text.

of meiosis, the DNA of chromosomes has been replicated during the prior interphase.

Leptonema During the **leptotene stage**, the interphase chromatin material begins to condense, and the chromosomes, although still extended, become visible. Along each chromosome are **chromomeres**, localized condensations that resemble beads on a string. Recent evidence suggests that a process called **homology search**, which precedes and is essential to the initial pairing of homologs, begins during leptonema.

Zygonema The chromosomes continue to shorten and thicken during the **zygotene stage**. During the process of homology search, homologous chromosomes undergo initial alignment with one another. This so-called rough pairing is complete by the end of zygonema. In yeast, homologs are separated by about 300 nm, and near the end of zygonema, structures referred to as lateral elements are visible between paired homologs. As meiosis proceeds, the overall length of the lateral elements increases and a more extensive ultrastructural component, the **synaptonemal complex**, begins to form between the homologs. We discuss this meiotic component later in the chapter.

At the completion of zygonema, the paired homologs are referred to as bivalents. Although both members of each bivalent have already replicated their DNA, it is not yet visually apparent that each member is a double structure. The number of bivalents in each species is equal to the haploid (n) number.

Pachynema In the transition from the zygotene to the **pachytene stage**, coiling and shortening of chromosomes continues, and further development of the synaptonemal complex occurs between the two members of each bivalent. This leads to a more intimate pairing referred to as **synapsis**. Compared to the rough-pairing characteristic of yeast zygonema, homologs are now separated by only 100 nm.

During pachynema, each homolog is first evident as a double structure, providing visual evidence of the earlier replication of the DNA of each chromosome. Thus, each bivalent contains four member chromatids. As in mitosis, replicates are called sister chromatids, while chromatids from maternal and paternal members of a homologous pair are called nonsister chromatids. The four-membered structure is also referred to as a tetrad, and each tetrad contains two pairs of sister chromatids.

Diplonema During the ensuing **diplotene stage**, it is even more apparent that each tetrad consists of two pairs of sister chromatids. Within each tetrad, each pair of sister chromatids begins to separate. However, one or more areas remain in contact where chromatids are intertwined. Each such area, called a **chiasma** (pl. **chiasmata**), is thought to represent a point where nonsister chromatids have undergone genetic exchange through the process referred to above as crossing over. Although the physical exchange between chromosome areas occurred during the previous pachytene

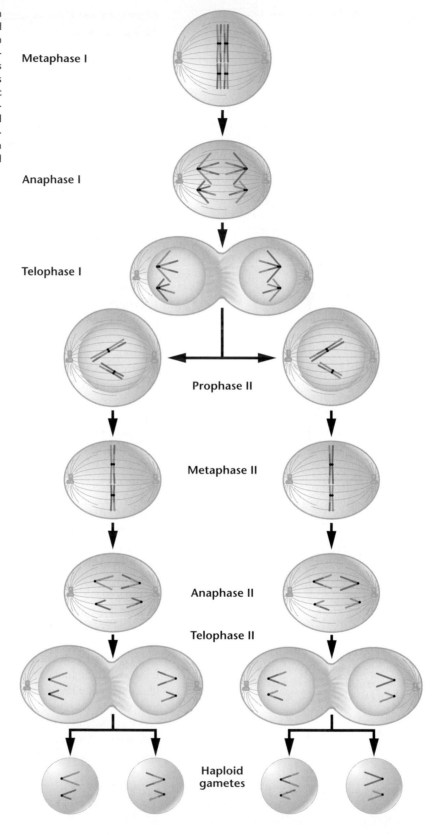

FIGURE 2–12 The major events in meiosis in an animal with a diploid number of 4, beginning with metaphase I. Note that the combination of chromosomes in the cells produced following telophase II is dependent on the random alignment of each tetrad and dyad on the equatorial plate during metaphase I and metaphase II. Several other combinations, which are not shown, can also be formed. The events depicted here are described in the text.

Metaphase I

Anaphase I

Telophase I

Prophase II

Metaphase II

Anaphase II

Telophase II

Haploid gametes

stage, the result of crossing over is visible only when the duplicated chromosomes begin to separate. Crossing over is an important source of genetic variability. As indicated earlier, new combinations of genetic material are formed during this process.

Diakinesis   The final stage of prophase I is **diakinesis**. The chromosomes pull farther apart, but nonsister chromatids remain loosely associated via the chiasmata. As separation proceeds, the chiasmata move toward the ends of the tetrad. This process, called **terminalization**, begins in late diplonema, and is completed during diakinesis. During this substage period of prophase I, the nucleolus and nuclear envelope break down, and the two centromeres of each tetrad attach to the recently formed spindle fibers. By the completion of prophase I, the centromeres of each tetrad structure are present on the equatorial plate of the cell.

Metaphase, Anaphase, and Telophase I

The remainder of the meiotic process is depicted in Figure 2–12. Following the first meiotic prophase, steps similar to those of mitosis occur. In the metaphase of the first division (**metaphase I**), the chromosomes have maximally shortened and thickened. The terminal chiasmata of each tetrad are visible and appear to be the only factor holding the nonsister chromatids together. Each tetrad interacts with spindle fibers, facilitating movement to the metaphase plate.

The alignment of each tetrad prior to this first anaphase is random. One-half of each tetrad is pulled to one or the other pole at random, and the other half then moves to the opposite pole. This random separation of dyads is the basis for the Mendelian postulate of **independent assortment**, which we discuss in Chapter 3. You may wish to revisit this discussion when you study this principle.

During the stages of meiosis I, a single centromere holds each pair of sister chromatids together. It does *not* divide. At **anaphase I**, one-half of each tetrad (one pair of sister chromatids—called a **dyad**) is pulled toward each pole of the dividing cell. This separation process is the physical basis of what we refer to as **disjunction**, the separation of chromosomes from one another. Occasionally, errors in meiosis occur and separation is not achieved, as we will see later in this chapter. The term **nondisjunction** describes such an error. At the completion of the normal anaphase I, a series of dyads equal to the haploid number is present at each pole.

If crossing over had not occurred in the first meiotic prophase, each dyad at each pole would consist solely of either paternal or maternal chromatids. However, the exchanges produced by crossing over create mosaic chromatids of paternal and maternal origin.

In many organisms, **telophase I** reveals a nuclear membrane forming around the dyads. Next, the nucleus enters into a short interphase period. In other cases, the cells go directly from the first anaphase into the second meiotic division. If interphase occurs, the chromosomes do not replicate since they already consist of two chromatids. In general, meiotic telophase is much shorter than the corresponding stage in mitosis.

The Second Meiotic Division

A second division, referred to as **meiosis II**, is essential if each gamete or spore is to receive only one chromatid from each original tetrad. The stages characterizing meiosis II are shown in the bottom half of Figure 2–12. During **prophase II**, each dyad is composed of one pair of sister chromatids attached by a common centromere. During **metaphase II**, the centromeres are positioned on the equatorial plate. When

they divide, **anaphase II** is initiated, and the sister chromatids of each dyad are pulled to opposite poles. Because the number of dyads is equal to the haploid number, **telophase II** reveals one member of each pair of homologous chromosomes present at each pole. Each chromosome is referred to as a **monad**. Following cytokinesis in telophase II, four haploid gametes may result from a single meiotic event. At the conclusion of meiosis, not only has the haploid state been achieved, but if crossing over has occurred, each monad is a combination of maternal and paternal genetic information. As a result, the offspring produced by any gamete will receive a mixture of genetic information originally present in his or her grandparents.

2.6 Spermatogenesis and Oogenesis

Although events that occur during the meiotic divisions are similar in all cells that participate in gametogenesis in most animal species, there are certain differences between the production of a male gamete (spermatogenesis) and a female gamete (oogenesis). Figure 2–13 summarizes these processes.

Spermatogenesis takes place in the testes, the male reproductive organs. The process begins with the expanded growth of an undifferentiated diploid germ cell called a **spermatogonium**. This cell enlarges to become a **primary spermatocyte**, which undergoes the first meiotic division. The products of this division, called **secondary spermatocytes**, contain a haploid number of dyads. The secondary spermatocytes then undergo the second meiotic division, and each of these cells produces two haploid **spermatids**. Spermatids go through a series of developmental changes, **spermiogenesis**, and become highly specialized, motile **spermatozoa** or **sperm**. All sperm cells produced during spermatogenesis receive equal amounts of genetic material and cytoplasm.

Spermatogenesis may be continuous or may occur periodically in mature male animals, with its onset determined by the nature of the species' reproductive cycle. Animals that reproduce year-round produce sperm continuously, whereas those whose breeding period is confined to a particular season produce sperm only during that time.

In animal **oogenesis**, the formation of **ova** (sing. **ovum**), or eggs, occurs in the ovaries, the female reproductive organs. The daughter cells resulting from the two meiotic divisions receive equal amounts of genetic material, but they do *not* receive equal amounts of cytoplasm. Instead, during each division, almost all the cytoplasm of the **primary oocyte**, itself derived from the **oogonium**, is concentrated in one of the two daughter cells. The concentration of cytoplasm is necessary because a major function of the mature ovum is to nourish the developing embryo following fertilization.

During the first meiotic anaphase in oogenesis, the tetrads of the primary oocyte separate, and the dyads move toward opposite poles. During the first telophase, the dyads present at one pole are pinched off with very little surrounding cytoplasm to form the **first polar body**. The other daughter cell produced by this first meiotic division contains most of the cytoplasm and is called the **secondary oocyte**. The first

FIGURE 2–13 Spermatogenesis and oogenesis in animal cells.

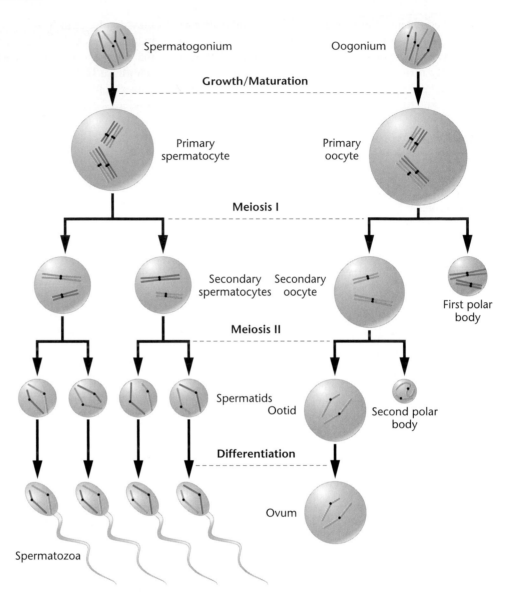

polar body may or may not divide again to produce two small haploid cells. The mature ovum will be produced from the secondary oocyte during the second meiotic division. During this division, the cytoplasm of the secondary oocyte again divides unequally, producing an **ootid** and a **second polar body**. The ootid then differentiates into the mature ovum.

Unlike the divisions of spermatogenesis, the two meiotic divisions of oogenesis may not be continuous. In some animal species, the two divisions may directly follow each other. In others, including humans, the first division of all oocytes begins in the embryonic ovary, but arrests in prophase I. Many years later, meiosis resumes in each oocyte just prior to its ovulation. The second division is completed only after fertilization.

2.7 The Significance of Meiosis

The process of meiosis is critical to the successful sexual reproduction of all diploid organisms. It is the mechanism by which the diploid amount of genetic information is reduced to the haploid amount. In animals, meiosis leads to the formation of gametes, while in plants haploid spores are produced, which in turn lead to the formation of haploid gametes.

Each diploid organism contains its genetic information in the form of homologous pairs of chromosomes. Each pair consists of one member derived from the maternal parent and one from the paternal parent. Following meiosis, haploid cells potentially contain either the paternal or maternal representative of each homologous pair of chromosomes. However, the process of crossing over, which occurs in the first meiotic prophase, reshuffles the genetic information. Crossing over occurs between the maternal and paternal members of each homologous pair, which then assort independently into gametes. This results in the great amounts of genetic variation in gametes.

It is important to touch briefly on the significant role that meiosis plays in the life cycles of fungi and plants. In many fungi, the predominant stage of the life cycle consists of hap-

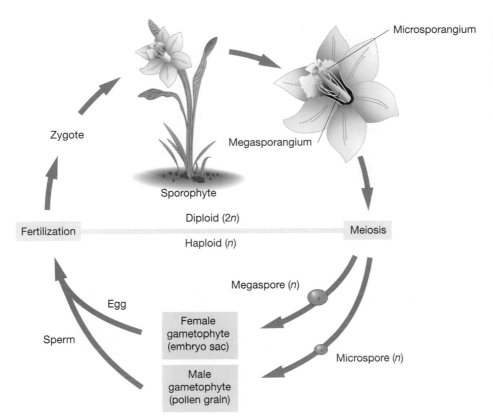

FIGURE 2–14 Alternation of generations between the diploid sporophyte (2n) and the haploid gametophyte (n) in a multicellular plant. The processes of meiosis and fertilization bridge the two phases of the life cycle. This is an angiosperm, where the sporophyte stage is the predominant phase.

loid vegetative cells. They arise through meiosis and proliferate by mitotic cell division. In multicellular plants, the life cycle alternates between the diploid **sporophyte stage** and the haploid **gametophyte stage** (Figure 2–14). While one or the other predominates in different plant groups during this "alternation of generations," the processes of meiosis and fertilization constitute the "bridge" between the sporophyte and gametophyte generations. Therefore, meiosis is an essential component of the life cycle of plants.

2.8   The Relationship Between Chromatin and Chromosomes

Thus far in this chapter, we have focused on mitotic and meiotic chromosomes, emphasizing their behavior during cell division and gamete formation. An interesting question is why chromosomes are invisible during interphase but present during the various stages of mitosis and meiosis. Studies using electron microscopy clearly show why chromosomes are visible only during division stages.

During interphase, only dispersed chromatin fibers are present in the nucleus [Figure 2.15(a)]. Once mitosis begins, however, the fibers coil and fold, condensing into typical mitotic chromosomes [Figure 2.15(b)]. If the fibers comprising the mitotic chromosome are loosened, areas of greatest spreading reveal individual fibers similar to those seen in interphase chromatin [Figure 2.15(c)]. Very few fiber

ends seem to be present, and in some cases, none can be seen. Instead, individual fibers always seem to loop back into the interior. Such fibers are obviously twisted and coiled around one another, forming the regular pattern of the mitotic chromosome. Starting in late telophase of mitosis and continuing during G1 of interphase, chromosomes then unwind to form the long fibers characteristic of chromatin, which consist of DNA and associated proteins, particularly proteins called histones. It is in this physical arrangement that DNA can most efficiently function during transcription and replication.

Electron microscopic observations of mitotic chromosomes in varying states of coiling led Ernest DuPraw to postulate the **folded-fiber model**, shown in Figure 2.15(d). During metaphase, each chromosome consists of two sister chromatids joined at the centromeric region. Each arm of the chromatid appears to consist of a single fiber wound much like a skein of yarn. The fiber is composed of tightly coiled double-stranded DNA and protein. An orderly coiling-twisting-condensing process appears to be involved in the transition of the interphase chromatin to the more condensed, mitotic chromosomes. It is estimated that during the transition from interphase to prophase, a 5000-fold contraction occurs in the length of DNA within the chromatin fiber!

After we have provided a more thorough description of DNA structure, we will return to this topic in Chapter 17 and explore the molecular nature of the chromatin fiber.

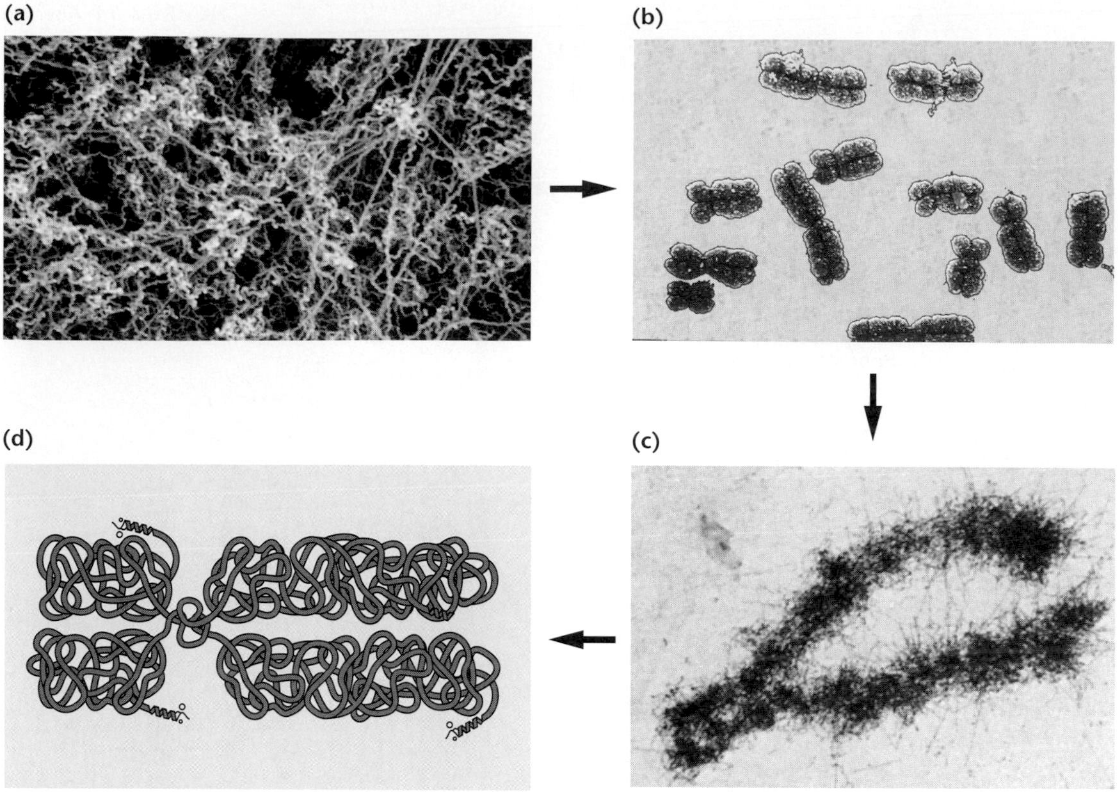

(a)

(b)

(d)

(c)

FIGURE 2–15 Comparison of (a) the chromatin fibers characteristic of the interphase nucleus with (b) and (c) metaphase chromosomes that are derived from chromatin during mitosis. Part (d) diagrams the mitotic chromosome and its various components, showing how chromatin is condensed into it. Parts (a) and (c) are transmission electron micrographs, while part (b) is a scanning electron micrograph. (*Biophoto Associates/Photo Researchers, Inc.*)

Chapter Summary

1. The structure of cells is elaborate and complex. Many components of cells are involved directly or indirectly with genetic processes.

2. In diploid organisms, chromosomes exist in homologous pairs. Each pair shares the same size, centromere placement, and gene sites. One member of each pair is derived from the maternal parent and one is derived from the paternal parent.

3. Mitosis and meiosis are mechanisms by which cells distribute genetic information contained in their chromosomes to progeny cells in a precise, orderly fashion.

4. Mitosis, or nuclear division, is part of the cell cycle and is the basis of cellular reproduction. Daughter cells are produced that are genetically identical to their progenitor cell.

5. Mitosis may be subdivided into discrete stages: prophase, prometaphase, metaphase, anaphase, and telophase. Condensation of chromatin into chromosome structures occurs during prophase. During prometaphase, chromosomes appear as double structures, each composed of a pair of sister chromatids. In metaphase, chromosomes line up on the equatorial plane of the cell. During anaphase, sister chromatids of each chromosome are pulled apart and directed toward opposite poles. Telophase completes daughter cell formation and is characterized by cytokinesis, the division of the cytoplasm.

6. The cell cycle is characteristic of all eukaryotes and is tightly regulated at three checkpoints: G1/S, G2/M, and M.

7. Meiosis, the underlying basis of sexual reproduction, results in the conversion of a diploid cell to a haploid gamete or spore. As a result of chromosome duplication and two subsequent divisions, each haploid cell receives one member of each homologous pair of chromosomes.

8. A major difference exists between meiosis in males and females. Spermatogenesis partitions cytoplasmic volume equally and produces four haploid sperm cells. Oogenesis, on the other hand, accumulates the cytoplasm in one egg cell and reduces the other haploid sets of genetic material to polar bodies. The extra cytoplasm contributes to zygote development following fertilization.

9. Meiosis results in extensive genetic variation by virtue of the exchange during crossing over between maternal and paternal chromatids and their random segregation into gametes. In addition, meiosis plays an important role in the life cycles of fungi and plants, serving as the bridge between alternating generations.

10. Mitotic chromosomes are produced as a result of the coiling and condensation of chromatin fibers characteristic of interphase.

Key Terms

AB antigen, 17

acrocentric, 19

allele, 20

anaphase, 22

anaphase I, 29

anaphase II, 29

apoptosis, 25

biparental inheritance, 20

bivalent, 26

cdc kinase, 24

cdc mutation (*cell division cycle* mutation), 24

Cdk protein (Cyclin-dependent kinase protein), 24

cell coat, 17

cell cycle, 21

cell furrow, 24

cell plate, 24

cell wall, 17

cellular organelle, 18

cellulose, 17

centriole, 18

centromere, 19

centrosome, 22

checkpoint, 24

chiasma (chiasmata), 27

chloroplast, 18

chromatid, 22

chromatin, 17

chromomere, 27

chromosome, 17

crossing over, 25

cyclin, 24

cytokinesis, 21

cytoplasm, 18

daughter chromosome, 24

diakinesis, 28

diploid number (*2n*), 19

diplotene stage, 27

disjunction, 29

dyad, 29

endoplasmic reticulum (ER), 18

endosymbiont hypothesis, 18

equational division, 26

first polar body, 29

folded-fiber model, 31

G0 stage, 21

G1 (gap I), 21

G1/S checkpoint, 24

G2 (gap II), 21

G2/M checkpoint, 24

gamete, 16

gametophyte stage, 31

genome, 20

histocompatibility antigen, 17

homologous chromosome, 19

homology search, 27

independent assortment, 29

interphase, 21

karyokinesis, 21

karyotype, 19

kinetechore, 23

leptotene stage, 27

locus (loci), 20

M checkpoint, 24

meiosis, 16

meiosis II, 29

metacentric, 19

metaphase, 22

metaphase I, 29

metaphase II, 29

metaphase plate, 23

middle lamella, 24

mitochondria, 18

mitosis, 16

MN antigen, 17

molecular motor, 24

monad, 26

nondisjunction, 29

nucleoid, 17

nucleolus, 17

nucleolus organizer region (NOR), 17

oogenesis, 29

oogonium, 29

ootid, 30

ovum (ova), 29

p arm, 19

p53 gene, 25

pachytene stage, 27

plasma membrane, 17

primary oocyte, 29

primary spermatocyte, 29

prokaryote, 17

prometaphase, 22

prophase, 22

prophase II, 29

q arm, 19

receptor molecule, 17

reductional division, 26

S phase, 21

second polar body, 30

secondary oocyte, 29

secondary spermatocyte, 29

sex-determining chromosome, 20

sister chromatids, 20

sperm, 29

spermatid, 29

spermatogenesis, 29

spermatogonium, 29

spermatozoa, 29

spermiogenesis, 29

spindle fibers, 22

spore, 16

sporophyte stage, 31

submetacentric, 19

synapsis, 27

synaptonemal complex, 27

telocentric, 19

telophase, 22

telophase I, 29

telophase II, 29

terminalization, 28

tetrad, 26

tumor-suppressor gene, 25

zygotene stage, 27

zygote, 20

Insights and Solutions

With this initial appearance of "Insights and Solutions," it is appropriate to describe its value to you as a student. This section precedes the "Problems and Discussion Questions" in each chapter; it provides sample problems and solutions that demonstrate approaches useful in genetic analysis. The insights you gain will help you arrive at correct solutions to ensuing problems.

1. In an organism with diploid number of 6, how many individual chromosomal structures will align on the metaphase plate during (a) mitosis, (b) meiosis I, and (c) meiosis II? Describe each configuration.

Solution: (a) In mitosis, where homologous chromosomes do not synapse, there will be 6 double structures, each consisting of a pair of sister chromatids. The number of structures is equivalent to the diploid number. (b) In meiosis I, the homologs have synapsed, reducing the number of structures to 3. Each is called a tetrad and consists of two pairs of sister chromatids. (c) In meiosis II, the same number of structures exist (3), but in this case they are called dyads. Each dyad is a pair of sister chromatids. When crossing over has occurred, each chromatid may contain parts of one of its nonsister chromatids, obtained during exchange in prophase I.

2. For the chromosomes shown in Figure 2–12, draw all possible alignment configurations that can occur during metaphase of meiosis I.

Solution: As shown in the diagram below, four configurations are possible when $n = 2$.

Case I
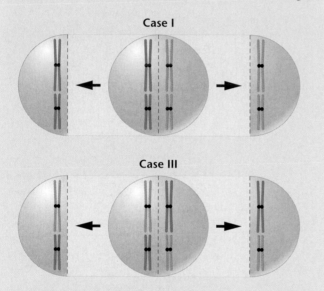

Case II
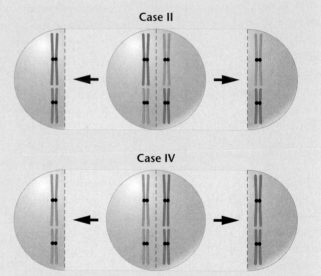

Case III

Case IV

3. Describe the composition of a meiotic tetrad as it exists during prophase I, assuming no crossover event has occurred. What impact would a single crossover event have on this structure?

Solution: Such a tetrad contains four chromatids, existing as two pairs. Members of each pair are replicas of one another and are called sister chromatids. They are held together by a common centromere. Members of one pair are maternally derived, whereas members of the other are paternally derived. Maternal and paternal members are called nonsister chromatids. A single crossover event has the effect of exchanging a portion of a maternal and a paternal chromatid, leading to a chiasma, where the two involved chromatids overlap physically in the tetrad. The process of exchange is referred to as crossing over.

Problems and Discussion Questions

1. What role do the following cellular components play in the storage, expression, or transmission of genetic information? (a) chromatin, (b) nucleolus, (c) ribosome, (d) mitochondrion, (e) centriole, (f) centromere.
2. Discuss the concepts of homologous chromosomes, diploidy, and haploidy. What characteristics are shared between two chromosomes considered to be homologous?
3. If two chromosomes of a species are the same length and have similar centromere placements yet are not homologous, what is different about them?
4. Describe the events that characterize each stage of mitosis.
5. If an organism has a diploid number of 16, how many chromatids are visible at the end of mitotic prophase? How many chromosomes are moving to each pole during anaphase of mitosis?
6. Describe how chromosomes are named on the basis of their centromere placement.
7. Contrast telophase in plant and animal mitosis.
8. Describe the phases of the cell cycle and the events that characterize each phase.
9. What "checkpoints" occur in the cell cycle? What is the role of each?

10. Describe the role and significance of the *p53* gene in humans. How does the process of apoptosis relate to your answer to the preceding question?

11. Examine Figure 2–13, which shows oogenesis in animal cells. Will the genotype of the second polar body (derived from meiosis II) always be identical to that of the ootid? Why or why not?

12. Contrast the end results of meiosis with those of mitosis.

13. Define and discuss these terms: (a) synapsis, (b) bivalents, (c) chiasmata, (d) crossing over, (e) chromomeres, (f) sister chromatids, (g) tetrads, (h) dyads, (i) monads.

14. An organism has a diploid number of 16 in a primary oocyte.
    (a) How many tetrads are present in the first meiotic prophase?
    (b) How many dyads are present in the second meiotic prophase?
    (c) How many monads migrate to each pole during the second meiotic anaphase?

15. Contrast spermatogenesis and oogenesis. What is the significance of the formation of polar bodies?

16. Explain why meiosis leads to significant genetic variation while mitosis does not.

17. A diploid cell contains three pairs of homologous chromosomes designated C1 C2, M1 M2, and S1 S2. No crossing over occurs. What possible combinations of chromosomes will be present (a) in two daughter cells following mitosis? (b) in the first meiotic metaphase? (c) in haploid cells following meiosis?

18. Considering the preceding problem, predict the number of different haploid cells that will occur if a fourth chromosome pair (W1 W2) is considered in addition to the C, M, and S chromosomes.

19. Describe the role of meiosis in the life cycle of a vascular plant.

20. Contrast the chromatin fiber with the mitotic chromosome. How are the two structures related?

21. Describe and distinguish between the bacterial chromosome and the eukaryotic chromosome in dividing cells.

22. You are given a metaphase chromosome preparation (a slide) from an unknown organism that contains 12 chromosomes. Two are clearly smaller than the rest, appearing identical in length and centromere placement. Describe all that you can about these chromosomes.

Selected Readings

Alberts, B., et al. 1994. *Molecular biology of the cell*, 3rd ed. New York: Garland.

Brachet, J., and Mirsky, A. E. 1961. *The cell: Meiosis and mitosis*, Vol. 3. Orlando, FL: Academic Press.

DuPraw, E. J. 1970. *DNA and chromosomes*. New York: Holt, Rinehart & Winston.

Glover, D. M., Gonzalez, C., and Raff, J. W. 1993. The centrosome. *Sci. Am.* (June) 268:62–68.

Golomb, H. M., and Bahr, G. F. 1971. Scanning electron microscopic observations of surface structures of isolated human chromosomes. *Science* 171:1024–26.

Hartwell, L. H., and Karstan, M. B. 1994. Cell cycle control and cancer. *Science* 266:1821–28.

Hartwell, L. H., and Weinert, T. A. 1989. Checkpoint controls that ensure the order of cell cycle events. *Science* 246:629–34.

Kleckner, N. 1996. Meiosis: How could it work? *Proc. Natl. Acad. Sci.* 93:8167–74.

Mazia, D. 1961. How cells divide. *Sci. Am.* (Jan.) 205:101–20.

———. 1974. The cell cycle. *Sci. Am.* (Jan.) 235:54–64.

McIntosh, J. R., and McDonald, K. L. 1989. The mitotic spindle. *Sci. Am.* (Oct.) 261:48–56.

Murray, A. W., and Kirschner, M. 1993. *The cell cycle: An introduction*. New York: Oxford Univ. Press.

Prescott, D. M., and Flexer, A. S. 1986. *Cancer, the misguided cell*, 2nd ed. Sunderland, MA: Sinauer.

Swanson, C. P., Merz, T., and Young, W. J. 1981. *Cytogenetics, the chromosome in division, inheritance, and evolution*, 2nd ed. Englewood Cliffs, NJ: Prentice-Hall.

Westergard, M., and von Wettstein, D. 1972. The synaptinemal complex. *Annu. Rev. Genet.* 6:71–110.

Wheatley, D. N. 1982. *The centriole: A central enigma of cell biology*. New York: Elsevier/North-Holland Biomedical.

Mendel's garden, as seen in the 1980s. *(Courtesy of Allan Gotthelf)*

3

Mendelian Genetics

CHAPTER CONCEPTS

Inherited characteristics are under the control of discrete particulate factors called genes that are transmitted from generation to generation on vehicles called chromosomes, according to rules first described by Gregor Mendel. The outcomes of crosses subject to these rules are affected by chance deviation and can be evaluated using statistical analysis. Human traits are initially studied using the pedigree method.

Although inheritance of biological traits has been recognized for thousands of years, the first significant insights into the mechanisms involved occurred about 130 years ago. In 1866, Gregor Johann Mendel published the results of a series of experiments that would lay the foundation for the formal discipline of genetics. Although Mendel's work went largely unnoticed until the turn of the century, in the ensuing years the concept of the gene as a distinct hereditary unit was established. Ways in which genes, as members of chromosomes, are transmitted to offspring and control traits were clarified. Research has continued unabated throughout the twentieth century. Indeed, studies in genetics, most recently at the molecular level, have remained continually at the forefront of biological research since the early 1900s.

When Mendel began his studies of inheritance using *Pisum sativum*, the garden pea, chromosomes and the role and mechanism of meiosis were totally unknown. Nevertheless, he determined that discrete **units of inheritance** exist and predicted their behavior during the formation of gametes. Subsequent investigators, with access to cytological data, were able to relate their observations of chromosome behavior during meiosis to Mendel's principles of inheritance. Once this correlation was made, Mendel's postulates were accepted as the basis for the study of what is known as **Mendelian**, or **transmission genetics**. These principles describe how parents transmit genes to offspring and were derived directly from Mendel's experimentation. Even today, they are the cornerstone of the study of inheritance. In this chapter we focus on the development of the Mendel's principles.

3.1 Gregor Johann Mendel

Johann Mendel was born in 1822 to a peasant family in the central European village of Heinzendorf. An excellent student in high school, he studied philosophy for several years afterward, and in 1843 was admitted to the Augustinian Monastery of St. Thomas in Brno, now part of the Czech Republic, taking the name of Gregor. In 1849, he was relieved of pastoral duties and accepted a teaching appointment that lasted several years. From 1851 to 1853, he attended the University of Vienna, where he studied physics and botany. In 1854, he returned to Brno, where, for the next 16 years, he taught physics and natural science. Mendel received support from the monastery for his studies and research throughout his life.

In 1856, Mendel performed his first set of hybridization experiments with the garden pea. The research phase of his career lasted until 1868, when he was elected abbot of the monastery. Although he retained his interest in genetics, his new responsibilities demanded most of his time. In 1884, Mendel died of a kidney disorder. The local newspaper paid him the following tribute: "His death deprives the poor of a benefactor, and mankind at large of a man of the noblest character, one who was a warm friend, a promoter of the natural sciences, and an exemplary priest."

Mendel first reported the results of some simple genetic crosses between certain strains of the garden pea in 1865. Although his was not the first attempt to provide experimental evidence pertaining to inheritance, Mendel's success where others failed can be attributed, at least in part, to his elegant model of experimental design and analysis.

Mendel showed remarkable insight into the methodology necessary for good experimental biology. First, he chose an organism that is easy to grow and to hybridize artificially. The pea plant is self-fertilizing in nature, but is easy to crossbreed experimentally. It reproduces well and grows to maturity in a single season. Mendel then followed seven visible features (unit characters), each represented by two contrasting forms or traits (Figure 3–1). For the character stem height, for example, he experimented with the traits *tall* and *dwarf*. He selected six other contrasting pairs of traits involving seed shape and color, pod shape and color, and pod and flower arrangement. From local seed merchants, Mendel obtained true-breeding strains, those in which each trait appeared unchanged generation after generation in self-fertilizing plants.

Several factors led to Mendel's success, in addition to his choice of a suitable organism. He restricted his examination to one or very few pairs of contrasting traits in each experiment. He also kept accurate quantitative records, a necessity in genetic experiments. From the analysis of his data, Mendel derived certain postulates that have become the principles of transmission genetics.

The results of Mendel's experiments went unappreciated until the turn of the century, well after his death. Once Mendel's publications were rediscovered by geneticists investigating the function and behavior of chromosomes, however, the implications of his postulates were immediately apparent. He had discovered the basis for the transmission of hereditary traits!

3.2 The Monohybrid Cross

Mendel's simplest crosses involved only one pair of contrasting traits. Each such breeding experiment is called a **monohybrid cross**. A monohybrid cross is made by mating individuals from two parent strains, each of which exhibits one of the two contrasting forms of the character under study. Initially, we examine the first generation of offspring of such a cross, and then we consider the offspring of **selfing** or **self-fertilizing** individuals from this first generation. The original parents are called the P_1 or **parental generation**, their offspring are the F_1 or **first filial generation**, and the individuals resulting from the selfing of the F_1 generation are the F_2 or **second filial generation**. We can, of course, continue to follow subsequent generations, if desired.

The cross between true-breeding pea plants with tall stems and dwarf stems is representative of Mendel's monohybrid crosses. *Tall* and *dwarf* are contrasting forms or traits of the character of stem height. Unless tall or dwarf plants are crossed together or with another strain, they will undergo self-fertilization and breed true, producing their respective trait generation after generation. However, when Mendel

Character	Contrasting traits		F₁ results	F₂ results	F₂ ratio
Seeds	round/wrinkled		all round	5474 round 1850 wrinkled	2.96:1
	yellow/green		all yellow	6022 yellow 2001 green	3.01:1
	full/constricted		all full	882 full 299 constricted	2.95:1
Pods	green/yellow		all green	428 green 152 yellow	2.82:1
	axial/terminal		all axial	651 axial 207 terminal	3.14:1
Flowers	violet/white		all violet	705 violet 224 white	3.15:1
Stem	tall/dwarf		all tall	787 tall 277 dwarf	2.84:1

FIGURE 3–1 A summary of the seven pairs of contrasting traits and the results of Mendel's seven monohybrid crosses of the garden pea (*Pisum sativum*, shown in the photograph). In each case, pollen derived from plants exhibiting one trait was used to fertilize the ova of plants exhibiting the other trait. In the F₁ generation, one of the two traits, (dominant) was exhibited by all plants. The contrasting trait (recessive) then reappeared in approximately 1/4 of the F₂ plants.

crossed tall plants with dwarf plants, the resulting F₁ generation consisted only of tall plants. When members of the F₁ generation were selfed, Mendel observed that 787 of 1064 F₂ plants were tall, while 277 of 1064 were dwarf. Note that in this cross (Figure 3–1) the dwarf trait disappears in the F₁, only to reappear in the F₂ generation.

Genetic data are usually expressed and analyzed as ratios. In this particular example, many identical P₁ crosses were made, and many F₁ plants—all tall—were produced. Of the 1064 F₂ offspring, 787 were tall and 277 were dwarf—a ratio of approximately 2.8:1.0, or about 3:1.

Mendel made similar crosses between pea plants exhibiting other pairs of contrasting traits. The results of these crosses are also shown in Figure 3–1. In every case, the outcome was similar to the tall/dwarf cross just described. All F₁ offspring

were identical to one of the parents. In the F₂ offspring, an approximate ratio of 3:1 was obtained. Three-fourths appeared like the F₁ plants, while one-fourth exhibited the contrasting trait, which had disappeared in the F₁ generation.

It is appropriate to point out one further aspect of the monohybrid crosses. In each, the F₁ and F₂ patterns of inheritance were similar regardless of which P₁ plant served as the source of pollen (sperm), and which served as the source of the ovum (egg). The crosses could be made either way—that is, pollination of dwarf plants by tall plants, or vice versa. These are called **reciprocal crosses**. Therefore, the results of Mendel's monohybrid crosses were not sex-dependent.

To explain these results, Mendel proposed the existence of particulate **unit factors** for each trait. He suggested that these factors serve as the basic units of heredity and are passed unchanged from generation to generation, determining various traits expressed by each individual plant. Using these general ideas, Mendel proceeded to hypothesize precisely how such factors could account for the results of the monohybrid crosses.

Mendel's First Three Postulates

Using the consistent pattern of results in the monohybrid crosses, Mendel derived the following three postulates or principles of inheritance.

1. UNIT FACTORS IN PAIRS

Genetic characters are controlled by unit factors that exist in pairs in individual organisms.

In the monohybrid cross involving tall and dwarf stems, a specific unit factor exists for each trait. Because the factors occur in pairs, three combinations are possible: two factors for tallness, two factors for dwarfness, or one for each factor. Every individual contains one of these three combinations, which determines stem height.

2. DOMINANCE/RECESSIVENESS

When two unlike unit factors responsible for a single character are present in a single individual, one unit factor is dominant to the other, which is said to be recessive.

In each monohybrid cross, the trait expressed in the F_1 generation is controlled by the dominant unit factor. The trait not expressed is controlled by the recessive unit factor. Note that this dominance/recessiveness relationship pertains only when unlike unit factors are present in pairs. The terms *dominant* and *recessive* are also used to designate the traits. In this case, tall stems are said to be dominant to the recessive dwarf stems.

3. SEGREGATION

During the formation of gametes, the paired unit factors separate or segregate randomly so that each gamete receives one or the other with equal likelihood.

If an individual contains a pair of like unit factors (e.g., both specific for tall), then all gametes receive one tall unit factor. If an individual contains unlike unit factors (e.g., one for tall and one for dwarf), then each gamete has a 50 percent probability of receiving either the tall or the dwarf unit factor.

These postulates provide a suitable explanation for the results of the monohybrid crosses. Let's use the tall/dwarf cross to illustrate. Mendel reasoned that P_1 tall plants contain identical paired unit factors, as do the P_1 dwarf plants. The gametes of tall plants all receive one tall unit factor as a result of segregation. Likewise, the gametes of dwarf plants all receive one dwarf unit factor. Following fertilization, all F_1 plants receive one unit factor from each parent, a tall factor from one and a dwarf factor from the other, reestablishing the paired relationship. Because tall is dominant to dwarf, all F_1 plants are tall.

When F_1 plants form gametes, the postulate of segregation demands that each gamete randomly receive either the tall or the dwarf unit factor. Following random fertilization events during F_1 selfing, four F_2 combinations result in equal frequency:

(1) tall/tall
(2) tall/dwarf
(3) dwarf/tall
(4) dwarf/dwarf

Combinations (1) and (4) result in tall and dwarf plants, respectively. According to the postulate of dominance/recessiveness, combinations (2) and (3) both yield tall plants. Therefore, the F_2 is predicted to consist of three-fourths tall and one-fourth dwarf, or a ratio of 3:1. This is approximately what Mendel observed in the cross between tall and dwarf plants. A similar pattern was observed in each of the other monohybrid crosses (Figure 3–1).

Modern Genetic Terminology

To illustrate the monohybrid cross and Mendel's first three postulates, we must first introduce several new terms as well as a set of symbols for the unit factors. Traits such as tall or dwarf are visible expressions of the information contained in unit factors. The physical appearance of a trait is called the **phenotype** of the individual.

Mendel's unit factors represent units of inheritance called **genes** by modern geneticists. For any given character, such as plant height, the phenotype is determined by alternative forms of a single gene called **alleles**. For example, the unit factors representing tall and dwarf are alleles determining the height of the pea plant.

By one convention, the first letter of the recessive trait is chosen to symbolize the character in question. The lowercase italic letter designates that allele for the recessive trait, and the uppercase italic letter designates the allele for the dominant trait. Therefore, we use *d* for the dwarf allele and *D* for the tall allele. When alleles are written in pairs to represent the two unit factors present in any individual (*DD*, *Dd*, or *dd*), these symbols are called the **genotype**. This term reflects the genetic makeup of an individual whether it is haploid or diploid. By reading the genotype, it is possible to know the phenotype of the individual: *DD* and *Dd* are tall, and *dd* is dwarf. When both alleles are the same (*DD* or *dd*), the individual is said to be **homozygous** or a **homozygote**; when the alleles are different (*Dd*), we use the term **heterozygous** or **heterozygote**. These symbols and terms are used in Figure 3–2 to illustrate the complete monohybrid cross.

Because he operated without the hindsight that modern geneticists enjoy, Mendel's analytical reasoning must be considered a truly outstanding scientific achievement. On the basis of rather simple but precisely executed breeding experiments, he not only proposed that discrete particulate units of heredity exist, he also explained how they are transmitted from one generation to the next!

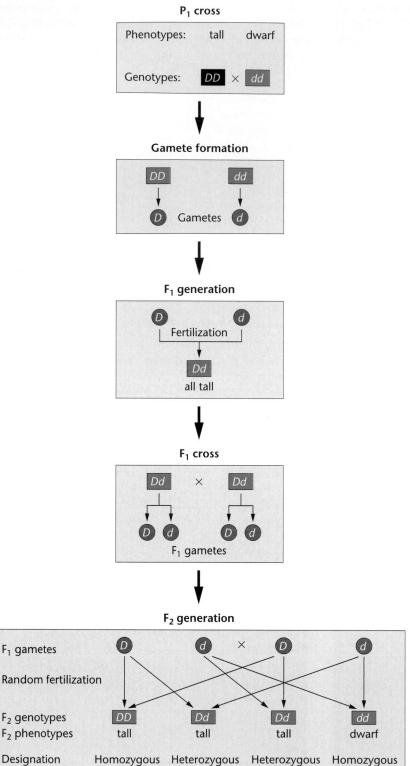

FIGURE 3–2 The monohybrid cross between tall and dwarf pea plants. The symbols *D* and *d* designate the tall and dwarf unit factors, respectively, in the genotypes of mature plants and gametes. Individuals are shown in rectangles, and gametes in circles.

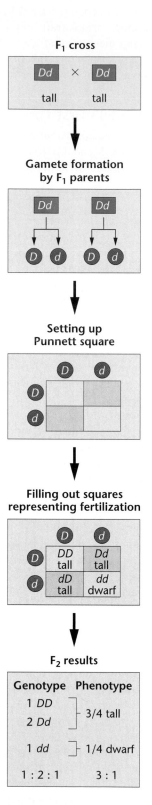

FIGURE 3–3 A Punnett square generates the F_2 ratio of the $F_1 \times F_1$ cross shown in Figure 3–2.

Test cross results

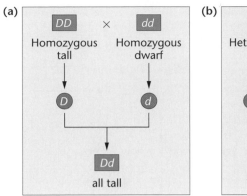

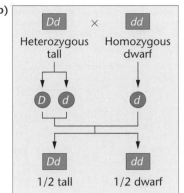

FIGURE 3–4 Test cross of a single character. In (a), the tall parent is homozygous. In (b), the tall parent is heterozygous. The genotype of each tall parent can be determined by examining the offspring when each is crossed to the homozygous recessive dwarf plant.

Punnett Squares

The genotypes and phenotypes resulting from the recombination of gametes during fertilization can be easily visualized by constructing a **Punnett square**, so named after the person who first devised this approach, Reginald C. Punnett. Figure 3–3 demonstrates this method of analysis for the $F_1 \times F_1$ monohybrid cross. Each of the possible gametes is assigned to a column or a row, with the vertical column representing those of the female parent and the horizontal row those of the male parent. After entering the gametes in rows and columns, the new generation is predicted by combining the male and female gametic information for each combination and entering the resulting genotypes in the boxes. This process thus lists all possible random fertilization events. The genotypes and phenotypes of all potential offspring are ascertained by reading the entries in the boxes.

The Punnett square method is particularly useful when you are first learning about genetics and how to solve problems. In Figure 3–3, note the ease with which the 3:1 phenotypic ratio and the 1:2:1 genotypic ratio is derived in the F_2 generation.

The Test Cross: One Character

Tall plants produced in the F_2 generation are predicted to be of either the *DD* or *Dd* genotype. You might ask if there is a way to distinguish the genotype. Mendel devised a rather simple method that is still used today in breeding plants and animals: the **test cross**. The organism expressing the dominant phenotype, but of unknown genotype, is crossed to a homozygous recessive individual. For example, as shown in Figure 3–4(a), if a tall plant of genotype *DD* is test-crossed to a dwarf plant, which must have the *dd* genotype, all offspring will be tall phenotypically and *Dd* genotypically. However, as shown in Figure 3–4(b), if a tall plant is *Dd* and it is crossed to a dwarf plant (*dd*), then one-half of the offspring will be tall (*Dd*) and the other half will be dwarf (*dd*). Therefore, a 1:1 tall/dwarf ratio demonstrates the heterozygous nature of the tall plant of unknown genotype. The test cross reinforced Mendel's conclusion that separate unit factors control the tall and dwarf traits.

3.3 The Dihybrid Cross

As a natural extension of the monohybrid cross, Mendel also designed experiments in which he examined two characters simultaneously. Such a cross, involving two pairs of contrasting traits, is a **dihybrid cross**. It is also called a **two-factor cross**. For example, if pea plants having yellow seeds that are also round are bred with those having green seeds that are also wrinkled, the results shown in Figure 3–5 will occur. The F_1 offspring will be all yellow and round. It is therefore apparent that yellow is dominant to green and that round is dominant to wrinkled. When the F_1 individuals are selfed, approximately 9/16 of the F_2 plants express yellow and round, 3/16 express yellow and wrinkled, 3/16 express green and round, and 1/16 express green and wrinkled.

A variation of this cross is also shown in Figure 3–5. Instead of crossing one P_1 parent with both dominant traits (yellow, round) and one with both recessive traits (green, wrinkled), plants with yellow, wrinkled seeds are crossed with those with green, round seeds. In spite of the change in the P_1 phenotypes, both the F_1 and F_2 results remain unchanged. It will become clear in the next section why this is so.

Mendel's Fourth Postulate: Independent Assortment

We can most easily understand the results of a dihybrid cross if we consider it theoretically as consisting of two monohybrid crosses conducted separately. Think of the two sets of traits as inherited independently of each other; that is, the chance of any plant having yellow or green seeds is not at all influenced by the chance that this plant will have round or wrinkled seeds. Thus, because yellow is dominant to green, all F_1 plants in the first theoretical cross would have yellow seeds. In the second theoretical cross, all F_1 plants would have round seeds because round is dominant to wrinkled. When Mendel examined the F_1 plants of the dihybrid cross, all were yellow and round, as we just predicted.

The predicted F_2 results of the first cross are 3/4 yellow and 1/4 green. Similarly, the second cross would yield 3/4 round and 1/4 wrinkled. Figure 3–5 shows that in the dihybrid cross, 12/16 F_2 plants are yellow while 4/16 are green, exhibiting the expected 3:1 (3/4:1/4) ratio. Similarly,

FIGURE 3–5 F_1 and F_2 results of Mendel's dihybrid crosses between yellow, round and green, wrinkled pea plants and between yellow, wrinkled and green, round pea plants.

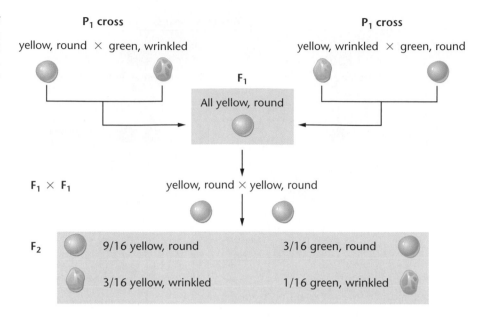

12/16 F_2 plants have round seeds while 4/16 have wrinkled seeds, again revealing the 3:1 (3/4:1/4) ratio.

Because it is evident that the two pairs of contrasting traits are inherited independently, we can predict the frequencies of all possible F_2 phenotypes by applying the **product law** of probabilities: *When two independent events occur simultaneously, the combined probability of the two outcomes is equal to the product of their individual probabilities of occurrence.* For example, the probability of an F_2 plant having yellow *and* round seeds is (3/4)(3/4), or 9/16, because 3/4 of all F_2 plants should be yellow and 3/4 of all F_2 plants should be round.

In a like manner, the probabilities of the other three F_2 phenotypes can be calculated: Yellow (3/4) and wrinkled (1/4) are predicted to be present together 3/16 of the time; green (1/4) and round (3/4) are predicted 3/16 of the time; and green (1/4) and wrinkled (1/4) are predicted 1/16 of the time. These calculations are shown in Figure 3–6. It is now apparent why the F_1 and F_2 results are identical whether the initial cross is yellow, round bred with green, wrinkled or if

yellow, wrinkled are bred with green, round. In both crosses, the F_1 genotype of all plants is identical. Each plant is heterozygous for both gene pairs. As a result, the F_2 generation is also identical in both crosses.

On the basis of similar results in numerous dihybrid crosses, Mendel proposed a fourth postulate called **independent assortment**: *During gamete formation, segregating pairs of unit factors assort independently of each other.* This postulate stipulates that segregation of any pair of unit factors occurs independently of all others. As a result of segregation, each gamete receives one member of every pair of unit factors. For one pair, whichever unit factor is received does not influence the outcome of segregation of any other pair. Thus, according to the postulate of independent assortment, all possible combinations of gametes are formed in equal frequency.

The Punnett square in Figure 3–7 shows how independent assortment works in the formation of the F_2 generation. Examine the formation of gametes by the F_1 plants. Segregation prescribes that every gamete receives either a G or g allele and a W or w allele. Independent assortment

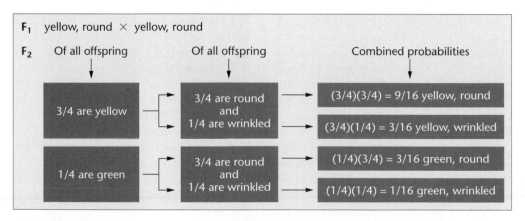

FIGURE 3–6 Computation of the combined probabilities of each F_2 phenotype for two independently inherited characters. The probability of each plant being yellow or green is independent of the probability of it being round or wrinkled.

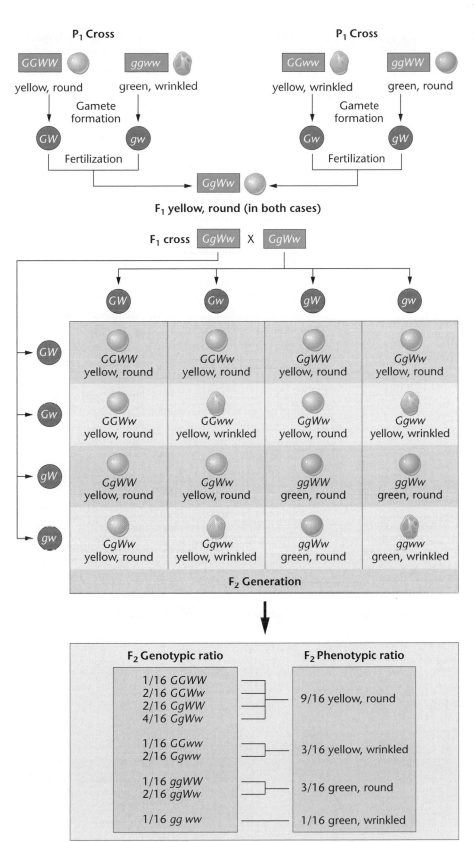

FIGURE 3–7 Analysis of the dihybrid crosses shown in Figure 3–5. The F₁ heterozygous plants are self-fertilized to produce an F₂ generation, which is computed using a Punnett square. Both the phenotypic and genotypic F₂ ratios are shown.

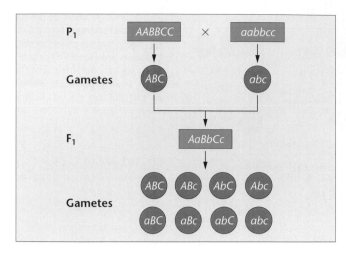

FIGURE 3–8 Formation of P₁ and F₁ gametes in a trihybrid cross.

stipulates that all four combinations (*GW*, *Gw*, *gW*, and *gw*) will be formed with equal probabilities.

In every F₁ × F₁ fertilization event, each zygote has an equal probability of receiving one of the four combinations from each parent. If many offspring are produced, 9/16 are yellow and round, 3/16 are yellow and wrinkled, 3/16 are green and round, and 1/16 are green and wrinkled, yielding what is designated as **Mendel's 9:3:3:1 dihybrid ratio**. This is an ideal ratio based on probability events involving segregation, independent assortment, and random fertilization. Because of deviation due strictly to chance, particularly if small numbers of offspring are produced, actual results will seldom match the ideal ratio exactly.

The Test Cross: Two Characters

The test cross can also be applied to individuals that express two dominant traits but whose genotypes are unknown. For example, the expression of the yellow, round phenotype in the F₂ generation just described may result from the *GGWW*, *GGWw*, *GgWW*, and *GgWw* genotypes. If an F₂ yellow, round plant is crossed with a homozygous recessive green, wrinkled plant (*ggww*), analysis of the offspring will indicate the actual genotype of that yellow, round plant. Each of the above genotypes will result in a different set of gametes,

and in a test cross, a different set of phenotypes in the resulting offspring. You should work out the results of each of the four crosses to make sure you understand this concept.

3.4   The Trihybrid Cross

Thus far, we have considered inheritance by individuals of up to two pairs of contrasting traits. Mendel demonstrated that the identical processes of segregation and independent assortment apply to three pairs of contrasting traits in what is called a **trihybrid cross**, also referred to as a **three-factor cross**.

Although a trihybrid cross is somewhat more complex than a dihybrid cross, its results are easily calculated if the principles of segregation and independent assortment are followed. For example, consider the cross shown in Figure 3–8, where the gene pairs of theoretical contrasting traits are represented by the symbols *A/a*, *B/b*, and *C/c*. In the cross between *AABBCC* and *aabbcc* individuals, all F₁ individuals are heterozygous for all three gene pairs. Their genotype, *AaBbCc*, results in the phenotypic expression of the dominant A, B, and C traits. When F₁ individuals serve as parents, each produces 8 different gametes in equal frequencies. At this point, we could construct a Punnett square with 64 separate boxes and read out the phenotypes. Because such a method is cumbersome in a cross involving so many factors, another method has been devised that is also used to calculate the predicted ratio.

The Forked-Line Method

It is much less difficult to consider each contrasting pair of traits separately and then to combine these results using the **forked-line method**, first shown in Figure 3–6. This method, also called a **branch diagram**, relies on the simple application of the laws of probability established for the dihybrid cross. Each gene pair is assumed to behave independently during gamete formation.

When the monohybrid cross *AA* × *aa* is made, we know that

1. All F₁ individuals have the genotype *Aa* and express the phenotype represented by the *A* allele, which is called the A phenotype in the following discussion.

FIGURE 3–9 Generation of the F₂ trihybrid phenotypic ratio using the forked-line method.

Generation of F₂ trihybrid phenotypes

A or a	B or b	C or c	Combined proportion		
3/4 A	3/4 B	3/4 C	(3/4)(3/4)(3/4) ABC	= 27/64	ABC
		1/4 c	(3/4)(3/4)(1/4) ABc	= 9/64	ABc
	1/4 b	3/4 C	(3/4)(1/4)(3/4) AbC	= 9/64	AbC
		1/4 c	(3/4)(1/4)(1/4) Abc	= 3/64	Abc
1/4 a	3/4 B	3/4 C	(1/4)(3/4)(3/4) aBC	= 9/64	aBC
		1/4 c	(1/4)(3/4)(1/4) aBc	= 3/64	aBc
	1/4 b	3/4 C	(1/4)(1/4)(3/4) abC	= 3/64	abC
		1/4 c	(1/4)(1/4)(1/4) abc	= 1/64	abc

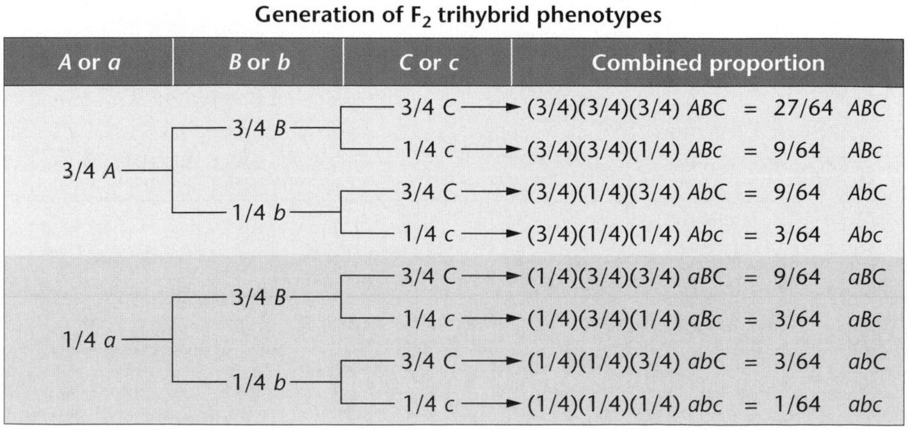

2. The F_2 generation consists of individuals with either the A phenotype or the a phenotype in the ratio of 3:1.

The same generalizations can be made for the *BB* × *bb* and *CC* × *cc* crosses. Thus, in the F_2 generation, 3/4 of all organisms express phenotype A, 3/4 express B, and 3/4 express C. Similarly, 1/4 of all organisms express phenotype a, 1/4 express b, and 1/4 express c. The proportions of organisms that express each phenotypic combination can be predicted by assuming that fertilization, following the independent assortment of these three gene pairs during gamete formation, is a random process. We simply apply the product law of probabilities once again. Figure 3–9 uses the forked-line method to calculate the phenotypic proportions of the F_2 generation. They fall into the trihybrid ratio 27:9:9:9:3:3:3:1. The same method can be used to solve crosses involving any number of gene pairs, *provided that all gene pairs assort independently from each other*. We shall see later that this is not always the case. However, it appeared to be true for all of Mendel's characters.

3.5 The Rediscovery of Mendel's Work

Mendel's work, initiated in 1856, was presented to the Brünn Society of Natural Science in 1865 and published the following year. However, while his findings were often cited and discussed, their significance went unappreciated for about 35 years! Many reasons have been suggested to explain why the significance of his research was not immediately recognized.

First of all, Mendel's adherence to mathematical analysis of probability events was an unusual approach in those days for biological studies. Perhaps his approach seemed foreign to his contemporaries. More important, his conclusions did not fit well with existing theories on the cause of variation among organisms. The source of natural variation intrigued students of evolutionary theory. These individuals, stimulated by the proposal developed by Charles Darwin and Alfred Russel Wallace, believed in **continuous variation**, whereby offspring were a blend of their parents' phenotypes. As we mentioned earlier, Mendel theorized that variation was due to discrete or particulate units, resulting in **discontinuous variation**. For example, Mendel proposed that the F_2 offspring of a dihybrid cross are merely expressing traits produced by new combinations of previously existing unit factors. As a result, Mendel's theories did not fit well with the evolutionists' preconceptions about causes of variation.

In the latter part of the nineteenth century, a remarkable observation set the scene for the rebirth of Mendel's work: Walter Flemming's discovery of chromosomes in the nuclei of salamander cells. In 1879, Flemming described the behavior of these threadlike structures during cell division. As a result of the findings of Flemming and many other cytologists, the presence of discrete units within the nucleus soon became an integral part of ideas about inheritance.

It was this mind set that prompted scientists to reexamine Mendel's findings.

In the early twentieth century, research led to renewed interest in Mendel's work. Hybridization experiments similar to Mendel's were performed independently by three botanists: Hugo de Vries, Karl Correns, and Erich Tschermak. De Vries's work focused on unit characters, and he demonstrated the principle of segregation in his experiments with several plant species. Apparently, he searched the existing literature and found that Mendel's work anticipated his own conclusions! Correns and Tschermak also reached conclusions similar to those of Mendel.

In 1902, two cytologists, Walter Sutton and Theodor Boveri, independently published papers linking their discoveries of the behavior of chromosomes during meiosis to the Mendelian principles of segregation and independent assortment. They pointed out that the separation of chromosomes during meiosis could serve as the cytological basis of these two postulates. Although they thought Mendel's unit factors were probably chromosomes rather than genes on chromosomes, their findings reestablished the importance of Mendel's work, which became the basis of ensuing genetic investigations. Sutton and Boveri are credited with initiating the **chromosomal theory of inheritance**, which was developed during the next two decades. We first discussed this theory in Chapter 1.

Unit Factors, Genes, and Homologous Chromosomes

Because the correlation between Sutton's and Boveri's observations and Mendelian principles is the foundation for the modern interpretation of transmission genetics, we examine this correlation in some depth before moving on to other topics.

As pointed out in Chapter 2, each species possesses a specific number of chromosomes in each somatic cell nucleus (except in gametes). For diploid organisms, this number is called the **diploid number ($2n$)** and is characteristic of that species. During the formation of gametes, this number is precisely halved (n), and when two gametes combine during fertilization, the diploid number is reestablished. During meiosis, however, the chromosome number is not reduced in a random manner. It was apparent to early cytologists that the diploid number of chromosomes is composed of homologous pairs identifiable by their morphological appearance and behavior. The gametes contain one member of each pair. The chromosome complement of a gamete is thus quite specific, and the number of chromosomes in each gamete is equal to the haploid number.

With this basic information, we can see the correlation between the behavior of unit factors and chromosomes and genes. Figure 3–10 lists three of Mendel's postulates (in the left column) and the chromosomal explanation of each (in the right column). Unit factors are really genes located on homologous pairs of chromosomes [Figure 3–10(a)]. Members of each pair of homologs separate, or segregate, during gamete formation [Figure 3–10(b)]. Two different alignments are possible, both of which are shown.

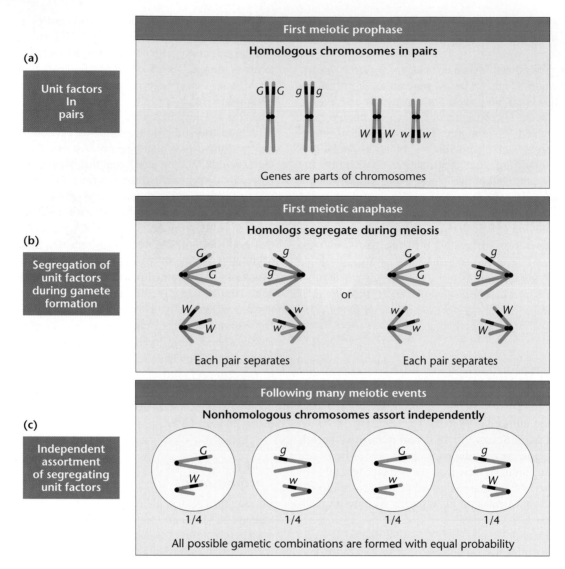

FIGURE 3–10 Correlation between the Mendelian postulates of (a) unit factors in pairs, (b) segregation, and (c) independent assortment, and the presence of genes located on homologous chromosomes and their behavior during meiosis.

To illustrate the principle of independent assortment, we must distinguish between members of any given homologous pair of chromosomes. One member of each pair is derived from the **maternal parent**, while the other comes from the **paternal parent**. We represent the different parental origins by different colors. As shown in Figure 3–10(c), following independent segregation of each pair of homologs, each gamete receives one member from each pair of chromosomes. All possible combinations are formed. If we add the symbols used in Mendel's dihybrid cross (G, g and W, w) to the diagram, we see why equal numbers of the four types of gametes are formed. The independent behavior of Mendel's pairs of unit factors (G and W in this example) is due to the fact that they are on separate pairs of homologous chromosomes.

From observations of the phenotypic diversity of living organisms, it is logical to assume that there are many more genes than chromosomes. Therefore, each homolog must carry genetic information for more than one trait. The currently accepted concept is that a chromosome is composed of a large number of linearly ordered, information-containing units called **genes**. Mendel's unit factors (which determine tall or dwarf stems, for example) actually constitute a pair of genes located on one pair of homologous chromosomes. The location on a given chromosome where any particular gene occurs is called its **locus** (pl. **loci**). The different forms taken by a given gene, called **alleles** (G or g), contain slightly different genetic information that determines the same character (stem length). Alleles are alternative forms of the same gene. Although we have only examined genes with two alternative alleles, most genes have more than two allelic forms. We discuss this concept of **multiple alleles** in Chapter 4.

We conclude this section by reviewing the criteria necessary to classify two chromosomes as a homologous pair:

1. During mitosis and meiosis, when chromosomes are visible as distinct figures, both members of a homologous pair are the same size and exhibit identical centromere locations.

2. During early stages of meiosis, homologous chromosomes form pairs, or synapse.

3. Although not generally microscopically visible, homologs contain identical, linearly ordered gene loci.

3.6 Independent Assortment and Genetic Variation

One major consequence of independent assortment is the production by an individual of genetically dissimilar gametes. Genetic variation results because the two members of any homologous pair of chromosomes are rarely, if ever, genetically identical. Therefore, because independent assortment leads to the production of all possible chromosome combinations, extensive genetic diversity results.

We have seen that the number of possible gametes, each with different chromosome compositions, is 2^n, where n equals the haploid number. Thus, if a species has a haploid number of 4, then 2^4 or 16 different gamete combinations can be formed as a result of independent assortment. Although this number is not high, consider the human species, where $n = 23$. If 2^{23} is calculated, we find that in excess of 8×10^6, or over 8 million, different types of gametes are represented. Because fertilization represents an event involving only one of approximately 8×10^6 possible gametes from each of two parents, each offspring represents only one of $(8 \times 10^6)^2$, or 64×10^{12}, potential genetic combinations! No wonder that, except for identical twins, each member of the human species demonstrates a distinctive appearance and individuality. This number of combinations is far greater than the number of humans who have ever lived on earth! Genetic variation resulting from independent assortment has been extremely important to the process of evolution in all organisms.

3.7 Probability and Genetic Events

Genetic ratios are most properly expressed as probabilities—for example, 3/4 tall:1/4 dwarf. These values predict the outcome of each fertilization event, such that the probability of each zygote having the genetic potential for becoming tall is 3/4, while the potential for becoming dwarf is 1/4. Probabilities range from 0, when an event is *certain not to occur*, to 1.0, when an event is *certain to occur*.

When two or more events occur independently but at the same time, we can calculate the probability of possible outcomes when they occur together. This is accomplished by applying the product law. As mentioned in our earlier discussion of independent assortment (see p. 42), the law states that the probability of two or more events occurring simultaneously is equal to the product of their individual probabilities. Two or more events are independent of one another if the outcome of each one does not affect the outcome of any of the others under consideration.

To illustrate the use of the product law, consider the possible results if you toss a penny (P) and a nickel (N) at the same time and examine all combinations of heads (H) and tails (T) that can occur. There are four possible outcomes:

$$(P_H:N_H) = (1/2)(1/2) = 1/4$$
$$(P_T:N_H) = (1/2)(1/2) = 1/4$$
$$(P_H:N_T) = (1/2)(1/2) = 1/4$$
$$(P_T:N_T) = (1/2)(1/2) = 1/4$$

The probability of obtaining a head or a tail in the toss of either coin is 1/2 and is unrelated to the outcome of the toss of the other coin. All four possible combinations are predicted to occur with equal probability.

If we want to calculate the probability where the possible outcomes of two events are independent of one another but can be accomplished in more than one way, we apply the **sum law**. For example, What is the probability of tossing our penny and nickel and obtaining one head and one tail? In such a case, we don't care whether it is the penny or the nickel that comes up heads, provided the other coin has the alternative outcome. As we have seen, there are two ways in which the desired outcome can be accomplished, each with a probability of 1/4. Thus, according to the sum law, the overall probability is equal to

$$(1/4) + (1/4) = 1/2$$

One-half of all such tosses are predicted to yield the desired outcome.

These simple probability laws will be useful throughout our discussions of transmission genetics, and as you solve genetics problems. In fact, we have already applied the product law, when we used the forked-line method to calculate the phenotypic results of Mendel's dihybrid and trihybrid crosses. When we wish to know the results of a cross, we need only calculate the probability of each possible outcome. The results of this calculation then allow us to predict the proportion of offspring expressing each phenotype or each genotype.

There is a very important point to remember when you deal with probability. Predictions of possible outcomes are based on large sample sizes. If we predict that 9/16 of the offspring of a dihybrid cross will express both dominant traits, it is very unlikely that, in a small sample, exactly 9 of every 16 offspring will express this phenotype. Instead, our prediction is that, of a large number of offspring, approximately 9/16 of them will do so. The deviation from the predicted ratio in smaller sample sizes is attributed to chance, a subject we examine in our discussion of statistics in the next section. As we shall see, the impact of deviation due strictly to chance diminishes as the sample size increases.

3.8 Evaluating Genetic Data: Chi-Square Analysis

Mendel's 3:1 monohybrid and 9:3:3:1 dihybrid ratios are hypothetical predictions based on the following assumptions: (1) Each allele is dominant or recessive; (2) segregation is operative; (3) independent assortment occurs; and (4) fertilization is random. The last three assumptions are influenced by chance events and are therefore subject to random fluctuation. This concept, called **chance deviation**, is most easily illustrated by tossing a single coin numerous times and recording the number of heads and tails observed. In each toss, there is a probability of 1/2 that a head will occur and a probability of 1/2 that a tail will occur. Therefore, the expected ratio of many tosses is 1:1. If a coin is tossed 1000 times, usually *about* 500 heads and 500 tails will be observed. Any reasonable fluctuation from this hypothetical ratio (e.g., 486 heads and 514 tails) is attributed to chance.

As the total number of tosses is reduced, the impact of chance deviation increases. For example, if a coin is tossed only 4 times, you wouldn't be too surprised if all 4 tosses result in only heads or only tails. For 1000 tosses, however, 1000 heads or 1000 tails would be most unexpected. In fact, you might believe that such a result would be impossible. Actually, all heads or all tails in 1000 tosses can be predicted to occur with a probability of only $(1/2)^{1000}$. Because $(1/2)^{20}$ is equivalent to less than one in a million times, an event occurring with a probability of only $(1/2)^{1000}$ is virtually impossible.

Two major points are significant here:

1. The outcomes of segregation, independent assortment, and fertilization, like coin tossing, are subject to random fluctuations from their predicted occurrences as a result of chance deviation.

2. As the sample size increases, the average deviation from the expected results decreases. Therefore, a larger sample size diminishes the impact of chance deviation on the final outcome.

In genetics, being able to evaluate observed deviation is a crucial skill. When we assume that data will fit a given ratio such as 1:1, 3:1, or 9:3:3:1, we establish what is called the **null hypothesis (H_0)**. It is so named because the hypothesis assumes that *no real difference* between the *measured values* (or ratio) and the *predicted values* (or ratio) exists. The apparent difference can be attributed purely to chance. The null hypothesis is evaluated using statistical analysis. On this basis, the null hypothesis may either (1) be rejected or (2) fail to be rejected. If it is rejected, the observed deviation from the expected is not attributed to chance alone. The null hypothesis and the underlying assumptions leading to it must be reexamined. If the null hypothesis fails to be rejected, any observed deviations are attributed to chance.

One of the simplest statistical tests devised to assess the null hypothesis is **chi-square (χ^2) analysis**. This test takes into account the observed deviation in each component of an expected ratio as well as the sample size and reduces them to a single numerical value. This value (χ^2) is then used to estimate how frequently the observed deviation can be expected to occur strictly as a result of chance. The formula used in chi-square analysis is

$$\chi^2 = \Sigma \frac{(o - e)^2}{e}$$

In this equation, o is the observed value for a given category and e is the expected value for that category. The Σ (the Greek letter sigma, the summation symbol) represents the sum of the calculated values for each category of the ratio. Because $(o - e)$ is the deviation (d) in each case, the equation reduces to

$$\chi^2 = \Sigma \frac{d^2}{e}$$

Table 3.1(a) shows the steps in the χ^2 calculation for the F_2 results of a hypothetical monohybrid cross. If you were analyzing these data, you would work from left to right, calculating and entering the appropriate numbers in each column. Regardless of whether the deviation d is positive or negative, it becomes positive after the number is squared. Table 3.1(b) analyzes the F_2 results of a hypothetical dihybrid cross. Make certain that you understand how each number was calculated in the dihybrid example.

The final step in chi-square analysis is to interpret the χ^2 value. To do so, you must initially determine the value of the **degrees of freedom (*df*)**, which is equal to $n - 1$, where n is the number of different categories into which each datum point may fall. For the 3:1 ratio, $n = 2$, so $df = 2 - 1 = 1$. For the 9:3:3:1 ratio, $df = 3$. Degrees of freedom must be taken into account because the greater the number of categories, the more deviation is expected as a result of chance.

Once you have determined the degrees of freedom, you can interpret the χ^2 value in terms of a corresponding **probability** value (***p***). Because this calculation is complex, we usually take the p value from a standard table or graph. Figure 3–11 shows a wide range of χ^2 and p values for various degrees of freedom in both a graph and a table. Let's use the graph to determine the p value. The caption for Figure 3–11(b) explains how to use the table.

To determine p, execute the following steps:

1. Locate the χ^2 value on the abscissa (the horizontal or X axis).

2. Draw a vertical line from this point up to the line on the graph representing the appropriate *df*.

3. Extend a horizontal line from this point to the left until it intersects the ordinate (the vertical or Y axis).

4. Estimate, by interpolation, the corresponding p value.

Using our first example (the monohybrid cross) in Table 3.1, we estimate the p value of 0.48 in this manner [Figure 3–11(a)]. For the dihybrid cross, use this method to see whether you can determine the value. The χ^2 value is 4.16 and $df = 3$. The approximate p value is 0.26. Using the table rather than the graph confirms that both p values are between 0.20 and 0.50. Examine the table to confirm this.

TABLE 3.1 Chi-square analysis

(a) Monohybrid Cross

Expected Ratio	Observed (o)	Expected (e)	Deviation (o − e)	Deviation² (d²)	d²/e
3/4	740	3/4 (1000) = 750	740 − 750 = −10	(−10)² = 100	100/750 = 0.13
1/4	260	1/4 (1000) = 250	260 − 250 = +10	(+10)² = 100	100/250 = 0.40
	Total = 1000				$\chi^2 = 0.53$
					$p = 0.48$

(b) Dihybrid Cross

Expected Ratio	o	e	o − e	d²	d²/e
9/16	587	567	+20	400	0.71
3/16	197	189	+8	64	0.34
3/16	168	189	−21	441	2.33
1/16	56	63	−7	49	0.78
	Total = 1008				$\chi^2 = 4.16$
					$p = 0.26$

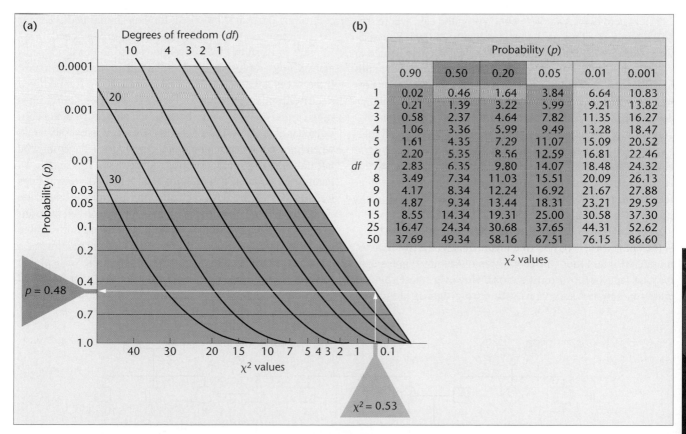

FIGURE 3–11 (a) Graph for converting χ^2 values to p values. (b) Table of χ^2 values for selected values of df and p. A χ^2 value greater than that shown at p = 0.05 justifies rejecting the null hypothesis. In our example, $\chi^2 = 0.53$ for 1 degree of freedom is converted to a p value between 0.20 and 0.50. The graph in (a) gives the more precise p value of 0.48 by interpolation. More darkly shaded values, in (a) and (b) justify failing to reject the null hypothesis.

Thus far, we have been concerned only with finding *p*. The most important aspect of χ^2 analysis is understanding what the *p* value means. We use the example of the dihybrid cross ($p = 0.26$) to illustrate. In these discussions, it is simplest to think of the *p* value as a percentage (e.g., $0.26 = 26\%$). In our example, the *p* value indicates that, if we repeat the same experiment many times, 26 percent of the trials would be expected to exhibit chance deviation as great as or greater than that seen in the initial trial. Conversely, 74 percent of the trials would show less deviation than initially observed as a result of chance.

The above discussion of the *p* values reveals that a hypothesis (a 9:3:3:1 ratio in this case) is never proved or disproved absolutely. Instead, a relative standard is set that enables us to either reject or fail to reject the null hypothesis. This standard is most often a *p* value of 0.05. When applied to chi-square analysis, a *p* value less than 0.05 means that the observed deviation in the set of results will be obtained by chance alone less than 5 percent of the time. Such a *p* value indicates that the difference between the observed and predicted results is substantial and thus enables us to reject the null hypothesis.

On the other hand, *p* values of 0.05 or greater (0.05 to 1.0) indicate that the observed deviation will be obtained by chance alone 5 percent or more of the time. The conclusion is not to reject the null hypothesis. Thus, the *p* value of 0.26, assessing the hypothesis that independent assortment accounts for the results, fails to be rejected. Therefore, the observed deviation can be reasonably attributed to chance.

A final note is relevant here concerning the case where the null hypothesis is rejected, that is, where $p \leq 0.05$. Suppose we are testing the null hypothesis that the data represented a 9:3:3:1 ratio, indicative of independent assortment. If the null hypothesis is rejected, what are alternative interpretations of the data? Researchers first reassess the assumptions that underlie the null hypothesis. We assumed that segregation operates faithfully for both gene pairs. We also assumed that fertilization is random, and that the viability of all gametes is equal irrespective of genotype—that is, that all gametes are equally likely to participate in fertilization. Finally, following fertilization, we assumed that all preadult stages and adult offspring are equally viable, regardless of their genotype.

An example will clarify this: Suppose our null hypothesis is that a dihybrid cross between fruit flies will result in 3/16 mutant wingless flies (the proportion of mutant zygotes that may, in fact, occur at fertilization). However, these mutant embryos may not survive as well during their preadult development, or as young adults, compared to flies whose genotype gives rise to wings. As a result, when the data are gathered, there are fewer than 3/16 wingless flies. Rejection of this null hypothesis alone is not cause for us to disregard the validity of the postulates of segregation and independent assortment, because other factors are operative.

3.9 Human Pedigrees

In the crosses discussed so far, one of the two traits for each character has been dominant to the other. Based on this observation, two significant questions arise:

1. Does the expression of all genes occur in this fashion?

2. Is it possible to ascertain the mode of inheritance of genes in organisms where designed crosses and the production of large numbers of offspring are not practical?

The answer to the first question is no. As we shall see in Chapter 4, many modes of inheritance exist that modify the monohybrid and dihybrid ratios observed by Mendel.

The answer to the second question is yes. The pattern of inheritance of a specific phenotype can be studied even in humans. The simplest way to study this pattern is to construct a family tree that shows the phenotype of the trait in question for each member. Such a family tree is called a **pedigree**. By analyzing the pedigree, we may be able to predict how the gene controlling the trait is inherited. If many similar pedigrees for the same trait are found, the prediction is strengthened.

Figure 3–12 shows the conventions used to construct pedigrees. Circles represent females, and squares designate males. If the sex is unknown, a diamond is used (II-2). If a pedigree traces only a single trait, as Figure 3–12 does, the circles, squares, and diamonds are shaded if the phenotype being considered is expressed. Those who fail to express a recessive trait, when known with certainty to be heterozygous, have only the left half of their square or circle shaded (see II-3 and II-4). Parents are connected by a horizontal line, and vertical lines lead to their offspring. All such offspring are called **sibs** and are connected by a horizontal **sibship line**. Sibs are placed from left to right according to birth order and are labeled with Arabic numerals. Each generation is indicated by a roman numeral.

Twins are indicated by diagonal lines from the vertical line connected to the sibship line. For **monozygotic** or **identical twins**, the diagonal lines are linked by a horizontal line

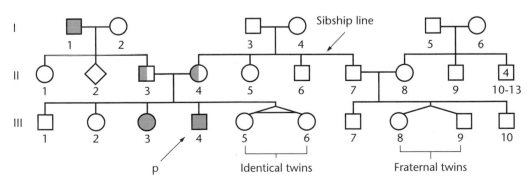

● **FIGURE 3–12** A representative pedigree for a single characteristic through three generations.

(III-5, 6). **Dizygotic** or **fraternal twins** lack this connecting line (III-8, 9). A number within one of the symbols (II-10–13) represents numerous sibs of the same or unknown phenotypes. The individual whose phenotype drew the attention of a physician or geneticist is called the **proband** and is indicated by an arrow connected to the designation **p** (III-4).

The pedigree shown in Figure 3–12 traces the pattern of inheritance of the human trait **albinism**. By analyzing the pedigree, we can see that albinism is a recessive trait.

The male parent of the first generation (I-1) is affected. Because none of his offspring show the disorder, we might conclude that the unaffected female parent (I-2) was a homozygous normal individual. Had she been heterozygous, one-half of the offspring would be expected to exhibit albinism. However, such a small sample (three offspring) prevents any certainty in the matter.

An unaffected second generation is characteristic of a rare recessive trait. If albinism were inherited as a dominant trait, individual II-3 would have to express the disorder in order to pass it to his offspring (III-3 and III-4). He does not. Inspection of the offspring constituting the third generation (row III) provides further support for the hypothesis that albinism is a recessive trait. If so, parents II-3 and II-4 are both heterozygous, and approximately one-fourth of their offspring should be affected. Two of the six offspring do show albinism. This deviation from the expected ratio is not unexpected in crosses with few offspring.

Pedigree analysis of many traits has been an extremely valuable research technique in human genetic studies. However, this approach does not usually provide the certainty in drawing conclusions that is afforded by designed crosses yielding large numbers of offspring. Nevertheless, when many independent pedigrees of the same trait or disorder are analyzed, consistent conclusions can often be drawn. Table 3.2 lists numerous human traits and classifies them according to their recessive or dominant expression. As we shall see in Chapter 4, the genes controlling some of these traits are located on the sex-determining chromosomes.

TABLE 3.2   Representative recessive and dominant human traits

Recessive Traits	*Dominant Traits*
Albinism	Achondroplasia
Alkaptonuria	Brachydactyly
Ataxia telangiectasia	Congenital stationary night blindness
Color blindness	Ehler–Danlos syndrome
Cystic fibrosis	Fascio-scapulo-humeral muscular dystrophy
Duchenne muscular dystrophy	Huntington disease
Galactosemia	Hypercholesterolemia
Hemophilia	Marfan syndrome
Lesch–Nyhan syndrome	Neurofibromatosis
Phenylketonuria	Phenylthiocarbamide tasting (PTC)
Sickle-cell anemia	Porphyria
Tay–Sachs disease	Widow's peak

How Mendel's Peas Become Wrinkled: A Molecular Explanation

Only recently, well over a hundred years after Mendel used wrinkled peas in his ground-breaking hybridization experiments, have we come to find out how the *wrinkled* gene makes peas wrinkled. The wild-type allele of the gene encodes a protein called **starch-branching enzyme (SBEI)**. This enzyme catalyzes the formation of highly branched starch molecules as the seed matures. Wrinkled peas, which result from the homozygous presence of the mutant form of the gene, lack the activity of this enzyme. The production of branch points is inhibited during the synthesis of starch within the seed. This in turn leads to the accumulation of more sucrose and to a higher water content while the seed develops. Osmotic pressure inside rises, which causes the loss of water internally and ultimately the wrinkled appearance of the seed during its maturation. In contrast, developing seeds that bear at least one copy of the normal gene (being either homozygous or heterozygous for the dominant allele) synthesize starch, and reach an osmotic balance that minimizes the loss of water. The end result is a smooth-textured outer coat.

The *SBEI* gene has been cloned and analyzed. Interestingly, the mutant gene contains a foreign sequence of some 800 base pairs that disrupts the normal coding sequence. This foreign segment closely resembles other such sequences, called **transposable elements**. These sequences have the ability to move from place to place in the genome of organisms. Transposable elements similar to this one have been found in maize (corn), parsley, and snapdragons. We cover transposable elements in greater detail in Chapter 14. Study of the *SBEI* gene provides greater insight into the relationship between genotypes and phenotypes.

Insights and Solutions

As a student, you will be asked to demonstrate your knowledge of transmission genetics by solving genetics problems. Success at this task represents not only comprehension of theory but its application to more practical genetic situations. Most students find problem solving in genetics to be challenging but rewarding. This section will provide you with basic insights into the reasoning essential to this process.

Genetics problems are in many ways similar to word problems in algebra. The approach taken is identical: (1) Analyze the problem carefully; (2) translate words into symbols, defining each one first; and (3) choose and apply a specific technique to solve the problem. The first two steps are critical. The third step is largely mechanical.

The simplest problems state all necessary information about the P_1 generation and ask you to find the expected ratios of the F_1 and F_2 genotypes and/or phenotypes. Always follow these steps when you encounter this type of problem:

1. Determine insofar as possible the genotypes of the individuals in the P_1 generation.

2. Determine what gametes may be formed by the P_1 parents.

3. Recombine gametes by the Punnett square or the forked-line methods, or, if the situation is very simple, by inspection. Read the F_1 phenotypes directly.

4. Repeat the process to obtain information about the F_2 generation.

Determining the genotypes from the given information requires that you understand the basic theory of transmission genetics. Consider this problem: *A recessive mutant allele, black, causes a very dark body in* Drosophila *when homozygous. The wild-type (normal) color is gray. What F_1 phenotypic ratio is predicted when a black female is crossed with a gray male whose father was black?*

To work this problem, you must understand dominance and recessiveness as well as the principle of segregation. Furthermore, you must use the information about the male parent's father. Here is one way to work this problem.

1. Because the female parent is black, she must be homozygous for the mutant allele (*bb*).

2. The male parent is gray; therefore, he must have at least one dominant allele (*B*). Because his father was black (*bb*) and he received one of the chromosomes bearing these alleles, the male parent must be heterozygous (*Bb*).

From here, the problem is simple:

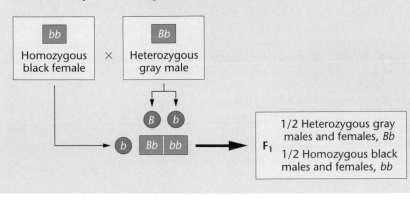

Apply this approach to the following problems.

1. In Mendel's work, he found that full pods are dominant to constricted pods while round seeds are dominant to wrinkled seeds. One of his crosses was between full, round plants and constricted, wrinkled plants. From this cross, he obtained an F_1 generation that was all full and round. In the F_2 generation, Mendel obtained his classic 9:3:3:1 ratio. Using this information, determine the expected F_1 and F_2 results of a cross between homozygous constricted, round and full, wrinkled plants.

Solution: First, define gene symbols for each pair of contrasting traits. Use the lowercase first letter of the recessive traits to designate those phenotypes and the uppercase first letter to designate the dominant traits. Thus, *C* and *c* indicate full and constricted, and *W* and *w* indicate the round and wrinkled phenotypes, respectively.

Now, determine the genotypes of the P_1 generation, form gametes, reconstitute the F_1 generation, and read off the phenotype(s):

P_1:	*ccWW*	×	*CCww*
	constricted, round		full, wrinkled

Gametes:	*cW*		*Cw*
F_1:		*CcWw*	
		full, round	

You can see immediately that the F_1 generation expresses both dominant phenotypes and is heterozygous for both gene pairs. Thus, you can expect that the F_2 generation will yield the classic Mendelian ratio of 9:3:3:1. Let's work it out anyway, just to confirm this, using the forked-line method. Because both gene pairs are heterozygous and can be expected to assort independently, we can predict the F_2 outcomes from each gene pair separately and then proceed with the forked-line method.

Every F_2 offspring is subject to the following probabilities:

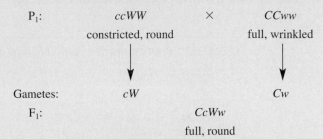

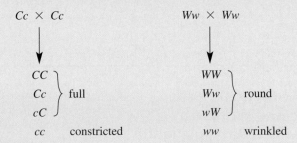

The forked-line method then allows us to confirm the 9:3:3:1 phenotypic ratio. Remember that this represents proportions of 9/16:3/16:3/16:1/16. Note that we are applying the product law as we compute the final probabilities:

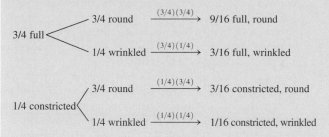

2. In another cross, involving parent plants of unknown genotype and phenotype, the following offspring were obtained. Determine the genotypes and phenotypes of the parents.

> *Offspring:* 3/8 full, round
>
> 3/8 full, wrinkled
>
> 1/8 constricted, round
>
> 1/8 constricted, wrinkled

Solution: This problem is more difficult and requires keener insight because you must work backward. The best approach is to consider the outcomes of pod shape separately from those of seed texture.

Of all plants, 6/8 (3/4) are full and 2/8 (1/4) are constricted. Of the various genotypic combinations that can serve as parents, which will give rise to a ratio of 3/4:1/4? Because this ratio is identical to Mendel's monohybrid F_2 results, we can propose that both unknown parents share the same genetic characteristic as the monohybrid F_1 parents; they must both be heterozygous for the genes controlling pod shape and thus are

$$Cc$$

Before accepting this hypothesis, let's consider the possible genotypic combinations that control seed texture. If we consider this characteristic alone, we see that the traits are expressed in a ratio of 4/8 (1/2) round:4/8 (1/2) wrinkled. To generate such a ratio, the parents cannot both be heterozygous, or their offspring would yield a 3/4:1/4 phenotypic ratio. They cannot both be homozygous, or all offspring would express a single phenotype. Thus, we are left with testing the hypothesis that one parent is homozygous and one is heterozygous for the alleles controlling texture. The potential case of $WW \times Ww$ does not work, since it also yields only a single phenotype. This leaves us with the potential case of $Ww \times ww$. Offspring in such a mating will yield 1/2 Ww (round):1/2 ww (wrinkled), exactly the outcome we are seeking.

Now, let's combine our hypotheses and predict the outcome of crossing. In our solution, we use a dash (–) to indicate that the second allele may be either dominant or recessive, since we are only predicting phenotypes.

3/4 C–
- 1/2 Ww ⟶ 3/8 C–Ww full, round
- 1/2 ww ⟶ 3/8 C–ww full, wrinkled

1/4 cc
- 1/2 Ww ⟶ 1/8 $ccWw$ constricted, round
- 1/2 ww ⟶ 1/8 $ccww$ constricted, wrinkled

As you can see, this cross produces offspring according to our initial information. Thus, we have solved the problem. Note that in this solution, we used genotypes in the forked-line method, in contrast to the use of phenotypes in the earlier solution.

3. Determine the probability that a plant of genotype $CcWw$ will be produced from parental plants of the genotypes $CcWw$ and $Ccww$.

Solution: Because the two gene pairs demonstrate straightforward dominance and recessiveness and assort independently during gamete formation, we need only calculate the individual probabilities of obtaining the two separate outcomes (Cc and Ww) and apply the product law to calculate the final probability:

$$Cc \times Cc \longrightarrow 1/4 \ CC:1/2 \ Cc:1/4 \ cc$$
$$Ww \times ww \longrightarrow 1/2 \ Ww:1/2 \ ww$$
$$p = (1/2 \ Cc)(1/2 \ Ww)=1/4 \ CcWw$$

4. In the laboratory, a genetics student crossed flies with normal, long wings with flies expressing the *dumpy* mutation (truncated wings), which she believed was a recessive trait. In the F_1 generation, all flies had long wings. The following results were obtained in the F_2 generation:

> 792 long-winged flies
>
> 208 dumpy-winged flies

The student tested the hypothesis that the dumpy wing is inherited as a recessive trait using χ^2 analysis of the F_2 data.

(a) What ratio was hypothesized?

(b) Did the χ^2 analysis support the hypothesis?

(c) What do the data suggest about the *dumpy* mutation?

Solution:

(a) The student hypothesized that the F_2 data (792:208) fit Mendel's 3:1 monohybrid ratio for recessive genes.

(b) The initial step in χ^2 analysis is to calculate the expected results (e) if the ratio is 3:1. Then we can compute deviation (d) and the remaining numbers.

Ratio	*o*	*e*	*d*	*d²*	*d²/e*
3/4	792	750	42	1764	2.35
1/4	208	250	−42	1764	7.06
Total =	1000				

$$\chi^2 = \Sigma \frac{d^2}{e}$$
$$= 2.35 + 7.06$$
$$= 9.41$$

We consult Figure 3–11 to determine the probability (p). This value helps us determine whether the deviations can be attributed to chance. There are two possible outcomes (n), so the degrees of freedom (df) = $n − 1$ or 1. The table in Figure 3–11 shows that p is a value between 0.01 and 0.001. The graph gives an estimate of about 0.001. Because $p < 0.05$, we reject the null hypothesis. The data do not fit a 3:1 ratio.

(c) When we accept Mendel's 3:1 ratio as a valid expression of the monohybrid cross, numerous assumptions are made. Examining our underlying assumptions may explain why the null hypothesis was rejected. We assumed that all genotypes are equally viable, that genotypes yielding long wings are equally likely to survive from fertilization through adulthood as the genotype yielding dumpy wings. Further study may reveal that dumpy flies are somewhat less viable than normal flies. As a result, we would expect less than 1/4 of the total offspring to express dumpy. This observation is borne out in the data, although we have not proven that this is the reason.

Chapter Summary

1. Over a century ago, Mendel studied inheritance patterns in the garden pea, establishing the principles of transmission genetics.

2. Mendel's postulates help describe the basis for the inheritance of phenotypic expression. He showed that unit factors, later called alleles, exist in pairs and exhibit a dominant/recessive relationship in determining the expression of traits.

3. Mendel postulated that unit factors (now called alleles) must segregate during gamete formation, such that each gamete receives only one of the two factors with equal probability.

4. Mendel's postulate of independent assortment states that each pair of unit factors segregates independently of other such pairs. As a result, all possible combinations of gametes will be formed with equal probability.

5. The discovery of chromosomes in the late 1800s and subsequent studies of their behavior during meiosis led to the rebirth of Mendel's work, linking the behavior of his unit factors to that of chromosomes during meiosis.

6. The Punnett square and the forked-line methods are used to predict the probabilities of phenotypes (and genotypes) from crosses involving two or more gene pairs.

7. Genetic ratios are expressed as probabilities. Thus, deriving outcomes of genetic crosses requires an understanding of the laws of probability.

8. Statistical analysis is used to test the validity of experimental outcomes. In genetics, variations from the expected ratios due to chance deviations can be anticipated.

9. Chi-square analysis allows us to assess the null hypothesis, namely, that there is no real difference between the expected and observed values. As such, it tests the probability of whether observed variations can be attributed to chance deviation.

10. Pedigree analysis is a method for studying the inheritance pattern of human traits over several generations. It frequently provides the basis for determining the mode of inheritance of human characteristics and disorders.

Key Terms

albinism, 51

allele, 39, 46

branch diagram, 44

chance deviation, 48

chi-square (χ^2) analysis, 48

chromosomal theory of inheritance, 45

continuous variation, 45

degrees of freedom (*df*), 48

dihybrid cross, 41

discontinuous variation, 45

diploid number (2*n*), 45

dizygotic twins, 51

dominance, 39

F_1 (first filial) generation, 37

F_2 (second filial) generation, 37

forked-line method, 44

fraternal twins, 51

gene, 39, 46

genotype, 39

heterozygote, 39

homozygote, 39

identical twins, 50

independent assortment, 42

locus, 46

maternal parent, 46

Mendelian genetics, 37

Mendel's 9:3:3:1 dihybrid ratio, 44

monohybrid cross, 37

monozygotic twins, 50

multiple alleles, 46

null hypothesis, 48

P_1 (parental) generation, 37

paternal parent, 46

pedigree, 50

phenotype, 39

probability, 48

proband, 51

product law, 42, 47

Punnett square, 41

recessive, 39

reciprocal cross, 38

segregation, 39

self-fertilization, 37

selfing, 37

sib, 50

sibship line, 50

starch-branching enzyme, 51

sum law, 47

test cross, 41

transmission genetics, 37

transposable elements, 51

trihybrid cross, 44

unit factors, 38

unit of inheritance, 37

Problems and Discussion Questions

When working genetics problems in this and succeeding chapters, always assume that members of the P_1 generation are homozygous, unless the information given, or the data, indicates otherwise.

1. In a cross between a black and a white guinea pig, all members of the F_1 generation are black. The F_2 generation is made up of approximately 3/4 black and 1/4 white guinea pigs. Diagram this cross, and show the genotypes and phenotypes.

2. Albinism in humans is inherited as a simple recessive trait. Determine the genotypes of the parents and offspring for the following families. When two alternative genotypes are possible, list both. (a) Two nonalbino (normal) parents have five children, four normal and one albino. (b) A normal male and an albino female have six children, all normal.

3. In a problem involving albinism (such as Problem 2), which of Mendel's postulates are demonstrated?

4. Why was the garden pea a good choice as an experimental organism in Mendel's work?

5. Pigeons exhibit a checkered or plain feather pattern. In a series of controlled matings, the following data were obtained:

	F₁ Progeny	
P₁ cross	Checkered	Plain
(a) checkered × checkered	36	0
(b) checkered × plain	38	0
(c) plain × plain	0	35

How are the checkered and plain patterns inherited? Predict the results of the $F_1 \times F_1$ mating from cross (b).

(Left photo: Joyce Photographics/Photo Researchers, Inc.; right photo: R. J. Erwin/Photo Researchers, Inc.)

6. Mendel crossed peas having round seeds and yellow cotyledons with peas having wrinkled seeds and green cotyledons. All the F_1 plants had round seeds with yellow cotyledons. Diagram this cross through the F_2 generation using both the Punnett square and forked-line methods.

7. Determine the genotypes of the parental plants by analyzing the phenotypes of the offspring from these crosses:

Parental Plants	Offspring
(a) round, yellow × round, yellow	3/4 round, yellow 1/4 wrinkled, yellow
(b) round, yellow × wrinkled, yellow	6/16 wrinkled, yellow 2/16 wrinkled, green 6/16 round, yellow 2/16 round, green
(c) round, yellow × wrinkled, green	1/4 round, yellow 1/4 round, green 1/4 wrinkled, yellow 1/4 wrinkled, green

8. Are any of the crosses in Problem 7 test crosses? If so, which one(s)?
9. Which of Mendel's postulates can be demonstrated in the crosses of Problem 7, but not in those in Problems 1 and 5? State this postulate.
10. Correlate Mendel's four postulates with what is now known about homologous chromosomes, genes, alleles, and the process of meiosis.
11. What is the basis for homology among chromosomes?
12. Distinguish between homozygosity and heterozygosity.
13. In *Drosophila*, gray body color is dominant to ebony body color, while long wings are dominant to vestigial wings. Work the following crosses through the F_2 generation and determine the genotypic and phenotypic ratios for each generation. Assume that the P_1 individuals are homozygous: (a) gray, long × ebony, vestigial; (b) gray, vestigial × ebony, long; (c) gray, long × gray, vestigial.

14. How many different types of gametes can be formed by individuals of the following genotypes? What are they in each case? (a) *AaBb*, (b) *AaBB*, (c) *AaBbCc*, (d) *AaBBcc*, (e) *AaBbcc*, and (f) *AaBbCcDdEe*?
15. Using the forked-line method, determine the genotypic and phenotypic ratios of these trihybrid crosses: (a) *AaBbCc × AaBBCC*, (b) *AaBBCc × aaBBCc*, and (c) *AaBbCc × AaBbCc*.
16. Mendel crossed peas with round, green seeds with ones with wrinkled, yellow seeds. All F_1 plants had seeds that were round and yellow. Predict the results of test-crossing these F_1 plants.
17. Shown are F_2 results of two of Mendel's monohybrid crosses. State a null hypothesis that you will test using χ^2 analysis. Calculate the χ^2 value and determine the p value for both crosses, then interpret the p values. Which cross shows a greater amount of deviation?

(a)	Full pods	882
	Constricted pods	299
(b)	Violet flowers	705
	White flowers	224

18. In one of Mendel's dihybrid crosses, he observed 315 round, yellow, 108 round, green, 101 wrinkled, yellow, and 32 wrinkled, green F_2 plants. Analyze these data using the chi-square test to see whether (a) they fit a 9:3:3:1 ratio; (b) the round:wrinkled traits fit a 3:1 ratio; (c) the yellow:green traits fit a 3:1 ratio.
19. A geneticist, in assessing data that fell into two phenotypic classes, observed values of 250:150. He decided to perform chi-square analysis using two different null hypotheses: (a) The data fit a 3:1 ratio; and (b) the data fit a 1:1 ratio. Calculate the χ^2 values for each hypothesis. What can you conclude about each hypothesis?
20. The basis for rejecting any null hypothesis is arbitrary. The researcher can set more or less stringent standards by deciding to raise or lower the critical p value used to reject or fail to reject the hypothesis. Would the use of a standard of $p = 0.10$ be more or less stringent in failing to reject the null hypothesis? Explain.
21. Consider three independently assorting gene pairs, *A/a*, *B/b*, and *C/c*, where each demonstrates typical dominance (*A*–, *B*–, *C*–), and recessiveness (*aa*, *bb*, *cc*). What is the probability of obtaining an offspring that is *AABbCc* from parents that are *AaBbCC* and *AABbCc*?
22. For the following pedigree, predict the mode of inheritance and the resulting genotypes of each individual. Assume that the alleles *A* and *a* control the expression of the trait.

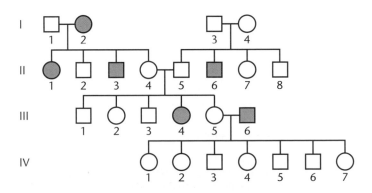

23. Which of Mendel's postulates are demonstrated by the pedigree in Problem 22? List and define these postulates.

24. The following pedigree follows the inheritance of myopia (near-sightedness) in humans. Predict whether the disorder is inherited as a dominant or a recessive trait. Based on your prediction, indicate the most probable genotype for each individual.

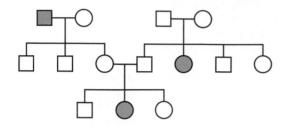

25. Draw all possible conclusions concerning the mode of inheritance of the trait denoted in each of the following limited pedigrees. (Each case is based on a different trait.)

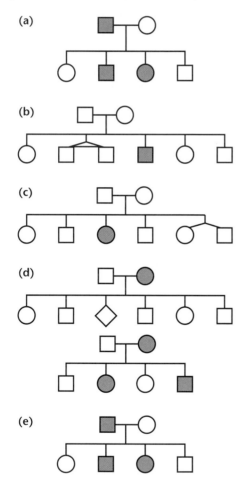

26. Two true-breeding pea plants are crossed. One parent is round, terminal, violet, constricted, while the other expresses the contrasting phenotypes of wrinkled, axial, white, full. The four pairs of contrasting traits are controlled by four genes, each located on a separate chromosome. In the F_1 generation, only round, axial, violet, and full are expressed. In the F_2 generation, all possible combinations of these traits are expressed in ratios consistent with Mendelian inheritance.
    (a) Based on the F_1 results, what conclusion can you draw about the inheritance of these traits?
    (b) In the F_2 results, which phenotype appears most frequently? Write a mathematical expression that predicts the frequency of occurrence of this phenotype.
    (c) Which F_2 phenotype is expected to occur least frequently? Write a mathematical expression that predicts this frequency.
    (d) In the F_2 generation, how often is either P_1 phenotype likely to occur?
    (e) If the F_1 plant is test-crossed, how many different phenotypes will be produced? How does this number compare to the number of different phenotypes in the F_2 generation discussed above?

27. Tay–Sachs disease (TSD) is an inborn error of metabolism that results in death usually by the age of 2. You are a genetic counselor, and you interview a phenotypically normal couple who consult you because the man had a female first cousin (on his father's side) who died from TSD and the woman had a maternal uncle with TSD. There are no other known cases in either family, and none of the matings were/are between related individuals. Assume that this trait is rare in this population.
    (a) Using standard pedigree symbols, draw a pedigree of these individuals' families showing the relevant individuals.
    (b) This couple asks you to calculate the probability that they both are heterozygous for the TSD allele.
    (c) They also want to know the probability that neither of them is heterozygous.
    (d) Finally, they ask you for the probability that one of them is heterozygous but the other is not.
    (*Hint:* The answers to b, c, and d should add up to 1.)

28. The wild-type (normal) fruit fly, *Drosophila melanogaster*, has straight wings and long bristles. Mutant strains have been isolated that have either curled wings or shaven bristles. The genes representing these two mutant traits are located on separate autosomes. Carefully examine the data from the five crosses below. (a) For each mutation, determine whether it is dominant or recessive. In each case, identify which crosses support your answer; and (b) define gene symbols, and for each cross, determine the genotypes of the parents.

				Number of Progeny			
	Cross		*straight wings, long bristles*	*straight wings, short bristles*	*curled wings, long bristles*	*curled wings, short bristles*
1	straight, short	× straight, short	30	90	10	30
2	straight, long	× straight, long	120	0	40	0
3	curled, long	× straight, short	40	40	40	40
4	straight, short	× straight, short	40	120	0	0
5	curled, short	× straight, short	20	60	20	60

Selected Readings

Carlson, E. A. 1987. *The gene: A critical history*. 2nd ed. Philadelphia: Saunders.

Cummings, M. R. 2000. *Human heredity: principles and issues*, 5th ed. Belmont, CA: Wadsworth.

Dunn, L. C. 1965. *A short history of genetics*. New York: McGraw-Hill.

Miller, J. A. 1984. Mendel's peas: A matter of genius or guile? *Science News* 125:108–09.

Olby, R. C. 1985. *Origins of Mendelism*, 2nd ed. London: Constable.

Orel, V. 1996. *Gregor Mendel. The first geneticist*. New York: Oxford University Press.

Peters, J., ed. 1959. *Classic papers in genetics*. Englewood Cliffs, NJ: Prentice-Hall.

Sokal, R. R., and Rohlf, F. J. 1987. *Introduction to biostatistics*, 2nd ed. New York: W. H. Freeman.

Soudek, D. 1984. Gregor Mendel and the people around him. *Am. J. Hum. Genet.* 36:495–98.

Stern, C. 1950. *The birth of genetics*. (Supplement to *Genetics* 35.)

Stern, C., and Sherwood, E. 1966. *The origin of genetics: a Mendel source book*. San Francisco: W. H. Freeman.

Stubbe, H. 1972. *History of genetics: from prehistoric times to rediscovery of Mendel's laws*. Cambridge, MA: MIT Press.

Sturtevant, A. H. 1965. *A history of genetics*. New York: Harper & Row.

Tschermak-Seysenegg, E. 1951. The rediscovery of Mendel's work. *J. Hered.* 42:163–72.

Voeller, B. R., ed. 1968. *The chromosome theory of inheritance: Classical papers in development and heredity*. New York: Appleton-Century-Crofts.

Welling, F. 1991. Historical study: Johann Gregor Mendel 1822–1884. *Am. J. Med. Genet.* 40:1–25.

Mice expressing the black, yellow, and agouti coat phenotypes. *(Tom Cerniglio/Oak Ridge National Laboratory)*

4

Modification of Mendelian Ratios

CHAPTER CONCEPTS

Specific phenotypes are often controlled by one or more gene pairs whose alleles exhibit modes of expression other than dominance and recessiveness. In all such cases, however, the Mendelian principles of segregation and independent assortment are operative during the distribution of the alleles into gametes.

In Chapter 3, we discussed the simplest principles of transmission genetics. We saw that genes are present on homologous chromosomes and that these chromosomes **segregate** from each other and **assort independently** with other segregating chromosomes during gamete formation. These two postulates are the fundamental principles of gene transmission from parent to offspring. However, when gene expression does not adhere to a simple dominant/recessive mode or when more than one pair of genes influences the expression of a single character, the classic 3:1 and 9:3:3:1 ratios are usually modified. Although more complex modes of inheritance result, the fundamental principles set down by Mendel still hold true in these situations.

In this chapter we restrict our initial discussion to the inheritance of traits that are under the control of only one set of genes. In diploid organisms, which have homologous pairs of chromosomes, two copies of each gene influence such traits. The copies need not be identical because alternative forms of genes, or alleles, occur within populations. How alleles influence a given phenotype is our primary focus. Then we consider how a single phenotype can be controlled by more than one set of genes, a situation sometimes described as gene interaction. Numerous examples are explored.

We also examine cases where genes are present on the X chromosome, illustrating **X-linkage**. Thus far, we have restricted our discussion to chromosomes other than the X and Y pair, referred to as **autosomes**. As we shall see, X-linkage provides yet another modification of Mendelian ratios. And finally, we entertain the ideas that the expression of a given phenotype depends on the cell or organism's overall environment and that phenotypic expression depends on more than just the genotype of an organism.

4.1 Potential Function of Alleles

Following the rediscovery of Mendel's work in the early 1900s, research focused on the many ways in which genes influence an individual's phenotype. This course of investigation, stemming from Mendel's findings, is called **neo-Mendelian genetics** (*neo* from the Greek word meaning since or new).

Each type of inheritance described in this chapter was investigated when observations of genetic data did not conform precisely to the expected Mendelian ratios. Hypotheses that modified and extended the Mendelian principles were proposed and tested with specifically designed crosses. Explanations were provided that were in accord with the principle that a phenotype is under the control of one or more genes located at specific loci on one or more pairs of homologous chromosomes.

To understand the various modes of inheritance, we must first examine the potential function of alleles. Alleles are alternative forms of the same gene. The allele that occurs most frequently in a population, the one that we arbitrarily designate as normal, is often referred to as the **wild-type allele**. This common allele is usually dominant—the allele for tall plants in the garden pea, for example—and its product is therefore functional in the cell. Wild-type alleles are responsible, of course, for the corresponding wild-type phenotype and are the standards against which all mutations at a particular locus are compared.

A mutant allele contains modified genetic information and often specifies an altered gene product. For example, in human populations, there are many known alleles of the gene that encodes the β-chain of human hemoglobin. All such alleles store information necessary for the synthesis of the β-chain polypeptide, but each allele specifies a slightly different form of the same molecule. Once manufactured, the product of an allele may or may not have its function altered.

The process of mutation is the source of new alleles. A new allele often leads to a change in the phenotype. A new phenotype results from a change in the functional activity of the cellular product controlled by that gene. Usually, the alteration or mutation is expressed as a loss of the specific wild-type function. For example, if a gene is responsible ultimately for the synthesis of a specific enzyme, a mutation in the gene may change the conformation of this enzyme and eliminate its affinity for the substrate. Such a mutation results in a total loss of function. Conversely, another organism may have a different mutation in the same gene, representing a different allele. The resulting enzyme may demonstrate a reduced or increased affinity for binding the substrate, or it may not have its affinity altered at all. Correspondingly, the functional capacity of the enzyme may be reduced, enhanced, or unchanged.

Although phenotypic traits can be affected by a single mutation, traits are often influenced by many gene products. In the case of enzymatic reactions, most are part of complex metabolic pathways. Therefore, phenotypic traits are often under the control of more than one gene and the allelic forms of each gene involved. In each of the many crosses discussed in the next few chapters, only one or a relatively few gene pairs are involved. Keep in mind that, in each cross, all genes that are not under consideration are assumed to have no effect on the inheritance patterns described.

4.2 Symbols for Alleles

In Chapter 3, we symbolized alleles for very simple Mendelian traits. The initial letter of the name of a recessive trait, lowercased and italic, denotes the recessive allele, and the same letter in uppercase refers to the dominant allele. Thus, for tall and dwarf, where dwarf is recessive, *D* and *d* represent the alleles responsible for these respective traits. Mendel used upper- and lowercase letters such as these to symbolize his unit factors.

Another useful system was developed in genetic studies of the fruit fly *Drosophila melanogaster* to discriminate between wild-type and mutant traits. This system uses the initial letter, or a combination of two or three letters, of the name of the mutant trait. If the trait is recessive, lowercase is used; if it is dominant, uppercase is used. The contrasting wild-type trait is denoted by the same letter, but with a superscript +. For example, *ebony* is a recessive body-color mutation in *Drosophila*. The normal wild-type body-color is gray. Using the above system, *ebony* is denoted by the symbol e, while gray is denoted by e^+. The responsible locus can be occupied by either the wild-type allele (e^+) or the mutant allele (e). A diploid fly may thus exhibit one of three possible genotypes:

e^+/e^+	gray homozygote (wild type)
e^+/e	gray heterozygote (wild type)
e /e	ebony homozygote (mutant)

The slash between the letters indicates that the two allele designations represent the same locus on two homologous chromosomes. If we instead consider a dominant wing mutation such as *Wrinkled* (*Wr*) wing in *Drosophila*, the three possible designations are Wr^+/Wr^+, Wr^+/Wr, and Wr/Wr. The latter two genotypes express the wrinkled-wing phenotype.

One advantage of this system is that further abbreviation can be used when convenient: The wild-type allele may simply be denoted by the $+$ symbol. Using *ebony* as an example, the designations of the three possible genotypes become

$+/+$	gray homozygote (wild type)
$+/e$	gray heterozygote (wild type)
e/e	ebony homozygote (mutant)

As we shall see in Chapter 8, this system is particularly useful when two or three genes are linked on the same chromosome and considered simultaneously. If no dominance exists, we simply use uppercase italic letters and superscripts to denote alternative alleles (e.g., R^1 and R^2, L^M and L^N, I^A and I^B). Their use will become apparent in ensuing sections of this chapter.

Two other points are important. First, although we have adopted a standard convention for assigning genetic symbols, many diverse systems of genetic nomenclature are used to identify genes in various organisms. Usually, the symbol selected reflects the function of the gene, or even a disorder caused by a mutant gene. For example, the *cdc* gene discussed in Chapter 2 refers to *cell division cycle* genes discovered in yeast. In bacteria, leu^- refers to a mutation that interrupts the biosynthesis of the amino acid leucine, where the wild-type gene is designated leu^+. The symbol *dnaA* represents a bacterial gene involved in DNA replication (and DnaA designates the protein made by that gene). In humans, capital letters are used to name genes: *BRCA1* represents a gene associated with susceptibility to breast cancer. Although these different systems may sometimes be confusing, they are all acceptable ways to symbolize genes.

4.3   Incomplete (Partial) Dominance

A cross between parents with contrasting traits may generate offspring with an intermediate phenotype. For example, if plants such as four-o'clocks or snapdragons with red flowers are crossed with white-flowered plants, the offspring have pink flowers. Because some red pigment is produced in the F_1 intermediate-colored pink flowers, neither red nor white flower color is dominant. Such a situation is known as **incomplete**, or **partial**, **dominance**.

If this phenotype is under the control of a single gene and two alleles where neither is dominant, the results of the F_1 (pink) $\times$ F_1 (pink) cross can be predicted. The resulting F_2 generation shown in Figure 4–1 confirms the hypothesis that only one pair of alleles determines these phenotypes. The genotypic ratio (1:2:1) of the F_2 generation is identical to that of Mendel's monohybrid cross. However, because neither allele is dominant, the phenotypic ratio is identical to the genotypic ratio. Note that because neither allele is recessive, we have chosen not to use upper- and lowercase letters as symbols. Instead, we denote the red and white alleles as R^1 and R^2. We could have used W^1 and W^2 or still other designations such as C^W and C^R, where C indicates "color" and the W and R superscripts indicate white and red.

Clear-cut cases of incomplete dominance, which result in intermediate expression of the overt phenotype, are relatively rare. However, even when complete dominance seems apparent, careful examination of the gene product, rather than the phenotype, often reveals an intermediate level of gene expression. An example is the human biochemical disorder **Tay–Sachs disease**, in which homozygous recessive indi-

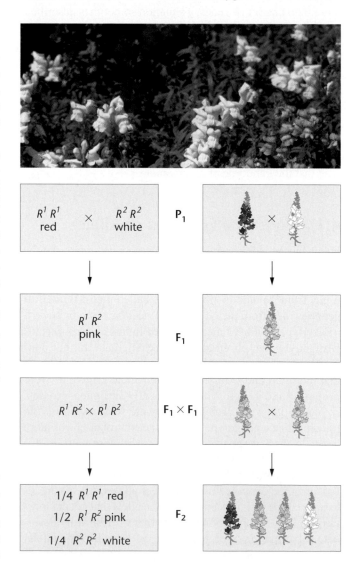

FIGURE 4–1 Incomplete dominance shown in the flower color of snapdragons. *(Photo: John D. Cunningham/Visuals Unlimited)*

viduals are severely affected with a fatal lipid storage disorder, and neonates die during their first 1–3 years of life. In afflicted individuals there is almost no activity of the responsible enzyme **hexosaminidase**, which is normally involved in lipid metabolism. Heterozygotes, with only a single copy of the mutant gene, are phenotypically normal, but express only about 50 percent of the enzyme activity found in homozygous normal individuals. Fortunately, this level of enzyme activity is adequate to achieve normal biochemical function. This situation is not uncommon in enzyme disorders.

4.4 Codominance

If two alleles of a single gene are responsible for producing two distinct, detectable gene products, a situation different from incomplete dominance or dominant/recessive arises. In such a case, the joint expression of both alleles in a heterozygote is called **codominance**. The **MN blood group** in humans illustrates this phenomenon and is characterized by a molecule called a glycoprotein, found on the surface of red blood cells. In the human population, two forms of this glycoprotein exist, designated M and N. An individual may exhibit either one or both of them.

The MN system is under the control of an autosomal locus found on chromosome 4 and two alleles designated L^M and L^N. Because humans are diploid, three combinations are possible, each resulting in a distinct blood type:

Genotype	Phenotype
$L^M L^M$	M
$L^M L^N$	MN
$L^N L^N$	N

As predicted, a mating between two MN parents may produce children of all three blood types:

$$L^M L^N \times L^M L^N$$
$$\downarrow$$

1/4	$L^M L^M$
1/2	$L^M L^N$
1/4	$L^N L^N$

This example shows that codominant inheritance is characterized by distinct expression of the gene products of both alleles. This characteristic distinguishes it from incomplete dominance, where heterozygotes express an intermediate, blended phenotype.

4.5 Multiple Alleles

Because the information stored in any gene is extensive, mutations can modify this information in many ways. Each change can potentially produce a different allele. Therefore, for any specific gene, the number of alleles within members of a population need not be restricted to only two. When three or more alleles of the same gene are found, **multiple alleles** are said to be present, creating a unique mode of inheritance. It is important to realize that *multiple alleles can be studied only in populations*. Any individual diploid organism has, at most, two homologous gene loci that may be occupied by different alleles of the same gene. However, among members of a species, many alternative forms of the same gene can exist.

The ABO Blood Group

The simplest case of multiple alleles is that in which three alternative alleles of one gene exist. This situation is illustrated by the **ABO blood group** in humans, discovered by Karl Landsteiner in the early 1900s. The ABO system, like the MN blood types, is characterized by the presence of antigens on the surface of red blood cells. The A and B antigens are distinct from the MN antigens and are under the control of a different gene, located on chromosome 9. As in the MN system, one combination of alleles in the ABO system exhibits a codominant mode of inheritance.

When individuals are tested using antisera that contain antibodies against the A or B antigen, four phenotypes are revealed. Each individual has either the A antigen (A phenotype), the B antigen (B phenotype), the A and B antigens (AB phenotype), or neither antigen (O phenotype). In 1924, it was hypothesized that these phenotypes were inherited as the result of three alleles of a single gene. This hypothesis was based on studies of the blood types of many different families.

Although different designations can be used, we use the symbols I^A, I^B, and I^O to distinguish these three alleles. The I designation stands for isoagglutinogen, another term for antigen. If we assume that the I^A and I^B alleles are responsible for the production of their respective A and B antigens and that I^O is an allele that does not produce any detectable A or B antigens, we can list the various genotypic possibilities and assign the appropriate phenotype to each:

Genotype	Antigen	Phenotype
$I^A I^A$	A	A
$I^A I^O$	A	
$I^B I^B$	B	B
$I^B I^O$	B	
$I^A I^B$	A, B	AB
$I^O I^O$	Neither	O

Note that in these assignments the I^A and I^B alleles behave dominantly to the I^O allele, but codominantly to each other. Our knowledge of human blood types has several practical applications, the most important of which is compatible blood transfusions.

The Bombay Phenotype

The biochemical basis of the ABO blood type system has now been carefully worked out. The A and B antigens are actually carbohydrate groups (sugars) that are bound to lipid molecules (fatty acids) protruding from the membrane of the red blood cell. The specificity of the A and B antigens is based on the terminal sugar of the carbohydrate group. Both

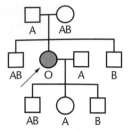

FIGURE 4–2 A partial pedigree of a woman displaying the Bombay phenotype. Functionally, her ABO blood group behaves as type O. Genetically, she is type B.

the A and B antigens are derived from a precursor molecule called the **H substance**, to which one or two terminal sugars is added.

In extremely rare instances, first recognized in a woman in Bombay, the H substance is incompletely formed. As a result, it is an inadequate substrate for the enzyme that normally adds either terminal sugar. This condition results in blood type O and is called the **Bombay phenotype**. It has been shown to be due to a rare recessive mutation, *h*, at a locus separate from that controlling the A and B antigens. Thus, even though an individual may contain the I^A and/or I^B alleles, if he or she displays the *hh* genotype, neither the A or B antigen can be added. This information helped explain why the woman in Bombay was typed as O even though one of her parents was type AB (and thus she should not have been type O), and why she was able to donate the I^B allele to her children (Figure 4–2).

The *white* Locus in *Drosophila*

Many other phenotypes in plants and animals are known to be controlled by multiple allelic inheritance. In *Drosophila*, for example, many alleles are known at practically every locus. The recessive mutation that causes white eyes, discovered by Thomas H. Morgan and Calvin Bridges in 1912, is one of over 100 alleles that can occupy this locus. In this allelic series, eye colors range from complete absence of pigment in the *white* allele, to deep ruby in the *white-satsuma*

TABLE 4.1 Some of the alleles present at the *white* locus of *Drosophila melanogaster* and their eye color phenotype

Allele	*Name*	*Eye Color*
w	*white*	pure white
w^a	*white-apricot*	yellowish orange
w^{bf}	*white-buff*	light buff
w^{bl}	*white-blood*	yellowish ruby
w^{cf}	*white-coffee*	deep ruby
w^e	*white-eosin*	yellowish pink
w^{mo}	*white-mottled orange*	light mottled orange
w^{sat}	*white-satsuma*	deep ruby
w^{sp}	*white-spotted*	fine grain, yellow mottling
w^t	*white-tinged*	light pink

allele, to orange in the *white-apricot* allele, to a buff color in the *white-buff* allele. These alleles are designated *w*, w^{sat}, w^a, and w^{bf}, respectively (Table 4.1). In each case, the total amount of pigment in these mutant eyes is reduced to less than 20 percent of that found in the brick-red, wild-type eye.

4.6 Lethal Alleles

Many gene products are essential to an organism's survival. Mutations resulting in the synthesis of a gene product that is nonfunctional can sometimes be tolerated in the heterozygous state; that is, one wild-type allele may be sufficient to produce enough of the essential product to allow survival. However, such a mutation behaves as a recessive **lethal allele**, and homozygous recessive individuals will not survive. The time of death will depend on when the product is essential. In mammals, for example, this might occur during development, early childhood, or even during adulthood.

In some cases, the allele responsible for a lethal effect when homozygous may also result in a distinctive mutant phenotype when present heterozygously. Such an allele is behaving as a recessive lethal but is dominant with respect to the phenotype. For example, a mutation that causes a yellow coat in mice was discovered in the early part of this century. The yellow coat varies from the normal agouti (wild-type) coat phenotype, as shown in Figure 4–3. Crosses between the various combinations of the two strains yield unusual results:

Crosses
A: agouti × agouti → all agouti
B: yellow × yellow → 2/3 yellow: 1/3 agouti
C: agouti × yellow → 1/2 yellow: 1/2 agouti

These results are explained on the basis of a single pair of alleles. With regard to coat color, the mutant *yellow* allele (A^Y) is dominant to the wild-type *agouti* allele (*A*), so heterozygous mice will have yellow coats. However, the *yellow* allele also behaves as a homozygous recessive lethal. When present in two copies, the mice die before birth. Thus, no homozygous yellow mice are ever recovered. The genetic basis for these three crosses is shown in Figure 4–3.

Other mutant genes are known to behave as dominant lethal alleles, where the presence of one copy of the allele results in the death of the individual. In humans, a disorder called **Huntington disease** (previously referred to as Huntington's chorea) is due to a dominant autosomal allele *H*, where the onset of the disease in heterozygotes (*Hh*) is delayed, usually well into adulthood. Affected individuals then undergo gradual nervous and motor degeneration until they die. This lethal disorder is particularly tragic because it has such a late onset, typically at about age 40. By that time, the affected individual may have produced a family. Each child has a 50 percent probability of inheriting the lethal allele, transmitting the allele to his or her offspring, and eventually developing the disorder. The American folk singer and composer Woody Guthrie died from this disease at age 39.

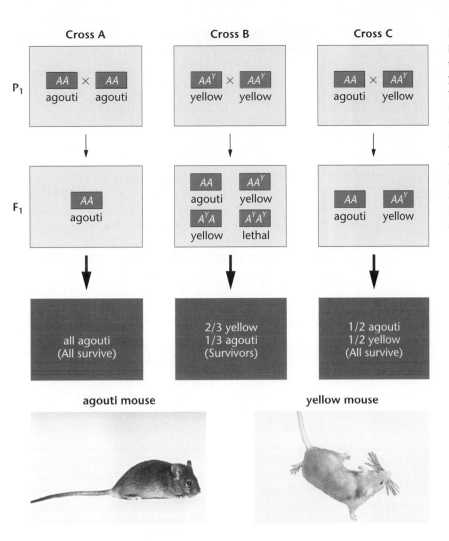

FIGURE 4–3 Inheritance patterns in three crosses involving the wild-type agouti allele (A) and the mutant yellow allele (A^Y) in the mouse. Note that the mutant allele behaves dominantly to the normal allele (A) in controlling coat color, but it also behaves as a homozygous lethal allele. The genotype A^YA^Y does not survive. *(Left photo: Courtesy of Stanton K. Short [The Jackson Laboratory, Bar Harbor, ME]); (Right photo: Tom Cerniglio/Oak Ridge National Laboratory)*

Dominant lethal alleles are rarely observed. For them to exist in a population, the affected individual must reproduce before lethality caused by the allele is expressed, as can occur in Huntington disease. If all affected individuals die before reaching reproductive age, the mutant gene will not be passed to future generations, and the mutation will disappear from the population unless it arises again as a result of a new mutation.

4.7 Combinations of Two Gene Pairs

Each example discussed so far modifies Mendel's 3:1 F_2 monohybrid ratio. Therefore, combining any two of these modes of inheritance in a dihybrid cross will likewise modify the classical 9:3:3:1 ratio. Having established the foundation for the modes of inheritance of incomplete dominance, codominance, multiple alleles, and lethal genes, we can now deal with the situation of two modes of inheritance occurring simultaneously. Mendel's principle of independent assortment applies to these situations, provided that the genes controlling each character are not linked on the same chromosome.

Consider, for example, a mating that occurs between two humans who are both heterozygous for the autosomal recessive gene that causes albinism and who are both of blood type AB.

What is the probability of a particular phenotypic combination occurring in each of their children? Albinism is inherited in the simple Mendelian fashion, and the blood types are determined by the series of three multiple alleles, I^A, I^B, and I^O. The solution to this problem is diagrammed in Figure 4–4, using the forked-line method. This dihybrid cross does not yield the classical four phenotypes in a 9:3:3:1 ratio; instead, six phenotypes occur in a 3:6:3:1:2:1 ratio, establishing the expected probability for each phenotype. This is just one of many variants of modified ratios possible when different modes of inheritance are combined.

4.8 Gene Interaction

Soon after Mendel's work was rediscovered, experimentation revealed that individual characteristics displaying discrete phenotypes are often under the control of more than one gene. This was a significant discovery because it revealed that genetic influence on the phenotype is much more complex than envisioned by Mendel. Instead of single genes controlling the development of individual parts of the plant and animal body, it soon became clear that phenotypic characters can be influenced by the interactions of many different genes and their products.

FIGURE 4–4 Calculation of the probabilities in a mating involving the ABO blood type and albinism in humans using the forked-line method.

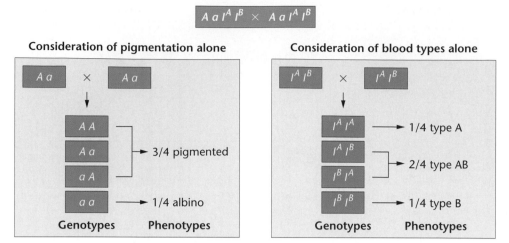

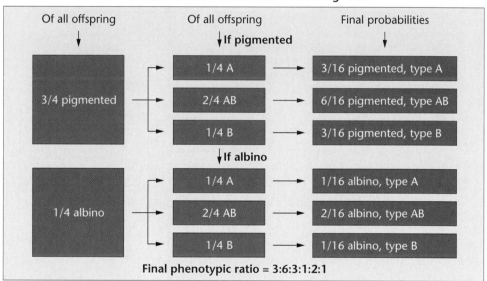

The term **gene interaction** is often used to describe the idea that several genes influence a particular characteristic. This does not mean, however, that two or more genes, or their products, necessarily interact directly with one another to influence a particular phenotype. Rather, the cellular function of numerous gene products contributes to the development of a common phenotype. For example, the development of an organ such as the compound eye of an insect is exceedingly complex and leads to a structure with multiple phenotypic manifestations—most simply described as a specific size, shape, texture, and color. The development of the eye can best be envisioned as a complex cascade of developmental events leading to its formation. This process exemplifies the developmental concept of **epigenesis** (first introduced in Chapter 1), whereby each ensuing step of development increases the complexity of this sensory organ and is under the control and influence of one or more genes.

Epistasis

Some of the best examples of gene interaction are those that show the phenomenon of **epistasis**. Derived from the Greek word for "stoppage," epistasis occurs when the expression of one gene or gene pair masks or modifies the expression of another gene or gene pair. The genes involved control the expression of the same general phenotypic characteristic, sometimes in an antagonistic manner, as when masking occurs. In other cases, however, the genes involved exert their influence on one another in a complementary, or cooperative, fashion.

For example, the homozygous presence of a recessive allele prevents or overrides the expression of other alleles at a second locus (or several other loci). In this case, the alleles at the first locus are said to be *epistatic* to those at the second locus, and the alleles at the second locus are *hypostatic* to those at the first locus. In another example, a single dominant allele at the first locus influences the expression of the alleles at a second gene locus. In a third case, two gene pairs complement one another such that at least one dominant allele at each locus is required to express a particular phenotype.

The Bombay phenotype discussed earlier is an example of a homozygous recessive condition at one locus masking the expression of a second locus (see Figure 4–2). There, we established that the homozygous condition (*hh*) masks the expression of the I^A and I^B alleles. Only individuals containing at least one *H* allele (designated *H–*) can form the A or B antigens. As a result, individuals whose genotypes include the *I^A* or *I^B* allele and who are also *hh* express the type O phenotype, regardless of their potential to make either antigen. An example of the outcome of matings between individuals heterozygous at both loci is shown in Figure 4–5. If many individuals of the genotype *I^AI^BHh* have children, the phenotypic ratio of 3 A: 6 AB: 3 B: 4 O is expected in their offspring.

It is important to note two things when examining this cross and the predicted phenotypic ratio:

1. A key distinction exists in this cross compared to the modified dihybrid cross shown in Figure 4–4: *only one characteristic—blood type—is being followed.* In the modified dihybrid cross in Figure 4–4, blood type *and* skin pigmentation are followed as separate phenotypic characteristics.

2. Even though only a single character was followed, the phenotypic ratio is expressed in 16ths. If we knew nothing about the H substance and the genes controlling it, we could still be confident that a second gene pair, other than that controlling the A and B antigens, is involved in the phenotypic expression. *When studying a single character, a ratio that is expressed in 16 parts (e.g., 3:6:3:4) suggests that two gene pairs are "interacting" during the expression of the phenotype under consideration.*

The study of gene interaction reveals a number of inheritance patterns that modify the classical Mendelian dihybrid F_2 ratio (9:3:3:1) in other ways. In several subsequent examples, epistasis has the effect of combining one or more of the four phenotypic categories in various ways. The generation of these four groups is reviewed in Figure 4–6, along with several modified ratios.

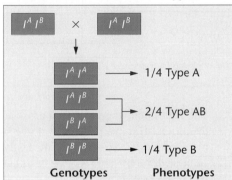
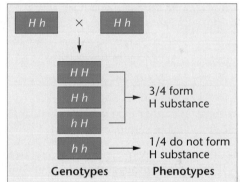
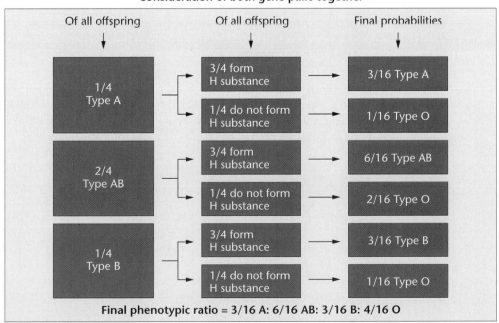

FIGURE 4–5 The outcome of a mating between individuals who are heterozygous at two genes determining their ABO blood type. Final phenotypes are calculated by considering both genes separately and then combining the results using the forked-line method.

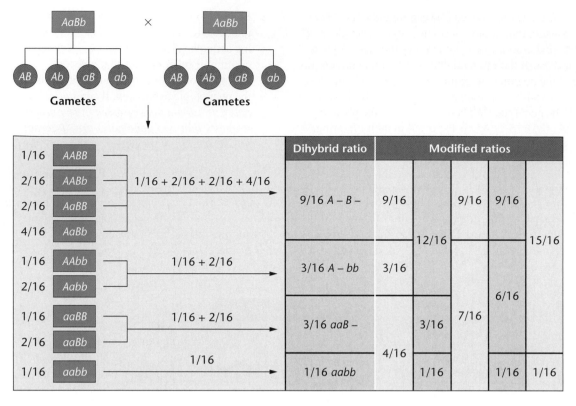

FIGURE 4–6   Generation of the various modified dihybrid ratios from the nine unique genotypes produced in a cross between individuals who are heterozygous at two genes.

As we discuss these and other examples, we will make several assumptions and adopt certain conventions:

1. In each case, distinct phenotypic classes are produced, each clearly discernible from all others. Such traits illustrate discontinuous variation, where phenotypic categories are discrete and qualitatively different from one another.

2. The genes considered in each cross are not linked and therefore assort independently of one another during gamete formation. So that you can easily compare the results of different crosses, we designate alleles as *A*, *a* and *B*, *b* in each case.

Case	Organism	Character	F₂ Phenotypes				Modified ratio
			9/16	3/16	3/16	1/16	
1	Mouse	Coat color	agouti	albino	black	albino	9:3:4
2	Squash	Color	white		yellow	green	12:3:1
3	Pea	Flower color	purple	white			9:7
4	Squash	Fruit shape	disc	sphere		long	9:6:1
5	Chicken	Color	white		colored	white	13:3
6	Mouse	Color	white-spotted	white	colored	white-spotted	10:3:3
7	Shepherd's purse	Seed capsule	triangular			ovoid	15:1
8	Flour beetle	Color	red / sooty / red / sooty	black	jet	black	6:3:3:4

FIGURE 4–7   The basis of modified dihybrid F₂ phenotypic ratios, resulting from crosses between doubly heterozygous F₁ individuals. The four groupings of the F₂ genotypes shown in Figure 4–6 and across the top of this figure are combined in various ways to produce these ratios.

3. When we assume that complete dominance exists between the alleles of any gene pair, such that *AA* and *Aa* or *BB* and *Bb* are equivalent in their genetic effects, we use the designations *A*– or *B*– for both combinations, where the dash (–) indicates that either allele may be present, without consequence to the phenotype.

4. All P_1 crosses involve homozygous individuals (e.g., *AABB* × *aabb*, *AAbb* × *aaBB*, or *aaBB* × *AAbb*). Therefore, each F_1 generation consists of only heterozygotes of genotype *AaBb*.

5. In each example, the F_2 generation produced from these heterozygous parents is our main focus of analysis. When two genes are involved (Figure 4–6), the F_2 genotypes fall into four categories: 9/16 *A–B–*, 3/16 *A–bb*, 3/16 *aaB–*, and 1/16 *aabb*. Because of dominance, all genotypes in each category are equivalent in their effect on the phenotype.

Our first case is the inheritance of coat color in mice (case 1 of Figure 4–7). Normal wild-type coat color is agouti, a grayish pattern formed by alternating bands of pigment on each hair. Agouti is dominant to black (nonagouti) hair, which is caused by a recessive mutation, *a*. Thus, *A*– results in agouti, while *aa* yields black coat color. When it is homozygous, a recessive mutation, *b*, at a separate locus, eliminates pigmentation altogether, yielding albino mice (*bb*), regardless of the genotype at the other locus. The presence of at least one *B* allele allows pigmentation to occur in much the same way that the *H* allele in humans allows the expression of the ABO blood types. In a cross between agouti (*AABB*) and albino (*aabb*), members of the F_1 are all *AaBb* and have agouti coat color. In the F_2 progeny of a cross between two F_1 heterozygotes, the following genotypes and phenotypes are observed:

F_1: *AaBb* × *AaBb*

↓

F_2 Ratio	Genotype	Phenotype	Final Phenotypic Ratio
9/16	*A– B–*	agouti	9/16 agouti
3/16	*A– bb*	albino	4/16 albino
3/16	*aa B–*	black	3/16 black
1/16	*aa bb*	albino	

We can envision gene interaction yielding the observed 9:3:4 F_2 ratio as a two-step process:

	Gene *B*		Gene *A*	
Precursor	↓	Black	↓	agouti
Molecule	⟶	Pigment	⟶	Pattern
(colorless)	*B*–		*A*–	

In the presence of a *B* allele, black pigment can be made from a colorless substance. In the presence of an *A* allele, the black pigment is deposited during the development of hair in a pattern that produces the agouti phenotype. If the *aa* genotype occurs, all of the hair remains black. If the *bb* genotype occurs, no black pigment is produced, regardless of the presence of the *A* or *a* alleles, and the mouse is albino. Therefore, the *bb* genotype masks or suppresses the expression of the *A* gene, demonstrating epistasis.

A second type of epistasis occurs when a dominant allele at one genetic locus masks the expression of the alleles at a second locus. For instance, case 2 of Figure 4–7 deals with the inheritance of fruit color in summer squash. Here, the dominant allele *A* results in white fruit color regardless of the genotype at a second locus, *B*. In the absence of the dominant *A* allele (the *aa* genotype), *BB* or *Bb* results in yellow color, while *bb* results in green color. Therefore, if two white-colored double heterozygotes (*AaBb*) are crossed, an interesting phenotypic ratio occurs because of this type of epistasis:

F_1: *AaBb* × *AaBb*

↓

F_2 Ratio	Genotype	Phenotype	Final Phenotypic Ratio
9/16	*A– B–*	white	12/16 white
3/16	*A– bb*	white	3/16 yellow
3/16	*aa B–*	yellow	1/16 green
1/16	*aa bb*	green	

Of the offspring, 9/16 are *A–B–* and are thus white. The 3/16 bearing the genotypes *A–bb* are also white. Of the remaining squash, 3/16 are yellow (*aaB–*), while 1/16 are green (*aabb*). When combined, the modified ratio of 12:3:1 occurs.

Our third example (case 3 of Figure 4–7), first discovered by William Bateson and Reginald Punnett (of Punnett square fame), is demonstrated in a cross between two strains of white-flowered sweet peas. Unexpectedly, the F_1 plants are all purple, and the F_2 occurs in a ratio of 9/16 purple to 7/16 white. The proposed explanation for these results suggests that the presence of at least one dominant allele of each of two gene pairs is essential for flowers to be purple. All other genotype combinations yield white flowers because the homozygous condition of *either* recessive allele masks the expression of the dominant allele at the other locus.

The cross is shown as follows:

P_1: *AAbb* × *aaBB*
white   white

↓

F_1: All *AaBb* (purple)

↓

F_2 Ratio	Genotype	Phenotype	Final Phenotypic Ratio
9/16	*A– B–*	purple	9/16 purple
3/16	*A– bb*	white	
3/16	*aa B–*	white	7/16 white
1/16	*aa bb*	white	

We can now envision how two gene pairs might yield such results:

	Gene *A*		Gene *B*	
Precursor	↓	Intermediate	↓	Final
Substance	⟶	Product	⟶	Product
(colorless)	*A–*	(colorless)	*B–*	(purple)

At least one dominant allele from each pair of genes is necessary to ensure both biochemical conversions to the final product, yielding purple flowers. In the cross above, this will occur in 9/16 of the F_2 offspring. All other plants (7/16) have flowers that remain white.

Novel Phenotypes

Other cases of gene interaction yield novel, or new, phenotypes in the F_2 generation, in addition to producing modified dihybrid ratios. Case 4 in Figure 4–7 depicts the inheritance of fruit shape in the summer squash *Cucurbita pepo*. When plants with disc-shaped fruit (*AABB*) are crossed to plants with long fruit (*aabb*), the F_1 generation all have disc fruit. However, in the F_2 progeny, fruit with a novel shape—sphere—appear, as well as fruit exhibiting the parental phenotypes. A variety of fruit shapes are shown in Figure 4–8.

The F_2 generation, with a modified 9:6:1 ratio, is generated as follows:

$$F_1: AaBb \times AaBb$$
$$\text{disc} \downarrow \text{disc}$$

F_2 Ratio	Genotype	Phenotype	Final Phenotypic Ratio
9/16	*A– B–*	disc	9/16 disc
3/16	*A– bb*	sphere	
3/16	*aa B–*	sphere	6/16 sphere
1/16	*aa bb*	long	1/16 long

In this example of gene interaction, both gene pairs influence fruit shape equivalently. A dominant allele at either locus ensures a sphere-shaped fruit. In the absence of dominant alleles, the fruit is long. However, if both dominant alleles (*A* and *B*) are present, the fruit displays a flattened, disc shape.

Other Modified Dihybrid Ratios

The remaining cases (5–8) in Figure 4–7 show additional modifications of the dihybrid ratio and provide still other examples of gene interactions. All eight cases have two things in common. First, in arriving at a suitable explanation of the inheritance pattern of each one, we have not violated the principles of segregation and independent assortment. Therefore, the added complexity of inheritance in these examples does not detract from the validity of Mendel's conclusions. Second, the F_2 phenotypic ratio in each example has been expressed in sixteenths. When similar observations are made in crosses where the inheritance pattern is unknown, it suggests to geneticists that two gene pairs are controlling the observed phenotypes. You should make the same infer-

FIGURE 4-8 Summer squash exhibiting various fruit-shape phenotypes, where disc (white), long (orange gooseneck), and sphere (bottom left) are apparent. *(Photo: Irene Vandermolen/Animals Animals/ Earth Scenes)*

ence in your analysis of genetics problems. Other observations on solving genetics problems are provided in Insights and Solutions at the conclusion of this chapter.

4.9   Complementation Analysis and Alleles

An interesting situation arises when two mutations, both of which produce a similar phenotype, are isolated independently. Suppose that two investigators, one in a genetics laboratory in the United States and the other in a genetics laboratory in Canada, independently isolate and establish a true-breeding strain of wingless *Drosophila* and demonstrate that each is due to a recessive mutation. We might assume that both strains contain mutations in the same gene. However, since we know that many genes are involved in the formation of wings, mutations in any one of them might inhibit wing formation during development. The experimental approach called **complementation analysis** allows us to determine whether two such mutations are in the same gene—that is, whether they are alleles or whether they represent mutations in separate genes.

Our analysis seeks to answer this simple question: *Are two mutations that yield similar phenotypes present in the same gene or in two different genes?* To find the answer, we cross the two mutant strains and analyze the F_1 generation. There are two alternative outcomes and interpretations of such a cross, as shown in Figure 4–9 and discussed below. The mutation isolated in the United States is designated m^{usa}, while the mutation isolated in Canada is designated m^{can}.

Case 1. *All offspring develop normal wings.*

Interpretation: The two recessive mutations are in separate genes and are not alleles of one another. Following the cross, all F_1 flies are heterozygous for both genes. *Complementation* is said to occur. Because each mutation

Case 1- Mutations are in
separate genes

Case 2- Mutations are in
different locations within
the same gene

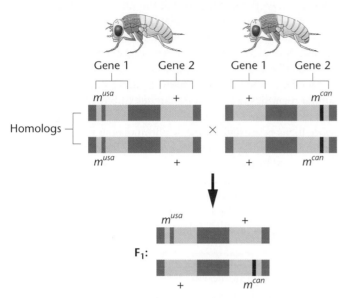

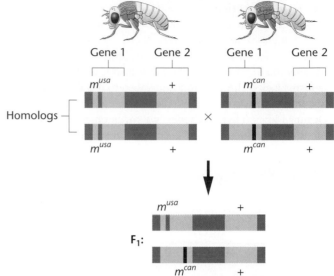

One normal copy of each gene is present.
Complementation occurs.

FLIES ARE WILD TYPE AND DEVELOP WINGS

Gene 1 is mutant in all cases, while Gene 2 is normal
No complementation occurs.

FLIES ARE MUTANT AND DO NOT DEVELOP WINGS

FIGURE 4–9 Complementation analysis of alternative outcomes of two wingless mutations in *Drosophila* (m^{usa} and m^{can}). In case 1, the mutations are not alleles of the same gene, while in case 2, the mutations are alleles of the same gene.

is in a separate gene and each F_1 fly is heterozygous at both loci, the normal products of both genes are produced (by the one normal copy of each gene). As a result, wings develop.

Case 2. *All offspring fail to develop wings.*

Interpretation: The two mutations affect the same gene and are alleles of one another. Complementation does *not* occur. Because the two mutations affect the same gene, the F_1 flies are homozygous for the two mutant alleles (the m^{usa} allele and the m^{can} allele). No normal product of the gene is produced. In the absence of this essential product, wings do not form.

Complementation analysis, as originally devised by the *Drosophila* geneticist Edward B. Lewis, is often called the **cis–trans** test. Borrowed from nomenclature used in organic chemistry, *cis* (along side of one another) refers to the case where the two mutations are on the same homolog. Likewise, *trans* (opposite one another) refers to the case where the mutations are on separate homologs. As shown in Figure 4–9, the *trans* configuration is critical in determining whether the two mutations are alleles of the same gene or not. If we determine that the two mutations are indeed alleles, their presence as heterozygotes in the *cis* configuration serves as an important control in complementation analyses. In this configuration, flies will develop wings.

Complementation analysis may be used to screen any number of individual mutations that result in the same phenotype. Such an analysis may reveal that only a single gene is involved, or that two or more genes are involved. All mutations determined to be present in any single gene are said to fall into the same **complementation group**, and they will complement mutations in all other groups.

If large numbers of mutations affecting the same trait are available and studied using complementation analysis, it is possible to predict the total number of genes involved in the determination of that trait. We return to complementation analysis in Chapter 14, when we discuss the inherited human disorder xeroderma pigmentosum. There, complementation analysis reveals that mutations in any of seven complementation groups (genes) lead to the disorder.

4.10 Genes on the X Chromosome

In many animal and some plant species, one of the sexes contains a pair of unlike chromosomes that are involved in sex determination. In many cases, these are designated as the X and Y chromosomes. For example, in both *Drosophila* and in all mammals, including humans, males contain an X and a Y chromosome, whereas females contain two X chromosomes.

The Y, while behaving as a homolog to the X during meiosis, contains only a few genes, none of which are homologous to those present on the X. As we shall see, this situation creates unique inheritance patterns. The term **X-linkage** describes the transmission and expression of the normal complement of genes located on the X chromosome.

We consider the role of the X and Y chromosomes in sex determination in Chapter 5. There, we also learn that some organisms lack a chromosome equivalent to the Y and that it is not always the case that the male has unlike sex-determining chromosomes. Here, however, we focus on how X-linkage modifies Mendelian ratios, the central theme of this chapter.

X-Linkage in *Drosophila*

One of the first cases of X-linkage was documented by Thomas H. Morgan around 1920 during his studies of the *white* mutation in the eyes of *Drosophila* (Figure 4–10). We use this case to illustrate X-linkage. The normal wild-type red eye color is dominant to white.

Morgan's work established that the inheritance pattern of the white-eye trait is clearly related to the sex of the parent carrying the mutant allele. Unlike the outcome of the typical monohybrid cross, reciprocal crosses between white- and red-eyed flies did not yield identical results. In contrast, in all of Mendel's monohybrid crosses, F_1 and F_2 data were very similar regardless of which P_1 parent exhibited the recessive mutant trait. Morgan's analysis led to the conclusion that the *white* locus is present on the X chromosome rather than on one of the autosomes. As such, both the gene and the trait are said to be **X-linked**.

Results of reciprocal crosses between white-eyed and red-eyed flies are shown in Figure 4–10. The obvious differences in phenotypic ratios in both the F_1 and F_2 generations are dependent on whether or not the P_1 white-eyed parent was male or female.

Morgan was able to correlate these observations with the difference found in the sex chromosome composition between male and female *Drosophila*. He hypothesized that the recessive allele for white eye is found on the X chromosome, but its corresponding locus is absent from the Y chromosome. Females thus have two available gene sites, one on each X chromosome, while males have only one available gene site on their single X chromosome.

Morgan's interpretation of X-linked inheritance, shown in Figure 4–11, provides a suitable theoretical explanation for his results. Since the Y chromosome lacks homology with most genes on the X chromosome, whatever alleles are present on the X chromosome of the males will be expressed directly in their phenotype. Because males cannot be either homozygous or heterozygous for X-linked genes, this condition is referred to as being **hemizygous**.

One result of X-linkage is the **crisscross pattern of inheritance**, whereby phenotypic traits controlled by recessive X-linked genes are passed from homozygous mothers to all sons. This pattern occurs because females exhibiting a recessive trait carry the mutant allele on both X chromosomes. Because male offspring receive one of their mother's two X chromosomes and are hemizygous for all alleles present on that X, all sons will express the same recessive X-linked traits as their mother.

Morgan's work has taken on great historical significance. By 1910, the correlation between Mendel's work and the behavior of chromosomes during meiosis had provided the basis for the **chromosome theory of inheritance**, as postulated by Sutton and Boveri (see Chapter 3). The work involving the X chromosome is considered to be the first solid experimental evidence in support of this theory. In the ensuing two decades, these findings inspired further research, the outcomes of which provided indisputable evidence in support of this theory.

X-Linked Inheritance in Humans

In humans, many genes and the respective traits controlled by them are recognized as being linked to the X chromosome, as shown in Table 4.2. These X-linked traits can be

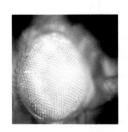 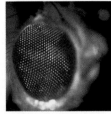

FIGURE 4–10 The F_1 and F_2 results of T. H. Morgan's reciprocal crosses involving the X-linked *white* mutation in *Drosophila melanogaster*. The actual F_2 data are shown in parentheses. The photographs show white eyes and the brick-red wild-type eye color. (*Photos: Carolina Biological Supply Co./Phototake NYC*)

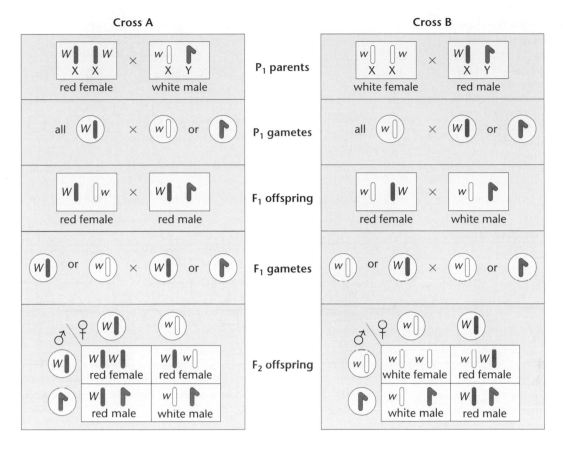

FIGURE 4–11 The chromosomal explanation of the results of the X-linked crosses shown in Figure 4–10.

easily identified in a **pedigree** because of the crisscross pattern of inheritance. A pedigree for one form of human **color blindness** is shown in Figure 4–12. The mother in generation I passes the trait to all her sons but to none of her daughters. If the offspring in generation II marry normal individuals, the color-blind sons will produce all normal male and female offspring (III-1, 2, and 3); the normal-visioned daughters will produce normal-visioned female offspring (III-4, 6, and 7), as well as color-blind (III-8) and normal-visioned (III-1 and 5) male offspring.

Because of the way in which X-linked genes are transmitted, unusual circumstances can be associated with recessive X-linked disorders, in comparison to recessive autosomal disorders. For example, if an X-linked disorder debilitates

TABLE 4.2 Human X-linked traits

Condition	Characteristics
Color blindness, deutan type	Insensitivity to green light.
Color blindness, protan type	Insensitivity to red light.
Fabry's disease	Deficiency of galactosidase A; heart and kidney defects, early death.
G-6-PD deficiency	Deficiency of glucose-6-phosphate dehydrogenase, severe anemic reaction following intake of primaquines in drugs and certain foods, including fava beans.
Hemophilia A	Classical form of clotting deficiency; deficiency of clotting factor VIII.
Hemophilia B	Christmas disease; deficiency of clotting factor IX.
Hunter syndrome	Mucopolysaccharide storage disease resulting from iduronate sulfatase enzyme deficiency; short stature, clawlike fingers, coarse facial features, slow mental deterioration, and deafness.
Ichthyosis	Deficiency of steroid sulfatase enzyme; scaly dry skin, particularly on extremities.
Lesch–Nyhan syndrome	Deficiency of hypoxanthine-guanine phosphoribosyl transferase enzyme (HGPRT) leading to motor and mental retardation, self-mutilation, and early death.
Muscular dystrophy (Duchenne type)	Progressive, life-shortening disorder characterized by muscle degeneration and weakness; sometimes associated with mental retardation; deficiency of the protein dystrophin.

FIGURE 4–12 (a) A human pedigree of the X-linked color blindness trait. (b) The most probable genotypes of each individual in the pedigree. The photograph is of an Ishihara color blindness chart. Red-green color blind individuals see a 3 rather than an 8 visualized by those with normal color vision. *(Photo: Mary Teresa Giancoli)*

Symbols

c = color blindness
C = normal vision
⸎ = Y chromosome

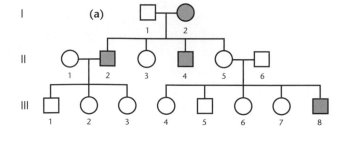

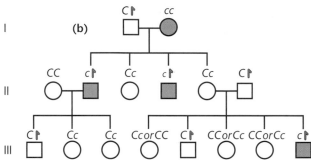

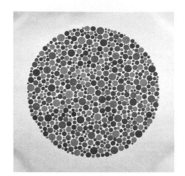

or is lethal to the affected individual prior to reproductive maturation, the disorder occurs exclusively in males. This is the case because the only sources of the lethal allele in the population are in heterozygous females who are "carriers" and do not express the disorder. They pass the allele to one-half of their sons, who develop the disorder because they are hemizygous but who rarely, if ever, reproduce. Heterozygous females also pass the allele to one-half of their daughters, who become carriers but do not develop the disorder. An example of such an X-linked disorder is Duchenne muscular dystrophy. The disease has an onset prior to age 6 and is often lethal prior to age 20. It normally occurs only in males.

4.11 Sex-Limited and Sex-Influenced Inheritance

In still other instances, inheritance is affected by the sex of an individual, though not necessarily by genes on the X chromosome. There are numerous examples in different organisms where the sex of the individual plays a determining role in the expression of certain phenotypes. In some cases, the expression of a specific phenotype is absolutely limited to one sex; in others, the sex of an individual influences the expression of a phenotype that is not limited to one sex or the other. This distinction differentiates **sex-limited inheritance** from **sex-influenced inheritance**.

In domestic fowl, tail and neck plumage is often distinctly different in males and females (Figure 4–13), demonstrating sex-limited inheritance. Cock feathering is longer, more curved, and pointed, while hen feathering is shorter and more rounded. Inheritance of these feather phenotypes is controlled by a single pair of autosomal alleles whose expression is modified by the individual's sex hormones.

As shown in the following chart, hen feathering is due to a dominant allele, H; but regardless of the homozygous presence of the recessive h allele, all females remain hen-feathered. Only in males does the hh genotype result in cock feathering.

Genotype	Phenotype	
	♀	♂
HH	Hen-feathered	Hen-feathered
Hh	Hen-feathered	Hen-feathered
hh	Hen-feathered	Cock-feathered

In certain breeds of fowl, the hen-feathering or cock-feathering allele has become fixed in the population. In the Leghorn breed, all individuals are of the *hh* genotype; as a result, males always differ from females in their plumage. Sebright bantams are all *HH*, resulting in no sexual distinction in feathering phenotypes.

FIGURE 4–13 Hen feathering (left) and cock feathering (right) in domestic fowl. The feathers in the hen are shorter and less curved. *(Photo: Hans Reinhard/Bruce Coleman, Inc.)*

FIGURE 4–14 Pattern baldness, a sex-influenced autosomal trait in humans. *(Photo: Debra P. Hershkowitz/Bruce Coleman, Inc.)*

Still another example of sex-limited inheritance involves the autosomal genes responsible for milk yield in dairy cattle. Regardless of the overall genotype that influences the quantity of milk production, those genes are obviously expressed only in females.

Cases of sex-influenced inheritance include **pattern baldness** in humans, horn formation in certain breeds of sheep (e.g., Dorsett Horn sheep), and certain coat patterns in cattle. In such cases, autosomal genes are responsible for the contrasting phenotypes displayed by both males and females, but the expression of these genes is dependent on the hormone constitution of the individual. Thus, the heterozygous genotype exhibits one phenotype in one sex and the contrasting one in the other. For example, pattern baldness in humans, where the hair is very thin on the top of the head (Figure 4–14), is inherited in the following way:

Genotype	Phenotype	
	♀	♂
BB	Bald	Bald
Bb	Not bald	Bald
bb	Not bald	Not bald

Even though females can display pattern baldness, this phenotype is much more prevalent in males. When females do inherit the *BB* genotype, the phenotype is much less pronounced than in males and is expressed later in life.

4.12 Phenotypic Expression

We conclude this chapter with the consideration of **phenotypic expression**. In Chapters 2 and 3, we assumed that the genotype of an organism is always directly expressed in its phenotype. For example, peas homozygous for the recessive *d* allele (*dd*) will always be dwarf. We discussed gene expression as though the genes operate in a closed, "black

box" system in which the presence or absence of functional products directly determines the collective phenotype of an individual. The situation is actually much more complex. Most gene products function within the internal milieu of the cell, and cells interact with one another in various ways. Further, the organism exists under diverse environmental influences. Thus, gene expression and the resultant phenotype are often modified through the interaction between an individual's particular genotype and the internal and external environment. In this final section, we deal with several important variables known to modify gene expression.

Penetrance and Expressivity

Some mutant genotypes are always expressed as a distinct phenotype, whereas others produce a proportion of individuals whose phenotypes cannot be distinguished from normal (wild type). The degree of expression of a particular trait can be studied quantitatively by determining the penetrance and expressivity of the genotype under investigation. The percentage of individuals that show at least some degree of expression of a mutant genotype defines the **penetrance** of the mutation. For example, the phenotypic expression of many mutant alleles in *Drosophila* can overlap with wild type. If 15 percent of mutant flies show the wild-type appearance, the mutant gene is said to have a penetrance of 85 percent.

By contrast, **expressivity** reflects the *range of expression* of the mutant genotype. Flies homozygous for the recessive mutant gene *eyeless* yield phenotypes that range from the presence of normal eyes to a partial reduction in size to the complete absence of one or both eyes (Figure 4–15). Although the average reduction of eye size is one-fourth to one-half, expressivity ranges from complete loss of both eyes to completely normal eyes.

Examples such as the expression of the *eyeless* gene provide the basis for experiments to determine the causes of phenotypic variation. If the laboratory environment is held constant and extensive variation is still observed, other genes may be influencing or modifying the *eyeless* phenotype. On the other hand, if the genetic background is not the cause of the phenotypic variation, environmental factors such as temperature, humidity, and nutrition may be involved. In the case of the *eyeless* phenotype, experiments have shown that both genetic background and environmental factors influence its expression.

Onset of Genetic Expression

Not all genetic traits are expressed at the same time during an organism's life span. In most cases, the age at which a gene is expressed corresponds to the normal sequence of growth and development. In humans, the prenatal, infant, preadult, and adult phases require different genetic information. As a result, many severe inherited disorders are often not manifested until after birth. In humans, for example, **Tay–Sachs disease**, inherited as an autosomal recessive, is a lethal lipid metabolism disease involving an abnormal enzyme, hexosaminidase A. Newborns appear phenotypically normal for the first few months. Then, developmental

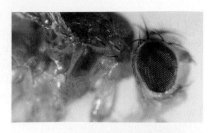

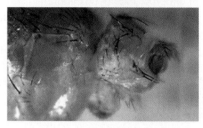

FIGURE 4–15 Variable penetrance as shown in the expression of the *eyeless* mutation in *Drosophila*. Gradations in phenotype range from wild type to partial reduction to eyeless. *(Top and bottom photos: Tanya Wolff/University of California at Berkeley. Middle photo: Joel C. Eisenberg, Ph.D., Dept. of Biochemistry, St. Louis University Medical Center)*

retardation, paralysis, and blindness ensue, and most affected children die by the age of 3.

The **Lesch–Nyhan syndrome**, inherited as an X-linked recessive disease, is characterized by abnormal nucleic acid metabolism (biochemical salvage of nitrogenous purine bases), leading to the accumulation of uric acid in blood and tissues, mental retardation, palsy, and self-mutilation of the lips and fingers. The disorder is due to a mutation in the gene encoding hypoxanthine-guanine phosphoribosyl transferase (HPRT). Newborns are normal for six to eight months prior to the onset of the first symptoms. Still another example involves **Duchenne muscular dystrophy (DMD)**, an X-linked recessive disorder associated with progressive muscular wasting. It is not usually diagnosed until the age of 3–5 years. Even with modern medical intervention, the disease is often fatal in the early twenties.

Perhaps the most variable of all inherited human disorders regarding age of onset is **Huntington disease**. Inherited as an autosomal dominant, Huntington disease affects the frontal lobes of the cerebral cortex where progressive cell death occurs over a period of more than a decade. Brain deterioration is accompanied by spastic uncontrolled movements, intellectual and emotional deterioration, and ultimately death. While onset has been reported at all ages, it most frequently occurs between ages 30 and 50, with a mean onset of 38 years.

Conditions such as these support the concept that the critical expression of normal genes varies throughout the life cycle of organisms, including humans. Gene products may play more essential roles at certain times. Further, it is likely that the internal physiological environment of an organism changes with age.

Genetic Anticipation

Interest in studying the genetic onset of phenotypic expression has intensified with the discovery of heritable disorders that exhibit a progressively earlier age of onset and an increased severity of the disorder in each successive generation. This general phenomenon is called **genetic anticipation**.

Myotonic dystrophy (DM), the most common type of adult muscular dystrophy, clearly illustrates genetic anticipation. Individuals afflicted with this autosomal dominant disorder exhibit extreme variation in the severity of symptoms. Mildly affected individuals develop cataracts as adults but have little or no muscular weakness. Severely affected individuals demonstrate more extensive myopathy and may be mentally retarded. In its most extreme form, the disease is fatal just after birth. A great deal of excitement was generated in 1989 when C. J. Howeler and colleagues confirmed the correlation of increased severity with earlier onset. They studied 61 parent–child pairs, and in 60 of the cases, age of onset was earlier in the child than in his or her affected parent.

In 1992, an explanation was put forward to explain both the molecular cause of the mutation responsible for DM as well as the basis of genetic anticipation. As we shall see later in the text, a particular region of the DM gene is repeated a variable number of times and is unstable. Normal individuals average about 5 copies of this region, minimally affected individuals reveal about 50 copies, and severely affected individuals demonstrate over 1000 copies. The most remarkable observation was that, in successive generations, the size of the repeated segment increases. Although it is not yet clear how the expansion in size affects onset and phenotypic expression, the correlation is extremely strong. Several other inherited human disorders, including the fragile-X syndrome, Kennedy disease, and Huntington disease, also reveal an association between the size of specific regions of the responsible gene and disease severity. We return to this general topic when we discuss the molecular explanation in detail in Chapter 14.

Genomic (Parental) Imprinting

Our final example of modification of the laws of Mendelian inheritance involves a variation of phenotypic expression that depends strictly on the parental origin of the chromosome carrying a particular gene, a phenomenon called **genomic** (or **parental**) **imprinting**. In some species, certain chromosomal regions and the genes contained within them somehow retain a memory, or an "imprint," of their parental origin that influences whether specific genes either are expressed or remain genetically silent—that is, are not expressed.

The imprinting step is thought to occur before or during gamete formation, leading to differentially marked genes (or chromosome regions) in sperm-forming versus egg-forming tissues. The process clearly differs from mutation because the imprint can be reversed in succeeding generations as genes pass from mother to son to granddaughter, and so on.

An example of imprinting involves the inactivation of one of the X chromosomes in mammalian females. In mice, prior to development of the embryo, imprinting occurs in tissues such that the X chromosome of paternal origin is genetically inactivated in all cells while the genes on the maternal X chromosome remain genetically active. As embryonic development is subsequently initiated, the imprint is "released," and random inactivation of either the paternal *or* the maternal X chromosome occurs.

In 1991, more specific information established that three specific mouse genes undergo imprinting. One is the gene encoding insulinlike growth factor II (*Igf2*). A mouse that carries two nonmutant alleles of this gene is normal in size, whereas a mouse that carries two mutant alleles lacks a growth factor and is a dwarf. The size of a heterozygous mouse (one allele normal and one mutant; Figure 4–16) depends on the parental origin of the normal allele. The mouse is normal in size if the normal allele came from the father, but dwarf if the normal allele came from the mother. From this we can deduce that the normal *Igf2* gene is imprinted to function poorly during the course of egg production in females but functions normally when it has passed through sperm-producing tissue in males.

Imprinting continues to depend on whether the gene passes through sperm-producing or egg-forming tissue leading to the next generation. For example, a heterozygous normal-sized male will donate to half his offspring a "normal-functioning" wild-type allele that counteracts the mutant allele received from the mother.

In humans, two distinct genetic disorders are thought to be caused by differential imprinting of the same region of chromosome 15. In both cases, the disorders *appear* to derive from an identical deletion of a specific region in one member of the chromosome 15 pair. The first disorder, **Prader–Willi syndrome (PWS)**, results when only an undeleted maternal chromosome remains. If only an undeleted paternal chromosome remains, an entirely different disorder, **Angelman syndrome (AS)**, results.

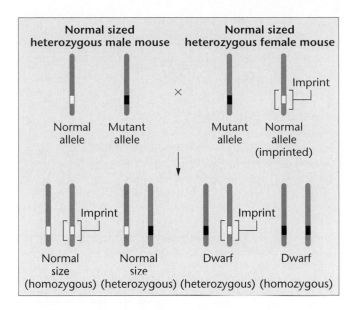

FIGURE 4–16 The effect of imprinting on the mouse *Igf2* gene, which produces dwarf mice in the homozygous condition. Heterozygous offspring that receive the normal allele from their father are normal in size. Heterozygotes that receive the normal allele from their mother, which has been imprinted, are dwarf.

The two conditions are clearly different phenotypically. PWS entails mental retardation as well as a severe eating disorder marked by an uncontrollable appetite, obesity, and diabetes. AS involves distinct behavioral manifestations as well as mental retardation. We can conclude that the involved region of chromosome 15 is imprinted differently in male and female gametes, and that both a maternal and paternal region are required for normal development.

Many questions about genomic imprinting remain unanswered. For example, though it appears that regions of chromosomes (rather than specific genes) are imprinted, it is not known how many genes are subject to imprinting. Current thinking suggests that methylation of certain pairs of nitrogenous bases that include cytosine may be involved. Such an explanation is in keeping with the knowledge that methylation of cytosine in DNA inhibits gene activity. Whatever the cause, this phenomenon is a fascinating topic and one that is under intense research investigation.

Genetics, Technology, and Society

Improving the Genetic Fate of Purebred Dogs

For dog lovers, nothing is quite so heartbreaking as watching a dog slowly go blind, standing by helplessly as he struggles to adapt to a life of perpetual darkness. That's what happens in progressive retinal atrophy (PRA), an inherited disorder first described in Gordon setters in 1911. Since then, PRA has been found in many other breeds of dogs, including Irish setters, border collies, Norwegian elkhounds, toy poodles, miniature schnauzers, cocker spaniels, and Siberian huskies.

The products of many genes are required for the development and maintenance of a healthy retina, and a defect in any one of them has the potential to cause retinal dysfunction. Decades of research have led to the identification of four such genes (*rcd1*, *rcd2*, *erd*, and *prcd*), and more are likely to be discovered. Different genes are mutated in different breeds—for example, *rcd1* in Irish setters and *prcd* in Labrador retrievers. In all dogs examined thus far, PRA shows a recessive pattern of inheritance.

Whichever mutation is responsible, PRA is more common in certain purebred dogs than in mixed breeds. The development of distinct breeds of dogs has involved an intensive selection for desirable attributes, whether a particular size, shape, color, or behavior. Since most desired characteristics are determined by recessive alleles, the fastest way to increase the homozygosity of these alleles and establish the characteristics in the population is to mate close relatives, which are likely to carry the same alleles. In the practice known as line breeding, for example, dogs may be mated to a cousin or a grandparent.

Unfortunately, the generations of inbreeding that have established favorable characteristics in pure breeds have also increased the homozygosity of certain harmful recessive alleles such as PRA, resulting in other inherited diseases. Many breeds are plagued with inherited hip dysplasia, which is particularly prevalent in German shepherds. Deafness and kidney disorders are common genetic maladies in Dalmatians. Over 300 such diseases have been characterized in purebred dogs, and most breeds of dogs are affected by one or more of them. By some estimates, fully 25 percent of purebred dogs are affected with one genetic ailment or another. Inbreeding is not the cause of these genetic diseases, but there is general agreement that breeding practices, if not executed properly, increase the frequency of disease.

Inbreeding is the likely explanation for the elevated frequency of PRA in certain breeds. Fortunately, advances in canine genetics are providing new tools for breeding healthy dogs. Since 1995, a genetic test has been available to identify mutations in the *rcd1* gene, which is responsible for the form of PRA that occurs in Irish setters. This test is now used to identify heterozygous carriers of *rcd1* mutations, dogs that show no symptoms of PRA but, if mated with other carriers, pass the trait on to about 25 percent of their offspring. Eliminating PRA carriers from breeding programs could, in theory, eradicate this condition from Irish setters in just a few generations.

It took many years to identify *rcd1* and other genes responsible for PRA. In the future, the isolation of genes underlying canine inherited disease should be faster, thanks to the Dog Genome Project, a collaborative effort involving scientists at the University of California, the University of Oregon, the Fred Hutchinson Cancer Research Center in Seattle, and other research centers. Its goal is the creation of a complete genetic map of the 39 chromosomes in the dog. In late 1997, preliminary stages were completed, paving the way for the development of a more comprehensive map. The identification of genes that cause inherited disease will allow genetic tests for diagnosing the disease before symptoms develop, as well as for ensuring that breeding animals are free of harmful recessive alleles.

The Dog Genome Project may also benefit humans beyond the reduction in disease in their canine companions. Since about 85 percent of the genes in the dog genome have equivalents in humans, identifying a disease-causing dog gene may short-cut the isolation of the corresponding gene in humans. For example, some forms of PRA in dogs appear to be equivalent to retinitis pigmentosum (RP) in humans, which afflicts about 1.5 million people worldwide. Despite much research, RP remains poorly understood, and current treatments only slow its progress. Understanding the genetic basis of PRA may lead to breakthroughs in the diagnosis and treatment of RP, potentially saving the sight of thousands of people every year. By contributing to the cure of human diseases, dogs may prove to be "man's best friend" in an entirely new way.

References

Berson, E. L. 1996. Retinitis pigmentosum: Unfolding its mystery. *Proc. Natl. Acad. Sci. USA*. 93:4526–28.

Ray, K., Baldwin, V., Acland, G., and Aquirre, G. 1995. Molecular diagnostic tests for ascertainment of genotype at the rod cone dysplasia (*rcd1* locus) in Irish setters. *Curr. Eye Res.* 14:243.

Smith, C. A. 1994. New hope for overcoming canine inherited disease. *Am. J. Vet. Med. Assoc.* 204:41–46.

Chapter Summary

1. Since Mendel's work was rediscovered, the study of transmission genetics has expanded to include many alternative modes of inheritance. In many cases, phenotypes can be influenced by two or more genes.

2. Incomplete or partial dominance is exhibited when an intermediate phenotypic expression of a trait occurs in an organism that is heterozygous for two alleles.

3. Codominance is exhibited when distinctive expression of two alleles occurs in a heterozygous organism.

4. The concept of multiple allelism applies to populations, since a diploid organism may host only two alleles at any given locus. Within a population, however, many alternative alleles of the same gene can occur.

5. Lethal mutations usually result in the inactivation or the lack of synthesis of gene products that are essential during development. Such mutations can be recessive or dominant. Some lethal genes, such as the one that causes Huntington disease, are not expressed until adulthood.

Insights and Solutions

Genetic problems take on added complexity if they involve two independent characters and multiple alleles, incomplete dominance, or epistasis. The most difficult types of problems are those that pioneering geneticists faced during laboratory or field studies. They had to determine the mode of inheritance by working backward from the observations of offspring to parents of unknown genotype.

1. Consider the problem of comb shape inheritance in chickens, where walnut, rose, pea, and single are the observed distinct phenotypes. *How is comb shape inherited, and what are the genotypes of the P_1 generation of each cross?* Use the following data to answer these questions.

Cross 1:	single × single	→ all single
Cross 2:	walnut × walnut	→ all walnut
Cross 3:	rose × pea	→ all walnut
Cross 4:	F_1 × F_1 of Cross 3	
	walnut × walnut	→ 93 walnut
		28 rose
		32 pea
		10 single

Solution: At first glance, this problem appears quite difficult. However, as with other seemingly difficult problems, applying a systematic approach and breaking the analysis into steps usually saves the day. The approach involves two steps. First, analyze the data carefully for any useful information. Once you identify something that is clearly helpful, follow an empirical approach; that is, formulate an hypothesis and, in a sense, test it against the given data. Look for a pattern of inheritance that is consistent with all cases.

(a) For example, this problem gives two immediately useful facts. First, in cross 1, P_1 singles breed true. Second, while P_1 walnut breeds true (cross 2), a walnut phenotype is also produced in crosses between rose and pea (cross 3). When these F_1 walnuts are crossed (cross 4), all four comb shapes are produced in a ratio that approximates 9:3:3:1. This observation immediately suggests a cross involving two gene pairs, because the resulting data display the same ratio as in Mendel's dihybrid crosses. Because only one trait is involved (comb shape),

epistasis may be occurring. This could serve as your working hypothesis, and you must now propose how the two gene pairs "interact" to produce each phenotype.

(b) If you call the allele pairs A, a and B, b, the prediction can be made that because walnut represents 9/16 in cross 4, A–B– will produce walnut. You might also hypothesize that in the case of cross 2, the genotypes are $AABB \times AABB$, where walnut was seen to breed true. (Recall that A– and B– mean AA or Aa and BB or Bb, respectively.)

(c) Because single is the phenotype representing 1/16 of the offspring of cross 4, you could predict that this phenotype is the result of the $aabb$ genotype. This is consistent with cross 1.

(d) Now you have only to determine the genotypes for rose and pea. The most logical prediction is that at least one dominant A or B allele combined with the double recessive condition of the other allele pair accounts for these phenotypes. For example,

$$A\text{–}bb \rightarrow \text{rose}$$
$$aa\ B\text{–} \rightarrow \text{pea}$$

If, in cross 3, $AAbb$ (rose) is crossed with $aaBB$ (pea), all offspring will be $AaBb$ (walnut). This is consistent with the data, and you must now only look at cross 4. We predict these walnut genotypes to be $AaBb$ (as above), and from the cross

$$AaBb \text{ (walnut)} \times AaBb \text{ (walnut)}$$

we expect

9/16 A–B– (walnut)

3/16 A–bb (rose)

3/16 aa B– (pea)

1/16 aa bb (single)

Our prediction is consistent with the information given. The initial hypothesis of the epistatic interaction of two gene pairs proves consistent throughout, and the problem is solved.

This example demonstrates the need for a basic theoretical knowledge of transmission genetics. Then, you must search for appropriate clues that enable you to proceed in a stepwise fashion toward a solution. Mastering problem solving requires practice, but will give you a great deal of satisfaction. Apply this general approach to the following problems.

Walnut

Pea

Rose

Single

(Single and pea photos: J. James Bitgood, University of Wisconsin Animal Sciences Dept., Madison, WI. Walnut and rose photos: Courtesy of Dr. Ralph Somes.)

2. In radishes, flower color may be red, purple, or white. The edible portion of the radish may be long or oval. When only flower color is studied, crossing red with white yields all purple. If these F_1 purples are interbred, no dominance is evident, and the F_2 generation consists of 1/4 red:1/2 purple:1/4 white. Regarding radish shape, long is dominant to oval in a normal Mendelian fashion.

(a) Determine the F_1 and F_2 phenotypes from a cross between a true-breeding red long radish and one that is white oval. (*Hint:* Be sure to define all gene symbols initially.)

Solution: This is a modified dihybrid cross in which the gene pair controlling color exhibits incomplete dominance. Shape is controlled conventionally. First, we establish gene symbols:

RR = red $\qquad$ Rr = purple $\qquad$ rr = white
$O–$ = long $\qquad$ oo = oval

P_1: $\quad RROO \quad \times \quad rroo$
(red long) $\qquad$ (white oval)
F_1: all $RrOo$ (purple long)
$F_1 \times F_1$: $RrOo \times RrOo$

F_2:

$$
\begin{cases}
1/4\ RR \begin{cases} 3/4\ O– \\ 1/4\ oo \end{cases} & \begin{matrix} 3/16\ RR\ O– & \text{red long} \\ 1/16\ RR\ oo & \text{red oval} \end{matrix} \\[1em]
2/4\ Rr \begin{cases} 3/4\ O– \\ 1/4\ oo \end{cases} & \begin{matrix} 6/16\ Rr\ O– & \text{purple long} \\ 2/16\ Rr\ oo & \text{purple oval} \end{matrix} \\[1em]
1/4\ rr \begin{cases} 3/4\ O– \\ 1/4\ oo \end{cases} & \begin{matrix} 3/16\ rr\ O– & \text{white long} \\ 1/16\ rr\ oo & \text{white oval} \end{matrix}
\end{cases}
$$

Note that to generate the F_2 results, we use the forked-line method. First, we consider the outcome of crossing F_1 parents for the color genes ($Rr \times Rr$). Then we consider the outcome of shape ($Oo \times Oo$).

(b) A red oval plant is crossed with a plant of unknown genotype and phenotype, yielding the following data. Determine the genotype and phenotype of the unknown plant.

Offspring: $\quad$ 103 red long:101 red oval:
98 purple long:100 purple oval

Solution: Since the two characters are inherited independently, consider them separately. The data indicate a 1/4:1/4:1/4:1/4 proportion. First, consider color:

P_1: $\quad$ red $\times$??? (unknown)
F_1: $\quad$ 204 red (1/2)
$\qquad\qquad$ 198 purple (1/2)

Because the red parent must be RR, the unknown parent must have a genotype of Rr to produce these results. It is thus purple. Now, consider shape:

P_1: $\quad$ oval $\times$??? (unknown)
F_1: $\quad$ 201 long (1/2)
$\qquad\qquad$ 201 oval (1/2)

Now, consider the oval and long characters. Because the oval plant must be oo, the unknown plant must have a genotype of Oo to produce these results. So it is long. The unknown plant is thus

$RrOo$ purple long

3. In humans, red-green color blindness is inherited as an X-linked recessive trait. A woman with normal vision, but whose father is color-blind, marries a male who has normal vision. Predict the color vision of their male and female offspring.

Solution: The female is heterozygous because she inherited an X chromosome with the mutant allele from her father. Her husband is normal. Therefore, the parental genotypes are

$Cc \times C\uparrow$ ($\uparrow$ is the Y chromosome)

All female offspring are normal (CC or Cc). One-half of the male children will be color-blind ($c\uparrow$) and the other half will have normal vision ($C\uparrow$).

Chapter Summary (cont.)

6. Mendel's classic F_2 ratio is often modified in instances where gene interaction results in discontinuous variation.

7. Epistasis may occur when two or more genes influence a single characteristic. Usually, the expression of one of the genes masks the expression of the other gene or genes.

8. Complementation analysis determines whether independently isolated mutations producing similar phenotypes are alleles of one another or whether they represent separate genes.

9. Genes located on the X chromosome display a unique mode of inheritance referred to as X-linkage.

10. Sex-limited and sex-influenced inheritance occur when the sex of the organism affects the phenotype controlled by a gene located on an autosome.

11. Phenotypic expression is not always the direct reflection of the genotype. Penetrance measures the percentage of organisms in a given population that exhibits evidence of the corresponding mutant phenotype. Expressivity, on the other hand, measures the range of phenotypic expression of a given genotype.

12. The time of onset of gene expression in organisms varies as the need for certain gene products occurs at different periods during development, growth, and aging.

13. Genetic anticipation refers to the phenomenon where the onset of phenotypic expression occurs earlier and becomes more severe in each ensuing generation.

14. Genomic imprinting is a process whereby a region of either the paternal or maternal chromosome is modified (marked or imprinted), thereby affecting phenotypic expression. Expression therefore depends on which parent contributes a mutant allele.

Key Terms

Problems and Discussion Questions

1. In shorthorn cattle, coat color may be red, white, or roan. Roan is an intermediate phenotype expressed as a mixture of red and white hairs. The following data are obtained from various crosses:

red × red	⟶	all red
white × white	⟶	all white
red × white	⟶	all roan
roan × roan	⟶	1/4 red: 1/2 roan: 1/4 white

   How is coat color inherited? What are the genotypes of parents and offspring for each cross?

2. Contrast incomplete dominance and codominance.

3. With regard to the ABO blood types in humans, determine the genotypes of the male parent and female parent:

   Male parent: Blood type B whose mother was type O
   Female parent: Blood type A whose father was type B
   Predict the blood types of the offspring that this couple may have and the expected proportion of each.

4. In foxes, two alleles of a single gene, *P* and *p*, may result in lethality (*PP*), platinum coat (*Pp*), or silver coat (*pp*). What ratio is obtained when platinum foxes are interbred? Is the *P* allele behaving dominantly or recessively in causing lethality? In causing platinum coat color?

5. Three gene pairs located on separate autosomes determine flower color and shape as well as plant height. The first pair exhibits incomplete dominance, where color can be red, pink (the heterozygote), or white. The second pair leads to personate (dominant) or peloric (recessive) flower shape, while the third gene pair produces either the dominant tall trait or the recessive dwarf trait. Homozygous plants that are red, personate, and tall are crossed with those that are white, peloric, and dwarf. Determine the F_1 genotype(s) and phenotype(s). If the F_1 plants are interbred, what proportion of the offspring will exhibit the same phenotype as the F_1 plants?

6. As in Problem 5, color may be red, white, or pink, and flower shape may be personate or peloric. For the following crosses, determine the P_1 and F_1 genotypes.

(a) red peloric × white personate	⟶	F_1: all pink personate
(b) red personate × white peloric	⟶	F_1: all pink personate
(c) pink personate × red peloric	⟶	F_1: 1/4 red personate
		1/4 red peloric
		1/4 pink personate
		1/4 pink peloric
(d) pink personate × white peloric	⟶	F_1: 1/4 white personate
		1/4 white peloric
		1/4 pink personate
		1/4 pink peloric

   What phenotype ratios would result from crossing the F_1 of (a) with the F_1 of (b)?

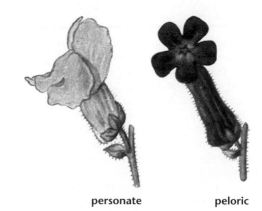

personate   **peloric**

(*Drawings of flowers: Erwin Bauer,* The Scientific Basics of Plant Cultivation: A Textbook for Farmers, Gardeners, and Foresters. *Gebruder Borntraeger, Berlin, 1924.*)

7. In some plants a red pigment, cyanidin, is synthesized from a colorless precursor. The addition of a hydroxyl group (—OH) to the cyanidin molecule causes it to become purple. In a cross between two randomly selected purple plants, the following results are obtained:

> 94 purple:31 red:43 colorless

How many genes are involved in determining these flower colors? Which genotypic combinations produce which phenotypes? Diagram the purple × purple cross.

8. In rats, the following genotypes of two independently assorting autosomal genes determine coat color:

A–B– (gray); A–bb (yellow); aaB– (black); aabb (cream)

A third gene pair on a separate autosome determines whether any color will be produced. The *CC* and *Cc* genotypes allow color according to the expression of the *A* and *B* alleles. However, the *cc* genotype results in albino rats regardless of the *A* and *B* alleles present. Determine the F_1 phenotypic ratio of the following crosses: (a) *AAbbCC × aaBBcc*; (b) *AaBBCC × AABbcc*; (c) *AaBbCc × AaBbcc*.

9. Given the inheritance pattern of coat color in rats described in Problem 8, predict the genotype and phenotype of the parents that produced the following F_1 offspring: (a) 9/16 gray:3/16 yellow:3/16 black:1/16 cream; (b) 9/16 gray:3/16 yellow:4/16 albino; (c) 27/64 gray:16/64 albino:9/64 yellow:9/64 black:3/64 cream.

10. A husband and wife have normal vision, although both of their fathers are red-green color-blind, which is inherited as an X-linked recessive condition. What is the probability that their first child will be (a) a normal son? (b) a normal daughter? (c) a color-blind son? (d) a color-blind daughter?

11. In humans, the ABO blood type is under the control of autosomal multiple alleles. Red-green color blindness is a recessive X-linked trait. If two parents who are both type A and have normal vision produce a son who is color-blind and type O, what is the probability that their next child will be a female who has normal vision and is type O?

12. In spotted cattle, the colored regions may be mahogany or red. If a red female and a mahogany male, both derived from separate true-breeding lines, are mated and the cross is carried to an F_2 generation, the following results are obtained:

> F_1: 1/2 mahogany males:1/2 red females
> F_2: 3/8 mahogany males:1/8 red males:
> 1/8 mahogany females:3/8 red females

When the reciprocal of the initial cross is performed (mahogany female and red male), identical results are obtained. Explain these results by postulating how the color is genetically determined. Diagram the crosses.

13. In cats, yellow coat color is determined by the *b* allele, and black coat color is determined by the *B* allele. The heterozygous condition results in a coat pattern known as tortoise shell. These genes are X-linked. What kinds of offspring would be expected from a cross of a black male and a tortoise-shell female? What are the chances of getting a tortoise-shell male?

14. In *Drosophila*, an X-linked recessive mutation, *scalloped (sd)* causes irregular wing margins. Diagram the F_1 and F_2 results if
   (a) A scalloped female is crossed with a normal male.
   (b) A scalloped male is crossed with a normal female.
   Compare these results to those that would be obtained if the *scalloped* gene is autosomal.

15. Another recessive mutation in *Drosophila*, *ebony (e)*, is on an autosome (chromosome 3) and causes darkening of the body compared with wild-type flies. What phenotypic F_1 and F_2 male and female ratios will result if a scalloped-winged female with normal body color is crossed with a normal-winged ebony male? Work this problem by both the Punnett square method and the forked-line method.

16. While *vermilion* is X-linked and causes the eye color to be bright red, *brown* is an autosomal recessive mutation that causes the eye to be brown. Flies carrying both mutations lose all pigmentation and are white-eyed. Predict the F_1 and F_2 results of the following crosses:
   (a) vermilion females × brown males
   (b) brown females × vermilion males
   (c) white females × wild males

17. In pigs, coat color may be sandy, red, or white. A geneticist spent several years mating true-breeding pigs of all different color combinations, even going so far as to obtain true-breeding lines from different parts of the country. For crosses 1 and 4 below, she encountered a major problem: her computer crashed and she lost the F_2 data. She nevertheless persevered, and using the limited data shown here, she was able to predict the mode of inheritance, the number of genes involved, and assign genotypes to each coat color. Based on the available data generated from the crosses shown, attempt to duplicate her analysis.

Cross	P_1	F_1	F_2
1	sandy × sandy	all red	data lost
2	red × sandy	all red	3/4 red: 1/4 sandy
3	sandy × white	all sandy	3/4 sandy: 1/4 white
4	white × red	all red	data lost

Once you have formulated a hypothesis to explain the mode of inheritance and assigned genotypes to the respective coat colors, predict the outcomes of the F_2 generations where the data were lost.

18. A geneticist from an alien planet that prohibits genetic research brought with him two true-breeding lines of frogs. One line croaks by *uttering* "rib-it rib-it" and has purple eyes. The other line croaks by *muttering* "knee-deep knee-deep" and has green eyes. He mated the two types, producing F_1 frogs, all of whom utter "rib-it" and have blue eyes. A large F_2 generation yielded the following ratio:

> 27/64 blue-eyed, rib-it utterer
> 12/64 green-eyed, rib-it utterer
> 9/64 blue-eyed, knee-deep mutterer
> 9/64 purple-eyed, rib-it utterer
> 4/64 green-eyed, knee-deep mutterer
> 3/64 purple-eyed, knee-deep mutterer

   (a) How many total gene pairs are involved in the inheritance of both eye color and croaking?
   (b) Of these, how many control eye color? croaking?
   (c) Assign gene symbols for all phenotypes and indicate the genotypes of the P_1, F_1, and F_2 frogs.
   (d) After many years, the frog geneticist isolated true-breeding lines of all six F_2 phenotypes. Indicate the F_1 and F_2 phenotypic ratios of a cross between a blue-eyed mutterer and a purple-eyed utterer.

19. In another cross, the frog geneticist (Problem 18) mated two purple-eyed utterers, with the results shown here. What were the genotypes of the parents?

> 9/16 purple-eyed utterers
> 3/16 purple-eyed mutterers
> 3/16 green-eyed utterers
> 1/16 green-eyed mutterers

20. In cattle, coats may be solid white, solid black, or black and white spotted. When true-breeding solid whites are mated with true-breeding solid blacks, the F_1 generation consists of all solid white individuals. Following many $F_1 \times F_1$ matings, the following ratio was observed in the F_2 generation:

> 12/16   solid white
> 3/16   black and white spotted
> 1/16   solid black

Explain the mode of inheritance governing coat color and pattern by determining how many gene pairs are involved and which genotypes yield which phenotypes. Is it possible to isolate a true-breeding strain of black and white spotted cattle? If so, what genotype would they have? If not, explain why not.

21. In the guinea pig, a locus controlling coat color may be occupied by any of four alleles: C (*black*), c^k (*sepia*), c^d (*cream*), or c^a (*albino*). A progressive order of dominance exists among these alleles when they are present heterozygously: $C > c^k > c^d > c^a$. In the following crosses, determine the genotype of each individual and predict the phenotypic ratios of the offspring.
    (a) sepia × cream, where both had an albino parent
    (b) sepia × cream, where the sepia individual had an albino parent and the cream individual had two sepia parents
    (c) sepia × cream, where the sepia individual had two full-color parents and the cream individual had two sepia parents
    (d) sepia × cream, where the sepia individual had two full color parents and the cream individual had two full color parents

22. In the parakeet, two autosomal genes (located on different chromosomes) control the production of feather pigment. Gene B controls the production of a blue pigment, and gene Y controls the production of a yellow pigment. Recessive mutations in each gene are known that result in the loss of synthesis of the respective pigment. Two green parakeets are mated and produce green, blue, yellow, and albino progeny.
    (a) Based on this information, explain the pattern of inheritance. Be sure to include the following in your answer:
       (i) the genotypes of the green parents
       (ii) the genotypes of all four phenotypic classes
       (iii) the fraction of total progeny that each phenotypic class represents
    (b) The parental (green) parakeets are the progeny of a cross between two true-breeding strains. What two types of crosses between true-breeding strains could have produced the green parents? Indicate the genotypes and phenotypes for each cross.

(Photo: Robert Pearcy/Animals Animals/Earth Scenes)

23. Consider the following three pedigrees, all involving a single human trait.

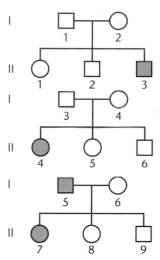

(a) Which sets of conditions if any, can be excluded?

  Conditions: dominant *and* X-linked
  dominant *and* autosomal
  recessive *and* X-linked
  recessive *and* autosomal

(b) For any set of conditions that you excluded, indicate the *single individual* in generation II (e.g., II-1, II-2, ...) that was instrumental in your decision to exclude that condition. If none were excluded, answer "none apply."

(c) Given your conclusions, indicate the *genotype* of the following individuals. If more than one possibility applies, list all possibilities. Use the symbols A and a for the genotypes.

  Individuals: II-1; II-6; II-9

24. Three autosomal recessive mutations in *Drosophila*, all with tan eye color (*r1*, *r2*, and *r3*) are independently isolated and subjected to complementation analysis. The results are shown here. Which, if any, are alleles of one another? Predict the results of the cross that is not shown—that is, *r2* × *r3*.

> Cross 1:   *r1* × *r2* ⟶ F₁: all wild-type eyes
>
> Cross 2:   *r1* × *r3* ⟶ F₁: all tan eyes

25. Labrador retrievers may be black, brown, or golden in color. While each color may breed true, many different outcomes occur if numerous litters are examined from a variety of matings, where the parents are not necessarily true-breeding. Shown here are just some of the many possibilities. Propose a mode of inheritance that is consistent with these data, and indicate the corresponding genotypes of the parents in each mating. Indicate as well the genotypes of dogs that breed true for each color.

(Photo: William H. Mullins/Photo Researchers, Inc.)

(a)	black × brown ⟶	all black
(b)	black × brown ⟶	1/2 black
		1/2 brown
(c)	black × brown ⟶	3/4 black
		1/4 golden
(d)	black × golden ⟶	all black
(e)	black × golden ⟶	4/8 golden
		3/8 black
		1/8 brown
(f)	black × golden ⟶	2/4 golden
		1/4 black
		1/4 brown
(g)	brown × brown ⟶	3/4 brown
		1/4 golden
(h)	black × black ⟶	9/16 black
		4/16 golden
		3/16 brown

Selected Readings

Bartolomei, M. S., and Tilghman, S. M. 1997. Genomic imprinting in mammals. *Annu. Rev. Genet.* 31:493–525.

Brink, R. A., ed. 1967. *Heritage from Mendel.* Madison: University of Wisconsin Press.

Bultman, S. J., Michaud, E. J., and Woychik, R. P. 1992. Molecular characterization of the mouse *agouti* locus. *Cell* 71:1195–1204.

Cattanach, B. M., and Jones, J. 1994. Genetic imprinting in the mouse: Implications for gene regulation. *J. Inherit. Metab. Dis.* 17:403–20.

Carlson, E. A. 1987. *The gene: A critical history*, 2nd ed. Philadelphia: Saunders.

Drayna, D., and White, R. 1985. The genetic linkage map of the human X chromosome. *Science* 230:753–58.

Dunn, L. C. 1966. *A short history of genetics.* New York: McGraw-Hill.

Feil, R., and Kelsey, G. 1997. Genomic imprinting: A chromatin connection. *Am. J. Hum. Genet.* 61:1213–19.

Foster, M. 1965. Mammalian pigment genetics. *Adv. Genet.* 13:311–39.

Grant, V. 1975. *Genetics of flowering plants.* New York: Columbia University Press.

Harper, P. S., et al. 1992. Anticipation in myotonic dystrophy: New light on an old problem. *Am. J. Hum. Genet.* 51:10–16.

Howeler, C. J., et al. 1989. Anticipation in myotonic dystrophy: Fact or fiction? *Brain* 112:779–97.

Mahedevan, M., et al. 1992. Myotonic dystrophy mutation: An unstable CTG repeat in the 3′ untranslated region of the gene. *Science* 255:1253–58.

McKusick, V. A. 1962 On the X chromosome of man. *Quart. Rev. Biol.* 37:69–175.

Morgan, T. H. 1910. Sex-limited inheritance in *Drosophila. Science* 32:120–22.

Peters, J. A., ed. 1959. *Classic papers in genetics.* Englewood Cliffs, NJ: Prentice-Hall.

Phillips, P.C. 1998. The language of gene interaction. *Genetics* 149:1167–71.

Race, R. R., and Sanger, R. 1975. *Blood groups in man*, 6th ed. Oxford, England: Blackwell.

Sapienza, C. 1990. Parental imprinting of genes. *Sci. Am.* (Oct.) 363:52–60.

Siracusa, L. D. 1994. The *agouti* gene: Turned on to yellow. *Cell* 10:423–28.

Vogel, F., and Motulsky, A. G. 1997. *Human genetics: Problems and approaches*, 3rd ed. New York: Springer-Verlag.

Watkins, M. W. 1966. Blood group substances. *Science* 152:172–81.

Yoshida, A. 1982. Biochemical genetics of the human blood group ABO system. *Am. J. Hum. Genet.* 34:1–14.

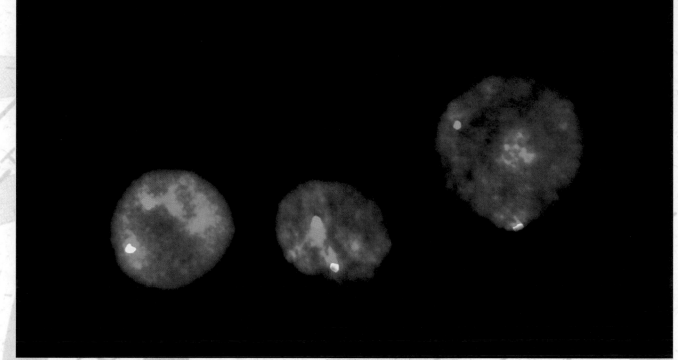

Demonstration of the X and Y chromosomes (the blue and the pink dots, respectively) in mammalian fetal cells using fluorescent in situ hybridization (FISH). *(James King-Holmes/Science Photo Library/Photo Researchers, Inc.)*

5

Sex Determination and Sex Chromosomes

CHAPTER CONCEPTS

Sexual differentiation plays an important role during the life cycle of various plants and animals. While a single pair of sex chromosomes (e.g., the X and the Y chromosome) often plays an important role in determining sexual maturation, genes present on these chromosomes as well as on autosomes serve as the underlying basis of sex determination. In humans, genes on the Y chromosome cause maleness, and in their absence, female development occurs. Mechanisms have developed to compensate for the dosage of genetic expression in organisms where one sex contains two X chromosomes, while the other has but a single X chromosome. In mammals, random inactivation of one of the X chromosomes is the compensatory mechanism. Still other modes of sex determination have evolved. Reptiles exemplify environmentally induced sex determination, where the temperature during the incubation of eggs is the critical factor.

In the biological world, a wide range of reproductive modes and life cycles are recognized. Asexual organisms exist where no evidence of sexual reproduction is evident. Other species alternate between short periods of sexual reproduction and prolonged periods of asexual reproduction. In most diploid eukaryotes, however, sexual reproduction is the only natural mechanism that results in new members of a species. Orderly transmission of genetic units from parents to offspring, and thus any phenotypic variability, relies on the processes of segregation and independent assortment that occur during meiosis. Meiosis produces haploid gametes so that, following fertilization, the resulting offspring maintain the diploid number of chromosomes characteristic of their species. Thus, meiosis ensures genetic constancy within members of the same species.

These events, which are involved in the perpetuation of all sexually reproducing organisms, ultimately depend on an efficient union of gametes during fertilization. In turn, successful mating between organisms, the basis for fertilization, relies on some form of sexual differentiation in organisms. Although not overtly apparent, this differentiation occurs as low on the evolutionary scale as bacteria and single-celled eukaryotic algae. In evolutionarily higher forms of life, the differentiation of the sexes is more evident as phenotypic dimorphism in the males and females of each species. The shield and spear ($\male$), the ancient symbols of iron and Mars, and the mirror ($\female$), the symbol of copper and Venus, represent the maleness and femaleness acquired by individuals.

While dissimilar, or **heteromorphic chromosomes**, such as the X-Y pair, often characterize one sex or the other, resulting in their label as **sex chromosomes**, genes rather than chromosomes determine sex. As we shall see, some of these genes are present on sex chromosomes, but others are autosomal. Extensive investigation reveals a wide variation in sex chromosome systems, even in closely related organisms, suggesting that mechanisms controlling sex determination have undergone rapid evolution in many instances.

In this chapter we first review several representative modes of sexual differentiation by examining the life cycle of three organisms often studied in genetics: the green alga *Chlamydomonas*; the maize plant, *Zea mays*; and the nematode (roundworm), *Caenorhabditis elegans* (most often referred to as *C. elegans*). These organisms contrast the different roles that sexual differentiation plays in the lives of diverse organisms. Then, we delve more deeply into what is known about the genetic basis for the determination of sexual differences, with a particular emphasis on two organisms: our own species, representative of mammals, and *Drosophila*, where pioneering sex-determining studies were performed.

5.1   Sexual Differentiation and Life Cycles

In multicellular organisms, it is important to distinguish between primary sexual differentiation, which involves only the gonads where gametes are produced, and secondary sexual differentiation, which involves the overall appearance of the organism, including clear differences in such organs as mammary glands and external genitalia. In plants and animals, the terms **unisexual**, **dioecious**, and **gonochoric** are equivalent; they refer to an individual containing only male *or* only female reproductive organs. Conversely, the terms **bisexual**, **monoecious**, and **hermaphroditic** refer to individuals containing both male *and* female reproductive organs, a common occurrence in both the plant and animal kingdoms. These organisms can produce fertile gametes of both sexes. The term **intersex** is usually reserved for individuals of intermediate sexual differentiation, who are most often sterile.

Chlamydomonas

The life cycle of the green alga *Chlamydomonas* (Figure 5–1) is representative of organisms that exhibit only infrequent periods of sexual reproduction. Such organisms spend most of their life cycle in the haploid phase, asexually producing daughter cells by mitotic divisions. However, under unfavorable nutrient conditions, such as nitrogen depletion, certain daughter cells function as gametes. Following fertilization, a diploid zygote, which can withstand the unfavorable environment, is formed. When conditions become more suitable, meiosis ensues and haploid vegetative cells are again produced.

In such species, there is little visible difference between the haploid vegetative cells that reproduce asexually and the haploid gametes that are involved in sexual reproduction. The two gametes that fuse during mating are morphologically indistinguishable and are called **isogametes**. Species producing them are said to be **isogamous**.

In 1954, Ruth Sager and Sam Granik demonstrated that gametes in *Chlamydomonas* can be subdivided into two mating types. Working with clones derived from single haploid cells, they showed that cells from a given clone mate with cells from some but not all other clones. When they tested mating abilities of large numbers of clones, all could be placed into one of two mating categories, either mt^+ or mt^-. "Plus" cells mate only with "minus" cells, and vice versa. Following fertilization and meiosis, the four haploid cells (zoospores) produced were found to consist of two plus types and two minus types.

Further experimentation established that plus and minus cells differ chemically. When extracts are prepared from cloned *Chlamydomonas* cells (or their flagella), and then added to cells of the opposite mating type, clumping or agglutination occurs. No such agglutination occurs if the extracts are added to cells of the mating type from which it was derived. These observations demonstrate that despite the morphological similarities between isogametes, they differentiate chemically. Therefore, in this alga, a primitive means of sex differentiation exists even though there is no morphological indication that such differentiation has occurred.

Maize (*Zea mays*)

As we first discussed in Chapter 2 (see Figure 2–14), life cycles of many plants alternate between the haploid gametophyte stage and the diploid sporophyte stage. The processes of meiosis and fertilization link the two phases during the life

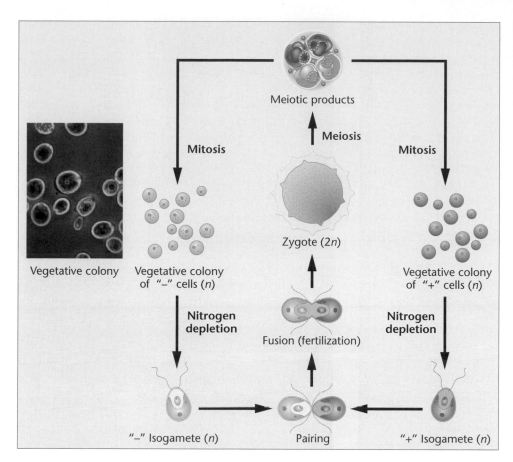

FIGURE 5–1 The life cycle of *Chlamydomonas.* Unfavorable conditions stimulate the formation of isogametes of opposite mating type that may fuse in fertilization. The resulting zygote undergoes meiosis, producing two haploid cells of each mating type. The photograph shows vegetative cells of this green alga. *(Photo: Biophoto Assoc./Photo Researchers, Inc.)*

Meiotic products

Meiosis

Mitosis

Mitosis

Zygote (2*n*)

Vegetative colony

Vegetative colony of "–" cells (*n*)

Vegetative colony of "+" cells (*n*)

Nitrogen depletion

Fusion (fertilization)

Nitrogen depletion

"–" Isogamete (*n*)

Pairing

"+" Isogamete (*n*)

cycle. Maize (*Zea mays*), familiar to you as corn, exemplifies a monoecious seed plant where the sporophyte phase and the morphological structures representing this stage predominate during the life cycle. Both male and female structures are present on the adult plant. Thus, sex determination occurs differently in different tissues of the same organism, as shown in the life cycle of this plant (Figure 5–2). The **stamens** (which collectively constitute the tassel), produce diploid microspore mother cells, each of which undergoes meiosis and gives rise to four haploid microspores. Each haploid microspore in turn develops into a mature male microgametophyte—the pollen grain—which contains two sperm nuclei.

Equivalent female diploid cells, known as megaspore mother cells, exist in the **pistil** of the sporophyte. Following meiosis, only one of the four haploid megaspores survives. It usually divides mitotically three times, producing a total of eight haploid nuclei enclosed in the embryo sac. Two of these nuclei unite near the center of the embryo sac, becoming the endosperm nuclei. At the micropyle end of the sac where the sperm enters, three nuclei remain: the oocyte nucleus and two synergids. The other three antipodal nuclei cluster at the opposite end of the embryo sac.

Pollination occurs when pollen grains make contact with the silks (or stigma) of the pistil and develop extensive pollen tubes that grow toward the embryo sac. When contact is made at the micropyle, the two sperm nuclei enter the embryo sac. One sperm nucleus unites with the haploid oocyte nucleus, and the other sperm nucleus unites with two endosperm nuclei. This process, known as double fertilization, results in the

diploid zygote nucleus and the triploid endosperm nucleus, respectively. Each ear of corn can contain as many as 1000 of these structures, each of which develops into a single kernel. Each kernel, if allowed to germinate, gives rise to a new plant, the sporophyte.

The mechanism of sex determination and differentiation in a monoecious plant like *Zea mays*, where the tissues that form both male and female gametes are of the same genetic constitution, was difficult to comprehend at first. However, the discovery of a large number of mutant genes that disrupt normal tassel and pistil formation supports the concept that normal products of these genes play an important role in sex determination by affecting the differentiation of male or female tissue in several ways.

For example, mutant genes that cause sex reversal provide valuable information. When homozygous, all mutations classified as *tassel seed* (*ts*) interfere with tassel production and induce the formation of female structures. Thus, a single gene can cause a normally monoecious plant to become functionally female. On the other hand, the recessive mutations *silkless* (*sk*) and *barren stalk* (*ba*) interfere with the development of the pistil, resulting in plants with only functional male reproductive organs.

Data gathered from studies of these and other mutants suggest that the products of many wild-type alleles of these genes interact in controlling sex determination. During development, certain cells are "determined" to become male or female structures. Following sexual differentiation into either male or female structures, male or female gametes are produced.

FIGURE 5–2 The life cycle of maize (*Zea mays*). The diploid sporophyte bears stamens and pistils that give rise to haploid microspores and megaspores, which develop into the pollen grain and the embryo sac that ultimately house the sperm and oocyte, respectively. Following fertilization, the embryo develops within the kernel and is nourished by the endosperm. Germination of the kernel gives rise to a new sporophyte (the mature corn plant), and the cycle repeats itself. *(Photo: Bill Beatty/Visuals Unlimited)*

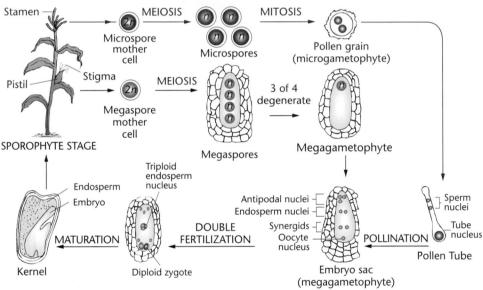

C. elegans

The roundworm *C. elegans* [Figure 5–3(a)] has become a popular organism in genetic studies, particularly during the investigation of the genetic control of development. Its usefulness is based on the fact that the adult consists of only about 1000 cells, the precise lineage of which can be traced back to specific embryonic origins. Among many interesting mutant phenotypes that have been studied, behavioral modifications are a favorite topic of inquiry.

There are two sexual phenotypes in these worms: males, which have only testes, and hermaphrodites, which contain both testes and ovaries. During larval development of hermaphrodites, testes form that produce sperm, which is stored. Ovaries are also produced, but oogenesis does not occur until the adult stage is reached several days later. The eggs that are then produced are fertilized by the stored sperm in the process of self-fertilization.

The outcome of this process is quite interesting [Figure 5–3(b)]. The vast majority of organisms that result, like the parental worm, are hermaphrodites; less than 1 percent of the offspring are males. As adults, they can mate with hermaphrodites, producing about half male and half hermaphrodite offspring.

The genetic signal that determines maleness rather than hermaphroditic development is provided by genes located on both the X chromosome and autosomes. *C. elegans* lacks a Y chromosome altogether. Hermaphrodites have two X chromosomes while males have only one X chromosome. It is believed that the ratio of X chromosomes to the number of sets of autosomes ultimately determines the sex of these worms. A ratio of 1.0 results in hermaphrodites and a ratio of 0.5 results in males. The absence of a heteromorphic Y chromosome is not uncommon in organisms.

(a)

(b)

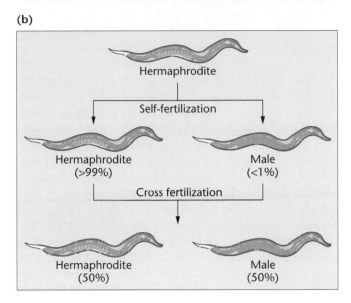

FIGURE 5–3 (a) Photomicrograph of an hermaphroditic nematode, *C. elegans.* (b) The outcomes of self-fertilization in a hermaphrodite and a mating of a hermaphrodite and a male worm. *(Photo: Dr. Maria Gallegos, University of California, San Francisco)*

5.2 X and Y Chromosomes: Early Studies

How sex is determined has long intrigued geneticists. In 1891, H. Henking identified a nuclear structure in the sperm of certain insects, which he labeled the X-body. Several years later, Clarance McClung showed that some grasshopper sperm contain an unusual genetic structure, which he called a **heterochromosome**, but the remainder lack such a structure. He mistakenly associated the presence of the heterochromosome with the production of male progeny. In 1906, Edmund B. Wilson clarified the findings of Henking and McClung when he demonstrated that female somatic cells in the insect *Protenor* contain 14 chromosomes, including 2 X chromosomes. During oogenesis, an even reduction occurs, producing gametes with 7 chromosomes, including 1 X. Male somatic cells, on the other hand, contain only 13 chromosomes, including a single X chromosome. During spermatogenesis, gametes are produced containing either 6 chromosomes, without an X, or 7 chromosomes, one of which is an X. Fertilization by X-bearing sperm results in female offspring, and fertilization by X-deficient sperm results in male offspring [Figure 5–4(a)].

The presence or absence of the X chromosome in male gametes provides an efficient mechanism for sex determination in this species and also produces a 1:1 sex ratio in the

(a) *Protenor* mode

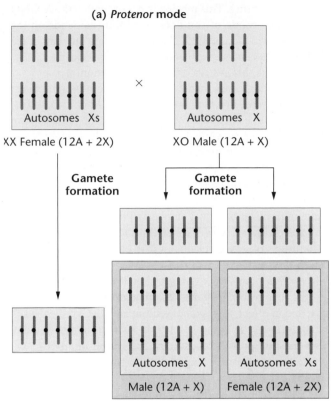

1:1 sex ratio

(b) *Lygaeus* mode

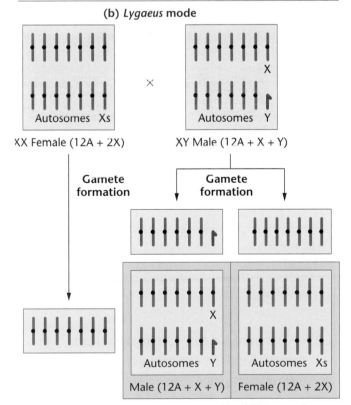

1:1 sex ratio

FIGURE 5–4 (a) The *Protenor* mode of sex determination where the heterogametic sex (the male in this example) is XO and produces gametes with or without the X chromosome. (b) The *Lygaeus* mode of sex determination, where the heterogametic sex (again, the male in this example) is XY and produces gametes with either an X or a Y chromosome. In both cases, the chromosome composition of the offspring determines its sex.

resulting offspring. This mechanism, now called the **XX/XO** or *Protenor* **mode of sex determination**, depends on the random distribution of the X chromosome into one-half of the male gametes during segregation. As we saw earlier, *C. elegans* exhibits this system of sex determination.

Wilson also experimented with the hemipteran insect *Lygaeus turicus*, in which both sexes have 14 chromosomes. Twelve of these are autosomes. In addition, the females have 2 X chromosomes, while the males have only a single X and a smaller heterochromosome labeled the **Y chromosome**. Females in this species produce only gametes of the (6A + X) constitution, but males produce two types of gametes in equal proportions: (6A + X) and (6A + Y). Therefore, following random fertilization, equal numbers of male and female progeny are produced with distinct chromosome complements. This mode of sex determination is called the *Lygaeus* or **XX/XY** type [Figure 5–4(b)].

In *Protenor* and *Lygaeus* insects, males produce unlike gametes. As a result, they are described as the **heterogametic** sex, and in effect, their gametes ultimately determine the sex of the progeny in those species. In such cases, the female, who has like sex chromosomes, is the **homogametic sex**, producing uniform gametes with regard to chromosome numbers and types.

The male is not always the heterogametic sex. In other organisms, the female produces unlike gametes, exhibiting either the *Protenor* (XX/XO) or *Lygaeus* (XX/XY) mode of sex determination. Examples include moths and butterflies, most birds, some fish, reptiles, amphibians, and at least one species of plants (*Fragaria orientalis*). To immediately distinguish situations in which the female is the heterogametic sex, some workers use the notation **ZZ/ZW**, where ZW is the heterogamous female, instead of the XX/XY notation.

The situation with fowl (chickens) demonstrates the difficulty in establishing which sex is heterogametic and whether the *Protenor* or *Lygaeus* mode is operative. While genetic evidence supported the hypothesis that the female is the heterogametic sex, the cytological identification of the sex chromosome was not accomplished until 1961 because of the large number of chromosomes (78) characteristic of chickens. When the sex chromosomes were finally identified, the female was shown to contain an unlike chromosome pair, including a heteromorphic chromosome (the W chromosome). Thus, in fowl, the female is indeed heterogametic and is characterized by the *Lygaeus* type of sex determination.

5.3   Chromosome Composition and Sex Determination in Humans

The first attempt to understand sex determination in our own species occurred almost 100 years ago and involved the examination of chromosomes present in dividing cells. Efforts were made to accurately determine the diploid chromosome number of humans, but because of the relatively large number of chromosomes, this proved to be quite difficult. In 1912, H. von Winiwarter counted 47 chromosomes in a spermato-

gonial metaphase preparation. It was believed that the sex-determining mechanism in humans was based on the presence of an extra chromosome in females, who were thought to have 48 chromosomes. However, in the 1920s, Theophilus Painter observed between 45 and 48 chromosomes in cells of testicular tissue and also discovered the small Y chromosome, which is now known to occur only in males. In his original paper, Painter favored 46 as the diploid number in humans, but he later concluded incorrectly that 48 was the chromosome number in both males and females.

For 30 years, this number was accepted. Then, in 1956, Joe Hin Tjio and Albert Levan discovered a better way to prepare chromosomes. The improved technique led to a strikingly clear demonstration of metaphase stages showing that 46 is indeed the human diploid number. Later that same year, C. E. Ford and John L. Hamerton, also working with testicular tissue, confirmed this finding. The familiar karyotype of humans (Figure 5–5) is based on Tjio and Levan's technique.

Within the normal 23 pairs of human chromosomes, one pair was shown to vary in configuration in males and females. These two chromosomes were designated the X and Y sex chromosomes. The human female has two X chromosomes, and the human male has one X and one Y chromosome.

We might believe that this observation is sufficient to conclude that the Y chromosome determines maleness. However, several other interpretations are possible. The Y could play no role in sex determination; the presence of two X chromosomes could cause femaleness; or maleness could result from the lack of a second X chromosome. The evidence that clarified which explanation was correct emerged in the study of variations in the human sex chromosome composition. As such investigations reveal, the Y chromosome does indeed determine maleness in humans.

Klinefelter and Turner Syndromes

About 1940, scientists identified two human abnormalities characterized by aberrant sexual development, **Klinefelter syndrome** and **Turner syndrome**.* Individuals with Klinefelter syndrome have genitalia and internal ducts that are usually male, but their testes are rudimentary and fail to produce sperm. They are generally tall and have long arms and legs and large hands and feet.

Although some masculine development does occur, feminine sexual development is not entirely suppressed. Slight enlargement of the breasts (gynecomastia) is common, and the hips are often rounded. This ambiguous sexual development can lead to abnormal social development. Intelligence is often below the normal range.

In Turner syndrome, the affected individual has female external genitalia and internal ducts, but the ovaries are rudimentary. Other characteristic abnormalities include short stature (usually under 5 feet), skin flaps on the back of the

---

*Although the possessive form of the names of most syndromes (eponyms) is sometimes used (e.g., Klinefelter's), the current preference is to use the nonpossessive form, which we have adopted for all human syndromes.

(a)

(b)

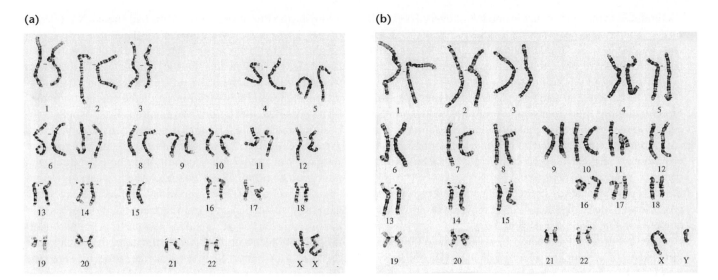

FIGURE 5–5 The traditional human karyotypes derived from a normal female and a normal male. Each contains 22 pairs of autosomes and two sex chromosomes. The female (a) contains two identical X chromosomes, while the male (b) contains one X and one Y chromosome (the Y is often referred to as a heterochromosome). *(Courtesy of the Greenwood Genetic Center, Greenwood, SC)*

neck, and underdeveloped breasts. A broad, shieldlike chest is sometimes noted. Intelligence is usually normal.

In 1959, the karyotypes of individuals with these syndromes were determined to be abnormal with respect to the sex chromosomes. Individuals with Klinefelter syndrome have more than one X chromosome. Most often they have an XXY complement in addition to 44 autosomes [Figure 5–6(a)]. People with this karyotype are designated **47,XXY**. Individuals with Turner syndrome are most often monosomic and have only 45 chromosomes, including just a single X chromosome. They are designated **45,X** [Figure 5–6(b)]. Note the convention used in designating these chromosome compositions. The number states the total number of chromosomes present,

and the information after the comma indicates the deviation from the normal diploid content. Both conditions result from **nondisjunction**, the failure of the X chromosomes to segregate properly during meiosis (see Figure 7–1).

The Klinefelter and Turner karyotypes and their corresponding sexual phenotypes allow us to conclude that the Y chromosome determines maleness in humans. In its absence, the sex of the individual is female, even if only a single X chromosome is present. The presence of the Y chromosome in the individual with Klinefelter syndrome is sufficient to determine maleness, even though its expression is not complete. Similarly, in the absence of a Y chromosome, as in the case of individuals with Turner syndrome, no masculinization occurs.

(a)

(b)

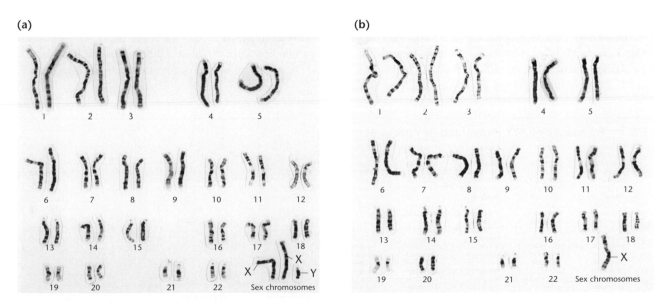

FIGURE 5–6 The karyotypes of individuals with (a) Klinefelter syndrome (47,XXY) and (b) Turner syndrome (45,X). *(Catherine B. Palmer, Dept. of Medical Genetics/Indiana University-Indianapolis)*

Klinefelter syndrome occurs in about 2 of every 1000 male births. The karyotypes 48,XXXY, 48,XXYY, 49,XXXXY, and 49,XXXXY are similar phenotypically to 47,XXY, but manifestations are often more severe in individuals with a greater number of X chromosomes.

Turner syndrome can also result from karyotypes other than 45,X, including individuals called mosaics whose somatic cells display two different genetic cell lines, each exhibiting a different karyotype. Such cell lines result from a mitotic error during early development, the most common chromosome combinations being 45,X/46,XY and 45,X/46,XX. Thus, an embryo that began life with a normal karyotype can give rise to an individual whose cells show a mixture of karyotypes and who exhibits this syndrome.

Turner syndrome is observed in about 1 in 2000 female births, a frequency much lower than that for Klinefelter syndrome. However, the vast majority of 45,X fetuses die *in utero* and are aborted spontaneously. Thus, the frequency of Turner syndrome may be much higher at conception.

47,XXX Syndrome

The presence of three X chromosomes along with a normal set of autosomes (**47,XXX**) results in female differentiation. This syndrome, which occurs in about 1 of 1200 female births, is highly variable in expression. Frequently, 47,XXX women are perfectly normal. In other cases, underdeveloped secondary sex characteristics, sterility, and mental retardation can occur. In rare instances, 48,XXXX and 49,XXXXX karyotypes have been reported. The syndromes associated with these karyotypes are similar to, but more pronounced than, the 47,XXX. Thus in many cases, the presence of additional X chromosomes appears to disrupt the delicate balance of genetic information essential to normal female development.

47,XYY Condition

Another human condition involving the sex chromosomes, **47,XYY**, has also been intensively investigated. Studies of this condition, where the only deviation from diploidy is the presence of an additional Y chromosome in an otherwise normal male karyotype, have led to a controversy initiated in 1965 by Patricia Jacobs. She found that 9 of 315 males in a Scottish maximum security prison had the 47,XYY karyotype. These males were significantly above average in height and had been incarcerated as a result of antisocial (nonviolent) criminal acts. Of the nine males studied, seven were of subnormal intelligence, and all suffered personality disorders. Several other studies produced similar findings.

The possible correlation between this chromosome composition and criminal behavior piqued considerable interest and extensive investigations of the phenotype and frequency of the 47,XYY condition in both criminal and noncriminal populations ensued. Above-average height (usually over 6 feet) and subnormal intelligence have been generally substantiated, and the frequency of males displaying this karyotype is indeed higher in penal and mental institutions compared with unincarcerated males (see Table 5.1). A particularly relevant question involves the characteristics displayed by XYY males who are not incarcerated. The only nearly constant association is that such individuals are over 6 feet tall.

A study that addressed this issue was initiated to identify 47,XYY individuals at birth and to follow their behavioral patterns during preadult and adult development. By 1974, the two investigators, Stanley Walzer and Park Gerald, had identified about 20 XYY newborns in 15,000 births at Boston Hospital for Women. However, they soon came under great pressure to abandon their research. Those opposed to the study argued that the investigation could not be justified and might cause great harm to those individuals who displayed this karyotype. The opponents argued that (1) no association between the additional Y chromosome and abnormal behavior had been previously established in the population at large, and (2) "labeling" these individuals in the study might become a self-fulfilling prophecy. That is, as a result of participation in the study, parents, relatives, and friends might treat individuals identified as 47,XYY differently, ultimately producing the expected antisocial behavior. Despite the support of a government funding agency and the faculty at Harvard Medical School, Walzer and Gerald abandoned the investigation in 1975.

More recently, it has become clear that many XYY males do not exhibit any form of antisocial behavior and lead normal lives. Therefore, we must conclude that there is a high, but not

TABLE 5.1   Frequency of XYY Individuals in Various Settings

Setting	Restriction	Number Studied	XYY Number	XYY Frequency
Control population	Newborns	28,366	29	0.10%
Mental-penal	No height restriction	4,239	82	1.93
Penal	No height restriction	5,805	26	0.44
Mental	No height restriction	2,562	8	0.31
Mental-penal	Height restriction	1,048	48	4.61
Penal	Height restriction	1,683	31	1.84
Mental	Height restriction	649	9	1.38

Source: Compiled from data presented in Hook, 1973, Tables 1–8. Copyright 1973 by the American Association for the Advancement of Science.

constant correlation between the extra Y chromosome and the predisposition of males to behavioral problems.

5.4 Sexual Differentiation in Humans

Once researchers established that, in humans, it is the Y chromosome that houses genetic information necessary for maleness, they attempted to pinpoint a specific gene or genes capable of providing the "signal" responsible for sex determination. Before we delve into this topic, it is useful to consider how sexual differentiation occurs in order to better comprehend how humans develop into sexually dimorphic males and females. During early development, every human embryo undergoes a period when it is potentially hermaphroditic. By the fifth week of gestation, gonadal primordia arise as a pair of ridges associated with each embryonic kidney. Primordial germ cells migrate to these ridges, where an outer cortex and inner medulla form. The **cortex** is capable of developing into an ovary, while the inner **medulla** may develop into a testis. In addition, two sets of undifferentiated male (Wolffian) and female (Mullerian) ducts exist in each embryo.

If the cells of the genital ridge have the XY constitution, development of the medullary region into a testis is initiated around the seventh week. However, in the absence of the Y chromosome, no male development occurs, and the cortex of the genital ridge subsequently forms ovarian tissue. Parallel development of the appropriate male or female duct system then occurs, and the other duct system degenerates. Substantial evidence indicates that in males, once testes differentiation is initiated, the embryonic testicular tissue secretes two hormones that are essential for continued male sexual differentiation.

In the absence of male development, as the 12th week of fetal development approaches, the oogonia within the ovaries begin meiosis and primary oocytes can be detected. By the 25th week of gestation, all oocytes become arrested in meiosis and remain dormant until puberty is reached some 10–15 years later. In males, on the other hand, primary spermatocytes are not produced until puberty is reached.

The Y Chromosome and Male Development

The Y chromosome, unlike the X, has long been thought to be nearly blank genetically. Currently, this is believed to be only partially true. Present on both ends of the chromosome are the so-called **pseudoautosomal regions (PARs)** that share homology with the X chromosome and synapse and recombine with it during meiosis. Such pairing is critical to the segregation of the X and Y during male gametogenesis. The remainder of the chromosome is referred to as the **NRY**, the **nonrecombining region of the Y**. As we shall see, some portions of the NRY are also homologous to genes on the X chromosome. The human Y chromosome is diagrammed in Figure 5–7.

While much of the chromosome is heterochromatic and genetically blank, the Y chromosome clearly carries genetic information absent from the X that controls male sexual development. In the absence of such genetic information, female development occurs. Thus, some region of the Y chromosome encodes a gene product that somehow triggers the undifferentiated gonadal tissue of the embryo to form testes. This product is called the **testis-determining factor (TDF)**. An extensive research effort is currently focused on its discovery.

It is now clear that a small part of the short arm of the human Y chromosome contains a gene called **SRY (sex-determining region Y)** that encodes TDF (Figure 5–7). Evidence proving that SRY is the responsible gene relies on the molecular geneticist's ability to identify the presence or absence of DNA sequences in rare individuals whose expected sex chromosome composition does not correspond to their sexual phenotype. For example, certain human males demonstrate two X and no Y chromosomes. Often, they have the region of the Y chromosome containing SRY attached to one of their X's. Some females have one X and one Y chromosome. Their Y is almost always missing the SRY region. These observations argue strongly in favor of the role of SRY in providing the primary signal for male development.

Further support of this conclusion involves an experiment using **transgenic mice**. Such animals are produced from fertilized eggs injected with foreign DNA that is subsequently incorporated into the genetic composition of the developing embryo. In normal mice, a chromosome region designated *Sry* has been identified that is comparable to SRY in humans. When DNA containing only mouse *Sry* is injected into normal XX mouse eggs, most of the offspring develop into males!

These studies suggest that the SRY gene determines maleness. Furthermore, it is present in all mammals examined so far, indicating that it has been conserved throughout the evolution of this diverse group of animals.

How specifically the product of this gene triggers the embryonic gonadal tissue to develop into testes rather than ovaries is a question under extensive investigation. A number of other autosomal genes are believed to be part of a cascade of genetic expression initiated by SRY. One such autosomal gene is SOX9 in humans. Others include WT1, found on human chromosome 11, originally identified as an

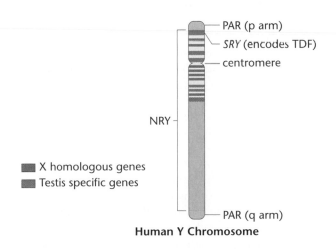

Human Y Chromosome

FIGURE 5–7 The various regions of the human Y chromosome.

oncogene associated with Wilms tumor, which affects the kidney and gonads. Another, SF1, is involved in the regulation of enzymes affecting steroid metabolism. In mice, this gene is initially active in both the male and female bisexual genital ridge, persisting until the point in development when testis formation is apparent. At that time, its expression persists in males, but is extinguished in females. The link between these various genes and sex determination brings us closer to the complete understanding of how males and females arise in humans.

We conclude this section on the Y chromosome by relating some recent findings by David Page and his colleagues that provide a more complete picture of this chromosome in humans (Figure 5–7). Using molecular probes, they have discovered 12 genes in the NRY, clearly separate from the pseudoautosomal regions. These are in addition to 8 other NRY genes found previously. The 12 genes sort into two groups. The first category consists of 5 NRY genes, each of which has a homolog on the X chromosome and is expressed in a wide range of tissues in addition to the testis. Genes in this group seem to encode information of general cellular function, thus their description as housekeeping genes. The 7 genes in the second group lack a functional homolog on the X chromosome and are expressed only in the testis. They appear to encode proteins specific to testis development and function.

These findings refute the so-called "wasteland" theory that depicts the human Y chromosome as genetically blank. They further provide a variety of genetic insights. For example, deletion of any of the testis-specific genes may account for infertility in otherwise normal males. Second, as we shall see in a subsequent section of this chapter, the group of Y-linked housekeeping genes must be accounted for in any mechanism of dosage compensation that equalizes the expression of genes present on the sex chromosomes.

5.5 The Sex Ratio in Humans

The presence of heteromorphic sex chromosomes in one sex of a species but not the other provides a potential mechanism for producing equal proportions of male and female offspring. This potential is premised on the segregation of the X and Y (or Z and W) chromosomes during meiosis, such that one-half of the gametes of the heterogametic sex receive one of the chromosomes and one-half receive the other one. Provided both types of gametes are equally successful in fertilization and the two sexes are equally viable during development, a 1.0 ratio of male and female offspring results.

Given the potential for producing equal numbers of both sexes, the actual proportion of male to female offspring has been investigated and is referred to as the **sex ratio**. We can assess it in two ways. The **primary sex ratio** reflects the proportion of males to females conceived in a population. The **secondary sex ratio** reflects the proportion of each sex that is born. The secondary sex ratio is much easier to determine, but has the disadvantage of not accounting for any disproportionate embryonic or fetal mortality.

When the secondary sex ratio in the human population was determined in 1969 using worldwide census data, it was found not to equal 1.0. For example, in the Caucasian population in the United States, the secondary ratio was a little less than 1.06, indicating that about 106 males are born for each 100 females. In 1995, this ratio dropped to slightly less than 1.05. In the African-American population in the United States, the ratio was 1.025. In other countries the excess of male births was even greater than reflected in these values. For example, in Korea, the secondary sex ratio was 1.15.

Despite these ratios, it is possible that the *primary sex ratio* is 1.0, and that it is altered between conception and birth. For the secondary ratio to exceed 1.0, then, prenatal female mortality would have to be greater than prenatal male mortality. However, this hypothesis has been examined and shown to be false. In fact, just the opposite occurs. In a Carnegie Institute study, reported in 1948, the sex of approximately 6000 embryos and fetuses recovered from miscarriages and abortions was determined, and fetal mortality was actually higher in males. On the basis of the data derived from that study, the primary sex ratio in U.S. Caucasians was estimated to be 1.079. More recent data have estimated that this figure is much higher—between 1.20 and 1.60, suggesting that many more males than females are conceived in the human population.

It is not clear why such a radical departure from the expected primary sex ratio of 1.0 occurs. To come up with a suitable explanation, we must examine the assumptions on which the theoretical ratio is based:

1. Because of segregation, males produce equal numbers of X- and Y-bearing sperm.

2. Each type of sperm has equivalent viability and motility in the female reproductive tract.

3. The egg surface is equally receptive to both X- and Y-bearing sperm.

While no direct experimental evidence contradicts any of these assumptions, the human Y chromosome is smaller than the X chromosome and therefore has less mass. Thus, it has been speculated that Y-bearing sperm are more motile than X-bearing sperm. If this is true, then the probability of a fertilization event leading to a male zygote is increased, providing one possible explanation for the observed primary ratio.

5.6 The X Chromosome and Dosage Compensation

The presence of two X chromosomes in normal human females and only one X in normal human males is unique compared with the equal numbers of autosomes present in the cells of both sexes. On theoretical grounds alone, it is possible to speculate that this disparity should create a "genetic dosage" problem between males and females for all X-linked genes. There is the potential for females to produce twice as much of each gene product for all X-linked genes. The additional X chromosomes in both males and females exhibiting the various syndromes discussed earlier in this chapter should compound this dosage problem even more. In this section, we describe research findings on X-linked gene expression that demonstrate a genetic mechanism allowing for **dosage compensation**.

Barr Bodies

Murray L. Barr and Ewart G. Bertram's experiments with female cats, and Keith Moore and Barr's subsequent study with humans, demonstrate a genetic mechanism in mammals that compensates for X chromosome dosage disparities. Barr and Bertram observed a darkly staining body in interphase nerve cells of female cats that was absent in similar cells of males. In humans this body can be easily demonstrated in female cells derived from the buccal mucosa or in fibroblasts but not in similar male cells (Figure 5–8). This highly condensed structure, about 1 μm in diameter, lies against the nuclear envelope of interphase cells. It stains positively in the Feulgen reaction for DNA.

Current experimental evidence demonstrates that this body, called a **sex chromatin body** or simply a **Barr body**, is an inactivated X chromosome. Susumo Ohno was the first to suggest that the Barr body arises from one of the two X chromosomes. This hypothesis is attractive because it provides a mechanism for dosage compensation. If one of the two X chromosomes is inactive in the cells of females, the dosage of genetic information that can be expressed in males and females is equivalent. Convincing but indirect evidence for this hypothesis comes from the study of the sex chromosome syndromes described earlier in this chapter. Regardless of how many X chromosomes exist, all but one of them appear to be inactivated and can be seen as Barr bodies. For example, no Barr body is seen in Turner 45,X females; one is seen in Klinefelter 47,XXY males; two in 47,XXX females; three in 48,XXXX females; and so on (Figure 5–9). Therefore, the number of Barr bodies follows an $N - 1$ rule, where N is the total number of X chromosomes present.

Although this mechanism of inactivating all but one X chromosome increases our understanding of dosage compensation, it further complicates our perception of other matters. Because one of the two X chromosomes is inactivated in normal human females, why then is the Turner 45,X individual not entirely normal? Why aren't females with the triplo-X and tetra-X karyotypes (47,XXX and 48,XXXX) normal? Further, in Klinefelter syndrome (47,XXY), X chromosome inactivation effectively renders such individuals 46,XY. Why aren't these males unaffected by the additional X chromosome in their nuclei?

One possible explanation is that chromosome inactivation does not normally occur in the very early developmental stages of those cells destined to form gonadal tissues. Another possible explanation is that not all of each X chromosome forming a Barr body is inactivated. If either hypothesis is correct, excessive expression of certain X-linked genes might still occur despite apparent inactivation of additional X chromosomes.

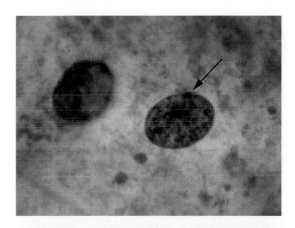

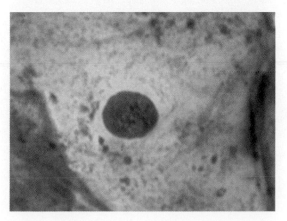

FIGURE 5–8 Photomicrographs comparing cheek epithelial cell nuclei from a male that fails to reveal Barr bodies (bottom) with a female that demonstrates Barr bodies (indicated by the arrow in the top image). This structure, also called a sex chromatin body, represents an inactivated X chromosome. *(Stuart Kenter Associates)*

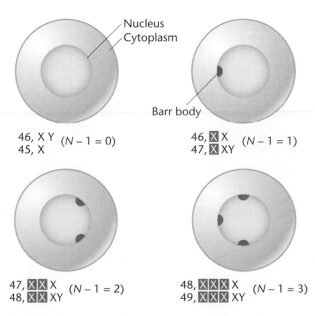

FIGURE 5–9 Barr body occurrence in various human karyotypes, where all X chromosomes except one ($N - 1$) are inactivated.

The Lyon Hypothesis

In mammalian females, one X chromosome is of maternal origin and the other is of paternal origin. Which one is inactivated? Is the inactivation random? Is the same chromosome inactive in all somatic cells? In 1961, Mary Lyon and Liane Russell independently proposed a hypothesis that answers these questions. They postulated that the inactivation of X chromosomes occurs randomly in somatic cells at a point early in embryonic development and that once inactivation has occurred, all progeny cells have the same X chromosome inactivated.

This explanation, which has come to be called the **Lyon hypothesis**, was initially based on observations of female mice heterozygous for X-linked coat color genes. The pigmentation of these heterozygous females was mottled, with large patches expressing the color allele on one X and other patches expressing the allele on the other X. Indeed, such a phenotypic pattern would result if different X chromosomes were inactive in adjacent patches of cells. Similar mosaic patterns occur in the black and yellow-orange patches of female tortoiseshell and calico cats (Figure 5–10). Such X-linked coat color patterns do not occur in male cats because all their cells contain the single maternal X chromosome and are therefore hemizygous for only one X-linked coat color allele.

The most direct evidence in support of the Lyon hypothesis comes from studies of gene expression in clones of human fibroblast cells. Individual cells are isolated following biopsy and cultured *in vitro*. If each culture is derived from a single cell, it is called a **clone**. The synthesis of the enzyme **glucose-6-phosphate dehydrogenase (G6PD)** is controlled by an X-linked gene. Numerous mutant alleles of this gene have been detected, and their gene products can be differentiated from the wild-type enzyme by their migration pattern in an electrophoretic field.

Fibroblasts have been taken from females heterozygous for different allelic forms of G6PD and studied. The Lyon hypothesis predicts that if inactivation of an X chromosome occurs randomly early in development and is permanent in all progeny cells, such a female should show two types of clones, each showing only one electrophoretic form of G6PD, in approximately equal proportions.

In 1963, Ronald Davidson and colleagues performed an experiment involving 14 clones from a single heterozygous female. Seven showed only one form of the enzyme, and 7 showed only the other form. What was most important was that none of the 14 showed both forms of the enzyme. Studies of G6PD mutants thus provide strong support for the random permanent inactivation of either the maternal or paternal X chromosome.

The Lyon hypothesis is generally accepted as valid; in fact, the inactivation of an X chromosome into a Barr body is sometimes referred to as **lyonization**. One extension of the hypothesis is that mammalian females are mosaics for all heterozygous X-linked alleles—some areas of the body express only the maternally derived alleles, and others express only the paternally derived alleles. Two especially interesting examples involve **red-green color blindness** and **anhidrotic ectodermal dysplasia**, both X-linked recessive disorders. In the former case, hemizygous males are fully color-blind in all retinal cells. However, heterozygous females display mosaic retinas with patches of defective color perception and surrounding areas with normal color perception. Males hemizygous for anhidrotic ectodermal dysplasia show absence of teeth, sparse hair growth, and lack of sweat glands. The skin of females heterozygous for this disorder reveals random patterns of tissue with and without sweat glands. In both examples, random inactivation of one or the other X chromosome early in the development of heterozygous females leads to these occurrences.

(a)

(b)

FIGURE 5–10 A calico cat (a), where the random distribution of orange and black patches demonstrates the Lyon hypothesis. The white patches are due to another gene, characterizing calicos from tortoiseshell cats (b), which lack the white patches. *(Left: Reed/Williams/Animals Animals/Earth Scenes; right: W. Layer/Okapia/Photo Researchers, Inc.)*

The Mechanism of Inactivation

The least understood aspect of the Lyon hypothesis is the mechanism of chromosome inactivation in mammals. How are almost all genes of an entire chromosome inactivated? Recent investigations are beginning to clarify this issue. A single region of the human X chromosome, called the **X-inactivation center (XIC)**, is the major control unit. Genetic expression of this region, located on the proximal end of the p arm, occurs only on the X chromosome that is inactivated. The constant association of expression of XIC and X chromosome inactivation supports the conclusion that this region is an important genetic component in the inactivation process.

A gene, *XIST* (X-inactive specific transcript), is now believed to represent the critical locus within the XIC. A comparable region (Xic) and gene (*Xist*) exist in the mouse, which has been the object of most investigation. Several interesting observations have been made regarding the RNA that is transcribed by the mouse *Xist* gene. First, the RNA product is quite large and lacks what is called an extended **open reading frame (ORF)**. An ORF includes the information necessary for translating the RNA product into a protein. Thus, the RNA is not translated, but instead plays a structural role in the nucleus, presumably in the mechanism of chromosome inactivation. This finding has led to the belief that the RNA products of *XIST* and *Xist* spread over and coat the X chromosome, producing some sort of molecular "cage" that entraps it, leading to its inactivation.

In 1996, a research group led by Graeme Penny provided convincing evidence that transcription of *Xist* is the critical event in chromosome inactivation. These researchers were able to introduce a targeted deletion (7 kb) into this gene. As a result, the chromosome bearing this mutation lost its ability to become inactivated. Several intriguing questions remain. First, in cells with more than two chromosomes, what sort of "counting" mechanism designates all but one X chromosome to be inactivated? Second, what "blocks" the Xic of the active chromosome, preventing transcription of *Xist*? Third, how is inactivation of the same X chromosome or chromosomes subsequently maintained in progeny cells, as the Lyon hypothesis calls for? The inactivation signal must somehow remain stable as cells proceed through mitosis. Whatever the answers to these questions, we have taken an exciting step toward understanding how dosage compensation is accomplished in mammals.

5.7   Chromosome Composition and Sex Determination in *Drosophila*

Because males and females in *Drosophila melanogaster* (and other *Drosophila* species) have the same general sex chromosome composition as humans (males are XY and females are XX), we might assume that the Y chromosome also causes maleness in these flies. However, the elegant work of Calvin Bridges in 1916 showed this not to be true. He studied flies with quite varied chromosome compositions, leading him to conclude that the Y chromosome is not involved in sex determination in this organism. Instead, Bridges proposed that both the X chromosomes and autosomes together play a critical role in sex determination. Recall that in the nematode, *C. elegans*, which lacks a Y chromosome, the sex chromosomes and autosomes are also critical to sex determination.

The most telling observation that differentiates the mechanism operating in *Drosophila* from that in humans involves the XXY and XO sex chromosome compositions. Contrary to what was later discovered in humans, Bridges found that the XXY flies are normal females, and the XO flies are sterile males. The presence of the Y chromosome in the XXY flies did not cause maleness, and its absence in the XO flies did not produce femaleness. From these data he concluded that the Y chromosome in *Drosophila* lacks male-determining factors but, since the XO males are sterile, it does contain genetic information essential to male fertility.

Bridges was able to clarify the mode of sex determination in *Drosophila* by studying the progeny of triploid females (3n), which have three copies each of the haploid complement of chromosomes. *Drosophila* has a haploid number of 4, thereby displaying three pairs of autosomes in addition to its pair of sex chromosomes. Triploid females apparently originate from rare diploid eggs fertilized by normal haploid sperm. Triploid females have heavy-set bodies, coarse bristles, and coarse eyes, and they can be fertile. Because of the odd number of each chromosome (3), during meiosis a wide range of chromosome complements is distributed into gametes that give rise to offspring with a variety of abnormal chromosome constitutions. A correlation among the sexual morphology, chromosome composition, and Bridges' interpretation is shown in Figure 5–11.

Bridges realized that the critical factor in determining sex is the ratio of X chromosomes to the number of haploid sets of autosomes (A) present. Normal (2X:2A) and triploid (3X:3A) females each have a ratio equal to 1.0, and both are fertile. As the ratio exceeds unity (3X:2A, or 1.5, for example), what was originally called a superfemale is produced. Because this female is rather weak and infertile and has lowered viability, this type is now more appropriately called a **metafemale**.

Normal (XY:2A) and sterile (XO:2A) males each have a ratio of 1:2, or 0.5. When the ratio decreases to 1:3, or 0.33, as in the case of an XY:3A male, infertile **metamales** result. Other flies recovered by Bridges in these studies contained an X:A ratio intermediate between 0.5 and 1.0. These flies were generally larger, and they exhibited a variety of morphological abnormalities and rudimentary bisexual gonads and genitalia. They were invariably sterile and expressed both male and female morphology, thus being designated as **intersexes**.

Bridges' results indicate that in *Drosophila*, factors that cause a fly to develop into a male are not localized on the sex chromosomes, but are instead found on the autosomes. Some female-determining factors, however, are localized on the X chromosomes. Thus, with respect to primary sex determination, male gametes containing one of each autosome plus a Y chromosome result in male offspring, not because of the presence of the Y but because of the lack of a second

Chromosome composition	Chromosome formulation	Ratio of X chromosomes to autosome sets	Sexual morphology
	3X/2A	1.5	Metafemale
	3X/3A	1.0	Female
	2X/2A	1.0	Female
	3X/4A	0.75	Intersex
	2X/3A	0.67	Intersex
	X/2A	0.50	Male
	XY/2A	0.50	Male
	XY/3A	0.33	Metamale

Normal diploid male

(IV)

(II)

(III)

(I)

X Y

2 sets of autosomes
+
X   Y

FIGURE 5–11 Chromosome compositions, the ratios of X chromosomes to sets of autosomes, and the resultant sexual morphology in *Drosophila melanogaster*. The normal diploid male chromosome composition is shown as a reference on the left (XY/2A).

X chromosome. This mode of sex determination is explained by the **genic balance theory**. Bridges proposed that a threshold for maleness is reached when the X:A ratio is 1:2 (X:2A), but that the presence of an additional X (XX:2A) alters this balance and results in female differentiation.

Numerous mutant genes have been identified that are involved in sex determination in *Drosophila*. The recessive autosomal gene *transformer* (*tra*), discovered over 50 years ago by Alfred H. Sturtevant, clearly demonstrates that a single autosomal gene can have a profound impact on sex determination. Females homozygous for *tra* are transformed into sterile males, but homozygous males are unaffected.

More recently, another gene, *Sex-lethal* (*Sxl*), has been shown to play a critical role, serving as a "master switch" in sex determination. Activation of the X-linked *Sxl* gene, which relies on a ratio of X chromosomes to sets of autosomes that equals 1.0, is essential to female development. In the absence

of activation, resulting, for example, from an X:A ratio of 0.5, male development occurs. It is interesting to note that mutations that inactivate the *Sxl* gene kill female embryos, but have no effect on male embryos, consistent with the role of the gene, as described earlier. While it is not yet exactly clear how this ratio influences the *Sxl* locus, we do have some insights into the question. The *Sxl* locus is part of a hierarchy of gene expression and exerts control over still other genes, including *tra* (discussed above), and *dsx* (*doublesex*), as well as others. The wild-type allele of *tra* is activated by the product of *Sxl* only in females, which in turn influences the expression of *dsx*. Depending on how the initial RNA transcript of *dsx* is processed (spliced), the resultant dsx protein activates either male- or female-specific genes required for sexual differentiation. Each step in this regulatory cascade requires a form of processing called **RNA splicing**, in which portions of the RNA are removed and the remaining fragments "spliced" back together prior to translation into a protein (see Chapter 15). In the case of the *Sxl* gene, its transcript can be spliced in several different ways, a phenomenon called **alternative splicing**. Two different RNA transcripts are produced in females and males, respectively. In potential females, the transcript is active and initiates a cascade of regulatory gene expression, ultimately leading to female differentiation. In potential males, the transcript is inactive, leading to different gene activity, whereby male differentiation occurs.

5.8 Temperature Variation and Sex Determination in Reptiles

We conclude this chapter by discussing several cases involving reptiles where the environment, specifically temperature, has a profound influence on sex determination. As we shall see, the investigations leading to this information may well come closer to revealing the true nature of the primary basis of sex determination than any finding previously discussed.

In many species of reptiles, sex is predetermined at conception by sex chromosome composition, as is the case in many organisms already considered in this chapter. For example, in many snakes, including vipers, a ZZ/ZW mode is in effect, where the female is the heterogamous sex. However, in boas and pythons, it is impossible to distinguish one sex chromosome from the other in either sex. In lizards, both the XX/XY and ZZ/ZW systems are found, depending on the species. In other reptilian species, including all crocodiles, most turtles, and some lizards, sex determination is achieved according to the incubation temperature of eggs during a critical period of embryonic development.

Interestingly, three distinct patterns of temperature-dependent sex determination emerge, as shown in Figure 5–12. In the first two, low temperatures yield 100% males and high temperatures yield 100% females (Case I), or just

the opposite occurs (Case II). In the third pattern (Case III), low *and* high temperatures yield 100% females, while intermediate temperatures yield various proportions of males. The third pattern is seen in various species of crocodiles, turtles, and lizards, although other members of these groups are known to exhibit the other patterns. Two observations are noteworthy. First, under certain temperatures in all three patterns, both male and female offspring result. Secondly, the pivotal temperature (T_P) is fairly narrow, usually less than 5°C, and sometimes only 1°C.

The central question raised by these observations is: What metabolic or physiological parameters are being affected by temperature that lead to the differentiation of one sex or the other? The answer is thought to involve steroids (mainly estrogens) and the enzymes involved in their synthesis. Studies clearly demonstrate that the effects of temperature on estrogens, androgens, and inhibitors of the enzymes controlling their synthesis are involved in the sexual differentiation of ovaries and testes. One enzyme in particular, **aromatase**, converts androgens (male hormones such as testosterone) to estrogens (female hormones such as estradiol). The activity of this enzyme is correlated with the pathway followed during gonadal differentiation activity and is high in developing ovaries and low in developing testes. Researchers in this field, including Claude Pieau and colleagues, have proposed that a thermosensitive factor mediates the transcription of the reptilian aromatase gene that leads to temperature-dependent sex determination. Several other genes are likely to be involved in this mediation.

The involvement of sex steroids in gonadal differentiation has also been documented in birds, fishes, and amphibians. Thus, sex-determining mechanisms involving estrogens seem to be characteristic of nonmammalian vertebrates. The regulation of such a system, while temperature dependent in many reptiles, appears to be controlled by sex chromosomes (XX/XY or ZZ/ZW) in many of these other organisms.

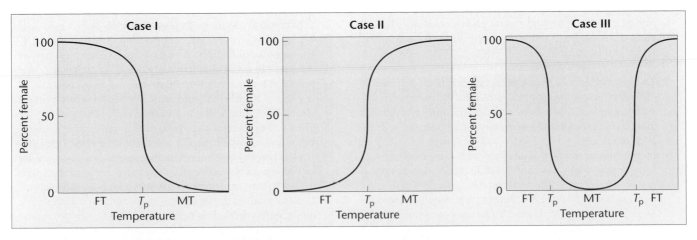

FIGURE 5–12 Three different patterns of temperature-dependent sex determination (TSD) in reptiles, as described in the text. The relative pivotal temperature T_p is crucial to sex determination during a critical point in embryonic development (FT = female-determining temperature; MT = male-determining temperature).

Male Sterility in Maize— Extrachromosomal Inheritance

While our discussion of sex determination in maize focuses on nuclear genes, which in mutant form cause either male or female sterility, another example of male sterility involves cytoplasmically transmitted factors. In the 1930s, Marcus Rhoades discovered a male-sterile strain of maize, and in a series of carefully designed crosses, established that sterility is inherited strictly through the female gamete, in contrast to the biparental inheritance that characterizes nuclear genes. Because all of the zygote's cytoplasm is derived from the female gamete, it is believed that this nonnuclear component is the source of the sterility factor(s). This exemplifies what we now call **extrachromosomal inheritance**, which is characterized by non-Mendelian inheritance patterns such as those observed by Rhoades.

Most cases of extrachromosomal inheritance are now attributed to the energy-transducing organelles found in the cytoplasm—mitochondria and chloro-

plasts. Examples include cytoplasmically transmitted mutations in *Chlamydomonas* (a green alga), *Neurospora* (a mold), *Saccharomyces* (a yeast), and even humans. As our knowledge of the molecular components of mitochondria and chloroplasts unfolded, primarily in the 1970s, the basis of this type of inheritance became clear. The relevant discovery was that both organelles contain their own genetic system, separate from that present in the nucleus, which includes a unique, circular DNA molecule. It contains genes that encode products related to the unique function of the organelle housing it. When mutations occur in those genes, they impact the function of mitochondria or chloroplasts, often causing far-reaching effects. Inheritance is through the parent that contributes the organelles to the zygote— this parent is the mother in heterogametic organisms.

In our example of male sterility in maize, the mitochondrion is the culprit. In both the case of Rhoades' male-sterile strain, and other male-sterile strains studied independently since his work, numerous genetic aberrations in mitochondrial function have been discovered. In some cases, small DNA molecules accessory to those

found normally in the organelle appear to be responsible. In other instances, mutations in the organelle DNA itself are sufficient to cause sterility. It is of interest to note that **mitochondrial DNA (mtDNA)**, discovered intitially through the use of electron microscopy, varies considerably in size in different organisms. While it can consist of as little as 10,000–20,000 bases (10–20 kb) in vertebrates, mtDNA in plants is much larger. For example, the garden pea *Pisum sativum* consists of 110 kb while the mustard plant *Arabidopsis* consists of 367 kb of DNA. Plant mitochondria may contain 10–20 copies of this DNA molecule.

Male sterility has actually played an important role during the development and agricultural use of hybrid strains of maize. By using male-sterile strains, self-fertilization is avoided, assuring pollination with the desired variety. However, in the 1970s, reliance on such strains led to a major problem. The strains then in use were more susceptible to a mutant form of the fungus causing southern leaf blight in maize, causing drastic losses in the corn crop in the United States. Fortunately, other strains were available to replace those susceptible to the blight.

Chapter Summary

1. In sexually reproducing organisms, meiosis, which both creates genetic variability and ensures genetic constancy, depends on fertilization. Fertilization ultimately relies on some form of sexual differentiation, which is achieved by a variety of sex-determining mechanisms.

2. The genetic basis of sexual differentiation is often related to different chromosome compositions in the two sexes. The heterogametic sex either lacks one chromosome or contains a unique heteromorphic chromosome, usually referred to as the Y chromosome.

3. In humans, the study of individuals with altered sex chromosome compositions has established that the Y chromosome is responsible for male differentiation. The absence of the Y chromosome leads to female differentiation. Similar studies in *Drosophila* have excluded the Y chromosome in such a role, instead demonstrating that a balance between the number of X chromosomes and sets of autosomes is the critical factor.

4. The primary sex ratio in humans substantially favors males at conception. During embryonic and fetal development, male

mortality is higher than that of females. The secondary sex ratio at birth still favors males by a small margin.

5. Dosage compensation mechanisms limit the expression of X-linked genes in females, who have two X chromosomes, as compared to males who have only one X. In mammals, compensation is achieved by the inactivation of either the maternal or paternal X early in development. This process results in the formation of Barr bodies in female somatic cells.

6. The Lyon hypothesis states that, early in development, inactivation is random between the maternal and paternal X. All subsequent progeny cells inactivate the same X as their progenitor cell. Mammalian females thus develop as genetic mosaics with respect to their expression of heterozygous X-linked alleles.

7. In many reptiles, the incubation temperature at a critical time during embryogenesis is often responsible for sex determination. Temperature influences the activity of enzymes involved in the metabolism of steroids related to sexual differentiation.

Insights and Solutions

1. In *Drosophila*, the X chromosomes can attach to one another ($\widehat{XX}$) such that they always segregate together. Some flies contain both an attached X chromosome and a Y chromosome.

(a) What sex would such a fly be? Explain why this is so.

Solution: The fly will be a female. The ratio of X chromosomes to sets of autosomes will be 1.0, leading to normal female development. The Y chromosome has no influence on sex determination in *Drosophila*.

(b) Given this answer, predict the sex of the offspring in a cross between this fly and a normal one of the opposite sex.

Solution: All flies will have two sets of autosomes, but each offspring will have one of the following sex chromosome compositions:

(1) $\widehat{XXX}$ → a metafemale with 3 X's (called a trisomic)

(2) $\widehat{XX}Y$ → a female like her mother

(3) X Y → a normal male

(4) Y Y → no development occurs

(c) If these offspring are allowed to interbreed, what will be the outcome?

Solution: A true-breeding stock will be created that maintains the attached X females generation after generation.

2. The X_g cell-surface antigen is coded for by a gene located on the X chromosome. No equivalent gene exists on the Y chromosome. Two codominant alleles of this gene have been identified: *Xg1* and *Xg2*. A woman of genotype *Xg2/Xg2* marries a man of genotype *Xg1/Y* and they produce a son with Klinefelter syndrome of genotype *Xg1/Xg2/Y*. Using proper genetic terminology, briefly explain how this individual was generated. In which parent and in which meiotic division did the mistake occur?

Solution: Because the son with Klinefelter syndromome is *Xg1 Xg2/Y*, he must have received both the *Xg1* allele and the Y chromosome from his father. Therefore, nondisjunction must have occurred during meiosis I in the father.

Key Terms

Problems and Discussion Questions

1. Define the terms heteromorphic and heterogamy as they relate to sex determination.
2. Contrast the *Protenor* and *Lygaeus* modes of sex determination.
3. Contrast the evidence explaining the different modes of sex determination in *Drosophila* and humans.
4. Devise a method of nondisjunction in human female gametes that would give rise to Klinefelter and Turner syndrome offspring following fertilization by a normal male gamete.
5. An insect species is discovered in which the heterogametic sex is unknown. An X-linked recessive mutation for *reduced wing* (*rw*) is discovered. Contrast the F_1 and F_2 generations from a cross between a female with reduced wings and a male with normal-sized wings when
   (a) The female is the heterogametic sex
   (b) The male is the heterogametic sex
   (c) Is it possible to distinguish between the *Protenor* and *Lygaeus* modes of sex determination based on the outcome of these crosses?
6. An attached X female fly, as described in the Insights and Solutions section of this chapter, expresses the recessive X-linked *white* eye phenotype. It is crossed with a male fly that expresses the X-linked recessive miniature wing phenotype. Determine the outcome of this cross regarding the sex, eye color, and wing size of the offspring.
7. It has been suggested that any male-determining genes contained on the Y chromosome in humans cannot be located in the limited region that synapses with the X chromosome during meiosis. What might be the outcome if such genes were located in this region?
8. What is a Barr body?
9. Indicate the expected number of Barr bodies in interphase cells of the following individuals: Klinefelter syndrome; Turner syndrome; and karyotypes 47,XYY, 47,XXX, and 48,XXXX.
10. Define the Lyon hypothesis.
11. Predict the potential effect of the Lyon hypothesis on the retina of a human female heterozygous for the X-linked red-green color-blindness trait.
12. Cat breeders are aware that kittens expressing the X-linked calico coat pattern are almost invariably females. Why?
13. What does the apparent need for dosage compensation mechanisms suggest about the expression of genetic information in normal diploid individuals?
14. The marine echiurid worm *Bonellia viridis* is an extreme example of the environment's influence on sex determination. Undifferentiated larvae either remain free-swimming and differentiate into females, or they settle on the proboscis of an adult female and become males. If larvae that have been on a female proboscis for a short period are removed and placed in seawater, they develop as intersexes. If larvae are forced to develop in an aquarium where pieces of proboscises have been placed, they develop into males. Contrast this mode of sexual differentiation with that of mammals. Suggest further experimentation to elucidate the mechanism of sex determination in *Bonellia*.
15. Discuss the possible reasons why the primary sex ratio in humans is as high as 1.40–1.60.
16. The X-linked dominant mutation in the mouse, *Testicular feminization* (*Tfm*), eliminates the normal response to the testicular hormone testosterone during sexual differentiation. An XY animal bearing the *Tfm* allele on the X chromosome develops testes, but no further male differentiation occurs. The external genitalia of such an animal are female. From this information, what might you conclude about the role of the *Tfm* gene product and the X and Y chromosomes in sex determination and sexual differentiation in mammals? Can you devise an experiment, assuming you can "genetically engineer" the chromosomes of mice, to test and confirm your explanation?
17. In the wasp, *Bracon hebetor*, a form of parthenogenesis (where unfertilized eggs initiate development) resulting in haploid organisms is not uncommon. All haploids are males. When offspring arise from fertilization, females almost invariably result. P. W. Whiting has shown that an X-linked gene with 9 multiple alleles (X_a, X_b, etc.) controls sex determination. Any homozygous or hemizygous condition results in males, and any heterozygous condition results in females. If an X_a/X_b female mates with an X_a male and lays 50 percent fertilized and 50 percent unfertilized eggs, what proportion of male and female offspring will result?

Selected Readings

Amory, J. K., et al. 2000. Klinefelter's syndrome. *Lancet* 356: 333–35.

Avner, P., and Heard, E. 2001. X-chromosome inactivation: counting, choice and initiation. *Nature Reviews Genetics* 2:59–67.

Barr, M. L. 1966. The significance of sex chromatin. *Int. Rev. Cytol.* 19:35–39.

Burgoyne, P. S. 1998. The mammalian Y chromosome: A new perspective. *Bioessays* 20:363–66.

Carrel, L., and Willard, H. F. 1998. Counting on *Xist. Nature Genetics* 19:211–12.

Court-Brown, W. M. 1968. Males with an XYY sex chromosome complement. *J. Med. Genet.* 5:341–59.

Davidson, R., Nitowski, H., and Childs, B. 1963. Demonstration of two populations of cells in human females heterozygous for glucose-6-phosphate dehydrogenase variants. *Proc. Natl. Acad. Sci. USA* 50:481–85.

Dellaporta, S. L., and Calderon-Urrea, A. 1994. The sex determination process in maize. *Science* 266:1501–05.

Erickson, J. D. 1976. The secondary sex ratio of the United States, 1969–71: Association with race, parental ages, birth order, paternal education and legitimacy. *Ann. Hum. Genet.* (London) 40:205–12.

Freije, D., et al. 1992. Identification of a second pseudoautosomal region near the Xq and Yq telomeres. *Science* 258:1784–87.

Gorman, M., Kuroda, M., and Baker, B. S. 1993. Regulation of sex-specific binding of maleness dosage compensation protein to the male X chromosome in *Drosophila*. Cell 72:39–49.

Haseltine, F. P., and Ohno, S. 1981. Mechanisms of gonadal differentiation. *Science* 211:1272–78.

Hodgkin, J. 1990. Sex determination compared in *Drosophila* and *Caenorhabditis*. *Nature* 344:721–28.

Hook, E. B. 1973. Behavioral implications of the humans XYY genotype. *Science* 179:139–50.

Irish, E. E. 1996. Regulation of sex determination in maize. *BioEssays* 18:363–69.

Jacobs, P. A., et al. 1974. A cytogenetic survey of 11,680 newborn infants. *Ann. Hum. Genet.* 37:359–76.

Koopman, P., et al. 1991. Male development of chromosomally female mice transgenic for *Sry*. *Nature* 351:117–21.

Lahn, B. T., and Page, D. C. 1997. Functional coherence of the human Y chromosome. *Science* 278:675–80.

Lucchesi, J. 1983. The relationship between gene dosage, gene expression, and sex in *Drosophila*. *Dev. Genet.* 3:275–82.

Lyon, M. F. 1961. Gene action in the X chromosome of the mouse (*Mus musculus L.*). *Nature* 190:372–73.

——————. 1972. X-chromosome inactivation and developmental patterns in mammals. *Biol. Rev.* 47:1–35.

——————. 1988. X-chromosome inactivation and the location and expression of X-linked genes. Am. J. Hum. Genet. 42:8–16.

——————. 1998. X-Chromosome inactivation spreads itself: effects in autosomes. *Am. J. Hum. Genet.* 63:17–19.

Marin, I., and Baker, S. S. 1998. The evolutionary dynamics of sex determination. *Science* 281:1990–94.

Marshall Graves, J. A. 1998. Interaction between *SRY* and *SOX* genes in mammalian sex determination. *BioEssays* 20:264–69.

McMillen, M. M. 1979. Differential mortality by sex in fetal and neonatal deaths. *Science* 204:89–91.

Page, D. C., et al. 1987. The sex-determining region of the human Y chromosome encodes a finger protein. *Cell* 51:1091–1104.

Penny, G. D., et al. 1996. Requirement for *Xist* in X chromosome inactivation. *Nature* 379:131–37.

Pieau, C. 1996. Temperature variation and sex determination in reptiles. *BioEssays* 18:19–26.

Reddy, K. S., and Sulcova, V. 1998. Pathogenetics of 45,X/46,XY gonadal mosaicism. *Cytogenet. Cell Genet.* 82:52–57.

Schafer, A. J. 1996. Sex determination in humans. *BioEssays* 18:955–63.

Vainio. S., et al. 1999. Female development in mammals is regulated by Wnt-4 signalling. *Nature* 397:405–09.

Westergaard, M. 1958. The mechanism of sex determination in dioecious flowering plants. *Adv. Genet.* 9:217–81.

Whiting, P. W. 1939. Multiple alleles in sex determination in *Habrobracon*. *J. Morphology* 66:323–55.

Witkin, H. A., et al. 1976. Criminality in XYY and XXY men. *Science* 193:547–55.

Phenotypic effect of the *fw2.2* transgene in the tomato. These fruits either result from the presence (+) or the absence (−) of the quantitative trait loci (QTL) allele causing small fruit.

6

Quantitative Genetics

CHAPTER CONCEPTS

Traits that exhibit quantitative phenotypic variation are often under the genetic control of alleles whose influence is additive in nature, resulting in continuous variation. Such traits can be analyzed and characterized using statistical methods, which also allow the assessment of the relative importance of genetic factors during phenotypic expression. Such calculations establish the heritability of traits in a population. In humans, twin studies provide a similar, but less precise, estimate. Recently developed techniques allow the localization within the genome of loci contributing to quantitative traits.

In Chapter 4, numerous examples of gene interaction were discussed. In each case, the resultant phenotypic variation was classified into distinct traits. Pea plants are tall or dwarf; squash shape is spherical, disc-shaped, or elongated; and fruit fly eye color is red or white. These phenotypes exemplify discontinuous variation, in which discrete phenotypic categories exist. Many other traits in a population demonstrate considerably more variation and are not as easily categorized into distinct classes. Such phenotypes are thus said to demonstrate continuous variation.

Traits exhibiting continuous variation are most often controlled by two or more genes that provide an additive component to the phenotype that can be quantified. In this chapter we examine patterns of inheritance and outline some statistical techniques used to study such traits. These patterns illustrate **quantitative**, or **polygenic**, **inheritance**. In addition, we consider how geneticists assess the relative importance of genetic versus environmental factors as they contribute to phenotypic variation, and we discuss an approach used to localize these "quantitative" genes within the genome.

6.1 Quantitative Inheritance

Throughout the 18th and 19th centuries, scientists studied traits that exhibited a continuous gradation of phenotypes. For example, Sir Francis Galton investigated the diameter of sweet peas. When plants with large peas were crossed to those with small peas, the F_1 plants contained peas that were all of an intermediate diameter. When the F_2 generation was examined, peas were of many sizes, as large or as small as the original parents, and many sizes in between!

Galton's study showed a pattern of inheritance encountered by other investigators, including at least one cross made by Mendel. The F_1 generations were intermediate blends of the parental phenotypes, and the F_2 generation exhibited more or less continuous variation of phenotypic expression. In each case, the traits under investigation behaved in a quantitative fashion expressed as size, height, weight, color, and so on.

Not surprisingly, these traits were difficult to study, and their mode of inheritance was not clarified until early in the 20th century. Because these results were exceptions to the patterns observed by Mendel in most of his crosses, they failed to support his hypotheses and no doubt delayed the acceptance of his work. Nevertheless, the genetic explanation of continuous variation serves as the foundation for our current understanding of the field of genetics called quantitative, or polygenic inheritance.

The Multiple-Factor Hypothesis

One of the first cases of continuous phenotypic variation was encountered by Josef Gottlieb Kolreuter when he crossed tall and dwarf tobacco plants. The plants of the F_1 generation were all intermediate in height. When the F_2 generation was examined, individuals showed continuous variation in height, ranging from tall to dwarf, like the original parents, including many heights in between. A critical observation involved the distribution of phenotypes in the second generation: The majority of the F_2 plants were intermediate like the F_1 plants, but only a few were as tall or dwarf as the P_1 parents. These distributions are shown in Figure 6–1. Note that the F_2 data

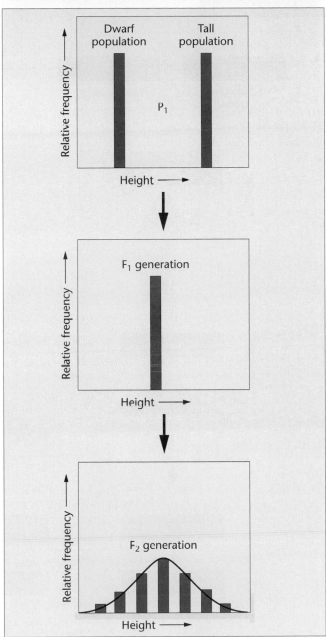

FIGURE 6–1 Histograms showing the relative frequency of individuals expressing various height phenotypes derived from Kolreuter's cross between dwarf and tall tobacco plants carried to the F_2 generation. The photograph shows a tobacco plant. *(Photo: Bildarchiv Okapia/Photo Researchers, Inc.)*

Corolla

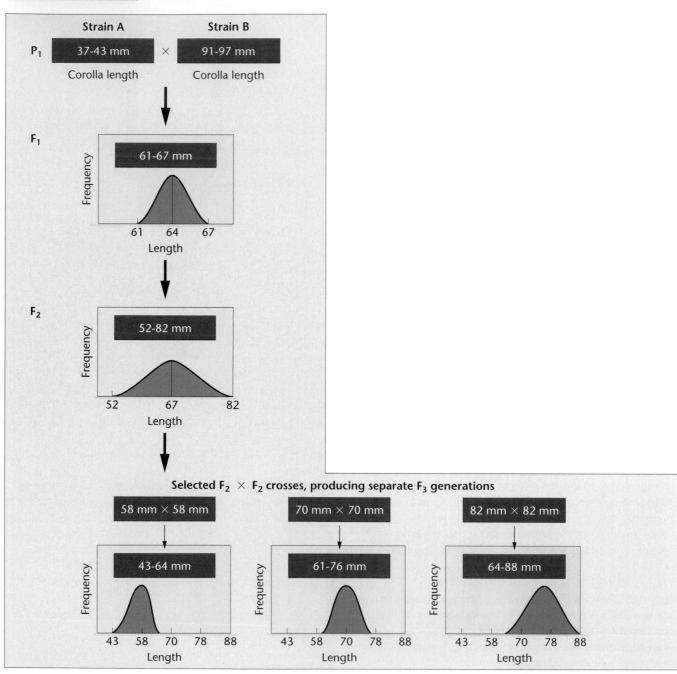

FIGURE 6–2 The F_1, F_2, and selected F_3 results of East's cross between two strains of *Nicotiana* with different corolla lengths. Plants of strain A vary from 37 to 43 mm, while plants of strain B vary from 91 to 97 mm. The photograph shows the flower and corolla of a tobacco plant. *(Photo: Norm Thomas/Photo Researchers, Inc.)*

demonstrate a normal distribution, as evidenced by the bell-shaped curve in the bottom histogram.

At the beginning of the 20th century, geneticists noted that many characters in different species had similar patterns of inheritance, such as height and stature in humans, seed size in the broad bean, grain color in wheat, and kernel number and ear length in corn. In each case, offspring in the succeeding generation seemed to be a blend of their parents' characteristics.

The issue of whether continuous variation could be accounted for in Mendelian terms caused considerable controversy in the early 1900s. William Bateson and Gudny Yule, who adhered to the Mendelian explanation of inheritance, suggested that a large number of factors or genes could account for the observed patterns. This proposal, called the **multiple-factor hypothesis**, implied that many factors or genes contribute to the phenotype in a *cumulative* or *quantitative* way. However, other geneticists argued that Mendel's unit factors could not account for the blending of parental phenotypes characteristic of these patterns of inheritance and were thus skeptical of these ideas.

By 1920, the conclusions of several critical sets of experiments largely resolved the controversy and demonstrated that Mendelian factors could account for continuous variation. In one experiment, Edward M. East crossed two strains of the tobacco plant *Nicotiana longiflora*. The fused inner petals of the flower, or corollas, of strain A were decidedly shorter than the corollas of strain B. With only minor variation, each strain was true-breeding. Thus, the differences between them were clearly under genetic control.

When plants from the two strains were crossed, the F_1, F_2, and selected F_3 data demonstrated a distinct pattern (Figure 6–2). The F_1 generation displayed corollas that were intermediate in length compared with the P_1 varieties, and showed only minor variability among individuals. While corolla lengths of the P_1 plants were about 40 mm and 94 mm, the F_1 generation contained plants with corollas that were all about 64 mm. In the F_2 generation, lengths varied much more, ranging from 52 mm to 82 mm. The majority of individuals resembled their F_1 parents, and as the deviation from this average increased, fewer and fewer plants were observed. When the data are plotted graphically (frequency vs. length), a bell-shaped curve results.

East further experimented with this population by selecting F_2 plants of various corolla lengths and allowing them to produce separate F_3 generations. Several are shown in Figure 6–2. In each case, a bell-shaped distribution was observed, with most individuals similar in height to the selected F_2 parents, but with considerable variation around this value.

East's experiments demonstrated that although the variation in corolla length seemed continuous, experimental crosses resulted in the segregation of distinct phenotypic classes as observed in the three independent F_3 categories. This key finding was the basis for the multiple-factor hypothesis, which explains how traits can deviate considerably in their expression.

Additive Alleles: The Basis of Continuous Variation

The multiple-factor hypothesis, suggested by the observations of East and others, embodies the following major points:

1. Characters that exhibit continuous variation can usually be quantified by measuring, weighing, counting, and so on.

2. Two or more pairs of genes, located throughout the genome, account for the hereditary influence on the phenotype in an *additive way*. Because many genes can be involved, inheritance of this type is often called *polygenic*.

3. Each gene locus may be occupied by either an **additive allele**, which contributes a set amount to the phenotype, or by a **nonadditive allele**, which does not contribute quantitatively to the phenotype.

4. The total effect on the phenotype of each additive allele, while small, is approximately equivalent to all other additive alleles at other gene sites.

5. Together, the genes controlling a single character produce substantial phenotypic variation.

6. Analysis of polygenic traits requires the study of large numbers of progeny from a population of organisms.

These points center around the concept that additive alleles at numerous loci control quantitative traits. To illustrate this, let's examine Herman Nilsson-Ehle's experiments involving grain color in wheat performed early in the 20th century. In one set of experiments, wheat with red grain was crossed to wheat with white grain (Figure 6–3). The F_1 generation demonstrated an intermediate color. In the F_2 generation, approximately 15/16 of the plants showed some degree of red grain, while 1/16 of the plants showed white grain. Because the ratio occurred in 16ths, he hypothesized that two gene pairs control the phenotype and, if so, they segregate independently from one another in a Mendelian fashion.

Careful examination of the F_2 generation revealed that grain color can be classified into four different shades of red. If two gene pairs are operative, each with one potential additive allele and one potential nonadditive allele, we can envision how the multiple-factor hypothesis accounts for this variation. In the P_1, both parents are homozygous; the red parent contains only additive alleles (uppercase letters), while the white parent contains only nonadditive alleles (lowercase letters). The F_1 generation, being heterozygous, contains only two additive alleles and expresses an intermediate phenotype. In the F_2, each offspring has either 4, 3, 2, 1, or 0 additive alleles (Figure 6–3). Wheat with no additive alleles (1/16) is white like one of the P_1 parents, while wheat with 4 additive alleles is red like the other P_1 parent. Plants with 3, 2, or 1 additive alleles constitute the other three categories of red color observed in the F_2 plants, with most (6/16) having 2 additive alleles like the F_1 plants.

FIGURE 6–3 How the multiple-factor hypothesis accounts for the 1:4:6:4:1 phenotypic ratio of grain color when all alleles designated by uppercase letters are additive and contribute an equal amount of pigment to the phenotype.

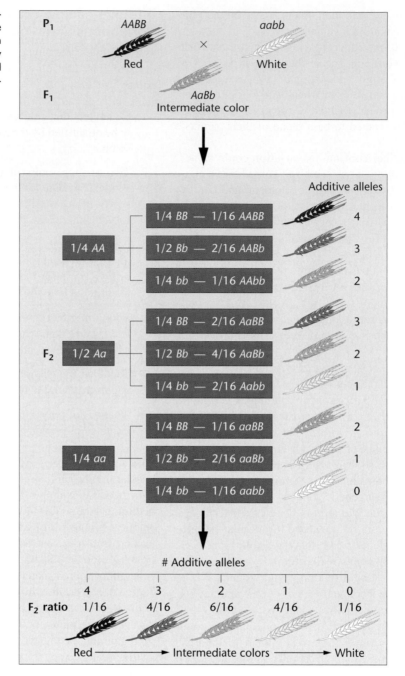

Therefore, continuous variation can be explained in a Mendelian fashion. Multiple-factor inheritance, where additive alleles influence the phenotype in a quantitative manner, results in such variation. As we saw in Nilsson-Ehle's initial cross, if two gene pairs are involved, only five F_2 phenotypic categories, in a 1:4:6:4:1 ratio, are expected. There is no reason why three, four, or more gene pairs cannot function in controlling various phenotypes. As greater numbers of gene pairs become involved, the number of classes increases and results in more complex ratios. The number of phenotypes and the expected F_2 ratios of crosses involving up to five gene pairs are shown in Figure 6–4.

Calculating the Number of Genes

When additive effects control polygenic traits, it is of interest to determine the number of genes involved. If the ratio (proportion) of F_2 individuals resembling *either* of the two most extreme phenotypes (the parental phenotypes) can be determined, then the number of gene pairs involved (*n*) can be calculated using the following simple formula:

$$\frac{1}{4^n} = \text{ratio of } F_2 \text{ individuals expressing either extreme phenotype}$$

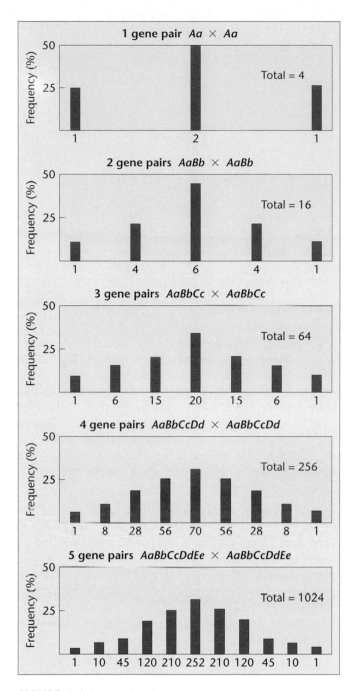

FIGURE 6–4 The results of crossing two heterozygotes when polygenic inheritance is operative with 1–5 gene pairs. Each histogram bar indicates a distinct phenotypic class from one extreme (left end) to the other extreme (right end). Each phenotype results from a different number of additive alleles.

TABLE 6.1 Determination of the Number of Gene Pairs (*n*) Involved in Polygenic Crosses

n	Individuals Expressing an Extreme Phenotype	Distinct F_2 Phenotypic Classes
1	1/4	3
2	1/16	5
3	1/64	7
4	1/256	9
5	1/1024	11

Table 6.1 lists the ratios and number of F_2 phenotypic classes produced in crosses involving up to five gene pairs.

For low numbers of gene pairs, it is sometimes easier to use the $(2n + 1)$ rule. If *n* equals the number of gene pairs, $2n + 1$ determines the total number of categories of possible phenotypes. When $n = 2$, $2n + 1 = 5$. That is, each phenotypic category can have 4, 3, 2, 1, or 0 additive alleles. When $n = 3$, $2n + 1 = 7$, and each phenotypic category can have 6, 5, 4, 3, 2, 1, or 0 additive alleles, and so on.

The Significance of Polygenic Inheritance

Polygenic inheritance is a significant concept because it appears to serve as the genetic basis for a vast number of traits involved in animal breeding and agriculture. For example, height, weight, and physical stature in animals, size and grain yield in crops, beef and milk production in cattle, and egg production in chickens are all thought to be under polygenic control. In most cases, it is important to note that the genotype, which is fixed at fertilization, establishes the potential range within which a particular phenotype falls. However, environmental factors determine how much of the potential will be realized. In the crosses described thus far, we have assumed an optimal environment, which minimizes variation from external sources.

Still other examples can be drawn from human genetic studies. Skin pigmentation, intelligence, various forms of behavior, obesity, and even the predisposition to certain diseases are all thought to be under the control of numerous genes. The latter two conditions, obesity and predisposition to disease (e.g., coronary heart disease) are good examples of **complex traits**. Unlike the cases of where continuous variation is observed, no clear Mendelian pattern of inheritance is observable. However, both conditions run in families such that sons and daughters of affected parents are much more likely to also be affected than are children of unaffected parents. In both cases, many genes are now known to be involved, but the environment substantially influences the ultimate expression of these traits. In such examples, when the combination of genetic and environmental factors are both substantial influences, **multifactorial** is sometimes used to describe the control of phenotypic expression.

In our previous example, the P_1 phenotypes represent these two extremes. In Figure 6–3, 1/16 of the F_2 are either red *or* white like the P_1 classes; this ratio can be substituted on the right side of the equation to solve for *n*:

$$\frac{1}{4^n} = \frac{1}{16}$$

$$\frac{1}{4^2} = \frac{1}{16}$$

$$n = 2$$

6.2   Analysis of Polygenic Traits

Analysis of a polygenic trait most often involves quantitative measurements, usually from many offspring generated from many crosses. The outcome can be expressed as a frequency diagram that often demonstrates a normal (bell-shaped) distribution (Figure 6–5) . While it is hoped that each series of crosses is representative of the population at large, variation in samples due strictly to chance can influence the data gathered. To assess the validity of the experimental data, statistical techniques are used. Such techniques were first devised by Galton early in this century to assess the inheritance of traits exhibiting continuous variation. Galton's efforts served as the initial basis of the field of study called **biometry**.

Statistical analysis serves three purposes:

1. Data can be mathematically reduced to provide a descriptive summary of the sample.

2. Data from a small but random sample can be used to infer information about groups larger than those from which the original data were obtained (statistical inference).

3. Two or more sets of experimental data can be compared to determine whether they represent significantly different populations of measurements.

Several statistical methods are useful in the analysis of traits that exhibit a normal distribution, including the mean, variance, standard deviation, and standard error of the mean.

The Mean

The distributions of the two sets of phenotypic measurements graphed in Figure 6–6 cluster around a central value. This clustering is called a central tendency, one measurement of which is the **mean** ($\overline{X}$). The mean is simply the arithmetic average of a set of measurements or data and is calculated as

$$\overline{X} = \frac{\Sigma X_i}{n}$$

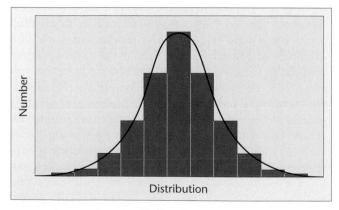

FIGURE 6–5 A normal frequency distribution characterized by a bell-shaped curve.

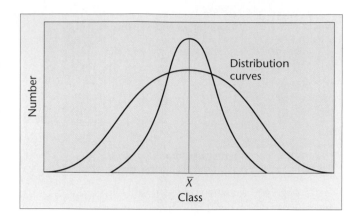

FIGURE 6–6 Two normal frequency distributions with the same mean but different amounts of variation.

where $\overline{X}$ is the mean, ΣX_i represents the sum of all individual values in the sample, and n is the number of individual values.

Although the mean provides a descriptive summary of the sample, it is of itself of limited value. As shown in Figure 6–6, a symmetrical distribution of values in the sample may, in one case, cluster near the mean. Another set of values may have the same mean, but be distributed widely around it. These contrasting conditions represent different types of variation within each sample called the **frequency distribution**. Whether due to chance or to one or more experimental variables, such variation creates the need for methods to describe sample measurements statistically.

Variance

As seen in Figure 6–6, the range and distribution of values on either side of the mean determines the shape of the distribution curve. The degree to which values within this distribution diverge from the mean is called the sample **variance** (s^2) and is used to estimate the variation present in an infinitely large population. The variance for a sample is calculated as

$$s^2 = \frac{\Sigma (X_i - \overline{X})^2}{n - 1}$$

where the sum (Σ) of the squared differences between each measured value (X_i) and the mean ($\overline{X}$) is divided by one less than the total sample size ($n - 1$). To avoid the numerous subtraction functions necessary to calculate s^2 for a large sample, we convert the equation to its algebraic equivalent:

$$s^2 = \frac{\Sigma X_i^2 - n\overline{X}^2}{n - 1}$$

The variance is a valuable measure of sample variability. As noted earlier, two distributions can have identical means ($\overline{X}$), yet vary considerably in their frequency distribution around the mean. The variance represents the average squared deviation of the measurements from the mean. Estimation of variance is particularly valuable in determin-

TABLE 6.2   Sample Inclusion for Various s Values

Multiples of s	Sample Included (%)
$\overline{X} \pm 1s$	68.3%
$\overline{X} \pm 1.96s$	95.0
$\overline{X} \pm 2s$	95.5
$\overline{X} \pm 3s$	99.7

ing the degree of genetic control of traits when the immediate environment also influences the phenotype.

Standard Deviation

Because the variance is a squared value, its unit of measurement is also squared (cm^2, mg^2, etc.). To express variation around the mean in the original units of measurement, we find the square root of the variance, a term called the **standard deviation** (s):

$$s = \sqrt{s^2}$$

Table 6.2 shows what percentage of the individual values within a normal distribution is included with different multiples of the standard deviation. The mean plus or minus one standard deviation ($\overline{X} \pm 1s$) includes 68 percent of all values in the sample. Over 95 percent of all values are found within two standard deviations ($\overline{X} \pm 2s$). As such, the standard deviation is an important descriptive tool. Furthermore, s can be interpreted as a probability. The $\overline{X} \pm 1s$ indicates a 68 percent probability that a measured value picked at random will fall within that range.

Standard Error of the Mean

To estimate how much the means of other similar samples drawn from the same population might vary, we calculate the **standard error of the mean** ($S_{\overline{X}}$):

$$S_{\overline{X}} = \frac{s}{\sqrt{n}}$$

where s is the standard deviation, and $\sqrt{n}$ is the square root of the sample size. The standard error of the mean measures the accuracy of the sample mean—that is, the variation of sample means in replications of the experiment. Because the standard error of the mean is computed by dividing s by $\sqrt{n}$, it is always a smaller value than the standard deviation.

Analysis of a Quantitative Character

To illustrate how biometric methods are used in quantitative analysis, we consider a simple example involving fruit weight in tomatoes. Let's assume that fruit weight is a quantitative character, and that one highly inbred strain (one that is highly homozygous) produces tomatoes averaging 18 oz. in weight, and another highly inbred strain produces fruit averaging 6 oz. in weight. These two varieties are crossed and produce an F_1 generation with weights ranging from 10 oz. to 14 oz. The F_2 population contains individuals that produce fruit ranging from 6 oz. to 18 oz. The results characterizing both generations are shown in Table 6.3.

The mean value for fruit weight in the F_1 generation is calculated as

$$\overline{X} = \frac{\Sigma X_i}{n} = \frac{626}{52} = 12.04$$

Similarly, the mean value for fruit weight in the F_2 generation is calculated as

$$\overline{X} = \frac{\Sigma X_i}{n} = \frac{872}{72} = 12.11$$

Average fruit weight is 12.04 oz. in the F_1 generation and 12.11 oz. in the F_2 generation. Although these mean values are similar, it is apparent from the frequency distributions (Table 6.3) that more variation is present in the F_2 generation. Fruit weight ranges from 6 oz. to 18 oz. in the F_2 generation, but only from 10 oz. to 14 oz. in the F_1 generation.

To quantify the amount of variation present in each generation, we calculate the variance (Table 6.4). As noted earlier, the sample variance is the sum of the squared differences between each value and the mean, divided by one less than the total number of observations. However, in the case where a number of observations (f) are grouped into representative classes (x), the variance is calculated according to the formula

$$s^2 = \frac{n \Sigma f(x^2) - (\Sigma f x)^2}{n(n - 1)}$$

As shown in Table 6.4, the variance is 1.29 for the F_1 generation, and 4.27 for the F_2 generation. When converted to the standard deviation ($s = \sqrt{s^2}$) the values become 1.13 and 2.06, respectively. Therefore, the distribution of tomato weight in the F_1 generation can be described as 12.04 $\pm$ 1.13 oz., and that in the F_2 generation can be described

TABLE 6.3   Distribution of F₁ and F₂ Progeny

		Weight (oz.)												
		6	7	8	9	10	11	12	13	14	15	16	17	18
Number of	F_1:					4	14	16	12	6				
Individuals	F_2:	1	1	2	0	9	13	17	14	7	4	3	0	1

TABLE 6.4 Calculation of Variance

	F_1				F_2		
x	f	$f(x)$	$f(x)^2$	x	f	$f(x)$	$f(x)^2$
6				6	1	6	36
7				7	1	7	49
8				8	2	16	128
9				9	0	0	0
10	4	40	400	10	9	90	900
11	14	154	1,694	11	13	143	1,573
12	16	192	2,304	12	17	204	2,448
13	12	156	2,028	13	14	182	2,366
14	6	84	1,176	14	7	98	1,372
15				15	4	60	900
16				16	3	48	768
17				17	0	0	0
18				18	1	18	324
	$n = \overline{52}$	$\Sigma fx = \overline{626}$	$\Sigma fx^2 = \overline{7,602}$		$n = \overline{72}$	$\Sigma fx = \overline{872}$	$\Sigma fx^2 = \overline{10,864}$

For F_1:

$$s^2 = \frac{52 \times 7602 - (626)^2}{52(52-1)}$$

$$= \frac{395,304 - 391,876}{2652}$$

$$= 1.29$$

For F_2:

$$s^2 = \frac{72 \times 10,864 - (872)^2}{72(72-1)}$$

$$= \frac{782,208 - 760,384}{5112}$$

$$= 4.27$$

as 12.11 ± 2.06 oz. This analysis indicates that the mean fruit weight of F_1 is nearly identical to that of F_2, but the F_2 shows greater variability in the distribution of weights than does F_1.

Observations about the inheritance of fruit weight in crosses between these two strains of tomatoes meet the expectations for polygenic traits. For the sake of this example, if we assume that each parental strain is homozygous for the additive or nonadditive alleles that control fruit weight, we can estimate the number of gene pairs involved in controlling fruit weight in these two strains of tomatoes. Since 1/72 of the F_2 offspring have a phenotype that overlaps one of the parental strains (72 total F_2 offspring, one weighs 6 oz., one weighs 18 oz.; see Table 6.3), using the formula $1/4^n = 1/72$ shows that n is between 3 and 4, indicative of the number of genes that control fruit weight in these tomato strains. If this experiment were repeated many times, with similar results, our confidence in this conclusion would be bolstered.

6.3 Heritability

Having just introduced several ways in which quantitative or continuous variation is measured and characterized in populations, we now consider how to assess the extent to which genetic factors contribute to such phenotypic variation. Often, much of the variation can be attributed to genetic factors, with the total environment having less influence. In other cases, the environment may have a greater influence on phe-

notypic variation within a population. The following discussion considers how geneticists attempt to define the impact of heredity versus environment on phenotypic variation.

Broad-Sense Heritability

If a trait can be measured quantitatively, experiments on many plants and animals can test the origin of variation. One approach uses inbred strains containing individuals of a relatively homogeneous (highly homozygous) genetic background. Experiments are then designed to test the effects of the range of prevailing environmental conditions on phenotypic variability. Variation observed *between* different inbred strains reared in a constant environment is due predominantly to genetic factors. Variation observed *among* members of the same inbred strain reared under different environmental conditions is due to nongenetic factors, which are generally categorized as "environmental."

The relative importance of genetic versus environmental factors can be formally assessed by examining the **heritability index** (H^2), which we calculate using an analysis of variance among individuals of a known genetic relationship. (In the ensuing discussion, we use the term V to designate variance, previously indicated as s^2.) This is an important approach for investigating organisms with long generation times. Also called **broad-sense heritability**, H^2 measures the degree to which **phenotypic variance** (V_P) is due to variation in genetic factors for a single population under the limits of environmental variation during the study. Note that the H^2 value *does not* determine the proportion of the total phenotype attributed to genetic factors. It *does* estimate the proportion of observed variation in the phenotype attributed to genetic factors, in comparison to environmental factors.

Phenotypic variance is due to the sum of three components: environmental variance (V_E), genetic variance (V_G), and variance resulting from the interaction of genetics and environment (V_{GE}). Therefore, phenotypic variance (V_P) is expressed as

$$V_P = V_E + V_G + V_{GE}$$

Because V_{GE} is often negligible, it's usually omitted. Therefore, the simpler equation is generally used:

$$V_P = V_E + V_G$$

Broad heritability expresses that proportion of variance due to the genetic component:

$$H^2 = \frac{V_G}{V_P}$$

An H^2 value approaching 1.0 indicates that environmental conditions have little impact on phenotypic variance in the population studied. An H^2 value close to 0 indicates that the environment is almost solely responsible for the observed phenotypic variation.

It is not possible to obtain an absolute H^2 value for any given character. If measured in a different population under a greater or lesser degree of environmental variability, H^2 might well be different for that character. For that reason, broad-sense heritability estimates are most useful in highly inbred strains or genetic clones such as artificially selected plant populations that are asexually propagated by cuttings. Furthermore, broad-sense heritability estimates are not very accurate in estimating the selection potential of quantitative traits since H^2 calculations take into account all forms of genetic variation, not simply additive genetic effects. Therefore, another type of calculation, narrow-sense heritability, has been devised that is useful for additive effects.

Narrow-Sense Heritability

Information regarding heritability is most useful in animal and plant breeding as a measure of potential response to selection. In this case, a different estimate of heritability must be used, based on a subcomponent of V_G referred to as **additive variance** (V_A):

$$V_G = V_A + V_D + V_I$$

Here V_A represents the additive variance that results from the average effect of additive components of genes. **Dominance variance**, V_D, is the deviation from additive components that results when phenotypic expression in heterozygotes is not precisely intermediate between the two homozygotes.

Interactive variance, V_I, is the deviation from additive components that occurs when two or more loci behave epistatically. Interactive variance is not associated with the average effect of V_A and is often negligible. Thus, it is often excluded from calculations.

When V_G is partitioned into V_A and V_D, a new assessment of heritability, h^2, or **narrow-sense heritability**, can be calculated. Thus h^2 values are useful in assessing selection potential in randomly breeding animal and plant populations:

$$h^2 = \frac{V_A}{V_P}$$

Because $V_P = V_E + V_G$ and $V_G = V_A + V_D$, we obtain:

$$h^2 = \frac{V_A}{V_E + V_A + V_D}$$

As we shall subsequently see, h^2 values are useful in predicting the phenotypes of offspring during selection procedures. The closer a value is to 1.0, the greater is our ability to make an accurate prediction, based on our knowledge of the parental phenotypes.

Artificial Selection

The process of selecting a specific group of organisms from an initially heterogeneous population for future breeding purposes is called **artificial selection**. A relatively high h^2 value predicts that selection will likely succeed in altering a population. As you can imagine based on our preceding discussion, measuring the components necessary to calculate h^2 is a complex task. A simplified approach involves measuring of the central tendencies (the means) of a trait from (1) a parental population exhibiting a bell-shaped distribution (M), (2) a "selected" segment of the parental population that expresses the most desirable quantitative phenotypes (M_1), and (3) the offspring (M_2) resulting from interbreeding the selected M_1 group. When this is accomplished, the following relationship of the three means and h^2 exists:

$$M_2 = M + h^2(M_1 - M)$$

Solving this equation for h^2 gives us

$$h^2 = \frac{M_2 - M}{M_1 - M}$$

Once these relationships are established, the equation can be further simplified by defining $M_2 - M$ as the **response** (R) and $M_1 - M$ as the **selection differential** (S), where h^2 reflects the ratio of the response observed to the total response possible. Thus,

$$h^2 = \frac{R}{S}$$

Now let's assess one case of selection. Suppose we measure the diameter of corn kernels in a population where the mean

FIGURE 6–7 Response of corn selected for high and low oil content over 76 generations. The numbers in parentheses at generations 9, 25, 52, and 76 for the "high oil" line indicate the calculation of heritability at these points in the continuing experiment.

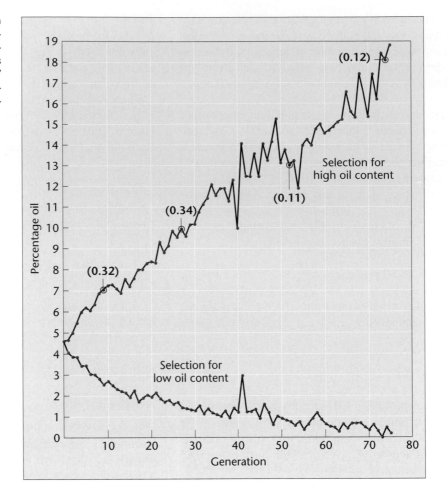

diameter (M) is much larger than desirable (20 mm), and from that population we select a group with the smallest diameters, for which the mean (M_1) is 10 mm. Plants that yielded this selected population are then interbred, and the progeny kernels yield a mean (M_2) of 13 mm; we then calculate h^2 to estimate the potential for artificial selection on kernel size:

$$h^2 = \frac{13 - 20}{10 - 20}$$
$$= \frac{-7}{-10}$$
$$= 0.70$$

On the basis of this calculation, we can conclude that the selection potential for kernel size is relatively high.

The longest-running artificial selection experiment known is still being conducted at the State Agricultural Laboratory in Illinois. Since 1896, corn has been selected for both high and low oil content. After 76 generations, selection continues to increase oil content (Figure 6–7). As selection has progressed, heritability of increased oil content has declined (see parenthetical values at generations 9, 25, 52, and 76 in Figure 6–7). The process will continue until all individuals in the population contain a uniform genotype for the additive alleles responsible for oil content. At that point, heritability will be reduced to zero and the response to artificial selection

will cease. Examination of the selection for low oil content shows that heritability is approaching this point.

Similar calculations involving artificial selection for other traits in a variety of organisms have established whether selection is an effective approach for obtaining populations exhibiting desirable phenotypes. Table 6.5 lists estimates of narrow heritability for a variety of traits in various organisms. These proportions are expressed as percentages. As you can see, heritability varies considerably among traits.

TABLE 6.5   Estimates of Heritability for Traits in Different Organisms

Trait	Heritability (h^2)
Mice	
Tail length	60%
Body weight	37
Litter size	15
Chickens	
Body weight	50
Egg production	20
Egg hatchability	15
Cattle	
Birth weight	51
Milk yield	44
Conception rate	3

In general, heritability is low for traits that are essential to an organism's survival, primarily because the genetic component has to be largely optimized during evolution. Egg production, litter size, and conception rate are examples where such physiological limitations on selection have already been established. Nevertheless, traits that are less critical to survival, such as body weight, tail length, and wing length, show higher heritability values. Narrow-sense heritability estimates are more valuable when they are based on data collected in many populations and environments and where a clear trend is established. Based on such estimates, selection techniques have led to vast improvements in the quality of animal and plant products.

Twin Studies in Humans

Traditional heritability studies are not possible in humans, for obvious reasons. However, human twins are very useful subjects for studying the heredity-versus-environment question. **Monozygotic** or **identical twins**, derived from the division and splitting of a single egg following fertilization, are identical in their genetic compositions. Although most identical twins are reared together and are exposed to very similar environments, some pairs are separated and raised in different settings. Thus, for particular traits, average similarities or differences in separated twins can be investigated. Such analyses are particularly useful because characteristics that remain similar in different environments are believed to have a strong genetic component. These data can then be compared with similar analyses of **dizygotic** or **fraternal twins**, who originate from two separate fertilization events. Dizygotic twins are no more genetically similar than other nontwin siblings, sharing (on average) one-half of their genes.

Another approach involves measuring **concordance** values of phenotypic expression in twin pairs raised together. These twins are said to be concordant for a given trait if both express it or neither expresses it. If one expresses the trait and the other does not, the pair is said to be **discordant**. Comparison of concordance values of MZ versus DZ twins reared together (Table 6.6) shows the potential value for heritability assessment.

These data must be examined very carefully before any conclusions are drawn. If the concordance value approaches 90–100 percent with monozygotic twins, we might be inclined to interpret this value as indicating a large genetic contribution to the expression of the trait. In some cases—for example, blood types and eye color—we know that this is indeed true. In the case of measles, however, a high concordance value merely indicates that the trait is almost always induced by a factor in the environment—in this case, a virus.

It is more meaningful to compare the *difference* between the concordance values of monozygotic and dizygotic twins. If these values are significantly higher for monozygotic twins than for dizygotic twins, a strong genetic component may be involved in the determination of the trait. We reach this conclusion because monozygotic twins, with identical genotypes, would be expected to show a greater concordance than genetically related, but not genetically identical, dizygotic twins. In the case of measles,

TABLE 6.6 **A Comparison of Concordance of Various Traits Between Monozygotic (MZ) and Dizygotic (DZ) Twins**

	Concordance	
Trait	*MZ*	*DZ*
Blood types	100%	66%
Eye color	99	28
Mental retardation	97	37
Measles	95	87
Haircolor	89	22
Handedness	79	77
Idiopathic epilepsy	72	15
Schizophrenia	69	10
Diabetes	65	18
Identical allergy	59	5
Cleft lip	42	5
Club foot	32	3
Mammary cancer	6	3

where concordance is high in both types of twins, the environment is assumed to contribute significantly.

Even though a particular trait may demonstrate considerable genetically based variation, it is often difficult to formulate a precise mode of inheritance using available data. In many cases, the trait is considered to be controlled by multiple-factor inheritance. However, when the environment is also exerting a partial influence, such a conclusion is particularly difficult to prove.

6.4 Mapping Quantitative Trait Loci

Because quantitative traits are influenced by numerous genes, geneticists would like to know where these genes are located in the genome. Are they linked on a single chromosome, or scattered throughout the genome among many chromosomes? As we shall see in subsequent chapters, locating, or "mapping," genes within the genome is an important step toward establishing the genetic identity of organisms. The initial approach is to localize genes controlling quantitative traits on a particular chromosome or chromosomes. In the context of this discussion, these genes are called **quantitative trait loci (QTLs)**. As we shall see, some rather ingenious methods have been devised to map these genes.

As an example of this type of analysis, we consider the phenotypic trait in *Drosophila* of resistance to the insecticide DDT, which has been shown to be under polygenic control. To find the loci responsible, strains selected for resistance to DDT and strains selected for sensitivity to DDT were crossed to flies carrying dominant genes that serve as markers on each of *Drosophila*'s four chromosomes. Following a variety of crosses, offspring were produced that contained many different combinations of marker chromosomes and chromosomes from either resistant or sensitive strains. As shown in Figure 6–8, flies with various chromosome combinations were then

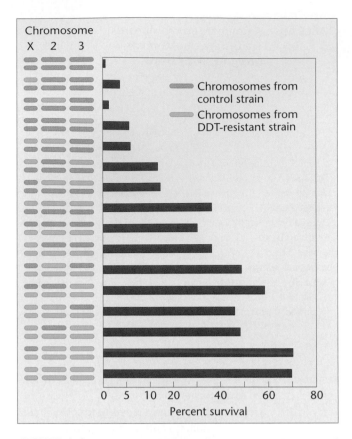

FIGURE 6–8 Survival rates of *Drosophila* carrying combinations of chromosomes from DDT-resistant and DDT-sensitive (control) strains when exposed to DDT. The results indicate that DDT resistance is polygenic, with genes on each of the major chromosomes making a major contribution. Chromosome 4 carries only a few *Drosophila* genes and was omitted from this analysis.

tested for resistance (their survival) when exposed to DDT. Results indicate that each chromosome in *Drosophila* contains genes that contribute to resistance. In other words, the loci bearing the genes that control DDT resistance are scattered throughout the genome.

We can now map the position of these genes more specifically along each chromosome in *Drosophila* because of the presence of molecular markers on each chromosome. The locations of QTLs are determined relative to the known positions of these markers. These markers, called **restriction fragment length polymorphisms (RFLPs)**, represent specific nuclease cleavage sites (see Chapter 16 for a detailed description of RFLPs). This approach enumerates and maps the loci responsible for quantitative traits.

RFLP markers are now available for many organisms of agricultural importance, making possible systematic mapping of QTLs. For example, hundreds of RFLP markers have been located in the tomato. They are spaced along all 12 chromosomes of this organism. Analysis is performed by crossing plants with extreme, but opposite, phenotypes and following the crosses through several generations. When both a marker and a phenotypic trait of interest are expressed together, they are said to *cosegregate*. Consistent cosegregation establishes the presence of a QTL at or near the RFLP marker along the chromosome. Whenever both an RFLP marker and

a QTL responsible for the trait under investigation are closely linked along a single chromosome, they are more likely to demonstrate an association throughout a pedigree than if they are not closely linked. When numerous QTLs are located, a genetic map is created for the involved genes.

RFLP analysis has resulted in extensive mapping of QTLs in the tomato, *Lycopersicon esculentum*. Many loci have been identified that are responsible for fruit weight, soluble solid content, and acidity. These loci are distributed on all 12 chromosomes representing the haploid genome of this plant. Several chromosomes contain loci for all three traits. Note that the RFLP method initially allows the identification of small chromosomal regions, not individual genes, although each locus may well house only a single gene.

Fruit weight in the tomato has been the focus of a highly successful research effort conducted by Steven Tanksley and his colleagues. Currently 28 QTLs responsible for phenotypic variation in fruit weight have been identified. One of these, *fw2.2*, has now been isolated, cloned, and transferred between plants, with most interesting results. While the cultivated tomato can weigh up to 1000 grams, the progenitor of the modern tomato is thought to have weighed only a few grams. Two distinct alleles of *fw2.2* are now recognized as a result of RFLP mapping studies. One allele is present in all wild small-fruited varieties of tomatoes investigated. The other allele is present in all domesticated large-fruited varieties. When the cloned allele from small-fruited varieties is transferred to a plant that normally produces large tomatoes, a remarkable result is achieved. The transformed plant produces fruits that are reduced in weight by about 30 percent. In the varieties employed in this study, this reduction averaged 17 grams, a significant phenotypic change caused by the action of a single gene.

Thus, for the first time, the genetic basis of quantitative variation is available for meaningful investigation. Tanksley's research group has now established that the *fw2.2* locus encodes a gene, *ORFX*, that is involved in the control of the number of carpels. Carpels are ovule-bearing units within the ovary of the flower that, following fertilization, develop into a fruit. The number of carpel units influences fruit size. That the allele for small fruit is partially dominant to the large fruit allele suggests that the genetic alteration between the two alleles is involved in the regulation of floral development, ultimately determining carpel number. This is in contrast to a genetic change that alters the sequence and structure of a protein that somehow imparts weight to the fruit. Of added interest to these findings is the observation that the *ORFX* gene is structurally similar to a human oncogene in the *ras* family (see Chapter 21) implicated in malignancy.

Mapping QTLs and defining the function of genes present in these regions in agriculturally important plants will greatly enhance programs designed to improve their yield. Knowledge gained from the study of so-called "quantitative genes" in plants will no doubt pave the way for investigations of similar genes in animals, including our own species. Since such loci are thought to be critical components of what have previously referred to as complex traits, our knowledge of how genes actually control phenotypes will be greatly extended.

Genetics, Technology, and Society

The Green Revolution Revisited

Of the greater than 6 billion people now living on Earth, about 750 million don't have enough to eat. And despite efforts to limit population growth, an additional 1 million people are expected to go hungry each year for the next several decades.

Will we be able to solve this problem? The past gives us some reasons to be optimistic. In the 1950s and 1960s, in the face of looming population increases, plant scientists around the world set about to increase the production of crop plants, including the three most important grains, wheat, rice, and maize. These efforts became known as the Green Revolution. The approach was three-pronged: (1) to increase the use of fertilizers, pesticides, and irrigation water, (2) to bring more land under cultivation, and (3) to develop improved varieties of crop plants by intensive plant breeding. While highly successful, in recent years the rate of increase in grain yields has slowed. If food production is to keep pace with the projected increase in the world's population, plant breeders will have to depend more and more on the genetic improvement of crop plants to provide higher yields. But is this possible? Are we approaching the theoretical limits of yield in important crop plants? Recent work with rice suggests that the answer to this question is a resounding no.

Rice ranks third in worldwide production, just behind wheat and maize. About 2 billion people, fully one-third of Earth's population, depend on rice for their basic nourishment. The majority of the world's rice crop is grown and consumed in Asia, but it is also a dietary staple in Africa and Central America. The Green Revolution for rice began in 1960 with the establishment of the International Rice Research Institute (IRRI), headquartered at Los Baños, Philippines. The goal was to breed rice with improved disease resistance and higher yield. Breeders were almost too successful: The first high-yield varieties were so top-heavy with grain that they tended to fall over. (Plant breeders call this "lodging.") To reduce lodging, IRRI breeders crossed a high-yield line with a dwarf native variety to create semi-dwarf lines, which were introduced to farmers in 1966. Due in large part to the adoption of the semi-dwarf lines, the world production of rice doubled in the next 25 years.

Rice breeders cannot afford to rest on their laurels, however, as the yield of modern rice varieties has not improved much in recent years. Predictions suggest that a 70 percent increase in the annual rice harvest may be necessary to keep pace with anticipated population growth during the next 30 years. Breeders are now looking to wild rice varieties for further crop improvement. Leading the way are Susan McCouch, Steven Tanksley, and their co-workers at Cornell University. To test the hypothesis that wild rice species carry genes that will improve the yield of cultivated varieties, they crossed cultivated rice (*Oryza sativa*) with a low-yield wild ancestor species (*Oryza rufipogon*) and then successively backcrossed the interspecific hybrid to cultivated rice for three generations. In theory, this would create lines whose genomes were about 95 percent from *O. sativa* and 5 percent from *O. rufipogon*. When testing these backcrossed lines for grain yield, they found that several of them outproduced cultivated rice by as much as 30 percent. These results demonstrated strikingly that even though wild rice relatives have low yields and appear to be inferior to cultivated rice, they still carry genes that will increase the yield of elite rice varieties. It will now be up to breeders to exploit the wild rice relatives.

But introducing favorable genes from a wild relative into a cultivated variety by conventional breeding is a long and involved process, often requiring a decade or more of crossing, selection, backcrossing, and more selection. Future improvements in cultivated species must be quicker if crop yields are to keep up with population growth. Fortunately, modern gene-mapping techniques are now leading to the identification of *quantitative trait loci* (QTL) that control complex traits such as yield and disease resistance. This makes possible a more direct approach to crop improvement, which has been termed the *advanced backcross QTL method*.

First, a cultivated variety is crossed with a wild relative, just as *O. sativa* was interbred with *O. rufipogon*. The hybrid is then backcrossed to the cultivated variety to generate lines that contain only a small fraction of the "wild" genome. The backcross lines with the best qualities (e.g., highest yield and most disease resistance) are selected, and the "wild QTL" responsible for the superior performance are identified using a detailed molecular linkage map. Once beneficial QTL are discovered by this method, they may be introduced into other cultivated varieties.

In order for this strategy to succeed, it is essential that wild crop relatives be preserved as a storehouse of potentially useful genes. Efforts were begun in the 1970s to protect the existing crop relatives of many plants in their natural habitats and to preserve them in seed banks. As the work of McCouch and Tanksley and others has shown, it is not possible to predict which wild varieties may be needed decades or even centuries from now to contribute their beneficial alleles to cultivated varieties. To prevent the loss of superior genes, the widest possible spectrum of wild species must be preserved, even those that have no obvious favorable characteristics.

Almost 60 years ago, the great Russian plant geneticist N. I. Vavilov suggested that wild relatives of crop plants could be the source of genes to improve agriculture. In the coming century, Vavilov's vision may finally be realized, as genes from long-neglected wild crop relatives, identified by new molecular methods, spark a revitalized Green Revolution.

References

MANN, C. 1997. Reseeding the green revolution. *Science* 277:1038–43.

RONALD, P. C. 1997. Making rice disease-resistant. *Sci. Am.* (Nov.) 277:98–105.

TANKSLEY, S. D., and MCCOUCH, S. R. 1997. Seed banks and molecular maps: Unlocking genetic potential from the wild. *Science* 277:1063–66.

XIAO, J. et al. 1996. Genes from wild rice improve yield. *Nature* 384:223–24.

Insights and Solutions

1. In a plant, height varies from 6 cm to 36 cm. When 6- and 36-cm plants are crossed, all F_1 plants are 21 cm. In the F_2 generation, a continuous range of heights is observed. Most are around 21 cm, and 3 of 200 are as short as the 6-cm P_1 parent.

(a) What mode of inheritance is demonstrated, and how many gene pairs are involved?

Solution: Polygenic inheritance where a continuous trait is involved and where alleles contribute additively to the phenotype is demonstrated. The 3/200 ratio of F_2 plants is the key to determining the number of gene pairs. This reduces to a ratio of 1/66.7, very close to 1/64. Using the formula $1/4^n = 1/64$ (where 1/64 is equal to the proportion of F_2 phenotypes as extreme as either P_1 parent), $n = 3$. Therefore, three gene pairs are involved.

(b) How much does each additive allele contribute to height?

Solution: The variation between the two extreme phenotypes is

$$36 - 6 = 30 \text{ cm}$$

Because there are six potential additive alleles (*AABBCC*), each contributes

$$30/6 = 5 \text{ cm}$$

to the base height of 6 cm, which results when no additive alleles (*aabbcc*) are part of the genotype.

(c) List all genotypes that give rise to 31-cm plants.

Solution: All genotypes that include 5 additive alleles will be 31 cm [(5 alleles $\times$ 5 cm) + 6 cm base height = 31 cm].

Therefore, the genotypes *AABBCc*, *AABbCC*, and *AaBBCC* will result in 31-cm plants.

2. The results shown below are recorded for ear length in corn:

Calculate the mean values for ear length for each parental strain and for the F_1 plants.

Solution: The mean values can be calculated as follows:

$$\overline{X} = \frac{\Sigma X_i}{n} \qquad P_A : \overline{X} = \frac{\Sigma X_i}{n} = \frac{378}{57} = 6.63$$

$$P_B : \overline{X} = \frac{\Sigma X_i}{n} = \frac{1697}{101} = 16.80$$

$$F_1 : \overline{X} = \frac{\Sigma X_i}{n} = \frac{836}{69} = 12.11$$

3. Compare the mean of the F_1 generation with that of each parental strain. What does this tell you about the type of gene action involved?

Solution: The F_1 mean (12.11) is almost midway between the parental means of 6.63 and 16.80. This indicates that the genes in question may be additive in effect.

4. The mean and variance of corolla length in two highly inbred strains of *Nicotiana* and their progeny are shown. One parent (P_1) has a short corolla and the other parent (P2) has a long corolla.

Strain	Mean (mm)	Variance (mm)
P_1 short	40.47	3.12
P_2 long	93.75	3.87
F_1 ($P_1 \times P_2$)	63.90	4.74
F_2 ($F_1 \times F_1$)	68.72	47.70

Calculate the heritability (H^2) of corolla length in this plant.

Solution: The formula for estimating heritability is $H^2 = V_G/V_P$, where V_G and V_P are the genetic and phenotypic components of variation, respectively. The main issue in this problem is obtaining some estimate of two components of phenotypic variation: genetic and environmental factors.

Recall that V_P is the combination of genetic and environmental variance. Because the two parental strains are true-breeding, they are assumed to be homozygous and the variances of 3.12 and 3.87 are considered to result from environmental influences. The average of these two values is 3.50. The F_1 generation is also genetically homogeneous and gives us an additional estimate of the environmental factors. By averaging with the parents,

$$\frac{3.50 + 4.74}{2} = 4.12$$

we obtain a relatively good idea of environmental impact on the phenotype. The phenotypic variance in F_2 is the sum of the genetic (V_G) and environmental (V_E) components. We have estimated the environmental input as 4.12, so $47.70 - 4.12$ gives us an estimate of V_G, which is 43.58. Heritability then becomes 43.58/47.70 or 0.91. This value indicates that about 91 percent of the variation in corolla length is due to genetic influences.

(Problem 2)—Length of Ear in cm

	5	6	7	8	9	10	11	12	13	14	15	16	17	18	19	20	21
Parent A	4	21	24	8													
Parent B									3	11	12	15	26	15	10	7	2
F_1					1	12	12	14	17	9	4						

Chapter Summary

1. Continuous variation is exhibited in crosses involving traits under polygenic control. Such traits are quantitative in nature and are inherited as a result of the cumulative impact of additive alleles.
2. Polygenic characteristics can be analyzed using statistical methods, which include the mean, the variance, the standard deviation, and the standard error of the mean. Such statistical analysis is descriptive, and can be used to make inferences about a population or to compare and contrast sets of data.
3. For many phenotypic characteristics, it is difficult to ascertain when variation is due to genetic or to environmental factors. Heritability, the estimate of the relative importance of genetic versus nongenetic factors in determining phenotypic variation in populations, can be calculated for many characters and is especially useful in selective breeding of commercially valuable plants and animals.
4. Studies involving twins are aimed at resolving the question of heredity versus environment in human traits. The degree of concordance of a trait is compared in monozygotic (identical) and dizygotic (fraternal) twins raised together or apart.
5. Loci bearing genes that control a quantitative trait are called quantitative trait loci (QTLs). Using either genetic or molecular markers, the location and distribution within the genome of QTLs can be ascertained.

Key Terms

additive allele, 105
additive variance, 111
artificial selection, 111
biometry, 108
broad-sense heritability, 110
complex traits, 107
concordance, 113
discordance, 113
dizygotic (fraternal) twins, 113
dominance variance, 111

frequency distribution, 108
heritability index, 110
interactive variance, 111
mean, 108
monozygotic (identical) twins, 113
multiple-factor hypothesis, 105
narrow-sense heritability, 111
nonadditive allele, 105
phenotypic variance, 110
polygenic inheritance, 103

quantitative inheritance, 103
quantitative trait loci (QTL), 113
restriction length fragment polymorphisms (RFLPs), 114
selection differential, 111
standard deviation, 109
standard error of the mean, 109
variance, 108

Problems and Discussion Questions

1. Distinguish between discontinuous and continuous variation. Which type relates to inheritance of a quantitative nature?
2. Define and discuss the following: (a) polygenes, (b) additive alleles, (c) multiple-factor hypothesis, (d) monozygotic versus dizygotic twins, (e) concordance versus discordance, and (f) heritability.
3. A dark red strain and a white strain of wheat are crossed and produce an intermediate, medium-red F_1. When the F_1 plants are interbred, an F_2 generation is produced in a ratio of 1 dark red:4 medium-dark red:6 medium red:4 light red:1 white. Further crosses reveal that the dark red and white F_2 plants are true-breeding.
   (a) Based on the ratio of offspring in F_2, how many genes are involved in the production of color?
   (b) How many additive alleles are needed to produce each possible phenotype?
   (c) Assign symbols to these alleles and list possible genotypes that give rise to the medium red and the light red phenotypes.
   (d) Predict the outcome of the F_1 and F_2 generations in a cross between a true-breeding medium red plant and a white plant.
4. Height in humans depends on the additive action of genes. Assume that this trait is controlled by the four loci R, S, T, and U and that environmental effects are negligible. Instead of additive versus nonadditive alleles, assume that additive and partially additive alleles exist. Additive alleles contribute two units and partially additive alleles contribute one unit to height.
   (a) Can two individuals of moderate height produce offspring that are much taller or shorter than either parent? If so, how?
   (b) If an individual with the minimum height specified by these genes marries an individual of intermediate or moderate height, will any of their children be taller than the tall parent? Why or why not?

5. An inbred strain of plants has a mean height of 24 cm. A second strain of the same species from a different geographical region also has a mean height of 24 cm. When plants from the two strains are crossed, the F_1 plants are the same height as the parent plants. However, the F_2 generation shows a wide range of heights; the majority are like the P_1 and F_1 plants, but approximately 4 of 1000 are only 12 cm high, and about 4 of 1000 are 36 cm high.
   (a) What mode of inheritance is occurring here?
   (b) How many gene pairs are involved?
   (c) How much does each gene contribute to plant height?
   (d) Indicate one possible set of genotypes for the original P_1 parents and the F_1 plants that could account for these results.
   (e) Indicate three possible genotypes that could account for 18-cm F_2 plants and three that could account for 33-cm F_2 plants.

6. Erma and Harvey were a compatible barnyard pair, but a curious sight. Harvey's tail was only 6 cm, while Erma's was 30 cm. Their F_1 piglet offspring all grew 18-cm tails. When inbred, an F_2 generation resulted in many piglets (Erma and Harvey's grandpigs), whose tails ranged in 4-cm intervals from 6 cm to 30 cm (6, 10, 14, 18, 22, 26, 30). Most had 18-cm tails, while 1/64 had 6-cm tails and 1/64 had 30-cm tails.
   (a) Explain how tail length is inherited by describing the mode of inheritance, indicating how many gene pairs are at work, and designating the genotypes of Harvey, Erma, and their 18-cm offspring.
   (b) If one of the 18-cm F_1 pigs is mated with a 6-cm F_2 pig, what phenotypic ratio would be predicted if many offspring resulted? Diagram the cross.

7. Discuss the use of monozygotic and dizygotic twins reared together and apart in assessing the genetic component responsible for phenotypic variation in humans.

8. In the following table, average differences in height and weight of monozygotic twins (reared together and apart), dizygotic twins, and siblings are compared. Draw as many conclusions as you can concerning the effects of genetics and the environment in influencing these human traits.

Trait	MZ Reared Together	MZ Reared Apart	DZ Reared Together	Sibs Reared Together
Height (cm)	1.7	1.8	4.4	4.5
Weight (kg)	1.9	4.5	4.5	4.7

9. List as many human traits as you can that are likely to be under the control of a polygenic mode of inheritance.

10. Corn plants from a test plot are measured, and the distribution of heights at 10-cm intervals is recorded:

Height (cm)	Plants (no.)
100	20
110	60
120	90
130	130
140	180
150	120
160	70
170	50
180	40

Calculate (a) the mean height, (b) the variance, (c) the standard deviation, and (d) the standard error of the mean. Plot a rough graph of plant height versus frequency. Do the values represent a normal distribution? Based on your calculations, how would you assess the variation within this population?

11. Contrast broad-sense heritability (H^2) and narrow-sense heritability (h^2). To what type of population is each calculation applicable? Which is useful in artificial selection procedures? Why?

12. The mean and variance of plant height of two highly inbred strains (P_1 and P_2) and their progeny (F_1 and F_2) are shown. Calculate the broad-sense heritability (H^2) of plant height in this species.

Strain	Mean (cm)	Variance
P_1	34.2	4.2
P_2	55.3	3.8
F_1	44.2	5.6
F_2	46.3	10.3

13. In a hypothetical study, vitamin A content and cholesterol content of eggs from a large population of chickens is investigated. The variances (V) are calculated, as shown:

Variance	Trait Vitamin A	Cholesterol
V_P	123.5	862.0
V_E	96.2	484.6
V_A	12.0	192.1
V_D	15.3	185.3

   (a) Calculate the narrow heritability (h^2) for both traits.
   (b) Which trait, if either, is likely to respond to selection?

14. In an assessment of learning in *Drosophila*, flies can be trained to avoid certain olfactory cues. In one population, a mean of 8.5 trials was required. A subgroup of this parental population that was trained most quickly (mean = 6.0 trials) was interbred and their progeny examined. These flies demonstrated a mean training value of 7.5 trials. Calculate and interpret narrow-sense heritability for olfactory learning in *Drosophila*.

15. A population of tomato plants with mean fruit weight of 60 g and an h^2 value of 0.3 is studied. Predict the results of artificial selection (the mean weight of the progeny) if tomato plants with an average fruit weight of 80 g are selected from the original population and interbred.

16. A mutant strain of *Drosophila* is isolated and shown to be resistant to an experimental insecticide, whereas normal (wild type) flies are sensitive to the chemical. Following a cross between resistant flies and sensitive flies, isolated populations are derived that have various combinations of chromosomes from the two strains. Each is tested for resistance, as shown at the top of the next column. Analyze the data and draw any appropriate conclusion about which chromosomes contain a gene responsible for inheritance of resistance to the insecticide.

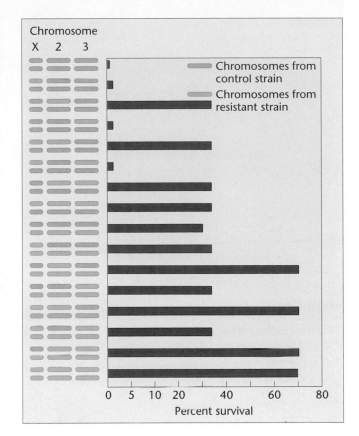

Selected Readings

Brink, R., ed. 1967. *Heritage from Mendel*. Madison: University of Wisconsin Press.

Crow, J. F. 1993. Francis Galton. Count and measure, measure and count. *Genetics* 135:1.

Dudley, J. W. 1977. 76 generations of selection for oil and protein percentage in maize. *In* Pollack, E., Kempthorne, O., and Bailey, T. (eds.), *Proceedings of the International Conference on Quantitative Genetics*, pp. 459–73. Ames: Iowa State University Press.

Falconer, D. 1989. *Introduction to quantitative genetics*, 3rd ed. Harlow, England: Longman.

Farber, S. 1980. *Identical twins reared apart*. New York: Basic Books.

Feldman, M. W., and Lewontin, R. C. 1975. The heritability hangup. *Science* 190:1163–66.

Fowler, C., and Mooney, P. 1990. *Shattering: Food, politics, and the loss of genetic diversity*. Tucson: University of Arizona Press.

Frary, A., et al. 2000. *fw2.2*: A quantitative trait locus key to the evolution of tomato fruit size. *Science* 289:85–88.

Haley, C. 1991. Use of DNA fingerprints for the detection of major genes for quantitative traits in domestic species. *Anim. Genet.* 22:259–77.

————. 1996. Livestock QTLs: Bringing home the bacon. *Trends Genet.* 11:488–90.

Lander, E., and Botstein, D. 1989. Mapping mendelian factors underlying quantitative traits using RFLP linkage maps. *Genetics* 121:185–99.

Lander, E., and Schork, N. 1994. Genetic dissection of complex traits. *Science* 265:2037–48.

Lewontin, R. C. 1974. The analysis of variance and the analysis of causes. *Am. J. Hum. Genet.* 26:400–11.

Lynch, M., and Walsh, B. 1998. *Genetics and analysis of quantitative traits*. Sunderland, MA: Sinauer Associates.

Macay, T. F. C. 2001. Quantitative trait loci in *Drosophila*. *Nature Reviews Genetics* 2:11–19.

Newman, H. H., Freeman, F. N., and Holzinger, K. T. 1937. *Twins: A study of heredity and environment*. Chicago: University of Chicago Press.

Paterson, A., Deverna, J., Lanini, B., and Tanksley, S. 1990. Fine mapping of quantitative traits loci using selected overlapping recombinant chromosomes in an interspecific cross of tomato. *Genetics* 124:735–42.

Plomin, R., McClearn, G., Gora-Maslak, G., and Neiderhiser, J. 1991. Use of recombinant inbred strains to detect quantitative trait loci associated with behavior. *Behav. Genet.* 21:99–116.

Stuber, C. W. 1996. Mapping and manipulating quantitative traits in maize. *Trends Genet.* 11:477–87.

Zar, J. H. 1996. *Biostatistical analysis*, 3rd ed. Upper Saddle River, NJ: Prentice Hall.

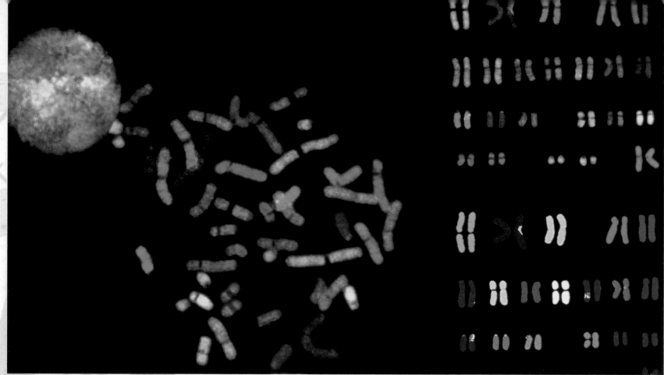

Spectral karyotyping of human chromosomes utilizing differentially labeled "painting" probes. *(Evelin Schrock, Stan du Manoir, and Tom Reid, National Institutes of Health)*

7

Chromosome Mutations: Variation in Number and Arrangement

CHAPTER CONCEPTS

Genetic information of diploid organisms is delicately balanced in both content and location within the genome. Chromosome mutation, a change in chromosome number or in the arrangement of a chromosome region within the genome, often results in phenotypic variation or disruption of development of an organism. Because the chromosome is the unit of transmission in meiosis, such variations can be passed to offspring in a predictable manner, resulting in many unique genetic outcomes.

Thus far, we have emphasized how mutations and the resulting alleles affect an organism's phenotype, and how traits are passed from parents to offspring according to Mendelian principles. In this chapter we look at phenotypic variation that results from more substantial changes than alterations of individual genes—modifications at the level of the chromosome.

Although most members of diploid species normally contain precisely two haploid chromosome sets, many known cases vary from this pattern. Modifications include a change in the total number of chromosomes, the deletion or duplication of genes or segments of a chromosome, and rearrangements of the genetic material either within or among chromosomes. Taken together, such changes are called **chromosome mutations** or **chromosome aberrations**, to distinguish them from gene mutations. Because, according to Mendelian laws, the chromosome is the unit of genetic transmission, chromosome aberrations are passed on to offspring in a predictable manner, resulting in many unique genetic outcomes.

Because the genetic component of an organism is delicately balanced, even minor alterations of either content or location of genetic information within the genome can result in some form of phenotypic variation. More substantial changes may be lethal, particularly in animals. Throughout the chapter we consider the many types of chromosomal aberrations, the phenotypic consequences for the organism that harbors an aberration, and the impact of the aberration on offspring of the affected individual. We will also discuss the role of chromosome aberrations in the evolutionary process.

7.1 Variation in Chromosome Number: An Overview

Before we embark on a discussion of the many types of variation involving numbers of chromosomes, it will be helpful to clarify the terminology that describes such changes. In the general condition known as **aneuploidy**, an organism gains or loses one or more chromosomes, but not a complete set. The loss of a single chromosome from an otherwise diploid genome is called *monosomy*. The gain of one chromosome results in *trisomy*.

Such changes are contrasted with the general condition of **euploidy**, where only complete haploid sets of chromosomes are present. If more than two sets are present, the term **polyploidy** applies. Organisms with three sets are specifically *triploid*; those with four sets are *tetraploid*, and so on. Table 7.1 gives you a framework to follow as we discuss aneuploid and euploid variation and the subsets within them.

7.2 Nondisjunction: The Origin of Aneuploidy

It is useful as we consider cases that include the gain or loss of chromosomes to examine how such aberrations originate. For instance, how do the syndromes arise where the number of sex-determining chromosomes in humans is altered, as described in Chapter 5? As you may recall, the gain (47,XXY) or loss (45,X) of an X chromosome from an otherwise diploid genome affects the phenotype, resulting in **Klinefelter** and **Turner syndrome**, respectively (see Figure 5–7). Human females are also known who contain extra X chromosomes (e.g., 47,XXX, 48,XXXX), and some males contain an extra Y chromosome (47,XYY).

Such chromosomal variation originates as the result of an error during meiosis. First introduced in Chapter 2, **nondisjunction**, the failure of paired homologues to disjoin during segregation, is a process where the normal distribution of chromosomes into gametes is disrupted. The results of nondisjunction during meiosis I and meiosis II for a single chromosome of a diploid organism are shown in Figure 7–1. As you can see, for the affected chromosome, abnormal gametes can form that contain either two members or none at all. Fertilizing these with a normal haploid gamete produces a zygote with either three members (trisomy) or only one member (monosomy) of this chromosome. As we shall see, nondisjunction leads to a variety of aneuploid conditons in humans and other organisms.

7.3 Monosomy

We turn now to a consideration of variations in the number of autosomes and the genetic consequence of such changes. The most common examples of aneuploidy where an organism has a chromosome number other than an exact multiple of the haploid set, are cases in which a single chromosome is either added to, or lost from, a normal diploid set. The loss of one chromosome produces a $2n - 1$ complement and is called **monosomy**.

Although monosomy for the X chromosome occurs in humans, as we have seen in 45,X Turner syndrome, monosomy for any of the autosomes is not usually tolerated in humans

TABLE 7.1 Terminology for Variation in Chromosome Numbers

Term	Explanation
Aneuploidy	$2n \pm x$ chromosomes
Monosomy	$2n - 1$
Trisomy	$2n + 1$
Tetrasomy, pentasomy, etc.	$2n + 2$, $2n + 3$, etc.
Euploidy	Multiples of n
Diploid	$2n$
Polyploidy	$3n, 4n, 5n, \ldots$
Triploidy	$3n$
Tetraploidy, pentaploidy, etc.	$4n, 5n$, etc.
Autopolyploidy	Multiples of the same genome
Allopolyploidy (Amphidiploidy)	Multiples of different genomes

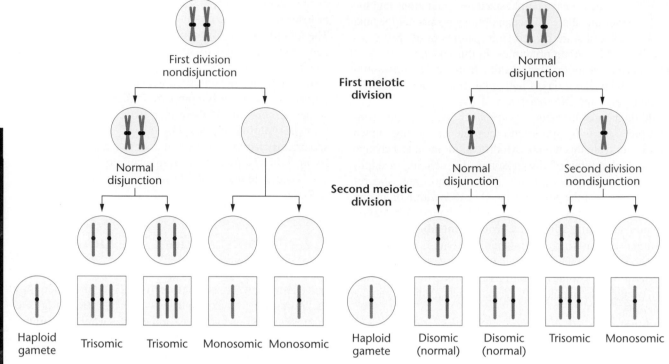

FIGURE 7–1 Nondisjunction during the first and second meiotic divisions. In both cases, some of the gametes formed either contain two members of a specific chromosome or lack that chromosome. Following fertilization by a gamete with a normal haploid content, monosomic, disomic (normal), or trisomic zygotes are produced.

or other animals. In *Drosophila*, flies that are monosomic for the very small chromosome 4—a condition referred to as **Haplo-IV**—develop more slowly, exhibit reduced body size, and have impaired viability. Monosomy for the larger chromosomes 2 and 3 is apparently lethal, because such flies have never been recovered.

The failure of monosomic individuals to survive in many animal species is at first quite puzzling, since at least a single copy of every gene is present in the remaining homolog. However, if just one of those genes is represented by a lethal allele, the unpaired chromosome condition leads to the death of the organism. This occurs because monosomy unmasks recessive lethals that are tolerated in heterozygotes carrying the corresponding wild-type alleles.

Aneuploidy is better tolerated in the plant kingdom. Monosomy for autosomal chromosomes has been observed in maize, tobacco, the evening primrose *Oenothera*, and the Jimson weed *Datura*, among other plants. Nevertheless, such monosomic plants are usually less viable than their diploid derivatives. Haploid pollen grains, which undergo extensive development before participating in fertilization, are particularly sensitive to the lack of one chromosome and are seldom viable.

Cri-du-Chat Syndrome

In humans, autosomal monosomy has not been reported beyond birth. Individuals with such chromosome complements are undoubtedly conceived, but none apparently survive embryonic and fetal development. There are, however, examples of survivors where only part of one chromosome is lost. These cases are sometimes referred to as **segmental deletions**. One such case was first reported by Jérôme LeJeune in 1963 when he described the clinical symptoms of the **cri-du-chat (cry of the cat) syndrome**. This syndrome is associated with the loss of part of the short arm of chromosome 5 (Figure 7–2). Thus, the genetic constitution may be designated as **46,5p–**, meaning that such an individual has all 46 chromosomes but that some of the *p* arm (the short or petite arm) of one member of the chromosome 5 pair is missing.

Infants with this syndrome exhibit anatomic malformations, including gastrointestinal and cardiac complications, and they are often mentally retarded. Abnormal development of the glottis and larynx is also characteristic of individuals with this syndrome. As a result, the infant has an unusual cry, one that is similar to the meowing of a cat, giving the syndrome its name.

Since 1963, hundreds of cases of cri-du-chat syndrome have been reported worldwide. An incidence of 1 in 50,000 live births has been estimated. The length of the short arm that is deleted varies somewhat; longer deletions appear to have a greater impact on the physical, psychomotor, and mental skill levels of those children who survive. Although the effects of the syndrome are severe, many individuals achieve a level of social development in the trainable range. Those who receive home care and early special schooling are ambulatory, develop self-care skills, and learn to communicate verbally.

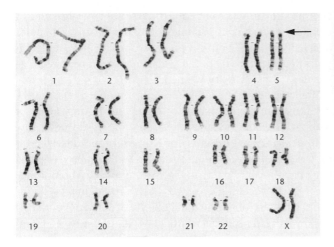

FIGURE 7–2 A representative karyotype and photograph of a child exhibiting cri-du-chat syndrome (46,5p–). In the karyotype, the arrow identifies the absence of a small piece of the short arm of one member of the chromosome 5 homologs. *(Left: Courtesy of the Greenwood Genetic Center, Greenwood, SC; right: Five P Minus Society)*

7.4 Trisomy

In general, the effects of trisomy parallel those of monosomy. However, the addition of an extra chromosome produces somewhat more viable individuals in both animal and plant species than does the loss of a chromosome. In animals, this is often true, provided that the chromosome involved is relatively small. However, the addition of a large autosome to the diploid complement in both *Drosophila* and humans has severe effects and is usually lethal during development.

In plants, trisomic individuals are viable, but their phenotype may be altered. A classical example involves the Jimson weed *Datura*, whose diploid number is 24. Twelve primary trisomic conditions are possible, and examples of each one have been recovered. Each trisomy alters the phenotype of the plant's capsule sufficiently to produce a unique phenotype. These capsule phenotypes were first thought to be caused by mutations in one or more genes.

Still another example is seen in the rice plant (*Oryza sativa*), which has a haploid number of 12. Trisomic strains for each chromosome have been isolated and studied. The plants of 11

of them can be distinguished from one another and from wild type. Trisomics for the longer chromosomes are the most distinctive and grow more slowly than the rest. This is in keeping with the belief that larger chromosomes cause greater genetic imbalance than smaller ones. In addition to growth rate, leaf structure, foliage, stems, grain morphology, and plant height vary between the various trisomies.

Down Syndrome

The only human autosomal trisomy in which a significant number of individuals survive longer than a year past birth was discovered in 1866 by Langdon Down. The condition is now known to result from trisomy of chromosome 21, one of the G group* (Figure 7–3), and is called **Down syndrome** or simply **trisomy 21** (and is designated **47,21+**). This trisomy is found in approximately 1 infant in every 800 live births.

---

*On the basis of size and centromere placement, human autosomal chromosomes are divided into seven groups: A (1–3); B (4–5); C (6–12); D (13–15); E (16–18); F (19–20); and G (21–22).

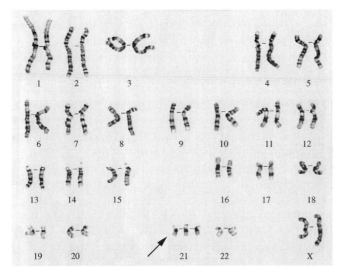

FIGURE 7–3 The karyotype and a photograph of a child with Down syndrome. In the karyotype, three members of the G-group chromosome 21 are present, creating the 47,21+ condition. *(Left: Courtesy of the Greenwood Genetic Center, Greenwood, SC; right: William McCoy/Rainbow)*

The overt phenotype of these individuals is so similar that they bear a striking resemblance to one another. They display a prominent epicanthic fold in the corner of the eye and are characteristically short. They may have flat faces, round heads; protruding, furrowed tongues, which cause the mouth to remain partially open; and short, broad hands with fingers showing characteristic palm and fingerprint patterns. Physical, psychomotor, and mental development is retarded, and poor muscle tone is characteristic.

Down children have a shortened life expectancy, and few survive to age 50. They are prone to respiratory disease and heart malformations and show an incidence of leukemia approximately 15 times as high as that of the normal population. However, careful medical scrutiny and treatment throughout their lives has extended their survival significantly. A striking observation is that death of older Down syndrome adults is frequently due to Alzheimer's disease.

The most characteristic origin of this trisomic condition is through nondisjunction of chromosome 21 during meiosis. Failure of paired homologs to disjoin during anaphase I or of chromatids to disjoin during anaphase II can result in male or female gametes with the $n + 1$ chromosome composition. Following fertilization with a normal gamete, the trisomic condition is created. Chromosome analysis has shown that while the additional chromosome can be derived from either the mother or father, the ovum is the source in 95 percent of the cases.

Before the development of techniques that distinguish paternal from maternal homologs, this conclusion was supported by other indirect evidence derived from studies of the age of mothers giving birth to Down infants. Figure 7–4 shows the relationship between maternal age and the incidence of Down syndrome newborns. While the frequency is about 1 in 1000 at maternal age 30, a tenfold increase to a frequency of 1 in 100 is noted at age 40. The frequency increases still further to about 1 in 50 at age 45. In spite of these statistics, it is important to point out that, overall, more than half of affected births occur to women who are under 35 years in age, primarily because there are so many more pregnancies in this group of women.

While the nondisjunctional event that produces Down syndrome seems more likely to occur during oogenesis in women between the ages of 35 and 45, we do not know with certainty why this is so. However, one observation may be relevant. In human females, all primary oocytes have been formed by birth. Therefore, once ovulation begins, each succeeding ovum has been arrested in meiosis for about a month longer than the one preceding it. Women 30 or 40 years old produce ova that are significantly older and arrested longer than those they ovulated 10 or 20 years previously. However, it is not yet known whether ovum age is the cause of the increased incidence of nondisjunction leading to Down syndrome.

These statistics pose a serious problem for the woman who becomes pregnant late in her reproductive years. **Genetic counseling** early in such pregnancies serves two purposes. First, it informs the parents about the probability that their child will be affected and educates them about Down syndrome. Although some individuals with Down syndrome must be institutionalized, others benefit greatly from special education programs and may be cared for at home. Further, these children are noted for their affectionate, loving natures. Second, a genetic counselor may recommend a prenatal diagnostic technique such as **amniocentesis** or **chorionic villus sampling (CVS)**. These techniques require the removal and culture of fetal cells. The karyotype of the fetus is then determined by cytogenetic analysis. If the fetus is diagnosed as having Down syndrome, a therapeutic abortion is one option the parents may consider.

Because Down syndrome appears to be caused by a random error—nondisjunction of chromosome 21 during maternal or paternal meiosis—the disorder is not expected to be inherited. Nevertheless, Down syndrome occasionally runs in families. This condition, familial Down syndrome, involves a **translocation** of chromosome 21, another type of chromosomal aberration, which we discuss later in this chapter.

Viability in Human Aneuploid Conditions

The reduced viability of individuals with recognized monosomic and trisomic conditions is evident. Only two other trisomies in humans survive to term. Both **Patau and Edwards syndromes, 47,13+** and **47,18+**, respectively, result in severe malformations and early lethality. Figure 7–5 illustrates the abnormal karyotype and the many defects characterizing Edwards infants.

Such observations lead us to believe that many other aneuploid conditions arise, but that the affected fetuses do not survive to term. This observation has been confirmed by karyotypic analysis of spontaneously aborted fetuses. These studies reveal some rather striking statistics. At least 15–20 percent of all conceptions terminate in spontaneous abortion (some estimates are considerably higher)! About 30 percent

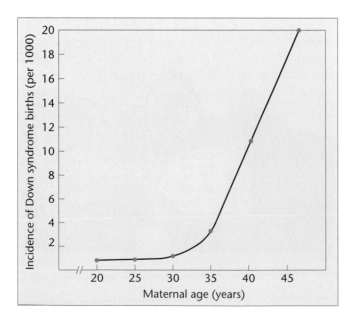

FIGURE 7–4 Incidence of Down syndrome births contrasted with maternal age.

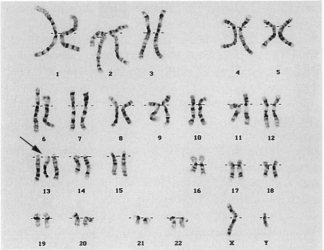

Mental retardation

Growth failure

Low set deformed ears

Deafness

Atrial septal defect

Ventricular septal defect

Abnormal polymorphonuclear granulocytes

Microcephaly

Cleft lip and palate

Polydactyly

Deformed finger nails

Kidney cysts

Double ureter

Umbilical hernia

Developmental uterine abnormalities

Cryptorchidism

FIGURE 7–5 The karyotype and phenotypic depiction of an infant with Patau syndrome, where three members of the D-group chromosome 13 are present, creating the 47,13+ condition.

of all spontaneously aborted fetuses demonstrate some form of chromosomal anomaly, and approximately 90 percent of all chromosomal anomalies are terminated prior to birth as a result of spontaneous abortion.

A large percentage of those demonstrating chromosomal abnormalities are aneuploids. The aneuploid with highest incidence among abortuses is the 45,X condition, which produces an infant with Turner syndrome if the fetus survives to term.

An extensive review of this subject by David H. Carr also reveals that a significant percentage of aborted fetuses are trisomic for one of the chromosome groups. Trisomies for

every human chromosome have been recovered. Monosomies were seldom found, however, even though nondisjunction should produce $n - 1$ gametes with a frequency equal to $n + 1$ gametes. This finding suggests that gametes lacking a single chromosome are functionally impaired to a serious degree or that the embryo dies so early in its development that recovery occurs infrequently. Various forms of polyploidy and other miscellaneous chromosomal anomalies were also found in Carr's study.

These observations support the hypothesis that normal embryonic development requires a precise diploid complement of chromosomes to maintain the delicate equilibrium in the expression of genetic information. The prenatal mortality of most aneuploids provides a barrier against the introduction of a general form of genetic anomalies into the human population.

7.5 Polyploidy and Its Origins

The term *polyploidy* describes instances in which more than two multiples of the haploid chromosome set are found. The naming of polyploids is based on the number of sets of chromosomes found: A triploid has $3n$ chromosomes; a tetraploid has $4n$; a pentaploid, $5n$; and so forth. Several general statements can be made about polyploidy. This condition is relatively infrequent in many animal species, but is well known in lizards, amphibians, and fish. It is much more common in plant species. Odd numbers of chromosome sets are not usually maintained reliably from generation to generation, because a polyploid organism with an uneven number of homologs often does not produce genetically balanced gametes. For this reason, triploids, pentaploids, and so on, are not usually found in plant species that depend solely on sexual reproduction for propagation.

Polyploidy originates in two ways: (1) The addition of one or more extra sets of chromosomes, identical to the normal haploid complement of the same species, resulting in **autopolyploidy**; and (2) the combination of chromosome sets from different species occurring as a consequence of hybridization, resulting in **allopolyploidy** (from the Greek word *allo*, meaning other or different). The distinction between auto- and allopolyploidy is based on the genetic origin of the extra chromosome sets, as shown in Figure 7–6.

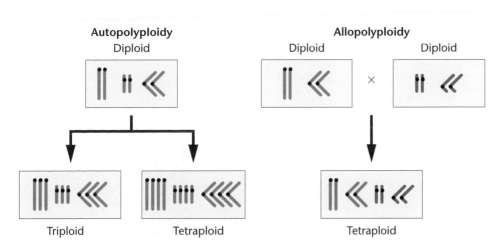

FIGURE 7–6 Contrasting chromosome origins of an autopolyploid versus an allopolyploid karyotype.

In our discussion of polyploidy, we use the following symbols to clarify the origin of additional chromosome sets. For example, if A represents the haploid set of chromosomes of any organism, then

$$A = a_1 + a_2 + a_3 + a_4 + \cdots + a_n$$

where a_1, a_2, and so on, are individual chromosomes and n is the haploid number. A normal diploid organism is represented simply as AA.

Autopolyploidy

In autopolyploidy, each additional set of chromosomes is identical to the parent species. Therefore, triploids are represented as AAA, tetraploids are $AAAA$, and so forth.

Autotriploids arise in several ways. A failure of all chromosomes to segregate during meiotic divisions can produce a diploid gamete. If such a gamete is fertilized by a haploid gamete, a zygote with three sets of chromosomes is produced. Or, rarely, two sperm may fertilize an ovum, resulting in a triploid zygote. Triploids are also produced under experimental conditions by crossing diploids with tetraploids. Diploid organisms produce gametes with n chromosomes, while tetraploids produce $2n$ gametes. Upon fertilization, the desired triploid is produced.

Because they have an even number of chromosomes, **autotetraploids** ($4n$) are theoretically more likely to be found in nature than are autotriploids. Unlike triploids, which often produce genetically unbalanced gametes with odd numbers of chromosomes, tetraploids are more likely to produce balanced gametes when involved in sexual reproduction.

How polyploidy arises naturally is of great interest. In theory, if chromosomes have replicated, but the parent cell never divides and instead reenters interphase, the chromosome number will be doubled. That this very likely occurs is supported by the observation that tetraploid cells can be produced experimentally from diploid cells. This is accomplished by applying cold or heat shock to meiotic cells or by applying colchicine to somatic cells undergoing mitosis. **Colchicine**, an alkyloid derived from the autumn crocus, interferes with spindle formation, and thus replicated chromosomes cannot separate at anaphase and do not migrate to the poles. When colchicine is removed, the cell can reenter interphase. When the paired sister chromatids separate and uncoil, the nucleus

contains twice the diploid number of chromosomes and is therefore $4n$. This process is shown in Figure 7–7.

In general, autopolyploids are larger than their diploid relatives. This increase seems to be due to larger cell size rather than greater cell number. Although autopolyploids do not contain new or unique information compared with the diploid relative, the flower and fruit of plants are often increased in size, making such varieties of greater horticultural or commercial value. Economically important triploid plants include several potato species of the genus *Solanum*, Winesap apples, commercial bananas, seedless watermelons, and the cultivated tiger lily *Lilium tigrinum*. These plants are propagated asexually. Diploid bananas contain hard seeds, but the commercial, triploid, "seedless" variety has edible seeds. Tetraploid alfalfa, coffee, peanuts, and McIntosh apples are also of economic value because they are either larger or grow more vigorously than do their diploid or triploid counterparts. The commercial strawberry is an octoploid.

Allopolyploidy

Polyploidy can also result from hybridizing two closely related species. If a haploid ovum from a species with chromosome sets AA is fertilized by sperm from a species with sets BB, the resulting hybrid is AB, where $A = a_1, a_2, a_3, ..., a_n$ and $B = b_1, b_2, b_3, ..., b_n$. The hybrid plant may be sterile because of its inability to produce viable gametes. Most often, this occurs when some or all of the a and b chromosomes are not homologous and therefore cannot synapse in meiosis. As a result, unbalanced genetic conditions result. If, however, the new AB genetic combination undergoes a natural or induced chromosomal doubling, two copies of all a chromosomes and two copies of all b chromosomes are now present, and they will pair during meiosis. As a result, a fertile $AABB$ tetraploid is produced. These events are shown in Figure 7–8. Since this polyploid contains the equivalent of four haploid genomes derived from separate species, such an organism is called an **allotetraploid**. In such cases, when both original species are known, an equivalent term, **amphidiploid**, is preferred in describing the allotetraploid.

Amphidiploid plants are often found in nature. Their reproductive success is based on their potential for forming balanced gametes. Since two homologs of each specific chromosome are present, meiosis occurs normally (Figure 7–8), and fertilization successfully propagates the plant sexually.

FIGURE 7–7 The potential involvement of colchicine in doubling the chromosome number, as occurs during the production of an autotetraploid. Two pairs of homologous chromosomes are followed. While each chromosome has replicated its DNA earlier during interphase, the chromosomes do not appear as double structures until late prophase. When anaphase fails to occur normally, the chromosome number doubles if the cell reenters interphase.

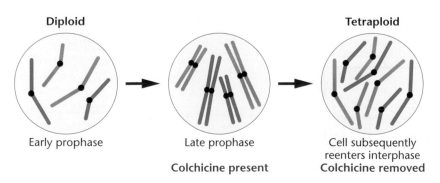

Diploid — Early prophase

Late prophase — **Colchicine present**

Tetraploid — Cell subsequently reenters interphase — **Colchicine removed**

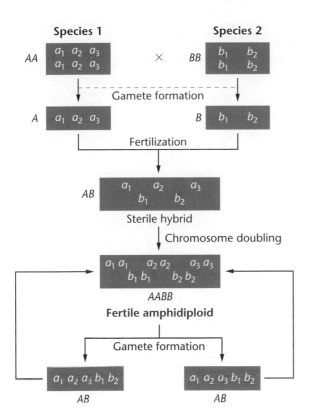

FIGURE 7–8 The origin and propagation of an amphidiploid. Species I contains genome *A* consisting of three distinct chromosomes, a_1, a_2, and a_3. Species 2 contains genome *B* consisting of two distinct chromosomes, b_1 and b_2. Following fertilization between members of the two species and chromosome doubling, a fertile amphidiploid containing two complete diploid genomes (*AABB*) is formed.

FIGURE 7–9 The pods of the amphidiploid form of *Gossypium*, the cultivated cotton plant. (*Ken Wagner/Phototake NYC*)

This discussion assumes the simplest situation, where none of the chromosomes in set *A* are homologous to those in set *B*. In amphidiploids formed from closely related species, some homology between *a* and *b* chromosomes is likely. Allopolyploids are rare in most animals because mating behavior is most often species-specific, and thus the initial step in hybridization is unlikely to occur.

A classical example of amphidiploidy in plants is the cultivated species of American cotton, *Gossypium* (Figure 7–9). This species has 26 pairs of chromosomes: 13 are large and 13 are much smaller. When it was discovered that Old World cotton had only 13 pairs of large chromosomes, allopolyploidy was suspected. After an examination of wild American cotton revealed 13 pairs of small chromosomes, this speculation was strengthened. J. O. Beasley reconstructed the origin of cultivated cotton experimentally by crossing the Old World strain with the wild American strain, then treating the hybrid with colchicine to double the chromosome number. The result of these treatments was a fertile amphidiploid variety of cotton. It contained 26 pairs of chromosomes and characteristics similar to the cultivated variety.

Amphidiploids often exhibit traits of both parental species. An interesting example, but one with no practical economic importance, is that of the hybrid formed between the radish *Raphanus sativus* and the cabbage *Brassica oleracea*. Both species have a haploid number of 9. The initial hybrid consists of 9 *Raphanus* and 9 *Brassica* chromosomes (9R + 9B). While hybrids are almost always sterile, some fertile amphidiploids (18R + 18B) have been produced. Unfortunately, the root of this plant is more like the cabbage and its shoot more like the radish. Had the converse occurred, the hybrid might have been of economic importance.

A much more successful commercial hybridization uses the grasses wheat and rye. Wheat (genus *Triticum*) has a basic haploid genome of 7 chromosomes. In addition to normal diploids ($2n = 14$), cultivated allopolyploids exist, including tetraploid ($4n = 28$) and hexaploid ($6n = 42$) species. Rye (genus *Secale*) also has a genome consisting of 7 chromosomes. The only cultivated species is the diploid plant ($2n = 14$).

Using the technique outlined in Figure 7–8, geneticists have produced various hybrids. When tetraploid wheat is crossed with diploid rye and the F1 plants treated with colchicine, a hexaploid variety ($6n = 42$) is derived. The hybrid, designated *Triticale* (see Figure 1–13), represents a new genus. Fertile hybrid varieties derived from various wheat and rye species can be crossed or backcrossed. These crosses have created many variations of the genus *Triticale*. The hybrid plants demonstrate characteristics of both wheat and rye. For example, certain hybrids combine the high protein content of wheat with the high content of the amino acid lysine of rye. The lysine content is low in wheat and thus is a limiting nutritional factor. Wheat is considered a high-yielding grain, whereas rye is noted for its versatility of growth in unfavorable environments. *Triticale* species, combining both traits, have the potential of significantly increasing grain production. Programs designed to improve crops through hybridization have long been underway in several underdeveloped countries.

7.6 Variation in Chromosome Structure and Arrangement: An Overview

The second general class of chromosome aberrations includes structural changes that delete, add, or rearrange substantial portions of one or more chromosomes. Included in this category are deletions and duplications of genes or part of a chromosome and rearrangements of genetic material in which a chromosome segment is either inverted, exchanged with a segment of a nonhomologous chromosome, or merely transferred to another chromosome. Before we discuss these aberrations, several general observations are pertinent.

In most instances, these changes are due to one or more breaks along the axis of a chromosome, followed by either the loss or rearrangement of some genetic material. Chromosomes can break spontaneously, but the rate of breakage may increase in cells exposed to chemicals or radiation. Although the ends of chromosomes, the telomeres, do not readily fuse with ends of "broken" chromosomes or with other telomeres, the ends produced at points of breakage are "sticky" and can rejoin other broken ends. If a breakage and rejoining event does not merely reestablish the original relationship, and if the alteration occurs in germ plasm, the gametes will contain the structural rearrangement, which is heritable.

If the aberration is found in one homolog but not the other, unusual but characteristic pairing configurations are formed during meiotic synapsis. These patterns are useful in identifying the type of change that has occurred. If no loss or gain of genetic material occurs, individuals bearing the aberration "heterozygously" are likely to be unaffected phenotypically. However, the unusual pairing arrangements often lead to gametes that are duplicated or deficient for some chromosomal regions. When this occurs, the offspring of "carriers" of certain aberrations often have an increased probability of demonstrating phenotypic manifestations.

7.7 Deletions

When a chromosome breaks in one or more places and a portion of it is lost, the missing piece is called a **deletion** (or a **deficiency**). The deletion can occur either near one end or from the interior of the chromosome. These are **terminal** or **intercalary deletions**, respectively [Figure 7–10(a) and (b)]. The portion of the chromosome retaining the centromere region is usually maintained when the cell divides, whereas the segment without the centromere is eventually lost in progeny cells following mitosis or meiosis. For synapsis to occur between a chromosome with a large intercalary deletion and a normal homolog, the unpaired region of the normal homolog must "buckle out" into a **deletion** or **compensation loop** [Figure 7–10(c)].

As noted in our discussion of the cri-du-chat syndrome, where only part of the short arm of chromosome 5 is lost, a deletion of a portion of a chromosome need not be very great before the effects become severe. If even more genetic in-

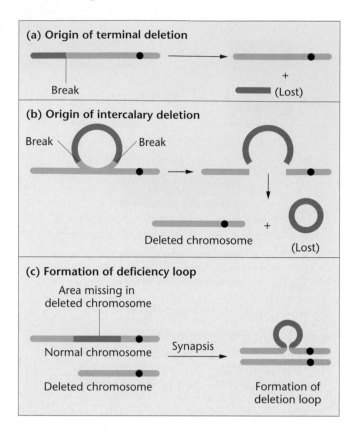

FIGURE 7–10 Origins of (a) a terminal and (b) an intercalary deletion. In (c), pairing occurs between a normal chromosome and one with an intercalary deletion by looping out the undeleted portion to form a deletion (or compensation) loop.

formation is lost as a result of a deletion, the aberration is often lethal. Such chromosome mutations never become available for study.

7.8 Duplications

When any part of the genetic material—a single locus or a large piece of a chromosome—is present more than once in the genome, it is called a duplication. As in deletions, pairing in heterozygotes can produce a compensation loop. Duplications arise as the result of unequal crossing over between synapsed chromosomes during meiosis (Figure 7–11) or through a replication error prior to meiosis. In the former case, both a duplication and a deletion are produced.

We consider three interesting aspects of duplications. First, they may result in gene redundancy. Second, as with deletions, duplications may produce phenotypic variation. Third, according to one convincing theory, duplications have also been an important source of genetic variability during evolution.

Gene Redundancy and Amplification— Ribosomal RNA Genes

Although many gene products are not needed in every cell of an organism, other gene products are known to be essential components of all cells. For example, ribosomal RNA must

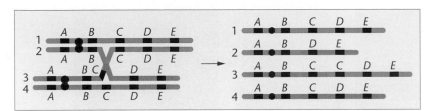

FIGURE 7–11 The origin of duplicated and deficient regions of chromosomes as a result of unequal crossing over. The tetrad on the left is mispaired during synapsis. A single crossover between chromatids 2 and 3 results in deficient and duplicated chromosomal regions (see chromosomes 2 and 3, respectively, on the right). The two chromosomes uninvolved in the crossover event remain normal in their gene sequence and content.

be present in abundance in order to support protein synthesis. The more metabolically active a cell is, the higher is the demand for this molecule. We might hypothesize that a single copy of the gene encoding rRNA is inadequate in many cells. Studies using the technique of molecular hybridization, which allows the determination of the percentage of the genome coding for specific RNA sequences, show that our hypothesis is correct! Indeed, multiple copies of genes code for rRNA. Such DNA is called **rDNA**, and the general phenomenon is called **gene redundancy**. For example, in the common intestinal bacterium *Escherichia coli* (*E. coli*), about 0.4 percent of the haploid genome consists of rDNA. This is equivalent to 5–10 copies of the gene. In *Drosophila melanogaster*, 0.3 percent of the haploid genome, equivalent to 130 copies, consists of rDNA. Although the presence of multiple copies of the same gene is not restricted to those coding for rRNA, we focus on them in this section.

In some cells, particularly oocytes, even the normal redundancy of rDNA is insufficient to provide adequate amounts of rRNA and ribosomes. Oocytes store abundant nutrients in the ooplasm for use by the embryo during early development. In addition, more ribosomes are included in the oocytes than in any other cell type. By considering how the amphibian *Xenopus laevis* acquires this abundance of ribosomes, we shall see a second way in which the amount of rRNA is increased. This phenomenon is called **gene amplification**.

The genes that code for rRNA are located in an area of the chromosome known as the **nucleolar organizer region (NOR)**. The NOR is intimately associated with the nucleolus, which is a processing center for ribosome production. Molecular hybridization analysis has shown that each NOR in the frog *Xenopus* contains the equivalent of 400 redundant gene copies coding for rRNA. Even this number of genes is apparently inadequate to synthesize the vast amount of ribosomes that must accumulate in the amphibian oocyte to support development following fertilization.

To further amplify the number of rRNA genes, the rDNA is selectively replicated, and each new set of genes is released from its template. Because each new copy is equivalent to an NOR, multiple small nucleoli are formed around each NOR in the oocyte. As many as 1500 of these "micronucleoli" have been observed in a single oocyte. If we multiply the number of micronucleoli (1500) by the number of gene copies in each NOR (400), we see that amplification in *Xenopus* oocytes can result in over half a million gene copies!

If each copy is transcribed only 20 times during the maturation of the oocyte, in theory, sufficient copies of rRNA are produced that well over 1 million ribosomes result.

The *Bar* Mutation in *Drosophila*

Duplications can cause phenotypic variation that might at first appear to be caused by a simple gene mutation. The *Bar* eye phenotype in *Drosophila* (Figure 7–12) is classic example. Instead of the normal oval eye shape, *Bar*-eyed flies have narrow, slitlike eyes. This phenotype is inherited in the same way as a dominant X-linked mutation.

In the early 1920s, Alfred H. Sturtevant and Thomas H. Morgan discovered and investigated this "mutation." Normal wild-type females (B^+/B^+) have about 800 facets in each eye. Heterozygous females (B/B^+) have about 350 facets, while homozygous females (B/B) average only about 70 facets. Females are occasionally recovered with even fewer facets and are designated as *double Bar* (B^D/B^+).

About ten years later, Calvin Bridges and Herman J. Muller compared the polytene X chromosome banding pattern of the *Bar* fly with that of the wild-type fly. Such chromosomes (see Chapter 17) contain specific banding patterns that have been well categorized into regions. As shown in Figure 7–12, their studies reveal that one copy of region 16A of the X chromosome is present in wild-type flies but that this region is duplicated in *Bar* flies and triplicated in *double Bar* flies. These observations provide evidence that the *Bar* phenotype is not the result of a simple chemical change in the gene, but is instead a duplication.

The Role of Gene Duplication in Evolution

During the study of evolution, it is intriguing to speculate on the possible mechanisms of genetic variation. The origin of unique gene products present in more recently evolved organisms but absent in ancestral forms is a topic of particular interest. In other words, how do "new" genes arise?

In 1970, Susumo Ohno published a provocative monograph, *Evolution by Gene Duplication*, in which he suggested that gene duplication is essential to the origin of new genes during evolution. Ohno's thesis is based on the supposition that the gene products of unique genes, present as only a single copy in the genome, are indispensable to the survival of members of any species during evolution. Therefore, unique genes are not free to accumulate mutations sufficient to alter their primary function and give rise to new genes.

Genotypes and Phenotypes

Genotype	Facet Number	Phenotype	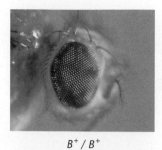 = 16A segments
B^+/B^+	779		
B/B^+	358		
B/B	68		
B^D/B^+	45		

B^+/B^+

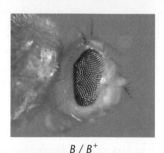

B/B^+

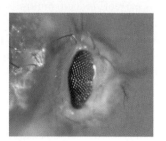

B/B

FIGURE 7–12 The duplication genotypes and resultant *Bar* eye phenotypes in *Drosophila*. Photographs show two *Bar* eye phenotypes and the wild type (B^+/B^+). *(Photos: Mary Lilly/Carnegie Institution of Washington)*

However, if an essential gene is duplicated in the germ line, major mutational changes in this extra copy will be tolerated in future generations because the original gene provides the genetic information for its essential function. The duplicated copy will be free to acquire many mutational changes over extended periods of time. Over short intervals, the new genetic information may be of no practical advantage. However, over long evolutionary periods, the duplicated gene may change sufficiently so that its product assumes a divergent role in the cell. The new function may impart an "adaptive" advantage to organisms, enhancing their fitness. Ohno has outlined a mechanism through which sustained genetic variability may have originated.

Ohno's thesis is supported by the discovery of genes that have a substantial amount of their DNA sequence in common, but whose gene products are distinct. For example, trypsin and chymotrypsin fit this description, as do myoglobin and hemoglobin. The DNA sequence is so similar (homologous) in each case that we may conclude that members of each pair of genes arose from a common an-

cestral gene through duplication. During evolution, the related genes diverged sufficiently that their products became unique.

Other support includes the presence of **gene families**, groups of contiguous genes whose products perform the same function. Again, members of a family show DNA sequence homology sufficient to conclude that they share a common origin. Examples are the various types of human hemoglobin polypeptide chains, as well as the immunologically important T-cell receptors and antigens encoded by the major histocompatibility complex.

7.9   Inversions

The **inversion**, another class of structural variation, is a type of chromosomal aberration in which a segment of a chromosome is turned around 180° within a chromosome. An inversion does not involve a loss of genetic information, but simply rearranges the linear gene sequence. An inversion requires breaks at two points along the length of the chromo-

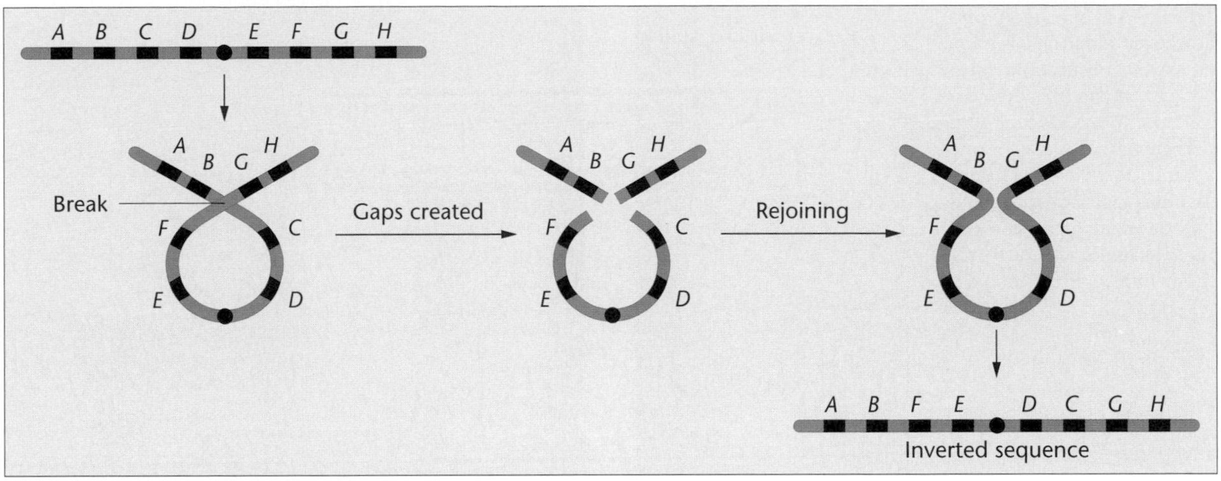

FIGURE 7–13   One possible origin of a pericentric inversion.

some and subsequent reinsertion of the inverted segment. Figure 7–13 illustrates how an inversion might arise. By forming a chromosomal loop prior to breakage, the newly created "sticky ends" are brought close together and rejoined.

The inverted segment may be short or quite long and may or may not include the centromere. If the centromere is not part of the rearranged chromosome segment, the inversion is said to be **paracentric**. If the centromere is part of the inverted segment, the term **pericentric** describes the inversion, which is the type shown in Figure 7–13.

Although inversions appear to have a minimal impact on individuals bearing them, their consequences are of great interest to geneticists. Organisms that are heterozygous for inversions may produce aberrant gametes that have a major impact on their offspring.

Consequences of Inversions During Gamete Formation

If only one member of a homologous pair of chromosomes has an inverted segment, normal linear synapsis during meiosis is not possible. Organisms with one inverted chromosome and one noninverted homolog are called **inversion heterozygotes**. Pairing between two such chromosomes in meiosis is accomplished only if they form an **inversion loop** (Figure 7–14).

If crossing over does not occur within the inverted segment of the inversion loop, the homologs will segregate, which results in two normal and two inverted chromatids that are distributed into gametes. However, if crossing over occurs within the inversion loop, abnormal chromatids are produced. The effect of a single exchange event within a paracentric inversion is diagrammed in Figure 7–14(a).

As in any meiotic tetrad, a single crossover between nonsister chromatids produces two parental chromatids and two recombinant chromatids. When the crossover occurs within a paracentric inversion, however, one recombinant chromatid is **dicentric** (two centromeres), and one recombinant chromatid is **acentric** (lacking a centromere). Both

contain duplications and deletions of chromosome segments as well. During anaphase, an acentric chromatid moves randomly to one pole or the other or may be lost, while a dicentric chromatid is pulled in two directions. This polarized movement produces **dicentric bridges** that are cytologically recognizable. A dicentric chromatid usually breaks at some point so that part of the chromatid goes into one gamete and part into another gamete during the reduction divisions. Therefore, gametes containing either recombinant chromatid are deficient in genetic material. When such a gamete participates in fertilization, the zygote most often develops abnormally, if at all.

A similar chromosomal imbalance is produced as a result of a crossover event between a chromatid bearing a pericentric inversion and its noninverted homolog, as shown in Figure 7–14(b). The recombinant chromatids that are directly involved in the exchange have duplications and deletions. In plants, gametes receiving such aberrant chromatids fail to develop normally, leading to aborted pollen or ovules. Thus, lethality occurs prior to fertilization, and inviable seeds result. In animals, the gametes have developed prior to the meiotic error, so fertilization is more likely to occur in spite of the chromosome error. However, the end result is the production of inviable embryos following fertilization. In both cases, viability is reduced.

Since viable offspring do not result in either plants or animals, it *appears* as if the inversion suppresses crossing over because offspring bearing crossover gametes are not recovered. Actually, in inversion heterozygotes, the inversion has the effect of *suppressing the recovery of crossover products* when chromosome exchange occurs within the inverted region. If crossing over always occurred within a paracentric or pericentric inversion, 50 percent of the gametes would be ineffective. The viability of the resulting zygotes is therefore greatly diminished. Furthermore, up to one-half of the viable gametes have the inverted chromosome, and the inversion will be perpetuated within the species. The cycle will be repeated continuously during meiosis in future generations.

FIGURE 7–14 The effects of a single crossover within an inversion loop in cases involving (a) a paracentric inversion, and (b) a pericentric inversion. In (a), two altered chromosomes are produced, one acentric and one dicentric. Both chromosomes also contain duplicated and deficient regions. In (b), two altered chromosomes are produced, both with duplicated and deficient regions.

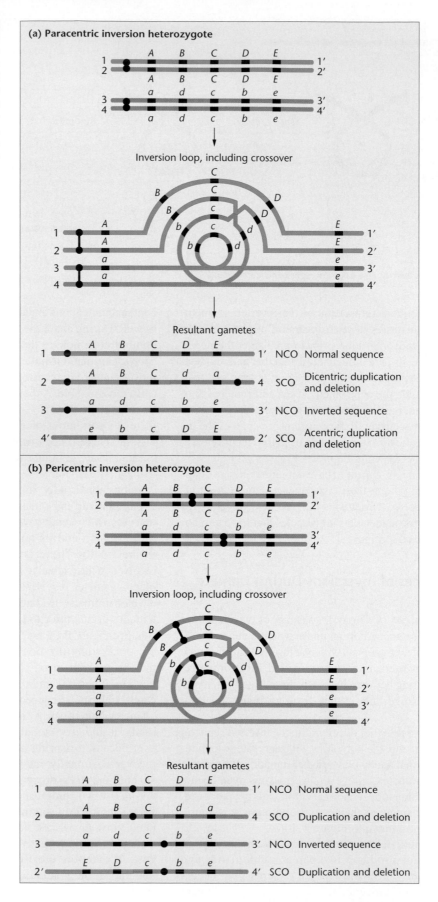

7.10   Translocations

Translocation, as the name implies, is the movement of a chromosomal segment to a new location in the genome. Reciprocal translocation, for example, involves the exchange of segments between two nonhomologous chromosomes. The least complex way for this event to occur is for two non-homologous chromosome arms to come close to each other so that an exchange is facilitated. Figure 7–15(a) shows a simple reciprocal translocation in which only two breaks are required. If the exchange includes internal chromosome segments, four breaks are required, two on each chromosome.

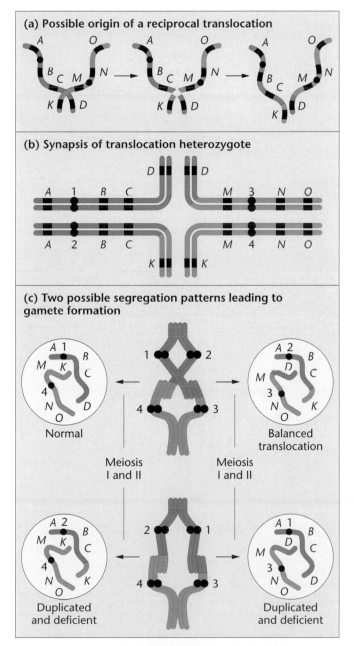

(a) Possible origin of a reciprocal translocation

(b) Synapsis of translocation heterozygote

(c) Two possible segregation patterns leading to gamete formation

Normal

Meiosis I and II

Balanced translocation

Meiosis I and II

Duplicated and deficient

Duplicated and deficient

FIGURE 7–15   (a) Possible origin of a reciprocal translocation. (b) Synaptic configuration formed during meiosis in an individual that is heterozygous for the translocation. (c) Two possible segregation patterns, one of which leads to a normal and a balanced gamete (called alternate segregation) and one that leads to gametes containing duplications and deficiencies (called adjacent segregation).

The genetic consequences of reciprocal translocations are, in several instances, similar to those of inversions. For example, genetic information is not lost or gained. Rather, there is only a rearrangement of genetic material. The presence of a translocation does not, therefore, directly alter the viability of individuals bearing it.

Homologs that are heterozygous for a reciprocal translocation undergo unorthodox synapsis during meiosis. As shown in Figure 7–15(b), pairing results in a crosslike configuration. As with inversions, genetically unbalanced gametes are also produced as a result of this unusual alignment during meiosis. In the case of translocations, however, aberrant gametes are not necessarily the result of crossing over. To see how unbalanced gametes are produced, focus on the homologous centromeres in Figure 7–15(b) and Figure 7–15(c). According to the principle of independent assortment, the chromosome containing centromere 1 migrates randomly toward one pole of the spindle during the first meiotic anaphase; it travels with *either* the chromosome having centromere 3 or the chromosome having centromere 4. The chromosome with centromere 2 moves to the other pole along with *either* the chromosome containing centromere 3 *or* centromere 4. This results in four potential meiotic products. The 1,4 combination contains chromosomes that are not involved in the translocation. The 2,3 combination, however, contains translocated chromosomes. These contain a complete complement of genetic information and are balanced. The other two potential products, the 1,3 and 2,4 combinations, contain chromosomes displaying duplicated and deleted segments.

When incorporated into gametes, the resultant meiotic products are genetically unbalanced. If they participate in fertilization, lethality often results. As few as 50 percent of the progeny of parents that are heterozygous for a reciprocal translocation survive. This condition, called **semisterility**, has an impact on the reproductive fitness of organisms, thus playing a role in evolution. Furthermore, in humans, such an unbalanced condition results in partial monosomy or trisomy, leading to a variety of birth defects.

Translocations in Humans: Familial Down Syndrome

Research conducted since 1959 has revealed numerous translocations in members of the human population. One common type of translocation involves breaks at the extreme ends of the short arms of two nonhomologous acrocentric chromosomes. These small segments are lost, and the larger segments fuse at their centromeric region. This type of translocation produces a new, large submetacentric or metacentric chromosome, often called a **Robertsonian translocation**.

One such translocation accounts for cases in which Down syndrome is inherited or familial. Earlier in this chapter we pointed out that most instances of Down syndrome are due to trisomy 21. This chromosome composition results from nondisjunction during meiosis in one parent. Trisomy accounts for over 95 percent of all cases of Down syndrome. In such instances, the chance of the same parents producing a second afflicted child is extremely low. However, in the remaining families with a Down child, the syndrome occurs in a much higher frequency over several generations.

Cytogenetic studies of the parents and their offspring from these unusual cases explain the cause of **familial Down syndrome**. Analysis reveals that one of the parents contains a 14/21 D/G translocation (Figure 7–16). That is, one parent has the majority of the G-group chromosome 21 translocated to one end of the D-group chromosome 14. This individual is phenotypically normal even though he or she has only 45 chromosomes. During meiosis, one-fourth of the individual's gametes have two copies of chromosome 21: a normal chromosome and a second copy translocated to chromosome 14. When such a gamete is fertilized by a standard haploid gamete, the resulting zygote has 46 chromosomes but three copies of chromosome 21. These individuals exhibit Down syndrome. Other potential surviving offspring contain either the standard diploid genome (with-

out a translocation) or the balanced translocation like the parent. Both cases result in normal individuals. Knowledge of translocations has allowed geneticists to resolve the seeming paradox of an inherited trisomic phenotype in an individual with an apparent diploid number of chromosomes.

7.11 Fragile Sites in Humans

We conclude this chapter with a brief discussion of the results of an intriguing discovery made around 1970 during observations of metaphase chromosomes prepared following human cell culture. In cells derived from certain individuals, a specific area along one of the chromosomes failed to stain, giving the appearance of a gap. In other individuals whose chromosomes displayed such morphology, the gaps appeared at other positions within the set of chromosomes. Such areas eventually became known as **fragile sites**, since they appeared

FIGURE 7–16 Chromosomal involvement in familial Down syndrome, as described in the text. The photograph shows the relevant chromosomes from a trisomy 21 offspring produced by a translocation carrier parent. *(Photo: Dr. Jorge Yunis)*

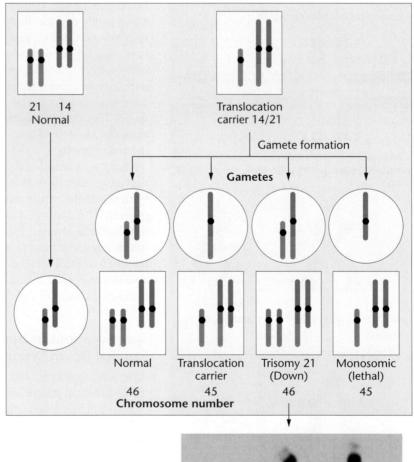

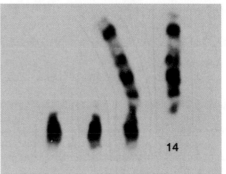

to be susceptible to chromosome breakage when cultured in the absence of certain chemicals such as folic acid, which is normally present in the culture medium. Fragile sites were at first considered curiosities, until a strong association was subsequently shown to exist between one of the sites and a form of mental retardation.

The cause of the fragility at these sites is unknown. Because they represent points along the chromosome that are susceptible to breakage, these sites may indicate regions where the chromatin is not tightly coiled. Note that even though almost all studies of fragile sites have been carried out *in vitro* using mitotically dividing cells, clear associations have been established between several of these sites and the corresponding altered phenotype, including mental retardation and cancer.

Fragile X Syndrome (Martin–Bell Syndrome)

Most fragile sites do not appear to be associated with any clinical syndrome. However, individuals bearing a folate-sensitive site on the X chromosome (Figure 7–17) exhibit the **fragile X syndrome** (or **Martin–Bell syndrome**), the most common form of inherited mental retardation. This syndrome affects about 1 in 1250 males and 1 in 2500 females. Because it is a dominant trait, females carrying only one fragile X chromosome can be mentally retarded. Fortunately, the trait is not fully expressed, as only about 30 percent of fragile X females are retarded, whereas about 80 percent of fragile X males are mentally retarded. In addition to mental retardation, affected males have characteristic long, narrow faces with protruding chins, enlarged ears, and increased testicular size.

A gene that spans the fragile site may be responsible for this syndrome. This gene, known as *FMR-1*, is one of a growing number of genes that have been discovered in which a sequence of three nucleotides is repeated many times, expanding the size of the gene. This phenomenon, called **trinucleotide repeats**, is also recognized in other human disorders, including Huntington disease. In *FMR-1*, the trinucleotide sequence CGG is repeated in an untranslated area adjacent to the coding sequence of the gene (called the "upstream" region). The number of repeats varies immensely within the human population, and a high number correlates directly with expression of fragile X syndrome. Normal individuals have between 6 and 54 repeats, whereas those with 55–200 repeats are considered "carriers" of the disorder. Above 200 repeats leads to expression of the syndrome.

It is thought that when the number of repeats reaches this level, the CGG regions of the gene become chemically modified so that the bases within and around the repeat are methylated, causing inactivation of the gene. The normal product of the gene is an RNA-binding protein known to be expressed in the brain. However, the relationship between the absence of this protein and fragile X syndrome is not yet clear.

From a genetic standpoint, perhaps the most interesting aspect of fragile X syndrome is the instability of the CGG repeats. An individual with 6–54 repeats transmits a gene containing the same number to his or her offspring. However, those with 55–200 repeats, while not at risk to develop the syndrome, may transmit to their offspring a gene with an increased number of repeats. The number of repeats continues to increase in future generations, demonstrating the phenomenon known as **genetic anticipation**, first introduced in Chapter 4. Once the threshold of 200 is exceeded, expression of the malady becomes more severe in each successive generation as the number of trinucleotide repeats increases.

While the mechanism that leads to the trinucleotide expansion has not yet been established, several factors are known that influence the instability. Most significant is the observation that expansion from the carrier status (55–200 repeats) to the syndrome status (over 200 repeats) occurs during the transmission of the gene by the maternal parent, but not by the paternal parent. Furthermore, several reports suggest that male offspring are more likely to receive the increased repeat size leading to the syndrome than are female offspring. Obviously, we have much to learn about the genetic basis of instability and expansion of DNA sequences.

Fragile Sites and Cancer

A second link between a fragile site and a human disorder was reported in 1996 by Carlo Croce, Kay Huebner, and their colleagues, who demonstrated an association between an autosomal fragile site and cancer. They showed that the gene *FHIT* (standing for *f*ragile *hi*stidine *t*riad), located within a well-defined fragile site on chromosome 3, is often altered in cells taken from tumors of individuals with lung cancer. A variety of mutations were found in cells derived from the tumors where the DNA had apparently been broken and incorrectly re-fused, resulting in deletions within the gene. In most cases, these mutations caused the *FHIT* gene to become inactivated.

This gene is part of the fragile region of the autosome designated *FRA3B*, which has been linked to other cancers, including the esophagus, colon, and stomach. The nature of the genetic alterations found in cancer cells suggests that the *FHIT* gene, because it is within a fragile region, may be highly susceptible to induced breaks in DNA. If these breaks are

FIGURE 7–17 A normal human X chromosome (left) contrasted with a fragile X chromosome (right). The "gap" region (near the bottom of the chromosome) is associated with the fragile X syndrome. (*Visuals Unlimited*)

incorrectly repaired, cancer-specific chromosome alterations may occur. Thus, this region of the chromosome appears to be particularly sensitive to carcinogen-induced damage, creating a susceptibility to cancer. It will be important to determine experimentally whether molecular polymorphism exists at this and other fragile sites within the human population, causing some individuals to be more susceptible to the effects of carcinogens than others.

Chapter Summary

1. Investigations into the uniqueness of each organism's chromosomal constitution is further enhancing our understanding of genetic variation. Alterations of the precise diploid content of chromosomes are called chromosomal aberrations or chromosomal mutations.

2. Deviations from the expected chromosomal number, or mutations in the structure of the chromosome, are inherited in predictable Mendelian fashion; they often result in inviable organisms or substantial changes in the phenotype.

3. Aneuploidy is the gain or loss of one or more chromosomes from the diploid content, resulting in conditions of monosomy, trisomy, tetrasomy, and so on. Studies of monosomic and trisomic disorders are increasing our understanding of the delicate genetic balance that enables normal development.

4. When complete sets of chromosomes are added to the diploid genome, polyploidy occurs. These sets can have identical or diverse genetic origin, creating either autopolyploidy or allopolyploidy, respectively.

5. Large segments of the chromosome can be modified by deletions or duplications. Deletions can produce serious conditions such as the cri-du-chat syndrome in humans, whereas duplications can be particularly important as a source of redundant or new genes.

6. Inversions and translocations, while altering the gene order along chromosomes, initially cause little or no loss of genetic information or deleterious effects. However, heterozygous combinations may cause genetically abnormal gametes following meiosis, with lethality often ensuing.

7. Fragile sites in human mitotic chromosomes have sparked research interest because one such site on the X chromosome is associated with the most common form of inherited mental retardation. Another fragile site, located on chromosome 3, has been linked to lung cancer.

Key Terms

acentric chromatid, 131
allopolyploidy, 125
allotetraploid, 126
amniocentesis, 124
amphidiploid, 126
aneuploidy, 121
autopolyploidy, 125
autotetraploid, 126
chorionic villus sampling (CVS), 124
chromosome aberration, 121
chromosome mutation, 121
colchicine, 126
compensation loop, 128
cri-du-chat (cry of the cat) syndrome, 122
deficiency, 128
deficiency loop, 128
deletion, 128
dicentric bridge, 131
dicentric chromatid, 131

Down syndrome, 123
Edwards syndrome (47,18+), 124
euploidy, 121
familial Down syndrome, 134
fragile site, 134
fragile X syndrome, 135
gene amplification, 129
gene family, 130
gene redundancy, 129
genetic anticipation, 135
genetic counseling, 124
Haplo-IV, 122
intercalary deletion, 128
inversion, 130
inversion heterozygote, 131
inversion loop, 131
Klinefelter syndrome, 121
Martin–Bell syndrome, 135
monosomy, 121

nondisjunction, 121
nucleolar organizer region (NOR), 129
paracentric inversion, 131
pericentric inversion, 131
Patau syndrome (47,13+), 124
polyploidy, 121
rDNA, 129
Robertsonian translocation, 133
segmental deletion, 122
semisterility, 133
terminal deletion, 128
translocation, 133
trinucleotide repeat, 135
trisomy 21, 123
Turner syndrome, 121
46,5p–, 122
47,21+, 123

Insights and Solutions

1. In a cross using maize involving three genes, *a*, *b*, and *c*, a heterozygote (*abc*/+++) is test-crossed to *abc*/*abc*. Even though the three genes are separated along the chromosome, thus predicting that crossover gametes and the resultant phenotype should be observed, only two phenotypes are recovered: *abc* and +++. Additionally, the cross produced significantly fewer viable plants than expected. Can you propose why no other phenotypes were recovered and why the viability was reduced?

Solution: One of the two chromosomes may contain an inversion that overlaps all three genes, effectively precluding the recovery of any "crossover" offspring. If this is a paracentric inversion and the genes are clearly separated (assuring that a significant number of crossovers occurs between them), then numerous acentric and dicentric chromosomes will form, resulting in the observed reduction in viability.

2. A male *Drosophila* from a wild-type stock is discovered to have only 7 chromosomes, whereas the normal 2*n* number is 8. Close examination reveals that one member of chromosome IV (the smallest chromosome) is attached to (translocated to) the distal end of chromosome II and is missing its centromere, thus accounting for the reduction in chromosome number.

(a) Diagram all members of chromosomes II and IV during synapsis in meiosis I.

Solution:

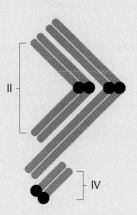

(b) If this male mates with a female with a normal chromosome composition who is homozygous for the recessive chromosome IV mutation *eyeless* (*ey*), what chromosome compositions will occur in the offspring regarding chromosomes II and IV?

Solution:

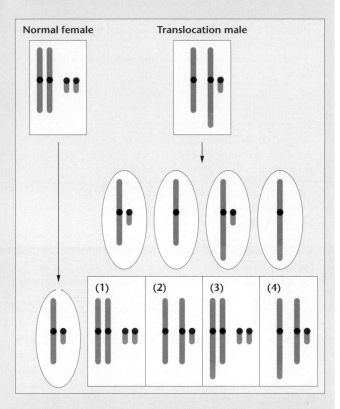

(c) Referring to the diagram in the solution to part (b), what phenotypic ratio will result regarding the presence of eyes, assuming all abnormal chromosome compositions survive?

Solution:

(1) normal (heterozygous)

(2) eyeless (monosomic, contains chromosome IV from mother)

(3) normal (heterozygous)

(4) normal (heterozygous)

The final ratio is 3/4 normal : 1/4 eyeless.

Problems and Discussion Questions

1. For a species with a diploid number of 18, indicate how many chromosomes will be present in the somatic nuclei of individuals that are haploid, triploid, tetraploid, trisomic, and monosomic.

2. Define the following pairs of terms, and distinguish between them:

   aneuploidy/euploidy
   monosomy/trisomy
   Patau syndrome/Edwards syndrome
   autopolyploidy/allopolyploidy
   autotetraploid/amphidiploid
   paracentric inversion/pericentric inversion

3. Contrast the relative survival times of individuals with Down, Patau, and Edwards syndromes. Speculate as to why such differences exist.

4. What evidence suggests that Down syndrome is more often the result of nondisjunction during oogenesis rather than during spermatogenesis?

5. What evidence indicates that humans with aneuploid karyotypes occur at conception, but are usually inviable?

6. Contrast the fertility of an allotetraploid with an autotriploid and an autotetraploid.

7. When two plants belonging to the same genus but different species are crossed, the F_1 hybrid is more viable and has more ornate flowers. Unfortunately, this hybrid is sterile and can only be propagated by vegetative cuttings. Explain the sterility of the hybrid. How might a horticulturist attempt to reverse its sterility?

8. Describe the origin of cultivated American cotton.

9. Predict how the synaptic configurations of homologous pairs of chromosomes might appear when one member is normal and the other member has sustained a deletion or duplication.

10. Inversions are said to "suppress crossing over." Is this terminology technically correct? If not, restate the description accurately.

11. Contrast the genetic composition of gametes derived from tetrads of inversion heterozygotes where crossing over occurs within a paracentric and pericentric inversion.

12. Discuss Ohno's hypothesis on the role of gene duplication in the process of evolution.

13. A human female with Turner syndrome also expresses the X-linked trait hemophilia, as did her father. Which of her parents underwent nondisjunction during meiosis, giving rise to the gamete responsible for the syndrome?

14. The primrose, *Primula kewensis*, has 36 chromosomes that are similar in appearance to the chromosomes in two related species, *Primula floribunda* ($2n = 18$) and *Primula verticillata* ($2n = 18$). How could *P. kewensis* arise from these species? How would you describe *P. kewensis* in genetic terms?

15. Certain varieties of chrysanthemums contain 18, 36, 54, 72, and 90 chromosomes; all are multiples of a basic set of 9 chromosomes. How would you describe these varieties genetically? What feature is shared by the karyotypes of each variety? A variety with 27 chromosomes has been discovered, but it is sterile. Why?

16. *Drosophila* may be monosomic for chromosome 4 yet remain fertile. Contrast the F_1 and F_2 results of the following crosses involving the recessive chromosome 4 trait, *bent* bristles:
   (a) monosomic IV, bent bristles × diploid, normal bristles
   (b) monosomic IV, normal bristles × diploid, bent bristles

17. Mendelian ratios are modified in crosses involving autotetraploids. Assume that one plant expresses the dominant trait green seeds and is homozygous (*WWWW*). This plant is crossed to one with white seeds that is also homozygous (*wwww*). If only one dominant allele is sufficient to produce green seeds, predict the F_1 and F_2 results of such a cross. Assume that synapsis between chromosome pairs is random during meiosis.

18. A couple, looking ahead to planning a family, are aware that through the past three generations on the husband's side a substantial number of stillbirths have occurred and several malformed babies were born who died early in childhood. The wife has studied genetics and urges her husband to visit a genetic counseling clinic, where a complete karyotype-banding analysis is performed. Although the tests show that he has a normal complement of 46 chromosomes, banding analysis reveals that one member of the chromosome 1 pair (in group A) contains an inversion covering 70 percent of its length. The homologue of chromosome 1 and all other chromosomes show the normal banding sequence.
   (a) How would you explain the high incidence of past stillbirths?
   (b) What can you predict about the probability of abnormality/normality of their future children?
   (c) Would you advise the woman that she will have to "wait out" each pregnancy to term to determine whether the fetus is normal? If not, what else can you suggest?

19. In a cross in *Drosophila*, a female heterozygous for the autosomally linked genes *a*, *b*, *c*, *d*, and *e* (*abcde*/+++++) is testcrossed to a male homozygous for all recessive alleles. Even though the distance between each of these loci is at least 3 map units, only four phenotypes are recovered, yielding the following data:

Phenotype	No. of Flies
+ + + + +	440
a b c d e	460
+ + + + e	48
a b c d +	52
	Total = 1000

Why are many expected crossover phenotypes missing? Can any of these loci be mapped from the data given here? If so, determine map distances.

20. A woman, seeking genetic counseling, is found to be heterozygous for a chromosomal rearrangement between the second and third chromosomes. Her chromosomes, compared to those in a normal karyotype, are diagrammed here:

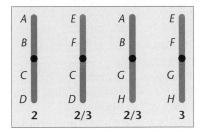

(a) What kind of chromosomal aberration is shown?

(b) Using a drawing, demonstrate how these chromosomes would pair during meiosis. Be sure to label the different segments of the chromosomes.

(c) This woman is phenotypically normal. Does this surprise you? Why or why not? Under what circumstances might you expect a phenotypic effect of such a rearrangement?

(d) This woman has had two miscarriages. She has come to you, an established genetic counselor, for advice. She raises the following questions.

- Is there a genetic explanation of her frequent miscarriages?
- Should she abandon her attempts to have a child of her own?
- If not, what is the chance that she could have a normal child?

Provide an informed response to her concerns.

Selected Readings

Antonarakis, S. E. 1998. Ten years of genomics, chromosome 21, and Down syndrome. *Genomics* 51:1–16.

Ashley-Koch, A. E., et al. 1997. Examination of factors associated with instability of the FMR1 CGG repeat. *Am. J. Hum. Genet.* 63:776–85.

Beasley, J. O. 1942. Meiotic chromosome behavior in species, species hybrids, haploids, and induced polyploids of *Gossypium*. *Genetics* 27:25–54.

Blakeslee, A. F. 1934. New jimson weeds from old chromosomes. *J. Hered.* 25:80–108.

Borgaonker, D. S. 1989. *Chromosome variation in man: A catalogue of chromosomal variants and anomalies*, 5th ed. New York: Alan R. Liss.

Boue, A. 1985. Cytogenetics of pregnancy wastage. *Adv. Hum. Genet.* 14:1–58.

Burgio, G. R., et. al., eds. 1981. *Trisomy 21*. New York: Springer-Verlag.

Carr, D. H. 1971. Genetic basis of abortion. *Annu. Rev. Genet.* 5:65–80.

Croce, C. M. 1996. The *FHIT* gene at 3p14.2 is abnormal in lung cancer. *Cell* 85:17–26.

Cummings, M. R. 2000. *Human heredity: Principles and issues.* 5th ed. Pacific Grove, CA: Brooks/Cole.

DeArce, M. A., and Kearns, A. 1984. The fragile X syndrome: The patients and their chromosomes. *J. Med. Genet.* 21:84–91.

Feldman, M., and Sears, E. R. 1981. The wild gene resources of wheat. *Sci. Am.* (Jan.) 244:102–12.

Gersh, M., et al. 1995. Evidence for a distinct region causing a cat-like cry in patients with 5p deletions. *Am. J. Hum. Genet.* 56:1404–10.

Gupta, P. K., and Priyadarshan, P. M. 1982. *Triticale:* Present status and future prospects. *Adv. Genet.* 21:256–346.

Hassold, T. J., et. al. 1980. Effect of maternal age on autosomal trisomies. *Ann. Hum. Genet. (London)* 44:29–36.

Hassold, T., and Jacobs, P. A. 1984. Trisomy in man. *Annu. Rev. Genet.* 18:69–98.

Hecht, F. 1988. Enigmatic fragile sites on human chromosomes. *Trends Genet.* 4:121–22.

Henikoff, S. 1994. A reconsideration of the mechanism of position effect. *Genetics.* 138:1–5.

Hulse, J. H., and Spurgeon, D. 1974. Triticale. *Sci. Am.* (Aug.) 231:72–81.

Jacobs, P. A., et al. 1974. A cytogenetic survey of 11,680 newborn infants. *Ann. Hum. Genet.* 37:359–76.

Kaiser, P. 1984. Pericentric inversions: Problems and significance for clinical genetics. *Hum. Genet.* 68:1–47.

Khush, G. S. 1973. *Cytogenetics of aneuploids.* Orlando, FL: Academic Press.

Khush, G. S., et al. 1984. Primary trisomics of rice: Origin, morphology, cytology and use in linkage mapping. *Genetics.* 107:141–63.

Lewis, E. B. 1950. The phenomenon of position effect. *Adv. Genet.* 3:73–115.

Lewis, W. H., ed. 1980. *Polyploidy: Biological relevance.* New York: Plenum Press.

Lynch, M., and Conery, J. S. 2000. The evolutionary fate and consequences of duplicate genes. *Science* 290:1151–54.

Lupski, J. R., Roth, J. R., and Weinstock, G. M. 1996. Chromosomal duplications in bacteria, fruit flies, and humans. *Am. J. Hum. Genet.* 58: 21–26.

Madan, K. 1995. Paracentric inversions: a review. *Hum. Genet.* 96:503–515.

Mantell, S. H., Mathews, J. A., and McKee, R. A. 1985. *Principles of plant biotechnology: An introduction to genetic engineering in plants.* Oxford: Blackwell.

Obe, G., and Basler, A. 1987. *Cytogenetics: Basic and applied aspects.* New York: Springer-Verlag.

Ohno, S. 1970. *Evolution by gene duplication.* New York: Springer-Verlag.

Oostra, B. A., and Verkerk, A. J. 1992. The fragile X syndrome: Isolation of the *FMR-1* gene and characterization of the fragile X mutation. *Chromosoma* 101:381–87.

Page, S. L., and Shaffer, L. G. 1997. Nonhomologous Robertsonian translocations form predominantly during female meiosis. *Nature Genetics* 15:231–32.

Patterson, D. 1987. The causes of Down syndrome. *Sci. Am.* (Aug.) 257:52–61.

Penny, G. D., et al. 1996. Requirement for *Xist* in X chromosome inactivation. *Nature* 379:131–37.

Schimke, R. T., ed. 1982. *Gene amplification*. Cold Spring Harbor, NY: Cold Spring Harbor Laboratory Press.

Shepard, J. F. 1982. The regeneration of potato plants from protoplasts. *Sci. Am.* (May) 246:154–166.

Shepard, J., et al. 1983. Genetic transfer in plants through interspecific protoplast fusion. *Science* 21:683–88.

Simmonds, N. W., ed. 1976. *Evolution of crop plants*. London: Longman.

Strickberger, M. W. 2000. *Evolution*, 3rd ed. Boston: Jones and Bartlett.

Sutherland, G. 1985. The enigma of the fragile X chromosome. *Trends Genet.* 1:108–11.

Taylor, A. I. 1968. Autosomal trisomy syndromes: A detailed study of 27 cases of Edwards syndrome and 27 cases of Patau syndrome. *J. Med. Genet.* 5:227–52.

Tjio, J. H., and Levan, A. 1956. The chromosome number of man. *Hereditas* 42:1–6.

Wilkins, L. E., Brown, J. X., and Wolf, B. 1980. Psychomotor development in 65 home-reared children with cri-du-chat syndrome. *J. Pediatr.* 97:401–5.

Yunis, J. J., ed. 1977. *New chromosomal syndromes*. Orlando, FL: Academic Press.

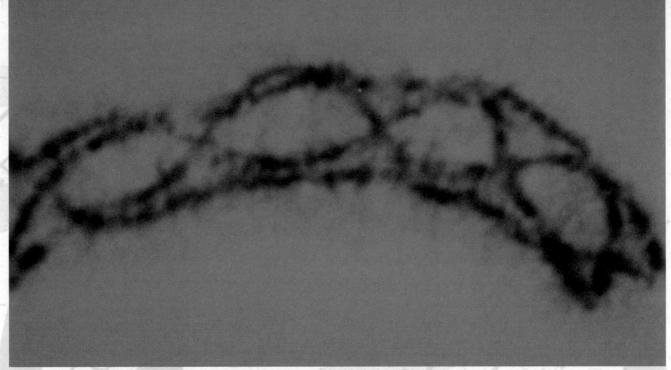

Chiasmata present between synapsed homologs during the first meiotic prophase. *(B. John/Cabisco/ Visuals Unlimited)*

8

Linkage and Chromosome Mapping in Eukaryotes

CHAPTER CONCEPTS

Many genes reside on each chromosome. Unless they are separated by crossing over, alleles at the loci on each homolog segregate as a unit during gamete formation. Recombinant gametes resulting from crossing over enhance genetic variability within a species and serve as the basis for constructing chromosome maps.

8.1 **Linkage Versus Independent Assortment**
The Linkage Ratio

8.2 **Incomplete Linkage, Crossing Over, and Chromosome Mapping**
Morgan and Crossing Over
Sturtevant and Mapping
Single Crossovers
Multiple Crossovers

8.3 **Mapping in *Drosophila* and Maize**
Three-Point Mapping in *Drosophila*
Determining the Gene Sequence
A Mapping Problem in Maize

8.4 **The Accuracy of Mapping Experiments**
Interference and the Coefficient of Coincidence

8.5 **The Genetic Map of *Drosophila***

8.6 **Somatic Cell Hybridization and Human Gene Mapping**

8.7 **Haploid Organisms in Linkage and Mapping Studies**
Gene-to-Centromere Mapping

8.8 **Other Aspects of Genetic Exchange**
Cytological Evidence for Crossing Over
Sister Chromatid Exchanges

8.9 **Did Mendel Encounter Linkage?**

Walter Sutton, along with Theodor Boveri, was instrumental in uniting the fields of cytology and genetics. As early as 1903, Sutton pointed out the likelihood that there must be many more "unit factors" than chromosomes in most organisms. Soon thereafter, genetic investigations revealed that certain genes segregate as if they were somehow joined or linked together. Further investigations showed that such genes are part of the same chromosome and are indeed transmitted as a single unit. We now know that most chromosomes contain a very large number of genes. Those that are part of the same chromosome are said to be *linked* and to demonstrate **linkage** in genetic crosses.

Because the chromosome, not the gene, is the unit of transmission during meiosis, linked genes are not free to undergo independent assortment. Instead, the alleles at all loci of one chromosome should, in theory, be transmitted as a unit during gamete formation. However, in many instances this does not occur. During the first meiotic prophase, when homologs are paired or synapsed, a reciprocal exchange of chromosome segments can take place. This event, called **crossing over**, results in the reshuffling, or recombination, of the alleles between homologs.

The degree of crossing over between any two loci on a single chromosome is proportional to the distance between them. Thus, depending on which loci are being studied, the percentage of recombinant gametes varies. This correlation allows us to construct **chromosome maps**, which give the relative locations of genes on chromosomes.

In this chapter we discuss linkage, crossing over, and chromosome mapping. We will conclude by entertaining the rather intriguing question of why Mendel, who studied seven genes in an organism with seven chromosomes, did not encounter linkage. Or did he?

8.1 Linkage Versus Independent Assortment

A simplified overview of the major theme of this chapter is given in Figure 8–1, which contrasts the meiotic consequences of (a) independent assortment, (b) linkage *without* crossing over, and (c) linkage *with* crossing over. In Figure 8–1(a) we see the results of independent assortment of two pairs of chromosomes, each containing one heterozygous gene pair. No linkage is exhibited. When a large number of meiotic events are observed, four genetically different gametes are formed in equal proportions. Each contains a different combination of alleles of the two genes.

Now let's compare these results with what occurs if the same genes are linked on the same chromosome. If no crossing over occurs between the two genes [Figure 8–1(b)], only two genetically different gametes are formed. Each gamete receives the alleles present on one homolog or the other, which transmits intact as the result of segregation. This case demonstrates **complete linkage**, which produces only **parental** or **noncrossover gametes**. The two parental gametes are formed in equal proportions. Though complete linkage between two genes seldom occurs, it is useful to consider the theoretical consequences of this concept.

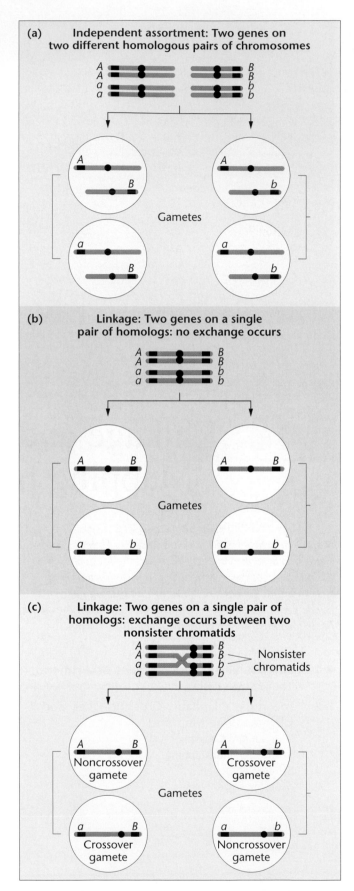

FIGURE 8–1 Results of gamete formation where two heterozygous genes are (a) on two different pairs of chromosomes; (b) on the same pair of homologs, but where no exchange occurs between them; and (c) on the same pair of homologs, where an exchange occurs between two nonsister chromatids.

Figure 8–1(c) shows the results of crossing over between two linked genes. As you can see, this crossover involves only two nonsister chromatids of the four chromatids present in the tetrad. This exchange generates two new allele combinations, called **recombinant** or **crossover gametes**. The two chromatids not involved in the exchange result in noncrossover gametes, like those in Figure 8–1(b). The frequency with which crossing over occurs between any two linked genes is generally proportional to the distance separating the respective loci along the chromosome. In theory, two randomly selected genes can be so close to each other that crossover events are too infrequent to easily detect. As shown in Figure 8–1(b), this circumstance, complete linkage, produces only parental gametes. On the other hand, if a small, but distinct distance separates two genes, few recombinant and many parental gametes will be formed. As the distance between the two genes increases, the proportion of recombinant gametes increases and that of the parental gametes decreases.

As we discuss later in this chapter, when the loci of two linked genes are far apart, the number of recombinant gametes approaches, but does not exceed, 50 percent. If 50 percent recombinants occurs, a 1:1:1:1 ratio of the four types (two parental and two recombinant gametes) results. In such a case, transmission of two linked genes is indistinguishable from that of two unlinked, independently assorting genes. That is, the proportion of the four possible genotypes is identical, as shown in Figure 8–1(a) and (c).

The Linkage Ratio

If complete linkage exists between two genes because of their close proximity, and organisms heterozygous at both loci are mated, an F_2 phenotypic ratio results that is unique, which we designate the **linkage ratio**. To illustrate this ratio, we consider a cross involving the closely linked recessive mutant genes *brown* (*bw*) eye and *heavy* (*hv*) wing vein in *Drosophila melanogaster* (Figure 8–2). The normal, wild-type alleles bw^+ and hv^+ are both dominant and result in red eyes and thin wing veins, respectively.

In this cross, flies with mutant brown eyes and normal thin veins are mated to flies with normal red eyes and mutant heavy veins. In more concise terms, brown-eyed flies are crossed with heavy-veined flies. If we extend the system of genetic symbols established in Chapter 4, linked genes are represented by placing their allele designations above and below a single or double horizontal line. Those above the line are located at loci on one homolog, and those below are located at the homologous loci on the other homolog. Thus, we represent the P_1 generation as follows:

$$P_1: \quad \frac{bw\ hv^+}{bw\ hv^+} \quad \times \quad \frac{bw^+\ hv}{bw^+\ hv}$$

brown, thin   red, heavy

Because the genes are located on an autosome, no distinction between males and females is necessary.

In the F_1 generation, each fly receives one chromosome of each pair from each parent; all flies are heterozygous for both gene pairs and exhibit the dominant traits of red eyes and thin veins:

$$F_1: \quad \frac{bw\ hv^+}{bw^+\ hv}$$

red, thin

As shown in Figure 8–2, when the F_1 generation is interbred, because of complete linkage, each F_1 individual forms only parental gametes. Following fertilization, the F_2 generation is produced in a 1:2:1 phenotypic and genotypic ratio. One-fourth of this generation shows brown eyes and thin veins; one-half shows both wild-type traits, namely, red eyes and thin veins; and one-fourth shows red eyes and heavy veins. In more concise terms, the ratio is 1 brown: 2 wild:1 heavy. Such a ratio is characteristic of complete linkage. Complete linkage is typically observed only when genes are very close together and the number of progeny is relatively small.

Figure 8–2 also gives the results of a test cross with the F_1 flies. Such a cross produces a 1:1 ratio of brown, thin and red, heavy flies. Had the genes controlling these traits been incompletely linked or located on separate autosomes, the test cross would have produced four phenotypes rather than two.

When large numbers of mutant genes present in any given species are investigated, genes located on the same chromosome will show evidence of linkage to one another. As a result, **linkage groups** can be established, one for each chromosome. In theory, the number of linkage groups should correspond to the haploid number of chromosomes. In diploid organisms in which large numbers of mutant genes are available for genetic study, this correlation has been confirmed.

8.2 Incomplete Linkage, Crossing Over, and Chromosome Mapping

It is highly improbable that two randomly selected genes linked on the same chromosome will be so close to one another along the chromosome that they demonstrate complete linkage. Instead, crosses involving two such genes almost always produce a percentage of offspring resulting from recombinant gametes. This percentage is variable and depends on the distance between the two genes along the chromosome. This phenomenon was first explained in 1911 by two *Drosophila* geneticists, Thomas H. Morgan and his undergraduate student, Alfred H. Sturtevant.

Morgan and Crossing Over

As you may recall from our discussion in Chapter 4, Morgan first discovered the phenomenon of X-linkage. In his studies, he investigated numerous *Drosophila* mutations located on the X chromosome. When he analyzed crosses involving only one trait, he deduced the mode of X-linked

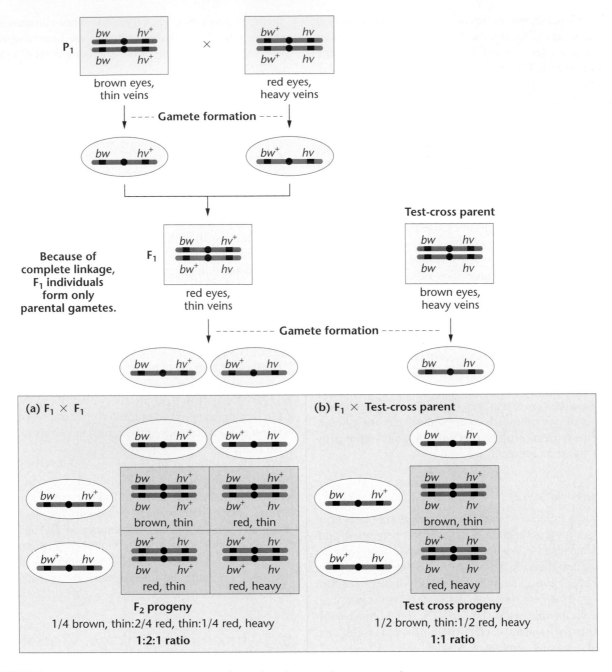

FIGURE 8–2 Results of a cross involving two genes located on the same chromosome where complete linkage is demonstrated. (a) The F₂ results of the cross. (b) The results of a test cross involving the F₁ progeny.

inheritance. However, when he crossed two X-linked genes, his results were at first puzzling. For example, as shown in cross A of Figure 8–3, he crossed female flies expressing mutant *yellow* body (*y*) and *white* eyes (*w*) with wild-type males (gray bodies and red eyes). The F₁ females were wild type, while the F₁ males expressed both mutant traits. In the F₂, 98.7 percent of the offspring showed the parental phenotypes—either yellow-bodied, white-eyed flies or wild-type flies (gray-bodied, red-eyed). The remaining 1.3 percent of the flies were either yellow-bodied with red eyes or gray-bodied with white eyes. It was as if the genes had somehow separated from each other during gamete formation in the F₁ flies.

When Morgan crossed other X-linked genes, the results were even more puzzling (cross B of Figure 8–3). The same basic pattern was observed, but the proportion of F₂ phenotypes differed; for example, in a cross involving white-eye, miniature-wing mutants, only 62.8 percent of the F₂ showed the parental phenotypes, while 37.2 percent of the offspring appeared as if the mutant genes had separated during gamete formation.

Morgan was faced with two questions: (1) What was the source of gene separation? and (2) Why did the frequency of the apparent separation vary depending on the genes being studied? The answer he proposed for the first question was based on his knowledge of earlier cytological observations

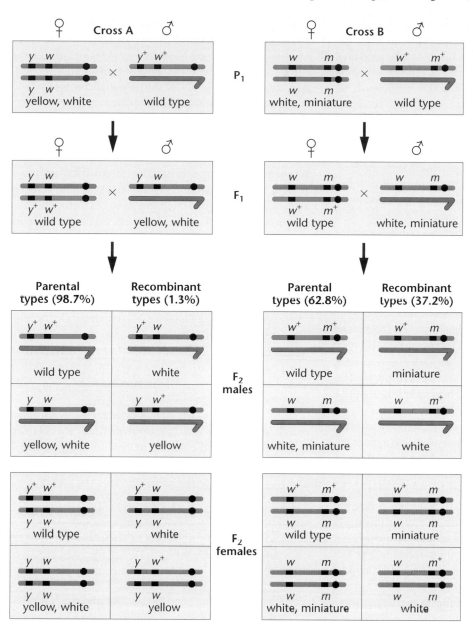

FIGURE 8–3 The F_1 and F_2 results of crosses involving the *yellow*-body, *white*-eye mutations and the *white*-eye, *miniature*-wing mutations. In cross A, 1.3 percent of the F_2 flies (males and females) demonstrate recombinant phenotypes, which express either *white* or *yellow*. In cross B, 37.2 percent of the F_2 flies (males and females) demonstrate recombinant phenotypes, which express either *miniature* or *white*.

made by F. A. Janssens and others. Janssens observed that synapsed homologous chromosomes in meiosis wrap around each other, creating **chiasmata** (sing., chiasma), where points of overlap are evident. Morgan proposed that these chiasmata could represent points of genetic exchange.

In the crosses shown in Figure 8–3, Morgan postulated that if an exchange occurs between the mutant genes on the two X chromosomes of the F_1 females, it leads to the observed results. He suggested that such exchanges led to 1.3 percent recombinant gametes in the *yellow-white* cross and 37.2 percent in the *white-miniature* cross. On the basis of this and other experiments, Morgan concluded that linked genes exist in a linear order along the chromosome and that a variable amount of exchange occurs between any two genes.

In his answer to the second question, Morgan proposed that two genes located relatively close to each other along a chromosome are less likely to have a chiasma form between them than if the two genes are farther apart on the chromosome.

Therefore, the closer two genes, the less likely a genetic exchange will occur between them. Morgan was the first to propose the term crossing over to describe the physical exchange leading to recombination.

Sturtevant and Mapping

Morgan's student, Alfred H. Sturtevant, was the first to realize that his mentor's proposal could be used to map the sequence of, as well as the distance between, linked genes. Sturtevant compiled further data on recombination between the genes represented by the *yellow*, *white*, and *miniature* mutants initially studied by Morgan. Frequencies of crossing over between each pair of these three genes were observed in separate crosses to be

(1)	*yellow, white*	0.5%
(2)	*white, miniature*	34.5%
(3)	*yellow, miniature*	35.4%

Because the sum of (1) and (2) approximately equals (3), Sturtevant suggested that the recombination frequencies between linked genes are additive. On this basis, he predicted that the order of the genes on the X chromosome is *yellow–white–miniature*. In arriving at this conclusion, he reasoned as follows: The *yellow* and *white* genes are apparently close to each other because the recombination frequency is low. However, both of these genes are quite far apart from *miniature* because the *white, miniature* and *yellow, miniature* combinations show large recombination frequencies. Because *miniature* shows more recombination with *yellow* than with *white* (35.4% vs. 34.5%), it follows that *white* is between the other two genes, not outside of them.

Sturtevant knew from Morgan's work that the frequency of exchange could be taken as an estimate of the distance between two genes or loci along the chromosome. He constructed a map of the three genes on the X chromosome, with 1 map unit equal to 1 percent recombination between two genes.* In the preceding example, the distance between *yellow* and *white* would thus be 0.5 map unit, and that between *yellow* and *miniature* would be 35.4 map units. It follows that the distance between *white* and *miniature* should be (35.4 − 0.5) or 34.9. This estimate is close to the actual frequency of recombination between *white* and *miniature* (34.5). The simple map for these three genes is shown in Figure 8–4.

In addition to these three genes, Sturtevant considered two other genes on the X chromosome and produced a more extensive map including all five genes. He and a colleague, Calvin Bridges, soon began a search for autosomal linkage in *Drosophila*. By 1923, they had clearly shown that linkage and crossing over are not restricted to X-linked genes, but can also be demonstrated with autosomes. During this work, they made another interesting observation. In *Drosophila*, crossing over was shown to occur only in females. The fact that no crossing over occurs in males made genetic mapping much less complex to analyze in *Drosophila*. However, crossing over does occur in both sexes in most other organisms.

Although many refinements in chromosome mapping have been developed since Sturtevant's initial work, his basic principles are accepted as correct. They are used to produce detailed chromosome maps of organisms for which large numbers of linked mutant genes are known. In addition to providing the basis for chromosome mapping, Sturtevant's

---

*In honor of Morgan's work, 1 map unit is now referred to as a centimorgan (cM).

findings are historically significant to the field of genetics. In 1910, the **chromosomal theory of inheritance** was still widely disputed. Even Morgan was skeptical of this theory before he conducted the bulk of his experimentation. Research has now firmly established that chromosomes contain genes in a linear order and that these genes are the equivalent of Mendel's unit factors.

Single Crossovers

Why should the relative distance between two loci influence the amount of recombination and crossing over observed between them? During meiosis, a limited number of crossover events occurs in each tetrad. These recombinant events occur randomly along the length of the tetrad. Therefore, the closer two loci reside along the axis of the chromosome, the less likely any single crossover event will occur between them. The same reasoning suggests that the farther apart two linked loci, the more likely a random crossover event will occur between them.

In Figure 8–5(a), a single crossover occurs between two nonsister chromatids, but not between the two loci; therefore, the crossover goes undetected because no recombinant gametes are produced. In Figure 8–5(b), where two loci are quite far apart, the crossover occurs between them, yielding recombinant gametes.

When a single crossover occurs between two nonsister chromatids, the other two chromatids of the tetrad are not involved in this exchange and enter the gamete unchanged. Even if a

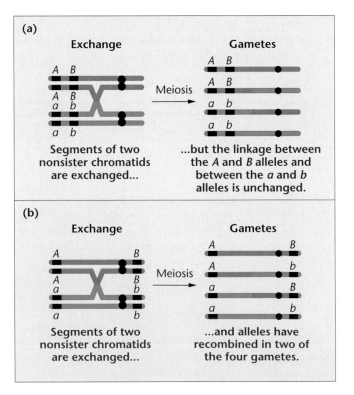

FIGURE 8–5 Two examples of a single crossover between two nonsister chromatids and the gametes subsequently produced. In (a), the exchange does not alter the linkage arrangement between the alleles of the two genes, only parental gametes form, and the exchange goes undetected. In (b), the exchange separates the alleles and results in recombinant gametes, which are detectable.

FIGURE 8–4 A map of the *yellow* (*y*), *white* (*w*), and *miniature* (*m*) genes on the X chromosome of *Drosophila melanogaster*. Each number represents the percentage of recombinant offspring produced in one of three crosses, each involving two different genes.

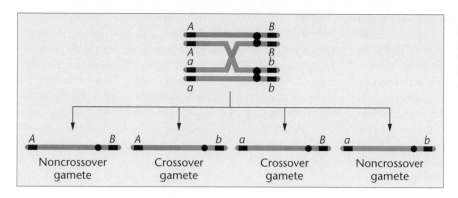

FIGURE 8–6 The consequences of a single exchange between two nonsister chromatids occurring in the tetrad stage. Two noncrossover (parental) and two crossover (recombinant) gametes are produced.

single crossover occurs 100 percent of the time between two linked genes, recombination is subsequently observed in only 50 percent of the potential gametes formed. This concept is diagrammed in Figure 8–6. Theoretically, if we consider only single exchanges and observe 20 percent recombinant gametes, crossing over actually occurs between these two loci in 40 percent of the tetrads. The general rule is that under these conditions, the percentage of tetrads involved in an exchange between two genes is twice as great as the percentage of recombinant gametes produced. Therefore, the theoretical limit of recombination due to crossing over is 50 percent.

When two linked genes are more than 50 map units apart, a crossover can theoretically be expected to occur between them in 100 percent of the tetrads. If this prediction were achieved, each tetrad would yield equal proportions of the four gametes shown in Figure 8–6, just as if the genes were on different chromosomes and assorting independently. For a variety of reasons, this theoretical limit is seldom achieved.

Multiple Crossovers

It is possible that in a single tetrad, two, three, or more exchanges will occur between nonsister chromatids as a result of several crossover events. Double exchanges of genetic material result from **double crossovers (DCO)**, as shown in Figure 8–7. For

a double exchange to be studied, three gene pairs must be investigated, each heterozygous for two alleles. Before we determine the frequency of recombination among all three loci, let's review some simple probability calculations.

As we have seen, the probability of a single exchange occurring between the A and B or the B and C genes relates directly to the distance between the respective loci. The closer A is to B and B is to C, the less likely a single exchange will occur between either of the two sets of loci. In the case of a double crossover, two separate and independent events or exchanges must occur simultaneously. The mathematical probability of two independent events occurring simultaneously is equal to the product of the individual probabilities. This is the product law we introduced in Chapter 3.

Suppose that crossover gametes resulting from single exchanges are recovered 20 percent of the time ($p = 0.20$) between A and B and 30 percent of the time ($p = 0.30$) between B and C. The probability of recovering a double-crossover gamete arising from two exchanges, between A and B and between B and C, is predicted to be $(0.20)(0.30) = 0.06$, or 6 percent. It is apparent from this calculation that the frequency of double-crossover gametes is always expected to be much lower than that of either single-crossover class of gametes.

If three genes are relatively close together along one chromosome, the expected frequency of double-crossover gametes

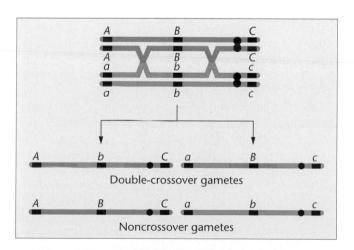

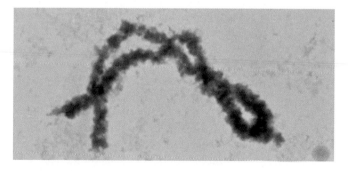

FIGURE 8–7 Consequences of a double exchange between two nonsister chromatids. Because the exchanges involve only two chromatids, two noncrossover gametes and two double-crossover gametes are produced. The photograph shows several chiasmata found in a tetrad isolated during the first meiotic prophase. *(Photo: Cabisco/Visuals Unlimited)*

is extremely low. For example, suppose the *A–B* distance in Figure 8–7 is 3 map units and the *B–C* distance is 2 map units. The expected double-crossover frequency is $(0.03)(0.02) = 0.0006$, or 0.06 percent. This translates to only 6 events in 10,000. Thus, in a mapping experiment such as this, where closely linked genes are involved, very large numbers of offspring are required in order to detect double-crossover events. In this example, it is unlikely that a double crossover will be observed even if 1000 offspring are examined. Thus, it is evident that if four or five genes are being mapped, even fewer triple and quadruple crossovers can be expected to occur.

8.3   Mapping in *Drosophila* and Maize

The information presented in the preceding section enables us to map three or more linked genes in a single cross. To illustrate the mapping process in its entirety, we examine two situations involving three linked genes in two quite different organisms.

Three-Point Mapping in *Drosophila*

In order to execute a successful mapping cross, three criteria must be met.

1. The genotype of the organism producing the crossover gametes must be heterozygous at all loci under consideration.

2. The cross must be constructed so that genotypes of all gametes can be determined accurately by observing the phenotypes of the resulting offspring. This is necessary because the gametes and their genotypes can never be observed directly. To overcome this problem, each phenotypic class must reflect the genotype of the gametes of the parents producing it.

3. A sufficient number of offspring must be produced in the mapping experiment to recover a representative sample of all crossover classes.

These criteria are met in the three-point mapping cross from *Drosophila melanogaster* shown in Figure 8–8. In this cross, three sex-linked recessive mutant genes—*yellow* (body color), *white* (eye color), and *echinus* (eye shape)—are considered. To diagram the cross, we must assume some theoretical sequence, even though we don't yet know if it is correct. In Figure 8–8, we initially assume the sequence of the three genes to be *y–w–ec*. If this is incorrect, our analysis will demonstrate this and reveal the correct sequence.

In the P_1 generation, males hemizygous for all three wild-type alleles are crossed to females that are homozygous for all three recessive mutant alleles. Therefore, the P_1 males are wild type with respect to body color, eye color, and eye shape. They are said to have a *wild-type phenotype*. The females, on the other hand, exhibit the three mutant traits—yellow body color, white eyes, and echinus eye shape.

This cross produces an F_1 generation consisting of females that are heterozygous at all three loci and males that, because

of the Y chromosome, are hemizygous for the three mutant alleles. Phenotypically, all F_1 females are wild type, while all F_1 males are *yellow*, *white*, and *echinus*. The genotype of the F_1 females fulfills the first criterion for mapping the three linked genes; that is, it is heterozygous at the three loci and can serve as the source of recombinant gametes generated by crossing over. Note that because of the genotypes of the P_1 parents, all three mutant alleles are on one homolog, and all three wild-type alleles are on the other homolog. *Other arrangements are possible.* For example, the heterozygous F_1 female might have the *y* and *ec* mutant alleles on one homolog and the *w* allele on the other. This would occur if, in the P_1 cross, one parent was *yellow* and *echinus* and the other parent was *white*.

In our cross, the second criterion is met by virtue of the gametes formed by the F_1 males. Every gamete contains either an X chromosome bearing the three mutant alleles or a Y chromosome, which is genetically inert for the three loci being considered. Whichever type participates in fertilization, the genotype of the gamete produced by the F_1 female will be expressed phenotypically in the F_2 male and female offspring derived from it. Thus, all F_1 noncrossover and crossover gametes can be detected by observing the F_2 phenotypes.

With these two criteria met, we can construct a chromosome map from the crosses shown in Figure 8–8. First, we determine which F_2 phenotypes correspond to the various noncrossover and crossover categories. Two of these can be determined immediately.

To determine the noncrossover F_2 phenotypes, we combine alleles present in the parental gametes formed by the F_1 female. Each such gamete contains an X chromosome *unaffected by crossing over*. As a result of segregation, approximately equal proportions of the two types of gametes, and subsequently the F_2 phenotypes, are produced. Because they derive from a heterozygote, the genotypes of the two parental gametes and the phenotypes of the two F_2 phenotypes complement one another. For example, if one is wild type, the other is completely mutant. This is the case in the cross under consideration. In other situations, if one chromosome shows one mutant allele or trait, the second chromosome shows the other two mutant traits, and so on. They are therefore called **reciprocal classes** of gametes and phenotypes.

The two noncrossover phenotypes are most easily recognized because *they exist in the greatest proportion*. Figure 8–8 shows that gametes (1) and (2) are present in the greatest numbers. Therefore, flies that express *yellow*, *white*, and *echinus* phenotypes and those that are normal (or wild type) for all three characters constitute the noncrossover category and represent 94.44 percent of the F_2 offspring.

The second category that can be easily detected is represented by the double-crossover phenotypes. Because of their low probability of occurrence, *they must be present in the least numbers*. Remember that this group represents two independent but simultaneous single-crossover events. Two reciprocal phenotypes can be identified: gamete (7), which shows the mutant traits *yellow* and *echinus* but normal eye color; and gamete (8), which shows the mutant trait *white* but normal body color and eye shape. Together these double-crossover phenotypes constitute only 0.06 percent of the F_2 offspring.

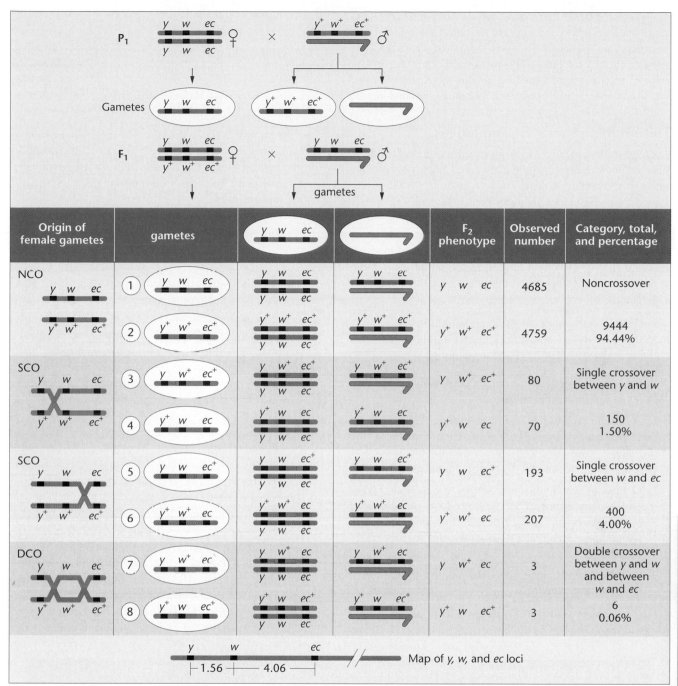

FIGURE 8–8 A three-point mapping cross involving the *yellow* (*y* or *y*⁺), *white* (*w* or *w*⁺), and *echinus* (*ec* or *ec*⁺) genes in *Drosophila melanogaster*. NCO, SCO, and DCO refer to noncrossover, single-crossover, and double-crossover groups, respectively. Because of the complexity of this and several of the ensuing figures, centromeres are not included on the chromosomes, and only two nonsister chromatids are shown initially. Crossing over always occurs in the tetrad stage.

The remaining four phenotypic classes represent two categories resulting from single crossovers. Gametes (3) and (4), reciprocal phenotypes produced by single-crossover events occurring between the *yellow* and *white* loci, are equal to 1.50 percent of the F₂ offspring. Gametes (5) and (6), constituting 4.00 percent of the F₂ offspring, represent the reciprocal phenotypes resulting from single-crossover events occurring between the *white* and *echinus* loci.

The map distances separating the three loci can now be calculated. The distance between *y* and *w*, or between *w* and *ec*, is equal to the percentage of all detectable exchanges occurring between them. For any two genes under consideration, this includes all appropriate single crossovers as well as all double crossovers. The latter are included because they represent two simultaneous single crossovers. For the *y* and *w* genes this includes gametes (3), (4), (7), and (8), totaling

1.50 percent + 0.06 percent, or 1.56 map units. Similarly, the distance between w and ec is equal to the percentage of offspring resulting from an exchange between these two loci: gametes (5), (6), (7), and (8), totaling 4.00 percent + 0.06 percent, or 4.06 map units. The map of these three loci on the X chromosome, based on these data, is shown at the bottom of Figure 8–8.

Determining the Gene Sequence

In the preceding example, the sequence (or order) of the three genes along the chromosome is assumed to be y–w–ec. Our analysis shows this sequence to be consistent with the data. However, in most mapping experiments the gene sequence is not known, and this constitutes another variable in the analysis. In our example, had the gene sequence been unknown, it could have been determined using a straightforward method.

This method is based on the fact that there are only three possible orders, each containing one of the three genes in between the other two:

(I) w–y–ec (y is in the middle)
(II) y–ec–w (ec is in the middle)
(III) y–w–ec (w is in the middle)

If you use the following steps during your analysis, you will be able to determine the gene order.

1. Assuming any one of the three orders, first determine the *arrangement of alleles* along each homolog of the heterozygous parent giving rise to noncrossover and crossover gametes (the F_1 female in our example).

2. Determine whether a double-crossover event occurring within that arrangement will produce the *observed double-crossover phenotypes*. Remember that these phenotypes occur least frequently and can be easily identified.

3. If this order does not produce the predicted phenotypes, try each of the other two orders. One does work!

These steps are shown in Figure 8–9, using the cross discussed above. The three possible orders are labeled I, II, and III, as shown previously. Either y, ec, or w must be in the middle.

1. Assuming that y is between w and ec, the arrangement of alleles along the homologs of the F_1 heterozygote is

$$\text{(I)} \quad \frac{w \quad y \quad ec}{w^+ \quad y^+ \quad ec^+}$$

We know this because of the way in which the P_1 generation was crossed. The P_1 female contributes an X chromosome bearing the w, y, and ec alleles, while the P_1 male contributes an X chromosome bearing the w^+, y^+, and ec^+ alleles.

2. A double crossover within the above arrangement yields the following gametes:

$$\underline{w \quad y^+ \quad ec} \text{ and } \underline{w^+ \quad y \quad ec^+}$$

Following fertilization, if y is in the middle, the F_2 double-crossover phenotypes will correspond to these gametic genotypes, yielding offspring that express the white, echinus phenotype and offspring that express the yellow phenotype. Instead, however, determination of the actual double-crossover phenotypes reveals them to be *yellow, echinus* flies and *white* flies. Therefore, our assumed order is incorrect.

3. If we consider the other orders, one with the ec/ec^+ alleles in the middle (II) or one with the w/w^+ alleles in the middle (III),

$$\text{(II)} \quad \frac{y \quad ec \quad w}{y^+ \quad ec^+ \quad w^+} \quad \text{or} \quad \text{(III)} \quad \frac{y \quad w \quad ec}{y^+ \quad w^+ \quad ec^+}$$

we see that arrangement II again provides *predicted* double-crossover phenotypes that *do not* correspond to the *actual* (observed) double-crossover phenotypes. The predicted phenotypes are *yellow, white* flies and *echinus* flies in the F_2 generation. Therefore, this order is also incorrect. However, arrangement III produces the observed phenotypes—*yellow, echinus* flies and *white* flies. Therefore, this order, with the w gene in the middle, is correct.

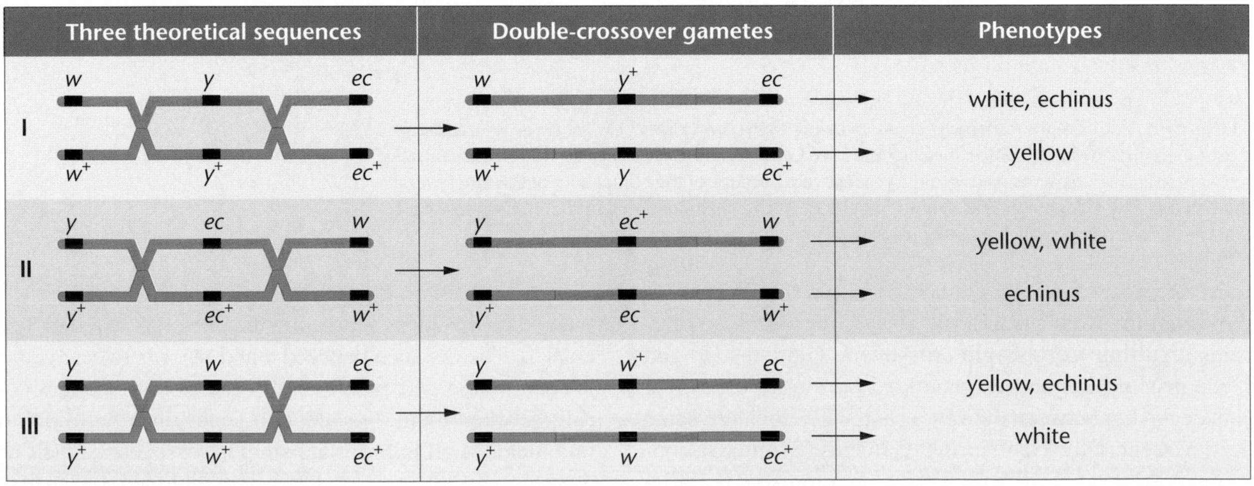

Three theoretical sequences	Double-crossover gametes	Phenotypes
I — w ... y ... ec / w^+ ... y^+ ... ec^+	w ... y^+ ... ec / w^+ ... y ... ec^+	white, echinus / yellow
II — y ... ec ... w / y^+ ... ec^+ ... w^+	y ... ec^+ ... w / y^+ ... ec ... w^+	yellow, white / echinus
III — y ... w ... ec / y^+ ... w^+ ... ec^+	y ... w^+ ... ec / y^+ ... w ... ec^+	yellow, echinus / white

FIGURE 8–9 The three possible sequences of the *white, yellow*, and *echinus* genes, the results of a double crossover in each case, and the resulting phenotypes produced in a test cross. For simplicity, the two noncrossover chromatids of each tetrad are omitted.

To summarize, this method is rather straightforward: First determine the arrangement of alleles on the homologs of the heterozygote yielding the crossover gametes. This is done by locating the reciprocal noncrossover phenotypes. Then, test each of three possible orders to determine which one yields the observed double-crossover phenotypes. The one that does so represents the correct order.

A Mapping Problem in Maize

Having established the basic principles of chromosome mapping, we now consider a related problem in maize (corn), in which the gene sequence and interlocus distances are unknown.

This analysis differs from the preceding example in several ways. First, the previous mapping cross involved X-linked genes. Here, we consider autosomal genes. Second, in the discussion of this cross, we change our use of symbols, as first suggested in Chapter 4. Instead of using the gene symbols and superscripts (e.g., bm^+, v^+, and pr^+), we simply use + to denote each wild-type allele. This system is easier to manipulate, but requires a better understanding of mapping procedures.

When we consider three autosomally linked genes in maize, the experimental cross must still meet the same three crite-ria established for the X-linked genes in *Drosophila*: (1) One parent must be heterozygous for all traits under consideration; (2) the gametic genotypes produced by the heterozygote must be apparent from observing the phenotypes of the offspring; and (3) a sufficient sample size must be available for complete analysis.

In maize, the recessive mutant genes *bm* (*brown* midrib), *v* (*virescent* seedling), and *pr* (*purple* aleurone) are linked on chromosome 5. Assume that a female plant is known to be heterozygous for all three traits. We do not know (1) the arrangement of the mutant alleles on the maternal and paternal homologs of this heterozygote, (2) the sequence of genes, or (3) the map distances between the genes. What genotype must the male plant have to allow successful mapping? To meet the second criterion, the male must be homozygous for all three recessive mutant alleles. Otherwise, offspring of this cross showing a given phenotype might represent more than one genotype, making accurate mapping impossible.

Figure 8–10 diagrams this cross. As shown, we know neither the arrangement of alleles nor the sequence of loci in the heterozygous female. Several possibilities are shown, but we have yet to determine which one is correct. In the test-cross male parent, *we don't know the sequence*, and so we

(a) Possible allele arrangements and gene sequences in a heterozygous female

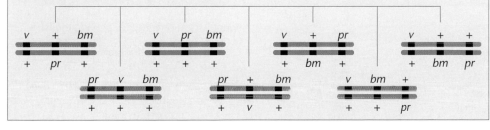

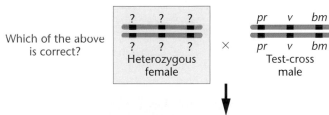

Which of the above is correct?

Heterozygous female × Test-cross male

FIGURE 8–10 (a) Some possible allele arrangements and gene sequences in a heterozygous female. The data from a three-point mapping cross, depicted in (b), where the female is test-crossed, determine which combination of arrangement and sequence is correct [see Figure 8–11(d)].

(b) Actual results of mapping cross

Phenotypes of offspring			Number	Total and percentage	Exchange classification
+	v	bm	230	467 42.1%	Noncrossover (NCO)
pr	+	+	237		
+	+	bm	82	161 14.5%	Single crossover (SCO)
pr	v	+	79		
+	v	+	200	395 35.6%	Single crossover (SCO)
pr	+	bm	195		
pr	v	bm	44	86 7.8%	Double crossover (DCO)
+	+	+	42		

must designate it randomly. Note that we initially place *v* in the middle. This may or may not be correct.

The offspring are arranged in groups of two for each pair of reciprocal phenotypic classes. The two members of each reciprocal class derive from either no crossing over (NCO), one of two possible single-crossover events (SCO), or a double crossover (DCO).

To solve this problem, it is helpful to refer to Figures 8–10 and 8–11 as you consider the following questions.

1. *What is the correct heterozygous arrangement of alleles in the female parent?*

    Determine the two noncrossover classes, those that occur with the highest frequency. In this case, they are + *v bm* and *pr* + +. Therefore, the alleles on the homologs of the female parent must be arranged as shown in Figure 8–11(a). These homologs segregate into gametes, unaffected by any recombination event. Any other arrangement of alleles will not yield the observed noncrossover classes. (Remember that + *v bm* is equivalent to *pr*$^+$ *v bm*, and that *pr* + + is equivalent to *pr v*$^+$ *bm*$^+$.)

2. *What is the correct sequence of genes?*

    We know that the arrangement of alleles is

    $$\frac{+ \quad v \quad bm}{pr \quad + \quad +}$$

    But is the assumed sequence correct? That is, will a double-crossover event yield the observed double-crossover phenotypes following fertilization? Simple observation shows that it will not [Figure 8–11(b)]. Now try the other two orders [Figure 8–11(c) and (d)], *keeping the same arrangement:*

    $$\frac{+ \quad bm \quad v}{pr \quad + \quad +} \quad \text{or} \quad \frac{v \quad + \quad bm}{+ \quad pr \quad +}$$

    Only the case on the right yields the observed double-crossover gametes [Figure 8–11(d)]. Therefore, the *pr* gene is in the middle. From this point on, work the problem using the above arrangement and sequence, with the *pr* locus in the middle.

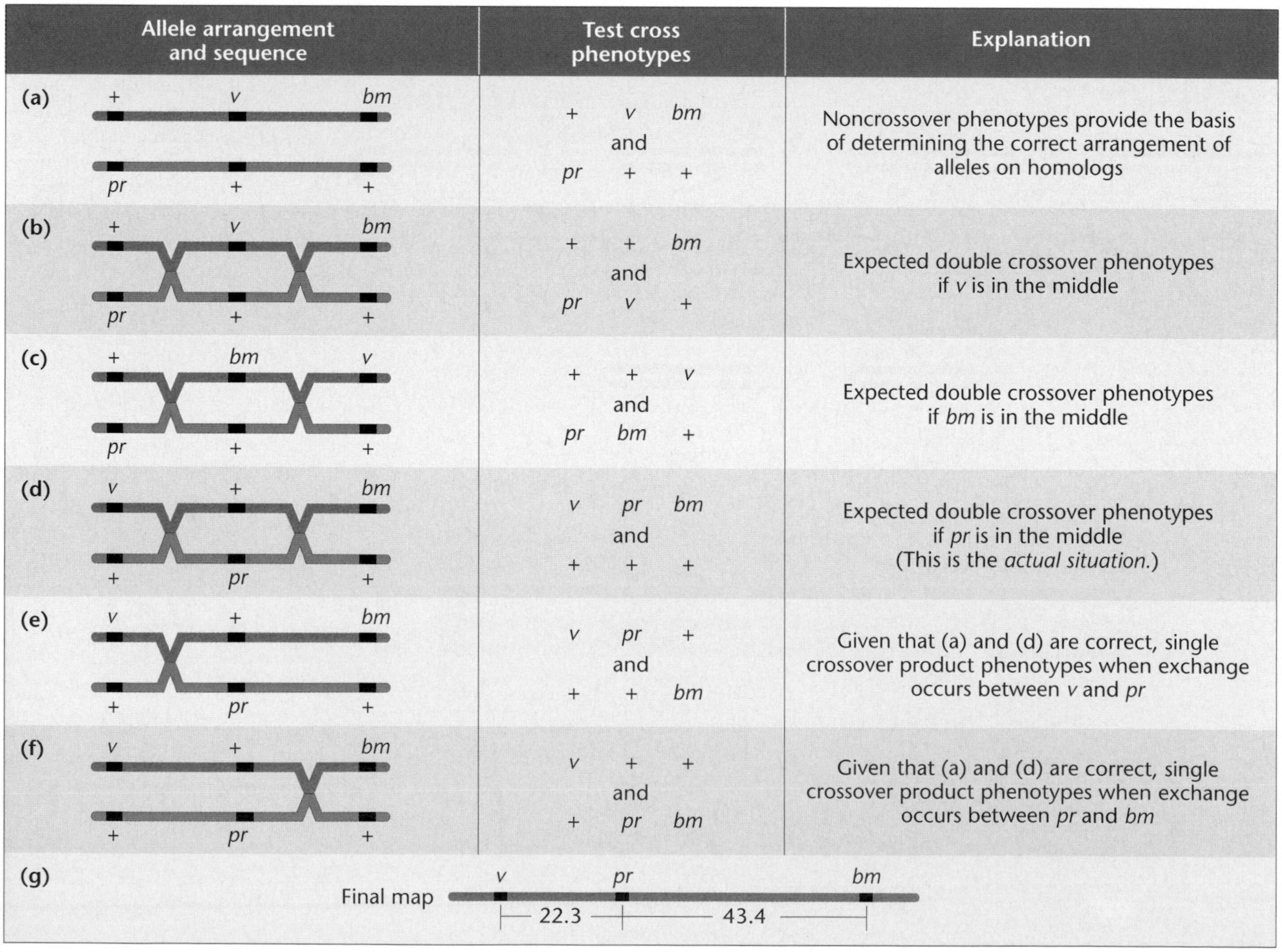

Allele arrangement and sequence			Test cross phenotypes	Explanation
(a) + *v bm* / *pr* + +			+ *v bm* and *pr* + +	Noncrossover phenotypes provide the basis of determining the correct arrangement of alleles on homologs
(b) + *v bm* / *pr* + +			+ + *bm* and *pr v* +	Expected double crossover phenotypes if *v* is in the middle
(c) + *bm v* / *pr* + +			+ + *v* and *pr bm* +	Expected double crossover phenotypes if *bm* is in the middle
(d) *v* + *bm* / + *pr* +			*v pr bm* and + + +	Expected double crossover phenotypes if *pr* is in the middle (This is the *actual situation*.)
(e) *v* + *bm* / + *pr* +			*v pr* + and + + *bm*	Given that (a) and (d) are correct, single crossover product phenotypes when exchange occurs between *v* and *pr*
(f) *v* + *bm* / + *pr* +			*v* + + and + *pr bm*	Given that (a) and (d) are correct, single crossover product phenotypes when exchange occurs between *pr* and *bm*
(g) Final map: *v* — 22.3 — *pr* — 43.4 — *bm*				

FIGURE 8–11 Producing a map of the three genes in the cross in Figure 8–10, where neither the arrangement of alleles nor the sequence of genes in the heterozygous female parent is known.

3. *What is the distance between each pair of genes?*

Having established the sequence of loci as *v–pr–bm*, we can now determine the distance between *v* and *pr* and between *pr* and *bm*. Remember that the map distance between two genes is calculated on the basis of all detectable recombination events occurring between them. This includes both single- and double-crossover events.

Figure 8–11(e) shows that the phenotypes <u>*v pr +*</u> and <u>*+ + bm*</u> result from single crossovers between the *v* and *pr* loci, accounting for 14.5 percent of the offspring. By adding the percentage of double crossovers (7.8%) to the number obtained for single crossovers, the total distance between the *v* and *pr* loci is calculated to be 22.3 map units.

Figure 8–11(f) shows that the phenotypes <u>*v + +*</u> and <u>*+ pr bm*</u> result from single crossovers between the *pr* and *bm* loci, totaling 35.6 percent. With the addition of the double-crossover classes (7.8%), the distance between *pr* and *bm* is calculated to be 43.4 map units. The final map for all three genes in this example is shown in Figure 8–11(g).

8.4 The Accuracy of Mapping Experiments

So far, we have assumed that crossover frequencies are directly proportional to the distance between any two loci along the chromosome. However, it is not always possible to detect all crossover events. A case in point is a double exchange that occurs between the two loci in question. As shown in Figure 8–12(a), if a double exchange occurs, the original arrangement of alleles on each nonsister homolog is recovered. Therefore, even though crossing over occurs, it is impossible to detect in these cases. This factor holds for all even-numbered exchanges between two loci.

Furthermore, as a result of complications posed by multiple strand exchanges, mapping determinations usually underestimate the actual distance between two genes. The farther apart two genes, the greater the probability that undetected crossovers will occur. While the discrepancy is minimal for two genes relatively close together, the degree of inaccuracy increases as the distance increases, as shown in the graph of recombination frequency versus map distance in Figure 8–12(b). The most accurate maps are constructed from experiments whose genes are relatively close together.

Interference and the Coefficient of Coincidence

As shown in the maize example, we can predict the expected frequency of multiple exchanges, such as double crossovers, once the distance between genes is established. For example, in the maize cross, the distance between *v* and *pr* is 22.3 map units, and the distance between *pr* and *bm* is 43.4 map units. If the two single crossovers that make up a double crossover occur independently of one another, we can calculate the expected frequency of double crossovers (DCO_{exp}):

$$DCO_{exp} = (0.223) \times (0.434) = 0.097 = 9.7 \text{ percent}$$

Most often in mapping experiments, the observed DCO frequency is less than the expected number of DCOs. In the maize cross, for example, only 7.8 percent DCOs are observed when 9.7 percent are expected. The phenomenon called **interference** (when a crossover event in one region of the chromosome inhibits a second event in nearby regions) causes this reduction.

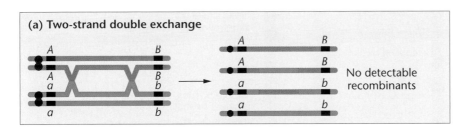

(a) Two-strand double exchange

No detectable recombinants

(b)

FIGURE 8–12 (a) A double crossover goes undetected because no rearrangement of alleles occurs. (b) The theoretical and actual percentage of recombinant chromatids versus map distance. The straight line shows the theoretical relationship if a direct correlation between recombination and map distance exists. The curved line is the actual relationship, which is derived from studies of *Drosophila*, *Neurospora*, and *Zea mays*.

To quantify the disparities that result from interference, we calculate the **coefficient of coincidence (C)**:

$$C = \frac{\text{Observed DCO}}{\text{Expected DCO}}$$

In the maize cross, we have

$$C = \frac{0.078}{0.097} = 0.804$$

Once we have found C, we can quantify interference (I) using the simple equation

$$I = 1 - C$$

In the maize cross, we have

$$I = 1.000 - 0.804 = 0.196$$

If interference is complete and no double crossovers occur, then $I = 1.0$. If fewer DCOs than expected occur, I is a positive number and positive interference has occurred. If more DCOs than expected occur, I is a negative number and negative interference has occurred. In the maize example, I is a positive number (0.196), indicating that 19.6 percent fewer double crossovers occurred than expected.

Positive interference is most often observed in eukaryotic systems. In general, the closer genes are to one another along the chromosome, the more positive interference occurs. In fact, in *Drosophila*, interference is often complete within a distance of 10 map units, and no multiple crossovers are re-

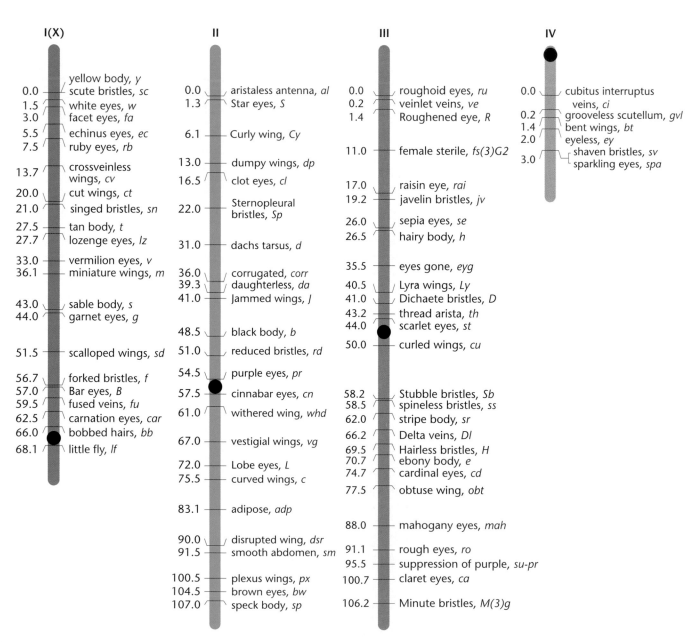

FIGURE 8–13 A partial genetic map of the four chromosomes of *Drosophila melanogaster*. The circle on each chromosome represents the position of the centromere.

covered. This observation suggests that physical constraints that prevent the formation of closely aligned chiasmata cause interference. This interpretation is consistent with the finding that interference decreases as the genes in question are located farther apart. In the maize cross in Figures 8–10 and 8–11, the three genes are relatively far apart, and 80 percent of the expected double crossovers are observed.

8.5 The Genetic Map of *Drosophila*

In organisms such as *Drosophila*, maize, and the mouse, where large numbers of mutations have been discovered and where mapping crosses are possible, extensive chromosome maps have been constructed. Figure 8–13 shows partial maps for the four chromosomes (I, II, III, and IV) of *Drosophila*. Virtually every morphological feature of the fruit fly has been subjected to mutations. The locus of each mutant gene is first localized to one of the four chromosomes (or linkage groups) and then mapped in relation to all other genes present on that chromosome. Based on cytological evidence, the relative lengths of these genetic maps correlate roughly with the relative physical lengths of these chromosomes.

8.6 Somatic Cell Hybridization and Human Gene Mapping

In humans, where neither designed matings nor large numbers of offspring are available, the earliest linkage studies were based on pedigree analysis. Attempts were made to establish whether a trait was X-linked or autosomal. For autosomal traits, geneticists have tried to distinguish whether pairs of traits demonstrate linkage or independent assortment. In this way, it was hoped that a human gene map could be created.

The difficulty arises, however, when two genes of interest are separated on a chromosome such that recombinant gametes are formed, obscuring linkage in a pedigree. In such cases, the demonstration of linkage is enhanced by an approach that relies on probability, called the **lod score method**. First devised by J. B. S. Haldane and C. A. Smith in 1947 and refined by Newton Morton in 1955, the lod score (for *log* of the *od*ds favoring linkage) assesses the probability that a particular pedigree involving two traits reflects linkage or not. First, the probability is calculated that family data (pedigrees) concerning two or more traits conform to the transmission of traits without linkage. Then the probability is calculated that the identical family data following these same traits result from linkage with a specified recombination frequency. The ratio of these probabilities expresses the "odds" for and against linkage.

Accuracy using the lod score method is limited by the extent of the family data, but nevertheless represents an important advance in assigning human genes to specific chromosomes and constructing preliminary human chromosome maps. The initial results were discouraging because of the method's inherent limitations and because of the relatively high haploid number of human chromosomes (23). By 1960, almost no linkage or mapping information had become available.

In the 1960s, a new technique, **somatic cell hybridization**, proved to be an immense aid in assigning human genes to their respective chromosomes. This technique, first discovered by Georges Barsky, relies on the fact that two cells in culture can be induced to fuse into a single hybrid cell. Barsky used two mouse cell lines, but it soon became evident that cells from different organisms could also be fused. When fusion occurs, an initial cell type called a **heterokaryon** is produced. The hybrid cell contains two nuclei in a common cytoplasm. Using the proper techniques, it is possible to fuse human and mouse cells, for example, and isolate the hybrids from the parental cells.

As the heterokaryons are cultured *in vitro*, two interesting changes occur. The nuclei eventually fuse, creating a **synkaryon**. Then, as culturing is continued for many generations, chromosomes from one of the two parental species are gradually lost. In the case of the human–mouse hybrid, human chromosomes are lost randomly until eventually the synkaryon has a full complement of mouse chromosomes and only a few human chromosomes. As we shall see, it is the preferential loss of human chromosomes (rather than mouse chromosomes) that makes possible the assignment of human genes to the chromosomes on which they reside.

The experimental rationale is straightforward. If a specific human gene product is synthesized in a synkaryon containing one to three human chromosomes, then the gene responsible for that product must reside on one of the human chromosomes remaining in the hybrid cell. On the other hand, if the human gene product is not synthesized in a synkaryon, the gene responsible is not present on those human chromosomes that remain in this hybrid cell. Ideally, a panel of 23 hybrid cell lines, each with just one unique human chromosome, would allow the immediate assignment of any human gene for which the product could be characterized.

In practice, a panel of cell lines, each with several remaining human chromosomes, is most often used. The correlation of the presence or absence of each chromosome with the presence or absence of each gene product is called **synteny testing**. Consider, for example, the hypothetical data provided in Figure 8–14, where four gene products (A, B, C, and D) are tested in relationship to eight human chromosomes. Let's analyze the gene that produces product A.

1. Product A is not produced by cell line 23, but chromosomes 1, 2, 3, and 4 are present in cell line 23. Therefore, we rule out the presence of gene A on those four chromosomes and conclude that it must be on chromosome 5, 6, 7, or 8.

2. Product A is produced by cell line 34, which contains chromosomes 5 and 6, but not 7 and 8. Therefore, gene A is on chromosome 5 or 6, but cannot be on 7 or 8 because they are absent, even though product A is produced.

3. Product A is also produced by cell line 41, which contains chromosome 5 but not 6. Using similar reasoning, gene A must be on chromosome 5.

Hybrid cell lines	Human chromosomes present								Gene products expressed			
	1	2	3	4	5	6	7	8	A	B	C	D
23	●	●	●	●					−	+	−	+
34	●	●			●	●			+	−	−	+
41	●		●		●		●		+	+	−	+

FIGURE 8–14   A hypothetical grid of data used in synteny testing to assign genes to their appropriate human chromosomes. Three somatic hybrid cell lines, designated 23, 34, and 41, have each been scored for the presence or absence of human chromosomes 1–8, as well as for their ability to produce the hypothetical human gene products A, B, C, and D.

Using a similar approach, gene *B* can be assigned to chromosome 3. You should perform this analysis to demonstrate for yourself that this is correct. Gene *C* presents a unique situation. The data indicate that it is not present on any of the first seven chromosomes (1–7). While it might be on chromosome 8, no direct evidence supports this conclusion. Other panels are needed. We leave gene *D* for you to analyze. On what chromosome does it reside?

Using this technique, literally hundreds of human genes have been assigned to one chromosome or another. Some of the assignments shown in Figure 8–15 were either derived or confirmed with the use of somatic cell hybridization techniques. To map genes for which the products have yet to be discovered, researchers have had to rely on other approaches. For example, by combining recombinant DNA technology with pedigree analysis, it has been possible to assign the genes responsible for **Huntington disease**, **cystic fibrosis**, and **neurofibromatosis** to their respective chromosomes, 4, 7, and 17. This approach is discussed in Chapter 19.

8.7   Haploid Organisms in Linkage and Mapping Studies

Many of the single-celled eukaryotes are haploid during the vegetative stages of their life cycle. The alga *Chlamydomonas* and the mold *Neurospora* demonstrate this genetic condition. These organisms do form reproductive cells that fuse during fertilization, producing a diploid zygote. However, this structure soon undergoes meiosis, resulting in haploid vegetative cells that then propagate by mitotic divisions. Small, haploid organisms have several important advantages in genetic studies compared with diploid eukaryotes. They can be cultured and manipulated in genetic crosses much more easily. In addition, a haploid organism contains only a single allele of each gene, which is expressed directly in the phenotype. This fact greatly simplifies genetic analysis. As a result, organisms such as *Chlamydomonas* and *Neurospora* serve as subjects of research investigations in many areas of genetics, including linkage and mapping studies.

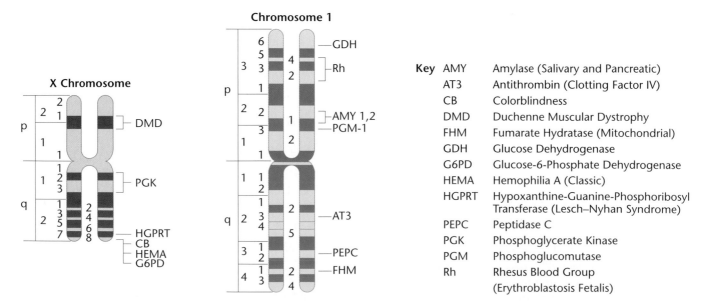

FIGURE 8–15   Representative regional gene assignments for human chromosome 1 and the X chromosome. Many assignments were initially derived using somatic-cell hybridization techniques.

In order to perform genetic experiments with such organisms, crosses are made, and following fertilization the meiotic structures are isolated. Because all four meiotic products give rise to spores in each structure, these structures are called **tetrads**. *The term "tetrad" has a different meaning here than when it was used earlier to describe a precise chromatid configuration in meiosis*. Individual tetrads are isolated, and the resultant cells are grown and analyzed separately from those of other tetrads. In the results we are about to describe, the data reflect the proportion of tetrads that show one combination of genotypes, the proportion that show another combination, and so on. Such experimentation is called **tetrad analysis**.

Gene-to-Centromere Mapping

When a single gene in *Neurospora* is analyzed (Figure 8–16), the data can be used to calculate the map distance between the gene and the centromere. This process is sometimes referred to as **mapping the centromere**. It is accomplished by experimentally determining the frequency of recombination using tetrad data. Note that once the four meiotic products of the tetrad are formed, a mitotic division occurs, resulting in eight ordered products (ascospores). If no crossover event occurs between the gene under study and the centromere, the pattern of ascospores (contained within an ascus, pl. asci) appears as shown in Figure 8–16(a) $(aaaa++++)$.*

This pattern represents **first-division segregation** because the two alleles are separated during the first meiotic division. However, a crossover event will alter this pattern, as shown in Figure 8–16(b) $(aa++aa++)$ and 8–16(c) $(++aaaa++)$. Actually, two other recombinant patterns occur, depending on the chromatid orientation during the second meiotic division: $++aa++aa$ and $aa++++aa$. The four latter patterns reflect **second-division segregation** because the two alleles are not separated until the second meiotic division. Usually, the ordered tetrad data are condensed to reflect the genotypes of the identical ascospore pairs. Six combinations are possible:

First-Division Segregation	*Second-Division Segregation*	
$aa++$	$a+a+$	$+aa+$
$++aa$	$+a+a$	$a++a$

To calculate the distance between the gene and the centromere, a large number of asci resulting from a controlled cross are counted. We then use these data to calculate the distance (d):

$$d = \frac{1/2 \text{ (second-division segregant asci)}}{\text{total asci scored}}$$

The distance (d) reflects the percentage of recombination and is only one-half the number of second-division segre-

gant asci. This is because crossing over in each occurs in only two of the four chromatids during meiosis.

To illustrate, we use a for albino and $+$ for wild type in *Neurospora*. In crosses between the two genetic types, suppose the following data are observed:

65 first-division segregants
70 second-division segregants

The distance between a and the centromere is

$$d = \frac{(1/2)(70)}{135} = 0.259$$

or about 26 map units.

As the distance increases to 50 map units, in theory all asci should reflect second-division segregation. However, numerous factors prevent this. As in diploid organisms, mapping accuracy based on crossover events is greatest when the gene and centromere are relatively close together.

We can also analyze haploid organisms in order to distinguish between linkage and independent assortment of two genes. Mapping distances between gene loci are calculated once linkage is established. As a result, detailed maps of organisms such as *Neurospora* and *Chlamydomonas* are now available.

8.8 Other Aspects of Genetic Exchange

Careful analysis of crossing over during gamete formation allows us to construct chromosome maps in both diploid and haploid organisms. However, we should not lose sight of the real biological significance of crossing over, which is to generate genetic variation in gametes and, subsequently, in the offspring derived from the resultant eggs and sperm. Because of the critical role of crossing over in generating variation, the study of genetic exchange is a key topic for study in genetics. Many important questions remain. For example, does crossing over involve an actual exchange of chromosome arms? Does exchange occur between paired sister chromatids during mitosis? We briefly consider possible answers to these questions.

Cytological Evidence for Crossing Over

Once genetic mapping was understood, it was of great interest to investigate the relationship between chiasmata observed in meiotic prophase I and crossing over. For example, are chiasmata visible manifestations of crossover events? If so, then crossing over in higher organisms appears to result from an actual physical exchange between homologous chromosomes. That this is the case was demonstrated independently in the 1930s by Harriet Creighton and Barbara McClintock in *Zea mays* and by Curt Stern in *Drosophila*.

Because the experiments are similar, we consider only one of them, the work with maize. Creighton and McClintock

---

*The pattern $(++++aaaa)$ can also be formed. However, it is indistinguishable from $(aaaa++++)$.

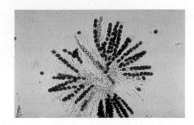

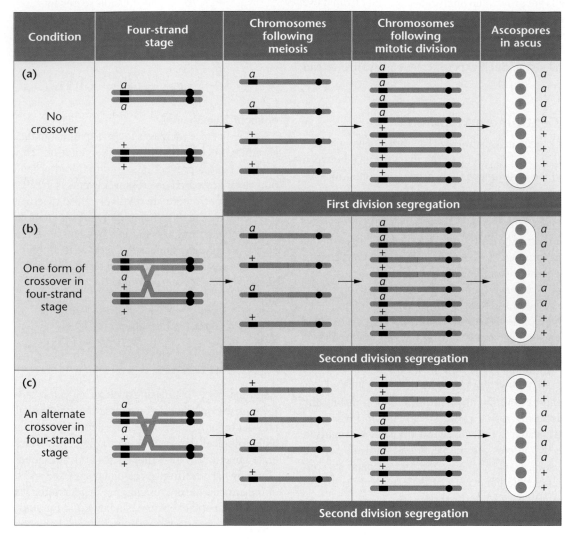

FIGURE 8–16 Three ways in which different ascospore patterns can be generated in *Neurospora*. Analysis of these patterns is the basis of gene-to-centromere mapping. The photograph shows a variety of ascospore arrangement within *Neurospora* asci. *(Photo: James W. Richardson/Visuals Unlimited)*

studied two linked genes on chromosome 9. At one locus, the alleles *colorless* (*c*) and *colored* (*C*) control endosperm coloration. At the other locus, the alleles *starchy* (*Wx*) and *waxy* (*wx*) control the carbohydrate characteristics of the endosperm. The maize plant studied is heterozygous at both loci. That one of the homologs contains two unique cytological markers is the key to the experiment. The markers consist of a densely stained knob at one end of the chromosome and a translocated piece of another chromosome (8) at the other end. The arrangements of alleles and cytological markers can be detected cytologically and are shown in Figure 8–17.

Creighton and McClintock crossed this plant to one homozygous for the color allele (*c*) and heterozygous for the

endosperm alleles. They obtained a variety of different phenotypes in the offspring, but they were most interested in a crossover result involving the chromosome with the unique cytological markers. They examined the chromosomes of this plant with the colorless, waxy phenotype (case I in Figure 8–17) for the presence of the cytological markers. If physical exchange between homologs accompanies genetic crossing over, the translocated chromosome will still be present, but the knob will not. This is exactly what happened! In a second plant (case II), the phenotype colored, starchy should result from either nonrecombinant gametes or from crossing over. Some of the cases then ought to contain chromosomes with the dense knob but

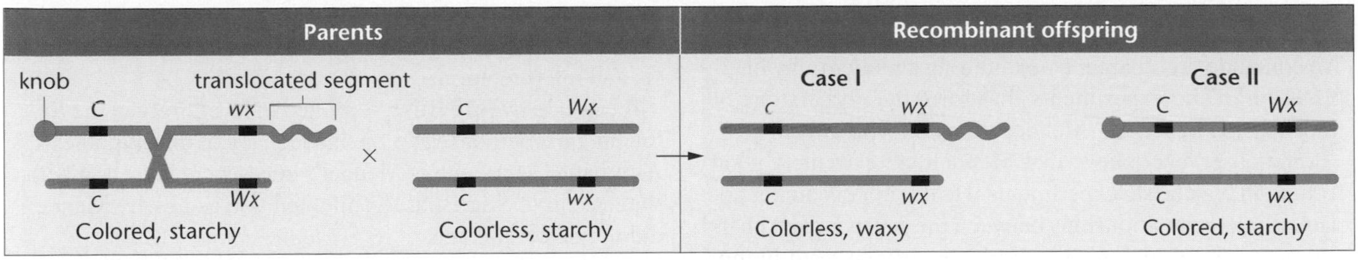

Parents		Recombinant offspring	
		Case I	**Case II**
knob translocated segment			
C *wx*	*c* *Wx*	*c* *wx*	*C* *Wx*
c *Wx*	*c* *wx*	*c* *wx*	*c* *wx*
Colored, starchy	Colorless, starchy	Colorless, waxy	Colored, starchy

FIGURE 8–17 The phenotypes and chromosome compositions of parents and recombinant offspring in Creighton and McClintock's experiment in maize. The knob and translocated segment are the cytological markers that established that crossing over involves an actual exchange of chromosome arms.

not the translocated chromosome. This condition was also found, and the conclusion that a physical exchange takes place was again supported. Along with Stern's findings with *Drosophila*, this work clearly established that crossing over has a cytological basis.

Sister Chromatid Exchanges

Knowing that crossing over occurs between synapsed homologs in meiosis, we might ask whether such a physical exchange occurs between homologs during mitosis. While homologous chromosomes do not usually pair up or synapse in somatic cells (*Drosophila* is an exception), each individual chromosome in prophase and metaphase of mitosis consists of two identical sister chromatids, joined at a common centromere. Surprisingly, several experimental approaches have demonstrated that reciprocal exchanges similar to crossing over occur between sister chromatids. While these **sister chromatid exchanges (SCEs)** do not produce new allelic combinations, evidence is accumulating that attaches significance to these events.

Identification and study of SCEs are facilitated by several modern staining techniques. In one technique, cells replicate for two generations in the presence of the thymidine analog **bromodeoxyuridine** (BUdR)*. Following two rounds of replication, each pair of sister chromatids has one member with one strand of DNA "labeled" with BUdR and one member with both strands labeled BUdR. Using a differential stain, chromatids with the analog in both strands stain less brightly than chromatids with it in only one strand. As a result, SCEs are readily detectable, if they occur. In Figure 8–18, numerous instances of SCE events are clearly evident. Because of the patchlike appearance, these sister chromatids are sometimes referred to as **harlequin chromosomes**.

While the significance of SCEs is still uncertain, several observations have led to great interest in this phenomenon. We know, for example, that agents that induce chromosome damage (viruses, X-rays, ultraviolet light, and certain chemical mutagens) increase the frequency of SCEs. The frequency of SCEs is also elevated in **Bloom syndrome**, a

human disorder caused by a mutation in the chromosome 15 *BLM* gene. This rare, recessively inherited disease is characterized by prenatal and postnatal retardation of growth, a great sensitivity of the facial skin to the sun, immune deficiency, a predisposition to malignant and benign tumors, and abnormal behavior patterns. The chromosomes from cultured leukocytes, bone marrow cells, and fibroblasts derived from homozygotes are very fragile and unstable compared to those of homozygous and heterozygous normal individuals. Increased breaks and rearrangements between nonhomologous chromosomes are observed in addition to excessive amounts of sister chromatid exchanges. Recent work by James German and colleagues suggest that the *BLM* gene encodes an enzyme called **DNA helicase**, which is best known for its role in DNA replication (see Chapter 11).

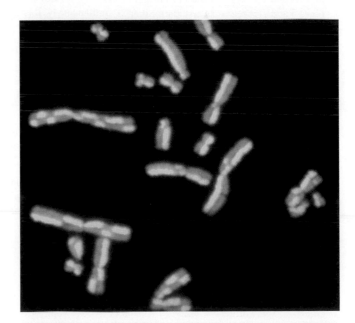

FIGURE 8–18 Light micrograph of sister chromatid exchanges (SCEs) in mitotic chromosomes. Sometimes called harlequin chromosomes because of their patchlike appearance, chromatids with the thymidine analog BUdR in both DNA strands fluoresce *less* brightly than do those with the analog in only one strand. These chromosomes were stained with 33258-Hoechst reagent and acridine orange and then viewed under fluorescence microscopy. *(Dr. Sheldon Wolff & Jody Bodycote/Laboratory of Radiology and Environmental Health, University of California, San Francisco)*

---

*The abbreviation BrdU is also used to denote bromodeoxyuridine.

8.9   Did Mendel Encounter Linkage?

We conclude this chapter by examining a modern-day interpretation of the experiments that form the cornerstone of transmission genetics—Mendel's crosses with garden peas.

Some observers believe that Mendel had extremely good fortune in his classic experiments. He did not encounter apparent linkage relationships between the seven mutant characters in any of his crosses. Had Mendel obtained highly variable data characteristic of linkage and crossing over, these unorthodox ratios might have hindered his successful analysis and interpretation.

The article by Stig Blixt, reprinted in its entirety in the following box, demonstrates the inadequacy of this hypothesis. As we shall see, some of Mendel's genes were indeed linked. We leave it to Stig Blixt to enlighten you as to why Mendel did not detect linkage.

Why Didn't Gregor Mendel Find Linkage?

It is quite often said that Mendel was very fortunate not to run into the complication of linkage during his experiments. He used seven genes, and the pea has only seven chromosomes. Some have said that had he taken just one more, he would have had problems. This, however, is a gross oversimplification. The actual situation, most probably, is shown in Table 1. This shows that Mendel worked with three genes in chromosome 4, two genes in chromosome 1, and one gene in each of chromosomes 5 and 7. It seems at first glance that, out of the 21 dihybrid combinations Mendel theoretically could have studied, no less than four (that is, *a–i*, *v–fa*, *v–le*, *fa–le*) ought to have resulted in linkages. As found, however, in hundreds of crosses and shown by the genetic map of the pea, *a* and *i* in chromosome 1 are so distantly located on the chromosome that no linkage is normally detected. The same is true for *v* and *le* on the one hand, and *fa* on the other, in chromosome 4. This leaves *v–le*, which ought to have shown linkage.

Mendel, however, seems not to have published this particular combination and thus, presumably, never made the appropriate cross to obtain both genes segregating simultaneously. It is therefore not so astonishing that Mendel did not run into the complication of linkage, although he did not avoid it by choosing one gene from each chromosome.

STIG BLIXT

Weibullsholm Plant Breeding Institute, Landskrona, Sweden, and Centro Energia Nucleate na Agricultura, Piracicaba, SP, Brazil.

Source: Reprinted by permission from *Nature*, Vol. 256, p. 206. Copyright 1975 Macmillan Magazines Limited.

TABLE 1   Relationship between Modern Genetic Terminology and Character Pairs Used by Mendel

Character Pair Used by Mendel	Alleles in Modern Terminology	Located in Chromosome
Seed color, yellow-green	I-i	1
Seed coat and flowers, colored-white	A-a	1
Mature pods, smooth expanded-wrinkled indented	V-v	4
Inflorescences, from leaf axis-umbellate in top of plant	Fa-fa	4
Plant height, 0.5–1 m	Le-le	4
Unripe pods, green-yellow	Gp-gp	5
Mature seeds, smooth-wrinkled	R-r	7

Chapter Summary

1. Genes located on the same chromosome are said to be linked. Alleles located on the same homolog, therefore, can be transmitted together during gamete formation. However, crossing over between homologs during meiosis results in the reshuffling of alleles and thereby contributes to genetic variability within gametes.

2. Early in this century, geneticists realized that crossing over provides an experimental basis for mapping the location of linked genes relative to one another along the chromosome.

3. Somatic cell hybridization techniques have made possible linkage and mapping analysis of human genes.

4. Mapping studies may also be performed with haploid organisms such as *Chlamydomonas* and *Neurospora*.
5. Cytological investigations of both maize and *Drosophila* reveal that crossing over involves a physical exchange of segments between nonsister chromatids.

6. An exchange of genetic material between sister chromatids can occur during mitosis as well. These events are referred to as sister chromatid exchanges (SCEs). An elevated frequency of such events is seen in the human disorder Bloom syndrome.
7. Evidence now suggests that several of the genes studied by Mendel are, in fact, linked. However, in such cases, the genes are sufficiently far apart to prevent the detection of linkage.

Insights and Solutions

1. In rabbits, *black* (*B*) is dominant to *brown* (*b*), while *full color* (*C*) is dominant to *chinchilla* (*c^{ch}*). The genes controlling these traits are linked. Rabbits that are heterozygous for both traits and express *black, full color*, are crossed to rabbits that express *brown, chinchilla*, with the following results:

> 31 *brown, chinchilla*
>
> 34 *black, full*
>
> 16 *brown, full*
>
> 19 *black, chinchilla*

Determine the arrangement of alleles in the heterozygous parents and the map distance between the two genes.

Solution: This is a two-point map problem, where the two reciprocal phenotypes most prevalent are the noncrossovers. The less frequent reciprocal phenotypes arise from a single crossover. The arrangement of alleles is derived from the noncrossover phenotypes because they enter gametes intact.

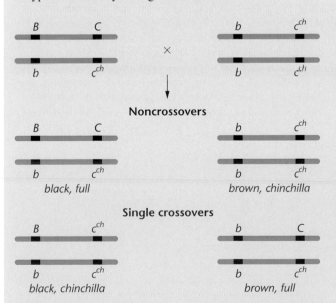

The single crossovers give rise to 35/100 offspring (35%). Therefore, the distance between the two genes is 35 map units (mu).

2. In *Drosophila*, *Lyra* (*Ly*) and *Stubble* (*Sb*) are dominant mutations located at locus 40 and 58, respectively, on chromosome 3. A recessive mutation with bright red eyes is discovered and

shown also to be on chromosome 3. A map is obtained by crossing a female who is heterozygous for all three mutations to a male that is homozygous for the bright red mutation (which we will call *br*). The following data are obtained:

Phenotype			Number
(1) *Ly*	*Sb*	*br*	404
(2) +	+	+	422
(3) *Ly*	+	+	18
(4) +	*Sb*	*br*	16
(5) *Ly*	+	*br*	75
(6) +	*Sb*	+	59
(7) *Ly*	*Sb*	+	4
(8) +	+	*br*	2
		Total =	1000

Determine the location of the *br* mutation on chromosome 3.

Solution: First, determine the arrangement of the alleles on the homologs of the heterozygous crossover parent (the female in this case). To do this, locate the most frequent reciprocal phenotypes, which arise from the noncrossover gametes. These are phenotypes (1) and (2). Each one represents the arrangement of alleles on one of the homologs. Therefore, the arrangement is

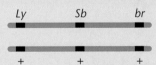

Second, find the correct sequence of the three loci along the chromosome. This is done by determining which sequence yields the observed double-crossover phenotypes, which are the least frequent reciprocal phenotypes (7 and 8).

If the sequence is correct as written, then a double crossover as depicted here,

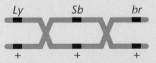

will yield *Ly* + *br* and + *Sb* + as phenotypes. Inspection shows that these categories (5 and 6) are actually single crossovers, not double crossovers. Therefore, the sequence, as written, is incorrect. Only two other sequences are possible. The *br* gene is either to the left of *Ly*, or it is between *Ly* and *Sb*:

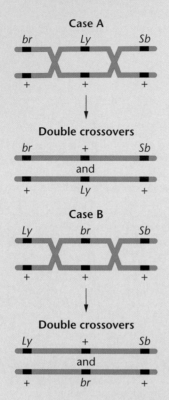

Case A

Double crossovers

Case B

Double crossovers

Comparison with the actual data shows that case B is correct. The double-crossover gametes yield flies that express *Ly* and *Sb* but not *br*, or express *br* but not *Ly* and *Sb*. Therefore, the correct arrangement and sequence is

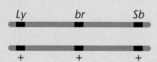

Once this sequence is found, determine the location of *br* relative to *Ly* and *Sb*. A single crossover between *Ly* and *br*, as shown here,

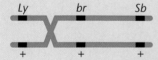

yields flies that are *Ly + +* and *+ br Sb* (categories 3 and 4). Therefore, the distance between the *Ly* and *br* loci is equal to

$$18 + 16 + 4 + 2/1000 = 40/1000 = 0.04 = 4 \text{ map units}$$

Remember to add the double crossovers because they represent two single crossovers occurring simultaneously. You need to know the frequency of all crossovers between *Ly* and *br*, so they must be included. Similarly, the distance between the *br* and *Sb* loci is derived mainly from single crossovers between them:

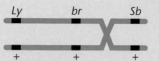

This event yields *Ly br +* and *+ + Sb* phenotypes (categories 5 and 6). Therefore, the distance equals

$$75 + 59 + 4 + 2/1000 = 140/1000 = 0.14 = 14 \text{ map units}$$

The final map shows that *br* is located at locus 44, since *Lyra* and *Stubble* are known:

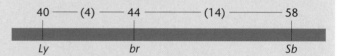

3. Refer to Figure 8–13, and predict what gene was discovered in Problem 2 (which we called *br*). Suggest an experimental cross that could confirm your prediction.

Solution: Inspection of Figure 8–13 reveals that the mutation *scarlet* (*st*) is present at locus 44.0, so it is reasonable to hypothesize that the bright-red eye mutation is an allele at the *scarlet* locus.

To test this hypothesis, you could perform complementation analysis (see Chapter 4) by crossing females expressing the bright red mutation with known *scarlet* males. If the two mutations are alleles, no complementation will occur, and all progeny will reveal a bright-red mutant eye phenotype. If complementation occurs, all progeny will express normal brick red (wild-type) eyes, since the bright red mutation and *scarlet* are at different loci (they are probably very close together). In such a case, all progeny will be heterozygous at both the bright eye and the *scarlet* loci, and will not express either mutation because they are both recessive. This type of complementation analysis is called an **allelism test**.

Key Terms

Problems and Discussion Questions

1. Why does more crossing over occur between two distantly linked genes than between two genes located very close together on the same chromosome?

2. Why are double-crossover events expected in lower frequency than single-crossover events?

3. What essential criteria must be met to execute a successful mapping cross?

4. The genes *dumpy* wing (*dp*), *clot* eye (*cl*), and *apterous* wing (*ap*) are linked on chromosome 2 of *Drosophila*. In a series of two-point mapping crosses, the following genetic distances are determined:

dp–ap	42
dp–cl	3
ap–cl	39

   What is the sequence of the three genes?

5. In maize, colored aleurone (in the kernels) is due to the dominant allele *R*. The recessive allele *r*, when homozygous, produces colorless aleurone. The plant color (not the kernel color) is controlled by the gene pair *Y* and *y*. The dominant *Y* gene results in green color, while the homozygous presence of the recessive *y* gene results in yellow color. In a test cross between a plant of unknown genotype and phenotype and a plant that is homozygous recessive for both traits, the following progeny are obtained:

Colored green	88
Colored yellow	12
Colorless green	8
Colorless yellow	92

   Explain these results determining the exact genotype and phenotype of the unknown plant, including the precise association of the two genes on the homologs and the distance between them.

6. Using two pairs of genes (*P/p* and *Z/z*), a test-cross parent (*ppzz*) is crossed to an organism of unknown genotype. Analysis of the data indicates that the gametes are produced in the following proportions:

   PZ, 42.4%   *Pz*, 6.9%   *pZ*, 7.1%   *pz*, 43.6%

   Draw all possible conclusions about the location of these genes.

7. Two different female *Drosophila* are isolated, each heterozygous for the autosomally linked genes *b* (*black* body), *d* (*dachs* tarsus), and *c* (*curved* wings). These genes are in the order *d–b–c*, with *b* closer to *d* than to *c*. Shown is the genotypic arrangement for each female along with the various gametes formed by both. Identify which categories are noncrossovers (NCOs), single crossovers (SCOs), and double crossovers (DCOs) in each case. Which category is likely to have the fewest gametes? Which will have the most gametes?

Female A			Female B		
d	*b*	+	*d*	+	+
+	+	*c*	+	*b*	*c*

   Gametes

Female A				Female B			
(1) *d b c*		(5) *d + +*		(1) *d b +*		(5) *d b c*	
(2) *+ + +*		(6) *+ b c*		(2) *+ + c*		(6) *+ + +*	
(3) *+ + c*		(7) *d + c*		(3) *d + c*		(7) *d + +*	
(4) *d b +*		(8) *+ b +*		(4) *+ b +*		(8) *+ b c*	

8. In *Drosophila*, females expressing the three X-linked recessive traits, *scute* (*sc*) bristles, *sable* body (*s*), and *vermilion* eyes (*v*) are crossed with wild-type males. In the F₁ generation, all females are wild type, while all males express all three mutant traits. The cross is carried to the F₂ generation, and 1000 offspring are counted, with the results shown here. No determination of sex has been made in the F₂ data.

Phenotype			Offspring
sc	*s*	*v*	314
+	+	+	280
+	*s*	*v*	150
sc	+	+	156
sc	+	*v*	46
+	*s*	+	30
sc	*s*	+	10
+	+	*v*	14

(a) Determine the genotypes of the P₁ and F₁ parents, using proper nomenclature.
(b) Determine the sequence of the three genes and the map distance between them.
(c) Are there more or fewer double crossovers than expected? Calculate the coefficient of coincidence. Does this represent positive or negative interference?

9. In *Drosophila*, *Dichaete* (*D*) is a chromosome 3 mutation with a dominant effect on wing shape. It is lethal when homozygous. The genes *ebony* (*e*) and *pink* (*p*) are chromosome 3 recessive mutations affecting body and eye color, respectively. Flies from a *Dichaete* stock are crossed to homozygous *ebony*, *pink* flies, and the F₁ progeny, with a *Dichaete* phenotype, are then backcrossed to the *ebony*, *pink* homozygotes. The results of this back cross are as follows:

Phenotype	Number
Dichaete	401
ebony, pink	389
Dichaete, ebony	84
pink	96
Dichaete, pink	2
ebony	3
Dichaete, ebony, pink	12
wild type	13

(a) Diagram this cross, showing the genotypes of the parents and offspring of both crosses.
(b) What is the sequence of and interlocus distance between these three genes?

10. In a cross in *Neurospora*, involving two alleles *B* and *b*, the following tetrad patterns are observed. Calculate the distance between the locus and the centromere.

Tetrad Pattern	Number
BBbb	36
bbBB	44
BbBb	4
bBbB	6
BbbB	3
bBBb	7

11. In a theoretical diploid organism, a female of genotype

$$\frac{a \quad b \quad c}{+ \quad + \quad +}$$

produces 100 meiotic tetrads. Of these, 68 show no crossover events. Of the remaining 32, 20 show a crossover between *a* and *b*, 10 show a crossover between *b* and *c*, and 2 show a double crossover between *a* and *b* and between *b* and *c*. Of the 400 meiotic products, how many of each of the 8 different genotypes will be produced? Assuming the order *a–b–c* and the allele arrangement shown above, what is the map distance between these loci?

12. In *Drosophila*, two mutations, *Stubble* (*Sb*) and *curled* (*cu*), are linked on chromosome 3. *Stubble* is a dominant gene that is lethal in the homozygous state, and *curled* is a recessive gene. If a female of genotype

$$\frac{Sb \quad cu}{+ \quad +}$$

is to be mated to detect recombinants among her offspring, what male genotype would you choose as a mate?

13. Another cross in *Drosophila* involves the recessive, X-linked genes *yellow* (*y*), *white* (*w*), and *cut* (*ct*). A female that is yellow-bodied and white-eyed with normal wings is crossed to a male whose eyes and body are normal but whose wings are cut. The F₁ females are wild type for all three traits, while the F₁ males express the yellow-body, white-eye traits. The cross is carried to an F₂, and only male offspring are tallied. A genetic map is constructed using the following data:

Phenotype			Male Offspring
y	+	*ct*	9
+	*w*	+	6
y	*w*	*ct*	90
+	+	+	95
+	+	*ct*	424
y	*w*	+	376
y	+	+	0
+	*w*	*ct*	0

(a) Diagram the genotypes of the F₁ parents.
(b) Construct a map, assuming that *white* is at locus 1.5 on the X chromosome.
(c) Do you expect double-crossover offspring?
(d) Can the F₂ female offspring be used to construct the map? Why or why not?

14. In a series of two-point map crosses involving five genes located on chromosome II in *Drosophila*, the following recombinant (single-crossover) frequencies are observed:

pr–adp	29
pr–vg	13
pr–c	21
pr–b	6
adp–b	35
adp–c	8
adp–vg	16
vg–b	19
vb–c	8
c–b	27

(a) If the *adp* gene is present near the end of the chromosome II (locus 83), construct a map of these genes.
(b) In another set of experiments, a sixth gene, *d*, is tested against *b* and *pr*:

$$d–b, \text{ } 17\% \qquad d–pr, \text{ } 23\%$$

Predict the results of two-point maps between *d* and *c*, *d* and *vg*, and *d* and *adp*.

15. An organism of genotype *AaBbCc* is test-crossed. The genotypes of the progeny are as follows:

20	*Aa Bb Cc*	20	*Aa Bb cc*
20	*aa bb Cc*	20	*aa bb cc*
5	*Aa bb Cc*	5	*Aa bb cc*
5	*aa Bb Cc*	5	*aa Bb cc*

(a) If these three genes are all assorting independently of each other, how many genotypic classes would you expect in the progeny of the testcross?

(b) If these three genes are so tightly linked that crossover never occurs, how many genotypic classes would you expect in the progeny of the test cross?

(c) What can you conclude from the actual data?

16. *Drosophila melanogaster* has one pair of sex chromosomes (XX or XY) and three autosomes, referred to as chromosomes 2, 3, and 4. A male fly with very short legs is discovered by a genetics student. Using this male, the student establishes a pure breeding stock of this mutant and finds that it is recessive. This mutant is then incorporated into a stock containing the recessive gene *black* (body color, located on chromosome 2) and the recessive gene *pink* (eye color, located on chromosome 3). A female from the homozygous black, pink, short stock is then mated to a wild-type male. The F_1 males of this cross are all wild type and are then backcrossed to the homozygous *b p sh* females. The F_2 results are shown in the following table. No other phenotypes are observed.

		Phenotype		
	Wild	*Pink**	*Black, Short*	*Black, Pink, Short*
Females	63	58	55	69
Males	59	65	51	60

*Pink indicates that the other two traits are wild type, and so on.

(a) Based on these results, the student assigns *short* to a linkage group (a chromosome). Which group is it? Give your step-by-step reasoning.

(b) The experiment is subsequently repeated making the reciprocal cross, F_1 females backcrossed to homozygous *b p sh* males, and 85 percent of the offspring fall into the above classes, but 15 percent of the offspring are equally divided among *b + p*, *b + +*, *+ sh p*, and *+ sh +* phenotypic males and females. How can these results be explained, and what information can you derive from the data?

Selected Readings

Allen, G. E. 1978. *Thomas Hunt Morgan: The man and his science.* Princeton, NJ: Princeton University Press.

Chaganti, R., Schonberg, S., and German, J. 1974. A manyfold increase in sister chromatid exchange in Bloom syndrome lymphocytes. *Proc. Natl. Acad. Sci.* 71:4508–12.

Creighton, H. S., and McClintock, B. 1931. A correlation of cytological and genetical crossing over in *Zea mays. Proc. Natl. Acad. Sci.* 17:492–97.

Douglas, L., and Novitski, E. 1977. What chance did Mendel's experiments give him of noticing linkage? *Heredity* 38:253–57.

Ellis, N. A., et al. 1995. The Bloom's syndrome gene product is homologous to RecQ helicases. *Cell* 83:655–66.

Ephrussi, B., and Weiss, M. C. 1969. Hybrid somatic cells. *Sci. Am.* (April) 220:26–35.

Latt, S. A. 1981. Sister chromatid exchange formation. *Annu. Rev. Genet.* 15:11–56.

Lindsley, D. L., and Grell, E. H. 1972. *Genetic variations of Drosophila melanogaster.* Washington, DC: Carnegie Institute of Washington.

Morgan, T. H. 1911. An attempt to analyze the constitution of the chromosomes on the basis of sex-linked inheritance in *Drosophila. J. Exp. Zool.* 11:365–414.

Morton, N. E. 1995. LODs—Past and present. *Genetics* 140:7–12.

Neuffer, M. G., Jones, L., and Zober, M. 1968. *The mutants of maize.* Madison, WI: Crop Science Society of America.

Perkins, D. 1962. Crossing over and interference in a multiply marked chromosome arm of *Neurospora. Genetics* 47:1253–74.

Ruddle, F. H., and Kucherlapati, R. S. 1974. Hybrid cells and human genes. *Sci. Am.* (July) 231:36–49.

Stahl, F. W. 1979. *Genetic recombination.* New York: W. H. Freeman.

Stern, C. 1936. Somatic crossing over and segregation in *Drosophila melanogaster. Genetics* 21:625–31.

Sturtevant, A. H. 1913. The linear arrangement of six sex-linked factors in *Drosophila,* as shown by their mode of association. *J. Exp. Zool.* 14:43–59.

————. 1965. *A history of genetics.* New York: Harper & Row.

Voeller, B. R., ed. 1968. *The chromosome theory of inheritance: Classical papers in development and heredity.* New York: Appleton-Century-Croft.

Wolff, S., ed. 1982. *Sister chromatid exchange.* New York: Wiley-Interscience.

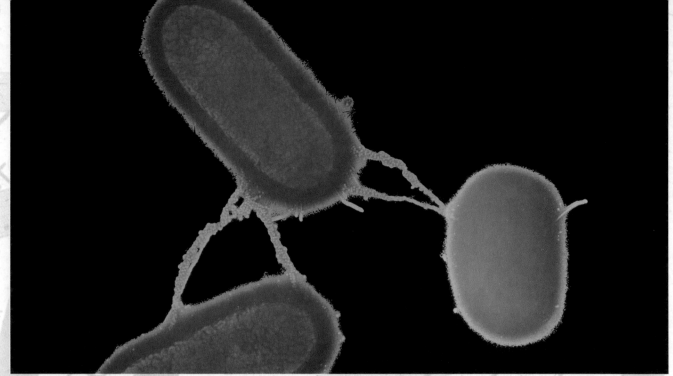

Transmission electron micrograph of conjugating *E. coli*. (*Dr. L. Caro/Science Photo Library/ Photo Researchers, Inc.*)

9

Mapping in Bacteria and Bacteriophages

CHAPTER CONCEPTS

Both bacteria and bacteriophages (bacterial viruses) demonstrate mechanisms by which genetic recombination occurs, processes that can serve as the basis for genetic mapping. As a result, both groups have been the subject of extensive analysis. Bacteria also contain extrachromosomal DNA in the form of plasmids, which can house a fertility factor that plays a critical role in genetic recombination. Both plasmids and bacteriophage DNA can be integrated into the bacterial chromosome, creating a number of interesting genetic scenarios.

In this chapter we shift from the consideration of transmission of genetic information in eukaryotes to a discussion of **bacteria** (prokaryotes) and **bacteriophages**, viruses that have bacteria as their host. The study of bacteria and bacteriophages has been essential to the accumulation of knowledge in many areas of genetic study. For example, much of what we know about molecular genetics, recombinational phenomena, and gene structure initially was derived from experimental work with these organisms. Furthermore, as we shall see in Chapters 16 and 19, our extensive knowledge of bacteria and their resident plasmids has made for their widespread use in DNA cloning and other recombinant DNA studies.

Bacteria and their viruses are especially useful research organisms in genetics for a number of reasons. First, they have extremely short reproductive cycles. Literally hundreds of generations, giving rise to billions of genetically identical bacteria or phages, can be produced in short periods of time. Furthermore, they can be studied in pure cultures. That is, a single species or mutant strain of bacteria or one type of virus can be isolated and investigated independently of other similar organisms.

In this chapter we focus on genetic recombination and chromosome mapping. Complex processes have evolved in bacteria and bacteriophages that facilitate genetic recombination within populations. As we shall see, these processes are the basis for the chromosome mapping analysis that forms the cornerstone of molecular genetic investigations of bacteria and the viruses that invade them.

9.1 Bacterial Mutation and Growth

It has long been known that pure cultures of bacteria give rise to cells that exhibit heritable variation, particularly with respect to growth under unique environmental conditions. Prior to 1943, the source of this variation was hotly debated. The majority of bacteriologists believed that environmental factors induced changes in certain bacteria that led to their adaptation to the new conditions. For example, strains of *Escherichia coli* are known to be sensitive to infection by the bacteriophage T1. Infection by the bacteriophage leads to the virus reproducing at the expense of the bacterial cell, from which new phages are released as the host cell is disrupted, or lysed. If a plate of *E. coli* is uniformly sprayed with T1, almost all cells are lysed. Rare *E. coli* cells, however, survive infection and are not lysed. If these cells are isolated and established in pure culture, all descendants are resistant to T1 infection. The **adaptation hypothesis**, put forth to explain this type of observation, implies that the interaction of the phage and bacterium is essential to the acquisition of immunity. In other words, the phage "induces" resistance in the bacteria.

On the other hand, the occurrence of **spontaneous mutations**, which occur in the presence or the absence of phage T1, suggested an alternative model to explain the origin of resistance in *E. coli*. In 1943, Salvador Luria and Max Delbrück presented the first convincing evidence that bacteria, like eukaryotic organisms, are capable of spontaneous mutation. This experiment, referred to as the **fluctuation test**, marks the initiation of modern bacterial genetic study. Spontaneous mutation is now considered the primary source of genetic variation in bacteria.

Mutant cells that arise spontaneously in otherwise pure cultures can be isolated and established independently from the parent strain by using established selection techniques. As a result, mutations for almost any desired characteristic can now be induced and isolated. Because bacteria and viruses usually contain only a single chromosome and are therefore haploid, all mutations are expressed directly in the descendants of mutant cells, adding to the ease with which these microorganisms can be studied.

Bacteria are grown in a liquid culture medium or in a Petri dish on a semisolid agar surface. If the nutrient components of the growth medium are simple and consist only of an organic carbon source (such as a glucose or lactose) and a variety of ions, including Na^+, K^+, Mg^{2+}, Ca^{2+}, and NH^{4+} present as inorganic salts, it is called **minimal medium**. To grow on such a medium, a bacterium must be able to synthesize all essential organic compounds (e.g., amino acids, purines, pyrimidines, sugars, vitamins, and fatty acids). A bacterium that can accomplish this remarkable biosynthetic feat—one that we ourselves cannot duplicate— is termed a **prototroph**. It is said to be wild type for all growth requirements. On the other hand, if a bacterium loses, through mutation, the ability to synthesize one or more organic components, it is said to be an **auxotroph**. For example, if it loses the ability to make histidine, then this amino acid must be added as a supplement to the minimal medium for growth to occur. The resulting bacterium is designated as a *his⁻* auxotroph, in contrast to its prototrophic *his⁺* counterpart.

To study mutant bacteria quantitatively, an inoculum of bacteria is placed in liquid culture medium. A characteristic growth pattern is exhibited (Figure 9–1). Initially, during the **lag phase**, growth is slow. Then, a period of rapid growth, called the **logarithmic (log) phase**, ensues. During this

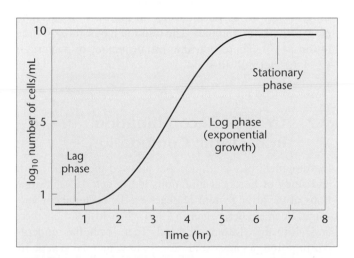

FIGURE 9–1 Typical bacterial population growth curve showing the initial lag phase, the subsequent log phase where exponential growth occurs, and the stationary phase that occurs when nutrients are exhausted.

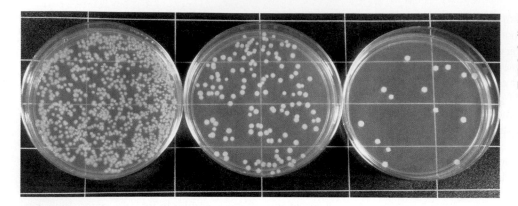

FIGURE 9–2 Results of the serial dilution technique and subsequent culture of bacteria. Each dilution varies by a factor of 10. Each colony is derived from a single bacterial cell. *(Michael G. Gabridge/Visuals Unlimited)*

phase, cells divide continually with a fixed time interval between cell divisions, resulting in exponential growth. When a cell density of about 10^9 cells/mL is reached, nutrients and oxygen become limiting and cells cease dividing; at this point, the cells enter the **stationary phase**. Because the doubling time during the log phase can be as short as 20 minutes, an initial inoculum of a few thousand cells easily achieves maximum cell density during an overnight incubation.

Cells grown in liquid medium can be quantified by transferring them to semisolid medium in a Petri dish. Following incubation and many divisions, each cell gives rise to a visible colony on the surface of the medium. If the number of colonies is too great to count, then a series of successive dilutions (a technique called serial dilution) of the original liquid culture is made and plated, until the colony number is reduced to the point where it can be counted (Figure 9–2). This technique allows the number of bacteria present in the original culture to be calculated.

As an example, let's assume that the three dishes in Figure 9–2 represent serial dilutions of 10^{-3}, 10^{-4}, and 10^{-5} (left to right). We need only select the dish in which the number of colonies can be counted accurately. Because each colony arose from a single bacterium, the number of colonies multiplied by the dilution factor represents the number of bacteria in each milliliter of the initial inoculum used to start the serial dilutions. In Figure 9–2, the rightmost dish has 15 colonies. The dilution factor for a 10^{-5} dilution is 10^5. Therefore, the initial number of bacteria is 15×10^5 per milliliter.

9.2 Genetic Recombination in Bacteria: Conjugation

Development of techniques that allowed the identification and study of bacterial mutations led to detailed investigations of the arrangement of genes on the bacterial chromosome. Such studies began in 1946 when Joshua Lederberg and Edward Tatum showed that bacteria undergo **conjugation**, a parasexual process in which the genetic information from one bacterium is transferred to and recombined with that of another bacterium (see the chapter opening photograph). Like meiotic crossing over in eukaryotes, genetic recombination in bacteria enabled the development of

methodology for chromosome mapping. Note that the term **genetic recombination**, as applied to bacteria and bacteriophages, leads to the *replacement* of one or more genes present in one strain with those from a genetically distinct strain. While this is somewhat different from our use of genetic recombination in eukaryotes, where the term describes crossing over that results in *reciprocal exchange events*, the overall effect is the same: Genetic information is transferred from one chromosome to another, resulting in an altered genotype. Two other phenomena that result in the transfer of genetic information from one bacterium to another, transformation and transduction, have helped us determine the arrangement of genes on the bacterial chromosome. We discuss these processes in later sections of this chapter.

Lederberg and Tatum's initial experiments were performed with two multiple-auxotroph strains of *E. coli* K12. Strain A required methionine and biotin in order to grow, while strain B required threonine, leucine, and thiamine (Figure 9–3). Neither strain would grow on minimal medium. The two strains were first grown separately in supplemented media, and cells from both were mixed and grown together for several more generations and then plated on minimal medium. Any bacterial cells that grew on minimal medium were prototrophs. It is highly improbable that any of the cells that contained two or three mutant genes underwent spontaneous mutation simultaneously at two or three independent locations, leading to wild-type cells. Therefore, the researchers assumed that any prototrophs recovered arose as a result of some form of genetic exchange and recombination between the two mutant strains.

In this experiment, prototrophs were recovered at a rate of $1/10^7$ (10^{-7}) cells plated. The controls for this experiment involved separate plating of cells from strains A and B on minimal medium. No prototrophs were recovered. Based on these observations, Lederberg and Tatum proposed that genetic exchange had occurred.

F$^+$ and F$^-$ Bacteria

Lederberg and Tatum's findings were soon followed by numerous experiments that elucidated the genetic basis of conjugation. It quickly became evident that different strains of bacteria are involved in a unidirectional transfer of genetic material. When cells serve as donors of parts of their chromosomes, they are designated as **F$^+$ cells** (F for "fertility").

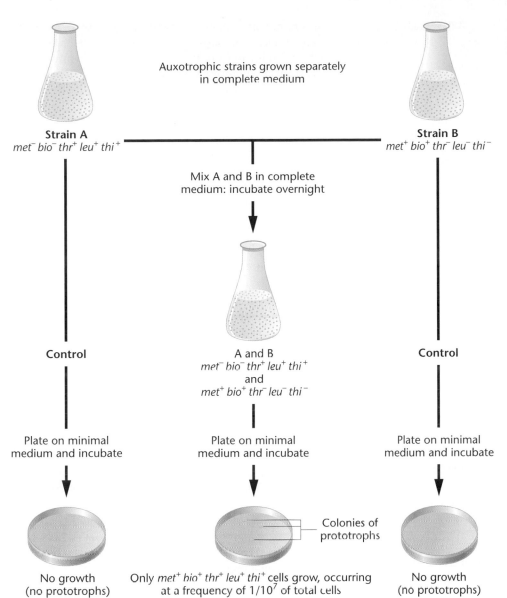

Auxotrophic strains grown separately
in complete medium

Strain A
met⁻ bio⁻ thr⁺ leu⁺ thi⁺

Strain B
met⁺ bio⁺ thr⁻ leu⁻ thi⁻

Mix A and B in complete
medium: incubate overnight

Control

A and B
met⁻ bio⁻ thr⁺ leu⁺ thi⁺
and
met⁺ bio⁺ thr⁻ leu⁻ thi⁻

Control

Plate on minimal
medium and incubate

Plate on minimal
medium and incubate

Plate on minimal
medium and incubate

Colonies of
prototrophs

No growth
(no prototrophs)

Only *met⁺ bio⁺ thr⁺ leu⁺ thi⁺* cells grow, occurring
at a frequency of 1/10⁷ of total cells

No growth
(no prototrophs)

FIGURE 9–3 Genetic recombination of two auxotrophic strains producing prototrophs. Neither auxotroph grows on minimal medium, but prototrophs do, suggesting that genetic recombination has occurred.

MEDIA TUTORIAL Bacterial Genetics

Recipient bacteria receive the donor chromosome material (now known to be DNA), and recombine it with part of their own chromosome. They are designated as **F⁻ cells**.

It was subsequently established that cell contact is essential for chromosome transfer to occur. Support for this concept was provided by Bernard Davis, who designed a U-tube for growing F⁺ and F⁻ cells (Figure 9–4). At the base of the tube is a sintered glass filter with a pore size that allows passage of the liquid medium, but that is too small to allow the passage of bacteria. The F⁺ cells are placed on one side of the filter, and F⁻ cells on the other side. The medium is moved back and forth across the filter so that the cells essentially share a common medium during bacterial incubation. Davis plated samples from both sides of the tube on minimal medium, but no prototrophs were found. He logically concluded that *physical contact between cells of the two strains is essential to genetic recombination*. We now know that this physical interaction is the initial step in the process of conjugation

and is established by a structure called the **F pilus** (or **sex pilus**, pl. pili). Bacteria often have many pili, which are tubular extensions of the cell. After contact is initiated between mating pairs (Figure 9–5), transfer of the chromosome begins.

Later evidence established that F⁺ cells contain a **fertility factor** (called the **F factor**) that confers the ability to donate part of their chromosome during conjugation. Experiments by Joshua and Esther Lederberg and by William Hayes and Luca Cavalli-Sforza show that certain conditions eliminate the F factor in otherwise fertile cells. However, if these "infertile" cells are then grown with fertile donor cells, the F factor is regained.

The conclusion that the F factor is a mobile element is further supported by the observation that, following conjugation and genetic recombination, recipient cells always become F⁺. Thus, in addition to the *rare* cases of gene transfer from the bacterial chromosome (genetic recombination), the F factor itself is passed to *all* recipient cells. On this basis,

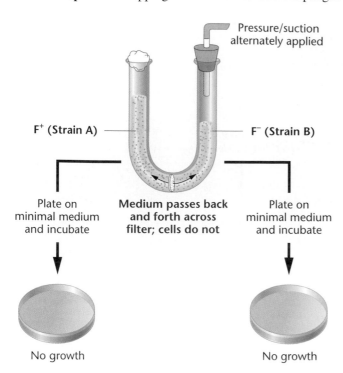

Pressure/suction alternately applied

F⁺ (Strain A) — F⁻ (Strain B)

Plate on minimal medium and incubate

Medium passes back and forth across filter; cells do not

Plate on minimal medium and incubate

No growth

No growth

FIGURE 9–4 When strain A and B auxotrophs are grown in a common medium but separated by a filter, no genetic recombination occurs and no prototrophs are produced. The apparatus shown is a Davis U-tube.

the initial crosses of Lederberg and Tatum (Figure 9–3) can be designated as follows:

STRAIN A		STRAIN B
F⁺	×	F⁻
DONOR		RECIPIENT

Isolation of the F factor confirmed these conclusions. Like the bacterial chromosome, though distinct from it, the F factor has been shown to consist of a circular, double-stranded DNA molecule, equivalent to about 2 percent of the bacteri-

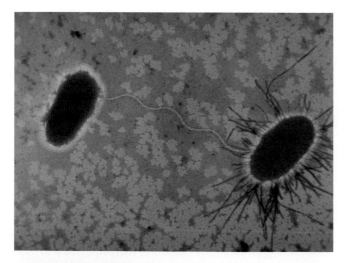

FIGURE 9–5 An electron micrograph of conjugation between an F⁺ *E. coli* cell and an F⁻ cell. The sex pilus linking them is clearly visible. (*Dennis Kunkel/Phototake NYC*)

al chromosome (about 100,000 nucleotide pairs). Contained in the F factor, among others, are 19 genes, the products of which are involved in the transfer of genetic information. These include those essential to the formation of the sex pilus.

As we soon shall see, the F factor is in reality an autonomous genetic unit referred to as a plasmid. However, in our historical coverage of its discovery, we will continue to refer to it as a "factor."

It is believed that the transfer of the F factor during conjugation involves separation of the two strands of its double helix and the movement of one of the two strands into the recipient cell. Both strands, one moving across the conjugation tube and one remaining in the donor cell, are replicated. The result is that both the donor *and* the recipient cells are F⁺. This process is diagrammed in Figure 9–6.

To summarize, an *E. coli* cell may or may not contain the F factor. When this factor is present, the cell is able to form a sex pilus and potentially serves as a donor of genetic information. During conjugation, a copy of the F factor is almost always transferred from the F⁺ cell to the F⁻ recipient, converting it to the F⁺ state. The question remained as to exactly why such a low proportion of cells involved in these matings (10^{-7}) also results in genetic recombination. The answer awaited further experimentation.

Hfr Bacteria and Chromosome Mapping

Subsequent discoveries not only clarified how genetic recombination occurs, but also defined a mechanism by which the *E. coli* chromosome could be mapped. We address chromosome mapping first.

In 1950, Cavalli-Sforza treated an F⁺ strain of *E. coli* K12 with nitrogen mustard, a chemical known to induce mutations. From these treated cells, he recovered a genetically altered strain of donor bacteria that underwent recombination at a rate of $1/10^4$ (or 10^{-4}), 1000 times more frequently than the original F⁺ strains. In 1953, Hayes isolated another strain that demonstrated an elevated frequency. Both strains were designated **Hfr**, for **high-frequency recombination**. Because Hfr cells behave as donors, they are a special class of F⁺ cells.

Another important difference was noted between Hfr strains and the original F⁺ strains. If the donor is from an Hfr strain, recipient cells, while sometimes displaying genetic recombination, never become Hfr; that is, they remain F⁻. In comparison, then,

$$F^+ \times F^- \rightarrow F^+ \quad \text{(low rate of recombination)}$$
$$Hfr \times F^- \rightarrow F^- \quad \text{(higher rate of recombination)}$$

Perhaps the most significant characteristic of Hfr strains is the *nature of recombination*. In any given strain, certain genes are more frequently recombined than others, and some not at all. This *nonrandom* pattern was shown to vary from Hfr strain to Hfr strain. While these results were puzzling, Hayes interpreted them to mean that some physiological alteration of the F factor had occurred, resulting in the production of Hfr strains of *E. coli*.

In the mid-1950s, experimentation by Ellie Wollman and François Jacob explained the difference between Hfr and F⁺

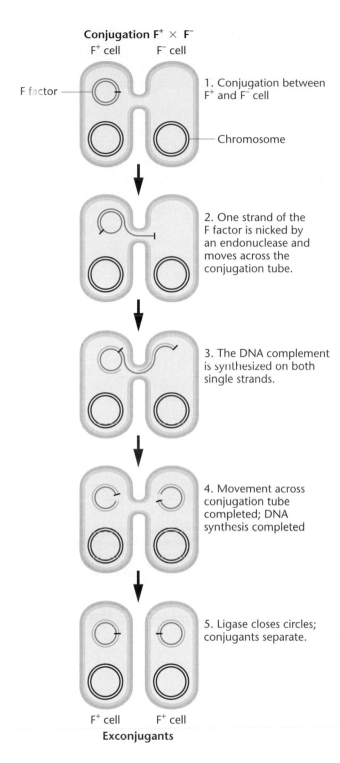

Conjugation F⁺ × F⁻
F⁺ cell   F⁻ cell

F factor

1. Conjugation between F⁺ and F⁻ cell

Chromosome

2. One strand of the F factor is nicked by an endonuclease and moves across the conjugation tube.

3. The DNA complement is synthesized on both single strands.

4. Movement across conjugation tube completed; DNA synthesis completed

5. Ligase closes circles; conjugants separate.

F⁺ cell   F⁺ cell
Exconjugants

FIGURE 9–6 An F⁺ × F⁻ mating demonstrating how the recipient F⁻ cell converts to F⁺. During conjugation, the DNA of the F factor replicates with one new copy entering the recipient cell, converting it to F⁺. The black bar added to the F factors follows their clockwise rotation during replication.

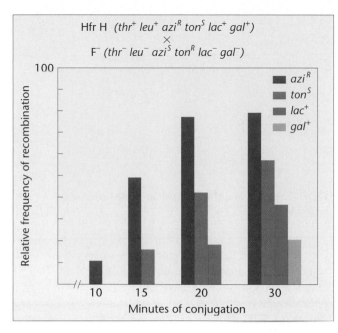

FIGURE 9–7 The progressive transfer during conjugation of various genes from a specific Hfr strain of *E. coli* to an F⁻ strain. Certain genes (*azi* and *ton*) transfer sooner than others and recombine more frequently. Others (*lac* and *gal*) take longer to transfer and recombine with a lower frequency. Others (*thr* and *leu*) are always transferred and are used in the initial screen for recombinants.

placed in a blender. The shear forces created in the blender separated conjugating bacteria so that the transfer of the chromosome was terminated effectively. The cells were then assayed for genetic recombination.

This process, called the **interrupted mating technique**, demonstrated that specific genes of a given Hfr strain were transferred and recombined sooner than others. Figure 9–7 illustrates this point. During the first 8 minutes after the two strains are mixed, no genetic recombination can be detected. At about 10 minutes, recombination of the *azi* gene can be detected, but no transfer of the *ton^S*, *lac⁺*, or *gal⁺* genes is noted. By 15 minutes, 70 percent of the recombinants are *azi⁺*; 30 percent are now also *ton^S*; but none is *lac⁺* or *gal⁺*. Within 20 minutes, the *lac⁺* is found among the recombinants; and within 25 minutes, *gal⁺* is also being transferred. Wollman and Jacob had demonstrated *an ordered transfer of genes* that correlated with the length of time conjugation proceeded.

It appeared that the chromosome of the Hfr bacterium was transferred linearly and that the gene order and distance between genes, as measured in minutes, could be predicted from such experiments (Figure 9–8). This information served as the basis for the first genetic map of the *E. coli* chromosome. "Minutes" in bacterial mapping are equivalent to "map units" in eukaryotes.

Wollman and Jacob then repeated the same type of experiment with other Hfr strains, obtaining similar results with one important difference. While genes were always transferred linearly with time, as in their original experiment, which genes entered first and which followed later seemed to vary from Hfr strain to Hfr strain [Figure 9–9(a)]. When

and showed how Hfr strains allow genetic mapping of the *E. coli* chromosome. In their experiments, Hfr and F⁻ strains with suitable marker genes were mixed and recombination of specific genes assayed at different times. To accomplish this, a culture containing a mixture of an Hfr and an F⁻ strain was first incubated and samples removed at various intervals and

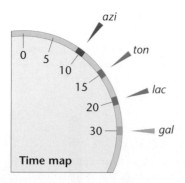

FIGURE 9–8 A time map of the genes studied in the experiment depicted in Figure 9–7.

they reexamined the rate of entry of genes, and thus the different genetic maps for each strain, a definite pattern emerged. The major difference between each strain was simply the point of origin (*O*) and the direction in which entry proceeded from that point [Figure 9–9(b)].

To explain these results, Wollman and Jacob postulated that the *E. coli* chromosome is circular (a closed circle, with no free ends). If the point of origin (*O*) varies from strain to strain, a different sequence of genes will be transferred in each case. But what determines *O*? They proposed that in various Hfr strains, the F factor integrates into the chromosome at different points. Its position determines the site of *O*. One such case of integration is shown in Figure 9–10 (step 1). During conjugation between an Hfr and an F⁻ cell, the position of the F factor determines the initial point of transfer

(steps 2 and 3). Those genes adjacent to *O* are transferred first. *The F factor becomes the last part that can be transferred* (step 4). However, conjugation rarely, if ever, lasts long enough to allow the entire chromosome to pass across the conjugation tube (step 5). This proposal explains why recipient cells, when mated with Hfr cells, remain F⁻.

Figure 9–10 also depicts the way in which the two strands making up a DNA molecule unwind during transfer, allowing for the entry of one strand of DNA into the recipient (step 3). Following replication, the entering DNA now has the potential to recombine with its homologous region of the host chromosome. The DNA strand that remains in the donor also undergoes replication.

The use of the interrupted mating technique with different Hfr strains allowed researchers to map the entire *E. coli* chromosome. Mapped in time units, strain K12 (or *E. coli* K12) is 100 minutes long. Over 900 genes have now been placed on the map.

Recombination in F⁺ × F⁻ Matings: A Reexamination

The preceding model helped geneticists better understand how genetic recombination occurs during the F⁺ × F⁻ matings. Recall that recombination occurs much less frequently in them than in Hfr × F⁻ matings, and that random gene transfer is involved. The current belief is that when F⁺ and F⁻ cells are mixed, conjugation occurs readily and that each F⁻ cell involved in conjugation with an F⁺ cell receives a copy of the F factor, *but that no genetic recombination oc-*

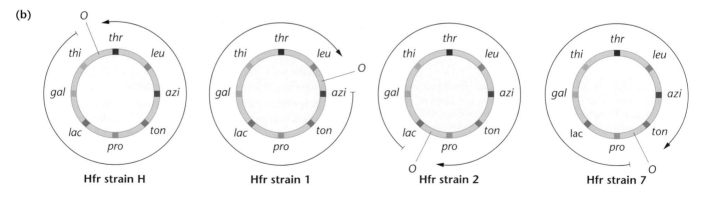

(a)

Hfr strain	Order of transfer (Earliest)												(Latest)
H	thr	–	leu	–	azi	–	ton	–	pro	–	lac	–	gal	–	thi
1	leu	–	thr	–	thi	–	gal	–	lac	–	pro	–	ton	–	azi
2	pro	–	ton	–	azi	–	leu	–	thr	–	thi	–	gal	–	lac
7	ton	–	azi	–	leu	–	thr	–	thi	–	gal	–	lac	–	pro

(b)

Hfr strain H   Hfr strain 1   Hfr strain 2   Hfr strain 7

FIGURE 9–9 (a) The order of gene transfer in four Hfr strains, suggesting that the *E. coli* chromosome is circular. (b) The point where transfer originates (*O*) is identified in each strain. Note that transfer proceeds in either direction, depending on the strain. The origin is determined by the point of integration into the chromosome of the F factor, and the direction of transfer is determined by the orientation of the F factor as it integrates.

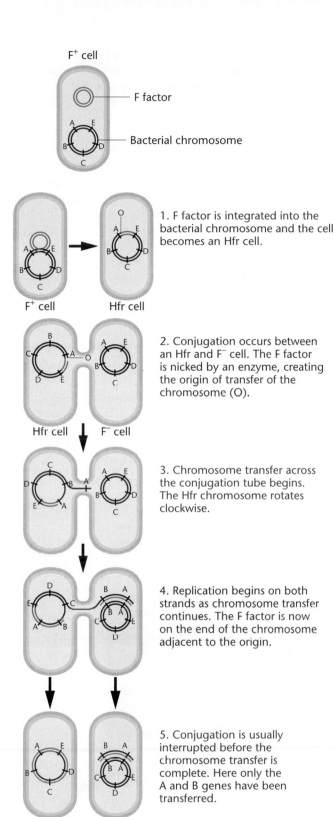

1. F factor is integrated into the bacterial chromosome and the cell becomes an Hfr cell.

2. Conjugation occurs between an Hfr and F⁻ cell. The F factor is nicked by an enzyme, creating the origin of transfer of the chromosome (O).

3. Chromosome transfer across the conjugation tube begins. The Hfr chromosome rotates clockwise.

4. Replication begins on both strands as chromosome transfer continues. The F factor is now on the end of the chromosome adjacent to the origin.

5. Conjugation is usually interrupted before the chromosome transfer is complete. Here only the A and B genes have been transferred.

F⁺ cell   F⁻ cell
Exconjugants

FIGURE 9–10 Conversion of F⁺ to an Hfr state occurs by integrating the F factor into the bacterial chromosome. The point of integration determines the origin (O) of transfer. During conjugation, an enzyme nicks the F factor, now integrated into the host chromosome, initiating transfer of the chromosome at that point. Conjugation is usually interrupted prior to complete transfer. Above, only the A and B genes are transferred to the F⁻ cell, which may recombine with the host chromosome.

curs. However, at an extremely low frequency in a population of F⁺ cells, the F factor integrates spontaneously from the cytoplasm to a random point in the bacterial chromosome, converting the F⁺ cell to the Hfr state as we saw in Figure 9–10. Therefore, in F⁺ × F⁻ crosses, the extremely low frequency of genetic recombination (10^{-7}) is attributed to the rare, newly formed Hfr cells, which then undergo conjugation with F⁻ cells. Because the point of integration of the F factor is random, which gene or genes are transferred by any newly formed Hfr donor will also appear to be random within the larger F⁺/F⁻ population. The recipient bacterium will appear as a recombinant, but will remain F⁻. If it subsequently undergoes conjugation with an F⁺ cell, it will be converted to F⁺.

The F′ State and Merozygotes

In 1959, during experiments with Hfr strains of *E. coli*, Edward Adelberg discovered that the F factor could lose its integrated status, causing the cell to revert to the F⁺ state (Figure 9–11, step 1). When this occurs, the F factor frequently carries several adjacent bacterial genes along with it (step 2). Adelberg labeled this condition F′ to distinguish it from F⁺ and Hfr. F′, like Hfr, is thus another special case of F⁺. This conversion is described as one from Hfr to F′.

The presence of bacterial genes within a cytoplasmic F factor creates an interesting situation. An F′ bacterium behaves like an F⁺ cell, initiating conjugation with F⁻ cells (Figure 9–11, step 3). When this occurs, the F factor, containing chromosomal genes, is transferred to the F⁻ cell (step 4). As a result, whatever chromosomal genes are part of the F factor are now present in duplicate in the recipient cell (step 5), because the recipient still has a complete chromosome. This creates a partially diploid cell called a **merozygote**. Pure cultures of F′ merozygotes can be established. They have been extremely useful in the study of bacterial genetics, particularly in genetic regulation.

9.3 Rec Proteins and Bacterial Recombination

Once researchers established that a unidirectional transfer of DNA occurs between bacteria, they were interested in determining how the actual recombination event occurs in the recipient cell. Just how does the donor DNA replace the comparable region in the recipient chromosome? As with many systems, the biochemical mechanism by which recombination occurs was deciphered through genetic studies. Major insights were gained as a result of isolating a group of mutations representing *rec* genes.

The first relevant observation in this case involved a series of mutant genes labeled *recA*, *recB*, *recC*, and *recD*. The first mutant gene, *recA*, diminished genetic recombination in bacteria 1000-fold, nearly eliminating it altogether. The other *rec* mutations reduced recombination by about 100 times. Clearly, the normal wild-type products of these genes play some essential role in the process of recombination.

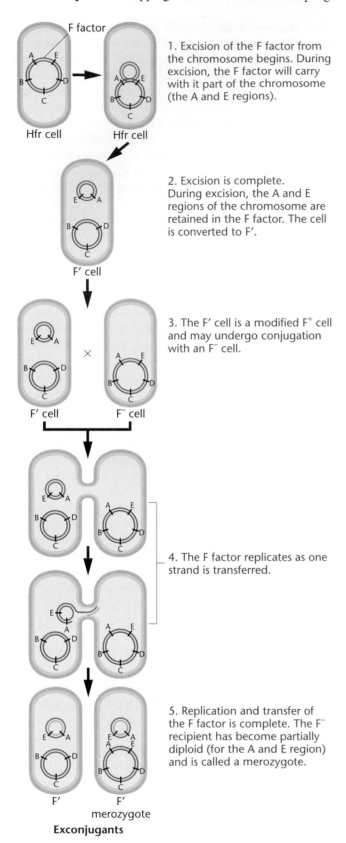

1. Excision of the F factor from the chromosome begins. During excision, the F factor will carry with it part of the chromosome (the A and E regions).

2. Excision is complete. During excision, the A and E regions of the chromosome are retained in the F factor. The cell is converted to F′.

3. The F′ cell is a modified F⁺ cell and may undergo conjugation with an F⁻ cell.

4. The F factor replicates as one strand is transferred.

5. Replication and transfer of the F factor is complete. The F⁻ recipient has become partially diploid (for the A and E region) and is called a merozygote.

FIGURE 9–11 Conversion of an Hfr bacterium to F′ and its subsequent mating with an F⁻ cell. The conversion occurs when the F factor loses its integrated status. During excision from the chromosome, it carries with it one or more chromosomal genes (A and E). Following conjugation with an F⁻ cell, the recipient cell becomes partially diploid and is called a merozygote. It also behaves as an F⁺ donor cell.

By looking for a functional gene product present in normal cells but missing in mutant cells, researchers subsequently isolated several gene products and showed that they play a role in genetic recombination. The first is called the **RecA protein**.* The second is a more complex protein called the **RecBCD protein**, an enzyme consisting of polypeptide subunits encoded by three other *rec* genes. The roles of these proteins have now been elucidated *in vitro*. This genetic research has extended our knowledge of the process of recombination considerably and underscores the value of isolating mutations, establishing their phenotypes, and determining the biological role of the normal, wild-type gene. Once we have thoroughly explored the topic of DNA structure, we examine the molecular role of the Rec proteins in recombination in Chapter 11.

9.4 F Factors and Plasmids

In the preceding sections we examined the extrachromosomal heredity unit called the F factor. When it exists autonomously in the bacterial cytoplasm, the F factor is composed of a double-stranded closed circle of DNA [Figure 9–12(a)]. These characteristics place the F factor in the more general category of genetic structures called **plasmids**. These structures contain one or more genes—often, quite a few. Their replication depends on the same enzymes that replicate the chromosome of the host cell, and they are distributed to daughter cells along with the host chromosome during cell division.

---

*Note that the names of bacterial genes use lowercase letters and are italicized, while the corresponding gene products begin with capital letters and are not italicized, as illustrated by the *recA* gene and RecA protein.

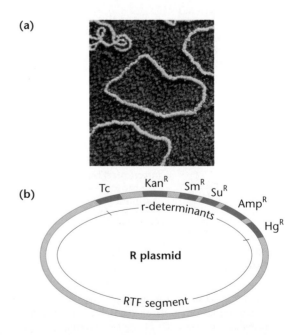

(a)

(b)

FIGURE 9–12 (a) Electron micrograph of a plasmid isolated from *E. coli*; (b) an R plasmid containing resistance transfer factors (RTFs) and multiple r-determinants (Tc, tetracycline; Kan, kanamycin; Sm, streptomycin; Su, sulfonamide; Amp, ampicillin; and Hg, mercury). *(Photo: K.G. Murti/Visuals Unlimited)*

Plasmids are generally classified according to the genetic information specified by their DNA. The F factor confers fertility and contains genes essential for sex pilus formation, upon which genetic recombination depends. Other examples of plasmids include the R and the Col plasmids.

Most **R plasmids** consist of two components: the **RTF (resistance transfer factor)** and one or more **r-determinants** [Figure 9–12(b)]. The RTF encodes genetic information essential to transferring the plasmid between bacteria, and the r-determinants are genes that confer resistance to antibiotics. While RTFs are similar to a variety of plasmids from different bacterial species, r-determinants, each of which is specific for resistance to one class of antibiotic, vary widely. Sometimes, a bacterial cell contains r-determinant plasmids but no RTF. Such a cell is resistant but cannot transfer the genetic information for resistance to recipient cells. The most commonly studied plasmids, however, contain the RTF as well as one or more r-determinants. Resistance to tetracycline, streptomycin, ampicillin, sulfonamide, kanamycin, and chloramphenicol are most frequently encountered. Sometimes these occur in a single plasmid, conferring multiple resistance to several antibiotics [Figure 9–12(b)]. Bacteria bearing such plasmids are of great medical significance, not only because of their multiple resistance, but because of the ease with which the plasmids can be transferred to other bacteria.

The **Col plasmid**, ColE1, derived from *E. coli*, is clearly distinct from R plasmids. It encodes one or more proteins that are highly toxic to bacterial strains that do not harbor the same plasmid. These proteins, called **colicins**, can kill neighboring bacteria. Bacteria that carry the plasmid are said to be colicinogenic. Present in 10–20 copies per cell, a gene in the plasmid encodes an immunity protein that protects the host cell from the toxin. Unlike an R plasmid, the Col plasmid is not usually transmissible to other cells.

9.5 Bacterial Transformation

Transformation also provides a mechanism for recombining genetic information in some bacteria. In transformation, small pieces of extracellular DNA are taken up by a living bacterium, ultimately leading to a stable genetic change in the recipient cell. We are interested in transformation in this chapter because, in those bacterial species where it occurs, the process can be used to map bacterial genes, although in a more limited way than conjugation. We return to the topic of transformation in Chapter 10 because of the central role the process played in experiments proving that DNA is the genetic material.

This process (Figure 9–13) consists of numerous steps divided into two main categories: (1) entry of DNA into a recipient cell, and (2) recombination of the donor DNA with its homologous region in the recipient chromosome. In a population of bacterial cells, only those in a particular physiological state, referred to as **competence**, take up DNA. Entry is thought to occur at a limited number of receptor sites on the surface of the bacterial cell. Passage across the cell wall and membrane is an active process that requires energy and specific transport molecules. This model is supported by the

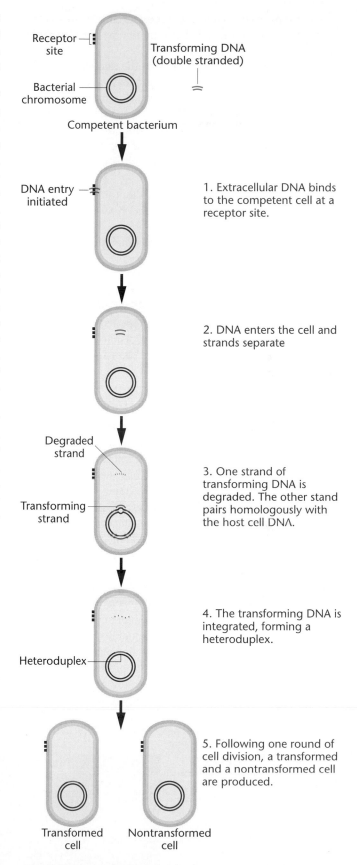

1. Extracellular DNA binds to the competent cell at a receptor site.

2. DNA enters the cell and strands separate

3. One strand of transforming DNA is degraded. The other stand pairs homologously with the host cell DNA.

4. The transforming DNA is integrated, forming a heteroduplex.

5. Following one round of cell division, a transformed and a nontransformed cell are produced.

FIGURE 9–13 Proposed steps for transforming a bacterial cell by exogenous DNA. Only one of the two entering DNA strands is involved in the transformation event, which is completed following cell division.

fact that substances that inhibit energy production or protein synthesis in the recipient cell also inhibit the transformation process.

During the process of entry, one of the two strands of the double helix is digested by nucleases, leaving only a single strand to participate in transformation (Figure 9–13, steps 2 and 3). The surviving strand of DNA then aligns with its complementary region of the bacterial chromosome. In a process involving several enzymes, the segment replaces its counterpart in the chromosome, which is excised and degraded (step 4).

For recombination to be detected, the transforming DNA must be derived from a different strain of bacteria that bears some genetic variation, such as a mutation. Once it is integrated into the chromosome, the recombinant region contains one host strand (present originally) and one mutant strand. Because these strands are from different sources, this helical region is referred to as a **heteroduplex**. Following one round of semiconservative replication, one chromosome is restored to its identical configuration, and the other contains the mutant gene. Following cell division, one nonmutant (untransformed) cell and one mutant (transformed) cell are produced (step 5).

Transformation and Linked Genes

For DNA to be effective in transformation, it must include between 10,000 and 20,000 nucleotide pairs, about 1/200 of the *E. coli* chromosome. This size is sufficient to encode several genes. Genes adjacent or very close to one another on the bacterial chromosome can be carried on a single segment of this size. Because of this fact, a single event can result in the **cotransformation** of several genes simultaneously. Genes that are close enough to each other to be contransformed are said to be *linked*. In contrast to the use of the term *linkage* in eukaryotes, which indicates all genes on a single chromosome, note that here linkage refers to the close proximity of genes.

If two genes are not linked, simultaneous transformation occurs only as a result of two independent events involving two distinct segments of DNA. As in double crossing over in eukaryotes, the probability of two independent events occurring simultaneously is equal to the product of the individual probabilities. Thus, the frequency of two unlinked genes being transformed simultaneously is much lower than if they are linked.

Subsequent studies have shown that, in addition to *Diplococcus pneumoniae*, a variety of bacteria readily undergo transformation (e.g., *Hemophilus influenzae*, *Bacillus subtilis*, *Shigella paradysenteriae*, and *E. coli*). Under certain conditions, studies have shown that relative distances between linked genes can be determined from transformation data. While analysis is more complex, such data are interpreted in a manner analogous to chromosome mapping in eukaryotes.

9.6 The Genetic Study of Bacteriophages

Bacteriophages, or **phages** as they are commonly known, are viruses that have bacteria as their hosts. During their reproduction, phages can be involved in still another mode of bacterial genetic recombination called transduction. To understand this process, we must consider the genetics of bacteriophages, which themselves also undergo recombination. In this section, we first examine the structure and life cycle of one type of bacteriophage. We then discuss how these phages are studied during their infection of bacteria. Finally, we contrast two possible modes of behavior once initial phage infection occurs. This information is background for our subsequent discussion of transduction and bacteriophage recombination. Furthermore, a great deal of genetic research has been done using bacteriophages as a model system, making them a worthy subject of discussion.

Phage T4: Structure and Life Cycle

Bacteriophage T4 is one of a group of related bacterial viruses referred to as T-even phages. It exhibits an intricate structure, as shown in Figure 9–14. Its genetic material, DNA, is contained within an icosahedral (a polyhedron with 20 faces) protein coat, together making up the head of the virus. The DNA is sufficient in quantity to encode more than 150 average-size genes. The head is connected to a tail that contains a collar and a contractile sheath surrounding a central core. Tail fibers, which protrude from the tail, contain binding sites in their tips that specifically recognize unique areas of the outer surface of the cell wall of the bacterial host, *E. coli*.

The life cycle of phage T4 (Figure 9–15) is initiated when the virus binds by adsorption to the bacterial host cell. Then,

FIGURE 9–14 The structure of bacteriophage T4 including an icosahedral head filled with DNA, a tail consisting of a collar, tube, sheath, base plate, and tail fibers. During assembly, the tail components are added to the head and then tail fibers are added. *(Photo: M. Wurtz/Biozentrum, University of Basel/Science Photo Library/Photo Researchers, Inc.)*

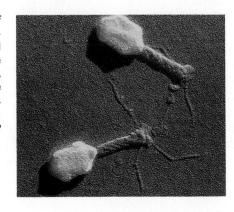

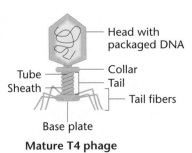

Head with packaged DNA
Tube
Sheath
Collar
Tail
Tail fibers
Base plate
Mature T4 phage

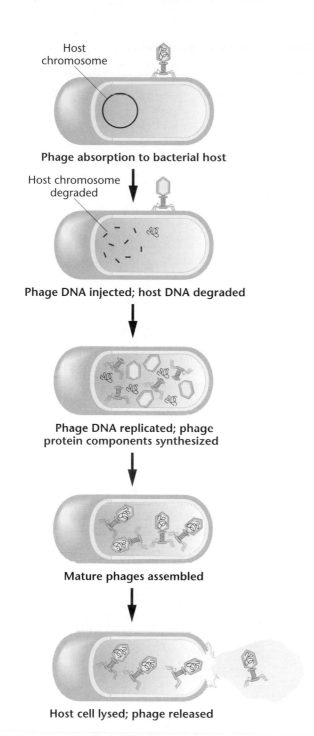

Host
chromosome

Phage absorption to bacterial host

Host chromosome
degraded

Phage DNA injected; host DNA degraded

**Phage DNA replicated; phage
protein components synthesized**

Mature phages assembled

Host cell lysed; phage released

FIGURE 9–15 Life cycle of bacteriophage T4.

an ATP-driven contraction of the tail sheath causes the central core to penetrate the cell wall. The DNA in the head is extruded, and it then moves across the cell membrane into the bacterial cytoplasm. Within minutes, all bacterial DNA, RNA, and protein synthesis is inhibited, and synthesis of viral molecules begins. At the same time, degradation of the host DNA is initiated.

A period of intensive viral gene activity characterizes infection. Initially, phage DNA replication occurs, leading to a pool of viral DNA molecules. Then, the components of the head, tail, and tail fibers are synthesized. The assembly of mature viruses is a complex process that has been well stud-

ied by William Wood, Robert Edgar, and others. Three sequential pathways occur: (1) DNA packaging as the viral heads are assembled, (2) tail assembly, and (3) tail fiber assscmbly. Once DNA is packaged into the head, it combines with the tail components, to which tail fibers are added. Total construction is a combination of self-assembly and enzyme-directed processes.

When approximately 200 viruses are constructed, the bacterial cell is ruptured by the action of lysozyme (a phage gene product), and the mature phages are released from the host cell. The 200 new phages infect other available bacterial cells, and the process repeats over and over again.

The Plaque Assay

Bacteriophages and other viruses have played a critical role in our understanding of molecular genetics. During infection of bacteria, enormous quantities of bacteriophages can be obtained for investigation. Often, over 10^{10} viruses are produced per milliliter of culture medium. Many genetic studies rely on the ability to quantify the number of phages produced following infection under specific culture conditions. The technique used routinely is called the **plaque assay**.

This assay is shown in Figure 9–16, where actual plaque morphology is also shown. A serial dilution of the original virally infected bacterial culture is first performed. Then, a 0.1-mL sample (an aliquot) from one or more dilutions is added to a small volume of melted nutrient agar (about 3 mL) into which a few drops of a healthy bacterial culture have been added. The solution is then poured evenly over a base of solid nutrient agar in a Petri dish and allowed to solidify before incubation. A clear area called a **plaque** occurs wherever a single virus initially infected one bacterium in the lawn that has grown up during incubation. The plaque represents clones of the single infecting bacteriophage, created as reproduction cycles are repeated. If the dilution factor is too low, the plaques are plentiful and they will fuse, lysing the entire lawn. This has occurred in the 10^{-3} dilution in Figure 9–16. On the other hand, if the dilution factor is increased, plaques can be counted and the density of viruses in the initial culture can be estimated:

$$\text{plaque number/mL} \times \text{dilution factor}$$

Using the results shown in Figure 9–16, 23 phage plaques are derived from the 0.1-mL aliquot of the 10^{-5} dilution. Therefore, we estimate that there are 230 phages/mL at this dilution. The initial phage density in the undiluted sample, where 23 plaques are observed from 0.1 mL, is calculated as

$$\text{initial phage density} = 230/\text{mL} \times 10^5 = 230 \times 10^5/\text{mL}$$

Because this figure is derived from the 10^{-5} dilution, we also estimate that there will be only 0.23 phage/0.1 mL in the 10^{-7} dilution. As a result, when 0.1 mL from this tube is assayed, it is predicted that no phage particles will be present. This possibility is borne out in Figure 9–16, where an intact

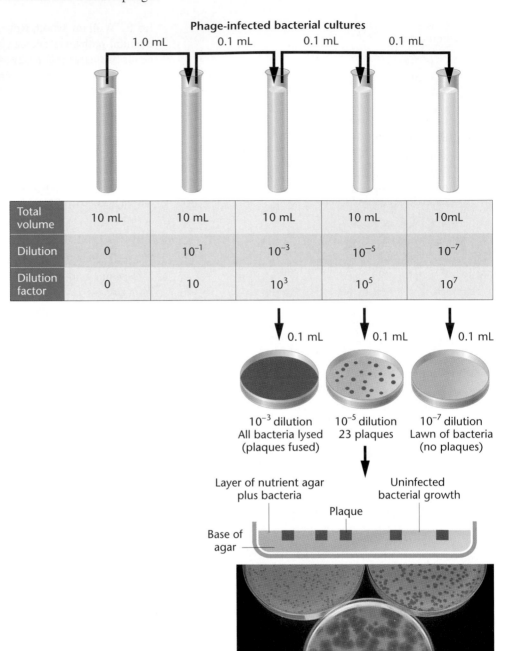

lawn of bacteria is depicted. The dilution factor is simply too great.

The use of the plaque assay is invaluable in mutational and recombinational studies of bacteriophages.

Lysogeny

The relationship between virus and bacterium does not always result in viral reproduction and lysis. As early as the 1920s, it was known that a virus can enter a bacterial cell and establish a symbiotic relationship with it. The precise molecular basis of this symbiosis is now well understood. Upon entry, the viral DNA, instead of replicating in the bacterial cytoplasm, is integrated into the bacterial chromosome, a step that characterizes the developmental stage referred to as **lysogeny**. Subsequently, each time the bacterial chromosome is replicated, the viral DNA is also replicated and passed to daughter bacterial cells following division. No new viruses are produced, and no lysis of the bacterial cell occurs. However, under certain stimuli, such as chemical or ultraviolet light treatment, the viral DNA loses its integrated status and initiates replication, phage reproduction, and lysis of the bacterium.

Several terms are used to describe this relationship. The viral DNA that integrates into the bacterial chromosome is a **prophage**. Viruses that either lyse the cell or behave as a prophage are **temperate**. Those that only lyse the cell are referred to as **virulent**. A bacterium harboring a prophage is **lysogenic**; that is, it is capable of being lysed as a result of induced viral reproduction. The viral DNA, which can replicate either in the bacterial cytoplasm or as part of the bacterial chromosome, is classified as an **episome**.

9.7 Transduction: Virus-Mediated Bacterial DNA Transfer

In 1952, Norton Zinder and Joshua Lederberg were investigating possible recombination in the bacterium *Salmonella typhimurium*. Although they recovered prototrophs from mixed cultures of two different auxotrophic strains, subsequent investigations revealed that recombination was occurring in a manner different from that attributable to the presence of an F factor, as in *E. coli*. What they discovered was a process of bacterial recombination mediated by bacteriophages and now called **transduction**.

The Lederberg–Zinder Experiment

Lederberg and Zinder mixed the *Salmonella* auxotrophic strains LA-22 and LA-2 together and, when the mixture was plated on minimal medium, they recovered prototrophic cells. The LA-22 strain was unable to synthesize the amino acids phenylalanine and tryptophan (phe^- trp^-), and LA-2 could not synthesize the amino acids methionine and histidine (met^- his^-). Prototrophs (phe^+ trp^+ met^+ his^+) were recovered at a rate of about $1/10^5$ (or 10^{-5}) cells.

Although these observations at first suggested that the recombination involved was the type observed earlier in conjugative strains of *E. coli*, experiments using the Davis U-tube soon showed otherwise (Figure 9–17). The two auxotrophic strains were separated by a sintered glass filter, thus preventing cell contact but allowing growth to occur in a common medium. Surprisingly, when samples were removed from both sides of the filter and plated independently on minimal medium, prototrophs were recovered only from the side of the tube containing LA-22 bacteria. Recall that if conjugation were responsible, the conditions in the Davis U-tube would be expected to prevent recombination altogether (see Figure 9–4).

Since LA-2 cells appeared to be the source of the new genetic information (phe^+ and trp^+), how that information crossed the filter from the LA-2 cells to the LA-22 cells, allowing recombination to occur, was a mystery. The unknown source was designated simply as a **filterable agent (FA)**.

Three subsequent observations were useful in identifying the FA:

1. The FA was produced by the LA-2 cells only when they were grown in association with LA-22 cells. If LA-2 cells were grown independently and that culture medium was then added to LA-22 cells, recombination did not occur. Therefore, LA-22 cells play some role in the production of FA by LA-2 cells and do so only when the two share common growth medium.

2. The presence of DNase, which enzymatically digests DNA, did not render the FA ineffective. Therefore, the FA is not naked DNA, ruling out transformation.

3. The FA could not pass across the filter of the Davis U-tube when the pore size was reduced below the size of bacteriophages.

Aided by these observations and aware that temperate phages can lysogenize *Salmonella*, researchers proposed that the genetic recombination event is mediated by bacteriophage P22, present initially as a prophage in the chromosome of the LA-22 *Salmonella* cells. It was hypothesized that rarely P22 prophages enter the vegetative or lytic phase, reproduce, and are released by the LA-22 cells. Such phages, being much smaller than a bacterium, then cross the filter of the U-tube and subsequently infect and lyse some of the LA-2 cells. In the process of lysis of LA-2, these P22 phages occasionally package in their heads a region of the LA-2 chromosome. If this region contains the phe^+ and trp^+ genes, and the phages subsequently pass back across the filter and infect LA-22 cells, these newly lysogenized cells will behave as prototrophs. This process of transduction, whereby bacterial recombination is mediated by bacteriophage P22, is diagrammed in Figure 9–18.

The Nature of Transduction

Further studies revealed the existence of transducing phages in other species of bacteria. For example, *E. coli* can be transduced by phages P1 and λ. *Bacillus subtilis* and *Pseudomonas aeruginosa* can be transduced by the phages SPO1 and F116, respectively. The details of several different modes of transduction have also been established. Even though the initial discovery of transduction involved a temperate phage and a lysogenized bacterium, the same process can occur during the normal lytic cycle. Sometimes a small piece of bacterial

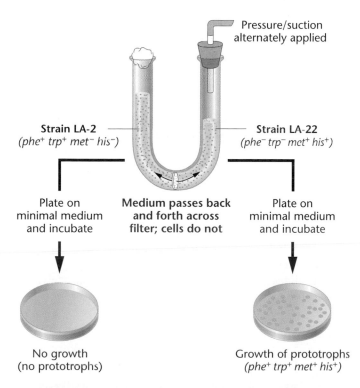

Strain LA-2
(phe^+ trp^+ met^- his^-)

Strain LA-22
(phe^- trp^- met^+ his^+)

Pressure/suction
alternately applied

**Medium passes back
and forth across
filter; cells do not**

Plate on
minimal medium
and incubate

Plate on
minimal medium
and incubate

No growth
(no prototrophs)

Growth of prototrophs
(phe^+ trp^+ met^+ his^+)

FIGURE 9–17 The Lederberg–Zinder experiment using *Salmonella*. After placing two auxotrophic strains on opposite sides of a Davis U-tube, Lederberg and Zinder recovered prototrophs from the side containing the LA-22 strain but not from the side containing the LA-2 strain. These initial observations led to the discovery of the phenomenon called transduction.

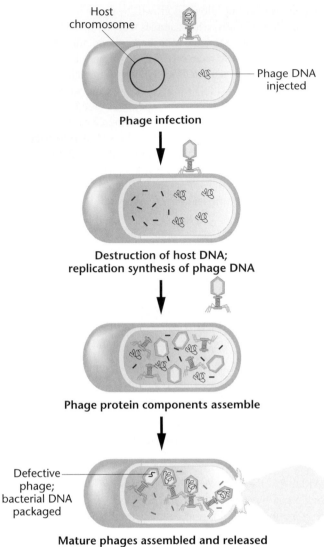

Host chromosome

Phage DNA injected

Phage infection

Destruction of host DNA; replication synthesis of phage DNA

Phage protein components assemble

Defective phage; bacterial DNA packaged

Mature phages assembled and released

Subsequent infection of another cell with defective phage; Bacterial DNA injected by phage

Transduction

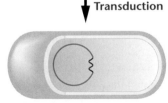

Integration of bacterial DNA into recipient chromosome

FIGURE 9–18 Generalized transduction.

DNA is packaged *along with* the viral chromosome so that the transducing phage contains both viral and bacterial DNA. In such cases, only a few bacterial genes are present in the transducing phage. However, when *only* bacterial DNA is packaged, regions as large as 1 percent of the bacterial chromosome can become enclosed in the viral head. In either case, the ability to infect is unrelated to the type of DNA in the phage head, making transduction possible.

When bacterial rather than viral DNA is injected into the bacterium, it either remains in the cytoplasm or recombines with the homologous region of the bacterial chromosome. If the bacterial DNA remains in the cytoplasm, it does not replicate but is transmitted to one progeny cell following each division. When this happens, only a single cell, partially diploid for the transduced genes, is produced—a phenomenon called **abortive transduction**. If the bacterial DNA recombines with its homologous region of the bacterial chromosome, the transduced genes are replicated as part of the chromosome and passed to all daughter cells. This process is called **complete transduction**.

Both abortive and complete transduction are subclasses of the broader category of **generalized transduction**, which is characterized by the random nature of DNA fragments and genes transduced. Each fragment of the bacterial chromosome has a finite but small chance of being packaged in the phage head. Most cases of generalized transduction are of the abortive type; some data suggest that complete transduction occurs 10–20 times less frequently.

Transduction and Mapping

Like transformation, generalized transduction was used in linkage and mapping studies of the bacterial chromosome. The fragment of bacterial DNA involved in a transduction event is large enough to include numerous genes. As a result, two genes that closely align (are linked) on the bacterial chromosome can be simultaneously transduced, a process called **cotransduction**. Two genes that are not close enough to one another along the chromosome to be included on a single DNA fragment require two independent events to be transduced into a single cell. Since this occurs with a much lower probability than cotransduction, linkage can be determined.

By concentrating on two or three linked genes, transduction studies can also determine the precise order of these genes. The closer linked genes are to each other, the greater the frequency of cotransduction. Mapping studies involving three closely aligned genes can thus be executed, and the analysis of such an experiment is predicated on the same rationale underlying other mapping techniques.

9.8 Intergenic Recombination and Mapping in Bacteriophages

Around 1947, several research teams demonstrated that genetic recombination can be detected in bacteriophages. This led to the discovery that gene mapping can be performed in these viruses. Such studies relied on finding numerous phage mutations that could be visualized or assayed. As in bacteria and eukaryotes, these mutations allow genes to be identified and followed in mapping experiments. Thus, before considering recombination and mapping in these bacterial viruses, we briefly introduce several of the mutations that were studied.

Bacteriophage Mutations

Phage mutations often affect the morphology of the plaques formed following lysis of bacterial cells. For example, in 1946, Alfred Hershey observed unusual T2 plaques on plates of *E. coli* strain B. Where the normal T2 plaques are small and have a clear center surrounded by a diffuse (nearly invisible) halo, the unusual plaques were larger and possessed a distinctive outer perimeter (Figure 9–19). When the viruses were isolated from these plaques and replated on *E. coli* B cells, the resulting plaque appearance was identical. Thus, the plaque phenotype was an inherited trait resulting from the reproduction of mutant phages. Hershey named the mutant *rapid lysis* (*r*) because the plaques were larger, apparently resulting from a more rapid or more efficient life cycle of the phage. It is now known that in wild-type phages, reproduction is inhibited once a particular-sized plaque has

TABLE 9.1	Some Mutant Types of T-Even Phages
Name	*Description*
minute	Small plaques
turbid	Turbid plaques on *E. coli* B
star	Irregular plaques
uv-sensitive	Alters UV sensitivity
acriflavin-resistant	Forms plaques on acriflavin agar
osmotic shock	Withstands rapid dilution into distilled water
lysozyme	Does not produce lysozyme
amber	Grows in *E. coli* K12, but not B
temperature-sensitive	Grows at 25°C, but not at 42°C

been formed. The *r* mutant T2 phages are able to overcome this inhibition, producing larger plaques.

Luria also discovered another bacteriophage mutation, *host range* (*h*). This mutation extends the range of bacterial hosts that the phage can infect. Although wild-type T2 phages can infect *E. coli* B, they normally cannot attach or be adsorbed to the surface of *E. coli* B-2. The *h* mutation, however, provides the basis for adsorption and subsequent infection of *E. coli* B-2. When grown on a mixture of *E. coli* B and B-2, the center of the *h* plaque appears much darker than the h^+ plaque (Figure 9–19).

Table 9.1 lists other types of mutations that have been isolated and studied in the T-even series of bacteriophages (e.g., T2, T4, T6). These mutations are important to the study of genetic phenomena in bacteriophages.

Intergenic Mapping

Genetic recombination in bacteriophages was discovered during **mixed infection experiments** in which two distinct mutant strains were allowed to *simultaneously* infect the same bacterial culture. These studies were designed so that the number of viral particles sufficiently exceeded the number of bacterial cells to ensure simultaneous infection of most cells by both viral strains. Because two loci are involved, recombination is referred to as intergenic.

For example, in one study using the T2/*E. coli* system, the parental viruses were of either the h^+r (wild-type host range, rapid lysis) or hr^+ (extended host range, normal lysis) genotype. If no recombination occurred, these two parental genotypes would be the only expected phage progeny. However, the recombinants h^+r^+ and hr were detected in addition to the parental genotypes (see Figure 9–19). As with eukaryotes, the percentage of recombinant plaques divided by the total number of plaques reflects the relative distance between the genes.

$$\text{recombinational frequency} = (h^+r^+ + hr)/\text{total plaques} \times 100$$

Sample data for the *h* and *r* loci are shown in Table 9.2.

Similar recombinational studies have been conducted with numerous mutant genes in a variety of bacteriophages. Data are analyzed in much the same way as they are in eukaryotic

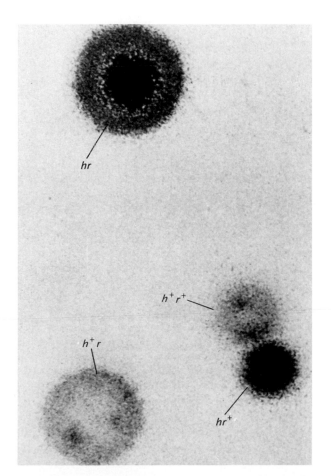

FIGURE 9–19 Plaque morphology phenotypes observed following simultaneous infection of *E. coli* by two strains of phage T2, h^+r and hr^+. In addition to the parental genotypes, recombinant plaques *hr* and h^+r^+ were also recovered. *(From Hershey & Chase, 1951)*

TABLE 9.2 **Results of a Cross Involving the**
h and r Genes in Phage T2 ($hr^+ \times h^+r$)

Genotype	Plaques	Designation
h r^+	42	Parental progeny
h^+ r	34	76%
h^+ r^+	12	Recombinants
h r	12	24%

Source: Data derived from Hershey and Rotman, 1949.

mapping experiments. Two- and three-point mapping crosses are possible, and the percentage of recombinants in the total number of phage progeny is calculated. This value is proportional to the relative distance between two genes along the DNA molecule constituting the chromosome.

Investigations into phage recombination support a model similar to that of eukaryotic crossing over—a breakage and reunion process between the viral chromosomes. A fairly clear picture of the dynamics of viral recombination is emerging. Following the early phase of infection, the chromosomes of the phages begin replication. As this stage progresses, a pool of chromosomes accumulates in the bacterial cytoplasm. If double infection by phages of two genotypes has occurred, then the pool of chromosomes initially consists of the two parental types. Genetic exchange between these two types will occur before, during, and after replication, producing recombinant chromosomes.

In the case of the h^+r and hr^+ example discussed here, recombinant h^+r^+ and hr chromosomes are produced. Each of these chromosomes can undergo replication, with new replicates exchanging with each other and with parental chromosomes. Furthermore, recombination is not restricted to exchanges between two chromosomes—three or more can be involved simultaneously. As phage development progresses, chromosomes are randomly removed from the pool and packed into the phage head, forming mature phage particles. Thus, a variety of parental and recombinant genotypes are represented in progeny phages.

Chapter Summary

1. Inherited phenotypic variation in bacteria results from spontaneous mutation.
2. Genetic recombination in bacteria can result from three different modes: conjugation, transformation, and transduction.
3. Conjugation is initiated by a bacterium housing a plasmid called the F factor. If the F factor is in the cytoplasm of a donor cell (F^+), the recipient F^- cell receives a copy of the F factor, converting it to the F^+ status.
4. If the F factor is integrated into the donor cell chromosome (Hfr), recombination is initiated with the recipient F^- cell, with genetic information flowing unidirectionally to it. Time mapping of the bacterial chromosome is based on the orientation and position of the F factor in the donor chromosome.
5. The products of a group of genes designated as *rec* are directly involved in recombination between the invading DNA and the recipient bacterial chromosome.
6. Plasmids, such as the F factor, are autonomously replicating DNA molecules found in the bacterial cytoplasm. Plasmids contain unique genes, such as those conferring antibiotic resistance, as well as those necessary for their transfer during conjugation.

7. In the phenomenon of transformation, which does not require cell contact, exogenous DNA enters the host chromosome of a recipient bacterial cell. Linkage mapping of closely aligned genes is performed using this process.
8. Bacteriophages (viruses that infect bacteria) demonstrate a defined life cycle during which they reproduce within the host cell. They are studied using the plaque assay.
9. Bacteriophages can be lytic, where they infect, reproduce, and lyse the host cell; or they can lysogenize the host cell, where they infect it, integrate their DNA into the host chromosome, and do not reproduce.
10. Transduction is virus-mediated bacterial DNA recombination. When a lysogenized bacterium subsequently reenters the lytic cycle, the new bacteriophages serve as the vehicles for the transfer of host (bacterial) DNA. In the process of generalized transduction, a random part of the bacterial chromosome is transferred.
11. Transduction is also used for bacterial linkage and mapping studies.

Genetics, Technology, and Society

Eradicating Cholera: Edible Vaccines

Almost lost amid the furor over the cloned sheep Dolly was the original purpose of the undertaking: to genetically engineer an animal that would produce a valuable pharmaceutical product, eventually leading to a herd of genetically identical drug-producing animals. The goal, in other words, was to be able to turn sheep, cattle, and other animals into living drug factories.

Meanwhile, almost unnoticed by the general public, the genetic engineering of plants to serve a similar role is much closer to reality. When foreign genes are inserted into the plant genome, transgenic plants are created that produce the foreign gene product. One of the most intriguing and potentially life-saving objectives is the genetic engineering of plants to produce vaccines against human diseases. Immunization would then involve eating foods altered to contain a protein that acts as an antigen and stimulates the production of antibodies to protect against bacterial or viral infection.

Plants have many advantages as vaccine-producers. Since plants can be grown in large numbers, plant-produced vaccines should be less expensive than conventional vaccines. Further, extensive purification and refrigerated transport and storage of vaccines will not be required. This is important in parts of the world where the supply of electricity is unreliable. Finally, since people would simply eat the vaccine-containing food, there would be no need for syringes and needles, or medical staff to give injections.

Leading the effort to develop edible vaccines in plants is Charles J. Arntzen, formerly at Texas A&M University and now president of the Boyce Thompson Institute for Plant Research at Cornell University. Arntzen's research team is focusing on intestinal diseases, especially cholera. Cholera may at first seem an odd target, since it is a disease that has not been a major public health problem in this country for over a century. But cholera remains a leading cause of death of infants and children throughout the Third World,

where basic sanitation is lacking and water supplies are often contaminated. For example, in July of 1994, 70,000 cases of cholera were reported among the Rwandans crowded into refugee camps in Goma, Zaire, leading to 12,000 fatalities. A cholera epidemic struck East Africa in the fall of 1997 killing nearly 3000. And after an absence of over 100 years, cholera reappeared in Latin America in 1991, spreading from Peru to Mexico and claiming more than 10,000 lives.

The causative agent of cholera is *Vibrio cholerae*, a curved, rod-shaped bacterium found mostly in rivers and oceans. Most strains of *V. cholerae* are harmless; only one strain, called O1, is pathogenic. Infection occurs when a person drinks water or eats food contaminated with this strain. Once in the digestive system, the bacteria colonize the small intestine and begin producing proteins called enterotoxins. The cholera enterotoxin binds to the surface of the mucosal cells lining the intestine, triggering the massive secretion of water and dissolved salts from these cells. This results in violent diarrhea, which, if untreated, is followed by severe dehydration, muscle cramps, lethargy, and often death.

The cholera enterotoxin consists of two polypeptides, called the A and B subunits, which individually have no effect. For the toxin to be active, one A subunit must be linked to five B subunits. Since it is the B subunit of this complex that binds to intestinal cells, Arntzen's group decided to use this polypeptide as their antigen, reasoning that antibodies against it would potentially prevent toxin binding and render the bacteria harmless.

Their test of this idea involved the B subunit of an *E. coli* enterotoxin, which is similar in structure and immunological properties to the cholera protein. The DNA coding sequence of the B-subunit gene was obtained, to which they attached a promoter that would prompt transcription in all tissues of the plant. They then introduced this hybrid gene into potato plants by means of *Agrobacterium*-mediated transformation. They chose the potato not only because methods for transformation and regeneration of this plant are fairly rou-

tine, but also so they could assay the effectiveness of the antigen in the edible part of the plant, the tuber. Analysis showed that the engineered plants expressed their new gene and produced the enterotoxin.

After feeding mice a few grams of these genetically engineered tubers that had produced the B subunit, they found that the mice began to produce specific antibodies against it and to secrete them into the small intestine. And, most critically, mice later fed purified enterotoxin were protected from its effects. They did not develop the symptoms of cholera. Clinical trials are now planned to test the efficacy of the potato-produced vaccine in humans.

In the meantime, the Arntzen group is also working towards producing edible vaccines in bananas, which have several advantages over potatoes. For one thing, bananas can be grown almost anywhere throughout the tropical or sub-tropical developing countries of the world, exactly where they are needed the most. And unlike potatoes, bananas are usually eaten raw, avoiding the potential inactivation of the antigenic proteins by cooking. Finally, bananas are well liked by infants and children, making this approach to immunization a more feasible one.

Procedures are now being perfected for the transformation of banana cells with genes encoding the cholera enterotoxin and the regeneration of transgenic plants. It will be some time before the engineered bananas are ready to test, however, since it takes three years to grow a banana crop. If all goes as planned, it may someday be possible to immunize all Third World children against cholera and other intestinal diseases, saving untold thousands of young lives.

References

Arntzen, C. J. 1997. Edible vaccines. *Public Health Rep.* 112: 190–197.

Haq, T. A., Mason, H. S., Clements, J. D., and Arntzen, C. J. 1995. Oral immunization with a recombinant bacterial antigen produced in transgenic plants. *Science* 268: 714–716.

Sanchez, J. L., and Taylor, D. N. 1997. Cholera. *Lancet* 349: 1825–1830.

Insights and Solutions

1. Time mapping is performed in a cross involving the genes *his*, *leu*, *mal*, and *xyl*. The recipient cells are auxotrophic for all four genes. After 25 minutes, mating is interrupted with the following results in recipient cells.

$$90\% \text{ are } xyl^+$$
$$80\% \text{ are } mal^+$$
$$20\% \text{ are } his^+$$
$$\text{none are } leu^+$$

What are the positions of these genes relative to the origin (O) of the F factor and to one another?

Solution: Because the *xyl* gene is transferred most frequently, it is closest to O (very close). The *mal* gene is next and reasonably close to *xyl*, followed by the *his* gene. The *leu* gene is well beyond these three, since no recombinants recovered include it. The diagram shows these relative locations along a piece of the circular chromosome:

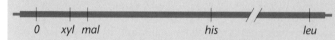

2. In four Hfr strains of bacteria, all derived from an original F^+ culture grown over several months, a group of hypothetical genes is studied and shown to transfer in the following orders:

Hfr Strain	Order of Transfer					
1	E	R	I	U	M	B
2	U	M	B	A	C	T
3	C	T	E	R	I	U
4	R	E	T	C	A	B

Assuming *B* is the first gene along the chromosome, determine the sequence of all genes shown. One strain creates an apparent dilemma. Which one is it? Explain why the dilemma is only apparent and not real.

Solution: The solution is found by overlapping the genes in each strain in the sequence in which they transfer:

```
        2  U  M  B  A  C  T
Strain: 3              C  T  E  R  I  U
        1                    E  R  I  U  M  B
```

Starting with *B*, the gene sequence is *BACTERIUM*.

Strain 4 creates a dilemma, which is resolved when we realize that the F factor is integrated in the opposite orientation; thus, the genes enter in the opposite sequence, starting with gene *R*.

$$\overrightarrow{MUIRETCAB}$$

3. Three strains of bacteria, each bearing a separate mutation, a^-, b^-, or c^- are the sources of donor DNA in a transformation experiment. Recipient cells are wild type for those genes, but express the mutant d^-.

(a) Based on the data, and assuming that the location of the *d* gene precedes the *a*, *b*, and *c* genes, propose a linkage map for the four genes.

DNA Donor	Recipient	Transformants	Frequency of Transformants
$a^-\ d^+$	$a^+\ d^-$	$a^+\ d^+$	0.21
$b^-\ d^+$	$b^+\ d^-$	$b^+\ d^+$	0.18
$c^-\ d^+$	$c^+\ d^-$	$c^+\ d^+$	0.63

Solution: These data reflect the relative distances between each of the *a*, *b*, *c* genes and the *d* gene. The *a* and *b* genes are about the same distance from the *d* gene and thus tightly linked to one another. The *c* gene is more distant. Assuming that the *d* gene precedes the others, the map looks like this:

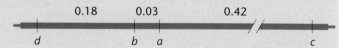

(b) If the donor DNA is wild type and the recipient cells are either a^-b^-, a^-c^-, or b^-c^-, in which case would wild-type transformants be expected most frequently?

Solution: Because the *a* and *b* genes are closely linked, they most likely cotransform in a single event. Thus, recipient cells of a^-b^- are most likely to convert to wild type.

Key Terms

Problems and Discussion Questions

1. Distinguish between the three modes of recombination in bacteria.
2. With respect to F^+ and F^- bacterial matings, answer the following questions: (a) How was it established that physical contact is necessary? (b) How was it established that chromosome transfer is unidirectional? (c) What is the basis of a bacterium being F^+ or Hfr?
3. List the major differences between (a) the $F^+ \times F^-$ and the Hfr $\times F^-$ bacterial crosses, and (b) F^+, F^-, Hfr, and F′ bacteria.
4. Describe the basis for chromosome mapping in the Hfr $\times F^-$ crosses.
5. When the interrupted mating technique is used with five different strains of Hfr bacteria, the following order of gene entry and recombination is observed. On the basis of these data, map the bacterial chromosome. Do the data support the concept of circularity?

Hfr Strain			Order		
1	T	C	H	R	O
2	H	R	O	M	B
3	M	O	R	H	C
4	M	B	A	K	T
5	C	T	K	A	B

6. Why are recombinants produced from an Hfr $\times F^-$ cross never F^+?
7. Explain the origin of F′ bacteria and merozygotes.
8. Describe the observations that led Zinder and Lederberg to conclude that the prototrophs recovered in their transduction experiments were not the result of F-mediated conjugation.
9. Define plaque, lysogeny, and prophage.
10. If a single bacteriophage infects one *E. coli* cell present on a lawn of bacteria and upon lysis yields 200 viable viruses, how many phages will exist in a single plaque if only three more lytic cycles occur?
11. A culture of an auxotrophic *leu*⁻ strain of bacteria is irradiated and incubated until it reaches the stationary phase. The control culture is not irradiated. These cultures are then serially diluted, and 0.1 mL of various dilutions is plated on minimal medium plus leucine and separately on minimal medium. The results, shown here are used to determine the spontaneous and X-ray-induced mutation rate of leu⁻ to leu⁺.

Culture Condition	Culture Medium		Dilution	Number of Colonies
Irradiated	(1) Minimal medium plus leucine		10^{-9}	24
	(2) Minimal medium		10^{-2}	12
Control	(1) Minimal medium plus leucine		10^{-9}	12
	(2) Minimal medium		10^{-1}	3

(a) Describe what is represented by each value. Which values should be approximately equal? Are they?
(b) Determine the induced and spontaneous mutation rate leading to prototrophic growth (leu⁻ to leu⁺).

12. In a transformation experiment, donor DNA is obtained from a prototrophic bacterial strain ($a^+b^+c^+$), and the recipient is auxotrophic for the three genes ($a^-b^-c^-$). The following data are obtained:

$u^+\ b^-\ c^-$	180
$a^-\ b^+\ c^-$	150
$u^+\ b^+\ c$	210
$a^-\ b^-\ c^+$	179
$a^+\ b^-\ c^+$	2
$a^-\ b^+\ c^+$	1
$a^+\ b^+\ c^+$	3

What general conclusions can you draw about the linkage relationships among the three genes?

13. Two theoretical phage strains ($a^-b^-c^-$ and $a^+b^+c^+$) are used to simultaneously infect a bacterial culture. Of 10,000 plaques scored, the following data are obtained:

$a^+\ b^+\ c^+$	4100	$a^-\ b^+\ c^-$	160
$a^-\ b^-\ c^-$	3990	$a^+\ b^-\ c^+$	140
$a^+\ b^-\ c^-$	740	$a^-\ b^-\ c^+$	90
$a^-\ b^+\ c^+$	670	$a^+\ b^+\ c^-$	110

Determine the genetic map of the three genes on the viral chromosome.

14. An Hfr strain is used to map three genes in an interrupted mating experiment. The cross is

$$\text{Hfr } a^+b^+c^+rif^S \times F^- \ a^-b^-c^-rif^R.$$

(No map order is implied in the listing of the alleles. The a^+ gene is required for the biosynthesis of nutrient A, the b^+ gene for nutrient B, and c^+ for nutrient C. The minus alleles indicate mutations rif is the antibiotic rifamycin.)

The cross is started at time 0, and at various times the mating mixture is plated on three types of plates. Each plate contains minimal medium *and* rifamycin *plus* specific supplements, which are indicated below. The results for each time point are shown as number of colonies growing on each plate.

Supplements Added to MM+ rifamycin	Time of Interruption			
	5 min	10 min	15 min	20 min
Nutrients A and B	0	0	4	21
Nutrients B and C	0	5	23	40
Nutrients A and C	4	25	60	82

(a) What genotype is being selected for on each type of medium?

(b) Map the *E. coli* chromosome and show the relative positions of genes a, b, and c approximately to scale (it takes about 90 minutes to transfer the F factor genes in). If the data do not give an unambiguous position for a specific gene, indicate so on your map. Also indicate the position and orientation of the F factor.

Selected Readings

Adelberg, E. A. 1960. *Papers on bacterial genetics*. Boston: Little, Brown.

Birge, E. A. 1988. *Bacterial and bacteriophage genetics—An introduction*. New York: Springer-Verlag.

Brock, T. 1990. *The emergence of bacterial genetics*. Cold Spring Harbor, NY: Cold Spring Harbor Laboratory Press.

Broda, P. 1979. *Plasmids*. New York: W. H. Freeman.

Bukhari, A. I., Shapiro, J. A., and Adhya, S. L., eds. 1977. *DNA insertion elements, plasmids, and episomes*. Cold Spring Harbor, NY: Cold Spring Harbor Laboratory.

Cairns, J., Stent, G. S., and Watson, J. D., eds. 1966. *Phage and the origins of molecular biology*. Cold Spring Harbor, NY: Cold Spring Harbor Laboratory.

Campbell, A. M. 1976. How viruses insert their DNA into the DNA of the host cell. *Sci. Am.* (Dec.) 235:102–13.

Fox, M.S. 1966. On the mechanism of integration of transforming deoxyribonucleate. *J. Gen. Physiol.* 49:183–96.

Hayes, W. 1968. *The genetics of bacteria and their viruses*, 2nd ed. New York: Wiley.

————. 1968. *The genetics of bacteria and their viruses*, 2nd ed. New York: Wiley.

Hershey, A. D., and Rotman, R. 1949. Genetic recombination between host range and plaque-type mutants of bacteriophage in single cells. *Genetics* 34:44–71.

Hotchkiss, R. D., and Marmur, J. 1954. Double marker transformations as evidence of linked factors in deoxyribonucleate transforming agents. *Proc. Natl. Acad. Sci. USA* 40:55–60.

Jacob, F., and Wollman, E. L. 1961. Viruses and genes. *Sci. Am.* (June) 204:92–106.

Lederberg, J. 1986. Forty years of genetic recombination in bacteria: A fortieth anniversary reminiscence. *Nature* 324:627–28.

Luria, S. E., and Delbruck, M. 1943. Mutations of bacteria from virus sensitivity to virus resistance. *Genetics* 28:491–511.

Lwoff, A. 1953. Lysogeny. *Bacteriol. Rev.* 17:269–337.

Miller, J. H. 1992. *A short course in bacterial genetics*. Cold Spring Harbor, NY: Cold Spring Harbor Press.

Morse, M. L., Lederberg, E. M., and Lederberg, J. 1956. Transduction in *Escherichia coli* K12. *Genetics* 41:141–56.

Novick, R. P. 1980. Plasmids. *Sci. Am.* (Dec.) 243:102–27.

Smith-Keary, P. F. 1989. *Molecular genetics of* Escherichia coli. New York: Guilford Press.

Stahl, F. W. 1987. Genetic recombination. *Sci. Am.* (Nov.) 256:91–101.

Stent, G. S. 1963. *Molecular biology of bacterial viruses*. New York: W. H. Freeman.

————. 1966. *Papers on bacterial viruses*, 2nd ed. Boston: Little, Brown.

Visconti, N., and Delbruck, M. 1953. The mechanism of genetic recombination in phage. *Genetics* 38:5–33.

Wollman, E. L., Jacob, F., and Hayes, W. 1956. Conjugation and genetic recombination in *Escherichia coli* K12. *Cold Spring Harb. Symp. Quant. Biol.* 21:141–62.

Zinder, N. D. 1958. Transduction in bacteria. *Sci. Am.* (Nov.) 199:38–46.

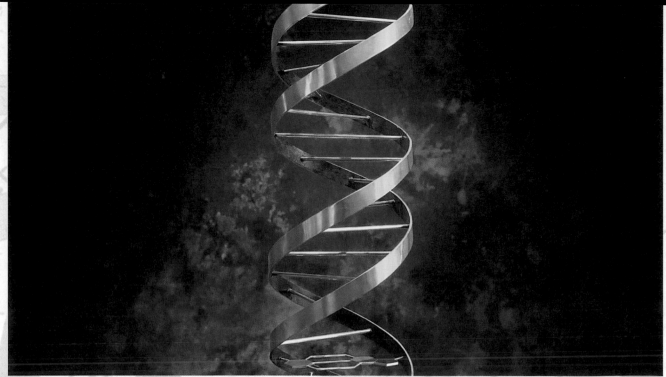

Bronze sculpture of double-helical DNA. *(Richard Megna/Fundamental Photographs)*

10

DNA Structure and Analysis

CHAPTER CONCEPTS

With few exceptions, the nucleic acid DNA serves as the genetic material in every living thing. The structure of DNA allows genetic information to be stored and expressed chemically within cells, as well as transmitting it to future generations. The molecule is a double-stranded helix united by hydrogen bonds formed between complementary nucleotides. In some viruses, RNA serves as the genetic material.

Earlier in the text, we discussed the existence of genes on chromosomes that control phenotypic traits and the way in which the chromosomes are transmitted through gametes to future offspring. Logically, genes must contain some sort of information, which, when passed to a new generation, influences the form and characteristics of the offspring; we refer to this as **genetic information**. We might also conclude that this same information in some way directs the many complex processes leading to the adult form.

Until 1944, it was not clear what chemical component of the chromosome makes up genes and constitutes the genetic material. Because chromosomes were known to have both a nucleic acid and a protein component, both were candidates. In 1944, however, there emerged direct experimental evidence that the nucleic acid DNA serves as the informational basis for heredity.

Once the importance of DNA in genetic processes was realized, work intensified with the hope of discerning not only the structural basis of this molecule but also the relationship of its structure to its function. Between 1944 and 1953, many scientists sought information that might answer the most significant and intriguing question in the history of biology: How does DNA serve as the genetic basis for the living process? Researchers believed the answer depended strongly on the chemical structure of the DNA molecule, given the complex but orderly functions ascribed to it.

These efforts were rewarded in 1953 when James Watson and Francis Crick set forth their hypothesis for the double-helical nature of DNA. The assumption that the molecule's functions would be clarified more easily once its general structure was determined proved to be correct. In this chapter we initially review the evidence that DNA is the genetic material and then discuss the elucidation of its structure.

10.1   Characteristics of the Genetic Material

For a molecule to serve as the genetic material, it must possess four major characteristics: **replication**, **storage of information**, **expression of information**, and **variation by mutation**. Replication of the genetic material is one facet of the cell cycle, a fundamental property of all living organisms. Once the genetic material of cells replicates and is doubled in amount, it must then be partitioned equally into daughter cells. During the formation of gametes, the genetic material is also replicated but is partitioned so that each cell gets only one-half of the original amount of genetic material—the process of *meiosis* discussed in Chapter 3. Although the products of mitosis and meiosis differ, these processes are both part of the more general phenomenon of cellular reproduction.

The characteristic of storage can be viewed as a repository of genetic information that may or may not be expressed. It is clear that while most cells contain a complete comple-

ment of DNA, at any point in time they express only part of its genetic potential. For example, bacteria "turn on" many genes in response to specific environmental conditions, and turn them off when such conditions change. In vertebrates, skin cells may display active melanin genes but never activate their hemoglobin genes; digestive cells activate many genes specific to their function, but do not activate their melanin genes.

Expression of the stored genetic information is the complex process of **information flow** within the cell (Figure 10–1). The initial event is the **transcription** of DNA, in which three types of RNA molecules are synthesized: **messenger RNA (mRNA)**, **transfer RNA (tRNA)**, and **ribosomal RNA (rRNA)**. Of these, mRNAs are translated into proteins. Each type of mRNA is the product of a specific gene and directs the synthesis of a different protein. **Translation** occurs in conjunction with ribosomes and involves tRNA, which adapts the chemical information in mRNA to the amino acids that make up proteins. Collectively, these processes form the **central dogma of molecular genetics**: "DNA makes RNA, which makes proteins."

The genetic material is also the source of variability among organisms through the process of mutation. If a mutation—a change in the chemical composition of DNA—occurs, the alteration is reflected during transcription and translation, affecting the specific protein. If a mutation is present in gametes, it will be passed to future generations and, with time, become distributed in the population. Genetic variation, which also includes rearrangements within and between chromosomes (see Chapter 7), provides the raw material for the process of evolution.

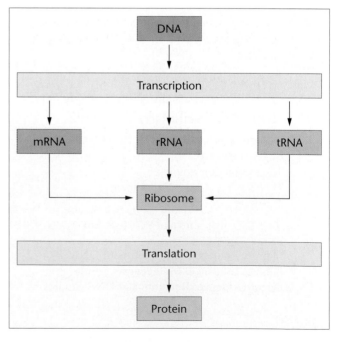

FIGURE 10–1 Simplified view of information flow involving DNA, RNA, and proteins within cells.

10.2   The Genetic Material: Early Studies

The idea that genetic material is physically transmitted from parent to offspring has been accepted for as long as the concept of inheritance has existed. Beginning in the late nineteenth century, research into the structure of biomolecules progressed considerably, setting the stage for describing the genetic material in chemical terms. Although proteins and nucleic acid were both major candidates for the role of the genetic material, until the 1940s, many geneticists favored proteins. This is not surprising, since a diversity of proteins was known to be abundant in cells, and much more was known about protein chemistry.

DNA was first studied in 1868 by a Swiss chemist, Friedrick Miescher. He isolated cell nuclei and derived an acid substance containing DNA that he called **nuclein**. As investigations progressed, however, DNA, which was shown to be present in chromosomes, seemed to lack the chemical diversity necessary to store extensive genetic information. This conclusion was based largely on Phoebus A. Levene's observations in 1910 that DNA contained approximately equal amounts of four quite similar molecules called **nucleotides**. Levene postulated incorrectly that identical groups of these four components were repeated over and over, which was the basis of his **tetranucleotide hypothesis** for DNA structure. Attention was thus directed away from DNA, favoring proteins, much by default. However, in the 1940s, Erwin Chargaff showed that Levene's proposal was incorrect when he demonstrated that most organisms do not contain precisely equal proportions of the four nucleotides. We shall see later that the structure of DNA accounts for Chargaff's observations.

10.3   Evidence Favoring DNA in Bacteria and Bacteriophages

The 1944 publication by Oswald Avery, Colin MacLeod, and Maclyn McCarty concerning the chemical nature of a "transforming principle" in bacteria was the event that led to the acceptance of DNA as the genetic material. Their work, along with subsequent findings of other research teams, constituted direct experimental proof that DNA, and not protein, is the biomolecule responsible for heredity. It marked the beginning of the era of molecular genetics, a period of discovery in biology that made biotechnology feasible and has moved us closer to understanding the basis of life. The impact of the initial findings on future research and thinking paralleled that of the publication of Darwin's theory of evolution and the subsequent rediscovery of Mendel's postulates of transmission genetics. Together, these events constituted the three great revolutions in biology.

Transformation Studies

The research that provided the foundation for Avery, MacLeod, and McCarty's work was initiated in 1927 by Frederick Griffith, a medical officer in the British Ministry of Health. He experimented with several different strains of the bacterium *Diplococcus pneumoniae.** Some were **virulent strains**, which cause pneumonia in certain vertebrates (notably humans and mice), while others were **avirulent strains**, which do not cause illness.

The difference in virulence depends on the existence of a polysaccharide capsule. Virulent strains have this capsule, whereas avirulent strains do not. The nonencapsulated bacteria are readily engulfed and destroyed by phagocytic cells in the animal's circulatory system. Virulent bacteria, which possess the polysaccharide coat, are not easily engulfed; they multiply and cause pneumonia.

The presence or absence of the capsule causes a visible difference between colonies of virulent and avirulent strains. Encapsulated bacteria form **smooth**, shiny-surfaced colonies (**S**) when grown on an agar culture plate; nonencapsulated strains produce **rough** colonies (**R**) (Figure 10–2). Thus, virulent and avirulent strains are easily distinguished by standard microbiological culture techniques.

Each strain of *Diplococcus* may be one of dozens of different types called serotypes. The specificity of the serotype is due to the detailed chemical structure of the polysaccharide constituent of the thick, slimy capsule. Serotypes are identified by immunological techniques and are usually designated by Roman numerals. Griffith used the avirulent type II*R* and the virulent type III*S* in his critical experiments. Table 10.1 summarizes the characteristics of these strains.

Griffith knew from the work of others that only living virulent cells produce pneumonia in mice. If heat-killed virulent bacteria are injected into mice, no pneumonia results, just as living avirulent bacteria fail to produce the disease. Griffith's critical experiment (Figure 10–2) involved an injection into mice of living II*R* (avirulent) cells combined with heat-killed III*S* (virulent) cells. Since neither cell type caused death in mice when injected alone, Griffith expected that the double injection would not kill the mice. But, after five days, mice that received both types of cells were all dead. Analysis of their blood revealed large numbers of living type III*S* bacteria!

As far as could be determined, these III*S* bacteria were identical to the III*S* strain from which the heat-killed cell preparation had been made. Control mice, injected only with living avirulent II*R* bacteria, did not develop pneumonia and remained healthy. This finding ruled out the possibility that the avirulent II*R* cells simply changed (or mutated) to virulent III*S* cells in the absence of the heat-killed III*S* fraction. Instead, some type of interaction took place between living II*R* and heat-killed III*S* cells.

Griffith concluded that the heat-killed III*S* bacteria somehow converted live avirulent II*R* cells into virulent III*S* cells. Calling the phenomenon **transformation**, he suggested that the **transforming principle** might be some part of the polysaccharide capsule or some compound required for capsule synthesis, although the capsule alone did not cause pneumonia. To use Griffith's term, the transforming principle from the dead III*S* cells served as a "pabulum" for the II*R* cells.

---

*Note that this organism is now designated *Streptococcus pneumonia*.

FIGURE 10–2 Griffith's transformation experiment. The photographs show bacterial colonies that contain cells with capsules (type IIIS) and without capsules (type IIR). *(Photos: Bruce Iverson)*

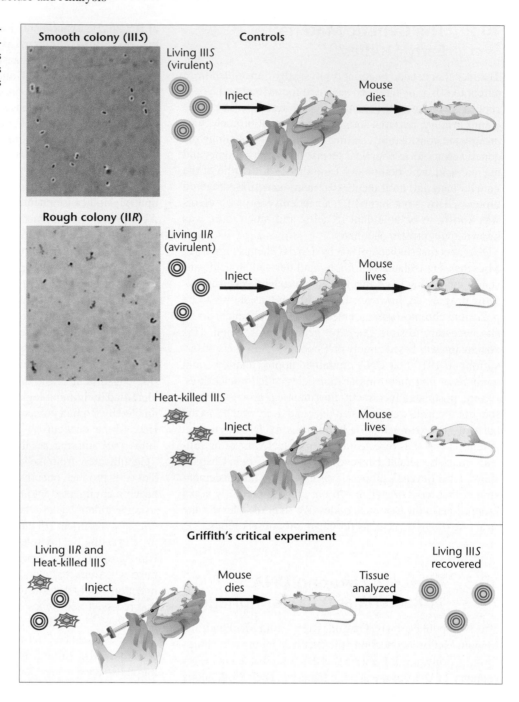

Griffith's work led other physicians and bacteriologists to explore the phenomenon of transformation. By 1931, Henry Dawson and his coworkers showed that transformation could occur *in vitro* (in a test tube containing only bacterial cells); that is, injection into mice was not necessary for transformation to occur. By 1933, Lionel J. Alloway had refined the *in vitro* experiments using extracts from S cells added to living R cells. The soluble filtrate from the heat-killed S cells was as effective in inducing transformation as were the intact cells! Alloway and others did not view transformation as a genetic event, but rather as a physiological modification of some sort. Nevertheless, the experimental evidence that a chemical substance was responsible for transformation was quite convincing.

Then, in 1944, after ten years of work, Avery, MacLeod, and McCarty published their results in what is now regarded as a classic paper in the field of molecular genetics. They reported that they had obtained the transforming principle in a highly purified state, and that beyond reasonable doubt it was DNA.

TABLE 10.1 **Strains of *Diplococcus pneumoniae* Used by Frederick Griffith in His Original Transformation Experiments**

Serotype	Colony Morphology	Capsule	Virulence
IIR	Rough	Absent	Avirulent
IIIS	Smooth	Present	Virulent

The details of their work are outlined in Figure 10–3. These researchers began their isolation procedure with large quantities (50–75 L) of liquid cultures of type III*S* virulent cells. The cells were centrifuged, collected, and heat-killed. Following various chemical treatments, a soluble filtrate was derived from these cells that retained the ability to induce transformation of type II*R* avirulent cells. Further testing clearly established that the transforming principle was DNA. The soluble filtrate was treated with a protein-digesting enzyme, called a protease, and an RNA-digesting enzyme, called ribonuclease. Such treatment destroyed the activity of any remaining protein and RNA. Nevertheless, transforming activity still remained. They concluded that neither protein nor RNA was responsible for transformation. The final confirmation came with experiments using crude samples of the DNA-digesting enzyme **deoxyribonuclease**, isolated from dog and rabbit sera. Digestion with this enzyme destroyed transforming activity present in the filtrate. Avery

and his coworkers were certain that the active transforming principle in these experiments was DNA!

The great amount of work, the confirmation and reconfirmation of the conclusions, and the logic of the experimental design involved in the research of these three scientists are truly impressive. Their conclusion in the 1944 publication, however, was stated very simply: "The evidence presented supports the belief that a nucleic acid of the desoxyribose* type is the fundamental unit of the transforming principle of *Pneumococcus* Type III."

They also immediately recognized the genetic and biochemical implications of their work. They suggested that the transforming principle interacts with the II*R* cell and gives rise to a coordinated series of enzymatic reactions that culminates in the synthesis of the type III*S* capsular polysaccharide. They

---

*Desoxyribose is now spelled deoxyribose.

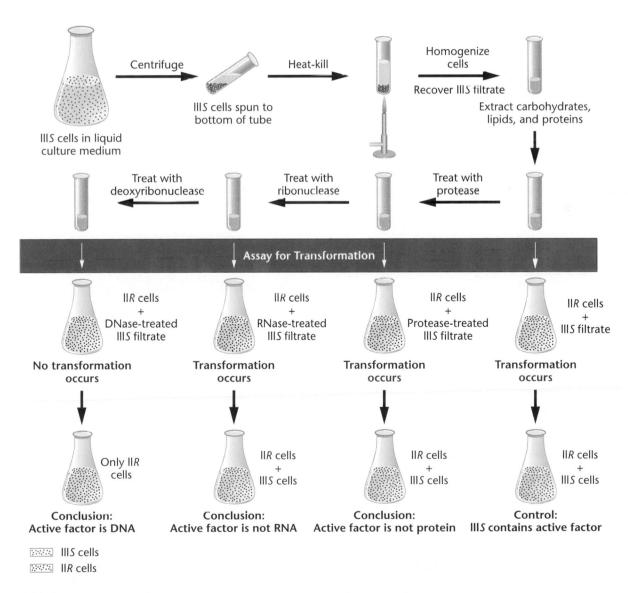

FIGURE 10–3 Summary of Avery, MacLeod, and McCarty's experiment, which demonstrated that DNA is the transforming principle.

emphasized that, once transformation occurs, the capsular polysaccharide is produced in successive generations. Transformation is therefore heritable, and the process affects the genetic material.

Transformation, previously introduced in Chapter 9, has been shown to occur in *Hemophilus influenzae*, *Bacillus subtilis*, *Shigella paradysenteriae*, and *Escherichia coli*, among many other microorganisms. Transformation of numerous genetic traits other than colony morphology has been demonstrated, including ones that resist antibiotics and that metabolize various nutrients. These observations further strengthened the belief that transformation by DNA is primarily a genetic event, rather than simply a physiological change. This idea is pursued again in the "Insights and Solutions" section at the end of this chapter.

The Hershey–Chase Experiment

The second major piece of evidence supporting DNA as the genetic material was provided during the study of the bacterium *Escherichia coli* and one of its infecting viruses, **bacteriophage T2**. Often referred to simply as a **phage**, the virus consists of a protein coat surrounding a core of DNA. Electron micrographs reveal the phage's external structure to be composed of a hexagonal head plus a tail. Figure 10–4 shows the life cycle of a T-even bacteriophage such as T2, as it was known in 1952. Briefly, the phage adsorbs to the bacterial cell and some component of the phage enters the bacterial cell. Following this infection step, the viral information "commandeers" the cellular machinery of the host and undergoes viral reproduction. In a reasonably short time, many new phages are constructed and the bacterial cell is lysed, releasing the progeny viruses.

In 1952, Alfred Hershey and Martha Chase published the results of experiments designed to clarify the events leading to phage reproduction. Several of the experiments clearly established the independent functions of phage protein and nucleic acid in the reproduction process of the bacterial cell. Hershey and Chase knew from existing data that

1. T2 phages consist of approximately 50 percent protein and 50 percent DNA.

2. Infection is initiated by adsorption of the phage by its tail fibers to the bacterial cell.

3. The production of new viruses occurs within the bacterial cell.

It appeared that some molecular component of the phage, DNA and/or protein, enters the bacterial cell and directs viral reproduction. Which was it?

Hershey and Chase used radioisotopes to follow the molecular components of phages during infection. Both ^{32}P and ^{35}S, radioactive forms of phosphorus and sulfur, were used. Because DNA contains phosphorus but not sulfur, ^{32}P effectively labels DNA. Because proteins contain sulfur but not phosphorus, ^{35}S labels protein. *This is a key point in the experiment.* If *E. coli* cells are first grown in the presence of ^{32}P or ^{35}S and then infected with T2 viruses, the progeny phage

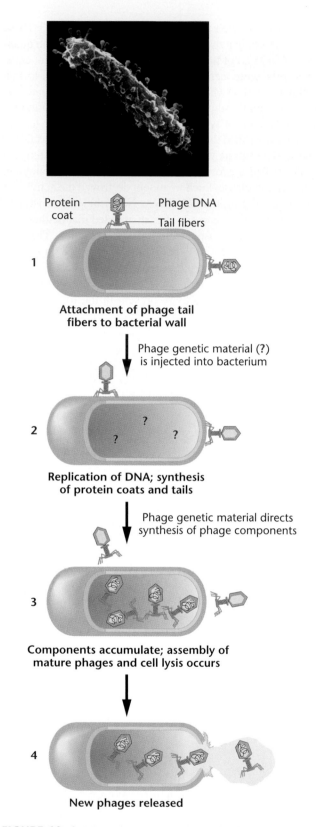

FIGURE 10–4 Life cycle of a T-even bacteriophage. The electron micrograph shows an *E. coli* cell during infection by numerous T2 phages (shown in blue). *(Photo: Oliver Meckes/MPI-Tubingen/Photo Researchers, Inc.)*

will have *either* a labeled DNA core *or* a labeled protein coat, respectively. These radioactive phages can be isolated and used to infect unlabeled bacteria (Figure 10–5).

When labeled phage and unlabeled bacteria are mixed, an adsorption complex forms as the phages attach their tail fibers to the bacterial wall. These complexes were isolated and sub-jected to a high shear force by placing them in a blender. This force strips off the attached phages, which can then be analyzed separately (Figure 10–5). By tracing the radioiso-topes, Hershey and Chase were able to demonstrate that most of the ^{32}P-labeled DNA had transferred into the bacterial cell following adsorption; on the other hand, most of the

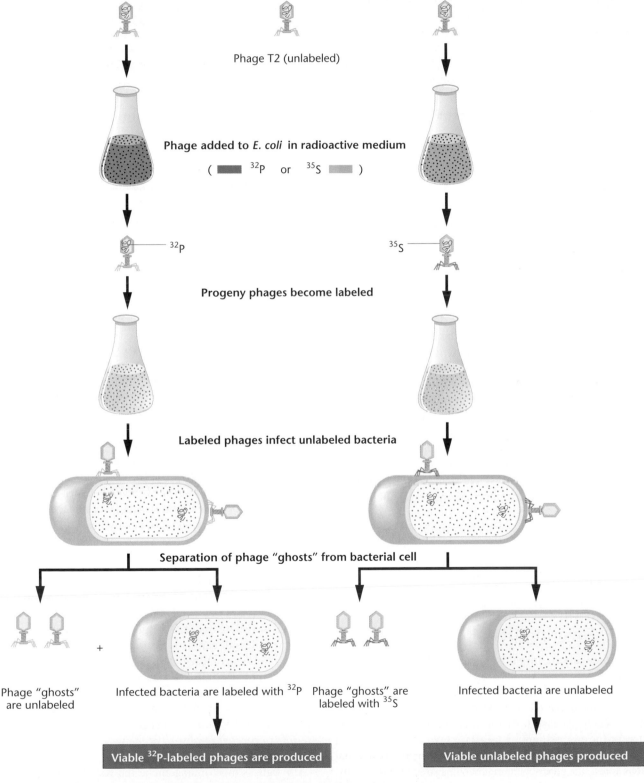

Phage T2 (unlabeled)

Phage added to *E. coli* in radioactive medium

(▇ ^{32}P or ^{35}S ▇)

^{32}P ^{35}S

Progeny phages become labeled

Labeled phages infect unlabeled bacteria

Separation of phage "ghosts" from bacterial cell

Phage "ghosts" are unlabeled   Infected bacteria are labeled with ^{32}P   Phage "ghosts" are labeled with ^{35}S   Infected bacteria are unlabeled

Viable ^{32}P-labeled phages are produced   **Viable unlabeled phages produced**

FIGURE 10–5 Summary of the Hershey–Chase experiment demonstrating that DNA, and not protein, is responsible for directing the reproduction of phage T2 during the infection of *E. coli*.

^{35}S-labeled protein remained outside the bacterial cell and was recovered in the phage "ghosts" (empty phage coats) after the blender treatment. Following this separation, the bacterial cells, which now contained viral DNA, were eventually lysed as new phages were produced. These progeny contained ^{32}P, but not ^{35}S.

Hershey and Chase interpreted these results as indicating that the protein of the phage coat remains outside the host cell and is not involved in the production of new phages. On the other hand, and most important, phage DNA enters the host cell and directs phage reproduction. Hershey and Chase had demonstrated the genetic material in phage T2 is DNA, not protein.

These experiments, along with those of Avery and his colleagues, provided convincing evidence that DNA is the molecule responsible for heredity. This conclusion has since served as the cornerstone of the field of molecular genetics.

Transfection Experiments

During the eight years following the publication of the Hershey–Chase experiment, additional research with bacterial viruses provided even more solid proof that DNA is the genetic material. In 1957, several reports demonstrated that if *E. coli* is treated with the enzyme lysozyme, the outer wall of the cell can be removed without destroying the bacterium. Enzymatically treated cells are naked, so to speak, and contain only the cell membrane as the outer boundary of the cell. Such structures are called **protoplasts** (or **spheroplasts**). John Spizizen and Dean Fraser reported independently that by using protoplasts, they were able to initiate phage multiplication with disrupted T2 particles. That is, provided protoplasts are used, a virus does not have to be intact for infection to occur.

Similar but refined experiments were reported in 1960 using only the DNA purified from bacteriophages. This process of infection by only the viral nucleic acid, called **transfection**, proves conclusively that phage DNA alone contains all the necessary information for producing mature viruses. Thus, the evidence that DNA serves as the genetic material in all organisms was further strengthened, even though all direct evidence had been obtained from bacterial and viral studies.

10.4 Indirect and Direct Evidence Favoring DNA in Eukaryotes

In 1950, eukaryotic organisms were not amenable to the types of experiments that in bacteria and viruses demonstrate that DNA is the genetic material. Nevertheless, it was generally assumed that the genetic material would be a universal substance and also serve this role in eukaryotes. Initially, support for this assumption relied on several circumstantial (indirect) observations that, taken together, indicated that DNA is also the genetic material in eukaryotes. Subsequently, direct evidence established unequivocally the central role of DNA in genetic processes.

TABLE 10.2 **DNA Content of Haploid Versus Diploid Cells of Various Species (in picograms)**

Organism	n	2n
Human	3.25	7.30
Chicken	1.26	2.49
Trout	2.67	5.79
Carp	1.65	3.49
Shad	0.91	1.97

Note: Sperm (*n*) and nucleated precursors to red blood cells (2*n*) were used to contrast ploidy levels.

Indirect Evidence: Distribution of DNA

The genetic material should be found where it functions—in the nucleus as part of chromosomes. Both DNA and protein fit this criterion. However, protein is also abundant in the cytoplasm, while DNA is not. Both mitochondria and chloroplasts are known to perform genetic functions, and DNA is also present in these organelles. Thus, DNA is found only where primary genetic function is known to occur. Protein, however, is found everywhere in the cell. These observations are consistent with the interpretation favoring DNA over protein as the genetic material.

Because it had been established earlier that chromosomes within the nucleus contain the genetic material, a correlation was expected between the ploidy (*n*, 2*n*, etc.) of cells and the quantity of the molecule that functions as the genetic material. Meaningful comparisons can be made between gametes (sperm and eggs) and somatic or body cells. The latter are recognized as being diploid (2*n*) and containing twice the number of chromosomes as gametes, which are haploid (*n*).

Table 10.2 compares the amount of DNA found in haploid sperm and the diploid nucleated precursors of red blood cells from a variety of organisms. The amount of DNA and the number of sets of chromosomes is closely correlated. No consistent correlation can be observed between gametes and diploid cells for proteins, thus again favoring DNA over proteins as the genetic material of eukaryotes.

Indirect Evidence: Mutagenesis

Ultraviolet (UV) light is one of a number of agents capable of inducing mutations in the genetic material. Simple organisms such as yeast and other fungi can be irradiated with various wavelengths of UV light, and the effectiveness of each wavelength can be measured by the number of mutations it induces. When the data are plotted, an **action spectrum** of UV light as a mutagenic agent is obtained. This action spectrum can then be compared with the **absorption spectrum** of any molecule suspected to be the genetic material (Figure 10–6). *The molecule serving as the genetic material is expected to absorb at the wavelengths found to be mutagenic.*

UV light is most mutagenic at the wavelength (λ) of 260 nanometers (nm). Both DNA and RNA absorb UV light most strongly at 260 nm. On the other hand, protein absorbs most

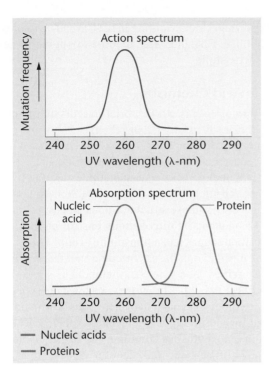

FIGURE 10–6 Comparison of the action spectrum, which determines the most effective mutagenic UV wavelength, and the absorption spectrum, which shows the range of wavelengths where nucleic acids and proteins absorb UV light.

strongly at 280 nm, yet no significant mutagenic effects are observed at this wavelength. This indirect evidence supports the idea that a nucleic acid is the genetic material and tends to exclude protein.

Direct Evidence: Recombinant DNA Studies

Although the circumstantial evidence described above does not constitute direct proof that DNA is the genetic material in eukaryotes, these observations spurred researchers to forge ahead, basing their work on this hypothesis. Today, there is no doubt of the validity of this conclusion. DNA *is* the genetic material in eukaryotes.

The strongest evidence is provided by molecular analysis utilizing **recombinant DNA technology**. In this procedure, segments of eukaryotic DNA corresponding to specific genes are isolated and literally spliced into bacterial DNA. Such a complex can be inserted into a bacterial cell and its genetic expression monitored. If a eukaryotic gene is introduced, the presence of the corresponding eukaryotic protein product demonstrates directly that this DNA is present and functional in the bacterial cell. This has been shown to be the case in countless instances. For example, the products of the human genes that specify insulin and the interferon are produced by bacteria following insertion of the human genes that encode these proteins. As the bacterium divides, the eukaryotic DNA replicates along with the host DNA and is distributed to the daughter cells,

which also express the human genes and synthesize the corresponding proteins.

The availability of vast amounts of DNA coding for specific genes, available as a result of recombinant DNA research, has led to other direct evidence that DNA serves as the genetic material. Work in the laboratory of Beatrice Mintz demonstrates that DNA encoding the human β-globin gene, when microinjected into a fertilized mouse egg, is later found to be present and expressed in adult mouse tissue and transmitted to and expressed in that mouse's progeny. These mice are examples of **transgenic animals**.

More recent work has introduced *rat* DNA encoding a growth hormone into fertilized *mouse* eggs. About one-third of the resultant mice grew to twice their normal size. This indicates that foreign DNA is present *and* functional in the experimental mice. Subsequent generations of mice inherited this genetic information and also grew to a large size.

We pursue the topic of recombinant DNA again later (see Chapters 16 and 19). The point to be made here is that in eukaryotes, DNA meets the requirement of expression of genetic information. Later we shall see exactly how DNA is stored, replicated, expressed, and mutated.

10.5 RNA as the Genetic Material in Some Viruses

Some viruses contain an RNA core rather than one composed of DNA. In these viruses, it appears that RNA serves as the genetic material—an exception to the general rule that DNA performs this function. In 1956, it was demonstrated that when purified RNA from **tobacco mosaic virus (TMV)** is spread on tobacco leaves, the characteristic lesions caused by viral infection subsequently appear on the leaves. It was concluded that RNA is the genetic material of this virus.

In 1965 and 1966, Norman Pace and Sol Spiegelman demonstrated further that RNA from the phage Qβ can be isolated and replicated *in vitro*. Replication depends on an enzyme, **RNA replicase**, which is isolated from host *E. coli* cells following normal infection. When the RNA replicated *in vitro* is added to *E. coli* protoplasts, infection and viral multiplication (transfection) occurs. Thus, RNA synthesized in a test tube serves as the genetic material in these phages by directing the production of all components necessary for viral replication.

Finally, one other group of RNA-containing viruses bears mentioning. These are the **retroviruses**, which replicate in an unusual way. Their RNA serves as a template for the synthesis of the complementary DNA molecule! The process, **reverse transcription**, occurs under the direction of an RNA-dependent DNA polymerase enzyme called **reverse transcriptase**. This DNA intermediate can be incorporated into the genome of the host cell, and when the host DNA is transcribed, copies of the original retroviral RNA chromosomes are produced. Retroviruses include the human immunodeficiency virus (HIV), which causes AIDS, as well as the RNA tumor viruses.

10.6 Structural Analysis of DNA

Having established that DNA is the genetic material in all living organisms (except certain viruses), we turn now to a consideration of the structure of this nucleic acid. In 1953, James Watson and Francis Crick proposed that the structure of DNA is in the form of a double helix. Their proposal was published in a short paper in the journal *Nature*, reprinted in its entirety (see p. 207). In a sense, this publication constituted the finish of a highly competitive scientific race to obtain what some consider to be the most significant finding in the history of biology. This "race," as recounted in Watson's book *The Double Helix*, demonstrates the human interaction, genius, frailty, and intensity involved in the scientific effort that eventually led to the elucidation of DNA structure.

The data available to Watson and Crick, crucial to the development of their proposal, came primarily from two sources: (1) base composition analysis of hydrolyzed samples of DNA and (2) X-ray diffraction studies of DNA. The analytical success of Watson and Crick can be attributed to model building that conformed to the existing data. If the correct solution to the structure of DNA is viewed as a puzzle, Watson and Crick, working in the Cavendish Laboratory in Cambridge, England, were the first to put the pieces together successfully.

Nucleic Acid Chemistry

Before turning to this work, a brief introduction to nucleic acid chemistry is in order. This chemical information was well known to Watson and Crick during their investigation and served as the basis of their model building.

DNA is a nucleic acid, and nucleotides are the building blocks of all nucleic acid molecules. Sometimes called mononucleotides, these structural units consist of three essential components: a **nitrogenous base**, a **pentose sugar** (a five-carbon sugar), and a **phosphate group**. There are two kinds of nitrogenous bases: the nine-membered, double-ringed **purines** and the six-membered, single-ringed **pyrimidines**. Two types of purines and three types of pyrimidines are commonly found in nucleic acids. The two purines are **adenine** and **guanine**, abbreviated **A** and **G**. The three pyrimidines are **cytosine**, **thymine**, and **uracil**, abbreviated **C**, **T**, and **U**. The chemical structures of A, G, C, T, and U are

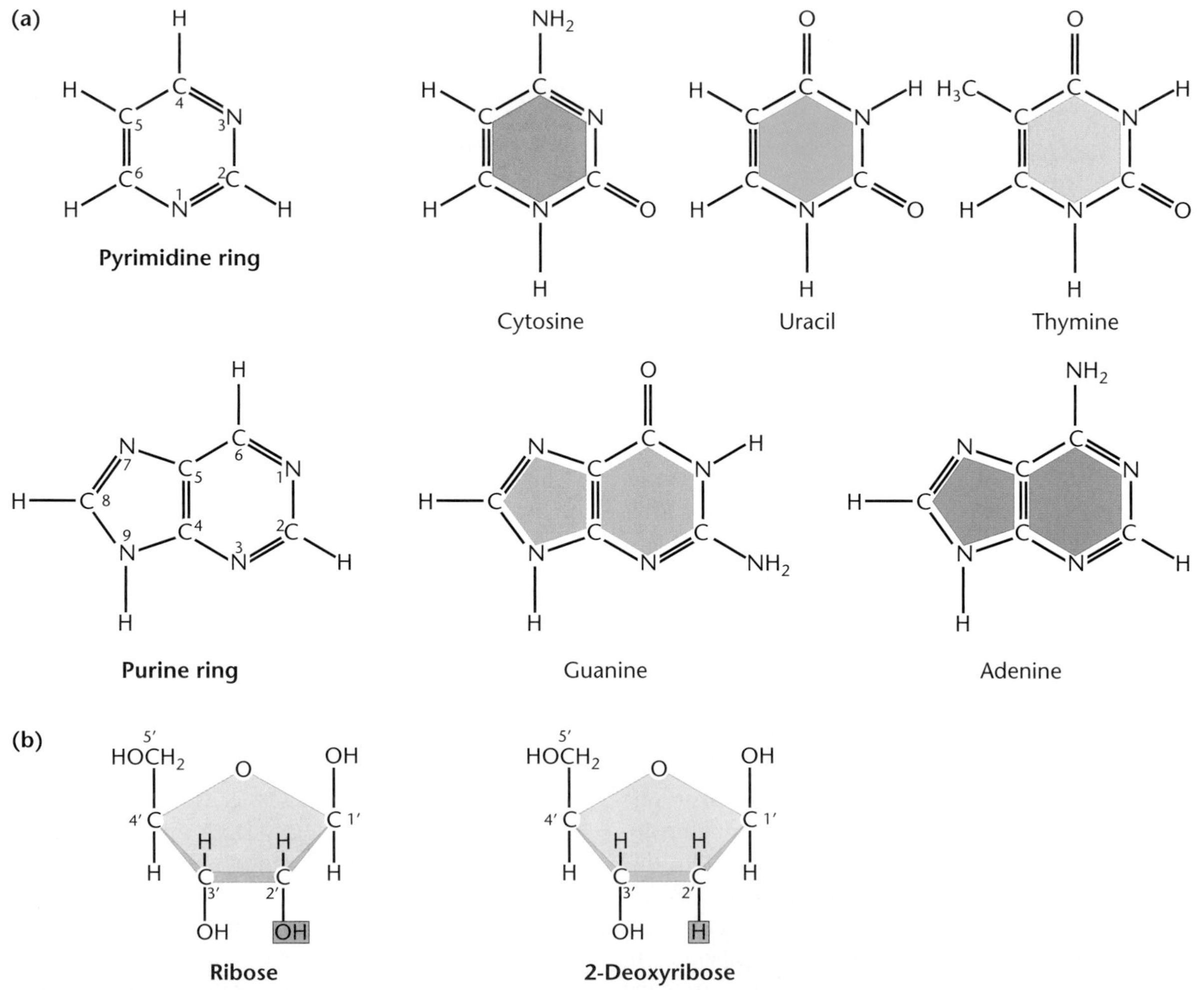

FIGURE 10–7 (a) Chemical structures of the pyrimidines and purines that serve as the nitrogenous bases in RNA and DNA. (b) Chemical ring structures of ribose and 2-deoxyribose, which serve as the pentose sugars in RNA and DNA, respectively.

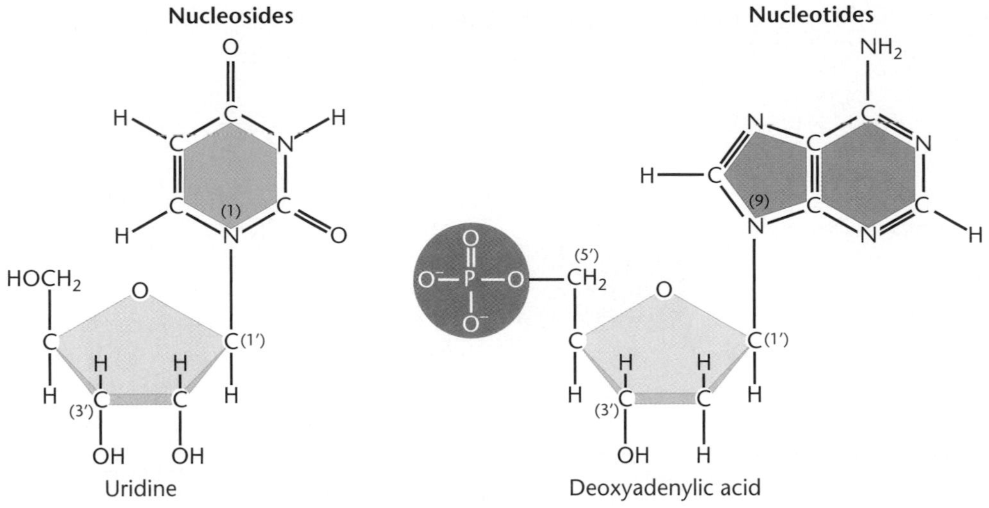

Nucleosides

Nucleotides

FIGURE 10–8 Structures and names of the nucleosides and nucleotides of RNA and DNA.

Uridine

Deoxyadenylic acid

Ribonucleosides	Ribonucleotides
Adenosine	Adenylic acid
Cytidine	Cytidylic acid
Guanosine	Guanylic acid
Uridine	Uridylic acid
Deoxyribonucleosides	**Deoxyribonucleotides**
Deoxyadenosine	Deoxyadenylic acid
Deoxycytidine	Deoxycytidylic acid
Deoxyguanosine	Deoxyguanylic acid
Deoxythymidine	Deoxythymidylic acid

shown in Figure 10–7(a). Both DNA and RNA contain A, C, and G; only DNA contains the base T, whereas only RNA contains the base U. Each nitrogen or carbon atom of the ring structures of purines and pyrimidines is designated by an unprimed number. Note that corresponding atoms in the purine and pyrimidine rings are numbered differently.

The pentose sugars found in nucleic acids give them their names. **Ribonucleic acids (RNA)** contain **ribose**, while **deoxyribonucleic acids (DNA)** contain **deoxyribose**. Figure 10–7(b) shows the structures for these two pentose sugars. Each carbon atom is distinguished by a number with a prime sign (C-1′, C-2′, etc.). As you can see, compared with ribose, deoxyribose has a hydrogen atom at the C-2′ position rather than a hydroxyl group. The presence of a hydroxyl group at the C-2′ position thus distinguishes RNA from DNA.

If a molecule is composed of a purine or pyrimidine base and a ribose or deoxyribose sugar, the chemical unit is called a **nucleoside**. If a phosphate group is added to the nucleoside, the molecule is now called a **nucleotide**. Nucleosides and nucleotides are named according to the specific nitrogenous base (A, T, G, C, or U) that is part of the molecule. The structure of a nucleotide and the nomenclature used in naming DNA nucleotides and nucleosides are as shown in Figure 10–8.

The bonding between the components of a nucleotide is highly specific. The C-1′ atom of the sugar is involved in the chemical linkage to the nitrogenous base. If the base is a purine, the N-9 atom is covalently bonded to the sugar. If the base is a pyrimidine, the bonding involves the N-1 atom. In a nucleotide, the phosphate group may be bonded to the C-2′, C-3′, or C-5′ atom of the sugar. The C-5′–phosphate configuration is shown in Figure 10–8. It is by far the most prevalent one in biological systems and the one found in DNA and RNA.

Nucleotides are also described by the term **nucleoside monophosphate (NMP)**. The addition of one or two phosphate groups results in **nucleoside diphosphates (NDP)** and **triphosphates (NTP)**, as shown in Figure 10–9. The triphosphate form is significant because it serves as the precursor molecule during nucleic acid synthesis within the cell. Additionally, the triphosphates adenosine triphosphate (ATP) and guanosine triphosphate (GTP) are important in the bioenergetics of cells because of the large amount of energy involved in adding or removing the terminal phosphate group. The hydrolysis of ATP or GTP to ADP or GDP and inorganic phosphate (P_i) is accompanied by the release of a large amount of energy in the cell. When these chemical conversions are coupled to other reactions, the energy produced can be used to drive them. As a result, ATP and GTP are involved in many cellular activities.

The linkage between two mononucleotides involves a phosphate group linked to two sugars. A **phosphodiester bond** forms, because phosphoric acid has been joined to two alcohols (the hydroxyl groups on the two sugars) by an ester linkage

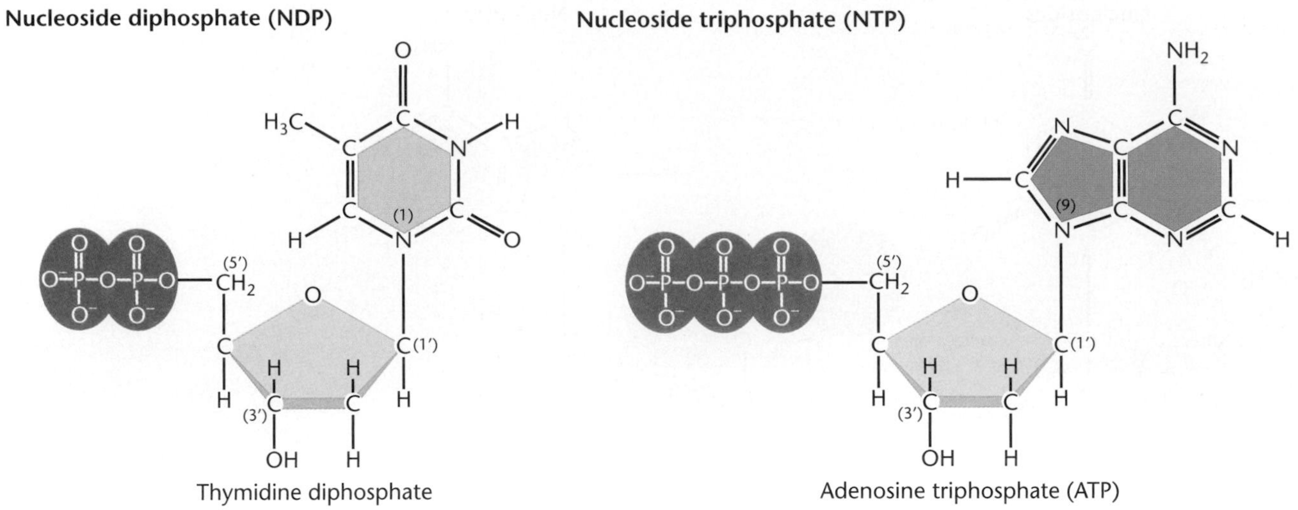

Nucleoside diphosphate (NDP)   **Nucleoside triphosphate (NTP)**

Thymidine diphosphate   Adenosine triphosphate (ATP)

FIGURE 10–9 Basic structures of nucleoside diphosphates and triphosphates, as illustrated by thymidine diphosphate and adenosine triphosphate.

on both sides. Figure 10–10 shows the resultant phosphodiester bond in DNA. Each bond has a C-3′ end and a C-5′ end. Two joined nucleotides form a dinucleotide, three nucleotides a trinucleotide, and so forth. Short chains consisting of approximately 20 nucleotides are called **oligonucleotides**. Still longer chains are referred to as **polynucleotides**.

Long polynucleotide chains account for the large molecular weight and explain the most important property of DNA—storage of vast quantities of genetic information. If each nu-

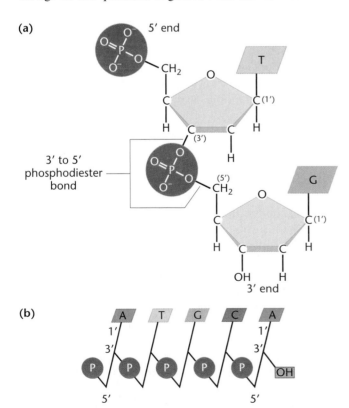

(a)

3′ to 5′
phosphodiester
bond

(b)

FIGURE 10–10 (a) Linkage of two nucleotides by the formation of a C-3′–C-5′ (3′–5′) phosphodiester bond, producing a dinucleotide. (b) Shorthand notation for a polynucleotide chain.

cleotide position in this long chain can be occupied by any one of four nucleotides, extraordinary variation is possible. For example, a polynucleotide that is only 1000 nucleotides in length can be arranged 4^{1000} different ways, each one different from all other possible sequences. This potential variation in molecular structure is essential if DNA is to store the vast amounts of chemical information necessary to direct cellular activities.

Base Composition Studies

Between 1949 and 1953, Erwin Chargaff and his colleagues used chromatographic methods to separate the four bases in DNA samples from various organisms. Quantitative methods were then used to determine the amounts of the four nitrogenous bases from each source. Table 10.3(a) lists some of Chargaff's original data. Parts (b) and (c) show more recently derived information that reinforces Chargaff's findings. As we shall see, Chargaff's data were critical to the successful model of DNA put forward by Watson and Crick.

On the basis of these data, the following conclusions may be drawn.

1. The amount of adenine residues is proportional to the amount of thymine residues in DNA (columns 1, 2, and 5). Also, the amount of guanine residues is proportional to the amount of cytosine residues (columns 3, 4, and 6).

2. Based on this proportionality, the sum of the purines (A + G) equals the sum of the pyrimidines (C + T), as shown in column 7.

3. The percentage of G + C does not necessarily equal the percentage of A + T. As you can see, this ratio varies greatly in different organisms, as shown in column 8.

These conclusions indicate a definite pattern of base composition of DNA molecules. The data provided the initial clue to "the puzzle." Additionally, they directly refuted Levene's tetranucleotide hypothesis, which stated that all four bases are present in equal amounts.

TABLE 10.3   DNA Base Composition Data

(a) Chargaff's data*

	Molar proportions[a]			
	1	*2*	*3*	*4*
Source	*A*	*T*	*G*	*C*
Ox thymus	26	25	21	16
Ox spleen	25	24	20	15
Yeast	24	25	14	13
Avian tubercle bacilli	12	11	28	26
Human sperm	29	31	18	18

(c) G + C content in several organisms

Organism	%G + C
Phage T2	36.0
Drosophila	45.0
Maize	49.1
Euglena	53.5
Neurospora	53.7

(b) Base compositions of DNAs from various sources

	Base composition				Base ratio		A + T/G + C ratio	
	1	*2*	*3*	*4*	*5*	*6*	*7*	*8*
Source	*A*	*T*	*G*	*C*	*A/T*	*G/C*	*(A + G)/(C + T)*	*(A + T)/(C + G)*
Human	30.9	29.4	19.9	19.8	1.05	1.00	1.04	1.52
Sea urchin	32.8	32.1	17.7	17.3	1.02	1.02	1.02	1.58
E. coli	24.7	23.6	26.0	25.7	1.04	1.01	1.03	0.93
Sarcina lutea	13.4	12.4	37.1	37.1	1.08	1.00	1.04	0.35
T7 bacteriophage	26.0	26.0	24.0	24.0	1.00	1.00	1.00	1.08

*Source: From Chargaff, 1950.
[a]Moles of nitrogenous constituent per mole of P (often, the recovery was less than 100 percent).

X-Ray Diffraction Analysis

When fibers of a DNA molecule are subjected to X-ray bombardment, these rays scatter according to the molecule's atomic structure. The pattern of scatter can be captured as spots on photographic film and analyzed, particularly for the overall shape of and regularities within the molecule. This process, **X-ray diffraction analysis**, was applied successfully to the study of protein structure by Linus Pauling and other chemists. The technique had been attempted on DNA as early as 1938 by William Astbury. By 1947, he had detected a periodicity within the structure of the molecule of 3.4 angstroms (Å)*, which suggested to him that the bases were stacked like coins on top of one another.

Between 1950 and 1953, Rosalind Franklin, working in the laboratory of Maurice Wilkins, obtained improved X-ray data from more purified samples of DNA (Figure 10–11). Her work confirmed the 3.4 Å periodicity seen by Astbury and suggested that the structure of DNA was some sort of helix. However, she did not propose a definitive model. Pauling had analyzed the work of Astbury and others and proposed incorrectly that DNA is a triple helix.

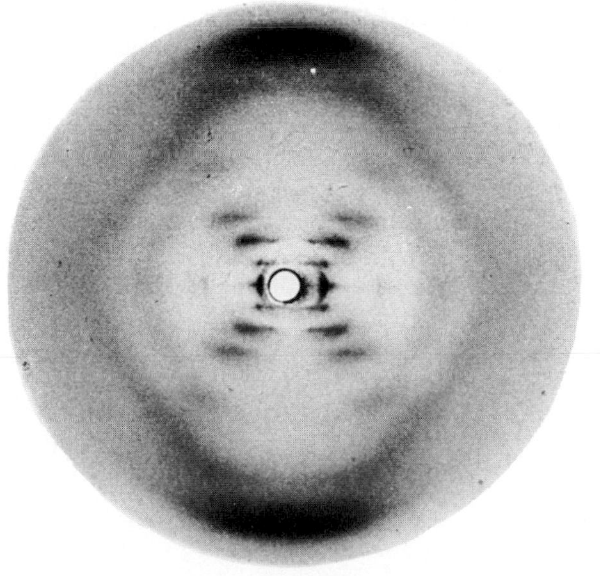

FIGURE 10–11 X-ray diffraction photograph of the B form of purified DNA fibers. The strong arcs on the periphery show closely spaced aspects of the molecule providing an estimate of the periodicity of nitrogenous bases, which are 3.4 Å apart. The inner cross pattern of spots shows the grosser aspect of the molecule, indicating its helical nature. (*M.H.F. Wilkens. Courtesy of Bio-Physics Department, King's College, London, England. Science Source/Photo Researchers, Inc.*)

---

*Today, measurement in nanometers (nm) is favored (1 nm = 10 Å).

10.7 The Watson–Crick Model

Watson and Crick published their analysis of DNA structure in 1953 (see pp. 207). By building models under the constraints of the information just discussed, they proposed the double-helical form of DNA shown in Figure 10–12(a). This model has the following major features.

1. Two long polynucleotide chains are coiled around a central axis, forming a right-handed double helix.

2. The two chains are antiparallel; that is, their C-5′-to-C-3′ orientations run in opposite directions.

3. The bases of both chains are flat structures, lying perpendicular to the axis; they are "stacked" on one another, 3.4 Å (0.34 nm) apart, and are located on the inside of the structure.

4. The nitrogenous bases of opposite chains are paired as the result of hydrogen bonds (described in the following discussion); in DNA, only A＝T and G≡C pairs are allowed.

5. Each complete turn of the helix is 34 Å (3.4 nm) long; thus, 10 bases exist per turn in each chain.

6. In any segment of the molecule, alternating larger **major grooves** and smaller **minor grooves** are apparent along the axis.

7. The double helix measures 20 Å (2.0 nm) in diameter.

The nature of base pairing (point 4 above) is the most genetically significant feature of the model. Before we discuss it in detail, several other important features warrant emphasis. First, the antiparallel nature of the two chains is a key part of the double-helix model. While one chain runs in the 5′-to-3′ orientation (what seems right side up to us), the other chain is in the 3′-to-5′ orientation (and thus appears upside down). This is illustrated in Figure 10–12(b and c). Given the constraints of the bond angles of the various nucleotide components, the double helix could not be constructed easily if both chains ran parallel to one another.

The key to the model proposed by Watson and Crick is the specificity of base pairing. Chargaff's data suggested that

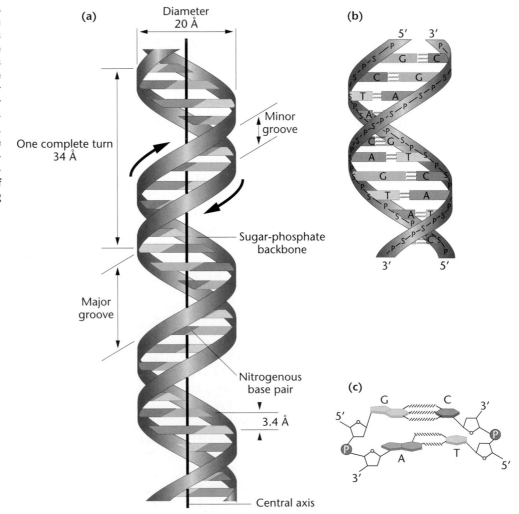

FIGURE 10–12 (a) The DNA double helix as proposed by Watson and Crick. The ribbonlike strands constitute the sugar–phosphate backbones, and the horizontal rungs constitute the nitrogenous base pairs, of which there are 10 per complete turn. The major and minor grooves are apparent. The solid vertical bar represents the central axis. (b) A detailed view depicting the bases, sugars, phophates, and hydrogen bonds of the helix. (c) A demonstration of the antiparallel nature of the helix and the horizontal stacking of the bases.

the amounts of A equaled T and that G equaled C. Watson and Crick realized that if A pairs with T and C pairs with G, thus accounting for these proportions, the members of each such base pair form hydrogen bonds [Figure 10–12(c)], providing the chemical stability necessary to hold the two chains together. Arranged in this way, both major and minor grooves become apparent along the axis. Further, a purine (A or G) opposite a pyrimidine (T or C) on each "rung of the spiral staircase" of the proposed helix accounts for the 20-Å (2-nm) diameter suggested by X-ray diffraction studies.

The specific A = T and G ≡ C base pairing is the basis for **complementarity**. This term describes the chemical affinity provided by hydrogen bonding between the bases. As we shall see, complementarity is very important in DNA replication and gene expression.

It is appropriate to inquire into the nature of a hydrogen bond, and whether is it strong enough to stabilize the helix. A **hydrogen bond** is a very weak electrostatic attraction between a covalently bonded hydrogen atom and an atom with an unshared electron pair. The hydrogen atom assumes a partial positive charge, while the unshared electron pair—characteristic of covalently bonded oxygen and nitrogen atoms—assumes a partial negative charge. These opposite charges are responsible for the weak chemical attractions. As oriented in the double helix, adenine forms two hydrogen bonds with thymine, and guanine forms three hydrogen bonds with cytosine. Although two or three hydrogen bonds taken alone are energetically very weak, 2000–3000 bonds in tandem (which would be found in two long polynucleotide chains) provide great stability to the helix.

Still another stabilizing factor is the arrangement of sugars and bases along the axis. In the Watson–Crick model, the **hydrophobic** (or "water-fearing") nitrogenous bases are stacked almost horizontally on the interior of the axis, thus shielded from water. The **hydrophilic** (or "water-loving") sugar–phosphate backbone is on the outside of the axis, where both components can interact with water. These molecular arrangements provide significant chemical stabilization to the helix.

A more recent and accurate analysis of the form of DNA that served as the basis for the Watson–Crick model reveals a minor structural difference. A precise measurement of the number of base pairs (bp) per turn has demonstrated a value of 10.4 bp rather than the 10.0 bp predicted by Watson and Crick. Where, in the classical model, each base pair is rotated 36° around the helical axis relative to the adjacent base pair, the new finding requires a rotation of 34.6°. Thus, there are slightly more than 10 bp per turn.

The Watson–Crick model had an immediate effect on the emerging discipline of molecular biology. Even in their initial 1953 article, the authors noted, "It has not escaped our notice that the specific pairing we have postulated immediately suggests a possible copying mechanism for the genetic material." Two months later, in a second article in *Nature*, Watson and Crick pursued this idea, suggesting a specific mode of replication of DNA–the **semiconservative model**. The second article alluded to two new concepts: (1) the stor-age of genetic information in the sequence of the bases, and (2) the mutations or genetic changes that would result from an alteration of the bases. These ideas have received vast amounts of experimental support since 1953 and are now universally accepted.

Watson and Crick's "synthesis" of ideas was highly significant with regard to subsequent studies of genetics and biology. The nature of the gene and its role in genetic mechanisms could now be viewed and studied in biochemical terms. Recognition of their work, along with that of Wilkins, led to their receiving the Nobel Prize in Physiology and Medicine in 1962. This was one of many such awards bestowed for their work in the field of genetics.

Alternative Forms of DNA

Under different conditions of isolation, several conformational forms of DNA have been recognized. At the time Watson and Crick performed their analysis, two forms—**A-DNA** and **B-DNA**—were known. Watson and Crick's analysis was based on X-ray studies of the B form by Franklin, which is present under aqueous, low-salt conditions and is believed to be the biologically significant conformation. While DNA studies around 1950 relied on the use of X-ray diffraction, more recent investigations have been performed using **single-crystal X-ray analysis**. The earlier studies achieved limited resolution of about 5 Å, but single crystals diffract X-rays at about 1 Å, near atomic resolution. As a result, every atom is "visible," and much greater structural detail is available during analysis.

Using these modern techniques, the A form of DNA has now been scrutinized. A-DNA is prevalent under high-salt or dehydration conditions. In comparison to B-DNA (Figure 10–13), A-DNA is slightly more compact, with 11 bp in each complete turn of the helix, which is 23 Å (2.3 nm) in diameter. While it is also a right-handed helix, the orientation of the bases is somewhat different. They are tilted and displaced laterally in relation to the axis of the helix. As a result, the appearance of the major and minor grooves is modified compared with those in B-DNA. It seems doubtful that A-DNA occurs under biological conditions.

Other forms of DNA (e.g., C-, D-, E- and most recently, P-DNA) are now known, but it is **Z-DNA** that has drawn the most attention. Discovered by Andrew Wang, Alexander Rich, and their colleagues in 1979 when they examined a small synthetic DNA fragment containing only C–G base pairs, Z-DNA takes on the rather remarkable configuration of a left-handed double helix (Figure 10–13). Like A- and B-DNA, Z-DNA consists of two antiparallel chains held together by Watson–Crick base pairs. Beyond these characteristics, Z-DNA is quite different. The left-handed helix is 18 Å (1.8 nm) in diameter, contains 12 bp per turn, and assumes a zigzag conformation (hence its name). The major groove present in B-DNA is nearly eliminated in Z-DNA.

Speculation abounds over the possibility that regions of Z-DNA exist in the chromosomes of living organisms. The

FIGURE 10–13 The top half of the figure shows computer-generated space-filling models of B-DNA (left), A-DNA (center), and Z-DNA (right). Below is an artist's depiction illustrating the orientation of the base pairs of B-DNA and A-DNA. (Note that in B-DNA the base pairs are perpendicular to the helix, while they are tilted and pulled away from the helix in A-DNA.) *(Photo: Ken Eward/Science Source/Photo Researchers, Inc.)*

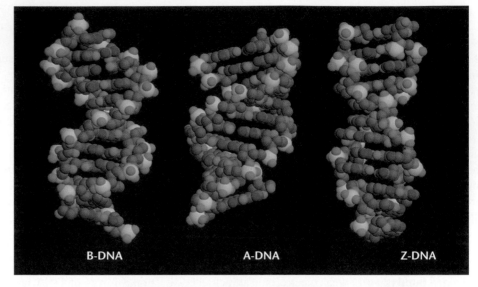

B-DNA   A-DNA   Z-DNA

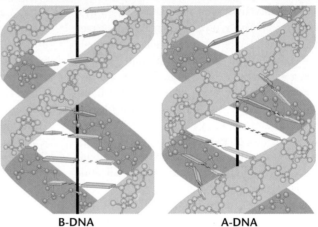

B-DNA   A-DNA

unique helical arrangement could provide an important recognition point for the interaction with other molecules. However, it is still not clear whether Z-DNA occurs *in vivo*.

10.8   The Structure of RNA

The second type of nucleic acid is ribonucleic acid, or RNA. The structure of these molecules resembles DNA, with several important exceptions. Although RNA also has as its building blocks nucleotides linked with polynucleotide chains, the sugar ribose replaces deoxyribose and the nitrogenous base uracil replaces thymine. Another important difference is that most RNA is single-stranded, although there are two important exceptions. First, RNA molecules sometimes fold back on themselves to form double-stranded regions of complementary base pairs. Second, some animal viruses that have RNA as their genetic material contain double-stranded helices.

As established earlier, three major classes of cellular RNA molecules function during the expression of genetic information: **ribosomal RNA (rRNA)**, **messenger RNA (mRNA)**, and **transfer RNA (tRNA)**. These molecules all originate as complementary copies of one of the two strands of DNA segments during the process of transcription. That is, their nucleotide sequence is complementary to the deoxyribonucleotide sequence of DNA, which served as the template for their synthesis. Because uracil replaces thymine in RNA, uracil is complementary to adenine during transcription and during RNA base pairing.

Table 10.4 characterizes the major forms of RNA found in prokaryotic and eukaryotic cells. Different RNAs are distinguished according to their sedimentation behavior in a centrifugal field and their size, as measured by the number of nucleotides each contains. Sedimentation behavior depends on a molecule's density, mass, and shape, and its measure is called the **Svedberg coefficient (S)**. While higher *S* values almost always designate molecules of greater molecular weight, the correlation is not direct; that is, a twofold increase in molecular weight does not lead to a twofold increase in *S*. This is because, in addition to a molecule's mass, the size and the shape of the molecule also impact its rate of sedimentation (*S*). As you can see, a wide variation exists in the size of the three classes of RNA.

Ribosomal RNA is generally the largest of these molecules (as is generally reflected in its *S* values) and usually constitutes about 80 percent of all RNA in the cell. Ribosomal RNAs are important structural components of **ribosomes**,

TABLE 10.4   RNA Characterization

RNA Class	% Total RNA*	Components (Svedberg Coefficient)	Eukaryotic (E) or Prokaryotic (P)	Number of Nucleotides
Ribosomal (rRNA)	80	5S	P and E	120
		5.8S	E	160
		16S	P	1542
		18S	E	1874
		23S	P	2904
		28S	E	4718
Transfer (tRNA)	15	4S	P and E	75–90
Messenger (mRNA)	5	varies	P and E	100–10,000

*In *E. coli*.

which function as nonspecific workbenches during the synthesis of proteins during the process of translation. The various forms of rRNA found in prokaryotes and eukaryotes differ distinctly in size.

Messenger RNA molecules carry genetic information from the DNA of the gene to the ribosome, where translation occurs. They vary considerably in size, which reflects the variation in the size of the protein encoded by the mRNA as well as the gene serving as the template for transcription of mRNA.

Transfer RNA, the smallest class of RNA molecules, carries amino acids to the ribosome during translation. Because more than one tRNA molecule interacts simultaneously with the ribosome, the molecule's smaller size facilitates these interactions.

We discuss the functions of the three classes of RNA in much greater detail in Chapter 12. In addition, as we proceed through the text, we will encounter other unique RNAs that perform various roles. For example, **small nuclear RNA (snRNA)** participates in processing mRNAs (Chapter 12). **Telomerase RNA** is involved in DNA replication at the ends of chromosomes (Chapter 11), and **antisense RNA** is involved in gene regulation (Chapter 15). Our purpose in this section has been to contrast the structure of DNA, which stores genetic information, with that of RNA, which most often functions in the expression of that information.

10.9 Hydrogen Bonds and the Analysis of Nucleic Acids

The unique nature of the hydrogen bond imparts an interesting and important set of qualities to the chemical behavior of nucleic acids under both laboratory and physiological conditions. For example, if DNA is isolated and subjected to slow heating, the double helix is denatured and unwinds. If a mixture of single strands that are complementary to each other are slowly cooled, they will reassociate, re-forming the helix. In the laboratory, these transformations can be "tracked" by monitoring the absorption of UV light (or optical density) at 260 nm (OD_{260}), using a spectrophotometer.

During unwinding, the viscosity of DNA decreases and UV absorption increases. A melting profile, in which OD_{260} is plotted against temperature, is shown for two DNA molecules in Figure 10–14. The midpoint of each curve is called the **melting temperature (T_m)**, where 50 percent of the strands are unwound. The molecule with a higher T_m has a higher percentage of $G \equiv C$ base pairs than $A = T$ base pairs compared to the molecule with the lower T_m, since $G \equiv C$ pairs share three hydrogen bonds compared to the two present between $A = T$ pairs.

Molecular Hybridization Techniques

The property of denaturation/renaturation of nucleic acids is the basis for one of the most powerful and useful techniques in molecular genetics—**molecular hybridization**. Provided that a reasonable degree of base complementarity exists, under the proper temperature conditions, two nucleic acid strands from different sources will rejoin. As a result,

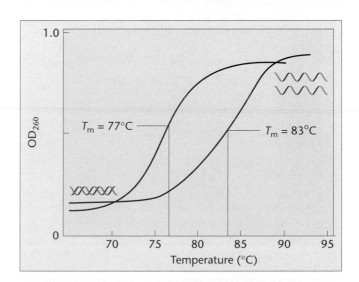

FIGURE 10–14 Increase in UV absorption vs. temperature (the hyperchromic effect) for two DNA molecules with different $G \equiv C$ contents. The molecule with a melting point (T_m) of 83°C has a greater $G \equiv C$ content than the molecule with a T_m of 77°C.

molecular hybridization is possible between DNA strands from different species and between DNA and RNA strands. For example, an RNA molecule will hybridize with the segment of DNA from which it was transcribed or with a DNA molecule from a different species, provided that its nucleotide sequence is nearly the same.

The technique can even be performed using the DNA present in cytological preparations as the "target" for hybrid formation. This process is called *in situ* **molecular hybridization**. Mitotic or interphase cells are first fixed to slides and then subjected to hybridization conditions. Single-stranded DNA or RNA is added, and hybridization is monitored. The nucleic acid that is added may either be radioactive or contain a fluorescent label to allow its detection. In the former case, the technique of autoradiography is used.

Figure 10–15 illustrates the use of a fluorescent label. A "probe," consisting of a short fragment of DNA that is complementary to DNA present in the chromosome's centromere regions, has been hybridized. Fluorescence occurs only in the centromere regions and thus identifies each one along its chromosome. Because fluorescence is used, the technique is known by the acronym **FISH (fluorescent *in situ* hybridization)**. The use of this technique to identify chromosomal locations housing specific genetic information has been a valuable addition to geneticists' repertoire of experimental techniques.

Reassociation Kinetics and Repetitive DNA

In one extension of molecular hybridization procedures, the *rate of reassociation* of complementary single DNA strands is analyzed. This technique, **reassociation kinetics**, was first refined and studied by Roy Britten and David Kohne.

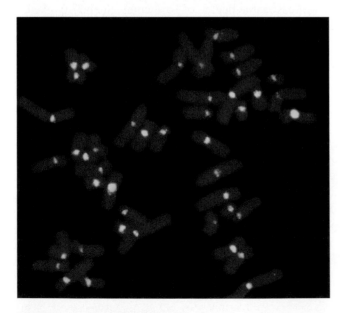

FIGURE 10–15 *In situ* hybridization of human metaphase chromosomes using a fluorescent technique (FISH). The probe, specific to centromeric DNA, produces a yellow fluorescence signal indicating hybridization. The red fluorescence is produced by propidium iodide counterstaining of chromosomal DNA. *(Ilse Light, John F. Connaughton/Ventana Medical Systems, Inc.)*

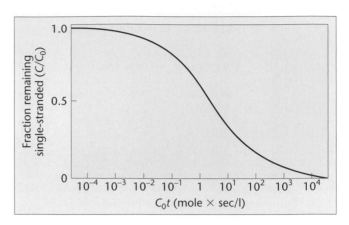

FIGURE 10–16 The ideal time course for reassociation of DNA (C/C_0) when, at time zero, all DNA consists of unique fragments of single-stranded complements. Note that the abscissa (C_0t) is plotted logarithmically.

The DNA used in such studies is first fragmented into small pieces by shearing forces introduced during its isolation. The resultant DNA fragments have an average size of several hundred base pairs. These fragments are then dissociated into single strands by heating. The temperature is then lowered, and reassociation is monitored. During reassociation, pieces of single-stranded DNA collide randomly. If they are complementary, a stable double strand is formed; if not, they separate and are free to encounter other DNA fragments. The process continues until all possible matches are made.

The results of one such experiment are presented in Figure 10–16. The percentage of reassociation of DNA fragments is plotted against a logarithmic scale of normalized time, a function referred to as C_0t. In this term, C_0 is the initial concentration of DNA single strands and t is the time.

A great deal of information can be obtained from studies that compare the reassociation of DNA of different organisms. For example, we can compare the point in the reaction when one-half of the DNA is present as double-stranded fragments. This point is called the **half reaction time**. Provided that all DNA fragments contain unique nucleotide sequences and all are about the same size, $C_0t_{1/2}$ varies directly with the total length of the DNA.

Figure 10–17 compares DNAs from three phage or bacterial sources, each with a different genome size. As genome size increases, the curves obtained have a similar shape but are shifted farther and farther to the right, indicative of an extended reassociation time. Reassociation occurs more slowly in larger genomes because it takes longer for initial matches if there are greater numbers of unique DNA fragments. This is so because collisions are random; the more sequences present, the greater the number of mismatches before all correct matchings occur.

When the kinetics of DNA reassociation in eukaryotic organisms with much larger genome sizes was first studied, a surprising observation was made. Rather than exhibiting a reduced rate of reassociation, the data revealed that *some* DNA segments reassociate even more rapidly than those

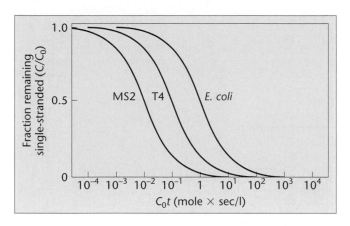

FIGURE 10–17 The reassociation rates (C/C_0) of DNA derived from phage MS2, phage T4, and *E. coli*. The genome of T4 is larger than MS2, and that of *E. coli* is larger than T4.

chromosomes. Careful study has shown that various levels of repetition exist. In some cases, short DNA sequences are repeated over a million times. In other cases, longer sequences are repeated only a few times, or intermediate levels of sequence redundancy are present. The discovery of repetitive DNA was one of the first clues that much of the DNA in eukaryotes is not contained in genes that encode proteins.

10.10 Electrophoresis of Nucleic Acids

We conclude this chapter by considering an essential technique in the analysis of nucleic acids, **electrophoresis**. This technique separates different-sized fragments of DNA and RNA chains and is invaluable in current research investigations in molecular genetics.

In general, electrophoresis separates molecules present in a mixture by causing them to migrate under the influence of an electric field. A sample is placed on a porous substance (a piece of filter paper or a semisolid gel), which is placed in a solution that conducts electricity. If two molecules have approximately the same shape and mass, the one with the greatest net charge migrates more rapidly toward the electrode of opposite polarity.

As electrophoretic technology developed from its initial application to protein separation, researchers discovered that using gels of varying pore sizes significantly improved the resolution of this research technique. This advance is particularly useful for mixtures of molecules with a similar charge:mass ratio but different sizes. For example, two polynucleotide chains of different lengths (e.g., 10 vs. 20 nucleotides) are both negatively charged based on the phosphate groups of the nucleotides. While both move to the positively charged pole (the anode), the charge:mass ratio is the same for each chain, and separation based strictly on the electric field is minimal. However, using a porous medium such as **polyacrylamide gels** or **agarose gels**, which can be prepared with various pore sizes, allows us to separate these two molecules.

In such cases, *the smaller molecules migrate at a faster rate through the gel than the larger molecules* (Figure 10–19). The key to separation is based on the matrix (pores) of the gel, which restricts migration of larger molecules more than it restricts smaller molecules. The resolving power is so great that polynucleotides that vary by even one nucleotide in length are clearly separated. Once electrophoresis is complete, bands representing the variously sized molecules are identified either by autoradiography (if a component of the molecule is radioactive) or by the use of a fluorescent dye that binds to nucleic acids.

Electrophoretic separation of nucleic acids is at the heart of a variety of commonly used research techniques discussed later in the text (Chapters 16 and 19). Of particular note are the various "blotting" techniques (e.g., Southern blots and Northern blots), as well as DNA sequencing methods.

derived from *E. coli*! The remaining DNA, as expected because of its greater size and complexity, takes longer to reassociate.

For example, Britten and Kohne examined DNA derived from calf thymus tissue. Based on their observations (Figure 10–18), they hypothesized correctly that the rapidly reassociating fraction might represent repetitive sequences present many times in the calf genome. This interpretation would explain why these segments reassociate so rapidly. Multiple copies of the same sequence are much more likely to make matches, thus reassociating more quickly than single copies. On the other hand, the remaining DNA segments consist of unique nucleotide sequences present only once in the genome. Because calf thymus DNA has many more of these unique sequences than *E. coli*, their reassociation takes longer.

Copies present many times in the genome are referred to collectively as **repetitive DNA**. Repetitive DNA is prevalent in eukaryotic genomes and is key to our understanding of how genetic information is organized in

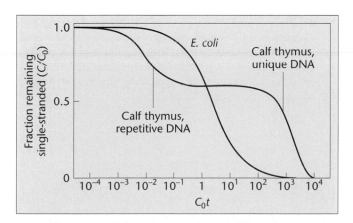

FIGURE 10–18 The C_0t curve of calf thymus DNA compared with *E. coli*. The repetitive fraction of calf DNA reassociates more quickly than that of *E. coli*, while the more complex unique calf DNA takes longer to reassociate than that of *E. coli*.

FIGURE 10–19 Electrophoretic separation of a mixture of DNA fragments that vary in length. The photograph at the bottom right shows an agarose gel with DNA bands. *(Photo: Dr. William S. Klug)*

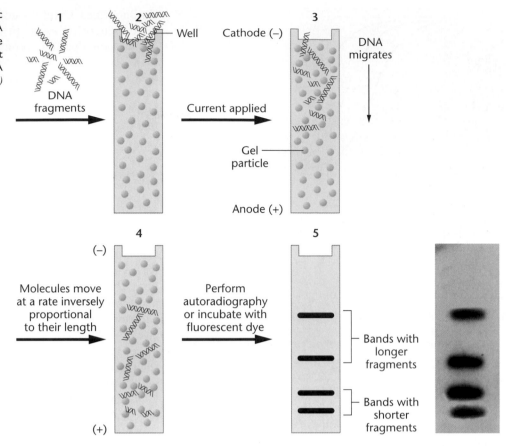

Chapter Summary

1. The existence of a genetic material capable of replication, storage, expression, and mutation is deducible from observed patterns of inheritance in organisms. Both proteins and nucleic acids were initially considered as possible candidates for the genetic material.

2. Transformation studies, as well as experiments using bacteria infected with bacteriophages, strongly suggested that DNA is the genetic material for bacteria and most viruses.

3. Initially, only circumstantial observations supported the hypothesis that DNA controls inheritance in eukaryotes. These included DNA distribution in the cell, quantitative analysis of DNA, and UV-induced mutagenesis. More recent recombinant DNA techniques, as well as experiments with transgenic mice, provide direct experimental evidence that the eukaryote genetic material is DNA.

4. In some viruses, RNA serves as the genetic material. These include bacteriophages as well as some plant and animal viruses.

5. By the 1950s, many scientists sought to determine the structure of DNA. These efforts culminated in 1953 with Watson and Crick's proposal. Based on base-pairing information and X-ray diffraction data, they constructed a model, the key features of which include two antiparallel polynucleotide chains held together in a right-handed double helix by the hydrogen bonds formed between complementary bases. To date, the basic tenets of this double helix have held true.

6. RNA varies from DNA by virtue of almost always being single-stranded, containing uracil rather than thymine, and having ribose rather than deoxyribose as its constituent sugar.

7. The structure of DNA lends itself to various forms of analysis, which have in turn led to studies of the functional aspects of the genetic machinery. Absorption of UV light, denaturation–reassociation, and electrophoresis procedures are important tools in the study of nucleic acids. Reassociation kinetics analysis enabled geneticists to postulate the existence of repetitive DNA in eukaryotes, where certain nucleotide sequences are present many times in the genome.

Molecular Structure of Nucleic Acids: A Structure for Deoxyribose Nucleic Acid

We wish to suggest a structure for the salt of deoxyribose nucleic acid (D. N. A.). This structure has novel features which are of considerable biological interest. A structure for nucleic acid has already been proposed by Pauling and Corey.[1] They kindly made their manuscript available to us in advance of publication.

Their model consists of three intertwined chains, with the phosphates near the fibre axis, and the bases on the outside. In our opinion, this structure is unsatisfactory for two reasons: (1) We believe that the material which gives the X-ray diagrams is the salt, not the free acid. Without the acidic hydrogen atoms it is not clear what forces would hold the structure together, especially as the negatively charged phosphates near the axis will repel each other. (2) Some of the van der Waals distances appear to be too small.

Another three-chain structure has also been suggested by Fraser (in the press). In his model the phosphates are on the outside and the bases on the inside, linked together by hydrogen bonds. This structure as described is rather ill-defined, and for this reason we shall not comment on it.

We wish to put forward a radically different structure for the salt of deoxyribose nucleic acid. This structure has two helical chains each coiled round the same axis. We have made the usual chemical assumptions, namely, that each chain consists of phosphate diester groups joining β-D-deoxyribofuranose residues with $3',5'$ linkages. The two chains (but not their bases) are related by a dyad perpendicular to the fibre axis. Both chains follow right-handed helices, but owing to the dyad the sequences of the atoms in the two chains run in opposite directions. Each chain loosely resembles Furberg's[2] model No. 1; that is, the bases are on the inside of the helix and the phosphates on the outside. The configuration of the sugar and the atoms near it is close to Furberg's "standard configuration," the sugar being roughly perpendicular to the attached base. There is a residue on each chain every 3.4 Å in the z-direction. We have assumed an angle of 36° between adjacent residues in the same chain, so that the structure repeats after 10 residues on each chain, that is, after 34 Å. The distance of a phosphate atom from the fibre axis is 10 Å. As the phosphates are on the outside, cations have easy access to them.

The structure is an open one, and its water content is rather high. At lower water content we would expect the bases to tilt so that the structure could become more compact.

The novel feature of the structure is the manner in which the two chains are held together by the purine and pyrimidine bases. The planes of the bases are perpendicular to the fibre axis. They are joined together in pairs, a single base from one chain being hydrogen-bonded to a single base from the other chain, so that the two lie side by side with identical z-co-ordinates. One of the pair must be a purine and the other a pyrimidine for bonding to occur. The hydrogen bonds are made as follows: purine position 1 to pyrimidine position 1; purine position 6 to pyrimidine position 6.

If it is assumed that the bases only occur in the structure in the most plausible tautomeric forms (that is, with the keto rather than the enol configuration) it is found that only specific pairs of bases can bond together. These pairs are: adenine (purine) with thymine (pyrimidine), and guanine (purine) with cytosine (pyrimidine).

In other words, if an adenine forms one member of a pair, on either chain, then on these assumptions the other member must be thymine; similarly for guanine and cytosine. The sequence of bases on a single chain does not appear to be restricted in any way. However, if only specific pairs of bases can be formed, it follows that if the sequence of bases on one chain is given, then the sequence on the other chain is automatically determined.

It has been found experimentally[3,4] that the ratio of the amounts of adenine to thymine, and the ratio of guanine to cytosine, are always very close to unity for deoxyribose nucleic acid.

It is probably impossible to build this structure with a ribose sugar in place of deoxyribose, as the extra oxygen atom would make too close a van der Waals contact.

The previously published X-ray data[5,6] on deoxyribose nucleic acid are insufficient for a rigorous test of our structure. So far as we can tell, it is roughly compatible with the experimental data, but it must be regarded as unproved until it has been checked against more exact results. Some of these are given in the following communications. We were not aware of the details of the results presented there when we devised our structure, which rests mainly though not entirely on published experimental data and stereochemical arguments.

It has not escaped our notice that the specific pairing we have postulated immediately suggests a possible copying mechanism for the genetic material. Full details of the structure, including the conditions assumed in building it, together with a set of co-ordinates for the atoms, will be published elsewhere.

We are much indebted to Dr. Jerry Donohue for constant advice and criticism, especially on interatomic distances. We have also been stimulated by a knowledge of the general nature of the unpublished experimental results and ideas of Dr. M. H. F. Wilkins, Dr. R. E. Franklin and their co-workers at King's College, London. One of us (J. D. W.) has been aided by a fellowship from the National Foundation for Infantile Paralysis.

J. D. Watson
F. H. C. Crick
Medical Research Council Unit for the Study of the Molecular Structure of Biological Systems, Cavendish Laboratory, Cambridge, England

[1]Pauling L., and Corey, R. B., *Nature*, 171, 346 (1953); *Proc. U.S. Nat. Acad. Sci.*, 39, 84 (1953).

[2]Furberg, S., *Acta Chem. Scand.*, 6, 634 (1952).

[3]Chargaff, E., for references see Zamenhof, S., Brawerman, G., and Chargaff, E., *Biochim. et Biophys. Acta*, 9, 402 (1952).

[4]Wyatt, G. R., *J. Gen. Physiol.*, 36, 201 (1952).

[5]Astbury, W. T., *Symp. Soc. Exp. Biol. 1, Nucleic Acid*, 66 (Camb. Univ. Press, 1947).

[6]Wilkins, M. H. F., and Randall, J. T., *Biochim. et Biophys. Acta*, 10, 192 (1953).

Genetics, Technology, and Society

Genetics and Society in the New Millennium

Since the beginning of human civilization, we have defined ourselves as the masters of the biological world. Civilization began when humans domesticated plants and animals and settled into societies—as recently as 12 millennia ago. Genetics, in the form of selective breeding, became the foundation of agricultural progress and contributed to the rise and fall of civilizations over thousands of years. Our ability to harness nature is reflected in religions and philosophies that place humans at the center of the universe, at the pinnacle of creation, above all other creatures.

Now, as we enter the 21st century, a revolution of biological thought has begun—a revolution that is changing both our mastery over living things, and our perception of ourselves.

The revolution began in 1953 with Watson and Crick's elucidation of the molecular structure of DNA. The structure of the DNA molecule immediately provided elegant solutions to age-old questions about the mechanisms of heredity, mutation, and evolution. Some of the greatest mysteries of life could suddenly be explained by the beauty and simplicity of a chemical that replicates and shuffles the code of life.

Over the next 30 years, DNA became the focus of laboratories around the world. Geneticists, biochemists, and molecular biologists quickly devised methods to purify, mutate, cut, and paste DNA in the test tube. DNA molecules from one organism were spliced into DNA molecules from another, and the chimeric molecules were introduced into bacteria or cells in culture. The nucleotide sequences of genes were defined and modified *in vitro*. Gene promoters, gene transcription units, and gene activities were assayed. The genetic traits of organisms such as bacteria, fungi, and fruit flies were modified by the removal or addition of genes from similar, or different, species. Genetic engineering had begun.

The DNA revolution has advanced at an explosive rate. In the 27 years since the first gene was cloned, scientists have discovered the genes that control hereditary diseases such as sickle-cell anemia, cystic fibrosis, and Tay-Sachs disease. Prenatal diagnosis for these and other genetic diseases is now available. Biotechnologists have genetically modified bacteria, plants, and animals to express proteins of agricultural and medical importance. Biotechnology now offers DNA forensic tests that have helped convict criminals, exonerate the innocent, and establish paternity. Scientists have even cloned mammals such as sheep and mice from adult somatic cells. DNA manipulation is now a powerful tool of medical research. Scientists are rapidly dissecting the molecular genetic mechanisms that control cell growth, aging, death, and cancer.

The effect that the DNA revolution has had on our views of ourselves and the world is reflected in everyday culture. Although scientists dismiss the idea that humans are simply the products of their genes, popular culture endows DNA with almost magical powers. In television sitcoms, magazines, and daily conversation, genes are said to explain personality, career choice, criminality, intelligence—even fashion preferences and political attitudes. Advertisements hijack the language of genetics in order to grant inanimate objects a "genealogy" or "genetic advantage." Popular culture speaks of DNA as an immortal force, with the ability to affect morality and fate. DNA is defined as the "essence of life" and the "immortal text," with the power to shape our future. Biological or genetic explanations for antisocial behavior appear to have more resonance for us than explanations involving social or economic factors.

But what of the future? Can we predict how DNA and genetics will help us shape the world in the new millennium? Although prophecy is certainly a risky business, some developments seem assured. The Human Genome Project has now decoded the entire human genome, ahead of schedule. This will lead to the identification of many genes that control normal and abnormal processes. In turn, this may enhance our ability to diagnose and predict genetic diseases. Scientists predict that the new millennium will bring us biotechnologies as complex as gene therapy and prenatal diagnosis and correction of genetic defects. Undoubtedly, the application of genetic engineering to agriculture will continue, as we manipulate plant genes for disease resistance, productivity, color, and flavor. The genetic engineering of farm animals is also likely to continue.

It is clear that the DNA revolution will have far-reaching practical consequences for humanity. It will also change how we think about ourselves and the world. As many more human genes are discovered, sequenced, and compared to those of other animals, it will become increasingly evident that we are closely related genetically to the rest of the animate world. The nucleotide sequence of our genome differs only about 1 percent from that of chimpanzees, and some of our genes are virtually identical to homologous genes in plants, animals, and bacteria. Will this knowledge alter our relationships with animals and with each other, as we realize the extent of our genetic kinship? As more genes are discovered that contribute to phenotypic traits as simple as eye color and as complicated as intelligence or sexual orientation, will this lead us to define ourselves more as genetic beings and less as creatures of free will or as the products of our environment?

As this millennium unfolds, we will inevitably be faced with the practical and philosophical consequences of the DNA revolution. Will society harness DNA for everyone's benefit, or will the new genetic knowledge be used as a vehicle for discrimination? At the same time that modern genetics grants us more dominion over life, will it paradoxically increase our feelings of powerlessness? Will our new genetic view of life increase our compassion for all life forms, or will it increase our perceived separation from the natural world? We will make our choices, and human history will proceed.

References:

Collins, F.S. et al. 1998. New goals for the U.S. human genome project: 1998–2003. Science 282: 682–89.

Nelkin, D., and Lindee, M.S. 1995. The DNA mystique, the gene as a cultural icon. New York: W.H. Freeman.

Wilkie, T. 1993. Perilous knowledge, the human genome project and its implications. London: Faber and Faber.

Key Terms

Insights and Solutions

In contrast to preceding chapters, this chapter does not emphasize genetic problem solving. Instead, it recounts some of the initial experimental analyses that launched the era of molecular genetics. Quite fittingly, then, our Insights and Solutions section shifts its emphasis to experimental rationale and analytical thinking, an approach that will continue through the remainder of the text whenever appropriate.

1. Based strictly on the transformation analysis of Avery, MacLeod, and McCarty, what objection might be made to the conclusion that DNA is the genetic material? What other conclusion might be considered?

 Solution: Based solely on their results, we could conclude that DNA is essential for transformation. However, DNA might have been a substance that caused capsular formation by converting nonencapsulated cells *directly* to ones with a capsule. That is, DNA may simply have played a catalytic role in capsular synthesis, leading to cells that display smooth type III colonies.

2. What observations argue against this objection?

 Solution: First, transformed cells pass the trait on to their progeny cells, thus supporting the conclusion that DNA is responsible for heredity, not for the direct production of polysaccharide coats. Second, subsequent transformation studies over the next five years showed that other traits, such as antibiotic resistance, could be transformed. Therefore, the transforming factor has a broad general effect, not one specific to polysaccharide synthesis.

3. If RNA were the universal genetic material, how would this have affected the Avery experiment and the Hershey–Chase experiment?

 Solution: In the Avery experiment, RNase rather than DNase would have eliminated transformation. Had this occurred, Avery and his colleagues would have concluded that RNA was the transforming factor. Hershey and Chase would have received identical results, since ^{32}P would also label RNA, but not protein.

4. Sea urchin DNA, which is double-stranded, contains 17.5 percent of its bases in the form of cytosine (C). What percentages of the other three bases are expected to be present in this DNA?

 Solution: The amount of C equals G, so guanine is also present at 17.5 percent. The remaining bases, A and T, are present in equal amounts and together they represent the remaining bases $(100 - 35)$. Therefore, A = T = 65/2 = 32.5 percent.

Problems and Discussion Questions

1. The functions ascribed to the genetic material are replication, expression, storage, and mutation. What does each of these terms mean?
2. Discuss the reasons why, prior to 1940, proteins were generally favored over DNA as the genetic material. What was the role of the tetranucleotide hypothesis in this controversy?
3. Contrast the various contributions made by Griffith with those of Avery and his coworkers to our understanding of transformation.
4. Why were ^{32}P and ^{35}S used in the Hershey–Chase experiment? Discuss the rationale and conclusions of this experiment.
5. What observations are consistent with DNA serving as the genetic material in eukaryotes? List and discuss. What direct evidence exists?
6. What are the exceptions to the general rule that DNA is the genetic material in all organisms? What evidence supports these exceptions?
7. Draw the chemical structure of the three components of a nucleotide and link them. What atoms are removed from the structures when the linkages are formed?
8. Adenine is also named 6-amino purine. How would you name the other four nitrogenous bases using this alternative system? ($=$O is oxy, and —CH_3 is methyl.)
9. Describe the various characteristics of the Watson–Crick double helix model for DNA.
10. What evidence did Watson and Crick have at their disposal in 1953? What was their approach in arriving at the structure of DNA?
11. Had Chargaff's data from a single source indicated the following, what might Watson and Crick have concluded?

Base	A	T	C	G
%	29	19	21	31

Why would this conclusion be contradictory to Wilkins' and Franklin's data?
12. List three main differences between DNA and RNA.
13. What is the chemical basis of molecular hybridization?
14. When the entire genome is utilized, why does unique sequence DNA take longer to reassociate than repetitive DNA sequences?
15. A genetics student is asked to draw the chemical structure of an adenine- and thymine-containing dinucleotide derived from DNA. The student made more than six major errors. His answer is shown here. One error (1) is circled and explained. Find five others. Circle them, label them 2–6, and briefly explain each, following the example.

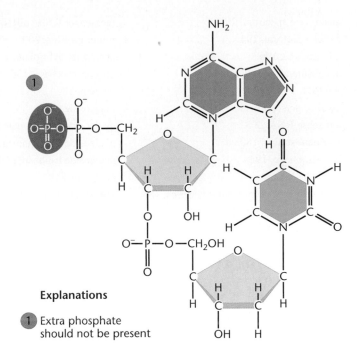

Explanations

1 Extra phosphate should not be present

16. Newsdate: March 1, 2005. A unique creature is discovered in outer space, and its genetic material is isolated and analyzed. In some ways, this material resembles DNA in its chemical make-up. It contains an abundance of the four-carbon sugar erythrose and a molar equivalent of phosphate groups. Additionally, it contains six nitrogenous bases: adenine (A), guanine (G), thymine (T), cytosine (C), hypoxanthine (H), and xanthine (X). These bases exist in the following relative proportions:

$$A = T = H \quad \text{and} \quad C = G = X$$

X-ray diffraction studies establish a regularity to the molecule and a constant diameter of about 30 Å. Together, these data suggest a model for the structure of this molecule. (a) Propose a general model and describe it briefly. (b) What base-pairing properties must exist for H and for X in the model? (c) Given the constant diameter of 30 Å, do you think that H and X are either (1) both purines or both pyrimidines, or (2) one is a purine and one is a pyrimidine?

17. A primitive eukaryote is discovered that displays a unique nucleic acid as its genetic material. Analysis reveals the following:
    (i) X-ray diffraction studies display a general pattern similar to DNA, but with somewhat different dimensions and more irregularity.
    (ii) A major hyperchromic shift is evident upon heating and monitoring UV absorption at 260 nm.
    (iii) Base composition analysis reveals four bases in the following proportions:
    Adenine—8%   Guanine—37%
    Xanthine—37%   Hypoxanthine—18%
    (iv) About 75% of the sugars are deoxyribose, while 25% are ribose.
    Attempt to solve the structure of this molecule by postulating a model that is consistent with the observations.

18. One of the most common spontaneous lesions that occurs in DNA under physiological conditions is the hydrolysis of the amino group of cytosine, converting it to uracil. What is the effect of a uracil replacing cytosine in the DNA?

19. In some organisms, cytosine is methylated at C-5 of the pyrimidine ring after it is incorporated into DNA. If a 5-methyl cytosine is hydrolyzed as described in Problem 18, what base is generated?

20. Compare the following curves, which represent reassociation kinetics. What can be said about the DNAs represented by curves A and B compared with the *E. coli* curve?

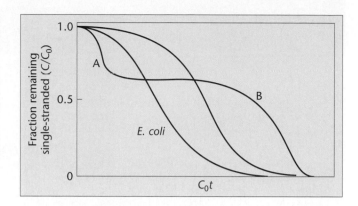

Selected Readings

Avery, O. T., MacLeod, C. M., and McCarty, M. 1944. Studies on the chemical nature of the substance inducing transformation of pneumococcal types. Induction of transformation by a desoxyribonucleic acid fraction isolated from pneumococcus type III. *J. Exp. Med.* 79:137–58. (Reprinted in Taylor, J. H. 1965. *Selected papers in molecular genetics.* Orlando, FL: Academic Press.)

Britten, R. J., and Kohne, D. E. 1970. Repeated segments of DNA. *Sci. Am.* (Apr.) 222:24–31.

Chargaff, E. 1950. Chemical specificity of nucleic acids and mechanism for their enzymatic degradation. *Experientia* 6:201–9.

Dawson, M. H. 1930. The transformation of pneumococcal types: I. The interconvertibility of type-specific *S. pneumococci. J. Exp. Med.* 51:123–47

DeRobertis, E. M., and Gurdon, J. B. 1979. Gene transplantation and the analysis of development. *Sci. Am.* (Dec.) 241:74–82.

Dickerson, R. E. 1983. The DNA helix and how it is read. *Sci. Am.* (June) 249:94–111.

Dickerson, R. E., et al. 1982. The anatomy of A-, B-, and Z-DNA. *Science* 216:475–85.

Dubos, R. J. 1976. *The professor, the institute and DNA: Oswald T. Avery, his life and scientific achievements.* New York: Rockefeller University Press.

Felsenfeld, G. 1985. DNA. *Sci. Am.* (Oct.) 253:58–78.

Fraenkel-Conrat, H., and Singer, B. 1957. Virus reconstruction: II. Combination of protein and nucleic acid from different strains. *Biochem. Biophys. Acta* 24:530–48 (Reprinted in Taylor, J.H. 1965. *Selected papers in molecular genetics,* Orlando, FL: Academic Press.)

Franklin, R. E., and Gosling, R. G. 1953. Molecular configuration in sodium thymonucleate. *Nature* 171:740–41.

Griffith, F. 1928. The significance of pneumococcal types. *J. Hyg.* 27:113–59.

Guthrie, G. D., and Sinsheimer, R. L. 1960. Infection of protoplasts of *Escherichia coli* by subviral particles. *J. Mol. Biol.* 2:297–305.

Hershey, A. D., and Chase, M. 1952. Independent functions of viral protein and nucleic acid and in growth of bacteriophage. *J. Gen. Physiol.* 36:39–56. (Reprinted in Taylor, J. H. 1965. *Selected papers in molecular genetics.* Orlando, FL: Academic Press.)

Judson, H. 1979. *The eighth day of creation: Makers of the revolution in biology.* New York: Simon & Schuster.

Levene, P. A., and Simms, H. S. 1926. Nucleic acid structure as determined by electrometric titration data. *J. Biol. Chem.* 70:327–41.

McCarty, M. 1980. Reminiscences of the early days of transformation. *Annu. Rev. Genet.* 14:1–16.

————. 1985. *The transforming principle: Discovering that genes are made of DNA.* New York: W. W. Norton.

Olby, R. 1974. *The path to the double helix.* Seattle: University of Washington Press.

Palmiter, R. D., and Brinster, R. L. 1985. Transgenic mice. *Cell* 41:343–45.

Pauling, L., and Corey, R. B. 1953. A proposed structure for the nucleic acids. *Proc. Natl. Acad. Sci. USA* 39:84–97.

Rich, A., Nordheim, A. and Wang, A. H.-J. 1984. The chemistry and biology of left-handed Z-DNA. *Annu Rev. Biochem.* 53:791–846.

Spizizen, J. 1957. Infection of protoplasts by disrupted T2 viruses. *Proc. Natl. Acad. Sci. USA* 43:694–701.

Stent, G. S., ed. 1981. *The double helix: Text, commentary review, and original papers.* New York: W. W. Norton.

Stewart, T. A., Wagner, E. F., and Mintz, B. 1982. Human β-globin gene sequences injected into mouse eggs, retained in adults, and transmitted to progeny. *Science* 217:1046–48.

Varmus, H. 1988. Retroviruses. *Science* 240:1427–35.

Watson, J. D. 1968. *The double helix.* New York: Atheneum.

Watson, J. D., and Crick, F. C. 1953a. Molecular structure of nucleic acids. A structure for deoxyribose nucleic acids. *Nature* 171:737–38.

Watson, J. D., and Crick, F. C. 1953b. Genetic implications of the structure of deoxyribose nucleic acid. *Nature* 171:964.

Weinberg, R. A. 1985. The molecules of life. *Sci. Am.* (Oct.) 253:48–57.

Wilkins, M. H. F., Stokes, A. R., and Wilson., H. R. 1953. Molecular structure of desoxypentose nucleic acids. *Nature* 171:738–40.

Yung, J. 1996. New FISH probes—The end in sight. *Nature Genetics* 14:10–12.

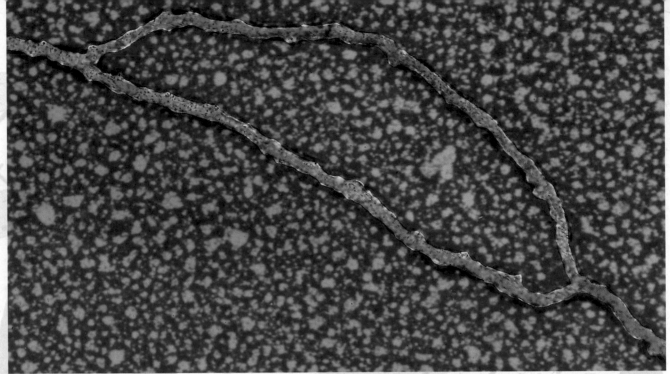

Transmission electron micrograph of human DNA from a HeLa cell, illustrating the replication bubble that characterizes DNA replication within a single replicon. *(Dr. Gopal Murti/Science Photo Library/Photo Researchers, Inc.)*

11

DNA—Replication and Synthesis

CHAPTER CONCEPTS

Semiconservative replication of DNA makes genetic continuity between parental and progeny cells possible, as predicted by the Watson–Crick model. Each strand of the parent helix serves as a template for its complement. DNA synthesis is a complex but orderly process orchestrated by a myriad of enzymes and other molecules. Together, they function with great fidelity to polymerize nucleotides into polynucleotide chains. Other enzymes interact with DNA, leading to genetic recombination.

ollowing Watson and Crick's proposal for the structure of DNA, scientists focused their attention on how this molecule replicates. Replication is an essential function of the genetic material and must be executed precisely if genetic continuity is to be maintained following cell division. This is an enormous and complex task. Consider for a moment that in the human genome, some 3 billion (10^9) base pairs exist within the 23 chromosomes. To duplicate a molecule of this size faithfully requires a mechanism of extreme precision. Even an error rate of only 10^{-6} (one in a million) will still create 3000 errors, obviously an excessive number during each replication cycle. While it is not error-free, an extremely accurate system of DNA replication has evolved in all organisms.

As Watson and Crick wrote in their 1953 paper, the model of the double helix provided their initial insight into how replication could occur. This mode, called semiconservative replication, is strongly supported from numerous studies of viruses, prokaryotes, and eukaryotes.

Once the general mode of replication was clarified, research to determine the precise details of DNA synthesis intensified. What has since been discovered is that numerous enzymes and other proteins are needed to copy a DNA helix. Because of the complexity of the chemical events during synthesis, this subject remains an extremely active area of research.

In this chapter we discuss the general mode of replication as well as the specific details of the synthesis of DNA. The research leading to this knowledge is yet another link in our understanding of life processes at the molecular level.

11.1 The Mode of DNA Replication

It was apparent to Watson and Crick that because of the arrangement and nature of the nitrogenous bases, each strand of a DNA double helix could serve as a template for the synthesis of its complement (Figure 11–1). They proposed that if the helix were unwound, each nucleotide along the two parent strands would have an affinity for its complementary nucleotide. As we learned in Chapter 10, complementarity is due to the hydrogen bonds that form. If thymidylic acid (T) were present, it would "attract" adenylic acid (A); if guanidylic acid (G) were present, it would "attract" cytidylic acid (C); likewise, A would attract T, and C would attract G. If these nucleotides were then linked covalently into polynucleotide chains along both templates, the result would be two new but identical double strands of DNA. Each replicated DNA molecule would consist of one "old" and one "new" strand, hence the reason for the name **semiconservative replication**.

Two other possible modes of replication also rely on the parental strands as a template (Figure 11–2). In **conservative replication**, synthesis of complementary polynucleotide chains occurs as described above. Following synthesis, however, the two newly created strands are brought together, and the parental strands reassociate. The original helix is thus "conserved."

In the second mode, called **dispersive replication**, the parental strands are seen to be dispersed into two new double helices following replication. Each strand would then consist of both old and new DNA. This mode would involve

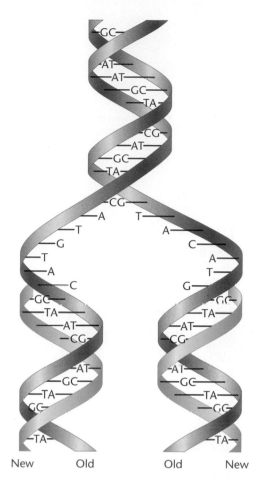

FIGURE 11–1 Generalized model of semiconservative replication of DNA. New synthesis is shown in blue.

New   Old   Old   New

cleavage of the parental strands during replication. It is the most complex of the three possibilities and is therefore least likely. It could not, however, be ruled out as an experimental model. Figure 11–2 shows the theoretical results of a single round of replication by the three modes.

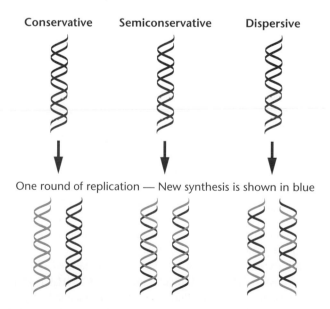

Conservative   **Semiconservative**   **Dispersive**

One round of replication — New synthesis is shown in blue

FIGURE 11–2 Results of one round of DNA replication for each of the three possible modes by which replication could be accomplished.

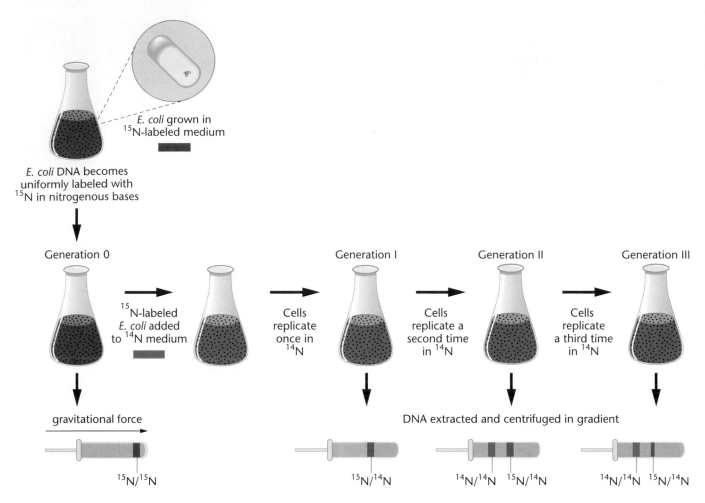

FIGURE 11–3 The Meselson–Stahl experiment.

The Meselson–Stahl Experiment

In 1958, Matthew Meselson and Franklin Stahl published the results of an experiment providing strong evidence that cells use semiconservative replication to produce new DNA molecules. *Escherichia coli* cells were grown for many generations in a medium where $^{15}NH_4Cl$ (ammonium chloride) was the only nitrogen source. A "heavy" isotope of nitrogen, ^{15}N contains one more neutron than the naturally occurring ^{14}N isotope. Unlike "radioactive" isotopes, ^{15}N is stable and thus does not decay (i.e., it is not radioactive). After many generations, all nitrogen-containing molecules, including the nitrogenous bases of DNA, contained the heavier isotope in the *E. coli* cells. DNA containing ^{15}N can be distinguished from ^{14}N-containing DNA by the use of sedimentation equilibrium centrifugation, in which centrifugation "forces" samples through a density gradient of a heavy metal salt such as cesium chloride. The more dense ^{15}N-DNA reaches equilibrium in the gradient at a point closer to the bottom (where the density is greater) than ^{14}N-DNA.

In this experiment (Figure 11–3), uniformly labeled ^{15}N cells were transferred to a medium containing only $^{14}NH_4Cl$. Thus, all new synthesis of DNA during replication contained only the "lighter" isotope of nitrogen. The time of transfer to the new medium was taken as time zero ($t = 0$). The *E. coli* cells were allowed to replicate over several generations, with cell samples removed after each replication cycle. DNA was isolated from each sample and subjected to sedimentation equilibrium centrifugation.

After one generation, the isolated DNA was present only in a single band of intermediate density—the expected result for semiconservative replication. Each replicated molecule was composed of one new ^{14}N-strand and one old ^{15}N-strand, as seen in Figure 11–4. This result effectively ruled out the conservative replication mode, in which two distinct bands are predicted.

After two cell divisions, DNA samples showed two density bands: One was intermediate and the other was lighter, corresponding to the ^{14}N position in the gradient. Similar results occurred after a third generation, except that the proportion of the ^{14}N-band increased. If replication were dispersive, all subsequent generations after $t = 0$ would demonstrate DNA of an intermediate density. In each subsequent generation the ratio $^{14}N/^{15}N$ would increase, and the hybrid band would become lighter and lighter, eventually approaching the ^{14}N-band. Since this result was not observed, the dispersive mode was ruled out. Thus, the results of the Meselson–Stahl experiment provided strong support for the semiconservative mode of DNA replication, as postulated by Watson and Crick.

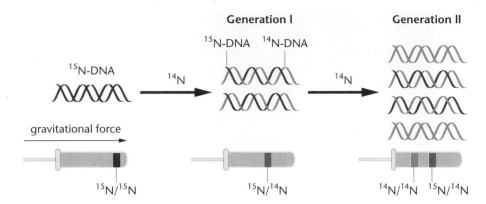

Generation I

^{15}N-DNA ^{14}N-DNA

Generation II

^{15}N-DNA

^{14}N

^{14}N

gravitational force

^{15}N/^{15}N ^{15}N/^{14}N ^{14}N/^{14}N ^{15}N/^{14}N

FIGURE 11–4 The expected results of two generations of semiconservative replication in the Meselson–Stahl experiment.

Semiconservative Replication in Eukaryotes

In 1957, the year before the work of Meselson and his colleagues was published, J. Herbert Taylor, Philip Woods, and Walter Hughes presented evidence that semiconservative replication also occurs in eukaryotic organisms. They experimented with root tips of the broad bean *Vicia faba*, which are an excellent source of dividing cells. These researchers examined the chromosomes of these cells following replication of DNA. They monitored the replication process by labeling DNA with ^{3}H-thymidine, a radioactive precursor of DNA, and then performing autoradiography.

Autoradiography is a cytological technique that pinpoints the location of an isotope in a cell. In this procedure, a photographic emulsion is placed over a section of cellular material (root tips in this experiment), and the preparation is stored in the dark. The slide is then developed, much as photographic film is processed. Because the radioisotope emits energy, the emulsion turns black at the approximate point of emission. The end result is the presence of dark spots or "grains" on the surface of the section, locating the newly synthesized DNA in the cell.

Root tips were grown for approximately one generation in the presence of the radioisotope and then placed in unlabeled medium, where cell division continued. At the conclusion of each generation, cultures were arrested at metaphase by the addition of colchicine (a chemical derived from the crocus plant, which poisons the mitotic spindle fibers), and chromosomes were examined by autoradiography. Figure 11–5 shows a single chromosome's replication over two division cycles as well as the distribution of grains. In this experiment, labeled thymidine was found only in association with chromatids that contained newly synthesized DNA.

The results are compatible with the semiconservative mode of replication. After the first replication cycle, radioactivity is detected over both sister chromatids. This finding is expected because each chromatid will contain one "new" radioactive DNA strand and one "old" unlabeled strand. After the second replication cycle, *which also takes place in unlabeled medium*, only one of the two new sister chromatids should be radioactive because half of the parent strands are unlabeled. With only the minor exceptions of **sister chromatid exchange** (see Chapter 8), this result was observed.

Together, the Meselson–Stahl and Taylor–Woods–Hughes experiments soon led to the general acceptance of the semiconservative mode of replication. The same conclusion has been reached in studies with other organisms. These experiments also strongly supported Watson and Crick's proposal for the double-helix model of DNA.

Origins, Forks, and Units of Replication

Semiconservative replication is the general mode by which DNA is duplicated. To enhance our understanding of this pattern, let's briefly consider a number of relevant issues. The first issue concerns the **origin of replication**. Where along the chromosome is DNA replication initiated? Is there only a single origin, or does DNA synthesis begin at more than one point? Is a point of origin random or is it located at a specific region along the chromosome? Second, once replication begins, does it proceed in a single direction or in both directions away from the origin? In other words, is replication **unidirectional** or **bidirectional**?

To address these issues, we introduce two terms. First, at the actual point along the chromosome where replication is occurring, the strands of the helix are unwound, creating a **replication fork**. Such a fork initially appears at the point of origin of synthesis and then moves along the DNA duplex as replication proceeds. If replication is bidirectional, two such forks will be present, migrating in opposite directions away from the origin. Second, the length of DNA that is replicated following one initiation event at a single origin is a unit called the **replicon**.

The evidence regarding the origin and direction of replication is clear. John Cairns tracked replication in *E. coli* using both radioisotopes and autoradiography. He demonstrated that replication is initiated in only one region. In *E. coli* this specific region, called *oriC*, has been mapped along the chromosome. It consists of 245 base pairs, though only a small number are essential to the initiation of DNA synthesis. Since in bacteriophages and bacteria DNA synthesis originates at a single point, the entire chromosome constitutes one replicon. The presence of only a single origin is characteristic of bacteria, which have only one circular chromosome.

Results put forward by other researchers, again relying on autoradiography, demonstrated that replication is

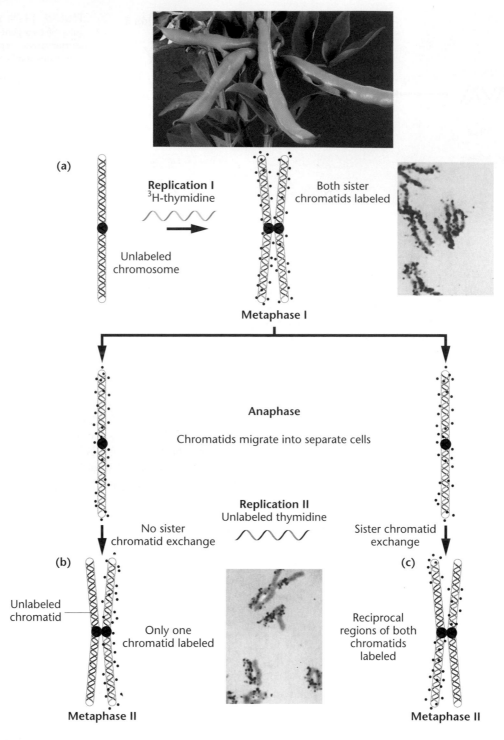

(a)

Replication I
³H-thymidine

Unlabeled chromosome

Both sister chromatids labeled

Metaphase I

Anaphase

Chromatids migrate into separate cells

Replication II
Unlabeled thymidine

No sister chromatid exchange

Sister chromatid exchange

(b)

Unlabeled chromatid

Only one chromatid labeled

Reciprocal regions of both chromatids labeled

(c)

Metaphase II

Metaphase II

FIGURE 11–5   The Taylor–Woods–Hughes experiment, demonstrating the semiconservative mode of replication of DNA in root tips of *Vicia faba*. A portion of the plant is shown in the top photograph. (a) An unlabeled chromosome proceeds through the cell cycle in the presence of ³H-thymidine. As it enters mitosis, both sister chromatids of the chromosome are labeled, as shown by autoradiography. After a second round of replication (b), this time in the absence of ³H-thymidine, only one chromatid of each chromosome is expected to be surrounded by grains. Except where a reciprocal exchange occurred between sister chromatids (c), the expectation was upheld. The micrographs are of the actual autoradiograms obtained in the experiment. *(Photos: Top from Walter H. Hodge/Peter Arnold, Inc.; right and bottom from "Molecular Genetics," Pt. 1, p. 74075, J. H. Taylor, ed. Reprinted by permission of Academic Press, Inc. [1963].)*

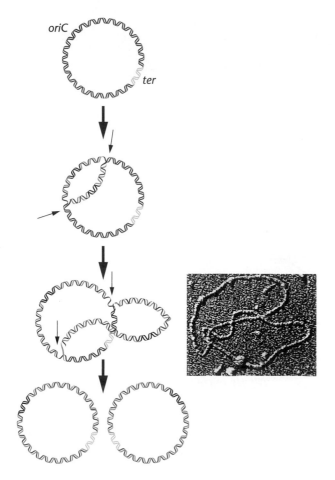

FIGURE 11–6 Bidirectional replication of the *E. coli* chromosome. The thin black arrows identify the advancing replication forks. The electron micrograph is of a bacterial chromosome in the process of replication, comparable to the figure next to it. *(Photo: Sundin and Varshavsky, Cell 25:659 [1981]. Courtesy of A. Vashavsky.)*

bidirectional, moving away from *oriC* in both directions (Figure 11–6). This creates two replication forks that migrate farther and farther apart as replication proceeds. These forks eventually merge as semiconservative replication of the entire chromosome is completed at a termination region, called *ter*.

11.2 Synthesis of DNA in Microorganisms

The determination that replication is semiconservative and bidirectional indicates only the *pattern* of DNA duplication and the association of the finished strands with each other once synthesis is complete. A more complex issue is how the actual *synthesis* of long complementary polynucleotide chains occurs on the template. As in most studies of molecular biology, this question was first approached by using microorganisms. Research began about the same time as the Meselson–Stahl work, and this topic remains an active area of investigation. What is most apparent in this research is the tremendous chemical complexity of the biological synthesis of DNA.

DNA Polymerase I

Studies of the enzymology of DNA replication were first reported by Arthur Kornberg and colleagues in 1957. They isolated an enzyme from *E. coli* that directed DNA synthesis in a cell-free (*in vitro*) system. The enzyme is now called **DNA polymerase I**, since it was the first of several to be isolated. Kornberg determined the following major requirements for *in vitro* DNA synthesis under the direction of the enzyme.

1. All four deoxyribonucleoside triphosphates (dATP, dCTP, dGTP, dTTP = dNTP)*

2. Template DNA

If any one of the four deoxyribonucleoside triphosphates was omitted from the reaction, no synthesis occurred. If derivatives of these precursor molecules other than the nucleoside triphosphate were used (nucleotides or nucleoside diphosphates), synthesis did not occur. If no template DNA was added, synthesis of DNA occurred but was reduced greatly. Template-dependent synthesis directed by Kornberg's enzyme appeared to be exactly the type required for semiconservative replication. The reaction is summarized in Figure 11–7. The enzyme has since been shown to consist of a single polypeptide containing 928 amino acids.

---

*dNTP designates the deoxyribose forms of the four nucleoside triphosphates; in a similar manner, dNMP refers to the monophosphate forms.

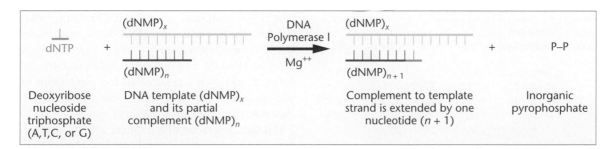

FIGURE 11–7 The chemical reaction catalyzed by DNA polymerase I. During each step, a single nucleotide is added to the growing complement of the DNA template using a nucleoside triphosphate as the substrate. The release of inorganic pyrophosphate drives the reaction energetically.

FIGURE 11–8 Demonstration of 5′-to-3′ synthesis of DNA.

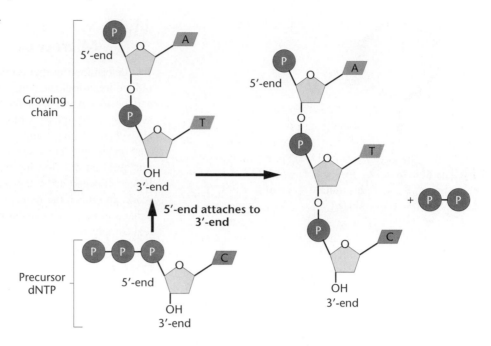

The way in which each nucleotide is added to the growing chain is a function of the specificity of DNA polymerase I. As shown in Figure 11–8, the precursor dNTP contains the three phosphate groups attached to the 5′-carbon of d-ribose. As the two terminal phosphates are cleaved during synthesis, the remaining phosphate attached to the 5′-carbon is covalently linked to the 3′-OH group of the d-ribose to which it is added. Thus, **chain elongation** occurs in the **5′-to-3′ direction** by the addition of one nucleotide at a time to the growing 3′-end. Each step provides a newly exposed 3′-OH group that can participate in the next addition of a nucleotide as DNA synthesis proceeds.

Having shown how DNA was synthesized, Kornberg sought to demonstrate the accuracy, or fidelity, with which the enzyme had replicated the DNA template. Because the nucleotide sequences of the template and the product could not be determined in 1957, he initially relied on several indirect methods.

One of Kornberg's approaches was to compare the nitrogenous base compositions of the DNA template with those of the recovered DNA product. Table 11.1 shows Kornberg's

base composition analysis of three DNA templates. These can be compared with the DNA product synthesized in each case. Within experimental error, the base composition of each product agreed with the template DNAs used. These data, along with other types of comparisons of template and product, suggested that the templates were replicated faithfully.

Synthesis of Biologically Active DNA

Despite Kornberg's extensive work, not all researchers were convinced that DNA polymerase I is the enzyme that replicates DNA within cells (*in vivo*). Their reservations involved observations that the *in vitro* rate of synthesis was much slower than the *in vivo* rate, that the enzyme was much more effective replicating single-stranded DNA than double-stranded DNA, and that the enzyme appeared to *degrade* DNA as well as to *synthesize* it; that is, it exhibits exonuclease activity.

Uncertain of the true cellular function of DNA polymerase I, Kornberg pursued another approach. He reasoned that if the enzyme could be used to synthesize **biologically active DNA** *in vitro*, then DNA polymerase I must be the major catalyzing force for DNA synthesis within the cell. The term *biological activity* means that the DNA synthesized supports metabolic activities and directs reproduction of the organism from which it was originally duplicated.

In 1967, Mehran Goulian, Kornberg, and Robert Sinsheimer showed that the DNA of the small bacteriophage φX174 could be completely copied by DNA polymerase I *in vitro*, and that the new product could be isolated and used to infect *E. coli*. This resulted in the production of mature phages under the direction of the synthetic DNA, thus demonstrating biological activity!

This demonstration of biological activity was viewed as a precise assessment of faithful copying. If even a single error had occurred to alter the base sequence of any of the 5386 nucleotides constituting the φX174 chromosome, the change

TABLE 11.1   **Base Composition of the DNA Template and the Product of Replication in Kornberg's Early Work**

Organism	Template or Product	%A	%T	%G	%C
T2	Template	32.7	33.0	16.8	17.5
	Product	33.2	32.1	17.2	17.5
E. coli	Template	25.0	24.3	24.5	26.2
	Product	26.1	25.1	24.3	24.5
Calf	Template	28.9	26.7	22.8	21.6
	Product	28.7	27.7	21.8	21.8

Source: Kornberg (1960).

might easily have caused a mutation that would prohibit the production of viable phages.

DNA Polymerases II and III

Although DNA synthesized under the direction of polymerase I demonstrated biological activity, a more serious reservation about the enzyme's true biological role was raised in 1969. Peter DeLucia and John Cairns reported the discovery of a mutant strain of *E. coli* that was deficient in polymerase I activity. The mutation was designated *polA1*. In the absence of the functional enzyme, this mutant strain of *E. coli* still duplicated its DNA and reproduced successfully! Other properties of the mutation led DeLucia and Cairns to conclude that in the absence of polymerase I, these cells are highly deficient in their ability to "repair" DNA. For example, the mutant strain is highly sensitive to ultraviolet light and radiation, both of which damage DNA and are therefore mutagenic. Nonmutant bacteria are able to repair a great deal of UV-induced damage.

These observations led to two conclusions:

1. At least one other enzyme that is responsible for replicating DNA *in vivo* is present in *E. coli* cells.

2. DNA polymerase I may serve a secondary function *in vivo*. This function is now believed by Kornberg and others to be critical to the *fidelity* of DNA synthesis, but this enzyme does not actually synthesize the entire complementary strand during replication.

To date, two other unique DNA polymerases have been isolated from cells lacking polymerase I activity and from normal cells that do contain polymerase I. Table 11.2 shows that the two enzymes, called **DNA polymerase II** and **III**, share several characteristics with DNA polymerase I. While none of the three *initiate* DNA synthesis on a template, all three can *elongate* an existing DNA strand, called a **primer**. As we shall see, RNA is also an adequate primer and is, in fact, used initially.

The DNA polymerase enzymes are all large complex proteins exhibiting a molecular weight in excess of 100,000 daltons. All three possess 3′–5′ **exonuclease activity**, which means that they have the potential to polymerize in one direction and then pause and excise nucleotides just added. As we discuss later in this chapter, this activity allows the enzymes to proofread newly synthesized DNA and to remove incorrect nucleotides, which can then be replaced.

DNA polymerase I also demonstrates 5′–3′ exonuclease activity. Thus, the enzyme can excise nucleotides starting at the

TABLE 11.2   Properties of Bacterial DNA Polymerases I, II, and III

Properties	I	II	III
Initiation of chain synthesis	−	−	−
5′–3′ polymerization	+	+	+
3′–5′ exonuclease activity	+	+	+
5′–3′ exonuclease activity	+	−	−
Molecules of polymerase/cell	400	?	15

TABLE 11.3   Subunits of the DNA Polymerase III Holoenzyme

Subunit	Function	Groupings
α ε θ	5′–3′ polymerization 3′–5′ exonuclease ??	"Core" enzyme: Elongates polynucleotide chain and proofreads
γ δ δ′ χ ψ	Loads enzyme on template (Serves as clamp loader)	γ complex
β	Sliding clamp structure (processivity factor)	
τ	Dimerizes core complex	

end where synthesis begins, and proceeding in the direction of synthesis, can remove the RNA primer. Two final observations probably explain why Kornberg isolated polymerase I and not polymerase III. Polymerase I is present in greater amounts than is polymerase III, and it is also much more stable.

What then are the roles of the three polymerases *in vivo*? Polymerase I is thought to be responsible for removing the primer and for the gap-filling synthesis. These gaps occur naturally as primers are removed. Its exonuclease activity also allows for proofreading during this process. Polymerase II appears to be involved in repairing DNA damaged by external forces, such as ultraviolet light. It is encoded by a gene that may be activated by disruption of DNA synthesis at the replication fork. Polymerase III is the enzyme responsible for the polymerization essential to replication. Its 3′–5′ exonuclease activity provides its proofreading function, as described.

We conclude this section by emphasizing the complexity of the DNA polymerase III molecule. Its active form, called a **holoenzyme**, consists of two sets (a dimer) of 10 separate polypeptide subunits (Table 11.3) and has a molecular weight in excess of 600,000 Da. The largest subunit, α, has a molecular weight of 140,000 Da. and, along with subunits ε and θ, constitutes the "core" enzyme responsible for the polymerization activity of the holoenzyme. The α subunit is responsible for nucleotide polymerization on the template strands, whereas the ε subunit of the core enzyme possesses the 3′–5′ exonuclease activity.

A second group of five subunits (γ, δ, δ′, χ, and ψ) forms what is called the γ complex, which "loads" the enzyme onto the template at the replication fork. This enzymatic function requires energy and is dependent on the hydrolysis of ATP. The β subunit prevents the core enzyme from falling off the template during polymerization. Finally, the τ subunit holds the two core polymerases together at the replication fork. The holoenzyme and several other proteins at the replication fork form a complex nearly as large as a ribosome known as a **replisome**. We consider the function of DNA polymerase III in more detail later in this chapter.

11.3   DNA Synthesis: A Model

We have thus far established that replication is semiconservative and bidirectional along a single replicon in bacteria and many viruses. Also, we know that synthesis is in the 5′-to-3′ mode under the direction of DNA polymerase III, creating two replication forks. These move in opposite directions away from the origin of synthesis. As you can see in the points noted below, many issues must still be resolved in order to gain a comprehensive understanding of DNA replication.

1. A mechanism must exist by which the helix is initially unwound (or denatured) and stabilized in this "open" configuration so that synthesis can proceed along both strands.

2. As unwinding and subsequent DNA synthesis proceeds, increased coiling creates tension farther down the helix, which must be reduced.

3. A primer of some sort must be synthesized so that polymerization can commence under the direction of DNA polymerase III. Surprizingly, RNA, not DNA, serves as this primer.

4. Once the RNA primers have been synthesized, DNA polymerase III commences synthesis of the complement of both strands of the parent molecule. Because the strands are antiparallel, continuous synthesis in the direction in which the replication fork moves is possible along only one of the two strands. On the other strand, synthesis is discontinuous in the opposite direction.

5. The RNA primers must be removed prior to completion of replication. The gaps that are temporarily created must be filled with DNA that is complementary to the template at each location.

6. The newly synthesized DNA strand that fills each temporary gap must be ligated to the adjacent strand of DNA.

As we consider these points, examine Figures 11–9, 11–10, and 11–11 to see how each issue is resolved. Figure 11–12 summarizes the model of DNA synthesis.

Unwinding the DNA Helix

As discussed earlier, DNA synthesis is initiated at a single origin along the circular chromosome of most bacteria and viruses. This region of the *E. coli* chromosome has been particularly well studied. Called *oriC*, it consists of 245 base pairs characterized by repeating sequences of 9 and 13 bases (called **9mers** and **13mers**). One particular protein (called **DnaA** because it is encoded by the gene *dnaA*) initiates unwinding of the helix. A number of subunits of the DnaA protein bind to each of several 9mers. This step is essential in facilitating the subsequent binding of **DnaB** and **DnaC** proteins that further open and destabilize the helix (Figure 11–9). Proteins such as these, which require the energy normally supplied by ATP hydrolysis to break hydrogen bonds and denature the double helix, are called **helicases**. Other proteins, called **single-stranded binding proteins (SSBPs)** stabilize this open conformation.

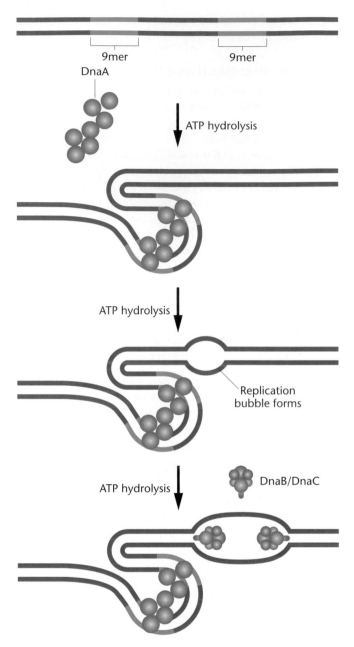

FIGURE 11–9 Helical unwinding of DNA during replication as accomplished by DnaA, DnaB, and DnaC proteins. Initial binding of many monomers of DnaA occurs at DNA sites that contain repeating sequences of nine nucleotides, called 9mers. Not illustrated are 13mers, which are also involved.

As unwinding proceeds, a coiling tension is created ahead of the replication fork, often producing **supercoiling**. In circular molecules, supercoiling takes the form of added twists and turns of the DNA, much like the coiling created in a rubberband by stretching it out and then twisting one end. Such supercoiling can be relaxed by **DNA gyrase**, a member of a larger group of enzymes referred to as **DNA topoisomerases**. The gyrase makes either single- or double-stranded "cuts" and also catalyzes localized movements that "undo" the twists and knots created during supercoiling. The strands are then resealed. These various reactions are driven by the energy released during ATP hydrolysis.

Together, the DNA, the polymerase complex, and associated enzymes make up an array of molecules that participate

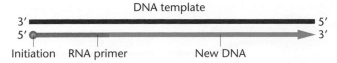

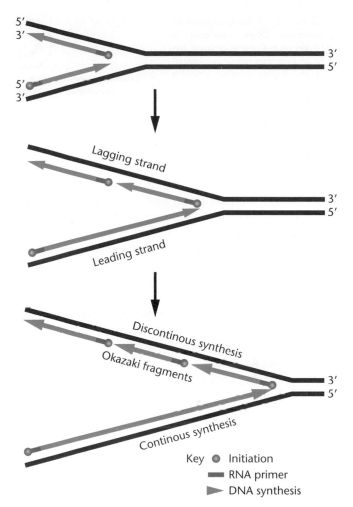

FIGURE 11–10 The initiation of DNA synthesis. A complementary RNA primer is first synthesized, to which DNA is added. All synthesis is in the 5′-to-3′ direction. Eventually, the RNA primer is replaced with DNA under the direction of DNA polymerase I.

Initiation of DNA Synthesis

Once a small portion of the helix is unwound, synthesis is initiated. As we have seen, DNA polymerase III requires a primer with a free 3′-end in order to elongate a polynucleotide chain. This prompted researchers to investigate how the first nucleotide could be added. Although no free 3′-hydroxyl group is initially present, it is now clear that, as pointed out above, RNA is the primer that initiates DNA synthesis.

A short segment of RNA, (about 5–15 nucleotides long), complementary to DNA, is first synthesized on the DNA template. Synthesis of the RNA is directed by a form of RNA polymerase called **primase**, which does not require a free 3′-end to initiate synthesis. It is to this short segment of RNA that DNA polymerase III begins to add 5′-deoxyribonucleotides, initiating DNA synthesis. Figure 11–10 shows how synthesis is initiated on a DNA template. At a later point, the RNA primer must be clipped out and replaced with DNA. This occurs under the direction of DNA polymerase I. Recognized in viruses, bacteria, and several eukaryotic organisms, RNA priming is a universal phenomenon during the initiation of DNA synthesis.

Continuous and Discontinuous DNA Synthesis

We must now reconsider the fact that the two strands of a double helix are **antiparallel** to each other; that is, one runs in the 5′–3′ direction, while the other runs in the opposite 3′–5′ direction. Because DNA polymerase III synthesizes DNA in only the 5′–3′ direction, synthesis along an advancing replication fork simultaneously occurs in one direction on one strand and in the opposite direction on the other. As a result, as the strands unwind and the replication fork progresses down the helix (Figure 11–11), only one strand can serve as a template for **continuous DNA synthesis**. This strand is called the **leading DNA strand**. As the fork progresses, many points of initiation are necessary on the opposite, or **lagging DNA strand**, resulting in **discontinuous DNA synthesis**.

Evidence supporting discontinuous DNA synthesis was first provided by Reiji Okazaki, Tuneko Okazaki, and their colleagues. They discovered that when bacteriophage DNA is replicated in *E. coli*, some of the newly formed DNA that is hydrogen-bonded to the template strand is present as small fragments containing 1000–2000 nucleotides. RNA primers are part of each such fragment. These pieces, called **Okazaki fragments**, convert into longer and longer DNA strands of higher molecular weight as synthesis proceeds.

FIGURE 11–11 Opposite polarity of DNA synthesis along the two strands, necessary because the two strands of DNA run antiparallel to one another and DNA polymerase III synthesizes only in one direction (5′ to 3′). On the lagging strand, synthesis must be discontinuous, resulting in the production of Okazaki fragments. On the leading strand, synthesis is continuous. RNA primers initiate synthesis on both strands.

Discontinuous synthesis of DNA requires enzymes that remove the RNA primer and that unite the Okazaki fragments into the continuous lagging strand. As we have noted, DNA polymerase I removes the primer and replaces the missing nucleotides. Joining the fragments appears to be the work of **DNA ligase**, which catalyzes the formation of the phosphodiester bond that closes the gap between the discontinuously synthesized strands. The evidence that DNA ligase performs this function during DNA synthesis is strengthened by observations of a ligase-deficient mutant strain (*lig*) of *E. coli* in which a large number of unjoined Okazaki fragments accumulate.

Concurrent Synthesis on the Leading and Lagging Strands

Given the model just discussed, you might well ask how DNA polymerase III synthesizes DNA on both the leading and lagging strands. Can both strands be replicated simultaneously at the same replication fork, or are the events distinct, involving two separate copies of the enzyme? Evidence suggests that both strands replicate simultaneously. As

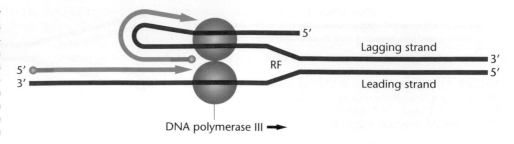

Figure 11–12 shows, if the lagging strand forms a loop, nucleotide polymerization occurs on both template strands under the direction of a dimer of the enzyme. After the synthesis of 100–200 base pairs, the monomer of the enzyme on the lagging strand encounters a completed Okazaki fragment, at which point it releases the lagging strand. A new loop is then formed with the lagging template strand, and the process repeats. Looping inverts the orientation of the template, but not the direction of actual synthesis on the lagging strand, which is always in the 5′-to-3′ direction.

Another important feature of the holoenzyme that facilitates synthesis at the replication fork is a dimer of the β subunit that forms a clamplike structure around the newly formed DNA duplex. This β-subunit clamp prevents the **core enzyme** (the α, ε, and θ subunits that are responsible for catalysis of nucleotide addition) from falling off the template as polymerization proceeds. Because the entire holoenzyme moves along the parent duplex, advancing the replication fork, the β-subunit dimer is often referred to as a sliding clamp.

Proofreading and Error Correction During DNA Replication

The underpinning of DNA replication is the synthesis of a new strand that is precisely complementary to the template strand at each nucleotide position. Although the action of DNA polymerases is very accurate, synthesis is not perfect and a noncomplementary nucleotide is occasionally inserted erroneously. To compensate for such inaccuracies, polymerases I and III both possess **3′–5′ exonuclease activity**, which enables them to detect and excise a mismatched nucleotide (in the 3′–5′ direction). Once the mismatched nucleotide is removed, 5′–3′ synthesis again proceeds. This process, **exonuclease proofreading**, increases the fidelity of synthesis. In the case of the holoenzyme form of DNA polymerase III, the epsilon (ε) subunit is directly involved in the proofreading step. In strains of *E. coli* where a mutation has rendered the ε subunit nonfunctional, the error rate (the mutation rate) during DNA synthesis is increased substantially.

11.4   A Coherent Model of DNA Synthesis

We can now combine the various aspects of DNA replication occurring at a single replication fork into a coherent model, as shown in Figure 11–13. At the advancing fork, a helicase is unwinding the double helix. Once unwound, single-stranded binding proteins associate with the strands, preventing the re-formation of the helix. In advance of the replication fork, DNA gyrase diminishes the tension created as the helix supercoils. Each half of the dimeric polymerase is a core enzyme bound to one template strand by a β-subunit sliding clamp. Continuous synthesis occurs on the leading strand, while the lagging strand must "loop" around for simultaneous synthesis to occur on both strands. Not shown in the figure, but essential to replication on the lagging strand, is the action of DNA polymerase I and DNA ligase, which replace the RNA primer with DNA and join the Okazaki fragments.

FIGURE 11–13 Summary of DNA synthesis at a single replication fork. Various enzymes and proteins essential to the process are shown.

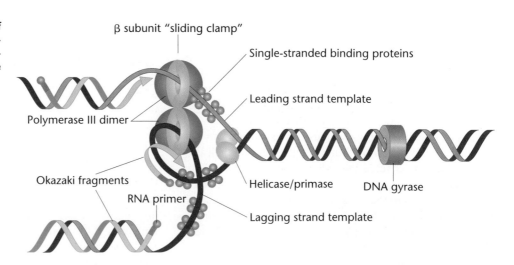

Because the investigation of DNA synthesis is an extremely active area of research, this model will no doubt be extended in the future. In the meantime, it gives us a summary of DNA synthesis against which we can interpret genetic phenomena.

11.5 Genetic Control of Replication

Much of what we know about DNA replication in viruses and bacteria is based on genetic analysis of the process. For example, we have already discussed studies involving the *polA1* mutation, which revealed that DNA polymerase I is not the major enzyme responsible for replication. Many other mutations interrupt or seriously impair some aspect of replication, such as the ligase-deficient and the proofreading-deficient mutations mentioned previously. Genetic analysis frequently uses **conditional mutations**, which are expressed under one condition but not under a different condition. For example, a **temperature-sensitive mutation** may not be expressed at a particular *permissive* temperature. When mutant cells are grown at a *non-permissive* (or *restrictive*) temperature, the mutation is expressed. The investigation of such temperature-sensitive mutants provides insights into the product and the associated function of the normal, nonmutant gene.

As shown in Table 11.4, a variety of genes in *E. coli* specify the subunits of polymerases I, II, and III and encode products involved in specification of the origin of synthesis, helix-unwinding and stabilization, initiation and priming, relaxation of supercoiling, repair, and ligation. The discovery of such a large group of genes attests to the complexity of the replication process, even in the relatively simple prokaryote. Given the enormous quantity of DNA that must be unerringly replicated in a very brief time, this level of complexity is not unexpected. As we see next, the process is even more involved and therefore more difficult to investigate in eukaryotes.

11.6 Eukaryotic DNA Synthesis

Research shows that eukaryotic DNA is replicated in a manner similar to that of bacteria. In both systems, double-stranded DNA unwinds at a replication origin, two replication forks form, and bidirectional synthesis of DNA occurs on the leading and lagging strand templates under the direction of DNA polymerase. Eukaryotic polymerases have the same fundamental requirements for DNA synthesis as do bacterial systems: four deoxyribonucleoside triphosphates, a template, and a primer. However, because eukaryotic cells contain much more DNA per cell and because this DNA is complexed with proteins, eukaryotes face many problems not encountered by bacteria. As we might expect, these complications make the process of DNA synthesis much more complex in eukaryotes and more difficult to study. However, a great deal is now known about the process.

Multiple Replication Origins

The most obvious difference between eukaryotic and prokaryotic DNA replication is that eukaryotic chromosomes contain multiple replication origins, in contrast to the single site that is part of the *E. coli* chromosome. The multiple origins are visible under the electron microscope (Figure 11–14). They are essential if the entire genome of a typical eukaryote is to be replicated in a reasonable time. Recall that (1) eukaryotes have much greater amounts of DNA than bacteria (e.g., yeast has 4 times as much and *Drosophila* has 100 times as much DNA as *E. coli*); and (2) the rate of synthesis by eukaryotic DNA polymerase is much slower—only about 50 nucleotides per second, a rate 20 times less than the comparable bacterial enzyme. Under these conditions, single-origin replication of a typical eukaryotic genome would take up to one month to complete! However, replication is accomplished in as little as 3 minutes in some eukaryotic organisms.

Many insights concerning the molecular nature of the multiple origins and the initiation of DNA synthesis at these sites are now available. Most information was originally derived from the study of yeast (e.g., *Saccharomyces cerevisiae*), which have between 250 and 400 replicons; subsequent studies used mammalian cells, which have as many as 25,000 replicons. The origins in yeast have been isolated and are

TABLE 11.4 Some of the Various *E. coli* Mutant Genes and Their Products or Role in Replication

Mutant Gene	Enzyme or Role
polA	DNA polymerase I
polB	DNA polymerase II
dnaE, N, Q, X, Z	DNA polymerase III subunits
dnaG	Primase
dnaA, I, P	Initiation
dnaB, C	Helicase at *oriC*
oriC	Origin of replication
gyrA, B	Gyrase subunits
lig	Ligase
rep	Helicase
ssb	Single-stranded binding proteins
rpoB	RNA polymerase subunit

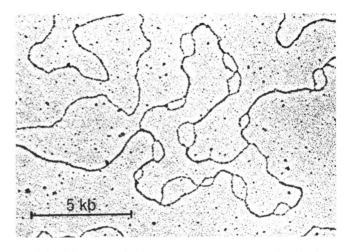

5 kb

FIGURE 11–14 A demonstration of the multiple origins of replication along a eukaryotic chromosome. Each origin is apparent as a replication bubble along the axis of the chromosome. (*H. J. Kreigstein and D.S. Hogness, "Proceedings National Academy of Sciences," 71, 136, 1974. [Fig. 2, p. 137].*)

called **autonomously replicating sequences (ARSs)**. They consist of a unit of 11 base pairs, flanked by other short sequences involved in efficient initiation. As we know from Chapter 2, DNA synthesis is restricted to the S phase of the eukaryotic cell cycle. Research has shown that the many origins are not all activated at once; instead clusters of 20–80 adjacent replicons are activated sequentially throughout the S phase until all DNA is replicated.

How the polymerase finds the ARS sequences among so much DNA is an obvious recognition problem. The solution involves a mechanism initiated prior to the S phase, where all ARS sequences are bound by a group of proteins (six in yeast); this mechanism forms a recognition complex that binds to DNA polymerase. Mutations in either the ARS sequence or in any of the genes encoding yeast's recognition-complex proteins prevent DNA synthesis.

Eukaryotic DNA Polymerases

The most complex aspect of eukaryotic replication is the array of polymerases involved in directing DNA synthesis. A total of six different forms of the enzyme have been isolated and studied. For the polymerases to access DNA, the topology of the helix must first be modified. As synthesis is triggered at each origin site, the double strands are opened up in an A $=$ T-rich region, which allows a helicase enzyme to enter that further unwinds the double-stranded DNA. Before polymerases begin synthesis, histone proteins complexed to the DNA (which form the characteristic nucleosomes of chromatin—see Chapter 17) also must be stripped away or otherwise modified. As DNA synthesis then proceeds, histones reassociate with the newly formed duplexes, reestablishing the characteristic nucleosome pattern (Figure 11–15). In eukaryotes, the synthesis of new histone proteins is tightly coupled to DNA synthesis during the S phase of the cell cycle.

Of the six known polymerases, three (Pol α, δ, and ε) are now considered essential to nuclear DNA replication in eukaryotic cells. Two others (Pol β and ξ) appear to be involved in DNA repair. The sixth form (Pol γ) is involved in the synthesis of mitochondrial DNA. Presumably, its replication function is limited to that organelle even though it is encoded by a nuclear gene. All but one of the enzyme's six forms

(the β form) consist of multiple subunits. Different subunits perform different functions during replication.

Pol α and δ may be the major forms of the enzyme involved in initiating nuclear DNA synthesis, so we concentrate our discussion on these. Initiation of synthesis involves the α polymerase. Two of the four subunits of the enzyme function as a primase in synthesizing RNA primers on both the leading and lagging template strands. Then, another subunit elongates the RNA primer by adding complementary deoxyribonucleotides, constituting the initial phase of DNA synthesis. Pol α is said to possess low **processivity**, a term that essentially reflects the length of DNA synthesized by an enzyme before it dissociates from the template. Thus, after a short DNA sequence is added to the RNA primer, an event known as **polymerase switching** occurs, whereby Pol α dissociates from the template and is replaced by Pol δ. This form of the enzyme possesses high processivity as well as 3′–5′ exonuclease activity, which gives it the potential to proofread. It is also capable of a 100-fold increase in the rate of synthesis in comparison to Pol α. Thus, under the direction of Pol δ, elongation and proofreading of the growing DNA strand occurs, as synthesis continues.

To accommodate the increased number of replicons, eukaryotic cells contain many more DNA polymerase molecules than do bacteria. While *E. coli* has about 15 copies of DNA polymerase III per cell, there may be up to 50,000 copies of the α form in animal cells. As pointed out earlier, the presence of greater numbers of smaller replicons in eukaryotes compensates for their slower rate of DNA synthesis as compared to bacteria. *E. coli* requires 20–40 minutes to replicate its chromosome, while *Drosophila*, with 40 times more DNA, accomplishes the same task in only 3 minutes during embryonic cell divisions.

11.7 DNA Replication, Telomeres, and Telomerase

A final difference that exists between prokaryotic and eukaryotic DNA synthesis involves the nature of the chromosomes. Unlike the closed, circular DNA of bacteria and most bacteriophages, eukaryotic chromosomes are linear. During replication they face a special problem at the "ends" of these linear molecules, called the **telomeres**. While synthesis proceeds normally to the end of the leading strand, a difficulty arises on the lagging strand as the RNA primer is removed (Figure 11–16). Normally, the newly created gap would be filled by adding a nucleotide to the existing 3′-OH group provided during discontinuous synthesis (to the right of gap b in Figure 11–16). However, there is no strand present to provide the 3′-OH group because this is the end of the chromosome. As a result, each successive round of synthesis theoretically shortens the chromosome by the length of the RNA primer. Because this is such a significant problem, we can suppose, at least for some cells, that a molecular solution would have developed early in evolution and be shared by all eukaryotes, and this is indeed the case.

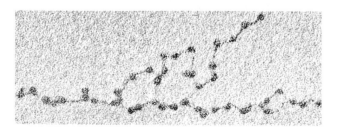

FIGURE 11–15 An electron micrograph of a eukaryotic replicating fork that demonstrates the presence of histone protein-containing nucleosomes on both branches. (*Dr. Harold Weintraub, Howard Hughes Medical Institute, Fred Hutchinson Cancer Center/Essential Molecular Biology" 2e, Freifelder & Malachinski, Jones & Bartlett, Fig. 7-24, p. 141.*)

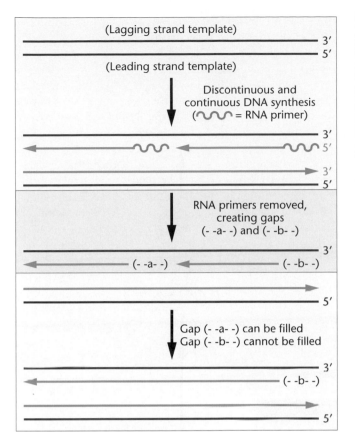

FIGURE 11–16 The difficulty encountered during the replication of the ends of linear chromosomes. A gap (- -b- -) is left following synthesis on the lagging strand.

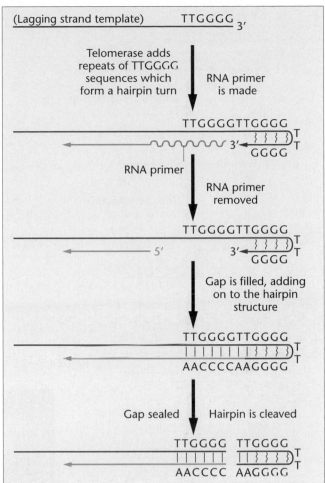

FIGURE 11–17 The predicted solution to the problem posed in Figure 11–16. The enzyme telomerase directs synthesis of the TTGGGG sequences, which results in the formation of a hairpin structure. The gap can now be filled, and following cleavage of the hairpin structure, this process averts the creation of a gap during replication of the ends of linear chromosomes.

Discovery of a unique eukaryotic enzyme, **telomerase**, has helped us understand how more complex organisms solve this problem. In the ciliated protozoan *Tetrahymena*, the many telomeres all terminate in the sequence 5′-TTGGGG-3′. Telomerase adds repeats of TTGGGG to the ends of molecules already containing this sequence, preventing the telomeric ends from shortening after each replication. As shown in Figure 11–17, telomerase adds several copies of the six-nucleotide repeat to the 3′-end of the lagging strand (using 5′–3′ synthesis). These repeats form a "hairpin loop," which is stabilized by unorthodox hydrogen bonding between opposite guanine residues (GG). This creates a free 3′-OH end that, following removal of the RNA primer, serves as a substrate for DNA polymerase I to fill the gap. The hairpin loop is then cleaved off, and the potential loss of DNA in each subsequent replication cycle is averted. This process also occurs in other eukaryotes under the direction of a similar enzyme.

Further investigation of the *Tetrahymena* telomerase enzyme by Elizabeth Blackburn and Carol Greider yielded an extraordinary finding. This enzyme adds the same TTGGGG sequence to DNA termini even if they lack this sequence. Thus, the TTGGGG sequence of the DNA substrate is not the signal for telomerase function. Blackburn and Greider have now established how the enzyme works. The enzyme is unique in that it contains within its molecular structure a short piece of RNA essential to its catalytic activity, making it a ribonucleoprotein. The RNA component encodes the se-

quences used by the enzyme as a template. The RNA contains 159 bases, including the sequence 5′-AACCCC-3′, which is complementary to the sequence whose synthesis it directs. Analogous enzyme functions have now been found in other single-celled organisms. The RNA-containing telomerase enzyme behaves in a manner analogous to the enzyme, reverse transcriptase, in that it synthesizes a DNA complement on an RNA template.

As we shall see in Chapter 17, telomeric DNA sequences have been highly conserved throughout evolution, reflecting the critical function of telomeres. In the essay at the end of this chapter, we shall see that telomere shortening has been linked to a molecular mechanism involved in the aging process of cells. In most eukaryotic somatic cells, telomerase is, in fact, not active, and thus with each cell division, the telomeres of each chromosome shorten. After many divisions, the telomere is seriously eroded and the cell loses the capacity for further division. Malignant cells, on the other hand, maintain telomerase activity and are immortalized.

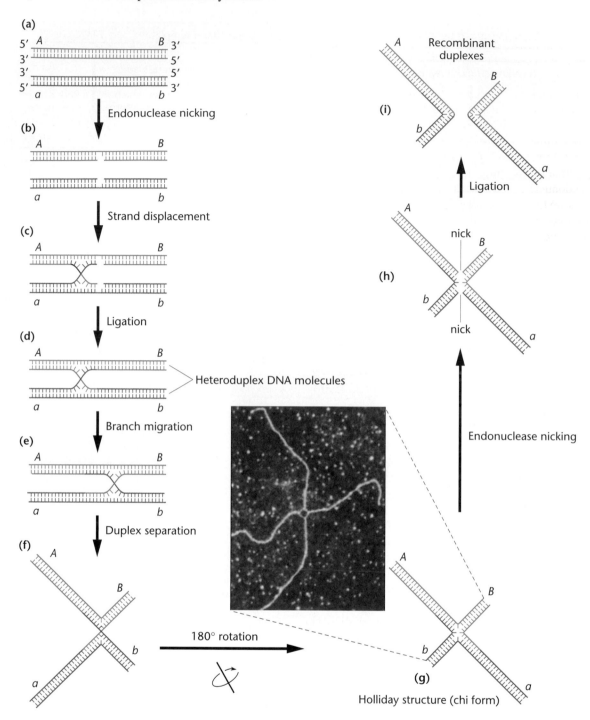

FIGURE 11–18 Model depicting how genetic recombination can occur as a result of the breakage and rejoining of heterologous DNA strands. Each stage is described in the text. The electron micrograph shows DNA in a chi-form structure similar to the diagram in (g); the DNA is an extended Holliday structure, derived from the *ColEI* plasmid of *E. coli. (Photo: David Dressler, Oxford University, England)*

11.8　DNA Recombination

We conclude this chapter by returning to a topic discussed in Chapter 8—**genetic recombination**. There, we pointed out that the process of crossing over depends on breakage and rejoining of the DNA strands between homologs. Now that we have discussed the chemistry and replication of DNA, it is appropriate to consider how recombi-

nation occurs at the molecular level. In general, the following information pertains to genetic exchange between any two homologous double-stranded DNA molecules, whether they be viral or bacterial chromosomes or eukaryotic homologs during meiosis. Genetic exchange at equivalent positions along two chromosomes with substantial DNA sequence homology is referred to as **general** or **homologous recombination**.

Several models attempt to explain crossing over, and they all share certain common features. First, all are based on the initial proposals put forth independently by Robin Holliday and Harold L. K. Whitehouse in 1964. They also depend on the complementarity between DNA strands for their precision of exchange. Finally, each model relies on a series of enzymatic processes to accomplish genetic recombination.

One such model is shown in Figure 11–18. It begins with two paired DNA duplexes or homologs (a), each of which has a single-stranded nick introduced (b) at an identical position by an endonuclease. The ends of the strands produced by these cuts are then displaced and subsequently pair with their complements on the opposite duplex (c). A ligase then seals the loose ends (d), creating hybrid duplexes called heteroduplex DNA molecules. The exchange creates a cross-bridged or Holliday structure. The position of the cross bridge then moves down the chromosome as a result of a process called branch migration (e). This occurs as a result of a zipperlike action as hydrogen bonds break and then re-form between complementary bases of the displaced strands of each duplex. This migration yields an increased length of heteroduplex DNA on both homologs.

If the duplexes separate (f) and the bottom portions rotate 180° (g), an intermediate planar structure called a **chi form** is created. If the two strands on opposite homologs previously uninvolved in the exchange are now nicked by an endonuclease (h) and ligation occurs (i), recombinant duplexes are created. Note that the arrangement of alleles is altered as a result of recombination.

Evidence supporting this model includes the electron microscopic visualization of chi-form planar molecules from bacteria where four duplex arms join at a single point of exchange [Figure 11–18(g)]. Further important evidence comes from the discovery in *E. coli* of the **RecA protein**. This molecule promotes the exchange of reciprocal single-stranded DNA molecules as occurs in step (c) of the model. RecA also enhances the hydrogen-bond formation during strand displacement, thus initiating heteroduplex formation. Finally, many other enzymes essential to the nicking and ligation process have also been discovered and investigated. The products of the *recB*, *recC*, and *recD* genes are thought to be involved in the nicking and unwinding of DNA. Numerous mutations that prevent genetic recombination have been found in viruses and bacteria. These mutations are thought to identify genes, whose products play an essential role in this process.

Gene Conversion

A modification of the preceding model has helped us better understand a unique genetic phenomenon known as **gene conversion**. Initially found in yeast by Carl Lindegren and in *Neurospora* by Mary Mitchell, gene conversion is characterized by a genetic exchange ratio involving two closely linked genes that is *nonreciprocal*. If we cross two *Neurospora* strains each bearing a separate mutation ($a+ \times b+$), a *reciprocal* recombination event between the genes yields spore pairs of the $+ +$ and *ab* genotypes. However, a nonreciprocal exchange yields one pair without the other. Working with pyridoxine mutants, Mitchell observed several asci containing spore patterns displaying the $+ +$ genotype, but not the reciprocal product (*ab*). Because the frequency of these events was higher than the predicted mutation rate and thus could not be accounted for by that phenomenon, they were called "gene conversions." They were so named because it appeared that one allele had somehow been "converted" to another during an event in which genetic exchange also occurred. Similar findings are apparent in the study of other fungi as well.

Gene conversion is now considered to be a consequence of the recombination process. One possible explanation interprets conversion as a mismatch of base pairs during heteroduplex formation, as shown in Figure 11–19. Mismatched regions of hybrid strands can be repaired by excising one

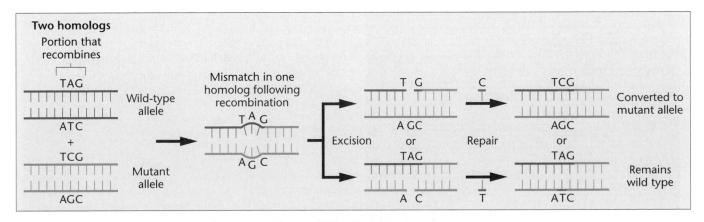

FIGURE 11–19 A proposed mechanism that accounts for the phenomenon of gene conversion. A base-pair mismatch occurs in one of the two homologs (bearing the mutant allele) during heteroduplex formation, which accompanies recombination in meiosis. During excision repair, one of the two mismatches is removed and the complement is synthesized. In one case, the mutant base pair is preserved. When it is subsequently included in a recombinant spore, the mutant genotype is maintained. In the other case, the mutant base pair is converted to the wild-type sequence. When included in a recombinant spore, the wild-type genotype is expressed, leading to a nonreciprocal exchange ratio.

Genetics, Technology, and Society

Telomerase: The Key to Immortality?

Humans, as do all multicellular organisms, grow old and die. As we age, our immune systems become less efficient, wound healing is impaired, and tissues and organs lose resilience. It has always been a mystery why we go through these age-related declines, and why each species has a characteristic finite lifespan. Why do we grow old? Can we reverse this march to mortality? Some recent discoveries suggest that the answers to these questions may lie at the ends of our chromosomes.

The study of human aging begins with a study of human cells growing in culture dishes. Like the organisms from which the cells are taken, cells in culture have a finite life span. This "replicative senescence" was noted over 30 years ago by Hayflick. He reported that normal human fibroblasts lose their ability to grow and divide after about 50 cell divisions. These senescent cells remain metabolically active, but can no longer proliferate. Eventually, they die. Although we don't know whether cellular senescence directly causes organismal aging, the evidence is suggestive. For example, cells from young people go through more divisions in culture than cells from older people; human fetal cells divide 60–80 times before undergoing senescence, whereas cells from older adults divide only 10–20 times. In addition, cells from species with short life spans stop growing after fewer divisions than cells from species with longer life spans; mouse cells divide 10–15 times in culture, but tortoise cells undergo over 100 divisions. Moreover, cells from patients with genetic premature aging syndromes (such as Werner syndrome) undergo fewer divisions in culture than cells from normal patients.

Another characteristic of aging cells is that their telomeres become shorter. *Telomeres* are the tips of linear chromosomes and consist of several thousand repeats of a short DNA sequence (TTAGGG in humans). Telomeres help preserve the structural integrity of chromosomes by protecting their ends from degrading or from fusing to other chromosomes. Telomeres are created and maintained by *telomerase*—a remarkable RNA-containing enzyme that adds telomeric DNA sequences onto the ends of linear chromosomes. Telomerase also solves the "end-replication" problem—which asserts that linear DNA molecules become shorter at each replication because DNA poly-

merase cannot synthesize new DNA at the 3'-ends of each parent strand. By adding numerous telomeric repeat sequences onto the 3'-ends of chromosomes, telomerase prevents the chromosomes from shrinking into oblivion. Unfortunately, normal somatic cells contain little if any telomerase. As a result, telomere length decreases by about 100 base pairs every time a normal cell divides. Telomere shortening may act as a clock that counts cell division and instructs the cell to stop dividing.

Could we gain perpetual youth and vitality by increasing our telomere lengths? A recent study suggests that it may be possible to reverse senescence by artificially increasing the amount of telomerase in our cells. When the investigators introduced cloned telomerase genes into normal human cells in culture, telomeres lengthened by thousands of base pairs and the cells continued to grow long past their senescence point. These observations confirm that telomere length acts as a cellular clock. In addition, they suggest that some of the atrophy of tissues that accompanies old age may someday be reversed by activating telomerase genes. However, before we rush out to buy telomerase pills, we must consider a possible consequence of cellular immortality—cancer.

Although normal cells undergo senescence after a specific number of cell divisions, cancer cells do not. It is thought that cancers arise after several genetic mutations accumulate in a cell. These mutations disrupt the normal checks and balances that control cell growth and division. It therefore seems logical that cancer cells would also stop the normal aging clock. If their telomeres became shorter after each cell division, tumor cells would eventually succumb to aging and cease growth. However, if they synthesized telomerase, they would arrest the ticking of the senescence clock and become immortal. In keeping with this idea, over 90% of human tumor cells contain telomerase activity, and have stable telomeres. The correlation between uncontrolled tumor cell growth and the presence of telomerase activity is so good that telomerase assays are being developed as diagnostic markers for cancer. Although there is currently some debate about whether the presence of telomerase is a prerequisite for, or simply a consequence of, cell transformation, it is possible that acquiring telomerase activity may be an important step in the development of a cancer cell. In support of this hypothe-

sis, recent studies show that the induction of telomerase activity, when combined with the inactivation of a tumor-suppressor gene (p16[INK4a]), results in cell immortalization—an essential step toward tumor development. Therefore, any attempt to increase telomerase activity in normal cells carries the risk of enhancing the development of tumors.

An attractive possibility is that telomerase may be an ideal target for anticancer drugs. Drugs that inhibit telomerase might destroy cancer cells by allowing their telomeres to shorten, thereby forcing the cells into senescence. Because most normal human cells do not express telomerase, such a therapy might be specific for tumor cells and hence less toxic than most current anticancer drugs. Such anti-telomerase anticancer drugs are currently under development by Geron Corporation. Although we don't yet know whether this approach will work in animals, it appears to work in cultured tumor cells. Tumor cells that are treated with an anti-telomerase agent lose telomeric sequences and die after about 25 cell divisions.

Before anti-telomerase drugs can be developed and used on humans, several questions must be answered. Is telomerase required by some normal human cells (such as lymphocytes and germ cells)? If so, anti-telomerase drugs may be unacceptably toxic. Could some cancer cells compensate for the loss of telomerase by using other telomere-lengthening mechanisms (such as recombination)? If so, anti-telomerase drugs may be doomed to failure. Even if we inhibit telomerase activity in tumor cells, could they undergo multiple divisions before reaching senescence and still damage the host?

Will telomerase allow us to both arrest cancers *and* reverse the descent into old age? Time will tell.

References

Bodnar, A. G., et al. 1998. Extension of life-span by introduction of telomerase into normal human cells. *Science* 279:349–52.

deLange, T. 1998. Telomeres and senescence: Ending the debate. *Science* 279:334–35.

Kiyono, T., et al. 1998. Both Rb/p16[INK4a] inactivation and telomerase activity are required to immortalize human epithelial cells. *Nature* 396:84–88.

strand and synthesizing the complement using the remaining strand as a template. Excision can occur in either strand, yielding two possible "corrections." One repairs the mismatched base pair and "converts" it to restore the original sequence. The other also corrects the mismatch, but does so by copying the altered strand, creating a base-pair substitution. Conversion may have the effect of creating identical alleles on the two homologs that were different initially.

In our example in Figure 11–19, suppose the G≡C pair on one homolog was responsible for the mutant allele, while the A=T pair was part of the wild-type gene sequence on the other homolog. Converting the G≡C pair to A=T changes the mutant allele to wild type, just as Mitchell originally observed.

Gene-conversion events help to explain other puzzling genetic phenomena in fungi. For example, when mutant and wild-type alleles of a single gene are crossed, asci should yield equal numbers of mutant and wild-type spores. However, exceptional asci with 3:1 or 1:3 ratios are sometimes observed. These ratios can be explained by gene conversion. The phenomenon has also been detected during mitotic events in fungi, as well as in studies of unique compound chromosomes in *Drosophila*.

Chapter Summary

1. In theory, three modes of DNA replication are possible: semiconservative, conservative, and dispersive. Though all three rely on base complementarity, semiconservative replication is the most straightforward and was predicted.

2. In 1958, Meselson and Stahl resolved this problem in favor of semiconservative replication in *E. coli*, showing that newly synthesized DNA consists of one old strand and one new strand. Taylor, Woods, and Hughes used root tips of the broad bean to demonstrate semiconservative replication in eukaryotes.

3. During the same period, Kornberg isolated the enzyme DNA polymerase I from *E. coli* and showed that it is capable of directing *in vitro* DNA synthesis, provided that a template and precursor nucleoside triphosphates are supplied.

4. The subsequent discovery of the *polA1* mutant strain of *E. coli*, capable of DNA replication in spite of its lack of polymerase I activity, cast doubt on this enzyme's *in vivo* replicative function. DNA polymerases II and III were then isolated. Polymerase III has been identified as the enzyme responsible for DNA replication *in vivo*.

5. During the process of DNA synthesis, the double helix unwinds, forming a replication fork where synthesis begins. Proteins stabilize the unwound helix and assist in relaxing the coiling tension created ahead of the replication activity.

6. Synthesis is initiated at specific sites along each template strand by the enzyme primase, which results in a short segment of RNA that provides a suitable 3'-end, upon which DNA polymerase III can begin polymerization.

7. Because of the antiparallel nature of the double helix, polymerase III synthesizes DNA continuously on the leading strand in a 5'-to-3' direction. On the opposite strand, called the lagging strand, synthesis results in short Okazaki fragments that are later joined by DNA ligase.

8. DNA polymerase I removes and replaces the RNA primer with DNA, which is joined to the adjacent polynucleotide by DNA ligase.

9. The isolation of numerous phage and bacterial mutant genes affecting many of the molecules involved in the replication of DNA has helped define the complex genetic control of the entire process.

10. DNA replication in eukaryotes is similar to, but more complex than, replication in prokaryotes. Multiple replication origins exist and multiple forms of DNA polymerase direct DNA synthesis.

11. Replication at the ends (telomeres) of linear molecules poses a special problem in eukaryotes that can be solved by a unique RNA-containing enzyme called telomerase.

12. Homologous recombination between genetic molecules relies on a series of enzymes that can cut, realign, and reseal DNA strands. The phenomenon of gene conversion may be best explained in terms of mismatch repair synthesis during these exchanges.

Key Terms

antiparallel, 221
autonomously replicating sequences (ARSs), 224
autoradiography, 215
bidirectional replication, 215
biologically active DNA, 218
chi form, 227
conditional mutation, 223
conservative replication, 213
continuous DNA synthesis, 221

core enzyme, 222
discontinuous DNA synthesis, 221
dispersive replication, 213
DnaA protein, 220
DnaB protein, 220
DnaC protein, 220
DNA gyrase, 220
DNA ligase, 221
DNA polymerase I, 217
DNA polymerase II, 219

DNA polymerase III, 219
DNA topoisomerase, 220
exonuclease activity, 219
exonuclease proofreading, 222
gene conversion, 227
general recombination, 226
genetic recombination, 226
helicase, 220
histone, 224
holoenzyme, 219

Insights and Solutions

1. Predict the theoretical results of conservative and dispersive models of DNA synthesis using the conditions of the Meselson–Stahl experiment. Follow the results through two generations of replication after cells have been shifted to ^{14}N-containing medium, using the following sedimentation pattern:

$$^{14}N/^{14}N \quad ^{15}N/^{14}N \quad ^{15}N/^{15}N$$

Solution:

Conservative replication

Dispersive replication

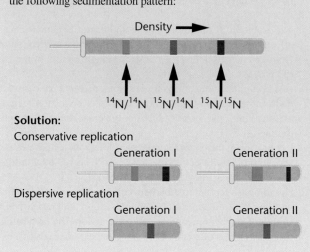

2. Mutations in the *dnaA* gene of *E. coli* are lethal and can only be studied following the isolation of conditional, temperature-sensitive mutations. Such mutant strains grow nicely and replicate their DNA at the permissive temperature of 18°C, but they do not grow or replicate their DNA at the restrictive temperature of 37°C. Two observations helped determine the function of the *dnaA* gene product. First, *in vitro* studies using DNA templates that have been unwound do not require the DnaA protein. Second, if intact cells are grown at 18°C and then shifted to 37°C, DNA synthesis continues at this temperature until one round of replication is completed, and DNA synthesis stops. What do these observations suggest about the role of the *dnaA* gene product?

Solution: These observations suggest that *in vivo* the DnaA protein is essential to the initiation of DNA synthesis. At 18°C (the permissive temperature), the mutation is not expressed and DNA synthesis begins. Following the shift to the restrictive temperature, DNA synthesis already initiated continues, but no new synthesis can begin. Because the DnaA protein is not required to synthesize unwound DNA, this observation suggests that the protein functions during initiation by interacting with the intact helix and somehow facilitating the localized denaturing necessary for synthesis to proceed. In fact, both conclusions are valid.

Problems and Discussion Questions

1. Compare conservative, semiconservative, and dispersive modes of DNA replication.
2. In the Meselson–Stahl experiment, which of the three modes of replication could be ruled out after one round of replication? After two rounds?
3. Predict the results of the Taylor–Woods–Hughes experiment if replication were (a) conservative, and (b) dispersive.
4. What are the requirements for the *in vitro* synthesis of DNA under the direction of DNA polymerase I?
5. What is meant by "biologically active" DNA?
6. Why was the phage φX174 chosen for the experiment demonstrating biological activity?
7. What was the significance of the *polA1* mutation during the study of DNA synthesis?
8. Summarize the properties and functions of polymerase I, II, and III.
9. List the proteins that unwind and stabilize DNA during *in vivo* DNA synthesis. How do they function?
10. Define and indicate the significance of (a) Okazaki fragments, (b) DNA ligase, and (c) primer RNA during DNA synthesis.
11. Outline the current model for DNA synthesis.
12. Why should DNA synthesis be more complex in eukaryotes than in bacteria? How is DNA synthesis similar in the two types of organisms?

13. Describe why DNA synthesis is "problematic" at the telomeres. What is unique about the enzyme that makes synthesis possible?
14. Define gene conversion and describe how this phenomenon is related to genetic recombination.
15. (a) Many of the gene products involved in DNA synthesis were initially defined by studying mutant *E. coli* strains that could not synthesize DNA. The α subunit of DNA pol III is responsible for the 5′-to-3′ polymerase (chain elongation) function and is coded for by the *dnaE* gene. What would be the resulting phenotype of a mutation in this gene that inactivates the α subunit? How is it possible to maintain such a strain?
    (b) The ε subunit of DNA pol III contains the 3′-to-5′ exonuclease activity and is coded for by the gene *dnaQ*. If this gene is mutated, what would be the resulting phenotype?

16. Suppose that *E. coli* synthesizes DNA at a rate of 100,000 nucleotides per minute and takes 40 minutes to replicate its chromosome.
    (a) How many base pairs are present in the entire *E. coli* chromosome?
    (b) What is the physical length of the chromosome in its helical configuration; that is, what is the circumference of the chromosome if it were opened into a circle?
17. The genome of the fruit fly *Drosophila melanogaster* consists of approximately 1.6×10^8 base pairs. DNA synthesis occurs at a rate of 30 base pairs per second. In the early embryo, the entire genome is replicated in 5 minutes. How many *bidirectional origins of synthesis* are required to accomplish this feat?

Selected Readings

Blackburn, E. H. 1991. Structure and function of telomeres. *Nature* 350:569–572.

Bodnar, A. G., et al. 1998. Extension of life-span by introduction of telomerase into normal human cells. *Science* 279:349–52.

Bramhill, D., and Kornberg, A. 1988. A model for initiation at origins of DNA replication. *Cell* 54:915–18.

DeLucia, P., and Cairns, J. 1969. Isolation of an *E. coli* strain with a mutation affecting DNA polymerase. *Nature* 224:1164–66.

Denhardt, D. T., and Faust, E. A. 1985. Eukaryotic DNA replication. *Bioessays* 2:148–53.

Diffley, J. F. X. 1996. Once and only once upon a time: Specifying and regulating origins of DNA replication in eukaryotic cells. *Genes and Development* 10:2819–30.

Dressler, D., and Potter, H. 1982. Molecular mechanisms in genetic recombination. *Annu. Rev. Biochem.* 51:727–61.

Greider, C. W., and Blackburn, E. H. 1989. A telomeric sequence in the RNA of Tetrahymena telomerase required for telomere repeat synthesis. *Nature* 337:331–36.

Greider, C. W. 1996. Telomeres, telomerase, and cancer. *Sci. Am.* (Feb.) 274:92–97.

———. 1998. Telomerase activity, cell proliferation, and cancer. *Proc. Natl. Acad. Sci. USA* 95:90–92.

Herendeen, D. R., and Kelly, T. J. 1996. DNA polymerase III: Running rings around the fork. *Cell* 84:5–8.

Hindges, R., and Hübscher, U. 1997. DNA polymerase essential for DNA transactions *Biol. Chem.* 378:345–62.

Holliday, R. 1964. A mechanism for gene conversion in fungi. *Genet. Res.* 5:282–304.

Huberman, J. C. 1987. Eukaryotic DNA replication: A complex picture partially clarified. *Cell* 48:7–8.

Kornberg, A. 1960. Biological synthesis of DNA. *Science* 131:1503–8.

———. 1974. *DNA synthesis*. New York: W. H. Freeman.

———. 1979. Aspects of DNA replication. *Cold Spring Harbor Symp. Quant. Biol.* 43:1–10.

Kornberg, A., and Baker, T. A. 1992. *DNA replication*, 2nd ed. New York: W. H. Freeman.

Lehman, I. R. 1974. DNA ligase: Structure, mechanism, and function. *Science* 186:790–97.

Lindegren, C. C. 1953. Gene conversion in *Saccharomyces*. *J. Genet.* 51:625–37.

Lodish, H., et al. 2000. *Molecular cell biology*, 4th ed. New York: W. H. Freeman.

Meselson, M., and Stahl, F. W. 1958. The replication of DNA in *Escherichia coli*. *Proc. Natl. Acad. Sci. USA* 44:671–82.

Mitchell, M. B. 1955. Aberrant recombination of pyridoxine mutants of *Neurospora*. *Proc. Natl. Acad. Sci. USA* 41:215–20.

Radding, C. M. 1978. Genetic recombination: Strand transfer and mismatch repair. *Annu. Rev. Biochem.* 47:847–80.

Radman, M., and Wagner, R. 1988. The high fidelity of DNA duplication. *Sci. Am.* (Aug.) 259:40–46.

Stahl, F. W. 1979. *Genetic recombination: Thinking about it in phage and fungi*. New York: W. H. Freeman.

———. 1987. Genetic recombination. *Sci. Am.* (Feb.) 256:90–101.

Taylor, J. H., Woods, P. S., and Hughes, W. C. 1957. The organization and duplication of chromosomes revealed by autoradiographic studies using tritium-labeled thymidine. *Proc. Natl. Acad. Sci. USA* 48:122–28.

Thömmes, P., and Hübscher, U. 1992. Eukaryotic DNA helicases: Essential enzymes for DNA transactions. *Chromosoma* 101:467–73.

Wang, J. C. 1982. DNA topoisomerases. *Sci. Am.* (July) 247:94–108.

Watson, J. D., et al. 1987. *Molecular biology of the gene, Vol. 1. General principles*, 4th ed. Menlo Park, CA: Benjamin/Cummings.

Whitehouse, H. L. K. 1982. *Genetic recombination: Understanding the mechanisms*. New York: Wiley.

Zyskind, J. W., and Smith, D. W. 1986. The bacterial origin of replication, *oriC*. *Cell* 46:489–90.

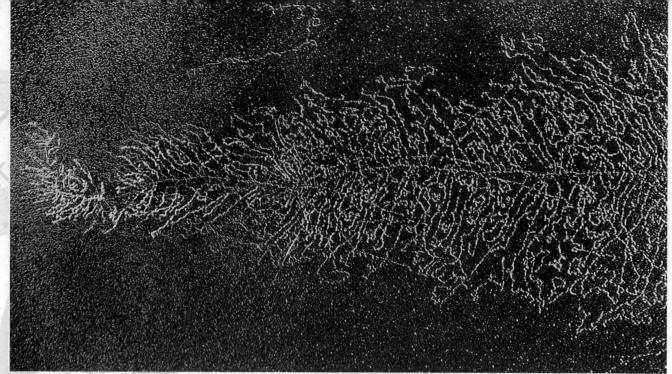

Electron micrograph visualizing the process of transcription. (*Prof. Oscar L. Miller/Science Photo Library/Photo Researchers, Inc.*)

12

The Genetic Code and Transcription

CHAPTER CONCEPTS

Genetic information, stored in DNA and transferred to RNA during the process of transcription, is present as three-letter code words. Using four different letters, corresponding to the four ribonucleotides in RNA, the 64 possible triplet codons specify the 20 different amino acids found in proteins, and provide signals that initiate and terminate protein synthesis. This unique language is the basis of life as we know it.

s we saw in Chapter 10, the structure of DNA consists of a linear sequence of deoxyribonucleotides. This sequence ultimately dictates the components constituting proteins, the end product of most genes. The central question is how such information stored as a nucleic acid can be decoded into a protein. Figure 12–1 gives a simple overview of how this transfer of information occurs. In the first step in gene expression, information present on one of the two strands of DNA (the template strand) is transferred into an RNA complement through the process of transcription. Once synthesized, this RNA acts as a "messenger" molecule, bearing the coded information—hence its name, messenger RNA (mRNA). Such RNAs then associate with ribosomes, where decoding into proteins occurs.

In this chapter we focus on the initial phases of gene expression by addressing two major questions. First, how is genetic information encoded? Second, how does the transfer from DNA to RNA occur, thus defining the process of transcription? As we shall see, ingenious analytical research established that the genetic code is written in units of three letters—ribonucleotides present in mRNA that reflect the stored information in genes. Each triplet code word directs the incorporation of a specific amino acid into a protein as it is synthesized. As we can predict based on our prior discussion of the replication of DNA, transcription is also a complex process dependent on a major polymerase enzyme and a cast of supporting proteins. We'll explore what is known

about transcription in bacteria and then contrast this prokaryotic model with the differences found in eukaryotes. Together, the information in this and the next chapter provides a comprehensive picture of molecular genetics, which serves as the most basic foundation for the understanding of living organisms. In Chapter 13, we address how translation occurs and discuss the structure and function of proteins.

12.1 Characteristics of the Genetic Code

Before we consider the various analytical approaches that led to our current understanding of the genetic code, let's summarize the general features that characterize it.

1. The genetic code is written in linear form, using the ribonucleotide bases that compose mRNA molecules as "letters." The ribonucleotide sequence is derived from the complementary nucleotide bases in DNA.

2. Each "word" within the mRNA contains three ribonucleotide letters. Each group of *three* ribonucleotides, called a **codon**, specifies *one* amino acid; the code is thus a **triplet**.

3. The code is **unambiguous**, meaning that each triplet specifies only a single amino acid.

4. The code is **degenerate**; that is, a given amino acid can be specified by more than one triplet codon. This is the case for 18 of the 20 amino acids.

5. The code contains "start" and "stop" signals, certain triplets that **initiate** and **terminate** translation.

6. No internal punctuation ("commas") is used in the code. Thus, the code is said to be **commaless**. Once translation of mRNA begins, the codons are read one after the other with no breaks between them.

7. The code is **nonoverlapping**. Once translation commences, any single ribonucleotide at a specific location within the mRNA is part of only one triplet.

8. The code is nearly **universal**. With only minor exceptions, a single coding dictionary is used by almost all viruses, prokaryotes, archaea, and eukaryotes.

12.2 Initial Insights into the Code

In the late 1950s, before it became clear that mRNA is the intermediate that transfers genetic information from DNA to proteins, it was thought that DNA itself might directly encode proteins during their synthesis. Because ribosomes had already been identified, the initial thinking was that information in DNA was transferred in the nucleus to the RNA of the ribosome, which served as the template for protein synthesis in the cytoplasm. This concept soon became untenable as accumulating evidence indicated that there was an unstable template intermediate. The RNA of ribosomes, on the other hand, was extremely stable. As a result, in 1961 François Jacob and Jacques Monod postulated the existence of **messenger RNA (mRNA)**.

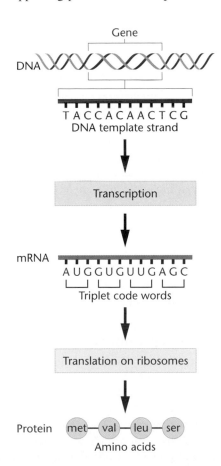

FIGURE 12–1 Flow of genetic information encoded in DNA to messenger RNA to protein.

Once mRNA was discovered, it was clear that even though genetic information is stored in DNA, the code that is translated into proteins resides in RNA. The central question then was how only four "letters"—the four nucleotides—could specify 20 "words"—the amino acids.

The Triplet Nature of the Code

In the early 1960s, Sidney Brenner argued on theoretical grounds that the code must be a triplet since three-letter words represent the minimal use of four letters to specify 20 amino acids. A code of four nucleotides, taken two at a time, for example, provides only 16 unique code words (4^2). A triplet code provides 64 words (4^3)—clearly more than the 20 needed—and it is much simpler than a four-letter code (4^4), which specifies 256 words.

Experimental evidence supporting the triplet nature of the code was subsequently derived from research by Francis Crick and his colleagues. Using phage T4, they studied **frameshift mutations**, which result from the addition or deletion of one or more nucleotides within a gene and subsequently the mRNA transcribed by it. The gain or the loss of one or more letters shifts the frame of reading during translation. Crick and his colleagues found that the gain or loss of one or two nucleotides caused such a mutation, but when three nucleotides were involved, the frame of reading was reestablished (Figure 12–2). This would not occur if the code consisted of anything other than a triplet. This work also suggested that most triplet codes are not blank, but rather encode amino acids, supporting the concept of a degenerate code.

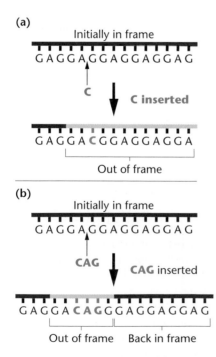

FIGURE 12–2 The effect of frameshift mutations on a DNA sequence repeating the triplet sequence GAG. (a) The insertion of a single nucleotide shifts all subsequent triplet reading frames. (b) The insertion of three nucleotides changes only two triplets, but the frame of reading is then reestablished to the original sequence.

12.3 Deciphering the Code

In 1961, Marshall Nirenberg and J. Heinrich Matthaei characterized the first specific coding sequences, which served as a cornerstone for the complete analysis of the genetic code. Their success, as well as that of others who made important contributions in deciphering the code, was dependent on the use of two experimental tools, an **_in vitro_ (cell-free) protein-synthesizing system**, and an enzyme, **polynucleotide phosphorylase**, which allowed the production of synthetic mRNAs. These mRNAs are templates for polypeptide synthesis in the cell-free system.

Cell-Free Polypeptide Synthesis

In the cell-free system, amino acids are incorporated into polypeptide chains. This _in vitro_ mixture, as might be expected, must contain the essential factors for protein synthesis in the cell: ribosomes, tRNAs, amino acids, and other molecules essential to translation. In order to follow (or trace) protein synthesis, one or more of the amino acids must be radioactive. Finally, an mRNA must be added, which serves as the template to be translated.

In 1961, mRNA had yet to be isolated. However, the use of the enzyme polynucleotide phosphorylase allowed the artificial synthesis of RNA templates, which could be added to the cell-free system. This enzyme, isolated from bacteria, catalyzes the reaction shown in Figure 12–3. Discovered in 1955 by Marianne Grunberg-Manago and Severo Ochoa, the enzyme functions metabolically in bacterial cells to degrade RNA. However, _in vitro_, with high concentrations of ribonucleoside diphosphates, the reaction can be "forced" in the opposite direction to synthesize RNA, as shown.

In contrast to RNA polymerase, polynucleotide phosphorylase requires no DNA template. As a result, each addition of a ribonucleotide is random, based on the relative concentration of the four ribonucleoside diphosphates added to the reaction mixtures. The probability of the insertion of a specific ribonucleotide is proportional to the availability of that molecule, relative to other available ribonucleotides. _This point is absolutely critical to understanding the work of Nirenberg and others in the ensuing discussion._

Taken together, the cell-free system for protein synthesis and the availability of synthetic mRNAs provided a means of deciphering the ribonucleotide composition of various triplets encoding specific amino acids.

The Use of Homopolymers

In their initial experiments, Nirenberg and Matthaei synthesized **RNA homopolymers**, each consisting of only one type of ribonucleotide. Therefore, the mRNA added to the _in vitro_ system was either UUUUUU..., AAAAAA..., CCCCCC..., or GGGGGG.... In testing each mRNA, they were able to determine which, if any, amino acids were incorporated into newly synthesized proteins. To do this, the researchers labeled one of the 20 amino acids added to the _in vitro_ system

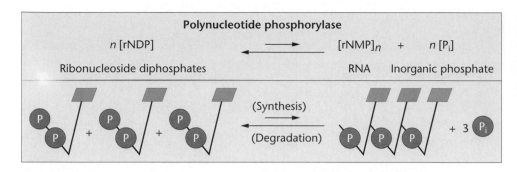

and conducted a series of experiments, each with a different amino acid made radioactive.

For example, consider the experiments using ^{14}C-phenylalanine (Table 12.1). Nirenberg and Matthaei concluded that the message poly U (polyuridylic acid) directs the incorporation of only phenylalanine into the homopolymer polyphenylalanine. Assuming a triplet code, they had determined the first specific codon assignment: UUU codes for phenylalanine. In related experiments, they quickly found that AAA codes for lysine and CCC codes for proline. Poly G was not an adequate template, probably because the molecule folds back on itself. Thus, the assignment for GGG had to await other approaches.

Note that the specific triplet codon assignments were possible only because homopolymers were used. The method yields only the composition of triplets, but since three Us, Cs, or As can have only one possible sequence (i.e., UUU, CCC, and AAA), the actual codon was identified.

Mixed Copolymers

With these techniques in hand, Nirenberg and Matthaei, and Ochoa and co-workers turned to the use of **RNA heteropolymers**. In this technique, two or more different ribonucleoside diphosphates are added in combination to form the artificial message. The researchers reasoned that if they knew the relative proportion of each type of ribonucleoside diphosphate, they could predict the frequency of any particular triplet codon occurring in the synthetic mRNA. If they then added the mRNA to the cell-free system and ascertained the percentage of any particular amino acid present in the new protein, they could analyze the results and predict the *composition* of triplets specifying specific amino acids.

This approach is shown in Figure 12–4 (page 236). Suppose that A and C are added in a ratio of 1A:5C. The insertion of

a ribonucleotide at any position along the RNA molecule during its synthesis is determined by the ratio of A:C. Therefore, there is a 1/6 chance for an A and a 5/6 chance for a C to occupy each position. On this basis, we can calculate the frequency of any given triplet appearing in the message.

For AAA, the frequency is $(1/6)^3$, or about 0.4 percent. For AAC, ACA, and CAA, the frequencies are identical— that is, $(1/6)^2(5/6)$, or about 2.3 percent for each. Together, all three 2A:1C triplets account for 6.9 percent of the total three-letter sequences. In the same way, each of three 1A:2C triplets accounts for $(1/6)(5/6)^2$, or 11.6 percent (or a total of 34.8%); CCC is represented by $(5/6)^3$, or 57.9 percent of the triplets.

By examining the percentages of any given amino acid incorporated into the protein synthesized under the direction of this message, we can propose probable base composition (Figure 12–4). Because proline appears 69 percent of the time we could propose that proline is encoded by CCC (57.9%) and one triplet consisting of 2C:1A (11.6%). Histidine, at 14 percent, is probably coded by one 2C:1A (11.6%) and one 1C:2A (2.3%). Threonine, at 12 percent, is likely coded by only one 2C:1A. Asparagine and glutamine each appear to be coded by one of the 1C:2A triplets, and lysine appears to be coded by AAA.

Using as many as all four ribonucleotides to construct the mRNA, the researchers conducted many similar experiments. Although determining the *composition* of the triplet code words for all 20 amino acids represented a significant breakthrough, the *specific sequences* of triplets were still unknown. Their determination awaited still other approaches.

The Triplet Binding Assay

It was not long before more advanced techniques were developed. In 1964, Nirenberg and Philip Leder developed the **triplet binding assay**, which led to specific assignments of triplets. The technique took advantage of the observation that ribosomes, when presented with an RNA sequence as short as three ribonucleotides, will bind to it and form a complex similar to that found *in vivo*. The triplet acts like a codon in mRNA, attracting the complementary sequence within tRNA (Figure 12–5, page 236). Such a triplet sequence in tRNA that is complementary to a codon of mRNA is known as an **anticodon**.

Although it was not yet feasible to chemically synthesize long stretches of RNA, triplets of known sequence could be synthesized in the laboratory to serve as templates. All that was needed was a method to determine which tRNA–amino

---

TABLE 12.1 Incorporation of ^{14}C-Phenylalanine into Protein

Artificial mRNA	Radioactivity (counts/min)
None	44
Poly U	39,800
Poly A	50
Poly C	38

Source: After Nirenberg and Matthaei (1961).

FIGURE 12–4 Results and interpretation of a mixed copolymer experiment where a ratio of 1A:5C is used (1/6A:5/6C).

Possible compositions	Probability of occurrence of any triplet	Possible triplets	Final %
3A	$(1/6)^3 = 1/216 = 0.4\%$	AAA	0.4
1C:2A	$(1/6)^2(5/6) = 5/216 = 2.3\%$	AAC ACA CAA	3 × 2.3 = 6.9
2C:1A	$(1/6)(5/6)^2 = 25/216 = 11.6\%$	ACC CAC CCA	3 × 11.6 = 34.8
3C	$(5/6)^3 = 125/216 = 57.9\%$	CCC	57.9
			100.0%

Chemical synthesis of message ⬇

━━━━━━━━━━━━━━━━━━━━━━━━━━━━━━━━━ RNA
C C C C C C C C A C C C C C C A A C C A C C C C C A C C C C C A C C C A A

Translation of message ⬇

Percentage of amino acids in protein		Probable base composition assignments
Lysine	<1%	AAA
Glutamine	2%	1C:2A
Asparagine	2%	1C:2A
Threonine	12%	2C:1A
Histidine	14%	2C:1A, 1C:2A
Proline	69%	CCC, 2C:1A

acid was bound to the triplet RNA–ribosome complex. The test system devised was quite simple. The amino acid to be tested was made radioactive, and a charged tRNA was produced. Because code compositions were known, researchers could narrow the decision as to which amino acids should be tested for each specific triplet.

The radioactively charged tRNA, the RNA triplet, and ribosomes are incubated together on a nitrocellulose filter, which retains the larger ribosomes but not the other smaller components, such as charged tRNA. If radioactivity is not retained on the filter, an incorrect amino acid has been test-

ed. If radioactivity remains on the filter, it is retained because the charged tRNA has bound to the triplet associated with the ribosome. In such a case, a specific codon assignment can be made.

Work proceeded in several laboratories, and in many cases clear-cut, unambiguous results were obtained. Table 12.2, for example, shows 26 triplets assigned to 9 amino acids. However, in some cases, the degree of triplet binding was inefficient, and assignments were not possible. Eventually, about 50 of the 64 triplets were assigned. These specific assignments of triplets to amino acids led to two major con-

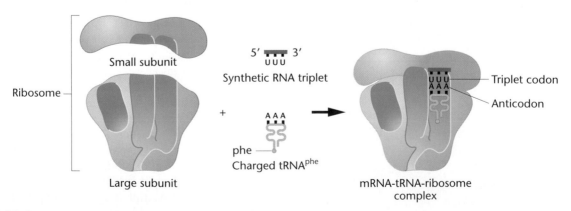

FIGURE 12–5 An example of the triplet binding assay. The UUU triplet acts as a codon, attracting the complementary tRNA^phe anticodon AAA.

TABLE 12.2   Amino Acid Assignments to Specific Trinucleotides Derived from the Triplet Binding Assay

Trinucleotides	Amino Acid	Trinucleotides	Amino Acid	Trinucleotides	Amino Acid
UGU UGC	Cysteine	CUC CUA CUG	Leucine	CCU CCC	Proline
GAA GAG	Glutamic acid	AAA AAG	Lysine	CCG CCA	Proline
AUU AUG AUA	Isoleucine	AUG	Methionine	UCU UCC	Serine
UUA UUG CUU	Leucine	UUU UUC	Phenylalanine	UCA UCG	Serine

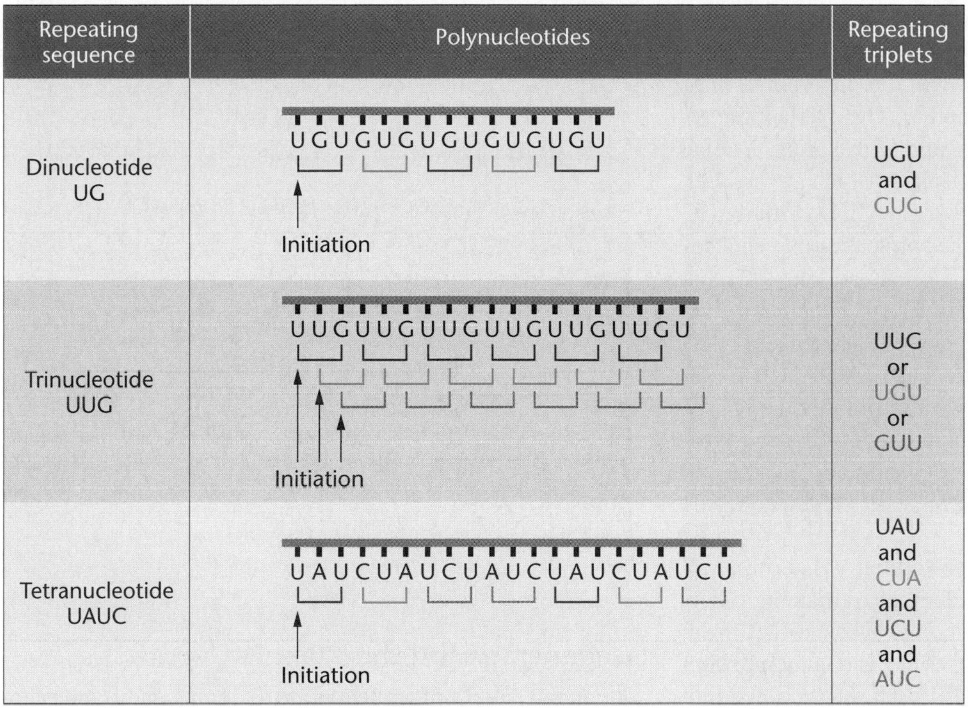

FIGURE 12–6 The conversion of di-, tri-, and tetranucleotides into repeating copolymers. The triplet codons produced in each case are shown.

clusions. First, the genetic code is degenerate; that is, one amino acid can be specified by more than one triplet. Second, the code is unambiguous; that is, a single triplet specifies only one amino acid. As we shall see later in this chapter, these conclusions have been upheld with only minor exceptions. The triplet binding technique was a major innovation in deciphering the genetic code.

Repeating Copolymers

Yet another innovative technique used to decipher the genetic code was developed in the early 1960s by Gobind Khorana, who chemically synthesized long RNA molecules consisting of short sequences repeated many times. First, he created shorter sequences (e.g., di-, tri-, and tetranucleotides), which were then replicated many times and finally joined enzymatically to form the long polynucleotides. As shown in Figure 12–6, a dinucleotide made in this way is converted to a mRNA with two repeating triplets. A trinucleotide is converted to a mRNA with three potential triplets, depending on the point at which initiation occurs, and a tetranucleotide creates four repeating triplets.

When these synthetic messages were added to a cell-free system, the predicted number of amino acids incorporated was upheld. Several examples are shown in Table 12.3.

TABLE 12.3   Amino Acids Incorporated Using Repeated Synthetic Copolymers of RNA

Repeating Copolymer	Codons Produced	Amino Acids in Polypeptides
UG	UGU	Cysteine
	GUG	Valine
AC	ACA	Threonine
	CAC	Histidine
UUC	UUC	Phenylalanine
	UCU	Serine
	CUU	Leucine
AUC	AUC	Isoleucine
	UCA	Serine
	CAU	Histidine
UAUC	UAU	Tyrosine
	CUA	Leucine
	UCU	Serine
	AUC	Isoleucine
GAUA	GAU	None
	AGA	
	UAG	
	AUA	

When such data were combined with those on composition assignment and triplet binding, specific assignments were possible.

One example of specific assignments made from such data demonstrates the value of Khorana's approach. Consider the following three experiments in concert with one another. The repeating trinucleotide sequence UUCUUCUUC... produces three possible triplets: UUC, UCU, and CUU, depending on the initiation point. When placed in a cell-free translation system, the polypeptides containing phenylalanine (phe), serine (ser), and leucine (leu) are produced. On the other hand, the repeating dinucleotide sequence UCUCUCUC... produces the triplets UCU and CUC with the incorporation of leucine and serine into the polypeptide. The overlapping results indicate that the triplets UCU and CUC specify leucine and serine but not which triplet specifies which amino acid. We can further conclude that *either* the CUU *or* the UUC triplet also encodes leucine or serine, while the other encodes phenylalanine.

To derive more specific information, let's examine the results of using the repeating tetranucleotide sequence UUAC, which produces the triplets UUA, UAC, ACU, and CUU. The CUU triplet is one of the two that interest us. Three amino acids are incorporated: leucine, threonine, and tyrosine. Because CUU must specify only serine or leucine, and because, of these two, only leucine appears, we can conclude that CUU specifies leucine.

Once this is established, we can logically determine all other assignments. Of the two triplet pairs remaining (UUC and UCU from the first experiment *and* UCU and CUC from the second experiment), whichever triplet is common to both must encode serine. This is UCU. By elimination, we find that UUC encodes phenylalanine and CUC encodes leucine. While the logic must be carefully followed, four specific triplets encoding three different amino acids have been assigned from these experiments.

From such interpretations, Khorana reaffirmed triplets that were already deciphered and filled in gaps left from other approaches. For example, the use of two tetranucleotide sequences, GAUA and GUAA, suggested that at least two triplets were termination signals. He reached this conclusion because neither of these sequences directed the incorporation of any amino acids into a polypeptide. Because there are no triplets common to both messages, he predicted that each repeating sequence contains at least one triplet that terminates protein synthesis. Table 12.3 lists the possible triplets for the poly-(GAUA) sequence, of which UAG is a termination codon.

12.4 The Coding Dictionary

The various techniques used to decipher the genetic code have yielded a dictionary of 61 triplet codon–amino acid assignments. The remaining three triplets are termination signals, not specifying any amino acid. Figure 12–7 designates the assignments in a particularly illustrative form first suggested by Francis Crick.

FIGURE 12–7 The coding dictionary. AUG encodes methionine, which initiates most polypeptide chains. All other amino acids except tryptophan, which is encoded only by UGG, are represented by two to six triplets. The triplets UAA, UAG, and UGA are termination signals and do not encode any amino acids.

Degeneracy and the Wobble Hypothesis

A general pattern of triplet codon assignments becomes apparent when we look at the genetic coding dictionary. Most evident is that the code is degenerate, as the early researchers predicted. That is, almost all amino acids are specified by two, three, or four different codons. Three amino acids (serine, arginine, and leucine) are each encoded by six different codons. Only tryptophan and methionine are encoded by single codons.

Also evident is the pattern of degeneracy. Most often in a set of codons specifying the same amino acid, the first two letters are the same, with only the third differing. Crick observed a pattern in the degeneracy at the third position, and in 1966, he postulated the **wobble hypothesis**.

Crick's hypothesis first predicted that the initial two ribonucleotides of triplet codes are more critical than the third member in attracting the correct tRNA. He postulated that hydrogen bonding at the third position of the codon–anticodon interaction is less constrained and need not adhere as specifically to the established base-pairing rules. The wobble hypothesis thus proposes a more flexible set of base-pairing rules at the third position of the codon (Table 12.4).

This relaxed base-pairing requirement, or "wobble," allows the anticodon of a single form of tRNA to pair with more than one triplet in mRNA. Consistent with the wobble hypothesis and the degeneracy of the code, U at the first position (the 5′-end) of the tRNA anticodon may pair with A or G at the third position (the 3′-end) of the mRNA codon, and

TABLE 12.4 Anticodon–Codon Base-Pairing Rules

Base at 1st position (5'-end) of tRNA	Base at 3rd position (3'-end) of mRNA
A	U
C	G
G	C or U
U	A or G
I	A, U or C

G may likewise pair with U or C. Inosine, one of the modified bases found in tRNA, may pair with C, U, or A. Applying these wobble rules, a minimum of about 30 different tRNA species is necessary to accommodate the 61 triplets specifying an amino acid. If nothing more, wobble can be considered a potential economy measure, provided that the fidelity of translation is not compromised. Current estimates are that 30–40 tRNA species are present in bacteria and up to 50 tRNA species in animal and plant cells.

Initiation and Termination

Initiation of protein synthesis is a highly specific process. In bacteria (in contrast to the *in vitro* experiments discussed earlier), the initial amino acid inserted into all polypeptide chains is a modified form of methionine—*N*-**formylmethionine** (**fmet**). Only one codon, AUG, codes for methionine, and it is sometimes called the **initiator codon**. However, when AUG appears internally in mRNA rather than at an initiating position, unformylated methionine is inserted into the polypeptide chain. Rarely, another triplet, GUG, specifies methionine during initiation, although it is not clear why this happens, since GUG normally encodes valine.

In bacteria, either the formyl group is removed from the initial methionine upon the completion of synthesis of a protein, or the entire formylmethionine residue is removed. In eukaryotes, methionine is also the initial amino acid during polypeptide synthesis. However, it is not formylated.

As mentioned in the preceding section, three other triplets (UAG, UAA, and UGA) serve as **termination codons**, punctuation signals that do not code for any amino acid. They are not recognized by a tRNA molecule, and translation terminates when they are encountered. Mutations that produce any of the three triplets internally in a gene also result in termination. Consequently, only a partial polypeptide is synthesized, since it is prematurely released

from the ribosome. When such a change occurs in the DNA, it is called a **nonsense mutation**.

12.5 Confirming the Code Using Phage MS2

All aspects of the genetic code discussed so far yield a fairly complete picture. The code is triplet in nature, degenerate, unambiguous, and commaless, but it contains punctuation with respect to start and stop signals. These individual principles have been confirmed by the detailed analysis of the RNA-containing bacteriophage MS2 by Walter Fiers and his co-workers.

MS2 is a bacteriophage that infects *E. coli*. Its nucleic acid (RNA) contains only about 3500 ribonucleotides, making up only three genes. These genes specify a coat protein, an RNA-directed replicase, and a maturation protein (the A protein). This simple system of a small genome and few gene products allowed Fiers and his colleagues to sequence the genes and their products. The amino acid sequence of the coat protein was completed in 1970, and the nucleotide sequence of the gene and several nucleotides on each of its ends were reported in 1972.

When the chemical nature of the gene and protein are compared, a colinear relationship is evident. That is, based on the coding dictionary, the linear sequence of nucleotides (and thus the sequence of triplet codons) corresponds precisely with the linear sequence of amino acids in the protein. Furthermore, the codon for the first amino acid is AUG, the common initiator codon; and the codon for the last amino acid is followed by two consecutive termination codons, UAA and UAG.

By 1976, the other two genes and their protein products were sequenced, providing similar confirmation. The analysis clearly shows that the genetic code, as established in bacterial systems, is identical in this virus. Other evidence suggests that the code is also identical in eukaryotes.

12.6 The Universality of the Genetic Code

Between 1960 and 1978, it was generally assumed that the genetic code would be found to be universal, applying equally to viruses, bacteria, archaea, and eukaryotes. Certainly, the nature of mRNA and the translation machinery seemed to be very similar in these organisms. For example, cell-free systems derived from bacteria can translate eukaryotic mRNAs. Poly U stimulates translation of polyphenylalanine in cell-free systems when the components are derived from eukaryotes. Many recent studies involving recombinant DNA technology (see Chapter 16) reveal that eukaryotic genes can be inserted into bacterial cells and transcribed and translated. Within eukaryotes, mRNAs from mice and rabbits have been injected into amphibian eggs and efficiently translated. For the many eukaryotic genes that have been sequenced, notably those for hemoglobin molecules, the amino acid sequence of the encoded proteins adheres to the coding dictionary established from bacterial studies.

However, several 1979 reports on the coding properties of DNA derived from mitochondria of yeast and humans (mtDNA) undermined the principle of the universality of the genetic language. Since then, mtDNA has been examined in many other organisms.

Cloned mtDNA fragments were sequenced and compared with the amino acid sequences of various mitochondrial proteins, revealing several exceptions to the coding dictionary (Table 12.5). Most surprising was that the codon UGA, normally specifying termination, specifies the insertion of tryptophan during translation in yeast and human mitochondria. In human mitochondria, AUA, which normally specifies isoleucine, directs the internal insertion of methionine. In yeast mitochondria, threonine is inserted instead of leucine when CUA is encountered in mRNA.

In 1985, several other exceptions to the standard coding dictionary were discovered in the bacterium *Mycoplasma capricolum*, and in the nuclear genes of the protozoan ciliates *Paramecium*, *Tetrahymena*, and *Stylonychia*. For example, as shown in Table 12.5, one alteration converts the termination codons (UGA) to tryptophan. Several others convert the normal termination codon to glutamine (gln). These changes are significant because both a prokaryote and several eukaryotes are involved, representing distinct species that have evolved over a long period of time.

Note the apparent pattern in several of the altered codon assignments. The change in coding capacity involves only a shift in recognition of the third, or wobble, position. For example, AUA specifies isoleucine in the cytoplasm and methionine in the mitochondrion. In the cytoplasm, methionine is specified by AUG. In a similar way, UGA calls for termination in the cytoplasm but for tryptophan in the mitochondrion. In the cytoplasm, tryptophan is specified by UGG. It has been suggested that such changes in codon recognition may represent an evolutionary trend toward reducing the number of tRNAs needed in mitochondria; only 22 tRNA species are encoded in human mitochondria, for example. However, until more examples are found, the differences

must be considered to be exceptions to the previously established general coding rules.

12.7 Transcription: DNA-Dependent RNA Synthesis

Even while the genetic code was being studied, it was quite clear that proteins were the end products of many genes. Thus, while some geneticists attempted to elucidate the code, other research efforts focused on the nature of genetic expression. The central question was how DNA, a nucleic acid, specifies a protein composed of amino acids.

The complex, multistep process begins with the transfer of genetic information stored in DNA to RNA. The process by which RNA molecules are synthesized on a DNA template is called **transcription**. It results in an mRNA molecule complementary to the gene sequence of one of the double helix's two strands. Each triplet codon in the mRNA is, in turn, complementary to the anticodon region of its corresponding tRNA as the amino acid is correctly inserted into the polypeptide chain during translation. The significance of transcription is enormous, for it is the initial step in the process of information flow within the cell. The idea that RNA is involved as an intermediate molecule in the process of information flow between DNA and protein is suggested by the following observations.

1. DNA is, for the most part, associated with chromosomes in the nucleus of the eukaryotic cell. However, protein synthesis occurs in association with ribosomes located outside the nucleus in the cytoplasm. Therefore, DNA does not appear to participate directly in protein synthesis.

2. RNA is synthesized in the nucleus of eukaryotic cells, where DNA is found, and is chemically similar to DNA.

3. Following its synthesis, most RNA migrates to the cytoplasm, where protein synthesis (translation) occurs.

4. The amount of RNA is generally proportional to the amount of protein in a cell.

TABLE 12.5 Exceptions to the Universal Code

Triplet	Normal Code Word	Altered Code Word	Source
UGA	termination	trp	Human and yeast mitochondria *Mycoplasma*
CUA	leu	thr	Yeast mitochondria
AUA	ile	met	Human mitochondria
AGA AGG	arg	termination	Human mitochondria
UAA	termination	gln	*Paramecium Tetrahymena Stylonychia*
UAG	termination	gln	*Paramecium*

Collectively, these observations suggest that genetic information, stored in DNA, is transferred to an RNA intermediate, which directs the synthesis of proteins. As with most new ideas in molecular genetics, the initial supporting experimental evidence was based on studies of bacteria and their phages. It was clearly established that (1) during initial infection RNA synthesis preceded phage protein synthesis; and (2) the RNA is complementary to phage DNA.

The results of these experiments agree with the concept of a messenger RNA (mRNA) being made on a DNA template and then directing the synthesis of specific proteins in association with ribosomes. This concept was formally proposed by François Jacob and Jacques Monod in 1961 as part of a model for gene regulation in bacteria. Since then, mRNA has been isolated and studied thoroughly. There is no longer any question about its role in genetic processes.

12.8 RNA Polymerase

To prove that RNA can be synthesized on a DNA template, it was necessary to demonstrate that there is an enzyme capable of directing this synthesis. By 1959, several investigators, including Samuel Weiss, had independently isolated such a molecule from rat liver. Called **RNA polymerase**, it has the same general substrate requirements as does DNA polymerase, the major exception being that the substrate nucleotides contain the ribose rather than the deoxyribose form of the sugar. Unlike DNA polymerase, no primer is required to initiate synthesis. The initial base remains as a nucleoside triphosphate (NTP). The overall reaction summarizing the synthesis of RNA on a DNA template can be expressed as

$$n(\text{NTP}) \xrightarrow[\text{enzyme}]{\text{DNA}} (\text{NMP})_n + n(\text{PP}_i)$$

As the equation reveals, nucleoside triphosphates (NTPs) serve as substrates for the enzyme, which catalyzes the polymerization of nucleoside monophosphates (NMPs), or nucleotides, into a polynucleotide chain $(\text{NMP})_n$. Nucleotides are linked during synthesis by 5′-to-3′ phosphodiester bonds (see Figure 10–10). The energy created by cleaving the triphosphate precursor into the monophosphate form drives the reaction, and inorganic phosphates (PP_i) are produced.

A second equation summarizes the sequential addition of each ribonucleotide as the process of transcription progresses:

$$(\text{NMP})_n + \text{NTP} \xrightarrow[\text{enzyme}]{\text{DNA}} (\text{NMP})_{n+1} + \text{PP}_i$$

As this equation shows, each transcription step involves the addition of one ribonucleotide (NMP) to the growing polyribonucleotide chain $(\text{NMP})_n$, using a nucleoside triphosphate (NTP) as the precursor.

RNA polymerase from *E. coli* has been extensively characterized and shown to consist of subunits designated α, β, β′, and σ. The active form of the enzyme, the **holoenzyme**, contains the subunits $\alpha_2\beta\beta'\sigma$ and has a molecular weight of almost 500,000 Da. Of these subunits, it is the β and β′ polypeptides that provide the catalytic basis and active site

for transcription. As we shall see, another subunit called the **σ (sigma) factor** plays a regulatory function involving the initiation of RNA transcription.

While there is but a single form of the enzyme in *E. coli*, there are several different sigma factors, creating variations of the polymerase holoenzyme. On the other hand, eukaryotes display three distinct forms of RNA polymerase, each consisting of a greater number of polypeptide subunits than does the bacterial form of the enzyme.

Promoters, Template Binding, and the σ Subunit

Transcription results in the synthesis of a single-stranded RNA molecule complementary to a region along only one strand of the DNA double helix. For simplicity, we'll call the transcribed DNA strand the **template strand** and its complement the **partner strand** [Figure 12–8(a), page 242].

The initial step is referred to as **template binding** [Figure 12–8(b)]. In bacteria, the site of this initial binding is established when the RNA polymerase σ subunit(s) recognizes specific DNA sequences called **promoters**. These regions are located in the 5′-region upstream from the point of initial transcription of a gene. It is believed that the enzyme "explores" a length of DNA until it recognizes the promoter region and binds to about 60 nucleotide pairs of the helix, 40 of which are upstream from the point of initial transcription. Once this occurs, the helix is denatured or unwound locally, making the DNA template accessible to the action of the enzyme.

The importance of promoter sequences cannot be overemphasized. They govern the efficiency of the initiation of transcription. In bacteria, both strong promoters and weak promoters have been discovered, leading to a variation of initiation from once every 1–2 seconds to only once every 10–20 minutes. Mutations in promoter sequences may severely reduce the initiation of gene expression. Because the interaction of promoters with RNA polymerase governs transcription, the nature of the binding between them is at the heart of discussions concerning genetic regulation, the subject of Chapter 15. While we will pursue information involving promoter–enzyme interactions in greater detail in those chapters, two points are appropriate here.

The first is the concept of **consensus sequences** of DNA. These are sequences that are similar (homologous) in different genes of the same organism or in one or more genes of related organisms. Their conservation throughout evolution attests to the critical nature of their role in biological processes. Two such sequences have been found in bacterial promoters. One, TATAAT, is located 10 nucleotides upstream from the site of initial transcription (the **−10 region**, or **Pribnow box**). The other, TTGACA, is located 35 nucleotides upstream (the **−35 region**). Mutations in both regions diminish transcription, often severely. In most eukaryotic genes studied, a consensus sequence comparable to that in the −10 region has been recognized. Because it is rich in adenine and thymine residues, it is called the **TATA box**.

(a) Transcription components

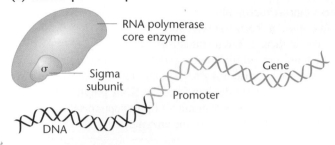

(b) Template binding and initiation of transcription

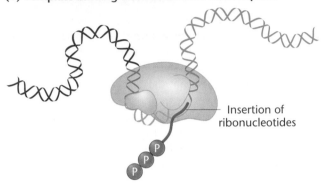

(c) Chain elongation

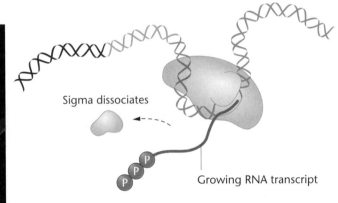

FIGURE 12–8 The early stages of transcription in prokaryotes, showing (a) the components of the process; (b) template binding at the −10 site involving the sigma subunit of RNA polymerase and subsequent initiation of RNA synthesis; and (c) chain elongation, after the sigma subunit has dissociated from the transcription complex and the enzyme moves along the DNA template.

The second point is that the degree of RNA polymerase binding to different promoters varies greatly, which causes the variable gene expression mentioned earlier. Currently, this is attributed to sequence variation in the promoters.

A final general point to be made involves the sigma (σ) subunit in bacteria. The major form is designated as σ^{70}, based on its molecular weight of 70 kilodaltons (kDa). While the promoters of most bacterial genes recognize this form, several alternative forms of RNA polymerase in *E. coli* have unique σ subunits associated with them (e.g., σ^{28}, σ^{32}, σ^{38}, and σ^{54}). These recognize different promoter sequences and provide specificity to the initiation of transcription.

Initiation and Elongation of RNA

Once it has recognized and bound to the promoter, RNA polymerase catalyzes **initiation**, the insertion of the first 5′-ribonucleoside triphosphate, which is complementary to the first nucleotide at the start site of the DNA template strand [Figure 12–8(b)]. As we noted, no primer is required. Subsequent ribonucleotide complements are inserted and linked by phosphodiester bonds as RNA polymerization proceeds. This process, called **chain elongation** [Figure 12–8(c)], continues in the 5′-to-3′ direction, creating a temporary DNA/RNA duplex whose chains run antiparallel to one another.

After a few ribonucleotides have been added to the growing RNA chain, the σ subunit dissociates from the holoenzyme, and elongation proceeds under the direction of the core enzyme. In *E. coli* this process proceeds at the rate of about 50 nucleotides/second at 37°C.

Eventually, the enzyme traverses the entire gene until it encounters a specific nucleotide sequence that acts as a termination signal. Such termination sequences, about 40 base pairs in length, are extremely important in prokaryotes because of the close proximity of the end of one gene and the upstream sequences of the adjacent gene. In some cases, the termination of synthesis is dependent on the **termination factor, rho (ρ)**. Rho is a large hexameric protein that physically interacts with the growing RNA transcript. At the point of termination, the transcribed RNA molecule is released from the DNA template, and the core polymerase enzyme dissociates. The synthesized RNA molecule is precisely complementary to a DNA sequence representing the template strand of a gene. Wherever an A, T, C, or G residue existed, a corresponding U, A, G, or C residue, respectively, is incorporated into the RNA molecule. Such RNA molecules ultimately provide the information leading to the synthesis of all proteins present in the cell.

It is important to note that in bacteria groups of genes whose products are related are often clustered along the chromosome. In many such cases, they are contiguous and all but the last gene lack the encoded signals for termination of transcription. The result is that during transcription a large mRNA is produced, encoding more than one protein. Since genes in bacteria are sometimes called cistrons (see Chapter 7), the RNA is called a **polycistronic mRNA**. Since the products of genes transcribed in this fashion are usually all needed at the same time, this is an efficient way to transcribe and, subsequently, to translate the needed genetic information. In eukaryotes, **monocistronic** mRNAs are the rule.

12.9   Transcription in Eukaryotes

Much of our knowledge of transcription derives from studies of prokaryotes. The general aspects of the mechanics of these processes are similar in eukaryotes, although there are several notable differences. We first summarize some of the major differences and then expand on a number of them in the following sections.

1. Transcription in eukaryotes occurs within the nucleus under the direction of three separate forms of RNA polymerase. Unlike prokaryotes, in eukaryotes the RNA transcript is not free to associate with ribosomes prior to the completion of transcription. For the mRNA to be translated, it must move out of the nucleus into the cytoplasm.

2. Initiation and regulation of transcription involve a more extensive interaction between upstream DNA sequences and protein factors involved in stimulating and initiating transcription. In addition to promoters, other control units called enhancers may be located in the 5' regulatory region upstream from the initiation point, but they have also been found within the gene or even in the 3' downstream region beyond the coding sequence.

3. Maturation of eukaryotic mRNA from the primary RNA transcript involves many complex stages referred to generally as "processing." An initial processing step involves the addition of a 5'-cap and a 3'-tail to most transcripts destined to become mRNAs. Other extensive modifications occur to the internal nucleotide sequence of eukaryotic RNA transcripts that eventually serve as mRNAs. The initial (or primary) transcripts are most often much larger than those that are eventually translated. Sometimes called **pre-mRNAs**, they are part of a group of molecules found only in the nucleus—a group referred to collectively as **heterogeneous nuclear RNA (hnRNA)**. Such RNA molecules are of variable but large size (up to 10^7 Da) and are complexed with proteins, forming **heterogeneous nuclear ribonucleoprotein particles (hnRNPs)**. Only about 25 percent of hnRNA molecules are converted to mRNA. In those that are converted, substantial amounts of the ribonucleotide sequence are excised, and the remaining segments are spliced back together prior to translation. This phenomenon has given rise to the concepts of split genes and **splicing** in eukaryotes.

In the remainder of this chapter we'll elaborate on these differences. We also return to them and other topics directly related to regulation of eukaryotic transcription in Chapter 15.

Initiation of Transcription in Eukaryotes

The recognition of certain highly specific DNA regions by RNA polymerase is the basis of orderly genetic function in all cells. Eukaryotic RNA polymerase exists in three unique forms, each of which transcribes different types of genes, as indicated in Table 12.6. Each enzyme is larger and more complex than

the single prokaryotic polymerase, consisting of two large subunits and 10–15 smaller subunits. In regard to the initial template binding step and promoter regions, most is known about polymerase II, which transcribes all mRNAs in eukaryotes.

At least three **cis-acting elements** of a eukaryotic gene aid the efficient initiation of transcription by polymerase II. Recall from our discussion of the *cis–trans* test in Chapter 4 that the term *cis* is drawn from organic chemistry nomenclature, meaning "next to" or on the same side as, in contrast to being "across from" or *trans*, to other functional groups. In molecular genetics, then, *cis*-elements are adjacent parts of the same DNA molecule.

One such element found within the promoter region is called the **Goldberg–Hogness** or **TATA box**, located about 30 nucleotide pairs upstream (-30) from the start point of transcription. The consensus sequence is a heptanucleotide consisting solely of A and T residues (TATAAAA). The sequence and function are analogous to that found in the -10 promoter region of prokaryotic genes. Because such a region is common to most eukaryotic genes, the TATA box is thought to be nonspecific and is only responsible for fixing the site of transcription initiation by facilitating denaturation of the helix. Such a conclusion is supported by the fact that $A = T$ base pairs are less stable than $G \equiv C$ pairs.

A second common *cis*-acting element is the **CAAT box**, located upstream within the promoter of many genes at about 80 nucleotides from the start of transcription (-80). It contains the consensus sequence GGCCAATCT. Still other upstream regulatory regions have been found, and most genes contain one or more of them. Along with the TATA and CAAT box, they influence the efficiency of the promoter. The location of each element is based on studies of deletions of particular regions of the promoter, each of which reduces the efficiency of transcription.

DNA regions called **enhancers** represent still another *cis*-acting element. Although their locations can vary, enhancers are often found even farther upstream than the regions already mentioned, or even downstream or within the gene. Thus, they can modulate transcription from a distance. Although they may not participate directly in template binding, they are essential to highly efficient initiation of transcription. We'll return to these elements in Chapter 15.

Complementing the *cis*-acting regulatory sequences are various **trans-acting factors** that facilitate template binding and, therefore, the initiation of transcription. These proteins are referred to as **transcription factors**. They are essential because RNA polymerase II cannot bind directly to eukaryotic promoter sites and initiate transcription without their presence. The transcription factors involved with human RNA polymerase II-binding are well characterized and designated **TFIIA**, **TFIIB**, and so on. One of these, **TFIID**, binds directly to the TATA-box sequence and is sometimes called the **TATA-binding protein (TBP)**. TFIID consists of about 10 polypeptide subunits. Once initial binding to DNA occurs, at least seven other transcription factors bind sequentially to TFIID, forming an extensive pre-initiation complex, which is then bound by RNA polymerase II.

Transcription factors with similar activity have been discovered in a variety of eukaryotes, including *Drosophila* and

---

TABLE 12.6 RNA Polymerases in Eukaryotes

Form	Product	Location
I	rRNA	Nucleolus
II	mRNA, snRNA	Nucleoplasm
III	5S rRNA	Nucleoplasm
	tRNA	Nucleoplasm

yeast. The nucleotide sequences in all organisms studied demonstrate a high degree of conservation. Transcription factors appear to supplant the role of the sigma factor in the prokaryotic enzyme. In Chapter 15, we return to a more in-depth consideration of the role of transcription factors in eukaryotic gene regulation, as well as the various DNA-binding domains that characterize them.

Heterogeneous Nuclear RNA and Its Processing: Caps and Tails

The genetic code is written in the ribonucleotide sequence of mRNA. This information originated, of course, in the template strand of DNA, where complementary sequences of deoxyribonucleotides exist. In bacteria, the relationship between DNA and RNA appears to be quite direct. The DNA base sequence is transcribed into an mRNA sequence, which

is then translated into an amino acid sequence according to the genetic code. By contrast, complex processing of mRNA in eukaryotes occurs before it is transported to the cytoplasm to participate in translation.

By 1970, accumulating evidence showed that eukaryotic mRNA is transcribed initially as a precursor molecule much larger than that which is translated. This notion was based on the observation by James Darnell and his co-workers of heterogeneous nuclear RNA (hnRNA) in mammalian nuclei that contained nucleotide sequences common to the smaller mRNA molecules present in the cytoplasm. They proposed that the initial transcript of a gene results in a large RNA molecule that must first be processed in the nucleus before it appears in the cytoplasm as a mature mRNA molecule. The various processing steps, discussed in the following sections, are summarized in Figure 12–9.

The initial **posttranscriptional modification** of eukaryotic RNA transcripts destined to become mRNAs involves the 5'-end of these molecules, where a **7-methylguanosine (7mG) cap** is added (Figure 12–9, step 2). The cap, which is added even before the initial transcript is complete, appears to be important to the subsequent processing within the nucleus, perhaps by protecting the 5'-end of the molecule from nuclease attack. Subsequently, this cap may be involved in the transport of mature mRNAs across the nuclear membrane into the cytoplasm. The cap is fairly complex and distinguished by a unique 5'–5' bonding between the cap and the initial ribonucleotide of the RNA. Some eukaryotes also contain a methyl group (CH_3) on the 2'-carbon of the ribose sugars of the first two ribonucleotides of the RNA.

Further insights into the processing of RNA transcripts during the maturation of mRNA came from the discovery that both hnRNAs and mRNAs contain at their 3'-end a stretch of as many as 250 adenylic acid residues. Such **poly-A sequences** are added after the 5'-7mG cap has been added. First, the 3'-end of the initial transcript is cleaved enzymatically at a point some 10–35 ribonucleotides from a highly conserved AAUAAA sequence (step 3). Then polyadenylation occurs by the sequential addition of adenylic acid residues (step 4). Poly A has now been found at the 3'-end of almost all mRNAs studied in a variety of eukaryotic organisms. The exceptions seem to be the products of histone genes.

While the AAUAAA sequence is not found on all eukaryotic transcripts, mutations in that sequence of those transcripts that do have it cannot add the poly-A tail. In the absence of this tail, the RNA transcripts are rapidly degraded. Therefore, both the 5'-cap and the 3'-poly-A tail are critical if an RNA transcript is to be further processed and transported to the cytoplasm.

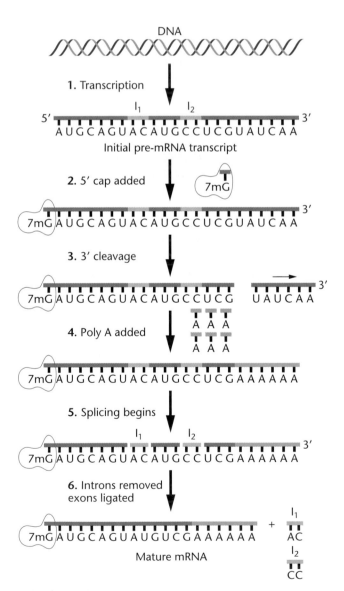

FIGURE 12–9 Posttranscriptional RNA processing in eukaryotes. Heterogeneous nuclear RNA (hnRNA) is converted to messenger (mRNA), which contains a 5'-cap and a 3'-poly-A tail, which then has introns spliced out.

12.10   Intervening Sequences and Split Genes

One of the most exciting breakthroughs in the history of molecular genetics occurred in 1977 when Susan Berget, Philip Sharp, and Richard Roberts presented direct evidence that the genes of animal viruses contain *internal* nucleotide sequences that are not expressed in the amino acid sequence of the proteins they encode. These internal DNA sequences are present in initial RNA transcripts, but they are removed before the mature mRNA is translated (Figure 12–9, steps 5 and 6). Such nucleotide segments are called **intervening sequences** (see I_1 and I_2 in Figure 12–9), and the genes containing them are known as **split genes**. Those DNA sequences that are not represented in the final mRNA product are also called **introns** ("int" for intervening), and those retained and expressed are called **exons** ("ex" for expressed). Splicing involves the removal of the ribonucleotide sequences present in introns as a result of an excision process and the rejoining of exons.

Similar discoveries were soon made in a variety of eukaryotes. Two approaches have been most fruitful. The first involves the molecular hybridization of purified, functionally mature mRNAs with DNA containing the genes specifying that message. Hybridization between nucleic acids that are not perfectly complementary results in **heteroduplexes**, in which introns present in the DNA but absent in the mRNA loop out and remain unpaired. Such structures can be visualized with the electron microscope, as shown in Figure 12–10. That heteroduplex contains seven loops (A–G) representing

seven introns whose sequences are present in DNA but not in the final mRNA.

The second approach provides more specific information (Figure 12–11). It involves a comparison of nucleotide sequences of DNA with those of mRNA and the correlation with amino acid sequences. Such an approach allows the precise identification of all intervening sequences.

Thus far, most eukaryotic genes have been shown to contain introns. One of the first so identified was the **β-globin gene** in mice and rabbits, studied independently by Philip Leder and Richard Flavell. The mouse gene contains an intron 550 nucleotides long, beginning immediately after the codon specifying the 104th amino acid. In the rabbit, there is an intron of 580 base pairs near the codon for the 110th amino acid. Additionally, a second intron of about 120 nucleotides exists earlier (upstream) in both genes. Similar introns have been found in the β-globin gene in all mammals examined.

The **ovalbumin gene** of chickens has been extensively characterized by Bert O'Malley in the United States and Pierre Chambon in France. As shown in Figure 12–11, the gene contains seven introns. In fact, as you can see, the majority of the gene's DNA sequence is "silent," being composed of introns. The initial RNA transcript is nearly three times the length of the mature mRNA. Compare the ovalbumin gene in Figures 12–10 and 12–11. Can you match the unpaired loops in Figure 12–10 with the sequence of introns specified in Figure 12–11?

The list of genes containing intervening sequences is long. In fact, few eukaryotic genes seem to be without introns. An extreme example of the number of introns in a single gene is that found in the gene coding for one subunit of collagen, the major connective tissue protein in vertebrates. The *pro-α-2(1) collagen* gene contains 50 introns. The precision of cutting and splicing that occurs must be extraordinary if errors are not to be introduced into the mature mRNA. What is equally noteworthy is a comparison of the size of genes with the size of the final mRNA once introns are removed. As shown in Table 12.7 (page 246), only about 15 percent of the collagen gene consists of exons that finally appear in

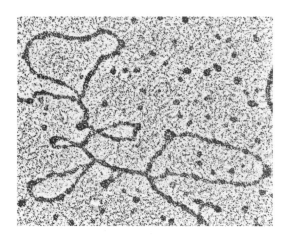

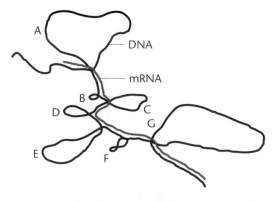

FIGURE 12–10 An electron micrograph and an interpretive drawing of the hybrid molecule (heteroduplex) formed between the template DNA strand of the chicken ovalbumin gene and the mature ovalbumin mRNA. Seven DNA introns, A–G, produce unpaired loops. (*Courtesy of Bert W. O'Malley, M.D*).

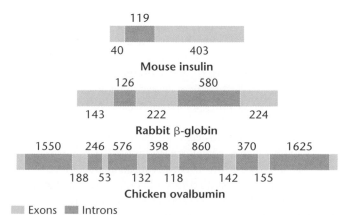

FIGURE 12–11 Intervening sequences in various eukaryotic genes. The numbers indicate the number of nucleotides present in various intron and exon regions.

TABLE 12.7 Contrasting Human Gene Size, mRNA Size, and the Number of Introns

Gene	Gene Size (kb)	mRNA size (kb)	Number of Introns
Insulin	1.7	0.4	2
Collagen [pro-α-2(1)]	38.0	5.0	50
Albumin	25.0	2.1	14
Phenylalanine hydroxylase	90.0	2.4	12
Dystrophin	2000.0	17.0	50

mRNA. For other proteins, an even more extreme picture emerges. Only about 8 percent of the albumin gene remains to be translated, and in the largest human gene known, dystrophin (which is the protein product absent in Duchenne muscular dystrophy), less than 1 percent of the gene sequence is retained in the mRNA. Several other human genes are also contrasted in Table 12.7.

While the vast majority of eukaryotic genes examined thus far contain introns, there are several exceptions. Notably, those coding for histones and for interferon do not appear to contain introns. It is not clear why or how the genes encoding these molecules have been maintained throughout evolution without acquiring the extraneous information characteristic of almost all other genes.

Splicing Mechanisms: Autocatalytic RNAs

The discovery of split genes led to intensive attempts to elucidate the mechanism by which introns of RNA are excised and exons are spliced back together, and a great deal of progress has been made. Interestingly, it appears that somewhat different mechanisms exist for different types of RNA, as well as for RNAs produced in mitochondria and chloroplasts.

Introns can be categorized into several groups based on their splicing mechanisms. In group I, represented by introns that are part of the primary transcript of rRNAs, no additional components are required for intron excision; the intron itself is the source of the enzymatic activity necessary for its own removal. This amazing discovery, which contradicted expectations, was made in 1982 by Thomas Cech and his colleagues during a study of the ciliate protozoan *Tetrahymena*. Reflecting their autocatalytic properties, RNAs that are capable of splicing themselves are sometimes called **ribozymes**.

The **self-excision process** is shown in Figure 12–12. Chemically, two nucleophilic reactions (called transesterification reactions) occur, the first involving an interaction between guanosine, which acts as a cofactor in the reaction, and the primary transcript [Figure 12–12(a)]. The 3'-OH group of guanosine is transferred to the nucleotide adjacent to the 5'-end of the intron. The second reaction involves the interaction of the newly acquired 3'-OH group on the left-hand exon and the phosphate on the 3'-end of the right intron [Figure 12–12(b)]. The intron

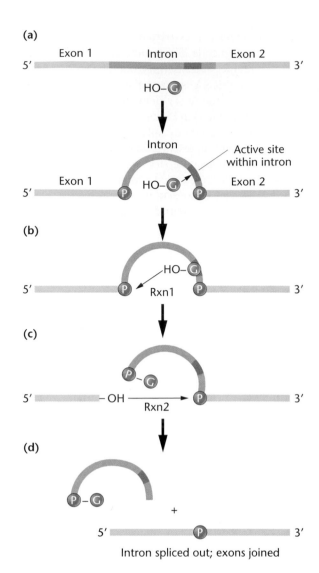

FIGURE 12–12 Splicing mechanism involved with group I introns removed from the primary transcript leading to rRNA. The process is one of self-excision involving two transesterification reactions.

is spliced out and the two exon regions are ligated, leading to the mature RNA [Figure 12–12(c)].

Self-excision of group I introns, as described, is now known to apply to pre-rRNAs from other protozoans. Self-excision also seems to govern the removal of introns present in the primary mRNA and tRNA transcripts produced in the mitochondria and chloroplasts. These are referred to as group II introns. Like group I molecules, splicing involves two autocatalytic reactions leading to the excision of introns. However, guanosine is not involved as a cofactor.

Splicing Mechanisms: The Spliceosome

Another major group of introns is found in nuclear-derived pre-mRNA transcripts. Compared to the other RNAs we've been discussing, introns in nuclear-derived mRNA can be much larger—up to 20,000 nucleotides—and they are more plentiful. Their removal appears to require a much more complex mechanism, which has been more difficult to define.

Nevertheless, many clues are now emerging and the model in Figure 12–13 illustrates the removal of one intron. You should refer to this figure during this discussion. First, the nucleotide sequences near the perimeters of this type of intron are often similar. Many begin at the 5′-end with a GU dinucleotide sequence and terminate at the 3′-end with an AG dinucleotide sequence. These, as well as other consensus sequences shared by introns, attract specific molecules that

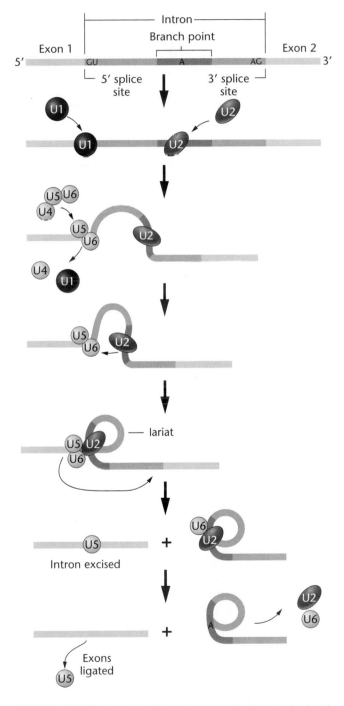

FIGURE 12–13 A model of the splicing mechanism involved with the removal of an intron from a pre-mRNA. Excision is dependent on various snRNAs (U1, U2, ..., U6) that combine with proteins to form snurps, which function as part of a large structure referred to as the spliceosome. The lariat structure in the intermediate stage is characteristic of this mechanism.

form a molecular complex essential to splicing. Such a complex, called a **spliceosome**, has been identified in extracts of yeast as well as mammalian cells. It is very large, being 40*S* in yeast and 60*S* in mammals. Perhaps the most essential component of spliceosomes is the unique set of **small nuclear RNAs (snRNAs)**. These RNAs are usually 100–200 nucleotides or less and are often complexed with proteins to form **small nuclear ribonucleoproteins** (**snRNPs**, or **snurps**). These are found only in the nucleus. Because they are rich in uridine residues, the snRNAs have been arbitrarily designated U1, U2, ..., U6.

The snRNA of U1 bears a nucleotide sequence that is homologous to the 5′-end of the intron. Base pairing resulting from this homology promotes binding that represents the initial step in the formation of the spliceosome. Following the addition of the other snurps (U2, U4, U5, and U6), splicing commences. As with group I splicing, two transesterification reactions are involved. The first involves the interaction of the 3′-OH group from an adenine (A) residue present within the region of the intron called the **branch point**. The A residue attacks the 5′-splice site, cutting the RNA chain. In a subsequent step involving several other snurps, an intermediate structure is formed and the second reaction ensues, linking the cut 5′-end of the intron to the A. This results in the formation of a characteristic loop structure called a lariat, which contains the excised intron. The exons are then ligated and the snurps released.

The processing involved in splicing represents a potential regulatory step during gene expression. For example, several cases are known where introns present in pre-mRNAs *derived from the same gene* are spliced *in more than one way*, thereby yielding different collections of exons in the mature mRNA. This process, referred to as **alternative splicing**, yields a group of mRNAs that, upon translation, result in a series of related proteins called **isoforms**. A growing number of examples have been found in organisms ranging from viruses to *Drosophila* to humans. Alternative splicing of pre-mRNAs provides the basis for producing related proteins from a single gene. We shall return to this topic in our discussion of the regulation of gene expression in eukaryotes (Chapter 15).

12.11 RNA Editing

In the late 1980s, still another quite unexpected form of post-transcriptional RNA processing was discovered. In this form, called **RNA editing**, the nucleotide sequence of a pre-mRNA is actually changed prior to translation. As a result, the ribonucleotide sequence of the mature RNA differs from the sequence encoded in the exons of the DNA from which the RNA was transcribed.

There are two main types of RNA editing—**substitution editing**, in which the identities of individual nucleotide bases are altered, and **insertion/deletion editing**, where nucleotides are added to or subtracted from the total number of bases. Substitution editing is used in some nuclear-derived eukaryotic RNAs and is prevalent in mitochondrial and

Genetics, Technology, and Society

Antisense Oligonucleotides: Attacking the Messenger

Standard chemotherapies for diseases such as cancer and AIDS are often accompanied by toxic side effects. Conventional therapeutic drugs target both normal and diseased cells, with diseased or infected cells being only slightly more susceptible than the patient's normal cells. Scientists have long wished for a magic bullet that could seek out and destroy the virus or cancer cell, leaving normal cells alive and healthy. Over the last decade, one particularly promising candidate for magic-bullet status has emerged—the *antisense oligonucleotide*.

Antisense therapies have arisen from an understanding of the molecular biology of gene expression. Gene expression is a two-step process. First, a single-stranded messenger RNA (mRNA) is copied from one strand of the duplex DNA molecule. Second, the mRNA is transported to the cytoplasm and complexed with ribosomes, and its genetic information is translated into the amino acid sequence of a polypeptide.

Normally, a gene is transcribed into RNA from only one strand of the DNA duplex. The resulting RNA is known as sense RNA. However, it is sometimes possible for the other DNA strand to be copied into RNA. The RNA produced by the transcription of the "wrong" strand of DNA is called antisense RNA. As with complementary strands of DNA molecules, complementary strands of RNA can form double-stranded molecules.

The formation of duplex structures between a sense and an antisense RNA may affect the sense RNA in at least two ways. The binding of antisense RNA to sense RNA may physically block ribosome binding or elongation, and hence may inhibit translation. Second, the binding of antisense RNA to sense RNA may trigger the degradation of the sense RNA because double-stranded RNA molecules are attacked by intracellular ribonucleases. In both cases, gene expression is blocked.

What makes the antisense approach so exciting is its potential specificity. Scientists can design antisense RNA (or DNA) molecules of known nucleotide sequence and can then synthesize large amounts of these antisense nucleic acids *in vitro*. Usually, oligonucleotides up to 20 nucleotides are used. It is theoretically possible to treat cells with these synthetic antisense oligonucleotides, to have the oligonucleotides enter the cell and bind precise target mRNAs, and to turn off the synthesis of one specific protein. If that protein is necessary for virus reproduction or cancer cell growth (but is not necessary in normal cells), the antisense oligonucleotide should have only therapeutic effects.

In the last few years, laboratory tests of antisense drugs have been so promising that several clinical trials are now in progress. In addition, one antisense drug (Vitravene©, ISIS Pharmaceuticals and CIBA Vision Corp.) has now been approved for sale in the United States. Vitravene is an antisense oligonucleotide used to treat cytomegalovirus-induced retinitis in AIDS patients. Cytomegalovirus (CMV) is a common virus that infects most people. Although it causes few problems in people with normal immune systems, it can cause serious symptoms in people with conditions that impair the immune system (such as AIDS). Up to 40 percent of AIDS patients develop retinitis or blindness as a result of CMV infections of the eye. Clinical trials indicate that the introduction of antisense CMV oligonucleotides into the eyes of AIDS patients with CMV-induced retinitis significantly delays the disease progression compared to untreated groups. It is not known how the oligonucleotides operate, but laboratory studies suggest that these antisense oligonucleotides may trigger the destruction of CMV mRNA and interfere with the absorption of the virus to the surface of host cells.

Antisense oligonucleotides may also act as anti-inflammatory drugs. Clinical trials are in progress to test antisense compounds in the treatment of asthma, rheumatoid arthritis, psoriasis, ulcerative colitis, and transplant rejection. One interesting antisense oligonucleotide (now in clinical trials conducted by ISIS and Boehringer Ingelheim) binds to the mRNA that encodes ICAM-1. ICAM-1 is a cell-surface glycoprotein that helps activate immune system and inflammatory cells. It is often overexpressed in tissues that suffer extreme inflammatory responses. Researchers hope that antisense oligonucleotides might temper inflammatory responses by reducing the expression of ICAM-1. The results of a recent clinical trial suggest that antisense ICAM-1 may be an effective treatment for Crohn's disease. Crohn's disease is a form of inflammatory bowel disease that affects about 200,000 people in the United States. It is a debilitating, chronic disease, and treatments such as steroids and immunosuppressive drugs are often ineffective and have toxic side effects. In one clinical trial, almost 50 percent of Crohn's patients went into remission after treatment with antisense ICAM-1, whereas none of the placebo group did so. In addition, one-third of the treated patients were well enough to stop steroid treatments by the end of the trial period. In this trial, the antisense ICAM-1 drugs appeared to be well-tolerated and safe.

Some interesting clinical trials are in progress to test antisense oligonucleotides as treatments for some types of cancer. These oligonucleotides are designed to reduce the synthesis of proteins that are either overexpressed in cancer cells or are present as mutant forms. These antisense drugs will be evaluated as treatments for tumors of the ovary, prostate, breast, brain, colon, and lung. Other antisense drugs in the pipeline are designed to attack the hepatitis B, hepatitis C, human papilloma, and AIDS viruses.

If antisense therapeutics pass the scrutiny of these scientific and clinical trials, we may indeed have acquired a magic molecular bullet to use in the battle against a variety of diseases.

References

Edgington, S. M. 1997. Antisense '97: A roundtable on the state of the industry. *Nature Biotechnology* 15:519–24.

Nyce, J. W. and Metzger, W. J. 1997. DNA antisense therapy for asthma in an animal model. *Nature* 385:721–25.

Roush, W. 1997. Antisense aims for a renaissance. *Science* 276:1192–93.

chloroplast RNAs transcribed in plants. *Physarum poly-cephalum*, a slime mold, uses both substitution and insertion/deletion editing for its mitochondrial mRNAs.

Trypanosoma, a parasite that causes African sleeping sickness, and its relatives use extensive insertion/deletion editing in mitochondrial RNAs. The number of uridines added to an individual transcript can make up more than 60 percent of the coding sequence, usually forming the initiation codon and placing the rest of the sequence into the proper reading frame. Insertion/deletion editing in trypanosomes is directed by **"gRNA" (guide RNA)** templates, which are also transcribed from the mitochondrial genome. These small RNAs are complementary to the edited region of the final, edited mRNAs. They base-pair with the pre-edited mRNAs to direct the editing machinery to make the correct changes.

The most well-studied examples of substitutional editing occur in mammalian nuclear-encoded mRNA transcripts. Apolipoprotein B (apo B) exists in both a long and a short form, although a single gene encodes both proteins. In human intestinal cells, apo B mRNA is edited by a single C-to-U change, which converts a CAA glutamine codon into a UAA stop codon and terminates the polypeptide at approximately half its genomically encoded length. The editing is performed by a complex of proteins that bind to a "mooring sequence" on the mRNA transcript just downstream of the editing site. In a second system, the synthesis of subunits constituting the glutamate receptor channels (GluR) in mammalian brain tissue is also affected by RNA editing. In this case, adenosine (A) to inosine (I) editing occurs in pre-mRNAs prior to their translation, where I is read as guanosine (G) during translation. A family of three ADAR (adenosine deaminase acting on RNA) enzymes are believed to be responsible for the editing of various sites within the glutamate channel subunits. The double-stranded RNAs required for editing by ADARs are provided by intron/exon pairing of the GluR mRNA transcripts. The editing changes alter the physiological parameters (solute permeability and desensitization response time) of the receptors containing the subunits.

Findings such as these in mammals have established that RNA editing provides still another important mechanism of posttranscriptional modification, and that this process is not restricted to small or asexually reproducing genomes such as mitochondria. Several new examples of RNA editing have been found each year since its discovery, and this trend is likely to continue. Further, the process has important implications for the regulation of genetic expression.

Chapter Summary

1. The genetic code, stored in DNA, is copied to RNA, where it is used to direct the synthesis of polypeptide chains. It is degenerate, unambiguous, nonoverlapping, and commaless.

2. The complete coding dictionary, determined using various experimental approaches, reveals that of the 64 possible codons, 61 encode the 20 amino acids found in proteins, while three triplets terminate translation. One of these 61 is the initiation codon and specifies methionine.

3. The observed pattern of degeneracy often involves only the third letter of a triplet series and led Francis Crick to propose the "wobble hypothesis."

4. Confirmation for the coding dictionary, including codons for initiation and termination, was obtained by comparing the complete nucleotide sequence of phage MS2 with the amino acid sequence of the corresponding proteins. Other findings support the belief that, with only minor exceptions, the code is universal for all organisms.

5. Transcription, the initial step in gene expression, describes the synthesis, under the direction of RNA polymerase, of a strand of RNA complementary to a DNA template.

6. The processes of transcription, like DNA replication, can be subdivided into the stages of initiation, elongation, and termination. Also like DNA replication, the process relies on base-pairing affinities between complementary nucleotides.

7. Initiation of transcription is dependent on an upstream (5′) DNA region called the promoter that represents the initial binding site for RNA polymerase. Promoters contain specific DNA sequences such as the TATA box that are essential to polymerase binding.

8. Transcription is more complex in eukaryotes than in prokaryotes. The primary transcript is a pre-mRNA that must be modified in various ways before it can be efficiently translated. Processing, which produces a mature mRNA, includes the addition of a 7mG cap and a poly-A tail, and the removal, through splicing, of intervening sequences, or introns. RNA editing of pre-mRNA prior to its translation also occurs in some systems.

Key Terms

Insights and Solutions

1. Calculate how many triplet codons would be possible had evolution seized on six bases (three complementary base pairs) rather than four bases within the structure of DNA. Would six bases accommodate a two-letter code, assuming 20 amino acids and start-and-stop codons?

Solution: Six things taken three at a time will produce $(6)^3$ or 216 triplet codes. If the code was a doublet, there would be $(6)^2$ or 36 two-letter codes, more than enough to accommodate 20 amino acids and start–stop punctuation.

2. In a heteropolymer experiment using 1/2C:1/4A:1/4G, how many different triplets will occur in the synthetic RNA molecule? How frequently will the most frequent triplet occur?

Solution: There will be $(3)^3$ or 27 triplets produced. The most frequent will be CCC, present $(1/2)^3$ or 1/8 of the time.

3. In a regular copolymer experiment, where UUAC is repeated over and over, how many different triplets will occur in the synthetic RNA, and how many amino acids will occur in the polypeptide when this RNA is translated? Be sure to consult Figure 12–7.

Solution: The synthetic RNA will repeat four triplets—UUA, UAC, ACU, and CUU—over and over.

Because both UUA and CUU encode leucine, while ACU and UAC encode threonine and tyrosine, respectively, the polypeptides synthesized under the directions of such an RNA contain three amino acids in the repeating sequence leu-leu-thr-tyr.

4. Actinomycin D inhibits DNA-dependent RNA synthesis. This antibiotic is added to a bacterial culture where a specific protein is being monitored. Compared to a control culture, where no antibiotic is added, translation of the protein declines over a period of 20 minutes, until no further protein is made. Explain these results.

Solution: The mRNA, which is the basis for the translation of the protein, has a lifetime of about 20 minutes. When actinomycin D is added, transcription is inhibited and no new mRNAs are made. Those already present support the translation of the protein for up to 20 minutes.

Problems and Discussion Questions

1. Crick, Barnett, Brenner, and Watts-Tobin, in their studies of frameshift mutations, found that either 3 (+)'s or 3 (−)'s restored the correct reading frame. If the code were a sextuplet (consisting of six nucleotides), would the reading frame be restored by either of these combinations?

2. In a mixed copolymer experiment using polynucleotide phosphorylase, 3/4 G:1/4 C is added to form the synthetic message. The amino acid composition of the ensuing protein is determined:

Glycine	36/64	(56%)
Alanine	12/64	(19%)
Arginine	12/64	(19%)
Proline	4/64	(6%)

From this information:

(a) Indicate the percentage (or fraction) of the time each possible triplet will occur in the message.

(b) Determine a complete set of base composition assignments for all amino acids present.

(c) Considering the wobble hypothesis, predict as many specific triplet assignments as possible.

3. In a mixed copolymer experiment, messengers are created with either 4/5 C:1/5 A or 4/5 A:1/5 C. These messages yield proteins with the following amino acid compositions. Using these data, predict the most specific coding composition for each amino acid.

4/5C:1/5A		4/5A:1/5C	
Proline	63.0%	Proline	3.5%
Histidine	13.0%	Histidine	3.0%
Threonine	16.0%	Threonine	16.6%
Glutamine	3.0%	Glutamine	13.0%
Asparagine	3.0%	Asparagine	13.0%
Lysine	0.5%	Lysine	50.0%
	98.5%		99.1%

4. In a coding experiment using repeating copolymers (as shown in Table 12.3), the following data are obtained:

Copolymer	Codons Produced	Amino Acids in Polypeptide
AG	AGA, GAG	Arg, Glu
AAG	AGA, AAG, GAA	Lys, Arg, Glu

AGG is known to code for arginine. Taking into account the wobble hypothesis, assign each of the four remaining different triplet codes to its correct amino acid.

5. In the triplet binding technique, radioactivity remains on the filter when the amino acid corresponding to the triplet is labeled. Explain the basis of this technique.

6. In studies of the amino acid sequence of wild-type and mutant forms of tryptophan synthetase in *E. coli*, the following changes have been observed:

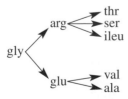

Determine a set of triplet codes in which a single nucleotide change produces each amino acid change.

7. Why doesn't polynucleotide phosphorylase synthesize RNA *in vivo*?

8. Refer to Table 12.1. Can you hypothesize why a mixture of poly U + poly A would not stimulate incorporation of ^{14}C-phenylalanine into protein?

9. Predict the amino acid sequence produced during translation by the following short theoretical mRNA sequences. Note that the second sequence was formed from the first by a deletion of only one nucleotide.

Sequence 1:	AUGCCGGAUUAUAGUUGA
Sequence 2:	AUGCCGGAUUAAGUUGA

What type of mutation gave rise to sequence 2?

10. A glycine residue exists at position −10 of the tryptophan synthetase enzyme of wild-type *E. coli*. If the codon specifying glycine is GGA, how many single-base substitutions will result in an amino acid substitution at position 210? What are they? How many will result if the wild-type codon is GGU?

11. Most proteins have more leucine than histidine residues, but more histidine than tryptophan residues. Correlate the number of codons for these three amino acids with this information.

12. Define the process of transcription. Where does this process fit into the central dogma of molecular genetics?

13. What was the initial evidence for the existence of mRNA?

14. Describe the structure of RNA polymerase in bacteria. What is the core enzyme? What is the role of the σ subunit?

15. In a written paragraph, describe the abbreviated chemical reactions that summarize RNA polymerase-directed transcription.

16. The following represent deoxyribonucleotide sequences derived from the template strand of DNA:

Sequence 1:	CTTTTTTGCCAT
Sequence 2:	ACATCAATAACT
Sequence 3:	TACAAGGGTTCT

(a) For each strand, determine the mRNA sequence that would be derived from transcription.

(b) Using Figure 12–7, determine the amino acid sequence that is encoded by these mRNAs.

(c) For sequence 1, what is the sequence of the partner strand?

17. Shown below are the amino acid sequences of the wild-type and three mutant forms of a short protein. Use this information to answer the following questions.

Wild type:	met-trp-tyr-arg-gly-ser-pro-thr
Mutant 1:	met-trp
Mutant 2:	met-trp-his-arg-gly-ser-pro-thr
Mutant 3:	met-cys-ile-val-val-val-gln-his

(a) Using Figure 12–7, predict the type of mutation that occurred leading to each altered protein.

(b) For each mutant protein, determine the specific ribonucleotide change that led to its synthesis.

(c) The wild-type RNA consists of nine triplets. What is the role of the ninth triplet?

(d) For the first eight wild-type triplets, which, if any, can you determine specifically from an analysis of the mutant proteins? Explain why or why not in each case.

(e) Another mutation (mutant 4) is isolated. Its amino acid sequence is unchanged, but mutant cells produce abnormally low amounts of the wild-type proteins. As specifically as you can, predict where in the gene this mutation exists.

Selected Readings

Barralle, F. E. 1983. The functional significance of leader and trailer sequences in eukaryotic mRNAs. *Int. Rev. Cytol.* 81:71–106.

Barrell, B. G., Air, G., and Hutchinson, C. 1976. Overlapping genes in bacteriophage φX174. *Nature* 264:34–40.

Barrell, B. G., Banker, A. T., and Drouin, J. 1979. A different genetic code in human mitochondria. *Nature* 282:189–94.

Birnstiel, M., Busslinger, M., and Strub, K. 1985. Transcription termination and 3′ processing: The end is in site. *Cell* 41:349–59.

Bonitz, S. G., et al. 1980. Codon recognition rules in yeast mitochondria. *Proc. Natl. Acad. Sci. USA* 77:3167–70.

Breitbart, R. E., and Nadal-Ginard, B. 1987. Developmentally induced, muscle-specific trans factors control the differential splicing of alternative and constitutive troponin T-exons. *Cell* 49:793–803.

Brenner, S. 1989. *Molecular biology: A selection of papers.* Orlando, FL: Academic Press.

Brenner, S., Jacob, F., and Meselson, M. 1961. An unstable intermediate carrying information from genes to ribosomes for protein synthesis. *Nature* 190:575–80.

Cattaneo, R. 1991. Different types of messenger RNA editing. *Annu. Rev. Genet.* 25:71–88.

Cech, T. R. 1986. RNA as an enzyme. *Sci. Am.* (Nov.) 255(5):64–75.

—————. 1987. The chemistry of self-splicing RNA and RNA enzymes. *Science* 236:1532–39.

Cech, T. R., Zaug, A. J., and Grabowski, P. J. 1981. *In vitro* splicing of the ribosomal RNA precursor of *Tetrahymena.* Involvement of a guanosine nucleotide in the excision of the intervening sequence. *Cell* 27:487–96.

Chambon, P. 1975. Eucaryotic nuclear RNA polymerases. *Annu. Rev. Biochem.* 44:613–38.

—————. 1981. Split genes. *Sci. Am.* (May) 244:60–71.

Cold Spring Harbor Laboratory. 1966. The genetic code. *Cold Spring Harbor Symp. Quant. Biol.* Vol. 31.

Crick, F. H. C. 1962. The genetic code. *Sci. Am.* (Oct.) 207:66–77.

—————. 1966a. The genetic code: III. *Sci. Am.* (Oct.) 215:55–63.

—————. 1966b. Codon-anticodon pairing: The wobble hypothesis. *J. Mol. Biol.* 19:548–55.

Crick, F. H. C., Barnett, L., Brenner, S., and Watts-Tobin, R. J. 1961. General nature of the genetic code for proteins. *Nature* 192:1227–32.

Darnell, J. E. 1983. The processing of RNA. *Sci. Am.* (Oct.) 249:90–100.

Dickerson, R. E. 1983. The DNA helix and how it is read. *Sci. Am.* (Dec.) 249:94–111.

Dugaiczk, A., et al. 1978. The natural ovalbumin gene contains seven intervening sequences. *Nature* 274:328–33.

Fiers, W., et al. 1976. Complete nucleotide sequence of bacteriophage MS2 RNA: Primary and secondary structure of the replicase gene. *Nature* 260:500–507.

Guthrie, C., and Patterson, B. 1988. Spliceosomal snRNAs. *Annu. Rev. Genet.* 22:387–419.

Hodges, P., and Scott, J. 1992. Apolipoprotein B mRNA editing: A new tier for the control of gene expression. *Trends Biochem. Sci.* 17:77–81.

Horton, H. R., et al. 2002. *Principles of biochemistry,* 3nd ed. Upper Saddle River, NJ: Prentice Hall.

Humphrey, T., and Proudfoot, N. J. 1988. A beginning to the biochemistry of polyadenylation. *Trends Genet.* 4:243–45.

Judson, H. F. 1979. *The eighth day of creation.* New York: Simon & Schuster.

Kable, M. L., et al. 1996. RNA editing: A mechanism for gRNA-specified uridylate insertion into precursor mRNA. *Science* 273:1189–95.

Khorana, H. G. 1967. Polynucleotide synthesis and the genetic code. *Harvey Lectures* 62:79–105.

Lodish, H., et al. 2000. Molecular cell biology. New York: W. H. Freeman.

Maniatis, T., and Reed, R. 1987. The role of small nuclear ribonucleoprotein particles in pre-mRNA splicing. *Nature* 325:673–78.

Nikolov, D. B., and Burley, S. K. 1997. RNA polymerase II transcription initiation: A structural view. *Proc. Natl. Acad. Sci. USA* 94:15–22.

Nilsen, T. W. 1994. RNA–RNA interactions in the spliceosome: Unraveling the ties that bind. *Cell* 78:1–4.

Nirenberg, M. W. 1963. The genetic code: II. *Sci. Am.* (March) 190:80–94.

Nirenberg, M. W., and Leder, P. 1964. RNA codewords and protein synthesis. *Science* 145:1399–1407.

Nirenberg, M. W., and Matthaei, H. 1961. The dependence of cell-free protein synthesis in *E. coli* upon naturally occurring or synthetic polyribosomes. *Proc. Natl. Acad. Sci. USA* 47:1588–1602.

O'Malley, B., et al. 1979. A comparison of the sequence organization of the chicken ovalbumin and ovomucoid genes. In *Eucaryotic gene regulation,* ed. R. Axel et al., pp. 281–99. Orlando, FL: Academic Press.

Padgett, R. A., et al. 1986. Splicing of messenger RNA precursors. *Annu. Rev. Biochem.* 55:1119–50.

Reed, R., and Maniatis, T. 1985. Intron sequences involved in lariat formation during pre-mRNA splicing. *Cell* 41:95–105.

Sharp, P. A. 1987. Splicing of messenger RNA precursors. *Science* 235:766–71.

—————. 1994. Nobel Lecture: Split genes and RNA splicing. *Cell* 77:805–15.

Sharp, P. A., and Eisenberg, D. 1987. The evolution of catalytic function. *Science* 238:729–30.

Steitz, J. A. 1988. Snurps. *Sci. Am.* (June) 258(6):56–63.

Volkin, E., Astrachan, L., and Countryman, J. L. 1958. Metabolism of RNA phosphorus in *E. coli* infected with bacteriophage T7. *Virology* 6:545–55.

Watson, J. D. 1963. Involvement of RNA in the synthesis of proteins. *Science* 140:17–26.

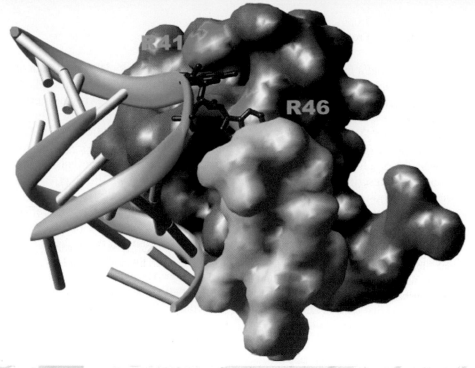

Model based on crystal structure analysis of the initiator factor IF1 bound to the 30S ribosomal subunit. *(Carter, et al. 2001. Science 291:500. From Figure 2D.)*

13
Translation and Proteins

CHAPTER CONCEPTS

Once genetic information has been transferred from DNA to RNA in the form of messenger RNA, it is ready to be translated into a polypeptide chain, the end product of most gene expression. This occurs during the process of translation, which occurs in association with the ribosome, the workbench of protein synthesis. Polypeptide chains achieve a three-dimensional conformation based on their primary amino acid sequences, often interacting with other such chains to create functional protein molecules. The function of any protein is closely tied to its structure, which can be disrupted by mutation, leading to a distinct phenotypic effect.

In Chapter 12, we established that a genetic code stores information in the form of triplet nucleotides in DNA and that this information is initially expressed through the process of transcription into a messenger RNA that is complementary to one strand of the DNA helix. However, the final product of gene expression, in almost all instances, is a polypeptide chain consisting of a linear series of amino acids whose sequence has been prescribed by the genetic code. In this chapter we examine how the information present in mRNA is processed to create polypeptides, which then fold into protein molecules. We also review the evidence that confirms that proteins are the end products of genes and briefly discuss the various levels of protein structure, diversity, and function. This information extends our understanding of gene expression and provides an important foundation for interpreting how the mutations that arise in DNA can result in the diverse phenotypic effects observed in organisms.

13.1 Translation: Components Essential to Protein Synthesis

Translation of mRNA is the biological polymerization of amino acids into polypeptide chains. This process, alluded to in our discussion in Chapter 12 of the genetic code, occurs only in association with ribosomes, which serve as nonspecific workbenches. The central question in translation is how triplet ribonucleotides of mRNA direct specific amino acids into their correct position in the polypeptide. This question was answered once **transfer RNA (tRNA)** was discovered. This class of molecules adapts specific triplet codons in mRNA to their correct amino acids. This adaptor role of tRNA was postulated by Francis Crick in 1957.

In association with a ribosome, mRNA presents a triplet codon that calls for a specific amino acid. A specific tRNA molecule contains within its nucleotide sequence three consecutive ribonucleotides complementary to the codon, called the **anticodon**, which can base-pair with the codon. Another region of this tRNA is covalently bonded to its corresponding amino acid.

Inside the ribosome, hydrogen bonding of tRNAs to mRNA holds the amino acids in proximity so that a peptide bond can be formed. As this process occurs over and over as mRNA runs through the ribosome, amino acids are polymerized into a polypeptide. Before we discuss the actual process of translation, let's first consider the structures of the ribosome and tRNA.

Ribosomal Structure

Because of its essential role in the expression of genetic information, the **ribosome** has been extensively analyzed. One bacterial cell contains about 10,000 of these structures, and a eukaryotic cell contains many times more. Electron microscopy reveals that the bacterial ribosome is about 250 μm at its largest diameter and consists of two subunits, one large and one small. Both subunits consist of one or more molecules of rRNA and an array of **ribosomal proteins**. When the

two subunits are associated with each other in a single ribosome, the structure is sometimes called a **monosome**.

The specific differences between prokaryotic and eukaryotic ribosomes are summarized in Figure 13–1. The subunit and rRNA components are most easily isolated and characterized on the basis of their sedimentation behavior in sucrose gradients (their rate of migration, abbreviated as S, as introduced in Chapter 10). In prokaryotes the monosome is a 70S particle, and in eukaryotes it is approximately 80S. Sedimentation coefficients, which reflect the variable rate of migration of different-size particles and molecules, are not additive. For example, the prokaryotic 70S monosome consists of a 50S and a 30S subunit, and the eukaryotic 80S monosome consists of a 60S and a 40S subunit.

The larger subunit in prokaryotes consists of a 23S RNA molecule, a 5S rRNA molecule, and 31 ribosomal proteins. In the eukaryotic equivalent, a 28S rRNA molecule is accompanied by a 5.8S and 5S rRNA molecule and about 50 proteins. The smaller prokaryotic subunits consist of a 16S rRNA component and 21 proteins. In the eukaryotic equivalent, an 18S rRNA component and about 33 proteins are found. The approximate molecular weights (MW) and number of nucleotides of these components are shown in Figure 13–1.

Regarding the components of the ribosome, it is now clear that the RNA molecules perform the all important catalytic functions associated with translation. The many proteins, whose functions were long a mystery, are thought to promote the binding of the various molecules involved in translation, and in general, to fine-tune the process. This conclusion is based on the observation that some of the catalytic functions in ribosomes still occur in experiments involving "ribosomal protein-depleted" ribosomes.

Molecular hybridization studies have established the degree of redundancy of the genes coding for the rRNA components. The *E. coli* genome contains seven copies of a single sequence that encodes all three components—23S, 16S, and 5S. The initial transcript of these genes produces a 30S RNA molecule that is enzymatically cleaved into these smaller components. Coupling of the genetic information encoding these three rRNA components ensures that, following multiple transcription events, equal quantities of all three will be present as ribosomes are assembled.

In eukaryotes, many more copies of a sequence encoding the 28S and 18S components are present. In *Drosophila*, approximately 120 copies per haploid genome are each transcribed into a molecule of about 34S. This is processed into the 28S, 18S, and 5.8S rRNA species. These are homologous to the three rRNA components of *E. coli*. In *X. laevis*, over 500 copies per haploid genome are present. In mammalian cells, the initial transcript is 45S. The rRNA genes, called **rDNA**, are part of the moderately repetitive DNA fraction and are present in clusters at various chromosomal sites.

Each cluster in eukaryotes consists of **tandem repeats**, with each unit separated by a noncoding **spacer DNA** sequence. In humans, these gene clusters have been localized near the ends of chromosomes 13, 14, 15, 21, and 22. The unique 5S rRNA component of eukaryotes is not part of this larger transcript. Instead, genes coding for this ribosomal component are dis-

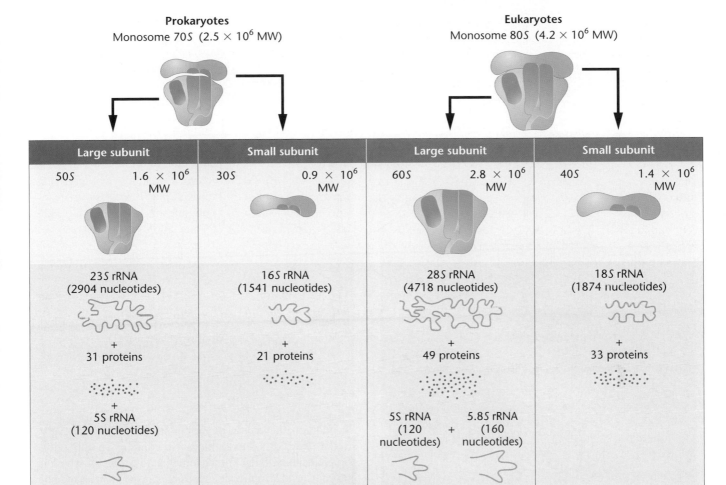

FIGURE 13–1 A comparison of the components in the prokaryotic and eukaryotic ribosome.

tinct and located separately. In humans, a gene cluster encoding them has been located on chromosome 1.

Despite the detailed knowledge available on the structure and genetic origin of the ribosomal components, a complete understanding of the function of these components has eluded geneticists. This is not surprising; the ribosome is perhaps the most intricate of all cellular structures. In bacteria, the monosome has a combined molecular weight of 2.5 million Da!

tRNA Structure

Because of their small size and stability in the cell, transfer RNAs (tRNAs) have been investigated extensively and are the best-characterized RNA molecules. They are composed of only 75–90 nucleotides, displaying a nearly identical structure in bacteria and eukaryotes. In both types of organisms, tRNAs are transcribed as larger precursors, which are cleaved into mature 4S tRNA molecules. In *E. coli*, for example, tRNA^{tyr} (the superscript identifies the specific tRNA and the cognate amino acid that binds to it) is composed of 77 nucleotides, yet its precursor contains 126 nucleotides.

In 1965, Robert Holley and his colleagues reported the complete sequence of tRNA^{ala} isolated from yeast. Of great interest was the finding that a number of nucleotides are

unique to tRNA. As shown in Figure 13–2 (page 255), each is a modification of one of the four nitrogenous bases expected in RNA (G, C, A, and U). These include inosinic acid, which contains the purine hypoxanthine, ribothymidylic acid, and pseudouridine, among others. These modified structures, variously referred to as *unusual*, *rare*, or *odd bases*, are created *following* transcription, illustrating the more general concept of posttranscriptional modification. In this case, the unmodified base is inserted during transcription and, subsequently, enzymatic reactions catalyze the chemical modifications to the base.

Holley's sequence analysis led him to propose the two-dimensional **cloverleaf model of tRNA**. It was known that tRNA demonstrates a secondary structure due to base pairing. Holley discovered that he could arrange the linear model in such a way that several stretches of base pairing would result. This arrangement created a series of paired stems and unpaired loops resembling the shape of a cloverleaf. Loops consistently contained modified bases that do not generally form base pairs. Holley's model is shown in Figure 13–3 (page 255).

Because the triplets GCU, GCC, and GCA specify alanine, Holley looked for an anticodon sequence complementary to one of these codons in his tRNA^{ala} molecule. He found it in the form of CGI (the 3′-to-5′ direction), in one loop of the cloverleaf. The nitrogenous base I (inosinic acid) can form

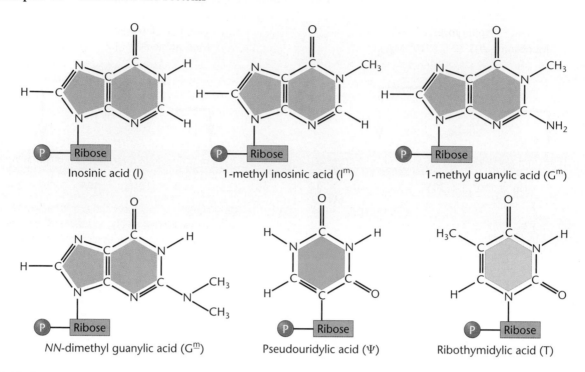

FIGURE 13–2 Unusual nitrogenous bases found in transfer RNA.

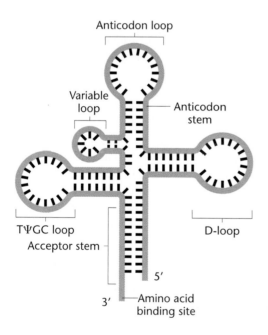

FIGURE 13–3 Holley's two-dimensional cloverleaf model of transfer RNA.

In addition, the lengths of various stems and loops are very similar. Each tRNA examined also contains an anticodon complementary to the known amino acid codon for which it is specific, and all anticodon loops are present in the same position of the cloverleaf.

Because the cloverleaf model was predicted strictly on the basis of nucleotide sequence, there was great interest in the

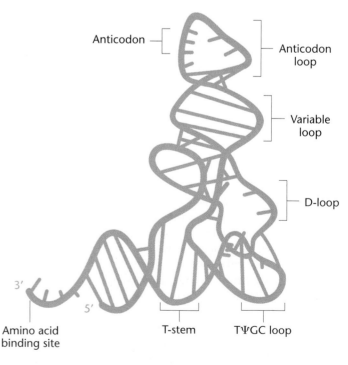

FIGURE 13–4 A three-dimensional model of transfer RNA.

hydrogen bonds with U, C, or A, the third members of the triplets. Thus, the **anticodon loop** was established.

Studies of other tRNA species reveal many constant features. First, at the 3′-end, all tRNAs contain the sequence **...pCpCpA-3′**. It is at this end of the molecule where the amino acid is covalently joined to the terminal adenosine residue. All tRNAs contain **5′-Gp...** at the other end of the molecule.

X-ray crystallographic examination of tRNA, which reveals a three-dimensional structure. By 1974, Alexander Rich and his colleagues in the United States, and J. Roberts, B. Clark, Aaron Klug, and their colleagues in England, had succeeded in crystallizing tRNA and performing X-ray crystallography at a resolution of 3 Å. At such resolution, the pattern formed by individual nucleotides is discernible.

As a result of these studies, a complete three-dimensional model of tRNA is now available (Figure 13–4). At one end of the molecule is the anticodon loop and stem, and at the other end is the 3'-acceptor region where the amino acid is bound. It has been speculated that the shapes of the intervening loops may be recognized by the specific enzymes responsible for adding the amino acid to tRNA, a subject to which we now turn our attention.

Charging tRNA

Before translation can proceed, the tRNA molecules must be chemically linked to their respective amino acids. This activation process, called **charging**, occurs under the direction of enzymes called **aminoacyl tRNA synthetases**. Because there are 20 different amino acids, there must be at least 20 different tRNA molecules and as many different enzymes. In theory, because there are 61 triplet codes, there could be the same number of specific tRNAs and enzymes. However, because of the ability of the third member of a triplet code to "wobble," it is now thought that there are at least 32 different tRNAs; it is also believed that there are only 20 synthetases, one for each amino acid, regardless of the greater number of corresponding tRNAs.

The charging process is outlined in Figure 13–5. In the initial step, the amino acid is converted to an activated form, reacting with ATP to create an **aminoacyladenylic acid**. A covalent linkage is formed between the 5'-phosphate group of ATP and the carboxyl end of the amino acid. This molecule remains associated with the synthetase enzyme, forming a complex that then reacts with a specific tRNA molecule. In this next step, the amino acid is transferred to the appropriate tRNA and bonded covalently to the adenine residue at the 3'-end. The charged tRNA may participate directly in protein synthesis. Aminoacyl tRNA synthetases are highly specific enzymes because they recognize only one amino acid and only a subset of corresponding tRNAs called **isoaccepting tRNAs**. This is a crucial point if fidelity of translation is to be maintained.

13.2 Translation: The Process

In a way similar to transcription, the process of translation can be best described by breaking it into discrete phases. We will consider three, each with its own set of illustrations (Figures 13–6, 13–7, and 13–8), but keep in mind that translation is a dynamic, continuous process. Correlate the following discussion with the step-by-step characterization in these figures. Many of the protein factors and their roles in translation are summarized in Table 13.1.

Initiation

Initiation of translation is depicted in Figure 13–6. Recall that the ribosome serves as a nonspecific workbench for the translation process. Most ribosomes, when they are not involved in translation, are dissociated into their large and small subunits. Initiation of translation in *E. coli* involves the small ribosome subunit, an mRNA molecule, a specific charged initiator tRNA, GTP, Mg^{++}, and at least three proteinaceous initiation factors (IFs) that enhance the binding affinity of the various translational components. In prokaryotes, the initiation codon of mRNA—AUG—calls for the modified amino acid **formylmethionine (fmet)**.

The small ribosomal subunit binds to several initiation factors, and this complex in turn binds to mRNA (step 1). In bacteria, this binding involves a sequence of up to six ribonucleotides (AGGAGG, not shown), which *precedes* the initial AUG start codon of mRNA. This sequence (containing only purines and called the **Shine–Dalgarno sequence**) base-pairs with a region of the 16S rRNA of the small ribosomal subunit, facilitating initiation.

Another initiation protein then enhances the binding of charged formylmethionyl–tRNA to the small subunit in

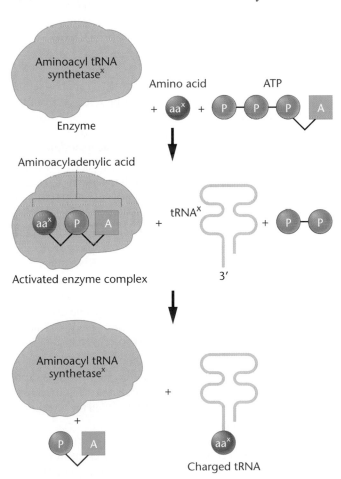

FIGURE 13–5 Steps involved in charging tRNA. The "x" denotes that for each amino acid only the corresponding specific tRNA and specific aminoacyl tRNA synthetase enzyme are involved in the charging process.

Translation components

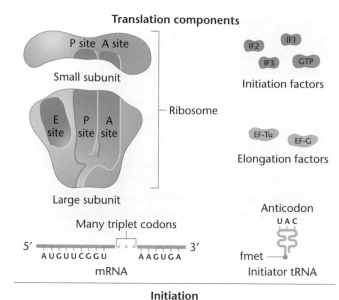

5′ ┬┬┬┬┬┬┬┬ ─── ┬┬┬┬┬┬ 3′
   A U G U U C G G U   A A G U G A

Many triplet codons

mRNA

Anticodon
U A C

fmet ──

Initiator tRNA

Initiation

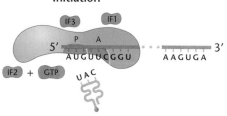

1. mRNA binds to small subunit along with initiation factors (IF1, 2, 3)

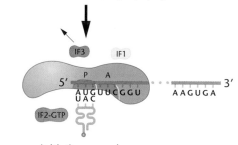

Initiation complex

2. Initiator tRNA^fmet binds to mRNA codon in P site; IF3 released

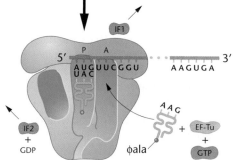

3. Large subunit binds to complex; IF1 and IF2 released; EF-Tu binds to tRNA, facilitating entry into A site

FIGURE 13–6 Initiation of translation. The components are depicted at the top of the figure.

response to the AUG triplet (step 2). This step "sets" the reading frame so that all subsequent groups of three ribonucleotides are translated accurately. This aggregate represents the **initiation complex**, which then combines with the large ribosomal subunit. In this process, a molecule of GTP is hydrolyzed, providing the required energy, and the initiation factors are released (step 3).

Elongation

The second phase of translation, elongation, is depicted in Figure 13–7. Once both subunits of the ribosome are assembled with the mRNA, binding sites for two charged tRNA molecules are formed. These are designated as the **P**, or **peptidyl**, and the **A**, or **aminoacyl**, **sites**. The charged initiator tRNA binds to the P site, provided that the AUG triplet of mRNA is in the corresponding position of the small subunit.

The increase of the growing polypeptide chain by one amino acid is called **elongation**. The sequence of the second triplet in mRNA dictates which charged tRNA molecule will become positioned at the A site (step 1). Once it is present, **peptidyl transferase** catalyzes the formation of the peptide bond, which links the two amino acids (step 2). This enzyme is part of the large subunit of the ribosome. At the same time, the covalent bond between the amino acid and the tRNA occupying the P site is hydrolyzed (broken). The product of this reaction is a dipeptide, which is attached to the 3′-end of tRNA still residing in the A site.

Before elongation can be repeated, the tRNA attached to the P site, which is now uncharged, must be released from the large subunit. The uncharged tRNA moves transiently through a third site on the ribosome called the **E site** (E stands for exit). The entire **mRNA–tRNA–aa₂–aa₁ complex** then shifts in the direction of the P site by a distance of three nucleotides (step 3). This event requires several protein elongation factors (EFs) as well as the energy derived from hydrolysis of GTP. The result is that the third triplet of mRNA is now in a position to accept another specific charged tRNA into the A site (step 4). One simple way to distinguish the two sites is to remember that, *following the shift*, the P site contains a tRNA attached to a peptide chain (P for peptide), whereas the A site contains a tRNA with an amino acid attached (A for amino acid).

The sequence of elongation is repeated over and over (steps 5 and 6). An additional amino acid is added to the growing polypeptide chain each time the mRNA advances through the ribosome. Once a polypeptide chain of reasonable size is assembled (about 30 amino acids), it begins to emerge from the base of the large subunit, as illustrated in step 6. A tunnel exists within the large subunit, through which the elongating polypeptide emerges.

As we have seen, the role of the small subunit during elongation is one of "decoding" the triplets present in mRNA, while that of the large subunit is peptide bond synthesis. The efficiency of the process is remarkably high; the observed error rate is only about 10^{-4}. An incorrect amino acid will occur only once in every 20 polypeptides of an average length of 500 amino acids! In *E. coli*, elongation occurs at a rate of about 15 amino acids per second at 37°C.

Elongation

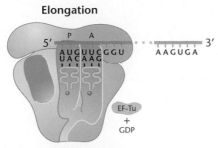

1. Second charged tRNA has entered A site, facilitated by EF-Tu; first elongation step commences

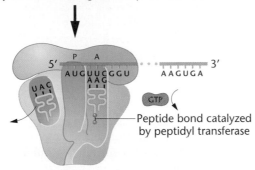

Peptide bond catalyzed by peptidyl transferase

2. Dipeptide bond forms; uncharged tRNA moves to E-site and then out of ribosome

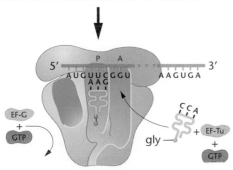

3. mRNA has shifted by 3 bases; EF-G facilitates the translocation step; first elongation step completed

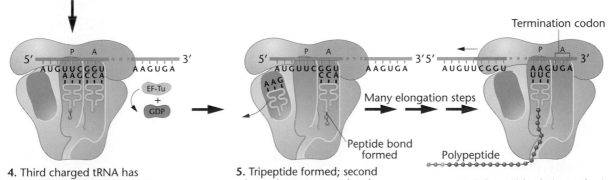

4. Third charged tRNA has entered A site, facilitated by EF-Tu; second elongation step begins

5. Tripeptide formed; second elongation step completed; uncharged tRNA moves to E site

Peptide bond formed

Many elongation steps

Termination codon

Polypeptide

6. Polypeptide chain synthesized and exiting ribosome

MEDIA TUTORIAL Translation

FIGURE 13–7 Elongation of the growing polypeptide chain during translation.

TABLE 13.1   Various Protein Factors Involved During Translation in *E. coli*

Process	*Factor*	*Role*
Initiation of translation	IF1	Stabilizes 30*S* subunit
	IF2	Binds fmet–tRNA to 30*S*–mRNA complex; binds to GTP and stimulates hydrolysis
	IF3	Binds 30*S* subunit to mRNA
Elongation of polypeptide	EF-Tu	Binds GTP; brings aminoacyl–tRNA to the A site of ribosome
	EF-Ts	Generates active EF-Tu
	EF-G	Stimulates translocation; GTP-dependent
Termination of translation and release of polypeptide	RF1	Catalyzes release of the polypeptide chain from tRNA and dissociation of the translocation complex; specific for UAA and UAG termination codons
	RF2	Behaves like RF1; specific for UGA and UAA codons
	RF3	Stimulates RF1 and RF2

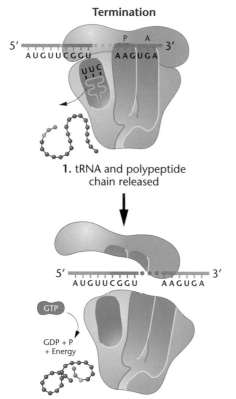

Termination

1. tRNA and polypeptide chain released

2. GTP-dependent termination factors activated; components separate; polypeptide folds into protein

FIGURE 13–8 Termination of the process of translation.

Termination

Termination, the third phase of translation, is depicted in Figure 13–8. Termination of protein synthesis is signaled by one or more of three triplet codes in the A site: UAG, UAA, or UGA. These codons do not specify an amino acid, nor do they call for a tRNA in the A site. These codons are called **stop codons**, **termination codons**, or **nonsense codons**. The finished polypeptide is therefore still attached to the terminal tRNA at the P site, and the A site is empty. The termination codon signals the action of **GTP-dependent release factors**, which cleave the polypeptide chain from the terminal tRNA, releasing it from the translation complex (step 1).

Once this cleavage occurs, the tRNA is released from the ribosome, which then dissociates into its subunits (step 2). If a termination codon should appear in the middle of an mRNA molecule as a result of mutation, the same process occurs, and the polypeptide chain is prematurely terminated.

Polyribosomes

As elongation proceeds and the initial portion of mRNA has passed through the ribosome, this mRNA is free to associate with another small subunit to form a second initiation complex. This process can be repeated several times with a single mRNA and results in what are called **polyribosomes** or just **polysomes**.

Polyribosomes can be isolated and analyzed following a gentle lysis of cells. In Figure 13–9(a) and (b), these complexes are seen under the electron microscope. In Figure 13–9(a), you can see mRNA (the thin line) between the individual ribosomes. The micrograph in Figure 13–9(b) is even more remarkable, for it shows the polypeptide chains emerging from the ribosomes during translation. The formation of polysome complexes represents an efficient use of the components available for protein synthesis during a unit of time. Using the analogy of a tape and tape recorder, in polysome complexes one tape (mRNA) would be played simultaneously by several recorders (the ribosomes), but at any given moment, each recording (the polypeptide being synthesized in each ribosome) would be at a different point of completion.

13.3   Translation in Eukaryotes

The general features of the model we just discussed were initially derived from investigations of the translation process in bacteria. As we saw, one main difference between translation in prokaryotes and eukaryotes is that in the latter, translation occurs on ribosomes that are larger and whose rRNA and protein components are more complex than those of prokaryotes (see Figure 13–1).

Several other differences are also important. Eukaryotic mRNAs are much longer-lived than their prokaryotic counterparts. Most exist for hours rather than minutes prior to

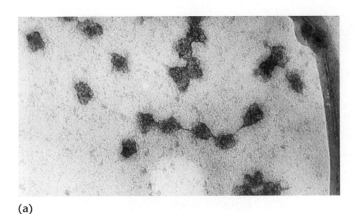

(a)

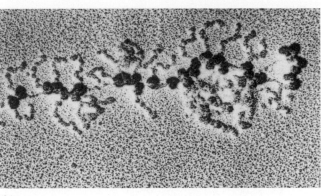

(b)

FIGURE 13–9 Polyribosomes visualized under the electron microscope. Those in (a) were derived from rabbit reticulocytes engaged in the translation of hemoglobin mRNA and in (b) from giant salivary gland cells of the midgefly, *Chironomus thummi*. In (b), the nascent polypeptide chain is apparent as it emerges from each ribosome. Its length increases as translation proceeds from left (5′) to right (3′) along the mRNA. *[a] from Rich, et al., 1963. Reproduced by permission of the Cold Spring Harbor Laboratory Press Symp. 38 (1963) Fig. 4D, p. 273, © 1964; [b] E. V. Kiseleva)*

their degradation by nucleases in the cell, remaining available much longer to orchestrate protein synthesis.

Several aspects that involve the initiation of translation are different in eukaryotes. First, as we discussed in our consideration of mRNA maturation, the 5′-end is "capped" with a 7-methylguanosine residue. The presence of this "cap," absent in prokaryotes, is essential to efficient translation, as RNAs lacking the cap are translated poorly. In addition, most eukaryotic mRNAs contain a short recognition sequence that surrounds the initiating AUG codon—5′-ACCAUGG. Named after Marilyn Kozak, who discovered it, this Kozak sequence appears to function during initiation in the same way that the Shine–Dalgarno sequence functions in prokaryotic mRNA. Both greatly facilitate the initial binding of mRNA to the small subunit of the ribosome.

Another difference is that the amino acid formylmethionine is not required to initiate eukaryotic translation. However, as in prokaryotes, the AUG triplet, which encodes methionine, is essential to the formation of the translational complex, and a unique transfer RNA (tRNA$_i^{met}$) is used during initiation.

Protein factors similar to those in prokaryotes guide the initiation, elongation, and termination of translation in eukaryotes. Many of these eukaryotic factors are clearly homologous to their counterparts in prokaryotes. However, a greater number of factors are usually required during each step, and some are more complex than in prokaryotes.

Finally, recall that in eukaryotes a large proportion of the ribosomes are found in association with the membranes that make up the endoplasmic reticulum (forming what is referred to as rough ER). Such membranes are absent from the cytoplasm of prokaryotic cells. This association in eukaryotes facilitates the secretion of newly synthesized proteins from the ribosomes directly into the channels of the endoplasmic reticulum. Recent studies using cryo-electron microscopy have established how this occurs. A tunnel in the large subunit of ribosomes begins near the point where the two subunits in-

terface and exits near the back of the large subunit. The location of this tunnel within the large subunit is the basis for the belief that it provides the conduit for the movement of the newly synthesized polypeptide chain out of the ribosome. In studies in yeast, newly synthesized polypeptides enter the ER through a membrane channel formed by a specific protein, Sec61. This channel is perfectly aligned with the exit point of the ribosomal tunnel. In prokaryotes, the polypeptides are released by the ribosome directly into the cytoplasm.

13.4 Proteins, Heredity, and Metabolism

Now let's consider how we know that proteins are the end products of genetic expression. The first insight into the role of proteins in genetic processes was provided by observations made by Sir Archibald Garrod and William Bateson early in the twentieth century. Garrod was born into an English family of medical scientists. His father was a physician with a strong interest in the chemical basis of rheumatoid arthritis, and his eldest brother was a leading zoologist in London. It is not surprising, then, that as a practicing physician, Garrod became interested in several human disorders that seemed to be inherited. Although he also studied albinism and cystinuria, we shall describe his investigation of the disorder **alkaptonuria**. Individuals afflicted with this disorder cannot metabolize the alkapton 2,5-dihydroxyphenylacetic acid, also known as homogentisic acid. As a result, an important metabolic pathway (Figure 13–10, page 262) is blocked. Homogentisic acid accumulates in cells and tissues and is excreted in the urine. The molecule's oxidation products are black and easily detectable in the diapers of newborns. The products tend to accumulate in cartilaginous areas, causing a darkening of the ears and nose. In joints, this deposition leads to a benign arthritic condition. This rare disease is not serious, but it persists throughout an individual's life.

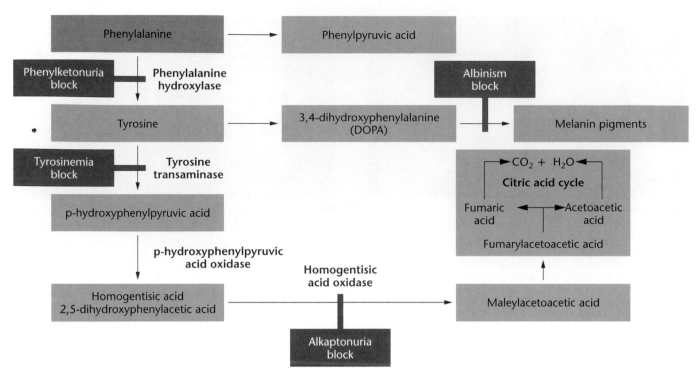

FIGURE 13–10 Metabolic pathway involving phenylalanine and tyrosine. Various metabolic blocks resulting from mutations lead to the disorders phenylketonuria, alkaptonuria, albinism, and tyrosinemia.

Garrod studied alkaptonuria by increasing dietary protein or adding to the diet the amino acids phenylalanine or tyrosine, both of which are chemically related to homogentisic acid. Under such condition, homogentisic acid levels increase in the urine of alkaptonurics but not in unaffected individuals. Garrod concluded that normal individuals can break down, or catabolize, this alkapton, but that afflicted individuals cannot. By studying the disorder's pattern of inheritance, Garrod further concluded that alkaptonuria is inherited as a simple recessive trait.

On the basis of these conclusions, Garrod hypothesized that hereditary information controls chemical reactions in the body, and that the inherited disorders he studied are the result of alternative modes of metabolism. While *genes* and *enzymes* were not familiar terms during Garrod's time, he used the corresponding concepts *unit factors* and *ferments*. Garrod published his initial observations in 1902.

Only a few geneticists, including Bateson, were familiar with or referred to Garrod's work. Garrod's ideas fit nicely with Bateson's belief that inherited conditions are caused by the lack of some critical substance. In 1909, Bateson published *Mendel's Principles of Heredity*, in which he linked ferments with heredity. However, for almost 30 years, most geneticists failed to see the relationship between genes and enzymes. Garrod and Bateson, like Mendel, were ahead of their time.

Phenylketonuria

The inherited human metabolic disorder, **phenylketonuria (PKU)**, results when another reaction in the pathway shown in Figure 13–10 is blocked. Described first in 1934, this dis-

order can result in mental retardation and is transmitted as an autosomal recessive disease. Afflicted individuals are unable to convert the amino acid phenylalanine to the amino acid tyrosine. These molecules differ by only a single hydroxyl group (OH), present in tyrosine, but absent in phenylalanine. The reaction is catalyzed by the enzyme **phenylalanine hydroxylase**, which is inactive in affected individuals and active at about a 30 percent level in heterozygotes. The enzyme functions in the liver. While the normal blood level of phenylalanine is about 1 mg/100 mL, phenylketonurics show a level as high as 50 mg/100 mL.

As phenylalanine accumulates, it can be converted to phenylpyruvic acid and, subsequently, to other derivatives. These are less efficiently resorbed by the kidney and tend to spill into the urine more quickly than phenylalanine. Both phenylalanine and its derivatives enter the cerebrospinal fluid, resulting in elevated levels in the brain. The presence of these substances during early development is thought to cause mental retardation.

As a result of early detection based on PKU screening of newborns, retardation can be prevented. When the condition is detected in the analysis of an infant's blood, a strict dietary regimen is instituted. A low-phenylalanine diet can reduce byproducts such as phenylpyruvic acid, and the abnormalities characterizing the disease can be diminished. The screening of newborns occurs routinely in all states of the United States. Phenylketonuria occurs in approximately 1 in 11,000 births.

Knowledge of inherited metabolic disorders such as alkaptonuria and phenylketonuria has caused a revolution in medical thinking and practice. Human disease, once thought to be attributed solely to the action of invading microorganisms, viruses, or parasites, clearly can have a genetic basis. We know now that literally thousands of medical conditions are caused by errors in metabolism that are the result of mutant genes. These human biochemical disorders include all classes of organic biomolecules.

13.5 The One-Gene:One-Enzyme Hypothesis

In two separate investigations beginning in 1933, George Beadle was to provide the first convincing experimental evidence that genes are directly responsible for the synthesis of enzymes. The first investigation, conducted in collaboration with Boris Ephrussi, involved *Drosophila* eye pigments. Together, they confirmed that mutant genes that alter the eye color of fruit flies can be linked to biochemical errors that, in all likelihood, involve the loss of enzyme function. Encouraged by these findings, Beadle then joined with Edward Tatum to investigate nutritional mutations in the pink bread mold *Neurospora crassa*. This investigation led to the **one-gene:one-enzyme hypothesis**.

Beadle and Tatum: *Neurospora* Mutants

In the early 1940s, Beadle and Tatum chose to work with *Neurospora* because much was known about its biochemistry, and mutations could be induced and isolated with relative ease. By inducing mutations, they produced strains that had genetic blocks of reactions essential to the growth of the organism.

Beadle and Tatum knew that this mold could manufacture nearly everything necessary for normal development. For example, using rudimentary carbon and nitrogen sources, this organism can synthesize 9 water-soluble vitamins, 20 amino acids, numerous carotenoid pigments, and all essential purines, and pyrimidines. Beadle and Tatum irradiated asexual conidia (spores) with X-rays to increase the frequency of mutations and allowed them to be grown on "complete" medium containing all the necessary growth factors (e.g., vitamins and amino acids, etc.). Under such growth conditions, a mutant strain unable to grow on minimal medium was able to grow by virtue of supplements present in the enriched complete medium. All the cultures were then transferred to minimal medium. If growth occurred on the minimal medium, the organisms were able to synthesize all the necessary growth factors themselves, and the researchers concluded that the culture did not contain a mutation. If no growth occurred, then they concluded that the culture contained a nutritional mutation, and the only task remaining was to determine its type. Both cases are shown in Figure 13–11(a).

Many thousands of individual spores derived by this procedure were isolated and grown on complete medium. In subsequent tests on minimal medium, many cultures failed to grow, indicating that a nutritional mutation had been induced. To identify the mutant type, the mutant strains were tested on a series of different minimal media [Figure 13–11(b) and (c)], each containing groups of supple-

ments, and subsequently on media containing single vitamins, amino acids, purines, or pyrimidines until one specific supplement that permitted growth was found. Beadle and Tatum reasoned that the supplement that restores growth is the molecule that the mutant strain cannot synthesize.

The first mutant strain isolated required vitamin B-6 (pyridoxine) in the medium, and the second one required vitamin B-1 (thiamine). Using the same procedure, Beadle and Tatum eventually isolated and studied hundreds of mutants deficient in the ability to synthesize other vitamins, amino acids, or other substances.

The findings derived from testing over 80,000 spores convinced Beadle and Tatum that genetics and biochemistry have much in common. It seemed likely that each nutritional mutation caused the loss of the enzymatic activity that facilitates an essential reaction in wild-type organisms. It also appeared that a mutation could be found for nearly any enzymatically controlled reaction. Beadle and Tatum had thus provided sound experimental evidence for the hypothesis that one gene specifies one enzyme, an idea alluded to over 30 years earlier by Garrod and Bateson. With modifications, this concept was to become a major principle of genetics.

Genes and Enzymes: Analysis of Biochemical Pathways

The one-gene:one-enzyme concept and its attendant methods have been used over the years to work out many details of metabolism in *Neurospora*, *Escherichia coli*, and a number of other microorganisms. One of the first metabolic pathways to be investigated in detail was that leading to the synthesis of the amino acid arginine in *Neurospora*. By studying seven mutant strains, each requiring arginine for growth (arg^-), Adrian Srb and Norman Horowitz ascertained a partial biochemical pathway that leads to the synthesis of this molecule. Their work demonstrates how genetic analysis can be used to establish biochemical information.

Srb and Horowitz tested each mutant strain's ability to reestablish growth if either citrulline or ornithine, two compounds with close chemical similarity to arginine, was used as a supplement to minimal medium. If either was able to substitute for arginine, they reasoned that it must be involved in the biosynthetic pathway of arginine. They found that both molecules could be substituted in one or more strains.

Of the seven mutant strains, four of them (*arg 4–7*) grew if supplied with either citrulline, ornithine, or arginine. Two of them (*arg 2* and *arg 3*) grew if supplied with citrulline or arginine. One strain (*arg 1*) would grow only if arginine were supplied; neither citrulline nor ornithine could substitute for it. From these experimental observations, the following pathway and metabolic blocks for each mutation were deduced:

$$\text{Precursor} \xrightarrow[\text{Enzyme A}]{arg\ 4\text{--}7} \text{Ornithine} \xrightarrow[\text{Enzyme B}]{arg\ 2\text{--}3} \text{Citrulline} \xrightarrow[\text{Enzyme C}]{arg\ 1} \text{Arginine}$$

The reasoning supporting these conclusions is based on the following logic. If mutants *arg 4* through *arg 7* can grow regardless of which of the three molecules is supplied as a supplement to minimal medium, the mutations preventing

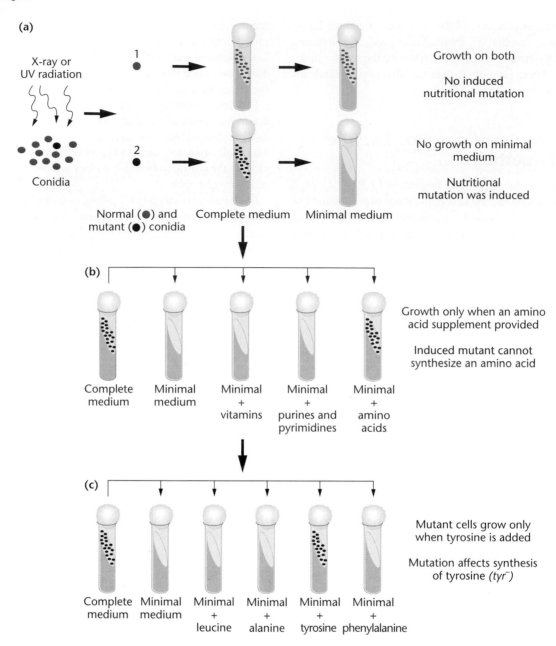

FIGURE 13–11 Induction, isolation, and characterization of a nutritional auxotrophic mutation in *Neurospora*. In (a), most conidia are not affected, but one conidium (shown in red) contains such a mutation. In (b) and (c), the precise nature of the mutation is established to involve the biosynthesis of tyrosine.

growth must cause a metabolic block that occurs *prior* to the involvement of ornithine, citrulline, or arginine in the pathway. When any one of these three molecules is added, its presence bypasses the block. As a result, both citrulline and ornithine appear to be involved in the biosynthesis of arginine. However, the sequence of their participation in the pathway cannot be determined on the basis of these data.

On the other hand, the *arg 2* and *3* mutations grow if supplied citrulline but not if they are supplied with only ornithine. Therefore, ornithine must occur in the pathway *prior* to the block. Its presence will not overcome the block. Citrulline, however, does overcome the block, so it must be

involved beyond the point of blockage. Therefore, the conversion of ornithine to citrulline represents the correct sequence in the pathway.

Finally, we can conclude that *arg 1* represents a mutation preventing the conversion of citrulline to arginine. Neither ornithine nor citrulline can overcome the metabolic block because both participate earlier in the pathway.

Taken together, these reasons support the sequence of biosynthesis outlined here. Since Srb and Horowitz's work in 1944, the detailed pathway has been worked out and the enzymes controlling each step characterized. The abbreviated metabolic pathway is shown in Figure 13–12.

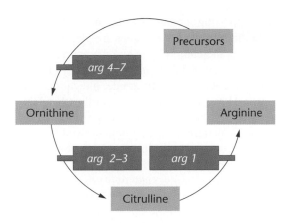

FIGURE 13–12 Abbreviated pathway resulting in the biosynthesis of arginine in *Neurospora*.

13.6 One-Gene:One-Polypeptide Chain

The concept of one-gene:one-enzyme developed in the early 1940s was not immediately accepted by all geneticists. This is not surprising because it was not yet clear how mutant enzymes could cause variation in many phenotypic traits. For example, *Drosophila* mutants demonstrate altered eye size, wing shape, wing vein pattern, and so on. Plants exhibit mutant varieties of seed texture, height, and fruit size. How an inactive, mutant enzyme could result in such phenotypes puzzled many geneticists.

Two factors soon modified the one-gene:one-enzyme hypothesis. First, while nearly all enzymes are proteins, not all proteins are enzymes. As the study of biochemical genetics proceeded, it became clear that all proteins are specified by the information stored in genes, leading to the more accurate phraseology **one-gene:one-protein**. Second, proteins often show a subunit structure consisting of two or more

polypeptide chains. This is the basis of the quaternary structure of proteins, which we discuss later in this chapter. Because each distinct polypeptide chain is encoded by a separate gene, a more modern statement of Beadle and Tatum's basic hypothesis is **one-gene:one-polypeptide chain**. These modifications of the original hypothesis became apparent during the analysis of hemoglobin structure in individuals afflicted with sickle-cell anemia.

Sickle-Cell Anemia

The first direct evidence that genes specify proteins other than enzymes came from the work on mutant hemoglobin molecules derived from humans afflicted with the disorder **sickle-cell anemia**. Affected individuals contain erythrocytes that, under low oxygen tension, become elongated and curved because of the polymerization of hemoglobin. The "sickle" shape of these erythrocytes is in contrast to the biconcave disc shape characteristic of those in normal individuals (Figure 13–13). Individuals with the disease suffer attacks when red blood cells aggregate in the venous side of capillary systems, where oxygen tension is very low. As a result, a variety of tissues are deprived of oxygen and suffer severe damage. When this occurs, an individual is said to experience a sickle-cell crisis. If untreated, a crisis can be fatal. The kidneys, muscles, joints, brain, gastrointestinal tract, and lungs can be affected.

In addition to suffering crises, these individuals are anemic because their erythrocytes are destroyed more rapidly than are normal red blood cells. Compensatory physiological mechanisms include increased red-cell production by bone marrow and accentuated heart action. These mechanisms lead to abnormal bone size and shape as well as dilation of the heart.

In 1949, James Neel and E. A. Beet demonstrated that the disease is inherited as a Mendelian trait. Pedigree analysis revealed three genotypes and phenotypes controlled by a single

(a)

(b)

FIGURE 13–13 A comparison of erythrocytes from normal individuals (a) and from individuals with sickle-cell anemia (b). *([a] Dennis Kunkel/Phototake NYC; [b] Francis Leroy/Biocosmos/ Science Photo Library/Photo Researchers, Inc.)*

pair of alleles, Hb^A and Hb^S. Normal and affected individuals result from the homozygous genotypes Hb^AHb^A and Hb^SHb^S, respectively. The red blood cells of heterozygotes, who exhibit the **sickle-cell trait** but not the disease, undergo much less sickling because over half of their hemoglobin is normal. Although largely unaffected, such heterozygotes are "carriers" of the defective gene, which is transmitted, on average, to 50 percent of their offspring.

In the same year, Linus Pauling and his co-workers provided the first insight into the molecular basis of the disease. They showed that hemoglobins isolated from diseased and normal individuals differ in their rates of electrophoretic migration. In this technique (see Chapter 10), charged molecules migrate in an electric field. If the net charge of two molecules is different, their rates of migration will be different. On this basis, Pauling and his colleagues concluded that a chemical difference exists between normal and sickle-cell hemoglobin. The two molecules are now designated **HbA** and **HbS**, respectively.

Figure 13–14(a) shows the migration pattern of hemoglobin derived from individuals of all three possible genotypes when subjected to **starch gel electrophoresis**. The gel pro-

vides the supporting medium for the molecules during migration. In this experiment, samples are placed at a point of origin between the cathode $(-)$ and the anode $(+)$, and an electric field is applied. The migration pattern reveals that all molecules move toward the anode, indicating a net negative charge. However, HbA migrates farther than HbS, suggesting that its net negative charge is greater. The electrophoretic pattern of hemoglobin derived from carriers reveals the presence of both HbA and HbS, confirming their heterozygous genotype.

Pauling's findings suggested two possibilities. It was known that hemoglobin consists of four nonproteinaceous, iron-containing heme groups and a globin portion containing four polypeptide chains. The alteration in net charge in HbS could be due, theoretically, to a chemical change in either component.

Work carried out between 1954 and 1957 by Vernon Ingram resolved this question. He demonstrated that the chemical change occurs in the primary structure of the globin portion of the hemoglobin molecule. Using the **fingerprinting technique**, Ingram showed that HbS differs in amino acid composition compared to HbA. Human adult hemoglobin con-

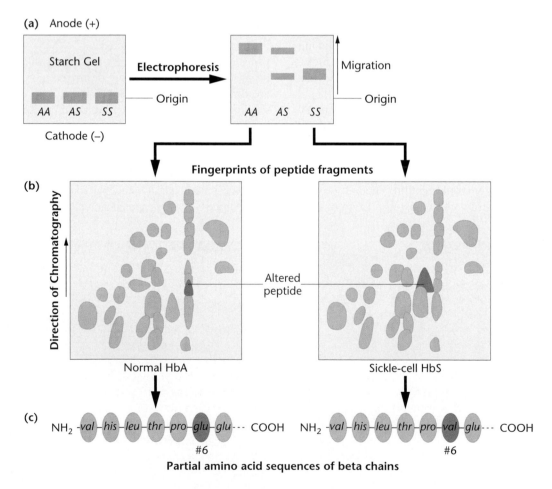

FIGURE 13–14 Investigation of hemoglobin derived from Hb^AHb^A and Hb^SHb^S individuals using electrophoresis, fingerprinting, and amino acid analysis. Hemoglobin from individuals with sickle-cell anemia (Hb^SHb^S) (a) migrates differently in an electrophoretic field, (b) shows an altered peptide in fingerprint analysis, and (c) shows an altered amino acid, valine, at the sixth position in the β chain. During electrophoresis, heterozygotes (Hb^AHb^S) reveal both forms of hemoglobin.

tains two identical α chains of 141 amino acids and two identical β chains of 146 amino acids in its quaternary structure.

The fingerprinting technique involves the enzymatic digestion of the protein into peptide fragments. The mixture is then placed on absorbent paper and exposed to an electric field, where migration occurs according to net charge. The paper is then turned at a right angle and placed in a solvent, where chromatographic action causes the migration of the peptides in the second direction. The end result is a two-dimensional separation of the peptide fragments into a distinctive pattern of spots or a "fingerprint." Ingram's work revealed that HbS and HbA differed by only a single peptide fragment [Figure 13–14(b)]. Further analysis then revealed a single amino acid change: Valine was substituted for glutamic acid at the sixth position of the β chain, accounting for the peptide difference [Figure 13–14(c)].

The significance of this discovery has been multifaceted. It clearly establishes that a single gene provides the genetic information for a single polypeptide chain. Studies of HbS also demonstrate that a mutation can affect the phenotype by directing a single amino acid substitution. Also, by providing the explanation for sickle-cell anemia, the concept of inherited **molecular disease** was firmly established. Finally, this work led to a thorough study of human hemoglobins, which has provided valuable genetic insights.

In the United States, sickle-cell anemia is found almost exclusively in the African-American population. It affects about 1 in every 625 African-American infants. Currently, about 50,000–75,000 individuals are afflicted. In about 1 of every 145 African-American married couples, both partners are heterozygous carriers. In these cases, each of their children has a 25 percent chance of having the disease.

13.7 Colinearity Between Genes and Proteins

Once it was established that genes specify the synthesis of polypeptide chains, the next logical question was how the genetic information contained in a gene's nucleotide sequence can be transferred to the amino acid sequence of a polypeptide chain. It seemed most likely that a **colinear relationship** would exist between the two molecules. That is, the order of nucleotides in the DNA of a gene would correlate directly with the order of amino acids in the corresponding polypeptide.

The initial experimental evidence in support of this concept was derived from studies by Charles Yanofsky of the *trpA* gene that encodes the A subunit of the enzyme **tryptophan synthetase** in *E. coli*. Yanofsky isolated many independent mutants that had lost the activity of the enzyme, mapped them, and established their location with respect to one another within the gene. He then determined where the amino acid substitution occurred in each mutant protein. When the two sets of data were compared, the colinear relationship was apparent. The location of each mutation in the *trpA* gene correlates with the position of the altered amino acid in the A polypeptide of tryptophan synthetase.

13.8 Protein Structure and Biological Diversity

Having established that the genetic information is stored in DNA and influences cellular activities through the proteins it encodes, we turn now to a brief discussion of protein structure. How can these molecules play such a critical role in determining the complexity of cellular activities? As we shall see, the fundamental aspects of the structure of proteins provide the basis for incredible complexity and diversity. At the outset, we should differentiate between the terms **polypeptides** and **proteins**. Both describe molecules composed of amino acids. The molecules differ, however, in their state of assembly and functional capacity. Polypeptides are the precursors of proteins. As assembled on the ribosome during translation, the molecule is called a *polypeptide*. When released from the ribosome following translation, a polypeptide folds up and assumes a higher order of structure. When this occurs, a three-dimensional conformation in space emerges. In many cases, several polypeptides interact to produce this conformation. When the final conformation is achieved, the molecule, now fully functional, is appropriately called a *protein*. It is the three-dimensional conformation that is essential to the function of the molecule.

The polypeptide chains of proteins, like nucleic acids, are linear nonbranched polymers. There are 20 amino acids that serve as the subunits (the building blocks) of proteins. Each amino acid has a **carboxyl group**, an **amino group**, and an **R (radical) group** (a side chain) bound covalently to a **central carbon atom**. The R group gives each amino acid its chemical identity. Figure 13–15 (page 268) shows the 20 R groups, which show a variety of configurations and can be divided into four main classes: (1) **nonpolar (hydrophobic)**, (2) **polar (hydrophilic)**, (3) **negatively charged**, and (4) **positively charged**. Because polypeptides are often long polymers, and because each position may be occupied by any one of the 20 amino acids with their unique chemical properties, enormous variation in chemical conformation and activity is possible. For example, if an average polypeptide is composed of 200 amino acids (molecular weight of about 20,000 Da), 20^{200} different molecules, each with a unique sequence, can be created using the 20 different building blocks.

Around 1900, German chemist Emil Fischer determined the manner in which the amino acids are bonded together. He showed that the amino group of one amino acid can react with the carboxyl group of another amino acid during a dehydration reaction, releasing a molecule of H_2O. The resulting covalent bond is known as a peptide bond (Figure 13–16, page 269). Two amino acids linked together constitute a dipeptide, three a tripeptide, and so on. Once 10 or more amino acids are linked by peptide bonds, the chain is referred to as a polypeptide. Generally, no matter how long a polypeptide is, it will contain a free amino group at one end (the N-terminus) and a free carboxyl group at the other end (the C-terminus).

Four levels of protein structure are recognized: primary, secondary, tertiary, and quaternary. The sequence of amino acids in the linear backbone of the polypeptide constitutes its

MEDIA TUTORIAL Mutation, Gene/Protein Colinearity

FIGURE 13–15 Chemical structures and designations of the 20 amino acids found in living organisms, divided into four major categories.

1. Nonpolar: Hydrophobic

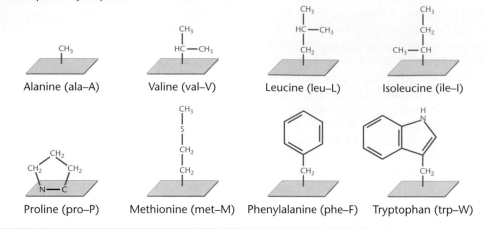

Alanine (ala–A)   Valine (val–V)   Leucine (leu–L)   Isoleucine (ile–I)

Proline (pro–P)   Methionine (met–M)   Phenylalanine (phe–F)   Tryptophan (trp–W)

2. Polar: Hydrophilic

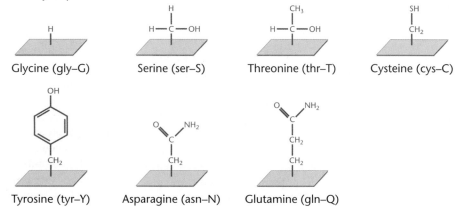

Glycine (gly–G)   Serine (ser–S)   Threonine (thr–T)   Cysteine (cys–C)

Tyrosine (tyr–Y)   Asparagine (asn–N)   Glutamine (gln–Q)

3. Polar: positively charged (basic)

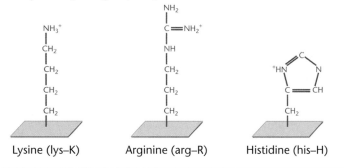

Lysine (lys–K)   Arginine (arg–R)   Histidine (his–H)

4. Polar: negatively charged (acidic)

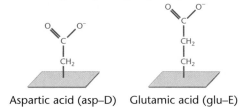

Aspartic acid (asp–D)   Glutamic acid (glu–E)

Amino group — Carboxyl group

Amino acid structure

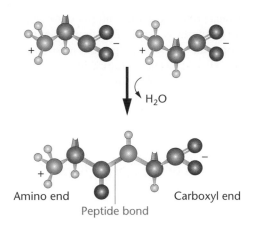

FIGURE 13–16 Peptide bond formation between two amino acids, resulting from a dehydration reaction.

primary structure. It is specified by the sequence of deoxyribonucleotides in DNA via an mRNA intermediate. The primary structure of a polypeptide helps determine the specific characteristics of the higher orders of organization as a protein is formed.

The **secondary structure** refers to a regular or repeating configuration in space assumed by amino acids closely aligned in the polypeptide chain. In 1951, Linus Pauling and Robert Corey predicted, on theoretical grounds, an **α helix** as one type of secondary structure. The α-helix model [Figure 13–17(a)] has since been confirmed by X-ray crystallographic studies. It is rodlike and has the greatest possible theoretical stability. The helix is composed of a spiral chain of amino acids stabilized by hydrogen bonds.

The side chains (the R groups) of amino acids extend outward from the helix, and each amino acid residue occupies a vertical distance of 1.5 Å in the helix. There are 3.6 residues per turn. While left-handed helices are theoretically possible, all proteins demonstrating an α helix are right-handed.

Also in 1951, Pauling and Corey proposed a second structure, the **β-pleated-sheet** configuration. In this model, a single polypeptide chain folds back on itself, or several chains run in either parallel or antiparallel fashion next to one another. Each such structure is stabilized by hydrogen bonds formed between atoms present on adjacent chains [Figure 13–17(b)]. A single zigzagging plane is formed in space with adjacent amino acids 3.5 Å apart.

As a general rule, most proteins demonstrate a mixture of α-helix and β-pleated-sheet structures. Globular proteins, most of which are round in shape and water soluble, usually contain a core of β-pleated-sheet structure as well as many areas with helical structures. The more rigid structural proteins, many of which are water insoluble, rely on more extensive β-pleated-sheet regions for their rigidity. For example, **fibroin**, the protein made by the silk moth, depends extensively on this form of secondary structure.

While the secondary structure describes the arrangement of amino acids within certain areas of a polypeptide chain, **tertiary protein structure** defines the three-dimensional conformation of the entire chain in space. Each protein twists and turns and loops around itself in a very specific fashion, characteristic of the specific protein. Three aspects of this level of structure are most important in determining this conformation and in stabilizing the molecule.

1. Covalent disulfide bonds form between closely aligned cysteine residues to make the unique amino acid cystine.

2. Nearly all of the polar hydrophilic R groups are located on the surface, where they can interact with water.

3. The nonpolar hydrophobic R groups are usually located on the inside of the molecule, where they interact with one another, avoiding interaction with water.

(a) Alpha helix   **(b) Beta-pleated sheet**

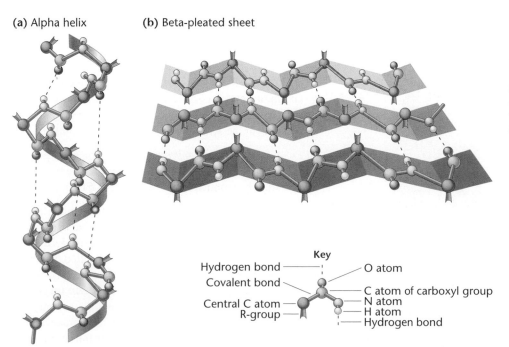

Key
Hydrogen bond ————— O atom
Covalent bond ————— C atom of carboxyl group
Central C atom ————— N atom
R-group ————— H atom
————— Hydrogen bond

FIGURE 13–17 (a) The right-handed α helix, which represents one form of secondary structure of a polypeptide chain. (b) The β-pleated-sheet configuration, an alternative form of secondary structure of polypeptide chains. For the sake of clarity, not all atoms are shown.

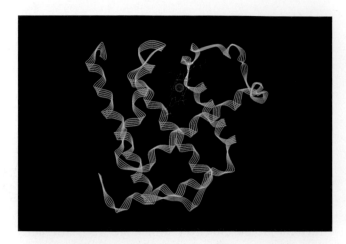

FIGURE 13–18 The tertiary level of protein structure in respiratory pigment myoglobin. The bound oxygen atom is shown in red. *(Horton, et al.* Principles of Biochemistry, © 2002. *Reprinted by permission of Prentice-Hall, Inc., Upper Saddle River, NJ)*

It is important to emphasize that the three-dimensional conformation achieved by any protein is a product of the primary structure of the polypeptide. Thus, the genetic code need only specify the sequence of amino acids in order to encode information that leads ultimately to the final assembly of proteins. The three stabilizing factors depend on the location of each amino acid relative to all others in the chain. As folding occurs, the most thermodynamically stable conformation possible results.

A model of the three-dimensional tertiary structure of the respiratory pigment **myoglobin** is shown in Figure 13–18. This level of organization is extremely important because the specific function of any protein is directly related to its three-dimensional conformation.

The **quaternary level of organization** applies only to proteins composed of more than one polypeptide chain, indicating the conformation of the various chains in relation to one another. This type of protein is called oligomeric, and each chain is called a protomer, or, less formally, a subunit.

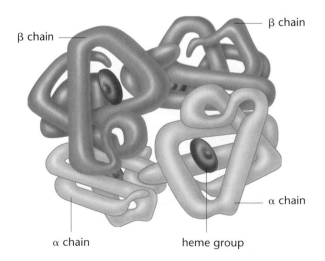

FIGURE 13–19 The quaternary level of protein structure as seen in hemoglobin. Four chains (2 α and 2 β) interact with four heme groups to form the functional molecule.

The individual protomers have conformations that fit together with other subunits in a specific complementary fashion. Hemoglobin, an oligomeric protein consisting of four polypeptide chains, has been studied in great detail. Its quaternary structure is shown in Figure 13–19. Most enzymes, including DNA and RNA polymerase, demonstrate quaternary structure.

13.9 Posttranslational Modification of Proteins

Before turning to a discussion of protein function, it is important to point out that polypeptide chains, like RNA transcripts, are often modified once they have been synthesized. This additional processing is broadly described as **posttranslational modification**. Although many of these alterations are detailed biochemical transformations and beyond the scope of this discussion, you should be aware that they occur and that they are critical to the functional capability of the final protein product. Several examples of posttranslational modification are presented below.

1. *The N-terminus and C-terminus amino acids are usually removed or modified.* For example, the initial N-terminal formylmethionine residue in bacterial polypeptides is usually removed enzymatically. Often, the amino group of the initial methionine residue is removed, and the amino group of the N-terminal residue is chemically modified (acetylated) in eukaryotic polypeptide chains.

2. *Individual amino acid residues are sometimes modified.* For example, phosphates may be added to the hydroxyl groups of certain amino acids, such as tyrosine. Modifications such as these create negatively charged residues that bond ionically with other molecules. The process of phosphorylation is extremely important in regulating many cellular activities and results from the action of enzymes called **kinases**. In other proteins, methyl groups may be added enzymatically.

3. *Carbohydrate side chains are sometimes attached.* These are added covalently, producing **glycoproteins**, an important category of molecules that includes many antigenic determinants, such as those specifying the antigens in the ABO blood-type system in humans.

4. *Polypeptide chains may be trimmed.* For example, insulin is first translated into a longer molecule that is enzymatically trimmed to its final form of 51 amino acids.

5. *Signal sequences are removed.* At the N-terminal end of some proteins, a sequence of up to 30 amino acids plays an important role in directing the protein to the location in the cell where it functions. This is called a **signal sequence**, and it determines the final destination of a protein in the cell. This process is called **protein targeting**. For example, proteins whose fate involves secretion or that are to become part of the plasma membrane are dependent on specific sequences for their initial transport

into the lumen of the endoplasmic reticulum. While the signal sequence of various proteins with a common destination might differ in their primary amino acid sequence, they share many chemical properties. For example, those destined for secretion all contain a string of up to 15 hydrophobic amino acids preceded by a positively charged amino acid at the N-terminus of the signal sequence. Once the polypeptides are transported, but prior to achieving their functional status as proteins, the signal sequence is enzymatically removed from these polypeptides.

6. *Polypeptide chains are often complexed with metals.* The tertiary and quaternary levels of protein structure often include and are dependent on metal atoms. The function of the protein is thus dependent on the molecular complex that includes both polypeptide chains and metal atoms. Hemoglobin, containing four iron atoms along with four polypeptide chains, is a good example.

These types of posttranslational modifications are obviously important in achieving the functional status specific to any given protein. Because the final three-dimensional structure of the molecule is intimately related to its specific function, how polypeptide chains ultimately fold into their final conformations is also an important topic. For many years, it was thought that protein folding was a spontaneous process whereby the molecule achieved maximum thermodynamic stability, based largely on the combined chemical properties inherent in the amino acid sequence of the polypeptide chain(s) composing the protein. However, numerous studies have shown that, for many proteins, folding is dependent upon members of a family of still other, ubiquitous proteins called **chaperones**. Chaperone proteins (sometimes called *molecular chaperones* or *chaperonins*) function to facilitate the folding of other proteins. While the mechanism by which chaperones function is not yet clear, like enzymes, they do not become part of the final product. Initially discovered in *Drosophila*, where they are called heat-shock proteins, chaperones have been discovered in a variety of organisms, including bacteria, animals, and plants.

13.10 Protein Function

The essence of life on the earth rests at the level of diverse cellular function. One can argue that DNA and RNA simply serve as vehicles to store and express genetic information. However, proteins are at the heart of cellular function. And it is the capability of cells to assume diverse structures and functions that distinguishes most eukaryotes from less evolutionarily advanced organisms such as bacteria. Therefore, an introductory understanding of protein function is critical to a complete view of genetic processes.

Proteins are the most abundant macromolecules found in cells. As the end products of genes, they play many diverse roles. For example, the respiratory pigments **hemoglobin** and **myoglobin** transport oxygen, which is essential for cellular metabolism. **Collagen** and **keratin** are structural proteins associated with the skin, connective tissue, and hair of organisms. **Actin** and **myosin** are contractile proteins, found in abundance in muscle tissue. Still other examples are the **immunoglobulins**, which function in the immune system of vertebrates; **transport proteins**, involved in movement of molecules across membranes; some of the **hormones** and their **receptors**, which regulate various types of chemical activity; and **histones**, which bind to DNA in eukaryotic organisms.

The largest group of proteins with a related function are the **enzymes**. Since we have referred to these molecules throughout this chapter, it may be useful to extend our discussion and include a more detailed description of their biological role. These molecules specialize in catalyzing chemical reactions within living cells. Enzymes increase the rate at which a chemical reaction reaches equilibrium, but they do not alter the end point of the chemical equilibrium. Their remarkable, highly specific catalytic properties largely determine the metabolic capacity of any cell type. The specific functions of many enzymes involved in the genetic and cellular processes of cells are described throughout the text.

Biological catalysis is a process whereby the **energy of activation** for a given reaction is lowered (Figure 13–20). The energy of activation is the increased kinetic energy state

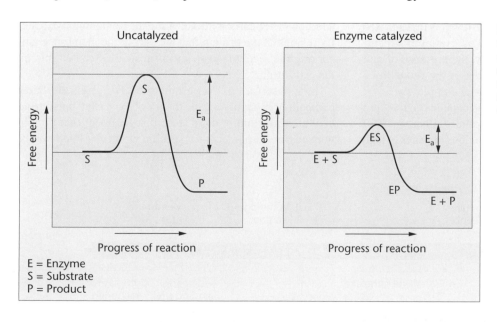

FIGURE 13–20 Energy requirements of an uncatalyzed versus an enzymatically catalyzed chemical reaction. The energy of activation (E_a) necessary to initiate the reaction is substantially lower as a result of catalysis.

that molecules must usually reach before they react with one another. While this state can be attained as a result of elevated temperatures, enzymes allow biological reactions to occur at lower physiological temperatures. In this way, enzymes make possible life as we know it.

The catalytic properties and specificity of an enzyme are determined by the chemical configuration of the molecule's **active site**. This site is associated with a crevice, a cleft, or a pit on the surface of the enzyme, which binds the reactants, or substrates, facilitating their interaction. Enzymatically catalyzed reactions control metabolic activities in the cell. Each reaction is either **catabolic** or **anabolic**. Catabolism is the degradation of large molecules into smaller, simpler ones with the release of chemical energy. Anabolism is the synthetic phase of metabolism, yielding the various components that make up nucleic acids, proteins, lipids, and carbohydrates.

13.11   Protein Domains and Exon Shuffling

We conclude our discussion of proteins by briefly discussing the important finding that regions made up of specific amino acid sequences are associated with specific functions in protein molecules. Such sequences, usually between 50 and 300 amino acids, constitute what are called **protein domains** and are represented by modular portions of the protein that fold into stable, unique conformations independently of the rest of the molecule. Different domains impart different functional capabilities. Some proteins contain only a single domain, while others contain two or more.

The significance of domains rests at the tertiary structure level of proteins. Each such modular unit can be a mixture of secondary structures, including both α helicies and β pleated sheets. The unique conformation that is assumed in a single domain imparts a specific function to the protein. For example, a domain may serve as the catalytic basis of an enzyme, or it may impart the capability to bind to a specific ligand as part of a membrane or another molecule. Thus, in the study of proteins, you will hear of *catalytic domains*, *DNA-binding domains*, and so on. The result is that a protein must be looked at as being composed of a series of structural and functional modules. Obviously, the presence of multiple domains in a single protein increases the versatility of each molecule and adds to its functional complexity.

An interesting proposal to explain the genetic origin of protein domains was put forward by Walter Gilbert in 1977. Gilbert suggested that the functional regions of genes in higher organisms consist of collections of exons originally present in ancestral genes that are brought together through recombination during the course of evolution. Referring to this process as **exon shuffling**, Gilbert proposed that exons, like protein domains, are also modular. Gilbert proposed that during evolution, exons may have been reshuffled between genes in eukaryotes such that different genes share similar domains.

Since 1977, a serious research effort has been directed toward the analysis of gene structure. In 1985, more direct evidence in favor of Gilbert's proposal of exon modules was presented. For example, the human gene encoding the membrane receptor for low-density lipoproteins (LDL) was isolated and sequenced. The LDL receptor protein is essential to the transport of plasma cholesterol into the cell. It mediates endocytosis and is expected to have numerous functional domains. These include the capability of this protein to bind specifically to the LDL substrates and to interact with other proteins at different levels of the membrane during transport across it. In addition, this receptor molecule is modified posttranslationally by the addition of a carbohydrate; a domain must exist that links to this carbohydrate.

Detailed analysis of the gene encoding this protein supports the concept of exon modules and their shuffling during evolution. The gene is quite large—45,000 nucleotides—and contains 18 exons. These represent only slightly less than 2600 nucleotides. These exons are related to the functional domains of the protein *and* appear to have been recruited from other genes during evolution.

Figure 13–21 shows these relationships. The first exon encodes a signal sequence that is removed from the protein before the LDL receptor becomes part of the membrane. The next five exons represent the domain specifying the binding site for cholesterol. This domain is made up of a sequence of 40 amino acids repeated seven times. The next domain consists of a sequence of 400 amino acids bearing a striking homology to the mouse peptide hormone epidermal growth factor (EGF). This region is encoded by eight exons and contains three repetitive sequences of 40 amino acids. A similar sequence is also found in three blood-clotting proteins. The fifteenth exon specifies the domain for the posttranslational addition of the carbohydrate, while the remaining two specify regions of the protein that are part of the membrane, anchoring the receptor to specific sites called coated pits on the cell surface.

These observations concerning the LDL exons are fairly compelling in support of the theory of exon shuffling during evolution. Certainly, there is no disagreement that protein domains are responsible for specific molecular interactions.

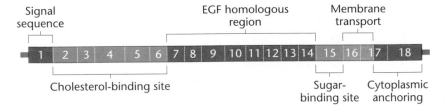

FIGURE 13–21 A comparison of the 18 exons making up the gene encoding the LDL receptor protein. The exons are organized into five functional domains and one signal sequence.

Genetics, Technology, and Society

Mad Cows and Heresies: The Prion Story

On March 20, 1996, the British government announced that a new brain disease had killed 10 young Britons, and that the victims might have caught the disease by eating infected beef. The disease, known as "mad cow disease" or bovine spongiform encephalopathy (BSE), slowly destroys brain cells and is always fatal. Recent studies confirm that BSE and the human disease, known as new variant Creutzfeldt–Jakob disease (nvCJD), are so similar at the molecular and pathological levels that they are very likely the same disease. Mad cow disease triggered political turmoil in Europe, a worldwide ban on British beef, and the near collapse of the $8.9-billion British beef industry. The European union demanded the slaughter and incineration of 4.7 million British cattle, a campaign that cost the government over $12 billion to compensate farmers, import milk, and buy cattle for new herds. BSE and nvCJD have recently appeared in France, several other European countries, and South Africa. At least 90 Europeans have now died of nvCJD, and epidemiologists estimate that between 10,000 and 500,000 cases of nvCJD may appear in the next two decades.

BSE and nvCJD are members of a group of neurological diseases known as spongiform encephalopathies. These diseases have long incubation times (up to 20 or 30 years), and lead to progressive neurodegeneration. Infected brain tissue resembles a sponge (hence "spongiform") and is riddled with proteinaceous deposits. Victims of the disease lose motor function, become demented, and eventually die. CJD cases arise spontaneously and randomly, at a rate of 1 per million per year worldwide. A less common form of CJD is inherited as an autosomal dominant condition. But the transmitted forms are the strangest. CJD can be transmitted through corneal or nervous tissue grafts or by injection of growth hormone derived from human pituitary glands. Kuru, a CJD-like disease of the Fore people of New Guinea, was transmitted from person to person via ritualistic cannibalism. Transmissible spongiform encephalopathies in animals include scrapie (sheep and goats), chronic wasting disease (deer and elk), and BSE. Like Kuru, BSE and scrapie are passed from animal to animal by ingestion of diseased animal remains, particularly neural tissue. The epidemic of BSE in Britain occurred because diseased cows and sheep were processed and fed to cattle as a protein supplement. In 1998, the British government banned the use of cows and sheep as feed for other cows and sheep, and the epidemic of BSE is subsiding. The European Union recently approved a temporary ban on the use of animal-containing feeds for all livestock, although the ban is nonbinding. U.S. and Canadian regulations still allow nonruminant animals to consume feed containing ruminant meat products, and ruminant animals to consume feed containing nonruminant animals as well as certain ruminant by-products such as blood, gelatin, and fat. Regulations may change, as recent studies suggest that BSE may spread between cows, pigs, chickens, and other livestock, or from cows to calves. The U.S. Department of Agriculture has banned importation of cattle or meat products from countries that have BSE. At present, no cases of BSE have been reported in the United States. Due to concerns that nvCJD may be spread by blood transfusions, Canada and the U.S. banned blood donations from persons who spent more than 6 months in Britain between 1980 and 1986.

For many years, the spongiform encephalopathies defied the best efforts of scientists to analyze them. The diseases are difficult to study, because they require injection of infected brain material into the brains of experimental animals and the diseases take months or years to develop. In addition, the infectious agent is apparently not a virus or bacterium, and infected animals do not develop antibodies against these mysterious agents. There are no treatments for the disease, and the only way to make a firm diagnosis is to examine brain tissue after death. The infectious material is unaffected by radiation or nucleases that damage nucleic acids; however, it is destroyed by some reagents that hydrolyze or modify proteins. In the early 1980s, American scientist Stanley Prusiner purified the infectious agent and concluded that it consists only of protein. He proposed that scrapie is spread by an infectious protein particle that he called a **prion**. His hypothesis was dismissed by most scientists, as the idea of an infectious agent that contains no DNA or RNA as genetic material was heretical. However, Prusiner and others presented evidence supporting the prion hypothesis, and the notion that the disease can be transmitted by an infectious particle that contains no genetic material has gained acceptance.

If prions are comprised only of protein, how do they cause disease? The answer may be as strange as the disease itself. The protein that makes up a prion (PrP) is a version of a normal protein that is synthesized in neurons and found in the brains of all adult animals. The difference between normal PrP and prion PrP lies in their protein secondary structures. Normal, noninfectious PrP folds into α-helices, whereas infectious prion PrP folds into β-sheets. When a normal PrP molecule contacts a prion PrP molecule, the normal protein is somehow unfolded and refolded into the abnormal PrP conformation. Once the normal PrP molecule has been transformed into an abnormal PrP molecule, it spreads its lethal conformation to neighboring normal PrP molecules, and the process takes off in a chain reaction. Normal PrP is a soluble protein that is easily destroyed by heat or enzymes that digest proteins. However, abnormal infectious PrP is insoluble in detergents, resists both heat and protease digestion, and is nearly indestructible. Hence, spongiform encephalopathies can be considered diseases of protein secondary structure.

Many urgent questions need to be addressed. How extensive is BSE contamination of the world's food supply? How many humans are infected with nvCJD but don't yet show symptoms? Can humans or animals act as asymptomatic carriers of prion diseases? Can prions exist in other parts of the body besides the brain and spinal cord, and if so, can CJD spread via blood transfusions, from mother to fetus, or by sterilized surgical instruments? Can we develop diagnostic tests and therapies for BSE and CJD? Are we near the end of the BSE story, or is it just beginning?

References

Almond, J., and Pattison, J. 1997. Human BSE. *Nature* 389:437–438.

Balter, M. 2000. Tracking the human fallout from "mad cow disease." *Science* 289:1452–1454.

Belay, E. D. 1999. Transmissible spongiform encephalopathies in humans. *Annu. Rev. Microbiol.* 53:283–314.

Prusiner, S. B. 1998. Prions. *Proc. Natl. Acad. Sci. U.S.A.* 95:15363–83.

Websites

The BSE Inquiry Report (released October 26, 2000).
http://www. bse.org.uk
The Official Mad Cow Disease Home Page (a compilation of news reports).
http://www.mad-cow.org

Chapter Summary

1. Translation describes the synthesis in cells of polypeptide chains, under the direction of mRNA and in association with ribosomes. This process ultimately converts the information stored in the genetic code of the DNA making up a gene into the corresponding sequence of amino acids making up the polypeptide.

2. Translation is a complex energy-requiring process that also depends on charged tRNA molecules and numerous protein factors. Transfer RNA (tRNA) serves as the adaptor molecule between an mRNA triplet and the appropriate amino acid.

3. The processes of translation, like transcription, can be subdivided into the stages of initiation, elongation, and termination. The process relies on base-pairing affinities between complementary nucleotides and is more complex in eukaryotes than in prokaryotes.

4. The first insight that proteins are the end products of genes was provided by the study of inherited metabolic disorders in humans early in the twentieth century by Garrod. Basic to his studies were inborn errors of metabolism leading to cystinuria, albinism, and alkaptonuria.

5. The investigation of nutritional requirements in *Neurospora* by Beadle and colleagues made it clear that mutations cause the loss of enzyme activity. Their work led to the concept of one-gene:one-enzyme.

6. The one-gene:one-enzyme hypothesis was later revised. Pauling and Ingram's investigations of hemoglobins from patients with sickle-cell anemia led to the discovery that one gene directs the synthesis of only one polypeptide chain.

7. The proposal suggesting that a gene's nucleotide sequence specifies in a colinear manner the sequence of amino acids in a polypeptide chain was confirmed by experiments involving mutations in the tryptophan synthetase gene in *E. coli*.

8. Proteins, the end products of genes, demonstrate four levels of structural organization that together provide the chemical basis for their three-dimensional conformation, which is the basis of the molecule's function.

9. Of the myriad functions performed by proteins, the most influential role is assumed by enzymes. These highly specific, cellular catalysts play a central role in the production of all classes of molecules in living systems.

10. Proteins consist of one or more functional domains, which are shared by different molecules. The origin of these domains may be the result of exon shuffling during evolution.

Key Terms

actin, 271
active site, 272
alkaptonuria, 261
alpha (α) helix, 269
amino group, 267
aminoacyl (A) site, 258
aminoacyl tRNA synthetases, 257
aminoacyladenylic acid, 257
anabolic, 272
anticodon, 254
anticodon loop, 256
β-pleated sheet, 269
biological catalysis, 271
carboxyl group, 267
catabolic, 272
central carbon atom, 267
chaperone, 271
charging (tRNA), 257
citrulline, 263
cloverleaf model of tRNA, 255
colinear relationship, 267
elongation, 258
endocytosis, 272
energy of activation, 271
enzyme, 271
exon shuffling, 272

fibroin, 269
fingerprinting technique, 266
formylmethionine (fmet), 257
globin portion (of hemoglobin), 266
glycoprotein, 270
GTP-dependent release factor, 260
HbA, 266
HbS, 266
heme group (of hemoglobin), 266
hemoglobin, 271
histone, 271
hormone, 271
immunoglobulin, 271
initiation complex, 258
isoaccepting tRNA, 257
keratin, 271
kinase, 270
LDL receptor protein, 272
molecular disease, 267
monosome, 254
myoglobin, 270
myosin, 271
nonsense codon, 260
one-gene:one-enzyme hypothesis, 263
one-gene:one-polypeptide chain hypothesis, 265

one-gene:one-protein hypothesis, 265
ornithine, 263
peptidyl (P) site, 258
peptidyl transferase, 258
phenylalanine hydroxylase, 262
phenylketonuria (PKU), 262
polypeptide, 265
polyribosome, 260
polysome, 260
posttranslational modification, 270
primary structure, 269
prion, 273
protein, 267
protein domain, 272
protein targeting, 270
quaternary protein structure, 265
R (radical) group, 267
receptors, 271
ribosomal proteins, 254
ribosome, 254
secondary structure, 269
Shine–Dalgarno sequence, 257
sickle-cell anemia, 265
sickle-cell trait, 266
signal sequence, 270
spacer DNA, 254

Insights and Solutions

1. The following growth responses are obtained using four mutant strains of *Neurospora* and the related compounds A, B, C, and D. None of the mutations grow on minimal medium. Draw all possible conclusions.

		Growth Product			
		C	**D**	**B**	**A**
	1	−	−	−	−
	2	−	+	+	+
Mutation	3	−	−	+	+
	4	−	−	+	−

Solution: First, nothing can be concluded about mutation 1, except that it lacks some essential growth factor, perhaps even unrelated to the biochemical pathway represented by mutations 2–4; nor can anything be concluded about compound C. If it is involved in the pathway, it is a product synthesized prior to the synthesis of A, B, and D.

We must now analyze these three compounds and the control of their synthesis by the enzymes encoded by genes 2, 3, and 4. Because product B allows growth in all three cases, it may be considered the "end product." It bypasses the block in all three instances. Using similar reasoning, product A precedes B in the pathway, since it allows a bypass in two of the three steps. Product D precedes B, yielding the more complete solution:

$$C(?) \longrightarrow D \longrightarrow A \longrightarrow B$$

Now determine which mutations control which steps. Since mutation 2 can be alleviated by products D, B, and A, it must control a step prior to all three products, perhaps the direct conversion to D, although we cannot be certain. Mutation 3 is alleviated by B and A, so its effect must precede them in the pathway. Thus, we will assign it as controlling the conversion of D to A. Likewise, we can assign mutation 4 to the conversion of A to B, leading to the more complete solution:

$$C(?) \xrightarrow{2(?)} D \xrightarrow{3} A \xrightarrow{4} B$$

Problems and Discussion Questions

1. List and describe the role of all of the molecular constituents present in a functional polyribosome.
2. Contrast the roles of tRNA and mRNA during translation and list all enzymes that participate in the transcription and translation process.
3. Francis Crick proposed the "adaptor hypothesis" for the function of tRNA. Why did he choose that description?
4. During translation, what molecule bears the codon? The anticodon?
5. The α chain of eukaryotic hemoglobin is composed of 141 amino acids. What is the minimum number of nucleotides in an mRNA coding for this polypeptide chain? Assuming that each nucleotide is 0.34 nm long in the mRNA, how many triplet codes can, at one time, occupy space in a ribosome that is 20 nm in diameter?
6. Summarize the steps involved in charging tRNAs with their appropriate amino acids.
7. For tRNA to carry out its role, each tRNA requires at least four specific recognition sites that must be inherent in its tertiary structure. What are they?
8. Discuss the potential difficulties involved in designing a diet to alleviate the symptoms of phenylketonuria.
9. Phenylketonurics cannot convert phenylalanine to tyrosine. Why don't these individuals exhibit a deficiency of tyrosine?
10. Phenylketonurics are often more lightly pigmented than are normal individuals. Can you suggest a reason why this is so?
11. The synthesis of flower pigments is known to be dependent on enzymatically controlled biosynthetic pathways. In the crosses shown here, postulate the role of mutant genes and their products in producing the observed phenotypes.

> (a) P_1: white strain A × white strain B
> F_1: all purple
> F_2: 9/16 purple: 7/16 white
> (b) P_1: white × pink
> F_1: all purple
> F_2: 9/16 purple: 3/16 pink: 4/16 white

12. A series of mutations in the bacterium *Salmonella typhimurium* results in the requirement of either tryptophan or some related molecule in order for growth to occur. From the data shown here, suggest a biosynthetic pathway for tryptophan.

	Growth Supplement				
Mutation	*Minimal Medium*	*Anthranilic Acid*	*Indole Glycerol Phosphate*	*Indole*	*Tryptophan*
trp-8	−	+	+	+	+
trp-2	−	−	+	+	+
trp-3	−	−	−	+	+
trp-1	−	−	−	−	+

13. The study of biochemical mutants in organisms such as *Neurospora* has demonstrated that some pathways are branched. The data shown here demonstrate the branched nature of the pathway resulting in the synthesis of thiamine. Why don't the data support a linear pathway? Can you postulate a pathway for the synthesis of thiamine in *Neurospora*?

	Growth Supplement			
Mutation	*Minimal Medium*	*Pyrimidine*	*Thiazole*	*Thiamine*
thi-1	−	−	+	+
thi-2	−	+	−	+
thi-3	−	−	−	+

14. Explain why the one-gene:one-enzyme hypothesis is not considered totally accurate today.

15. Why is an alteration of electrophoretic mobility interpreted as a change in the primary structure of the protein under study?

16. Using sickle-cell anemia as a basis, describe what is meant by a molecular or genetic disease. What are the similarities and dissimilarities between this type of a disorder and a disease caused by an invading microorganism?

17. Contrast the contributions of Pauling and Ingram to our understanding of the genetic basis for sickle-cell anemia.

18. Hemoglobins from two individuals are compared by electrophoresis and by fingerprinting. Electrophoresis reveals no difference in migration, but fingerprinting shows an amino acid difference. How is this possible?

19. Describe what colinearity means. Of what significance is the concept of colinearity in the study of genetics?

20. Define and compare the four levels of protein organization.

21. List as many different categories of protein functions as you can. Wherever possible, give an example of each category.

22. How does an enzyme function? Why are enzymes essential for living organisms on Earth?

23. Shown here are several amino acid substitutions in the α and β chains of human hemoglobin. Using the coding dictionary (Figure 12–7), determine how many of them can occur as a result of a single nucleotide change.

Hb Type	*Normal Amino Acid*	*Substituted Amino Acid*
HbJ Toronto	ala	asp (α-5)
HbJ Oxford	gly	asp (α-15)
Hb Mexico	gln	glu (α-54)
Hb Bethesda	tyr	his (β-145)
Hb Sydney	val	ala (β-67)
HbM Saskatoon	his	tyr (β-63)

24. In 1962, F. Chapeville and others reported an experiment in which they isolated radioactive ^{14}C-cysteinyl–tRNA^{cys} (charged tRNA^{cys} + cysteine). They then removed the sulfur group from the cysteine, creating alanyl–tRNA^{cys} (charged tRNA^{cys} + alanine). When alanyl–tRNA^{cys} was added to a synthetic mRNA calling for cysteine but not alanine, a polypeptide chain was synthesized containing alanine. What can you conclude from this experiment? [*Reference:* Chapeville et al., *Proc. Natl. Acad. Sci. USA* 48:1086–93 (1962).]

25. Deep in a previously unexplored South American rain forest, a species of plants is discovered with true-breeding varieties whose flowers are either pink, rose, orange, or purple. A very astute plant geneticist makes a single cross, carried to the F_2 generation, as shown here. Based solely on these data, he proposes both a mode of inheritance for flower pigmentation and a biochemical pathway for the synthesis of these pigments.

Carefully study the data. Create your own hypothesis to explain the mode of inheritance. Then propose a biochemical pathway consistent with your hypothesis. How could you test the hypothesis by making other crosses?

P_1: purple × pink
F_1: all purple
F_2: 27/64 purple
16/64 pink
12/64 rose
9/64 orange

Selected Readings

Anfinsen, C. B. 1973. Principles that govern the folding of protein chains. *Science* 181:223–30.

Bartholome, K. 1979. Genetics and biochemistry of phenylketonuria—Present state. *Hum. Genet.* 51:241–45.

Bateson, W. 1909. *Mendel's principles of heredity.* Cambridge: Cambridge University Press.

Beadle, G. W. 1945. Genetics and metabolism in *Neurospora*. *Physiol. Rev.* 25:643.

Beadle, G. W., and Tatum, E. L. 1941. Genetic control of biochemical reactions in *Neurospora*. *Proc. Natl. Acad. Sci. USA* 27:499–506.

Beet, E. A. 1949. The genetics of the sickle-cell trait in a Bantu tribe. *Ann. Eugenics* 14:279–84.

Boyer, P. D., ed. 1974. *The enzymes*, Vol 10, 3rd ed. Orlando, FL: Academic Press.

Bray, D. 1995. Protein molecules as computational elements in living cells. *Nature* 376:307–12.

Chapeville, F., et al. 1962. On the role of soluble ribonucleic acid in coding for amino acids. *Proc. Natl. Acad. Sci. USA* 48:1086–93.

Cigan, A. M., Feng, L., and Donahue, T. F. 1988. tRNA^{met} functions in directing the scanning ribosome to the start site of translation. *Science* 242:93–98.

Dahlberg, A. E. 1989. The functional role of ribosomal RNA in protein synthesis. *Cell* 57:525–29.

Dickerson, R. E., and Geis, I. 1983. *Hemoglobin: Structure, function, evolution, and pathology*. Menlo Park, CA: Benjamin/Cummings.

Doolittle, R. F. 1985. Proteins. *Sci. Am.* (Oct.) 253:88–99.

Ezzell, C. 1994. Evolutions: Molecular chaperones and protein folding. *J. NIH Res.* 6:103.

Frank, J. 1998. How the ribosome works. *Amer. Scient.* 86:428–39.

Garrod, A. E. 1902. The incidence of alkaptonuria: A study in chemical individuality. *Lancet* 2:1616–20.

————. 1909. *Inborn errors of metabolism*. London: Oxford University Press. (Reprinted 1963, Oxford University Press, London.)

Garrod, S. C. 1989. Family influences on A. E. Garrod's thinking. *J. Inher. Metab. Dis.* 12:2–8.

Horton, H. R., et al. 2002. *Principles of biochemistry*, 3rd ed. Upper Saddle River, NJ: Prentice-Hall.

Ingram, V. M. 1957. Gene mutations in human hemoglobin: The chemical difference between normal and sickle cell hemoglobin. *Nature* 180:326–28.

Koshland, D. E. 1973. Protein shape and control. *Sci. Am.* (Oct.) 229:52–64.

LaDu, B. N., Zannoni, V. G., Laster, L., and Seegmiller, J. E. 1958. The nature of the defect in tyrosine metabolism in alkaptonuria. *J. Biol. Chem.* 230:251.

Lake, J. A. 1981. The ribosome. *Sci. Am.* (Aug.) 245:84–97.

Maniatis, T., et al. 1980. The molecular genetics of human hemoglobins. *Annu. Rev. Genet.* 14:145–78.

Moore, P. B. 1988. The ribosome returns. *Nature* 331:223–27.

Murayama, M. 1966. Molecular mechanism of red cell sickling. *Science* 153:145–49.

Neel, J. V. 1949. The inheritance of sickle-cell anemia. *Science* 110:64–66.

Nirenberg, M. W., and Leder, P. 1964. RNA code words and protein synthesis. *Science* 145:1399–1407.

Noller, H. F. 1973. Assembly of bacterial ribosomes. *Science* 179:864–73.

Nomura, M. 1984. The control of ribosome synthesis. *Sci. Am.* (Jan.) 250:102–14.

Pauling, L., Itano, H. A., Singer, S. J., and Wells, I. C. 1949. Sickle-cell anemia: A molecular disease. *Science* 110:543–48.

Porce, B. T., and Garrett, R. A. 1999. Ribosomal mechanics, antibiotics, and GTP hydrolysis. *Cell* 97:423–26.

Rich, A., and Houkim, S. 1978. The three-dimensional structure of transfer RNA. *Sci. Am.* (Jan.) 238:52–62.

Rich, A., Warner, J. R., and Goodman, H. M. 1963. The structure and function of polyribosomes. *Cold Spring Harbor Symp. Quant. Biol.* 28:269–85.

Richards, F. M. 1991. The protein folding problem. *Sci. Am.* (Jan.) 264:54–63.

Rould, M. A., et al. 1989. Structure of *E. coli* glutaminyl-tRNA synthetase complexed with tRNA^{gln} and ATP at 2.8 Å resolution. *Science* 246:1135–42.

Saks, M. E., Sampson, J. R., and Abelson, J. N. 1994. The transfer RNA identity problem: A search for rules. *Science* 263:191–97.

Scott-Moncrieff, R. 1936. A biochemical survey of some Mendelian factors for flower colour. *J. Genet.* 32:117–70.

Scriver, C. R., and Clow, C. L. 1980. Phenylketonuria and other phenylalanine hydroxylation mutants in man. *Annu. Rev. Genet.* 14:179–202.

Sharp, P. A., and Eisenberg, D. 1987. The evolution of catalytic function. *Science* 238:729–30.

Srb, A. M., and Horowitz, N. H. 1944. The ornithine cycle in *Neurospora* and its genetic control. *J. Biol. Chem.* 154:129–39.

Uchino, T., et al. 1995. Molecular basis of phenotypic variation in patients with argininemia. *Hum. Genet.* 96:255–60.

Warner, J., and Rich, A. 1964. The number of soluble RNA molecules on reticulocyte polyribosomes. *Proc. Natl. Acad. Sci. USA* 51:1134–41.

Yanofsky, C., Drapeau, G., Guest, J., and Carlton, B. 1967. The complete amino acid sequence of the tryptophan synthetase A protein and its colinear relationship with the genetic map of the A gene. *Proc. Natl. Acad. Sci. USA* 57:296–98.

Ziegler, I. 1961. Genetic aspects of ommochrome and pterin pigments. *Adv. Genet.* 10:349–403.

Zubay, G. L., and Marmur, J., eds. 1973. *Papers in biochemical genetics*, 2nd ed. New York: Holt, Rinehart & Winston.

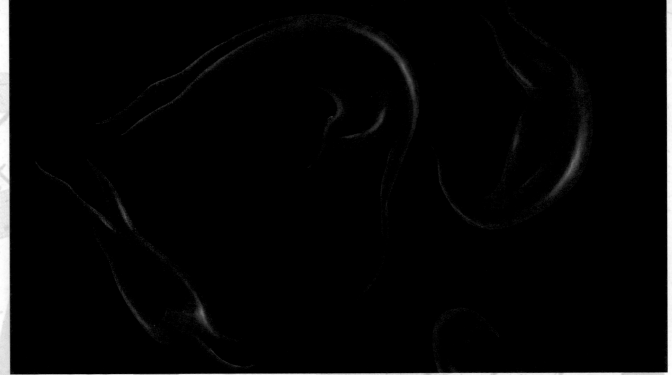

Mutant erythrocytes derived from an individual with sickle-cell anemia. *(Francis Leroy/Biocosmos/ Science Photo Library/Photo Researchers, Inc.)*

14

Gene Mutation, DNA Repair, and Transposable Elements

CHAPTER CONCEPTS

Aside from chromosomal mutations, the major basis of diversity among organisms, even those that are closely related, is genetic variation at the level of the gene. The origins of such variation are gene mutations, whereby coding sequences are altered as a result of substitution, addition, or deletion of one or more bases within those sequences. Such mutations are subject to elaborate repair mechanisms, make the study of genetics possible, and are the raw material on which evolution relies.

In Chapter 10, we defined the four characteristics or functions ascribed to the genetic information: replication, storage, expression, and variation by mutation. In a sense, mutation is a failure to store the genetic information faithfully. If a change occurs in the stored information, it may be reflected in the expression of that information and will be propagated following replication. Historically, the term *mutation* includes both chromosomal changes and changes within single genes. We discussed the former alterations in Chapter 7, referring to them collectively as **chromosomal aberrations**. In this chapter we are concerned with **gene mutations**. A change may be a simple substitution of one nucleotide or may involve the insertion or deletion of one or more nucleotides within the normal sequence of DNA.

Mutations form the basis for genetic studies. The resulting phenotypic variability provides the basis for geneticists to identify and study the genes that control the traits that have been modified. Without the phenotypic variability that mutations provide, genetic analysis would be impossible. For example, if all pea plants displayed a uniform phenotype, Mendel would have had no basis for his experimentation. Because of the importance of mutations, great attention has been given to their origin, induction, and classification.

Certain organisms lend themselves to induction of mutations that can be detected easily and studied throughout reasonably short life cycles. Viruses, bacteria, fungi, fruit flies, other invertebrates, certain plants, and mice fit these criteria. Thus, these organisms have been widely used to study mutation and mutagenesis, and through other studies they have also contributed to more general aspects of genetic knowledge.

14.1 Classification of Mutations

Mutations are classifed by various schemes. These schemes are not mutually exclusive, but instead depend simply on which aspects of mutation are being investigated or discussed. In this section we describe several distinctions that are used to classify mutations.

Spontaneous Versus Induced Mutations

All mutations are described as either *spontaneous* or *induced*. Although these two categories overlap to some degree, **spontaneous mutations** are those that just happen in nature. No specific agents are associated with their occurrence, and they are generally assumed to be random changes in the nucleotide sequences of genes. Most such mutations are linked to normal chemical processes in the organism that alter the structure of the nitrogenous bases that are part of the existing genes. Most spontaneous mutations are thought to occur during the enzymatic process of DNA replication, an idea that we discuss later in this chapter. Once an error is present in the genetic code, it may be reflected in the amino acid composition of the specified protein. If the changed amino acid is present in a part of the molecule critical to the structure or biochemical activity, a functional alteration can result.

In contrast to such spontaneous events, those that result from the influence of any artificial factor are considered to be **induced mutations**. It is generally agreed that any natural phenomenon that heightens chemical reactivity in cells will induce mutations. For example, radiation from cosmic and mineral sources and ultraviolet radiation from the sun are energy sources that most organisms are exposed to and, as such, may be factors that cause spontaneous mutations. The earliest demonstration of the artificial induction of mutation occurred in 1927, when Hermann J. Muller reported that X rays could cause mutations in *Drosophila*. In 1928, Lewis J. Stadler reported that X rays had the same effect on barley. In addition to various forms of radiation, a wide spectrum of chemical agents is also known to be mutagenic, as we shall see later in this chapter.

Gametic Versus Somatic Mutations

When we consider the effects of mutation in eukaryotic organisms, it is important to distinguish whether the change occurs in somatic cells or in gametes. Mutations arising in somatic cells are not transmitted to future generations. Mutations occurring in somatic cells that create recessive autosomal alleles are rarely of any consequence to the organism. The expression of most such mutations is likely to be masked by the dominant allele. Somatic mutations will have a greater impact if they are dominant or if they are X-linked, since such mutations are most likely to be immediately expressed. Similarly, the impact will be more noticeable if such somatic mutations occur early in development, when undifferentiated cells give rise to several differentiated tissues or organs. Mutations occurring in adult tissues are often masked by the thousands upon thousands of nonmutant cells performing the normal function.

Mutations in gametes or gamete-forming tissues are part of the germ line and are of greater concern because they are transmitted to offspring. **Dominant autosomal mutations** will be expressed phenotypically in the first generation. **X-linked recessive mutations** arising in the gametes of a heterogametic female may be expressed in hemizygous male offspring. This will occur provided that the male offspring receives the affected X chromosome. Because of heterozygosity, the occurrence of an **autosomal recessive mutation** in the gametes of either males or females (even one resulting in a lethal allele) may go unnoticed for many generations, until the resultant allele has become widespread in the population. The new allele will become evident only when a chance mating brings two copies of it together in the homozygous condition.

Other Categories of Mutation

Various types of mutations are classified on the basis of their effect on the organism. Note that a single mutation may well fall into more than one category. The most easily observed mutations are those affecting a **morphological trait**. For example, all of Mendel's pea characters and many genetic variations encountered in the study of *Drosophila* fit this designation. They cause obvious changes in morphology.

A second broad category of mutations includes those that exhibit **nutritional** or **biochemical variations** in phenotype. In bacteria and fungi, the inability to synthesize a particular amino acid or vitamin is an example of a typical nutritional mutation. In humans, sickle-cell anemia and hemophilia are examples of biochemical mutations. While such mutations in these organisms are not visible and do not always affect specific morphological characters, they can have a more general effect on the well-being and survival of the affected individual.

A third category consists of mutations that affect behavior patterns of an organism. For example, mating behavior or circadian rhythms of animals may be altered. The primary effect of **behavior mutations** is often difficult to discern. For example, the mating behavior of a fruit fly can be impaired if it cannot beat its wings. However, the defect may be in (1) the flight muscles, (2) the nerves leading to them, or (3) the brain, where the nerve impulses that initiate wing movements originate. The study of behavior and the genetic factors influencing it has benefited immensely from investigations of behavior mutations.

Another type of mutation can affect the regulation of genes. A regulatory gene can produce a product that controls the transcription of another gene. In other instances, a region of DNA either close to, or far from a gene can modulate its activity. In either case, **regulatory mutations** can disrupt normal regulatory processes and permanently activate or inactivate a gene. Our knowledge of genetic regulation depends on the study of mutations that disrupt this process.

Still another group consists of **lethal mutations**. Nutritional and biochemical mutations can also fall into this category. A mutant bacterium that cannot synthesize a specific amino acid it needs will be unable to grow and divide if plated on a medium lacking that amino acid. Various human biochemical disorders, such as Tay–Sachs disease and Huntington disease, are lethal at different points in the life cycle of humans.

Finally, any of these categories can exist as **conditional mutations**. Even though a mutation is present in the genome of an organism, it may not be evident under certain conditions. Among the best examples are **temperature-sensitive mutations**, found in a variety of organisms. At certain "permissive" temperatures, a mutant gene product functions normally, only to lose its functional capability at a different, "restrictive" temperature. When shifted to a restrictive temperature, the impact of the mutation becomes apparent, even lethal, and is amenable to investigation. The study of conditional mutations has been extremely important in experimental genetics, particularly in understanding the function of genes essential to the viability of organisms.

14.2 Detection of Mutation

Before geneticists can study the mutational process directly or obtain mutant organisms for genetic investigations, they must be able to detect mutations. The ease and efficiency of detecting mutations in a particular organism has generally determined the organism's usefulness in genetic studies. In this section we use several examples to show how mutations are detected.

Detection in Bacteria and Fungi

Detection of mutations is most efficient in haploid microorganisms such as bacteria and fungi. Detection depends on a selection system by which mutant cells can be isolated easily from nonmutant cells. The general principles are similar in bacteria and fungi. To illustrate, we describe how nutritional mutations in the fungus *Neurospora crassa* are detected.

Neurospora is a pink mold that normally grows on bread. It can also be cultured in the laboratory. This eukaryotic mold is haploid in the vegetative phase of its life cycle. Thus, mutations can be detected without the complications generated by heterozygosity in diploid organisms. Wild-type *Neurospora* grows on a **minimal culture medium** of glucose, a few inorganic acids and salts, a nitrogen source such as ammonium nitrate, and the vitamin biotin. Induced nutritional mutants will not grow on minimal medium, but will grow on a supplemented or **complete medium** that also contains numerous amino acids, vitamins, nucleic acid derivatives, and so forth. Microorganisms that are nutritional wild types (requiring only minimal medium) are called **prototrophs**, while those mutants that require a specific supplement to the minimal medium are called **auxotrophs**.

Nutritional mutants can be detected and isolated by their failure to grow on minimal medium and their ability to grow on complete medium. The mutant cells cannot synthesize some essential compound that is absent in minimal medium but present in complete medium. Once a nutritional mutant is detected and isolated, the missing compound is determined by attempts to grow the mutant strain in a series of tubes, each containing minimal medium supplemented with a single compound. The auxotrophic mutation can be defined in this way.

In Chapter 13, this detection technique was used by Beadle and Tatum during their "one-gene:one-enzyme" work, where we discussed it in the context of isolation of mutant *Neurospora* strains (see Figure 13–11). Similar techniques have been useful in the analysis of microorganisms (Chapter 9). These have been particularly critical during the study of molecular genetics.

Detection in *Drosophila*

Muller, in his studies demonstrating that X-rays are mutagenic, developed a number of detection systems in *Drosophila melanogaster*. These systems are used to estimate both spontaneous and induced rates of X-linked and autosomal recessive lethal mutations. We shall consider the **attached-X procedure**, which assesses X-linked mutations.

Attached-X females have two X chromosomes attached to a single centromere and one Y chromosome, in addition to the normal diploid complement of autosomes. When attached-X females are mated to males with normal sex chromosomes (XY), four types of progeny result: triplo-X females that die, viable attached-X females, YY males that

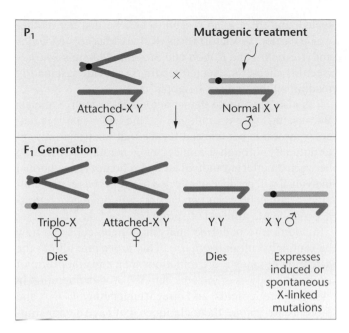

FIGURE 14–1 The attached-X method for detection of induced morphological mutations in *Drosophila*.

also die, and viable XY males. Figure 14–1 shows how P₁ males that have been treated with a mutagenic agent produce F₁ male offspring that express any induced X-linked recessive mutation.

The same approach works equally well whether or not a mutagenic agent has been used. When agents are not used, spontaneous mutations are expressed in the first generation. In detection techniques devised for recessive autosomal lethals in *Drosophila*, dominant marker mutations are followed through a series of three generations. Although these techniques are more cumbersome to perform, they are also fairly efficient.

Detection in Plants

Genetic variation in plants is extensive. Mendel's peas were the basis for the fundamental postulates of transmission genetics, and subsequent studies of plants have enhanced our understanding of gene interaction, polygenic inheritance, linkage, sex determination, chromosome rearrangements, and polyploidy. Many variations are detected simply by vi-

sual observation, but techniques also exist for detecting biochemical mutations in plants.

One technique involves the analysis of a plant's biochemical composition. For example, the isolation of proteins from maize endosperm, hydrolysis of the proteins, and determination of the amino acid composition have revealed that the *opaque*-2 mutant strain contains significantly more lysine than do other, nonmutant lines. Because maize protein is usually low in lysine, this mutation significantly improves the nutritional value of the plant. Once this mutant was discovered, plant geneticists and other specialists analyzed the amino acid compositions of various strains of other grain crops, including rice, wheat, barley, and millet. The results of such analyses are useful in combating malnutrition diseases resulting from inadequate protein or the lack of essential amino acids in the diet.

Other detection techniques involve the tissue culture of plant cell lines in defined medium. The plant cells are treated like microorganisms, and their resistance to herbicides or disease toxins can be determined by adding these compounds to the culture medium. The techniques associated with conditional lethal mutants can be used on plant cells in tissue culture and then applied to the genetics of higher plants, providing a means of detection that would not be possible in an intact plant.

Detection in Humans

Because humans are obviously not suitable experimental organisms, the techniques that have been developed for detecting of mutations in organisms such as *Drosophila* are not available to human geneticists. To determine the genetic basis for any human characteristic or disorder, geneticists first analyze a pedigree that traces the family history as far back as possible. If a trait is shown to be inherited, it is possible to predict whether the mutant allele is behaving as a dominant or a recessive and whether it is X-linked or autosomal.

Dominant mutations are the simplest to detect. If they are present on the X chromosome, affected fathers pass the phenotypic trait to all their daughters. If dominant mutations are autosomal, approximately 50 percent of the offspring of an affected heterozygous individual are expected to show the trait. Figure 14–2 shows a pedigree illustrating the initial occurrence of an autosomal dominant allele for cataracts of the eye. The parents in generation I were unaffected, but one of

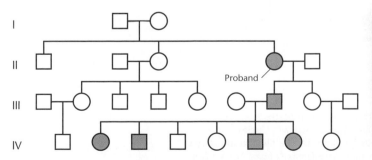

FIGURE 14–2 A hypothetical pedigree of inherited cataract of the eye in humans. *(Photo: Sue Ford/Science Photo Library/Photo Researchers, Inc.)*

three offspring (generation II) developed cataracts. This female, the proband, produced two children, of which the male child was affected. Of his six offspring, four of six were affected (generation IV). These observations are consistent with, but do not prove, an autosomal dominant mode of inheritance. However, the high percentage of affected offspring in generation IV favors this conclusion. Also, the unaffected daughter in this generation argues against X-linkage, because she received her X chromosome from her affected father. This conclusion is sound, provided that the mutant allele is completely penetrant.

X-linked recessive mutations can also be detected by pedigree analysis, as discussed in Chapter 4. The most famous case of an X-linked mutation in humans is that of **hemophilia**, which was found in the descendants of Queen Victoria. The recessive mutation for hemophilia has occurred many times in human populations, but the political consequences of the mutation that occurred in the royal family were sweeping. Inspection of the pedigree in Figure 14–3 leaves little doubt that Victoria was heterozygous (*Hh*) for the trait. Her father was not af-

fected, and there is no reason to believe that her mother was a carrier, as Victoria was. Robert Massie's *Nicholas and Alexandra* and Robert and Suzanne Massie's *Journey* (see this chapter's selected readings) provide fascinating reading on the topic of hemophilia.

It is also possible to detect autosomal recessive alleles. Because this type of mutation is "hidden" when it is heterozygous, it is not unusual for the trait to appear only intermittently through a number of generations. A mating between an affected individual and a homozygous normal individual will produce unaffected heterozygous carrier children. Matings between two carriers will produce, on average, one-fourth affected offspring.

In addition to pedigree analysis, human cells can now be routinely cultured *in vitro*. This procedure allows the detection of many more mutations than any other form of analysis. Analysis of enzyme activity, protein migration in electrophoretic fields, and direct sequencing of DNA and proteins are among the techniques that have demonstrated wide genetic variation between individuals in human populations.

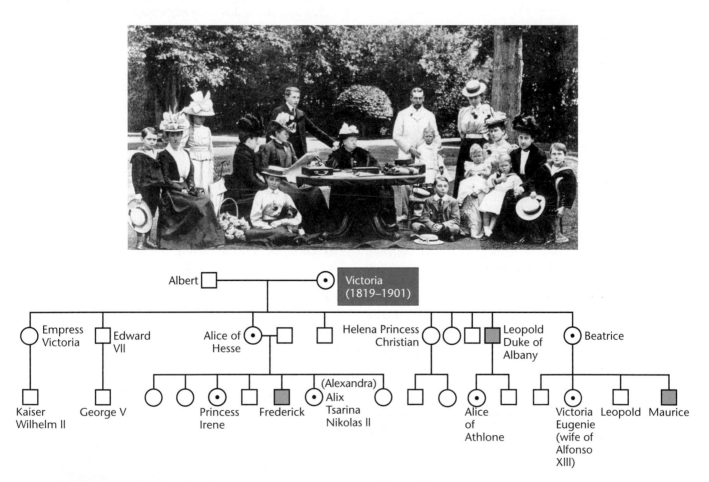

FIGURE 14–3 A partial pedigree of hemophilia in the British royal family descended from Queen Victoria. The pedigree is typical of the transmission of X-linked recessive traits. Circles with a dot in them indicate presumed female carriers heterozygous for the trait. The photograph shows Queen Victoria (seated front center) and some of her immediate family. (*Photo: Mary Evans Picture Library/Photo Researchers, Inc.*)

14.3 Spontaneous Mutation Rate

The types of detection systems we have been discussing allow geneticists to estimate mutation rates. However, we can only ascertain induced mutation when the induced rate clearly exceeds the spontaneous mutation rate for the organism under study. Determining the rate of spontaneous mutation provides the baseline for measuring the rate of experimentally induced mutation.

Examination of the spontaneous rate in a variety of organisms reveals many interesting points. First, the rate is exceedingly low for all organisms studied. Second, the rate is seen to vary considerably in different organisms. Third, even within the same species, the spontaneous mutation rate varies from gene to gene.

Viral and bacterial genes undergo spontaneous mutation on an average of about 1 in 100 million (10^{-8}) cell divisions. While *Neurospora* exhibits a similar rate, maize, *Drosophila*, and humans demonstrate a rate several orders of magnitude higher. The genes studied in these groups average between 1/1,000,000 and 1/100,000 (10^{-6} to 10^{-5}) mutations per gamete formed. Mouse genes are still another order of magnitude higher in their spontaneous mutation rate, 1/100,000 to 1/10,000 (10^{-5} to 10^{-4}) It is not clear why such a large variation occurs in mutation rate. The variation might reflect the relative efficiency of enzyme systems whose function is to repair errors created during replication. We discuss repair systems later in this chapter.

14.4 The Molecular Basis of Mutation

Even though we are aware that the gene is a reasonably complex genetic unit, particularly in eukaryotes, we will use a *simplified* definition of a gene in the following description of the molecular basis of mutation. In this context, it is easiest to consider a gene as a linear sequence of nucleotide pairs representing stored chemical information. Because the genetic code is a triplet, each sequence of three nucleotides specifies a single amino acid in the corresponding polypeptide. Any change that disrupts these sequences or the coded

information provides sufficient basis for a mutation. The least complex change is the substitution of a single nucleotide. In Figure 14–4, such a change is compared with our own written language, using three-letter words to be consistent with the genetic code. A change of one letter alters the meaning of the sentence "THE CAT SAW THE DOG," to "THE CAT SAW THE HOG" or "THE BAT SAW THE DOG," creating what is called *missense*. These are analogies to what are most appropriately referred to as **base substitutions** or **point mutations**. The mutation has changed the sense of the information into various forms of missense.

Two other more formal terms are used to describe nucleotide substitutions. If a purine replaces a purine or a pyrimidine replaces a pyrimidine, a **transition** has occurred. If a purine and a pyrimidine are interchanged, a **transversion** has occurred.

A second type of change that can occur in the nucleotide sequence is the insertion or deletion of a single nucleotide at any point along the gene. As illustrated in Figure 14–4, the remainder of the three-letter (code) words become garbled, creating much more extensive missense. These examples are called **frameshift mutations** because the frame of reading has been altered.

The analogy in Figure 14–4 demonstrates that insertions and deletions have the potential to change all subsequent triplets in a gene. It is probable that, of the 64 possible triplets, one of many altered triplets will be either UAA, UAG, or UGA. These are termination codons (see Chapter 12). When one is encountered during translation, polypeptide synthesis is terminated. Obviously, the results of frameshift mutations can be very severe!

Tautomeric Shifts

In 1953, immediately after they proposed a molecular structure of DNA, Watson and Crick published a paper in which they discussed the genetic implications of this structure. They recognized that the purines and pyrimidines found in DNA could exist in **tautomeric forms**; that is, each can exist in alternative chemical forms, differing by only a single proton shift in the molecule. Watson and Crick suggested that **tautomeric shifts** could result in base-pair changes or mutations.

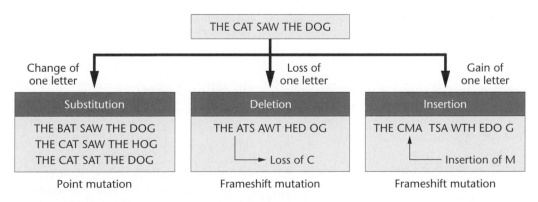

FIGURE 14–4 The impact of the substitution, deletion, and insertion of one letter in a sentence composed of three-letter words as analogies to point and frameshift mutations.

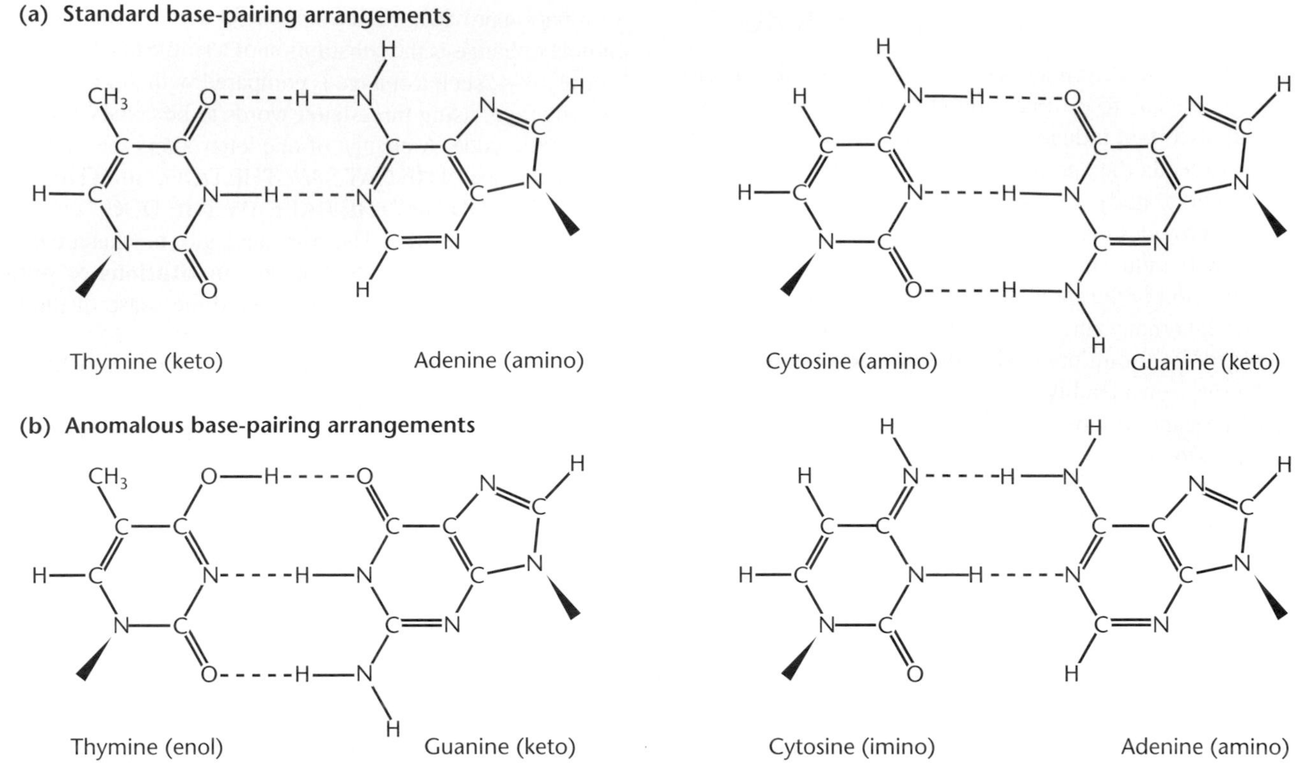

(a) Standard base-pairing arrangements

Thymine (keto)   Adenine (amino)   Cytosine (amino)   Guanine (keto)

(b) Anomalous base-pairing arrangements

Thymine (enol)   Guanine (keto)   Cytosine (imino)   Adenine (amino)

FIGURE 14–5 The standard base-pairing relationships compared with two anomalous arrangements occurring as a result of tautomeric shifts. The dense arrowhead indicates the point of bonding to the pentose sugar.

The most stable tautomers of the nitrogenous bases result in the standard hydrogen bonds that serve as the basis of the double helical model of DNA. The less frequently occurring tautomers are capable of hydrogen bonding with noncomplementary bases. However, the pairing is always between a pyrimidine and a purine. Figure 14–5 compares the normal base-pairing relationships with the rare unorthodox pairings. The biologically important unstable tautomers involve keto–enol pairs for thymine and guanine, and amino–imino pairs for cytosine and adenine.

The effect leading to mutation occurs during DNA replication when a rare tautomer in the template strand matches with a noncomplementary base. In the next round of replication, the "mismatched" members of the base pair are separated, and each specifies its normal complementary base. The end result is a transition mutation (see Figure 14–6).

Base Analogs

Base analogs, which are mutagenic chemicals, are molecules that can substitute for purines or pyrimidines during nucleic acid biosynthesis. The halogenated derivative of uracil in the number-5 position of the pyrimidine ring **5-bromouracil (5-BU)*** is a good example. Figure 14–7

(page 286) compares the structure of this thymine analog with the structure of thymine. The presence of the bromine atom in place of the methyl group increases the probability that a tautomeric shift will occur. If 5-BU is incorporated into DNA in place of thymine and a tautomeric shift to the enol form occurs, 5-BU base-pairs with guanine. After one round of replication, an $A = T$ to $G \equiv C$ transition results. There are other base analogs that are mutagenic. One, **2-amino purine (2-AP)**, can serve successfully as an analog of adenine. In addition to its base-pairing affinity with thymine, 2-AP can also base-pair with cytosine. As such, transitions from $A = T$ to $G \equiv C$ can result following replication.

Because of the specificity by which base analogs such as 2-AP induce transition mutations, base analogs can also be used to induce reversion to the wild-type nucleotide sequence. This alteration is called **reverse mutation**. The process also occurs spontaneously, but at a much lower rate.

Alkylating Agents

The sulfur-containing mustard gases were one of the first groups of chemical mutagens discovered. This discovery was made in studies involving chemical warfare during World War I. Mustard gases are **alkylating agents**; that is, they donate an alkyl group such as CH_3— or CH_3—CH_2— to amino or keto groups in nucleotides. For example, **ethylmethane sulfonate (EMS)**, commonly used as an experimental mutagen, alkylates the keto group in the number 6 position of guanine and

---

*If 5-BU is chemically linked to d-ribose, the nucleoside analog bromodeoxyuridine (BUdR) is formed.

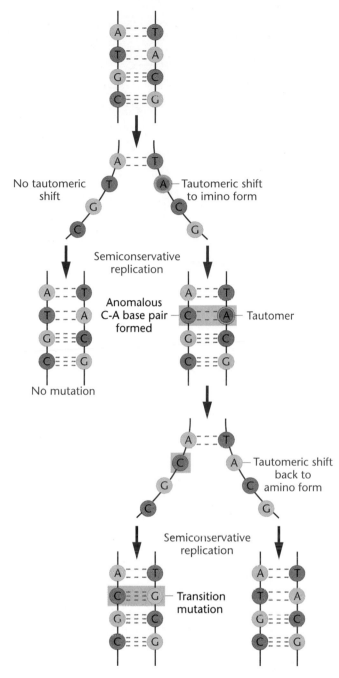

FIGURE 14–6 Formation of a T══A to a C≡≡G transition mutation as a result of a tautomeric shift in adenine.

in the number 4 position of thymine (Figure 14–8, page 286). As with base analogs, base-pairing affinities are altered and transition mutations result. In the case of **6-ethylguanine**, this molecule acts like a base analog of adenine, causing it to pair with thymine.

Acridine Dyes and Frameshift Mutations

Other chemical mutagens cause frameshift mutations, as originally depicted in Figure 14–4. These result from adding or removing one or more base pairs in the polynucleotide sequence of the gene. Inductions of frameshift mutations have

been studied in detail with a group of aromatic molecules known as **acridine dyes**. Proflavin, the most widely studied acridine mutagen, and acridine orange are examples. Acridine dyes are of about the same dimension as a nitrogenous base pair and are known to intercalate or wedge between purines and pyrimidines of intact DNA. Intercalation of acridine dyes induces contortions in the DNA helix, causing deletions and insertions.

One model suggests that the resultant frameshift mutations are generated at gaps produced in DNA during replication, repair, or recombination. During these events, there is the possibility of slippage and improper base pairing of one strand with the other. The model suggests that intercalation of the acridine into an improperly base-paired region can extend the existence of these slippage structures. If so, the probability increases that the mispaired configuration will exist when synthesis and rejoining occurs, thereby resulting in an addition or deletion of one or more bases from one of the strands.

Apurinic Sites and Deamination

Still another type of mutation involves the spontaneous loss of one of the nitrogenous bases in an intact double-helical DNA molecule. Most frequently, such an event involves either guanine or adenine. These sites, created by the "breaking" of the glycosidic bond linking the 1'-C of d-ribose and the number 9 position of the purine ring, are called **apurinic sites*** (**AP sites**). It has been estimated that thousands of such spontaneous lesions are formed daily in the DNA of mammalian cells in culture.

The absence of a nitrogenous base at an AP site will alter the genetic code if the strand involved is transcribed and translated. If replication occurs, the AP site is an inadequate template and can cause replication to stall. If a nucleotide is inserted, it is frequently incorrect, causing still another mutation! Fortunately, as we shall soon see, cells contain repair systems that often counteract and correct this type of lesion.

Another type of lesion is known to be the source of some mutations. In the process of **deamination**, an amino group is converted to a keto group in cytosine and adenine. In these two cases, cytosine is converted to uracil and adenine is changed to hypoxanthine.

The major effect of these changes is to alter the base-pairing specificities of these two molecules during DNA replication. For example, cytosine normally pairs with guanine. Following its conversion to uracil, which pairs with adenine, the original G≡≡C pair is converted to an A══U pair and, following an additional replication, to an A══T pair. When adenine is deaminated, an original A══T pair is converted to a G≡≡C pair because hypoxanthine pairs naturally with cytosine. Nitrous acid is a known mutagen that is capable of inducing deamination of bases in DNA.

---

*Apyrimidinic sites, where a pyrimidine has been lost, occur as well, and are also referred to as AP sites.

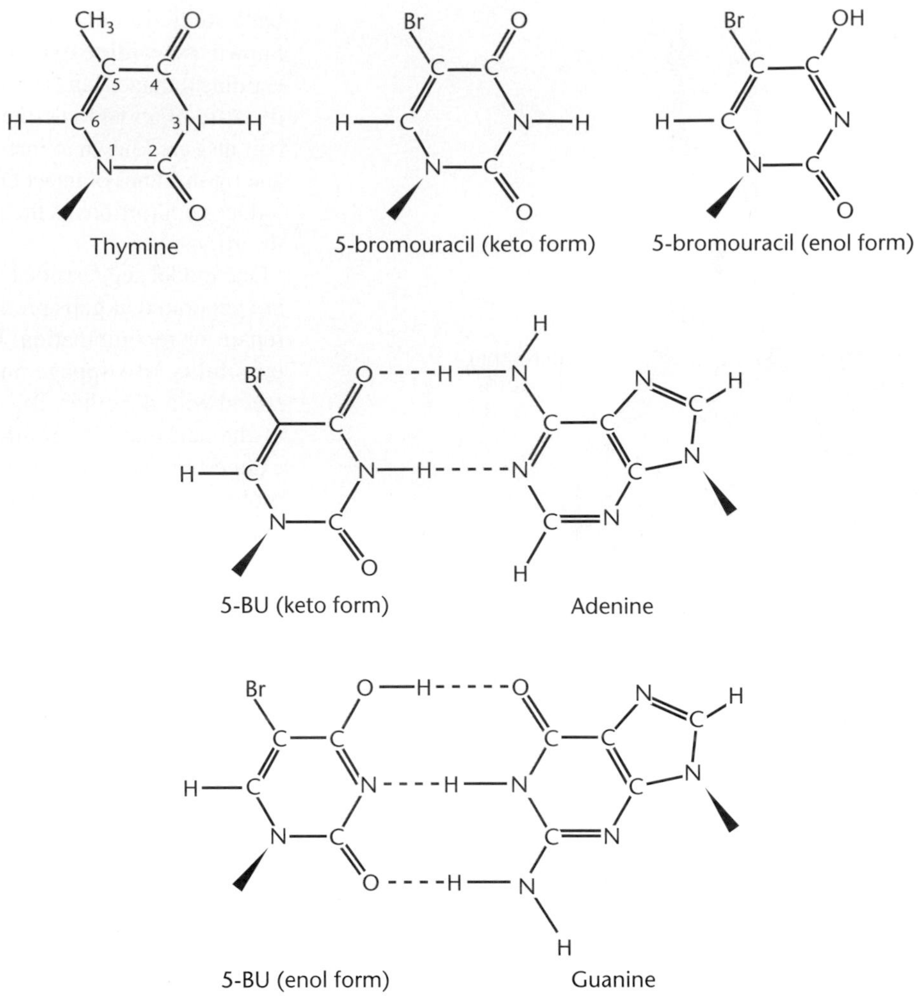

FIGURE 14–7 Similarity of 5-bromouracil (5-BU) structure to thymine structure. In the common keto form, 5-BU base-pairs normally with adenine, behaving as an analog. In the rare enol form, it pairs anomalously with guanine.

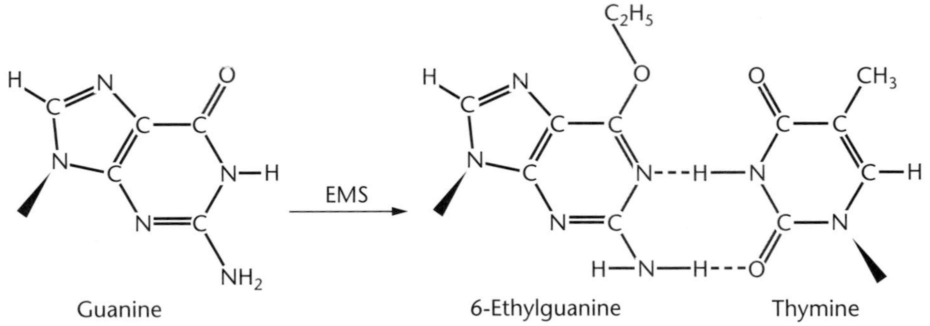

Thymine

5-bromouracil (keto form)

5-bromouracil (enol form)

5-BU (keto form)   Adenine

5-BU (enol form)   Guanine

FIGURE 14–8 Conversion of guanine to 6-ethylguanine by the alkylating agent ethylmethane sulfonate (EMS). The 6-ethylguanine base-pairs with thymine.

Guanine   6-Ethylguanine   Thymine

14.5   Ultraviolet and High Energy Radiation

All energy on the earth consists of a series of electromagnetic components of varying wavelength (Figure 14–9). Referred to as the **electromagnetic spectrum**, the energy associated with the various components varies inversely with wavelength. Everything longer than, and including visible light, is benign when it interacts with most organic molecules. However, everything of shorter wavelength than visible light, which is inherently more energetic, has a disruptive impact on organic molecules, such as those comprising living tissue. For example, in Chapter 10, we

emphasized the fact that purines and pyrimidines absorb **ultraviolet (UV) radiation** most intensely at a wavelength of about 260 nm, a property that has been useful in the detection and analysis of nucleic acids. In 1934, as a result of studies involving *Drosophila* eggs, it was discovered that UV radiation is mutagenic, and by 1960, several studies concerning the *in vitro* effect on the components of nucleic acids had been completed. The major effect of UV radiation is to induce the formation of **pyrimidine dimers**, particularly between two thymine residues (Figure 14–10). While cytosine–cytosine and thymine–cytosine dimers can also be formed, they are less prevalent. The dimers distort the DNA conformation and inhibit normal replication. As

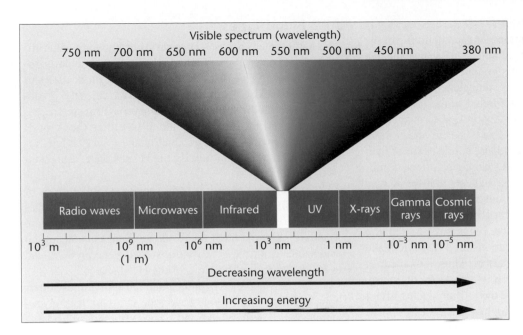

a result, errors can be introduced in the base sequence of DNA during replication, and when extensive, these are responsible, at least in part, for the killing effects of UV radiation on microorganisms.

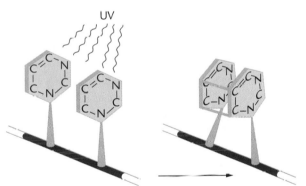

Dimer formed between adjacent thymidine residues along a DNA strand

FIGURE 14–10 Induction of a thymine dimer by UV radiation, leading to distortion of the DNA. The atoms of the pyrimidine ring are shown to illustrate the formation of the covalent cross-links.

As we shall see later in this chapter, several mechanisms have evolved to correct UV-induced lesions. We illustrate the essential nature of such "repair" processes in humans by discussing the inherited disorder xeroderma pigmentosum, where mutation has inactivated one UV-repair mechanism. Such individuals have been described as "children of the night," since they cannot risk exposure to the ultraviolet component of sunlight without the risk of epidermal malignancy.

Ionizing Radiation

As shown in Figure 14–9, X-rays, gamma rays, and cosmic rays have even shorter wavelengths than does UV radiation and are therefore more energetic. As a result, they are strong enough to penetrate deeply into tissues, causing ionization of the molecules encountered along the way. Hermann J. Muller and Lewis J. Stadler established in the 1920s that these sources of **ionizing radiation** are mutagenic. Since that time, the effects of ionizing radiation, particularly X-rays, have been studied intensely.

As X-rays penetrate cells, electrons are ejected from the atoms of molecules encountered by the radiation. Thus, stable molecules and atoms are transformed into free radicals and reactive ions. The trail of ions left along the path of a high-energy ray can initiate a variety of chemical reactions. These reactions can directly or indirectly affect the genetic material, altering the purines and pyrimidines in DNA and resulting in point mutations. Ionizing radiation is also capable of breaking phosphodiester bonds, disrupting the integrity of chromosomes and producing a variety of aberrations.

Figure 14–11 graphs the percentage of induced X-linked recessive lethal mutations versus the dose of X rays

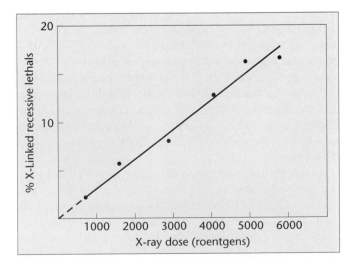

FIGURE 14–11 Plot of the percentage of X-linked recessive mutations induced by increasing doses of X-rays. If extrapolated, the graph intersects the zero axis.

administered. A linear relationship is evident between X-ray dose and the induction of mutation; for each doubling of the dose, twice as many mutations are induced. Because the line intersects near the zero axis, this graph suggests that even very small doses of irradiation are mutagenic.

These observations can be interpreted in the form of the **target theory**, first proposed in 1924 by J. A. Crowther and F. Dessauer. The theory proposes that there are one or more sites, or targets, within cells and that a single event of irradiation at one site will bring about a damaging effect, or mutation. In effect, the target theory suggests that the X-rays interact directly with the genetic material.

Two other observations concerning irradiation effects are of particular interest. First, in some organisms studied, the *intensity of the dose* (dose rate) administered seems to make little difference in the mutagenic effect. Thus, a total exposure of 100 roentgens (a **roentgen** is a measure of energy dose) seems to produce the same overall mutagenic effect whether it is administered in a single acute dose or spread out over time in several smaller chronic doses. *Drosophila* has shown this response. In mammals such as mice and humans, however, this appears not to be the case, suggesting that repair of the damage may occur during the intervals between irradiation.

Second, certain portions of the cell cycle are more susceptible to radiation effects. As we mentioned, X-rays can break chromosomes, resulting in terminal or intercalary deletions, translocations, and general chromosome fragmentation. Damage occurs most readily when chromosomes are greatly condensed in mitosis. This property is one reason why radiation is used to treat human malignancy. Because tumorous cells undergo division more often than their nonmalignant counterparts, they are more susceptible to the immobilizing effect of radiation.

14.6   Mutations in Humans: Case Studies

So far we have been discussing the molecular basis of mutation largely in terms of nucleic acid chemistry. We know that various types of nucleotide conversions or frameshift mutations occur, primarily because our analyses of amino acid sequences of many proteins within the populations of a species show substantial diversity. This diversity, which has arisen during evolution, is a reflection of changes in the triplet codons following substitution, insertion, or deletion of one or more nucleotides in the DNA sequences constituting genes.

As our ability to analyze DNA more directly increases, we are better able to look specifically at the actual nucleotide sequence of genes and to gain greater insights into mutation. Several techniques capable of accurate, rapid sequencing of DNA (see Chapter 16) have greatly extended our knowledge of molecular genetics. In this section we examine the results of a number of studies that have investigated the actual gene sequence of various mutations that affect humans.

ABO Blood Types

The ABO system is based on a series of antigenic determinants found on erythrocytes and other cells, particularly epithelial types. As we discussed in Chapter 4, three alleles of a single gene exist, the product of which modifies the H substance. The modification involves glycosyltransferase activity, converting the H substance to either the A or B antigen, as a result of the product of the I^A or I^B allele, respectively; or failing to modify the H substance, as a result of the I^O allele.

The responsible gene has been sequenced in 14 cases of varying ABO status. When the DNAs of the I^A and I^B alleles were compared, four consistent nucleotide substitutions were found. It is assumed that the resulting changes in the amino acid sequence of the glycosyltransferase gene product lead to the different modifications of the H substance.

The I^O allele situation is unique and intriguing. Individuals homozygous for this allele have type O blood, lack glycosyltransferase activity, and fail to modify the H substance. Analysis of the DNA of this allele shows one consistent change that is unique compared to the sequence of the other alleles, the deletion of a single nucleotide early in the coding sequence, causing a frameshift mutation. A complete messenger RNA is transcribed, but at translation, the frame of reading shifts at the point of deletion and continues out of frame for about 100 nucleotides before a stop codon is encountered. At this point, the polypeptide chain terminates prematurely, resulting in a nonfunctional product.

These findings provide a direct molecular explanation of the ABO allele system and the basis for the biosynthesis of the corresponding antigens. The molecular basis for the antigenic phenotypes is clearly the result of structural alterations, or mutations, of the nucleotide sequence of the gene encoding the glycosyltransferase enzyme.

Muscular Dystrophy

Muscular dystrophy is characterized by severe progressive muscular degeneration, or myopathy, resulting in the death of the affected individuals by early adulthood. Because the condition is recessive and X-linked, and because affected males die before they can reproduce, females are rarely affected by the disorder. The incidence of 1/3500 live male births makes muscular dystrophy one of the most common life-shortening hereditary disorders known. Two related forms exist. **Duchenne muscular dystrophy (DMD)** is more common and more severe than the allelic form, **Becker muscular dystrophy (BMD)**.

The region containing the gene, which consists of over 2 million bp, has been analyzed extensively. In normal (unaffected) individuals, transcription results in a messenger RNA containing about 14,000 bases (14 kb) that is translated into the protein dystrophin, consisting of 3685 amino acids. This protein can be detected in most cases of the less severe BMD, but is rarely found in DMD. This has led to the hypothesis that most mutations causing BMD do not alter the reading frame, but that most DMD mutations change the reading frame early in the gene, resulting in the premature termination of dystrophin translation. This hypothesis is consistent with the observed differences in severity of these two forms of the disorder.

In an extensive analysis of the DNA of 194 patients (160 DMD and 34 BMD), J. T. Den Dunnen and associates found that 128 of these mutations (65%) consisted of substantial deletions or insertions. Of 115 deletions, 17 occurred in BMD cases, and of 13 insertions, 1 occurred in BMD, with the remainder in DMD cases. In most cases, the results were consistent with the "reading frame" hypothesis: With few exceptions, DMD mutations changed the frame of reading, whereas BMD mutations usually did not alter the reading frame.

Perhaps the most noteworthy of Den Dunnen's findings is the high percentage of deleterious mutations that represent the deletion or insertion of nucleotides within the gene. This observation reflects the fact that a mutation caused by a random single-nucleotide substitution within a gene is more likely to be tolerated without the devastating effect of muscular dystrophy than the addition or loss of numerous nucleotides that can alter the frame of reading. There are three reasons for this.

1. A nucleotide substitution may not change the encoded amino acid, since the code is degenerate.

2. If an amino acid substitution does result, the change may not be present at a location within the protein that is critical to its function.

3. Even if the altered amino acid is present at a critical region, it may still have little or no effect on the function of the protein. For example, an amino acid might be changed to another with nearly identical chemical properties or to one with very similar recognition properties, such as shape.

As a result, single-base substitutions sometimes have little or no effect on protein function, or they may simply reduce the efficiency but not eliminate the functional capacity of the gene product. As more mutant genes are analyzed directly, our picture of mutation will become increasingly clear.

Trinucleotide Repeats in Fragile-X Syndrome, Myotonic Dystrophy, and Huntington Disease

Beginning about 1990, the molecular analysis of the DNA representing the genes responsible for a number of inherited human disorders provided a remarkable set of observations. Mutant genes were characterized by an expansion of a simple **trinucleotide repeat sequence**, usually from fewer than 15 copies in normal individuals to a large number in affected individuals. For example, in each of the three separate genes responsible for the autosomal disorders **myotonic dystrophy** and **Huntington disease** and the X-linked **fragile-X syndrome**, each gene was found to contain a unique trinucleotide DNA sequence repeated many times. While repeated sequences are also present in the nonmutant (normal) allele of each gene, the mutations are characterized by a significant variable increase in the number of times the trinucleotide is repeated. Table 14.1 summarizes the various cases that are discussed below, with particular emphasis on the size of the repeats in normal and diseased individuals.

Recall that in Chapter 4 we discussed the onset of the expression of various phenotypes. In several cases, a correlation has been found between the number of repeats and the manifestation of mutant phenotypes. The greater the number of repeats, the earlier disease onset occurs. Further, the number of repeats can increase in each subsequent generation. This general phenomenon, **genetic anticipation**, represents a unique form of mutation related to an instability of specific regions of the three different genes. Other normal genes in humans, as well as many other species, also contain trinucleotide repeats; however, the repeats do not balloon in size, and the genes maintain normal function.

In Chapter 7, we discussed fragile-X syndrome. The responsible gene, *FMR-1*, can have several hundred to several thousand copies of the trinucleotide sequence CGG. Individuals with up to 50 copies are normal and do not display the mental retardation associated with the syndrome. Individuals with 50–200 copies are considered "carriers." Although they are normal, their offspring may contain even more copies and express the syndrome.

Myotonic dystrophy, the most common form of adult muscular dystrophy, is due to a mutant gene that contains multiple copies of the sequence CTG. Fewer than 35 copies result in normal gene expression. Above this number, symptoms range from mild myopathy and cataracts to severe dystrophy, retardation, and even death in early childhood. Both the severity and onset are directly correlated with the size of the repeated sequence.

More recently, the gene responsible for Huntington disease has been found to demonstrate a similar pattern. The trinucleotide CAG repeat, present 10–35 times in normal individuals, exists in significantly increased numbers in diseased individuals. Much earlier onset occurs when a very large number of copies are present. Interestingly, in still another

---

TABLE 14.1 Summary of Trinucleotide-Repeat Disorders

	Trinucleotide Repeat	*Number (normal)*	*Number (affected)*	*Inheritance Pattern*
Huntington disease	CAG	6–35	36–120	autosomal dominant
Myotonic dystrophy	CTG	5–35	>200	autosomal dominant
Fragile-X syndrome	CGG	6–50	>200	X-linked dominant
Spinobulbar muscular atrophy	CAG	10–30	35–60	X-linked dominant

disorder, **spinobulbar muscular atrophy (Kennedy disease)**, the gene also contains repeated copies of the CAG triplet. However, diseased individuals usually have only 35–60 copies of the sequence.

The role of such repeated sequences in normal and mutant genes remains a mystery. Their locations within the gene vary in each case. In Huntington and Kennedy diseases, the repeat lies within the coding portion of the gene. This is not the case in the other two disorders, however. In the gene responsible for fragile-X syndrome, the repeat is upstream (5′) in an area that is most often involved in regulating gene expression. In the case of myotonic dystrophy, the repeat is downstream (the 3′-end).

The mechanism by which the repeated sequence expands from generation to generation is of great interest. How such an instability during DNA replication affects only specific areas of certain genes is currently an important research topic. This instability seems to be more prevalent in humans than in many other organisms.

14.7 Detection of Mutagenicity: The Ames Test

There is great concern about the possible mutagenic properties of any chemical that enters the human body, whether through the skin, digestive system, or respiratory tract. For example, great attention has been given to residual materials of air and water pollution, food preservatives and additives, artificial sweeteners, herbicides, pesticides, and pharmaceutical products. As we have seen, mutagenicity can be tested in various organisms, including *Drosophila*, mice, and cultured mammalian cells. The most common test, which involves bacteria, was devised by Bruce Ames.

The Ames test (Figure 14–12) uses any of several strains of the bacterium *Salmonella typhimurium* that were selected for sensitivity and specificity for mutagenesis. One strain is used to detect base-pair substitutions, and the other three detect various frameshift mutations. Each mutant strain is unable to synthesize histidine, and therefore requires histidine for growth (*his⁻*). The assay measures the frequency of reverse mutation, which yields wild-type (*his⁺*) bacteria. Greater sensitivity to mutagens occurs because these strains bear other mutations that eliminate both the DNA excision-repair system (discussed later in this chapter) and the lipopolysaccharide barrier that coats and protects the surface of the bacteria.

It is very interesting to note that many substances entering the human body are relatively innocuous until activated metabolically, usually in the liver, to a more chemically reactive product. Thus, the Ames test includes a step in which the test compound is incubated *in vitro* in the presence of a mammalian liver extract. Or, test compounds can be injected into a mouse, where they are modified by liver enzymes and then recovered.

FIGURE 14–12 The Ames test, which screens potential compounds for mutagenicity.

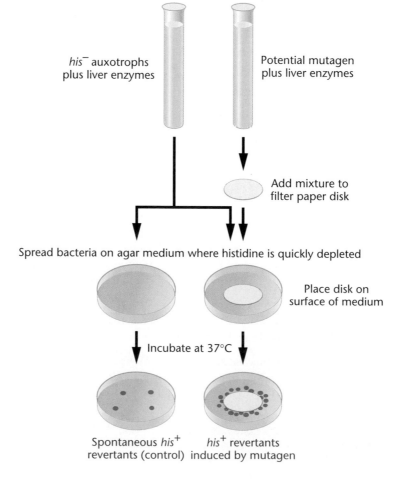

his⁻ auxotrophs plus liver enzymes

Potential mutagen plus liver enzymes

Add mixture to filter paper disk

Spread bacteria on agar medium where histidine is quickly depleted

Place disk on surface of medium

Incubate at 37°C

Spontaneous *his⁺* revertants (control)

his⁺ revertants induced by mutagen

In the initial use of Ames testing in the 1970s, a large number of known carcinogens were examined and over 80 percent were shown to be strong mutagens! This is not surprising; the transformation of cells to the malignant state undoubtedly occurs as a result of some alteration of DNA. Although a positive response as a mutagen does not prove that a compound is carcinogenic, the Ames test is useful as a preliminary screening device. It is used extensively in conjunction with the industrial and pharmaceutical development of chemical compounds.

14.8 Counteracting DNA Damage: Repair Systems

In previous sections of this chapter we established that replicating and nonreplicating DNA molecules are vulnerable to various forms of errors, which induce lesions that lead to gene mutations. Living systems have evolved a variety of very elaborate repair systems that are able to counteract many of the forms of DNA damage that lead to mutation. As we shall see, such repair systems are essential to the maintenance of the genetic integrity of organisms, and as such, to the survival of organisms on the earth.

Photoreactivation Repair: Reversal of UV Damage in Prokaryotes

As shown in Figure 14–10, UV light is mutagenic as a result of the creation of pyrimidine dimers. The study of the mutagenicity of UV radiation paved the way for the discovery

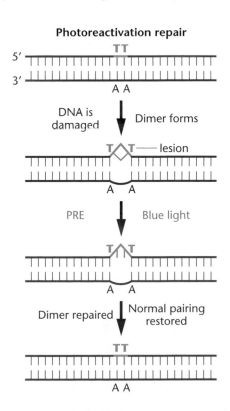

FIGURE 14–13 Damaged DNA repaired by photoreactivation repair. The bond creating the thymine dimer is cleaved by the photoreactivation enzyme (PRE), which must be activated by blue light.

of many forms of natural repair of DNA damage. The first relevant discovery concerning UV repair in bacteria was made in 1949 when Albert Kelner observed the phenomenon of **photoreactivation repair**. He showed that the UV-induced damage to *E. coli* DNA could be partially reversed if, following irradiation, the cells were exposed briefly to light in the blue range of the visible spectrum. The photoreactivation repair process was subsequently shown to be temperature dependent, suggesting that the light-induced mechanisms involve an enzymatically controlled chemical reaction. Visible light appears to induce the repair process of the DNA damaged by UV radiation.

Further studies of photoreactivation have revealed that the process is dependent on the activity of a protein called the **photoreactivation enzyme (PRE)**. This molecule can be isolated from extracts of *E. coli* cells. The enzyme's mode of action is to cleave the bonds between thymine dimers, thus reversing the effect of UV radiation on DNA (Figure 14–13). Although the enzyme will associate with a dimer in the dark, it must absorb a photon of light to cleave the dimer. In spite of the potential for diminishing UV-induced mutations, photoreactivation repair is not essential in *E. coli*, since a null mutation in the gene coding for the PRE is not lethal. In addition, the enzyme cannot be detected in humans and other eukaryotes, who must rely on other repair mechanisms to reverse the effects of UV radiation.

Excision Repair in Prokaryotes and Eukaryotes

Investigations in the early 1960s suggested that, in addition to PRE, a *light-independent* repair system exists in prokaryotes such as *E. coli* that repairs damage to DNA caused by both exogenous and endogenous agents. The basic mechanism of this type of repair, referred to as **excision repair**, has been conserved throughout evolution, occurring in all prokaryotic and eukaryotic organisms. The process consists of three basic steps.

1. The distortion or error is recognized and enzymatically clipped out by a nuclease. This "excision" may affect only a single base, a single nucleotide, or include several nucleotides adjacent to the error as well, leaving a gap in the helix.

2. DNA polymerase I fills this gap by inserting d-ribonucleotides complementary to those on the intact strand. The enzyme adds these bases to the 3′-OH end of the clipped DNA.

3. The joining enzyme DNA ligase seals the final "nick" that remains at the 3′-OH end of the last base inserted, closing the gap.

DNA polymerase I, the enzyme discovered by Arthur Kornberg, was once assumed to be the universal DNA-replication enzyme (see Chapter 11). However, it was the discovery of the *polA1* mutation that demonstrated the role of this enzyme in repairing UV-induced lesions in *E. coli*. Cells carrying the *polA1* mutation lack functional polymerase I, replicate their DNA normally, and are unusually sensitive to UV radiation. Apparently, such cells are unable to fill the gap created by the excision of the thymine dimers.

There are now known to be two types of excision repair: base excision repair and nucleotide excision repair. **Base excision repair (BER)** is involved in correcting damage to nitrogenous bases created by the spontaneous hydrolysis of DNA bases, as well as by agents that chemically alter them. As shown in Figure 14–14, the first step of the BER pathway in *E. coli* involves the recognition of the chemically altered base by **DNA glycosylases**, which are specific to different types of DNA damage. For example, the enzyme uracil-DNA glycosylase recognizes the unusual presence of uracil when it is part of a nucleotide in DNA. The enzyme first cuts the glycosidic bond between the base and the sugar, creating an apyrimidinic (AP) site. Such a sugar with a missing base is then recognized by an enzyme called **AP endonuclease**. The endonuclease makes a cut in the sugar backbone at the AP site. This creates a distortion in the DNA helix that is recognized by the excision-repair system, which is then activated, leading ultimately to the correction of the error.

Although much has been learned about glycosylases in *E. coli*, very little is known about the DNA glycosylases in eukaryotes. In addition, unlike the nucleotide excision pathway we are going to discuss next, there is no human or animal disease model available that has a defective BER pathway, so it is difficult to assess the importance of BER in eukaryotes.

While base excision repair recognizes and replaces modified bases in DNA, the **nucleotide excision repair (NER)** pathway repairs "bulky" lesions in DNA that alter or distort the regular DNA double helix, such as the UV-induced pyrimidine dimers discussed previously. The NER pathway, diagrammed in Figure 14–15, was first discovered in *E. coli* by Paul Howard-Flanders and co-workers who were able to isolate several independent mutants that demonstrated sensitivity to UV radiation. One group of genes was designated *uvr* (ultraviolet repair) and included the *uvrA*, *uvrB*, and *uvrC* mutations. In the NER pathway, the *uvr* gene products are involved in recognizing and clipping out the lesions in the DNA. The repair is then completed by DNA polymerase I and DNA ligase.

Xeroderma Pigmentosum and Nucleotide Excision Repair

The mechanism of nucleotide excision repair (NER), which has been elucidated in eukaryotes, is much more complicated than prokaryotic NER because it involves many more proteins. Much of what is known about the system in humans has resulted from detailed studies of individuals with **xeroderma**

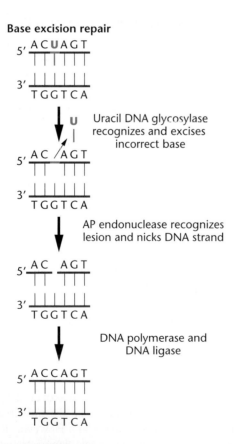

Base excision repair

FIGURE 14–14 Base excision repair (BER) accomplished by uracil DNA glycosylase, AP endonuclease, DNA polymerase, and DNA ligase. Uracil is recognized as a noncomplementary base, excised, and replaced with the complementary base (C).

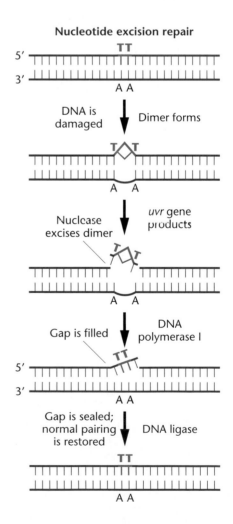

Nucleotide excision repair

FIGURE 14–15 Nucleotide excision repair (NER) of a UV-induced thymine dimer. In actuality, 12 bases are excised in prokaryotes and 28 bases are excised in eukaryotes during repair.

pigmentosum (XP), a rare genetic disorder that predisposes individuals to severe skin abnormalities. These individuals have lost their ability to undergo NER. As a result, individuals suffering from XP who are exposed to the ultraviolet radiation present in sunlight exhibit reactions that range from initial freckling and skin ulceration to the subsequent development of skin cancer. Figure 14–16 contrasts two XP individuals, one of whom has been detected early and protected from sunlight.

The condition is very severe and can be lethal, although early detection and protection from sunlight can arrest it. Because sunlight contains UV radiation, a causal relationship had been predicted between thymine dimer production and XP. Thus, the ability to repair UV-induced lesions was investigated in human fibroblast cultures derived from XP and normal individuals. (Fibroblasts are undifferentiated connective tissue cells.) The results suggested that the XP phenotype may be caused by more than one mutant gene.

In 1968, James Cleaver showed that cells from XP patients were deficient in **unscheduled DNA synthesis** (DNA synthesis other than that occurring during chromosome replication), which is elicited in normal cells by UV radiation. Since this type of synthesis is thought to represent the activity of the excision-repair system, the suggestion was that XP cells are deficient in excision repair.

The link between xeroderma pigmentosum and inadequate excision repair has been strengthened by the use of **somatic cell hybridization** studies, a technique that we first introduced in Chapter 8. Cultured fibroblast cells from any two unrelated XP patients can be induced to fuse, forming a heterokaryon where the two nuclei share a common cytoplasm. After fusion, excision repair, as assayed by unscheduled DNA synthesis, may be reestablished in the heterokaryon. When this occurs, the two variants are said to demonstrate **complementation**. Alone, neither cell type demonstrates excision repair, but together in a heterokaryon the process occurs. In genetic terms, this is strong evidence that in the two patients from whom the cells were derived, different affected genes led to the disease. Complementation occurs because the heterokaryon has at least one normal copy of each gene.

Based on many studies, patients have been divided into seven complementation groups, suggesting that at least seven different genes may be involved in excision repair. These genes or their protein products have now been identified. The genes are found in disparate regions of the genome, and a homologous gene for each has been identified in yeast. Approximately 20 percent of XP patients do not fall into any of the seven groups. They manifest similar symptoms, but their fibroblasts do not demonstrate defective excision repair. There is some evidence that they are less efficient in normal DNA replication.

Proofreading and Mismatch Repair

Other types of repair systems have evolved to correct other types of errors in DNA. For example, as we pointed out in our discussion of DNA synthesis in Chapter 11, DNA polymerase III possesses a **proofreading** function. During polymerization, when an incorrect nucleotide is inserted, the enzyme complex has the potential to recognize the error and "reverse" it by cutting out the incorrect nucleotide and replacing it. In bacterial systems, proofreading is thought to increase fidelity during synthesis by two orders of magnitude. If initial mismatches occur in $1/10^5$ nucleotide pairs (a rate of 10^{-5}), proofreading decreases final mismatches to $1/10^7$ (a rate of 10^{-7}).

To cope with those errors that remain after proofreading, still another mechanism, called **mismatch repair**, may be activated. Proposed over 20 years ago by Robin Holliday, the molecular basis of this process is now well established. As in other DNA lesions, the alteration or mismatch must be detected, the incorrect nucleotide(s) must be removed, and the replacement with the correct nucleotide(s) must occur. But a special problem exists with correction of a mismatch. How does the repair system recognize which strand is correct (the template during replication) and which contains the mismatched base (the newly synthesized strand)? How the repair system discriminates and recognizes the "new" nucleotide puzzled geneticists for decades. If the mismatch is recognized, but no discrimination occurred and

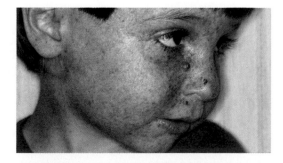
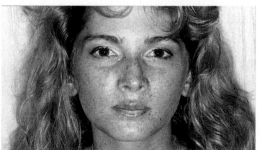

FIGURE 14–16 Two individuals with xeroderma pigmentosum. The 4-year-old boy on the left shows marked skin lesions induced by sunlight. Mottled redness (erythema) and irregular pigment changes that are a response to cellular injury are apparent. Two nodular cancers are present on his nose. The 18-year-old girl on the right has been carefully protected from sunlight since the diagnosis of xeroderma pigmentosum in infancy. Several cancers have been removed and she has worked as a successful model. *(W. Clark Lambert, M.D., Ph.D., University of Medicine & Dentistry of New Jersey)*

excision was random, half the time the strand bearing the correct base would be clipped out. The concept of strand discrimination by a repair enzyme is thus a critical step.

At least in some bacteria, including *E. coli*, this process has been elucidated and is based on the process of **DNA methylation**. These bacteria contain an enzyme, **adenine methylase**, that recognizes the DNA sequence

5′ ... GATC ... 3′

3′ ... CTAG ... 5′

as a substrate. Upon recognition, a methyl group is added to each of the adenine residues. This modification is stable throughout the cell cycle.

Following a further round of replication, the newly synthesized strands remain temporarily unmethylated. It is at this point that the repair enzyme recognizes the mismatch and preferentially binds to the unmethylated strand. A nick is made by an endonuclease protein, either 5′ or 3′, to the mismatch on the unmethylated strand. The nicked DNA strand is then unwound, degraded, and replaced until the mismatch is reached and excised. The proteins involved in this process are slightly different depending on which side of the mismatch the nick was made. A series of *E. coli* gene products, MutH, L, S, and U are involved in the discrimination step. Mutations in each result in strains deficient in mismatch repair. While the mechanism described here is based on studies of *E. coli*, similar mechanisms involving homologous proteins are known to exist in yeast and mammals.

Post-Replication Repair and the SOS Repair System

Still other types of repair have been discovered, illustrating the diversity of such mechanisms that have evolved to overcome DNA damage. One system, called **post-replication repair**, was discovered in an excision-defective strain of *E. coli* and first proposed by Miroslav Radman. It is a system that responds *after* damaged DNA has escaped repair and failed to be completely replicated, thus its name. As shown in Figure 14–17, when DNA bearing a lesion of some sort (such as a pyrimidine dimer) is being replicated, DNA polymerase at first stalls at the lesion and then skips over it, leaving a gap along the newly synthesized strand. To counteract this, the RecA protein directs a recombinational exchange process whereby this gap is filled as a result of the insertion of a segment initially present on the undamaged complementary strand. This event creates a gap on the "donor" strand, which can then be filled by repair synthesis as replication proceeds. Because a recombinational event is involved, post-replication is also referred to as *homologous recombination repair*, a more general category.

Still another repair pathway in *E. coli* is called the **SOS repair system**, which also responds to damaged DNA, but in a different way. Instead of skipping over a lesion during DNA synthesis, as in post-replication repair, the system creates conditions that allow DNA polymerase to continue repli-

Post replication repair

1. Lesion present in DNA unwound prior to replication

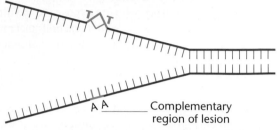

2. Replication skips over lesion and continues

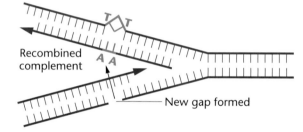

3. Undamaged complementary region of parental strand is recombined

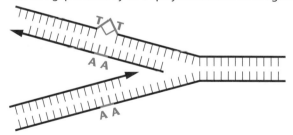

4. New gap is filled by DNA polymerase and DNA ligase

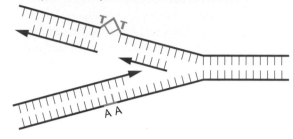

FIGURE 14–17 Post-replication repair that occurs if DNA replication has skipped over a lesion, such as a thymine dimer. Through the process of recombination, the correct complementary sequence is recruited from the parental strand and inserted into the gap opposite the lesion. The new gap created is filled by DNA polymerase and DNA ligase.

cation across the lesion. While no gap is created, the fidelity of replication is compromised, leading to the description of the system as being *error prone*. Phil Hanawalt and Paul Howard-Flanders, among others, have established that as many as 20 different genes, including *lexA*, *recA*, and *uvr*, may be induced by DNA damage, so that their products become involved in the SOS response.

Double-Strand Break Repair in Mammals

Thus far, we have only discussed repair pathways that deal with damage *within* a particular strand of DNA. We conclude our discussion of DNA repair by considering what happens when both strands of the DNA helix are cleaved, for example, as a result of exposure to ionizing radiation. This causes what are called double-strand breaks. While such damage occurs in bacteria as well, we will focus our discussion on eukaryotic cells. Fortunately, a specialized process, the **DNA double-strand break repair (DSB repair)** pathway, is activated and is responsible for ultimately reannealing the two DNA segments. Recently, interest has grown in DNA DSB repair because defects in this pathway are associated with X-ray hypersensitivity and immune deficiency. Such defects may also underlie familial disposition to breast and ovarian cancer.

Similar to post-replication repair, one pathway involved in double-strand break repair is referred to as **homologous recombinational repair** because damaged DNA is actually recombined and replaced with homologous undamaged DNA. This is necessary because, when both strands are broken, there is no undamaged parental strand available to use as the source of the complementary DNA sequence during repair. Therefore, the genetic information present in the homologous region of either a sister or nonsister homolog is "recruited" to replace the damaged double-stranded break. The undamaged homologous region is actually recombined into the damaged DNA molecule. While the exact mechanism is not yet understood, the complex is able to restore the integrity of the broken double helix.

14.9 Site-Directed Mutagenesis

This section introduces a useful experimental technique, **site-directed mutagenesis**, that allows researchers to introduce a designed mutation at a prescribed site within a gene of interest. The technique relies on the availability of a cloned gene and utilizes a number of manipulations involving recombinant DNA technology, which we introduce in Chapter 16. However, the underlying principles are based on information previously presented.

The goal of the technique is to alter one or more specific nucleotides within a gene in order to change a specific triplet codon. Upon transcription and translation, this change causes the insertion of a "mutant" amino acid into the protein encoded by the original gene. Such designed mutations are particularly useful in studying the effects of genetic change on protein function. Various approaches can be used to accomplish the goal; most are variations on the theme we are about to describe (Figure 14–18).

The initial step (step 1) is to determine the nucleotide sequence of the gene being studied. This can be accomplished by DNA sequencing techniques, or it can be predicted if the amino acid sequence of the protein is known, by utilizing our knowledge of the genetic code.

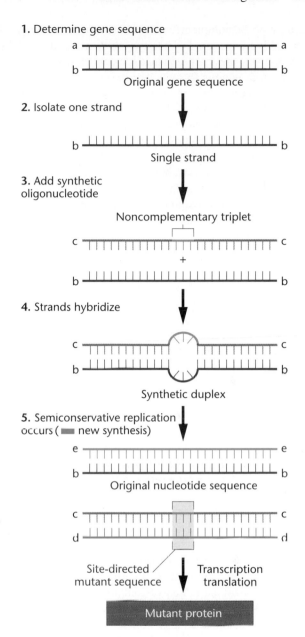

FIGURE 14–18 Site-directed mutagenesis. A single strand of DNA from a gene of interest is initially isolated. This is hybridized with a synthetic oligonucleotide containing a triplet altered so as to encode an amino acid of choice. Following semiconservative replication, a different complementary base pair is present in one of the new duplexes. Upon transcription and translation, a mutant protein, "designed" in the laboratory, will be produced.

The next step is to isolate one of the two complementary strands from the DNA of known sequence and decide which nucleotides are to be changed (step 2). Then, a small piece of DNA, a synthetic oligonucleotide, is chemically synthesized, which is complementary to that region at all points except in the triplet sequence or sequences that are to be altered (step 3). This sequence includes the triplet encoding the amino acid that will change in the protein.

As shown in step 4, this short piece of DNA is hybridized with the original parent strand, forming a partial duplex because of the complementarity along most of its length. If DNA polymerase and DNA ligase are then added to this hybrid complex, the short sequence is extended so that a duplex of the entire gene is formed. The two strands are perfectly complementary except at the point of alteration.

If this DNA is replicated (step 5), two types of duplexes are formed; one is like the original, unaltered gene and the other contains the newly designed sequence. Recombinant DNA technology makes it possible not only to complete the above manipulations, but also to allow the altered gene to be expressed so that large amounts of the desired protein are available for study. Or, as we shall see in the next section, the altered gene can be introduced into an organism and studied.

14.10 Knockout Genes and Transgenes

The ability to assess the function of specific genes has been enhanced tremendously with the development of techniques that allow experimental analysis *in vivo*. While a gene bearing a site-directed mutation can be investigated *in vitro*, the insertion of that gene into a living organism allows for the assessment of its function. If the insertion process involves the *replacement* of the comparable gene of the organism from which it originated, the technique is referred to as **gene knockout**. The genetically altered organism is called a **knockout organism**, for example, a "knockout mouse."

While specific gene replacement is difficult to achieve, it has been accomplished and is now fairly routine in research involving yeast and mice. In mice, the application of gene knockout techniques has been particularly fruitful in studies involving the genetic control of early development and behavior. Most often, a "loss-of-function" mutation replaces the normal gene, and the effects are investigated.

Furthermore, knockout mice now serve as models for studying human genetic disorders. In such cases, the comparable mouse gene that causes the human disorder is isolated, subjected to directed mutagenesis, and used to replace the nonmutant mouse gene. Cystic fibrosis and Duchenne muscular dystrophy have been investigated in this way. Applications of this general technology are discussed in more detail in Chapter 19.

If a gene is inserted into an organism *in addition* to its normal copies, it is called a **transgene**, and the organism is called a **transgenic organism** (e.g., a "transgenic plant"). In such cases, the gene may have undergone directed mutagenesis, or it may be a foreign gene isolated from another organism. This technology has been used more extensively than gene knockout, since it is easier to accomplish. This is so because specific replacement is not required.

The study of transgenic organisms is performed routinely in plants, *Drosophila*, mice, and a variety of other organisms. Of particular note is the potential provided in agricultural studies, as well as gene therapy techniques in our own species.

14.11 Transposable Genetic Elements

We conclude this chapter by discussing the phenomenon of **transposable genetic elements**, sometimes referred to as **transposons**, genetic units that can move or be "transposed" within the genome. It is appropriate to discuss transposable elements here because the movement of genetic units from one place in the genome to another often disrupts genetic function and results in phenotypic variation. As such, the impact of the transposition of genetic units often fits into a broad definition of mutation.

Transposable elements were first studied in maize by Barbara McClintock almost 50 years ago. However, even in the 1950s and 1960s the idea that genetic information was *not* fixed within the genome of an organism was slow to find acceptance. Such a notion was quite alien to the classical interpretation of genes on chromosomes. It was not until other transposable elements were discovered, and their molecular basis revealed, that the phenomenon was found to be nearly universal.

Insertion Sequences

Although the presence of transposable elements in maize had been predicted earlier, the first observation at the molecular level did not come until the early 1970s. A number of independent researchers, including Peter Starlinger and James Shapiro, visualized a unique class of mutations affecting different genes in various bacterial strains. For example, the expression of a cluster of related genes involving galactose metabolism in *E. coli* was repressed as a result of one such mutation; the phenotypic effect was heritable, but was found not to be caused by a base-pair change characteristic of conventional gene mutations. Instead, it was shown that a short specific DNA segment had been inserted into the bacterial chromosome at the beginning of the galactose gene cluster. When this segment was spontaneously excised from the bacterial chromosome, wild-type function was restored. Such a segment came to be called an **insertion sequence (IS)**.

It was subsequently revealed that several other distinct DNA segments could behave in a similar fashion, inserting into the chromosome and affecting gene function. These DNA segments are relatively short, not exceeding 2000 bp (2 kb). The first insertion sequence to be characterized in *E. coli*, IS1, is about 800 bp long; IS2, 3, 4, and 5 are about 1250–1400 bp in length.

The analysis of the DNA sequences of most IS units reveals a feature important to their mobility. The nucleotide sequences contain **inverted terminal repeats (ITRs)** of one another; that is, the final sequence at the 5'-end of one strand is the same as the final sequence at the 5'-end of the other strand, except that the sequences run in opposite directions (Figure 14–19). Although the figure shows the terminal-repeating unit to consist of only a few nucleotides, many more are actually involved. For example, in *E. coli* the IS1 termini contain about 20 nucleotide pairs, IS2 and IS3 about 40 pairs, and IS4 about 18 pairs. It seems likely that these ter-

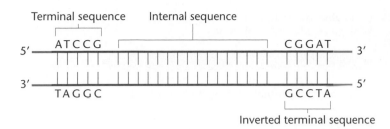

Terminal sequence Internal sequence

5'— ATCCG ————————————————— CGGAT —3'

3'— TAGGC ————————————————— GCCTA —5'

Inverted terminal sequence

FIGURE 14–19 An insertion sequence (IS). The terminal sequences are perfect inverted repeats of one another.

minal sequences are an integral part of the mechanism of the insertion of IS units into DNA. That insertion of IS units is more likely to occur at certain DNA regions than at others suggests that IS termini can recognize certain target sequences in the DNA during the process of insertion.

Careful investigation has revealed IS units in the wild-type *E. coli* chromosome as well as in other autonomous segments of bacterial DNA, called plasmids. Thus, their presence does not always result in mutation. In the *E. coli* chromosome, five or more copies of IS1, IS2, and IS3 are present. The exact number of each varies, depending on the strain examined.

Bacterial Transposons

In addition to their potential mutational effects, IS units play an even more significant role in the formation and movement of the larger **transposon (Tn) elements**. Transposons in bacteria consist of IS units that contain within their internal DNA sequence genes whose functions are unrelated to the insertion process. Like IS units, Tn elements are mobile in both bacterial and viral chromosomes and in plasmids. The Tn elements provide a mechanism for movement of genetic information from place to place both within and between organisms. Transposons were first discovered to move between DNA molecules as a result of observations of antibiotic-resistant bacteria. In the mid-1960s, Susumu Mitsuhashi first suggested that genes responsible for resistance to several antibiotics were mobile and could move between bacterial plasmids and chromosomes.

Transposons have become the focus of increased interest, particularly because they have been found in organisms other than bacteria. Bacteriophages that demonstrate the ability to insert their genetic material into the host chromosome behave in a similar fashion. The bacteriophage mu, consisting of over 35,000 nucleotides, can insert its DNA at various places in the *E. coli* chromosome. Like IS units, if insertion occurs within a gene, mutant behavior at that locus results. Transposons have also been discovered in higher organisms, including yeast, corn, *Drosophila*, and humans.

The *Ac-Ds* System in Maize

With our knowledge of insertion sequences and transposons in bacteria, it is no surprise that mobile genetic units also exist in eukaryotes. They are, in fact, more widespread, and, like bacteria, they often have the effect of altering the expression of genetic information.

About 20 years before the discovery of transposons in prokaryotic organisms, Barbara McClintock analyzed the genetic behavior of two mutations, *Dissociation* (*Ds*) and *Activator* (*Ac*), in corn plants (maize). She examined the phenotypes of the maize kernels resulting from genes expressed in either the endosperm or aleurone layers (Figure 14–20) and correlated her observations with a cytological examination of the maize chromosomes. Initially, McClintock determined that *Ds* is located on chromosome 9. If *Ac* is also present in the genome, *Ds* induces breakage at a point on the chromosome adjacent to its own location. If breakage occurs in somatic cells during their development, progeny cells often lose part of chromosome 9, causing a variety of phenotypic effects.

Subsequent analysis suggested to McClintock that both the *Ds* and *Ac* genes are sometimes transposed to different chromosomal locations. While *Ds* moves only if *Ac* is also present, *Ac* is capable of autonomous movement. Where *Ds* comes to reside determines its genetic effect; that is, it might cause chromosome breakage or it might inhibit gene expression. In cells where expression is inhibited, *Ds* might move again, releasing this inhibition. In these cases, the *Ds* element is believed to insert into a gene and subsequently to depart from it, causing changes in gene expression.

Figure 14–21 (page 298) shows the sort of movements and effects of the *Ds* and *Ac* elements described. In McClintock's original observation, when the *Ds* element jumped out of chromosome 9, this excision event restored normal gene function. In such cells, pigment synthesis was restored. McClintock concluded that the *Ds* and *Ac* genes are **transposable controlling elements**.

It was not until many years later that anything comparable to the controlling elements in maize was recognized in other

FIGURE 14–20 Corn kernel showing spots of colored aleurone produced by genetic transposition involving the *Ac–Ds* system. (*Dr. Nina Fedoroff/Penn State University*)

FIGURE 14-21 Consequences of the influence of the *Activator* (*Ac*) element on the *Dissociation* (*Ds*) element. In (b), *Ds* is transposed to a region adjacent to a theoretical gene *W*. Subsequent chromosomal breakage is induced, the *W*-bearing segment is lost, and mutant gene expression occurs. In (c), *Ds* is transposed to a region within the *W* gene, causing immediate mutant expression. *Ds* may also "jump" out of the *W* gene, with the accompanying restoration of *W* gene activity and its wild-type expression.

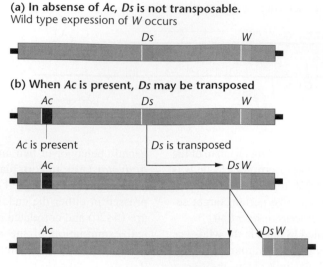

(a) In absense of *Ac*, *Ds* is not transposable.
Wild type expression of *W* occurs

(b) When *Ac* is present, *Ds* may be transposed

Ac is present   *Ds* is transposed

Chromosome breaks and fragment is lost
Expression of *W* ceases, producing mutant effect

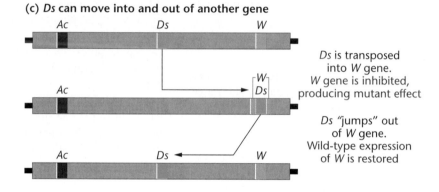

(c) *Ds* can move into and out of another gene

Ds is transposed into *W* gene. *W* gene is inhibited, producing mutant effect

Ds "jumps" out of *W* gene. Wild-type expression of *W* is restored

organisms. When bacterial insertion sequences and transposons were discovered, many parallels were evident. Transposons and insertion sequences were seen to move into and out of chromosomes, to insert at different positions, and to affect gene expression at the point of insertion.

Several *Ac* and *Ds* elements have now been isolated and carefully analyzed, and the relationship between the two elements has been clarified (Figure 14–22). The first *Ac* element sequence is 4563 bp long and strikingly similar to one of the known bacterial transposons. This sequence contains two 11-bp imperfect terminal-inverted repeats, two **open reading frames (ORFs)**, and three noncoding regions. Open reading frames contain initiation and termination sequences and are considered to encode genetic products. The first *Ds* element studied (*Ds-a*) is nearly identical in structure to *Ac* except for a 194-bp segment that has been deleted from the largest open reading frame. There is some evidence that this gene encodes a **transposase enzyme**, essential to transposition of both the *Ac* and *Ds* elements. The deletion of part of this gene in the *Ds-a* element explains its dependence on the *Ac* element for transposition. Several other *Ds* elements have also been sequenced, and each reveals an even larger deletion in the same region. In each case, however, the terminal repeats are retained and seem to be essential for transposition, provided that a functional transposase enzyme is supplied by the gene in the *Ac* element.

Although the validity of Barbara McClintock's proposed mobile elements was questioned following her initial observations, molecular analysis has since verified her conclusions. For her work, Barbara McClintock was awarded the Nobel Prize in 1983.

Copia and P Elements in *Drosophila*

Transposable elements have been discovered in still other eukaryotic organisms, notably in yeast, *Drosophila*, and primates, including humans. In 1975, David Hogness and his colleagues, David Finnigan, Gerald Rubin, and Michael Young, identified a class of genes in *Drosophila melanogaster* that they designated as **copia**. These genes transcribe "copious" amounts of RNA (hence their name). Present up to 30 times in the genome of cells, *copia* genes are nearly identical in nucleotide sequence. Mapping studies show that they are transposable to different chromosomal locations and are dispersed throughout the genome.

Copia genes appear to be only one of approximately 30 families of transposable elements in *Drosophila*, each of which is present 20–50 times in the genome. Together, these families constitute about 5 percent of the *Drosophila* genome and over half of the middle repetitive DNA of this organism. One estimate projects that 50 percent of all visible mutations

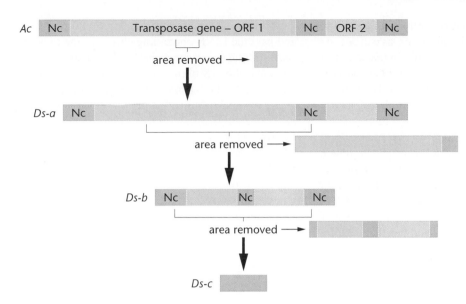

FIGURE 14–22 A comparison of the structure of an *Ac* element with three *Ds* elements, all of which have been isolated and sequenced. The imperfect and inverted repeats are at the ends of the *Ac* element. The transposase gene is in an open reading frame (ORF 1). No function has yet been assigned to ORF 2. Noncoding regions are designated by Nc. As this scheme shows, *Ds-a* appears to be simply an *Ac* element containing a small deletion in the gene encoding the transposase enzyme.

in *Drosophila* are the result of the insertion of transposons into otherwise wild-type genes!

Despite the variability in DNA sequence between the members of different families, they share a common structural organization thought to be related to the insertion and excision processes of transposition. Each *copia* gene consists of approximately 5000–8000 bp of DNA, including a long family-specific **direct terminal repeat (DTR)** sequence of 276 bp at each end. Within each repeat is a short **inverted terminal repeat (ITR)** of 17 bp. These features are shown in Figure 14–23. The DTR sequences are found in other transposons in other organisms but are not universal. However, the shorter ITR sequences are considered universal.

Insertion of *copia*, as with other transposable elements, appears to be dependent on ITR sequences and appears to occur at specific target sites in the genome. In general, eukaryotic transposons are strikingly similar to one another and share many features with those in bacteria.

Still another interesting category of transposable elements in *Drosophila* is the family called **P elements**. These were discovered while studying the phenomenon of **hybrid dysgenesis**, a condition that causes sterility, elevated mutation rate, and chromosome rearrangement in the offspring of crosses between certain strains of fruit flies.

Hybrid dysgenesis is caused by high rates of P-element transposition in the germ line, where these mobile DNA elements insert themselves into or near genes, thereby causing mutations. P elements are 2.9 kb long, with 31-bp terminal inverted repeats. The elements encode at least two proteins, one of which is the transposase enzyme that is required for transposition. The transposase is expressed only in the germ line, accounting for the tissue specificity of P-element transposition.

Mutations can arise from several kinds of insertional events. If a P element inserts into the coding region of a gene, it can destroy the normal gene product. If it inserts into the promoter region of a gene, it can affect the level of expression of the gene. Insertions into introns can affect splicing or cause premature termination of transcription.

Transposable Elements in Humans

The final class of transposable elements that we discuss is represented by the ***Alu* family** of **short interspersed elements (SINEs),** which are characteristic of moderately repetitive DNA in mammals. Human DNA contains some 500,000 copies of this 200- to 300-bp sequence in the genome. Originally detected using reassociation kinetic analysis (see Chapter 10), these transposable elements received their name because they contain specific nucleotide sequences that are cleaved by a restriction endonuclease called *Alu*I.

The *Alu* family is considered to be a significant set of elements in the genome based on the fact that they have been found in the DNA of all the primates and rodents studied. Within the *Alu* elements in mammals a specific 40-bp DNA sequence has been conserved. Finally, *Alu* sequences are represented in some nuclear transcripts.

The *Alu* elements are considered transposable based on several lines of evidence. The most important is the fact that their 200- to 300-bp sequence is flanked on either side by direct repeat sequences consisting of 7–20 bp, paralleling bacterial insertion sequences. These flanking regions are related to the insertion process during transposition. Secondly, clustered regions of *Alu* sequences vary in the DNA of both normal and diseased individuals and in different tissues of the same individual. Additionally, the sequences have been found extrachromosomally.

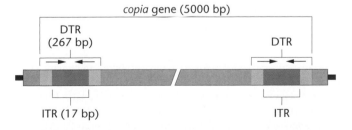

FIGURE 14–23 Structural organization of a *copia* transposable element in *Drosophila melanogaster*, showing the terminal repeats.

The potential mobility and mutagenic effects of these and other elements have far-reaching implications, as can be seen in a recent example of a transposon "caught in the act." The case involves a male child with hemophilia. One cause of hemophilia is a defect in blood-clotting factor VIII, the product of an X-linked gene. Haig Kazazian and his colleagues found a transposable element much longer than *Alu* sequences inserted at two points within the gene. Also present elsewhere in the genome, this sequence was shown to be a type of repetitive DNA referred to as a **long interspersed element (LINE)**. Both LINEs and SINEs are discussed in more detail in Chapter 17.

Researchers were very interested in determining if one of the mother's X chromosomes also contained this specific LINE. If so, the unaffected mother would be heterozygous and pass the LINE-containing chromosome to her son. The startling finding was that the LINE sequence is *not* present on either of her X chromosomes, but *was* detected on chromosome 22 of both parents. This suggests that this mobile element may have "moved" from one chromosome to another in the gamete-forming cells of the mother, prior to being transmitted to the son.

Many questions remain concerning this and other transposable elements. What is their origin? Were they once some sort of retrovirus? Exactly how do they move, and what has been their role during evolution? These and other questions will intrigue researchers for many years to come.

Genetics, Technology, and Society

Chernobyl's Legacy

On April 26, 1986, Reactor 4 of the Chernobyl Nuclear Power Station exploded, and ejected massive amounts of radioactive material into the surrounding countryside and throughout the northern hemisphere. The accident killed 31 emergency workers and caused acute radiation sickness in over 200 others. In the nine days following the initial explosion, as the reactor's temperature approached meltdown, radioactive fission products were discharged into the environment. High levels of radioactive iodine, xenon, strontium, and cesium were released into the atmosphere, along with over 5000 tons of vaporized boron carbide, dolomite, clay, sand, and lead. The plume of fallout traveled in a northwesterly direction through central Europe, reaching Finland and Sweden three days after the initial explosion. By May 5, the radioactive cloud reached the United Kingdom and remnants of the cloud arrived in North America one week later. Millions of people in Europe and the former Soviet Union were exposed to measurable amounts of radioactivity. People living within 30 km of Chernobyl were exposed to high levels of radioactivity prior to their evacuation 36 hours following the accident. Equally high were the exposures of some of the 600,000 military and civilian workers who were sent to Chernobyl to decontaminate the area and to encase the shattered reactor in a sarcophagus.

Chernobyl was the world's largest accidental release of radioactive material. In addition, this disaster inflicted enormous economic and personal burdens on the people of the former Soviet Union. The question that remains is whether Chernobyl's radioactive pollution directly threatens the long-term health of millions of people.

More is known about the effects of ionizing radiation on human health than any other agent (except perhaps cigarette smoke). Radiation refers to the energy that is emitted from a source, and includes heat, light, radio waves, microwaves, X-rays, and gamma rays from radioactive elements. Ionizing radiation (such as X-rays and gamma rays) is a subset of radiation that possesses sufficient energy to eject electrons from atoms. Ionizing radiation can damage any cellular component, alter nucleotides, and induce strand breaks in DNA. These DNA lesions can lead to mutations or chromosomal translocations. It is this DNA-damaging capacity of ionizing radiation that makes it a mutagen.

High doses of ionizing radiation increase the risk of developing certain cancers. The survivors of the atomic bomb blasts of Hiroshima and Nagasaki showed an increased incidence of leukemia within 2 years of the bombing. Breast cancers increased 10 years after exposure, as did cancers of the lung, thyroid, colon, ovary, stomach, and nervous system. However, the incidence of other cancers, such as those of the gall bladder, pancreas, uterus, and bone, did not increase due to radiation exposure. As ionizing radiation induces DNA damage, the offspring of A-bomb survivors were expected to show increases in birth defects, yet these increases were not detected.

The problem in extrapolating from Hiroshima to Chernobyl is the difference in dose. In A-bomb survivors, the cancer rate increased among those exposed to at least 200 mSv (mSv = millisievert, a unit dose of absorbed radiation). It is estimated that people living in the most contaminated areas close to Chernobyl may have received about 50 mSv, and some cleanup workers were exposed to approximately 250 mSv. Outside the Chernobyl area, radiation doses were estimated at 0.4–0.9 mSv in Germany and Finland, 0.01 mSv in the United Kingdom and 0.0006 mSv in the United States.

In order to put these numbers in context, the average dose for medical diagnostic procedures (such as chest and dental X-rays) is 0.39 mSv per year. A person's radiation exposure from natural background sources (cosmic rays, rocks, and radon gas) is about 2–3 mSv per year. Smokers expose themselves to an average extra dose of about 2.8 mSv per year from the intake of naturally occurring radioactive materials in tobacco smoke. A whole-body X-ray dose of 3000 mSv or more can be fatal.

It appears that mutation rates amongst plants and animals exposed to Chernobyl's radioactive waste may be two- to tenfold higher than normal. Also, Chernobyl cleanup workers exhibit a 25 percent higher mutation frequency at the *HPRT* locus. Mutation rates at mini-satellites among children born in polluted areas are twofold higher than those from controls in the United Kingdom. But do these increased mutation rates translate into health effects?

In the 115,000 people evacuated from Chernobyl, it was estimated that 26 leukemias might appear above the 25–30 that would occur spontaneously. It was also estimated that up to 17,000 additional cancers might occur in Europe, above the 123 million that would occur normally. At present, however, there have been no detectable increases in leukemias or solid tumors in contaminated areas of the former Soviet Union, Finland, or Sweden, or in the 600,000 Chernobyl cleanup workers.

Despite the lack of detectable increases in leukemias and solid tumors, one type of cancer has increased in the Chernobyl area. The rate of childhood thyroid cancer has reached over 100 cases per million children per year, whereas the normal rates would be expected to be between 0.5 and 3 cases per million children per year. To date, approximately 1800 children have been diagnosed with thyroid cancer. The regions with the greatest contamination appear to have the highest rates of thyroid cancer. Although epidemiologists debate whether these increases are entirely due to

radiation or to the increased reporting of cases, the scale of the increase and the fact that radioactive isotopes of iodine made up a significant portion of the Chernobyl fallout, make the link between thyroid cancer and Chernobyl fallout a plausible one.

The most profound effects of Chernobyl have been psychological. A recent study of Chernobyl cleanup workers found no increases in cancers, including leukemias or thyroid cancers. However, there was a 50 percent increase in suicides and a detectable increase in smoking- and alcohol-related disease. This mirrors other studies showing that 45 percent of people living within 300 km of Chernobyl believe that they have a radiation-induced illness. These people may be correct, as health effects such as depression, sleeps disturbance, hypertension, and altered perception have been documented. People suffer from the stress of relocation, the loss of personal and real property, and the feeling of a lack of control over their radiation exposure or their future. Posttraumatic stress may be a greater threat to health than the actual radiation exposure from the accident. Without solid information about their real risks, people feel that they live in constant danger and are simply awaiting the results of a cancer "lottery." Even if cancer rates and genetic defects do not increase dramatically, the indirect health effects from the Chernobyl Nuclear Power

Plant explosion have been, and continue to be, immense.

References

Anspaugh, L. R., Catlin, R. J. and Goldman, M. 1988. The global impact of the Chernobyl reactor accident. *Science* 242:1513–19.

Ginzberg, H. M. 1993. The psychological consequences of the Chernobyl accident—Findings from the International Atomic Energy Agency study. *Public Health Rep.* 108:184–92.

Jacob, P., et al. 2000. Thyroid cancer risk in Belarus after the Chernobyl accident: Comparison with external exposures. *Radiation and Environmental Biophysics* 39: 25–31.

Rahu, M., et al. 1997. The Estonian study of Chernobyl cleanup workers: II. Incidence of cancer and mortality. *Radiation Res.* 147;653–57.

Williams, D. 1994. Chernobyl, eight years on. *Nature* 371:556.

Website

UNSCEAR focuses on Chernobyl accident in General Assembly Report. June 6, 2000. Press release of the United Nations Information Service.
www.un.org/ha/chernobyl/unsceare.htm

Chapter Summary

1. The phenomenon of mutation not only provides the basis for most of the inherent variation present in living organisms, but also serves as the working tool of the geneticist in studying and understanding the nature of genetic processes.

2. Mutations are distinguished by the tissues affected. Somatic mutations are those that may affect the individual but are not heritable. Mutations arising in gametes may produce new alleles that can be passed on to offspring and can enter the gene pool.

3. Another classification of mutations relies on their effect. Morphological mutations, for example, may be detected visibly. Other types include biochemical, lethal, conditional, and regulatory mutations, groups that are not mutually exclusive.

4. Spontaneous mutations may arise naturally as a result of rare chemical rearrangements of atoms, or tautomeric shifts, and as the result of errors occurring during DNA replication. Although spontaneous mutations are very rare, their rate of occurrence may be increased experimentally by a variety of mutagenic agents.

5. Mutagenic agents such as base analogs as well as alkylating and deaminating agents cause chemical changes in nucleotides that alter their base-pairing affinities. As a result, base substi-

tution mutations arise following DNA replication. Mutagenicity of chemicals can be assessed using the Ames test.

6. Frameshift mutations, induced specifically by acridine dyes, arise when the addition or deletion of one or more nucleotides (but not multiples of three) occurs.

7. Direct analysis of DNA from individuals with specific ABO blood types and muscular dystrophy has been informative. Complete loss of function, as in the case in blood type O and the Duchenne form of muscular dystrophy, occurs when deletions or duplications of nucleotides have shifted the reading frame during translation.

8. Another form of mutation, discovered in several human disorders, includes unstable trinucleotide units that balloon in size during DNA replication and that are inherited in this form through successive generations. Increasing numbers of repeats often correlate with early onset and severity of disease.

9. Ultraviolet light and high-energy radiation from gamma, cosmic, or X-ray sources are also potent mutagenic agents. UV light induces the formation of pyrimidine dimers in DNA, while high-energy radiation causes the ionization of molecules in its path and is more penetrating than UV.

10. Various forms of repair of DNA lesions, such as pyrimidine dimers, have been discovered, including photoreactivation, excision repair, mismatch repair, proofreading, and various forms of recombinational repair, including double-stranded break repair. The significance of repair mechanisms is apparent when, in humans, excision repair is lost due to mutation. The result is the severe human disorder xeroderma pigmentosum.

11. Site-directed mutagenesis is a technique that allows researchers to create specific alterations in the nucleotide sequence of the DNA of genes.

12. The insertion sequences in bacteria and other transposable elements in eukaryotes are mobile genetic units characterizing the genomes of all organisms. They have a profound effect on genetic expression, thus serving as a distinct category of mutagenic agent.

Key Terms

acridine dye, 285

adenine methylase, 294

alkylating agent, 284

Alu family, 299

AP endonuclease, 292

apurinic site (AP site), 285

attached-X procedure, 280

autosomal recessive mutation, 279

auxotroph, 280

base analog, 284

base excision repair (BER), 292

base substitution, 283

Becker muscular dystrophy (BMD), 288

behavior mutations, 280

biochemical variation, 280

chromosomal aberration, 279

complementation, 293

complete medium, 280

conditional mutation, 280

copia, 298

deamination, 285

direct terminal repeat (DTR), 299

DNA glycosylase, 292

DNA methylation, 294

dominant autosomal mutation, 279

double-strand break repair (DSB repair), 295

Duchenne muscular dystrophy (DMD), 288

electromagnetic spectrum, 286

ethylmethane sulfonate (EMS), 284

excision repair, 291

fragile-X syndrome, 289

frameshift mutation, 283

gene knockout, 296

gene mutation, 279

genetic anticipation, 289

hemophilia, 282

homologous recombinational repair, 295

Huntington disease, 289

hybrid dysgenesis, 299

induced mutation, 279

insertion sequence (IS), 296

inverted terminal repeat (ITR), 296

ionizing radiation, 287

knockout organism, 296

lethal mutation, 280

long interspersed element (LINE), 300

minimal culture medium, 280

mismatch repair, 293

morphological trait, 279

myotonic dystrophy, 289

nucleotide excision repair (NER), 292

nutritional variation, 280

open reading frame (ORF), 298

P element, 299

photoreactivation enzyme (PRE), 291

photoreactivation repair, 291

point mutation, 283

post-replication repair, 294

proofreading, 293

prototroph, 280

pyrimidine dimer, 286

regulatory mutation, 280

reverse mutation, 284

roentgen, 288

short interspersed element (SINE), 299

site-directed mutagenesis, 295

somatic cell hybridization, 293

SOS repair system, 294

spinobulbar muscular atrophy (Kennedy disease), 290

spontaneous mutation, 279

target theory, 288

tautomeric shift, 283

tautomerism, 283

temperature-sensitive mutation, 280

transgenic organism, 296

transition, 283

transposable genetic element, 296

transposase enzyme, 298

transposon (Tn) element, 297

transversion, 283

trinucleotide repeat sequence, 289

ultraviolet (UV) radiation, 286

unscheduled DNA synthesis, 293

xeroderma pigmentosum (XP), 292

X-linked recessive mutation, 279

2-amino purine (2-AP), 284

5-bromouracil (5-BU), 284

6-ethylguanine, 285

Insights and Solutions

1. How could you isolate a mutant strain of bacterial cells that is resistant to penicillin, an antibiotic that inhibits cell wall synthesis?

Solution: Grow a culture of bacterial cells in liquid medium and plate the cells on agar medium to which penicillin has been added. Only penicillin-resistant cells will reproduce and form colonies. Each colony will, in all likelihood, represent a cloned group of cells with the identical mutation. Isolate members of each colony. To enhance the chance of such a mutation arising, you might want to add a mutagen to the liquid culture.

2. The base analog 2-amino purine (2-AP) substitutes for adenine during DNA replication, but it may base-pair with cytosine. The base analog 5-bromouracil (5-BU) substitutes for thymine, but it may base-pair with guanine. Follow the double-stranded trinucleotide sequence shown here through three rounds of replication, assuming that in the first round, both analogs are present and become incorporated wherever possible. In the second and third round of replication, they are removed. What final sequences occur?

Solution:

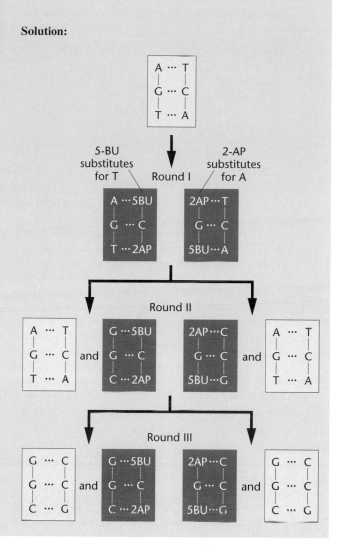

Problems and Discussion Questions

1. What is the difference between a chromosomal mutation and a gene mutation? Between a somatic and a gametic mutation?
2. Why do you suppose that a random mutation is more likely to be deleterious than beneficial?
3. Most mutations in a diploid organism are recessive. Why?
4. In *Drosophila*, induced mutations on chromosome 2 that are recessive lethals may be detected using a second chromosomal stock, *Curly, Lobe/Plum (Cy L/Pm)*. These alleles are all dominant and lethal in the homozygous condition. Detection is performed by crossing *Cy L/Pm* females to wild-type males that have been subjected to a mutagen. Three generations are required. In the F_1, *Cy L* males are selected and individually backcrossed to *Cy L/Pm* females. In the F_2 of each cross, flies expressing *Cy* and *L* are mated to produce a series of F_3 generations. Diagram these crosses and predict how an F_3 culture will vary if a recessive lethal was induced in the original test male

compared to the case where no lethal mutation resulted. In which F_3 flies would a recessive morphological mutation be expressed?
5. Describe tautomerism and the way in which this chemical event may lead to mutation.
6. Acridine dyes induce frameshift mutations. Is such a mutation likely to be more detrimental than a point mutation, where a single pyrimidine or purine has been substituted? If so, why?
7. Contrast the various types of DNA repair mechanisms known to counteract the effects of UV light and other DNA damage. What is the role of visible light in photoreactivation?
8. Mammography is an accurate screening technique for the early detection of breast cancer in humans. Because this technique uses X-rays diagnostically, it has been highly controversial. Can you explain why? What reasons justify the use of X-rays for this type of medical diagnosis?

9. Presented below are theoretical findings from studies of heterokaryons formed from human xeroderma pigmentosum cell strains.

	XP1	*XP2*	*XP3*	*XP4*	*XP5*	*XP6*	*XP7*
XP1	0						
XP2	0	0					
XP3	0	0	0				
XP4	+	+	+	0			
XP5	+	+	+	+	0		
XP6	+	+	+	+	0	0	
XP7	+	+	+	+	0	0	0

Note: + = complementing; 0 = noncomplementing.

These data represent the occurrence of unscheduled DNA synthesis in the fused heterokaryon when neither of the strains alone showed synthesis. What does unscheduled DNA synthesis represent? Which strains fall into the same complementation groups? How many groups are revealed based on these limited data? How do we interpret the presence of these complementation groups?

10. Why are X-rays a more potent mutagen than UV radiation?

11. Speculate as to how insertion sequences and other transposable elements disrupt genetic expression when inserted into wild-type genes.

12. Demonstrate your insights into both chromosomal and gene mutation by projecting yourself as one of the team of geneticists who immediately launched the study of the genetic effects of high-energy radiation on the surviving Japanese population following the atomic bomb attacks at Hiroshima and Nagasaki in 1945. Outline a comprehensive short-term and long-term study that would address this topic. Be sure to include stategies for considering the effects on both somatic and germ line tissues.

13. If the human genome contains 30,000 genes, and the mutation rate at each of these loci is 10^{-5} per gamete formed, what is the average number of new mutations that exist in each individual? If the current population is 4.3 billion people, how many newly arisen mutations exist in the current populace?

14. Compose a short essay that relates the molecular basis of fragile-X syndrome, myotonic dystrophy, and Huntington disease to the severity of the disorders, as well as to the phenomenon called genetic anticipation.

15. Cystic fibrosis (CF) is a severe autosomal recessive disorder in humans that results from a chloride ion–channel defect in epithelial cells. Over 500 sequence alterations have been identified in the 24 exons of the responsible gene (*CFTR* for cystic fibrosis transmembrane regulator), including dozens of different missense mutations and frameshift mutations, as well as numerous splice-site defects. Although all affected CF individuals demonstrate chronic obstructive lung disease, there is variation in pancreatic enzyme insufficiency (PI). Speculate on which types of observed mutations are likely to give rise to less severe symptoms of CF, including only minor PI. Of the 300 sequence alterations, a number of them found within the exon regions of the *CFTR* gene do not give rise to cystic fibrosis. Taking into account your accumulated knowledge of the genetic code, gene expression, protein function, and mutation, explain to a freshman biology major how this might be.

Selected Readings

Ames, B. N., McCann, J., and Yamasaki, E. 1975. Method for detecting carcinogens and mutagens with the *Salmonella*/mammalian microsome mutagenicity test. *Mut. Res.* 31:347–64.

Auerbach, C. 1978. Forty years of mutation research: A pilgrim's progress. *Heredity* 40:177–87.

Bates, G., and Lehrach, H. 1994. Trinucleotide repeat expansions and human genetic disease. *BioEssays* 16:277–83.

Beadle, G. W., and Tatum, E. L. 1945. *Neurospora* II. Methods of producing and detecting mutations concerned with nutritional requirements. *Am. J. Bot.* 32:678–86.

Berg, D., and Howe, M., eds. 1989. *Mobile* DNA. Washington, DC: American Society of Microbiology.

Carter, P. 1986. Site-directed mutagenesis. *Biochem.* J. 237:1–7.

Cleaver, J. E. 1968. Defective repair replication of DNA in xeroderma pigmentosum. *Nature* 218:652–56.

————, and Karentz, D. 1986. DNA repair in man: Regulation by a multiple gene family and its association with human disease. *Bioessays* 6:122–27.

Cohen, S. N., and Shapiro, J. A. 1980. Transposable genetic elements. *Sci. Am.* (Feb.) 242:40–49.

Deering, R. A. 1962. Ultraviolet radiation and nucleic acids. *Sci. Am.* (Dec.) 207:135–44.

Den Dunnen, J. T., et al. 1989. Topography of the Duchenne muscular dystrophy (DMD) gene. *Am. J. Hum. Genet.* 45:835–47.

Devoret, R. 1979. Bacterial tests for potential carcinogens. *Sci. Am.* (Aug.) 241:40–49.

Doring, H. P., and Starlinger, P. 1984. Barbara McClintock's controlling elements: Now at the DNA level. *Cell* 39:253–59.

Drake, J. W. 1991. A constant rate of spontaneous mutation in DNA-based microbes. *Proc. Natl. Acad. Sci. USA* 88:7160–64.

Drake, J. W., et al. 1975. Environmental mutagenic hazards. *Science* 187:505–14.

Drake, J. W., Glickman, B. W., and Ripley, L. S. 1983. Updating the theory of mutation. *Am. Sci.* 71:621–30.

Eyre-Walker, A., and Keightley, P. D. 1999. High genomic deleterious mutation rates in hominids. *Nature* 397:344–47.

Federoff, N. V. 1984. Transposable genetic elements in maize. *Sci. Am.* (June) 250:85–98.

Friedberg, E. C., Walker, G. C., and Siede, W. 1995. *DNA repair and mutagenesis*. Washington, DC: ASM Press.

Hanawalt, P. C., and Haynes, R. H. 1967. The repair of DNA. *Sci. Am.* (Feb.) 216:36–43.

Haseltine, W. A. 1983. Ultraviolet light repair and mutagenesis revisited. *Cell* 33:13–17.

Hendricikson, E. A. 1997. Cell-cycle regulation of mammalian DNA double-strand break repair. *Am. J. Hum. Genet.* 61:795–800.

Howard-Flanders, P. 1981. Inducible repair of DNA. *Sci. Am.* (Nov.) 245:72–80.

Jiricny, J. 1998. Eukaryotic mismatch repair: An update. *Mutation Research* 409:107–21.

Kelner, A. 1951. Revival by light. *Sci. Am.* (May) 184:22–25.

Knudson, A. G. 1979. Our load of mutations and its burden of disease. *Am. J. Hum. Genet.* 31:401–13.

Kraemer, F. H., et al. 1975. Genetic heterogeneity in xeroderma pigmentosum: Complementation groups and their relationship to DNA repair rates. *Proc. Natl. Acad. Sci. USA* 72:59–63.

Little, J. W., and Mount, D. W. 1982. The SOS regulatory system of *E. coli. Cell* 29:11–22.

Massie, R. 1967. *Nicholas and Alexandra.* New York: Atheneum.

—————, and Massie, S. 1975. *Journey.* New York: Knopf.

McCann, J., Choi, E., Yamasaki, E., and Ames, B. 1975. Detection of carcinogens as mutagens in the *Salmonella*/microsome test: Assay of 300 chemicals. *Proc. Natl. Acad. Sci. USA* 72:5135–39.

McClintock, B. 1956. Controlling elements and the gene. *Cold Spring Harbor Symp. Quant. Biol.* 21:197–216.

Macdonald, M. E., et al. 1993. A novel gene containing a trinucleotide repeat that is expanded and unstable in Huntington's disease chromosome. *Cell* 72:971–80.

McKusick, V. A. 1965. The royal hemophilia. *Sci. Am.* (Aug.) 213:88–95.

Miki, Y. 1998. Retrotransposal integration of mobile genetic elements in human disease. *J. Human Genet.* 43:77–84.

Muller, H. J. 1927. Artificial transmutation of the gene. *Science* 66:84–87.

—————. 1955. Radiation and human mutation. *Sci. Am.* (Nov.) 193:58–68.

O'Hare, K. 1985. The mechanism and control of P element transposition in *Drosophila. Trends Genet.* 1:250–54.

Osanna, N., Peterson, K. R., and Mount, D. W. 1986. Genetics of DNA repair in bacteria. *Trends Genet.* 2:55–58.

Radman, M., and Wagner, R. 1988. The high fidelity of DNA duplication. *Sci. Am.* (Aug.) 259(2):40–46.

Sherratt, D. J. (ed.)1995. *Mobile genetic elements.* New York: Oxford University Press.

Shortle, D., DiMario, D., and Nathans, D. 1981. Directed mutagenesis. *Annu. Rev. Genet.* 15:265–94.

Sigurbjornsson, B. 1971. Induced mutations in plants. *Sci. Am.* (Jan.) 224:86–95.

Spradling, A. C., Stern, D. M., Kiss, I., Roote, J., Laverty, T., and Rubin, G.M. 1995. Gene disruptions using P transposable elements: An integral component of the *Drosophila* Genome Project. *Proc. Natl. Acad. Sci. USA* 92:10824–30.

Stadler, L. J. 1928. Mutations in barley induced by X-rays and radium. *Science* 66:84–87.

Sutherland, B. M. 1981. Photoreactivation. *Bioscience* 31:439–44.

Tomlin, N. V., and Aprelikova, O. N. 1989. Uracil DNA glycosylases and DNA uracil repair. *Int. Rev. Cytol.* 114:81–124.

Vogel, F. 1992. Risk calculations for hereditary effects of ionizing radiation in humans. *Hum. Genet.* 89:127–46.

Wells, R. D. 1994. Molecular basis of genetic instability of triplet repeats. *J. Biol. Chem.* 271: 2875–78.

Wills, C. 1970. Genetic load. *Sci. Am.* (Mar.) 222:98–107.

Yamamoto, F., et al. 1990. Molecular genetic basis of the histoblood group ABO system. *Nature* 345:229–33.

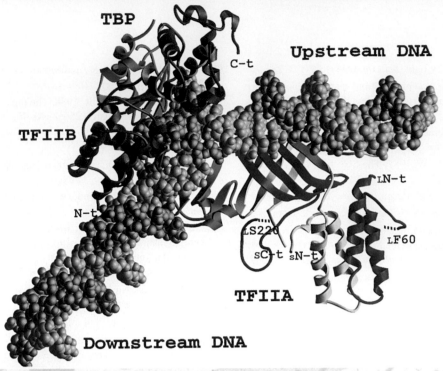

TBP

C-t

Upstream DNA

TFIIB

LN-t

N-t

LS220

LF60

sC-t sN-t

TFIIA

Downstream DNA

Molecular complex formed between TATA box-binding protein (TBP), transcription factors TFIIA and TFIIB, and TATA box DNA during the initiation of transcription. *(Yale University and the Howard Hughes Medical Institute, Dr. Paul Sigler)*

15

Regulation of Gene Expression

CHAPTER CONCEPTS

Expression of genetic information is dependent on regulatory mechanisms that either activate or repress the transcription of genes. Transcription is modulated by the interaction of various regulatory molecules with DNA sequences, most often located upstream from affected genes. Genetic regulation in eukaryotes also occurs during post-transcriptional events.

In previous chapters we established how DNA is organized into genes, how genes store genetic information, and how this information is expressed. We now consider one of the most fundamental issues in molecular genetics: *How is genetic expression regulated?* A functional bacterial cell contains thousands of proteins present at widely different concentrations, yet each is encoded by a single gene. Synthesis of bacterial gene products changes dramatically in response to environmental conditions. Bacterial cells normally synthesize the enzymes to metabolize lactose only when it is present in the environment and other, more readily metabolized carbon sources are not. In the absence of lactose, the enzymes needed to metabolize this sugar are also absent. These observations lend support to the idea that gene action is regulated.

In eukaryotic organisms, cells of the pancreas do not make retinal pigment, and retinal cells do not make insulin. In multicellular eukaryotes, genetic regulation is at the heart of cellular differentiation (Chapter 20). Differential gene expression serves as the basis for phenotypic specialization at the cellular and tissue levels in both plants and animals.

In this chapter we discuss examples of gene regulation in bacteria, bacteriophages, and eukaryotes. Pivotal to this discussion is our knowledge that all cells in an organism contain a complete set of genes characteristic of that species. Regulation is *not* accomplished by eliminating unused genetic information; instead, mechanisms have evolved to control the expression of genes. Some of these mechanisms, particularly in bacterial systems and their phages, have been extensively characterized.

15.1 Genetic Regulation in Prokaryotes: An Overview

Regulation of gene expression has been studied extensively in prokaryotes, particularly in *Escherichia coli*. Highly efficient mechanisms have evolved that turn genes on and off, depending on the cell's metabolic needs in particular environments. Detailed analysis of proteins in *E. coli* has shown that for the more than 4000 polypeptide chains encoded by the genome, a vast range of concentration of gene products exists. Some proteins may be present in as few as 5–10 molecules per cell, whereas others, such as ribosomal proteins and the many proteins involved in the glycolytic pathway, are present in as many as 100,000 copies per cell.

The idea that microorganisms regulate the synthesis of gene products can be illustrated by considering the utilization of lactose (a galactose-glucose–containing disaccharide) as a carbon source. When it is present in the growth medium, many bacteria and yeast produce enzymes specific to lactose metabolism. When lactose is absent, the enzymes are not manufactured. These organisms thus "adapt" to their environment, producing certain enzymes only when specific chemical substrates are present. Such enzymes are said to be **inducible**, reflecting the role of the substrate, which serves as the **inducer** of enzyme production. In contrast, other enzymes that are produced continuously, regardless of the chemical makeup of the environment, are described as **constitutive**.

Studies have also revealed cases where the presence of a specific molecule causes inhibition of genetic expression. This is often the case for molecules that are the end products of biosynthetic pathways. Amino acids can be synthesized by bacterial cells, but if the amino acids are present in the growth medium, they can be taken up and used. In such cases, it is inefficient for the cell to produce the enzymes necessary for the synthesis of those amino acids, and transcription of mRNA for the appropriate biosynthetic enzymes is repressed. This is an example of a **repressible system** of gene regulation.

Regulation, whether it is inducible or repressible, may be under either **negative** or **positive control**. Under negative control, genetic expression occurs *unless it is shut off by some form of a regulator molecule*. In contrast, under positive control, transcription occurs *only if a regulator molecule directly stimulates RNA production*. In theory, either type of control can govern inducible or repressible systems. Our discussion in the ensuing sections of this chapter will clarify these contrasting systems of regulation. For the enzymes involved in lactose and tryptophan, negative control is operative.

15.2 Lactose Metabolism in *E. coli*: An Inducible System

Beginning in 1946 Jacques Monod, and continuing through the next decade, Joshua Lederberg, François Jacob, and Andre L'woff amassed genetic and biochemical evidence involving lactose metabolism. These researchers and others provided insights into how gene activity is repressed when lactose is absent but induced when it is available. In the presence of lactose, the concentration of the enzymes responsible for its metabolism increases rapidly from a few molecules to thousands per cell. The enzymes responsible for lactose metabolism are thus *inducible*, and lactose serves as the *inducer*.

In prokaryotes, genes that code for enzymes with related functions (e.g., genes involved with lactose metabolism) tend to be organized in clusters, and they are often under the coordinated genetic control of a single regulatory unit. The site of this unit is almost always linked upstream to the gene cluster it controls and is known as a **cis-acting element**. Interactions at the site of such an element involve binding molecules that control transcription of the gene cluster. Such molecules are called **trans-acting elements**. Actions at the regulatory site determine whether the genes are expressed, and thus whether the corresponding enzymes or other protein products are present. Binding a *trans*-acting element at a *cis*-acting site can regulate the gene cluster either negatively (by turning the genes off) or positively (by turning genes in the cluster on). In this section we discuss how such bacterial gene clusters are coordinately regulated.

Paramount to the understanding of how gene expression is controlled in this system was the discovery of a regulatory gene and a regulatory site that are part of the gene cluster.

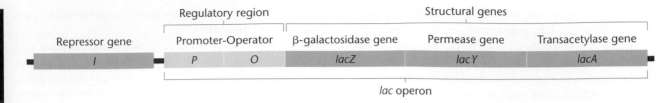

Repressor gene	Regulatory region		Structural genes		
	Promoter-Operator		β-galactosidase gene	Permease gene	Transacetylase gene
I	*P*	*O*	*lacZ*	*lacY*	*lacA*

lac operon

FIGURE 15–1 A simplified overview of the genes and regulatory units involved in the control of lactose metabolism (this region of DNA is not drawn to scale).

Neither of these elements encodes enzymes necessary for lactose metabolism; that is the function of the three genes in the cluster. As shown in Figure 15–1, the three structural genes and the adjacent regulatory site constitute the **lactose,** or **lac operon.** Together, the entire gene cluster functions in an integrated fashion to provide a rapid response to the presence or absence of lactose.

Structural Genes

There are three structural genes in the *lac* operon. The *lacZ* gene encodes **β-galactosidase,** an enzyme that converts the disaccharide lactose to the monosaccharides glucose and galactose (Figure 15–2). This conversion is essential if lactose is to serve as the primary energy source in glycolysis. The second gene, *lacY,* specifies the primary structure of **permease,** an enzyme that facilitates the entry of lactose into the bacterial cell. The third gene, *lacA,* codes for the enzyme **transacetylase.** While its physiological role is not completely clear, there is some thought that it helps remove toxic by-products of lactose digestion from the cell.

To study the genes coding for these three enzymes, researchers isolated numerous mutations in order to eliminate the function of one or the other enzyme. Such *lac⁻* mutants were first isolated and studied by Joshua Lederberg. Mutant cells that fail to produce active β-galactosidase (*lacZ⁻*) or permease (*lacY⁻*) cannot use lactose as an energy source. Mutations also were found in the transacetylase gene. Mapping studies by Lederberg established that all three genes are closely linked or contiguous to one another in the order *Z–Y–A* (see Figure 15–1).

Knowledge of their close linkage led to the discovery that all three genes are transcribed as a single unit, resulting in a **polycistronic mRNA** (Figure 15–3). Thus, the regulation of the *lac* genes is coordinated because a single message serves as the basis for translation of all three gene products.

The Discovery of Regulatory Mutations

How does lactose activate structural genes and induce the synthesis of the related enzymes? A partial answer comes from the discovery and study of **gratuitous inducers,** chemical analogs of lactose such as the sulfur analog **iso-propylthiogalactoside (IPTG),** shown in Figure 15–4. Gratuitous inducers behave like natural inducers, but they do not serve as substrates for the enzymes that are subse-

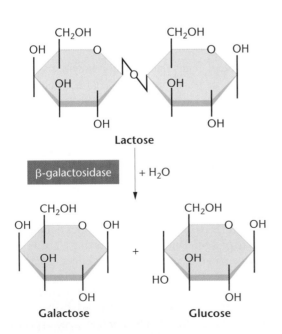

FIGURE 15–2 The catabolic conversion of the disaccharide lactose into its monosaccharide units, galactose and glucose.

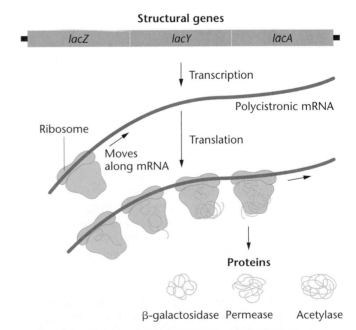

FIGURE 15–3 The structural genes of the *lac* operon are transcribed into a single polycistronic mRNA, which is translated simultaneously by several ribosomes into the three enzymes encoded by the operon.

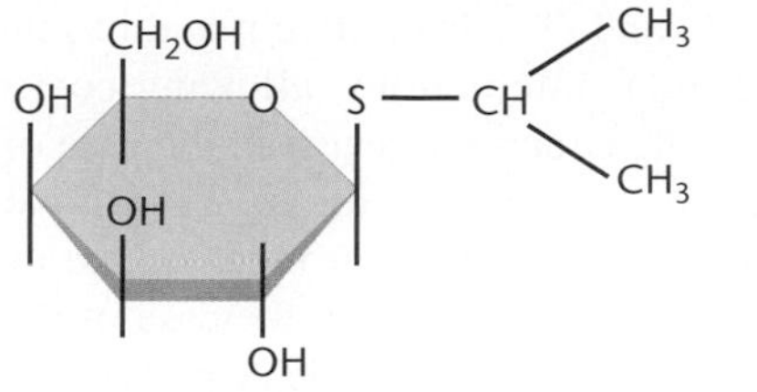

CH_2OH

$$CH_2OH$$

FIGURE 15–4 The gratuitous inducer isopropylthiogalactoside (IPTG).

quently synthesized. Their discovery provides strong evidence that the primary induction event does not depend on the interaction between the inducer and the enzyme.

What, then, is the role of lactose in induction? The answer to this question required the study of another class of mutations called **constitutive mutants**. In this type of mutation, the enzymes are produced regardless of the presence or absence of lactose. Maps of the first type of constitutive mutation, *lacI⁻*, showed that it is located at a site on the DNA close to, but distinct from, the structural genes. The *lacI* gene is appropriately called a **repressor gene**. We shall soon see

why this name is appropriate. A second set of constitutive mutations producing identical effects was found in a region immediately adjacent to the structural genes. This class of mutations, designated *lacO^C*, identifies the **operator region** of the operon. Because inducibility has been eliminated in both types of constitutive mutation (the enzymes are continually produced), clearly regulation has been disrupted by genetic changes.

The Operon Model: Negative Control

Around 1960, Jacob and Monod proposed the **operon model**, a scheme involving negative control, whereby a group of genes is regulated and expressed together as a unit. As we saw in Figure 15–1, the *lac* operon they proposed consists of the *Z*, *Y*, and *A* structural genes as well as the adjacent sequences of DNA referred to as the operator region. They argued that the *lacI* gene regulates the transcription of the structural genes by producing a **repressor molecule**, and that the repressor is **allosteric**, meaning that the molecule reversibly interacts with another molecule, causing both a conformational change in three-dimensional shape and a change in chemical activity. Figure 15–5 illustrates the

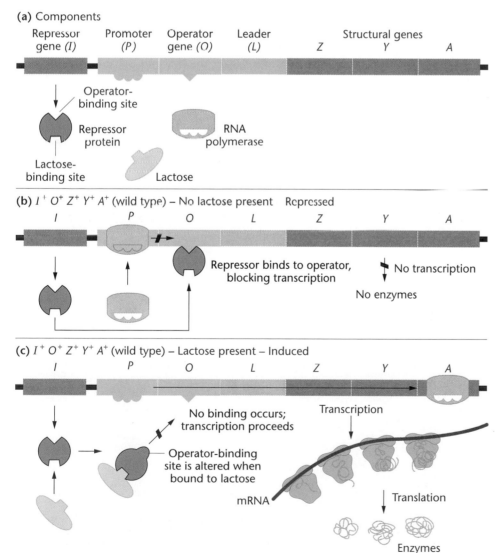

FIGURE 15–5 The components of the wild-type *lac* operon and the response in the absence and the presence of lactose, as described in the text.

(a) Components

Repressor gene *(I)* · Promoter *(P)* · Operator gene *(O)* · Leader *(L)* · Structural genes *Z* · *Y* · *A*

Operator-binding site
Repressor protein
Lactose-binding site
RNA polymerase
Lactose

(b) *I⁺ O⁺ Z⁺ Y⁺ A⁺* (wild type) – No lactose present   Repressed

I · *P* · *O* · *L* · *Z* · *Y* · *A*

Repressor binds to operator, blocking transcription
No transcription
No enzymes

(c) *I⁺ O⁺ Z⁺ Y⁺ A⁺* (wild type) – Lactose present – Induced

I · *P* · *O* · *L* · *Z* · *Y* · *A*

No binding occurs; transcription proceeds
Operator-binding site is altered when bound to lactose
Transcription
mRNA
Translation
Enzymes

components of the *lac* operon as well as the action of the *lac* repressor in the presence and absence of lactose. You should refer to it as you read the following section.

Jacob and Monod suggested that the repressor normally interacts with the DNA sequence of the operator region. When it does so, it inhibits the action of RNA polymerase, effectively repressing the transcription of the structural genes [Figure 15–5(b)]. However, when lactose is present, it binds to the repressor and causes an allosteric conformational change. This change alters the binding site of the repressor, rendering it incapable of interacting with operator DNA [Figure 15–5(c)]. In the absence of the repressor–operator interaction, RNA polymerase transcribes the structural genes, and the enzymes necessary for lactose metabolism are produced. Since transcription occurs only when the repressor fails to bind to the operator region, negative control is exerted.

The operon model uses these potential molecular interactions to explain the efficient regulation of the structural genes. In the absence of lactose, the enzymes encoded by the genes are not needed and are repressed. When lactose is present, it indirectly induces the activation of the genes by binding with the repressor.* If all lactose is metabolized, none is available to bind to the repressor, which is again free to bind to operator DNA and repress transcription.

---

*Technically, allolactose, an isomer of lactose, is the inducer. Allolactose is produced during the initial step in the metabolism of lactose by β-galactosidase.

Both the *I⁻* and *Oᶜ* constitutive mutations interfere with these molecular interactions, allowing continuous transcription of the structural genes. In the case of the *I⁻* mutant, seen in Figure 15–6(a), the repressor protein is altered and cannot bind to the operator region, so the structural genes are always turned on. In the case of the *Oᶜ* mutant [Figure 15–6(b)], the nucleotide sequence of the operator DNA is altered and will not bind with a normal repressor molecule. The result is the same: Structural genes are always transcribed.

Genetic Proof of the Operon Model

The operon model is a good one because it leads to three major predictions that can be tested to determine its validity. The major predictions to be tested are that (1) the *I* gene produces a diffusible cellular product, (2) the *O* region is involved in regulation but does not produce a product, and (3) the *O* region must be adjacent to the structural genes in order to regulate transcription.

The construction of partially diploid bacteria allows us to assess these assumptions, particularly those that predict *trans*-acting regulatory elements. For example, as introduced in Chapter 9, the F plasmid may contain chromosomal genes, in which case it is designated F⁺. When an F⁻ cell acquires such a plasmid, it now contains its own chromosome plus one or more additional genes present in the plasmid, creating a host cell, called a **merozygote**, that is diploid for those

FIGURE 15–6 The response of the *lac* operon in the absence of lactose when a cell bears either the *I⁻* or the *Oᶜ* mutation.

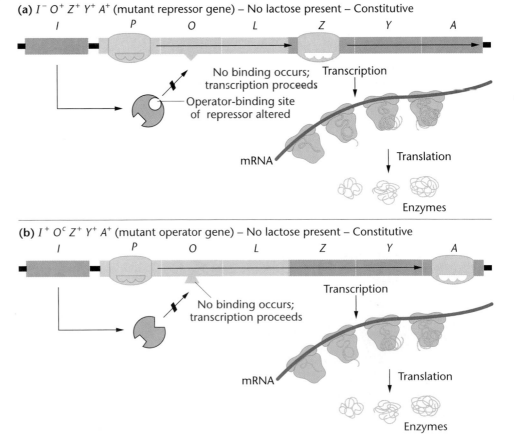

(a) *I⁻ O⁺ Z⁺ Y⁺ A⁺* (mutant repressor gene) – No lactose present – Constitutive

(b) *I⁺ Oᶜ Z⁺ Y⁺ A⁺* (mutant operator gene) – No lactose present – Constitutive

TABLE 15.1 A Comparison of Gene Activity (+ or −) in the Presence or Absence of Lactose for Various *E. coli* Genotypes

| | | Presence of β-Galactosidase Activity | |
	Genotype	Lactose Present	Lactose Absent
A.	$I^+O^+Z^+$	+	−
	$I^+O^+Z^-$	−	−
	$I^-O^+Z^+$	+	+
	$I^+O^CZ^+$	+	+
B.	$I^-O^+Z^+/F'I^+$	+	−
	$I^+O^CZ^+/F'O^+$	+	+
C.	$I^+O^+Z^+/F'I^-$	+	−
	$I^+O^+Z^+/F'O^C$	+	−
D.	$I^SO^+Z^+$	−	−
	$I^SO^+Z^+/F'I^+$	−	−

Note: In parts B to D, most genotypes are partially diploid, containing an F factor plus attached genes (F′).

genes. The use of such a plasmid makes it possible, for example, to introduce an I^+ gene into a host cell whose genotype is I^-, or to introduce an O^+ gene into a host cell of genotype O^C. The Jacob–Monod operon model predicts how regulation should be affected in such cells. Adding an I^+ gene to an I^- cell should restore inducibility, because a normal repressor, which is a *trans*-acting factor, would again be produced. Adding an O^+ gene to an O^C cell should have no effect on constitutive enzyme production, since regulation depends on an O^+ gene immediately adjacent to the structural genes; that is, O^+ is a *cis*-acting regulator.

Results of these experiments are shown in Table 15.1, where Z represents the structural genes. The inserted genes are listed after the designation F′. In both cases described above, the Jacob–Monod model is upheld (part B of Table 15.1). Part C shows the reverse experiments, where either an I^- gene or an O^C region is added to cells of normal inducible genotypes. As the model predicts, inducibility is maintained in these partial diploids.

Another prediction of the operon model is that certain mutations in the I gene should have the opposite effect of I^-. That is, instead of being constitutive by failing to interact with the operator, mutant repressor molecules should be produced that cannot interact with the inducer, lactose. As a result, the repressor would always bind to the operator sequence, and the structural genes would be permanently repressed (Figure 15–7). If this were the case, the presence of an additional I^+ gene would have little or no effect on repression.

In fact, such a mutation, I^S, was discovered where the operon is "superrepressed," as shown in part D of Table 15.1. An additional I^+ gene does not effectively relieve repression of gene activity. These observations again support the operon model for gene regulation.

Isolation of the *lac* Repressor

Although the operon theory of Jacob and Monod succeeded in explaining many aspects of genetic regulation in prokaryotes, the nature of the repressor molecule was not known when their landmark paper was published in 1961. While they had assumed that the allosteric repressor was a protein, RNA was also a candidate because activity of the molecule required the ability to bind to DNA. Despite many attempts to isolate and characterize the hypothetical repressor molecule, no direct chemical evidence was immediately forthcoming. A single *E. coli* cell contains no more than 10 or so copies of the *lac* repressor; direct chemical identification of 10 molecules in a population of millions of proteins and RNAs in a single cell presented a tremendous challenge.

In 1966, Walter Gilbert and Benno Müller-Hill reported the isolation of the *lac* repressor in partially purified form. To achieve the isolation, they used a regulator quantity (I^q) mutant strain that contains about 10 times as much repressor as do wild-type *E. coli* cells. Also instrumental in their success were the use of the gratuitous inducer, IPTG, which binds to the repressor, and the technique of **equilibrium dialysis**. In this technique, extracts of I^q cells are placed in a dialysis bag and allowed to attain equilibrium with an external solution of radioactive IPTG, which is small enough to diffuse freely in and out of the bag. At equilibrium, the concentration of IPTG is higher inside the bag than in the external solution, indicating that an IPTG-binding material is present in the cell

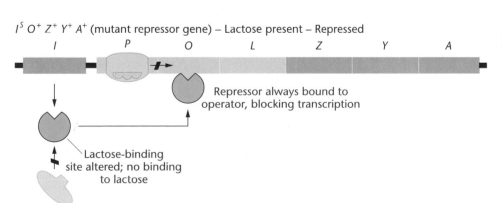

$I^S\ O^+\ Z^+\ Y^+\ A^+$ (mutant repressor gene) – Lactose present – Repressed

Repressor always bound to operator, blocking transcription

Lactose-binding site altered; no binding to lactose

FIGURE 15–7 The response of the *lac* operon in the presence of lactose in a cell bearing the I^S mutation.

extract and that this material is too large to diffuse across the wall of the bag.

Ultimately the IPTG-binding material was purified and shown to have various characteristics of a protein. In contrast, extracts of I^- constitutive cells having no *lac* repressor activity did not exhibit IPTG-binding activity, strongly suggesting that the isolated protein was the repressor molecule.

The CAP Protein: Positive Control of the *lac* Operon

As is apparent from the preceding discussion of the *lac* operon, the role of β-galactosidase is to cleave lactose into its components glucose and galactose. Then, for galactose to be used by the cell, it is converted to glucose. What if the cell finds itself in an environment that contains an ample amount of lactose and glucose? It is not energetically efficient for a cell to be "induced" by lactose to make β-galactosidase, since what it really needs, glucose, is already present. As we shall see, still another molecular component, called the **catabolite-activating protein (CAP)**, is involved in effectively repressing the expression of the *lac* operon when glucose is present. This inhibition, called **catabolite repression**, re-

flects the greater simplicity with which glucose may be metabolized in comparison to lactose. The cell "prefers" glucose, and if it is available, the *lac* operon is not activated, even when lactose is present.

To understand CAP and its role in regulation, let's backtrack for a moment. When the *lac* repressor is bound to the inducer, the *lac* operon is activated and RNA polymerase transcribes the structural genes. As we have learned in Chapter 12, transcription is initiated as a result of the binding that occurs between RNA polymerase and the nucleotide sequence of the **promoter region**, found upstream (5′) from the initial coding sequences. Within the *lac* operon, the promoter is found between the *I* gene and the operator region (*O*) (see Figure 15–1). Careful examination has revealed that polymerase binding is never very efficient unless CAP is also present to facilitate the process.

The mechanism is summarized in Figure 15–8. In the absence of glucose, and under inducible conditions, CAP exerts **positive control** by binding to the CAP site, facilitating RNA polymerase binding at the promoter, and thus transcription. Therefore, for maximal transcription, the repressor must be bound by lactose (so as not to repress operon expression), and CAP must be bound to the CAP-binding site.

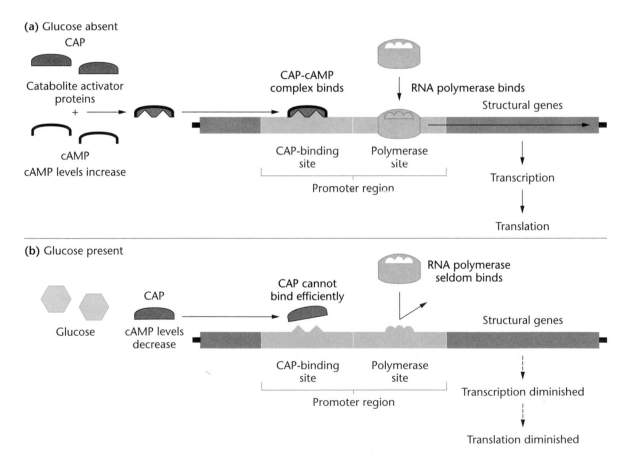

FIGURE 15–8 Catabolite repression. (a) In the absence of glucose, cAMP levels increase, resulting in the formation of a CAP–cAMP complex, which binds to the CAP site of the promoter, stimulating transcription. (b) In the presence of glucose, cAMP levels decrease, CAP–cAMP complexes are not formed, and transcription is not stimulated.

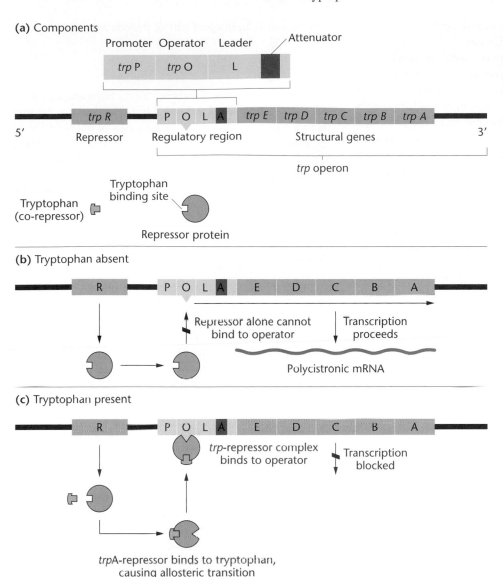

(a) Components

Promoter Operator Leader Attenuator

| trp P | trp O | L | |

| trp R | | P | O | L | A | trp E | trp D | trp C | trp B | trp A | |

5′   Repressor   Regulatory region   Structural genes   3′

trp operon

Tryptophan
binding site

Tryptophan
(co-repressor)

Repressor protein

(b) Tryptophan absent

| R | | P | O | L | A | E | D | C | B | A | |

Repressor alone cannot
bind to operator   Transcription
proceeds

Polycistronic mRNA

(c) Tryptophan present

| R | | P | O | L | A | E | D | C | B | A | |

trp-repressor complex
binds to operator   Transcription
blocked

*trp*A-repressor binds to tryptophan,
causing allosteric transition

FIGURE 15–9 (a) The components involved in the regulation of the tryptophan operon. Regulatory conditions are depicted involving either (b) activation or (c) repression of the structural genes. In the absence of tryptophan, an inactive repressor is made that cannot bind to the operator (*O*), thus allowing transcription to proceed. In the presence of tryptophan, it binds to the repressor, causing an allosteric transition to occur. This complex binds to the operator region, leading to repression of the operon.

This leads to the central question about CAP. What role does glucose play in inhibiting CAP binding when it is present? The answer involves still another molecule, **cyclic adenosine monophosphate (cAMP)**, on which CAP binding is dependent. In order to bind to the promoter, CAP must be linked to cAMP. The level of cAMP is itself dependent on an enzyme **adenyl cyclase**, which catalyzes the conversion of ATP to cAMP* (Figure 15–9).

The role of glucose in catabolite repression is now clear. It inhibits the activity of adenyl cyclase, causing a decline in the level of cAMP in the cell. Under this condition, CAP cannot form the CAP–cAMP complex essential to the positive control of transcription. Regulation of the *lac* operon by catabolite repression results in efficient energy use because the presence of glucose will override the need for the metabolism

of lactose, should it also be available to the cell. Catabolite repression involving CAP has also been observed for other inducible operons, including those controlling the metabolism of galactose and arabinose.

15.3 Tryptophan Metabolism in *E. coli*: A Repressible Gene System

Although the process of induction had been known for some time, it was not until 1953 that Monod and colleagues discovered a repressible operon. Wild-type *E. coli* are capable of producing the enzymes necessary for the biosynthesis of amino acids as well as other essential macromolecules. Focusing his studies on the amino acid tryptophan and the enzyme **tryptophan synthetase**, Monod discovered that if tryptophan is present in sufficient quantity in the growth medium, the enzymes necessary for its synthesis are not produced. Energetically, repression of the genes involved in the

---

*Because of its involvement with cAMP, CAP is also called cyclic AMP receptor protein (CRP), and the gene encoding the protein is named *crp*. Because the protein was first named CAP, we will adhere to the initial nomenclature.

production of these enzymes is highly economical for the cell when ample tryptophan is present.

Further investigation showed that a series of enzymes encoded by five contiguous genes on the *E. coli* chromosome is involved in tryptophan synthesis. These genes are part of an operon, and in the presence of tryptophan, all are coordinately repressed and none of the enzymes are produced. Because of the great similarity between this repression and the induction of enzymes for lactose metabolism, Jacob and Monod proposed a model of gene regulation analogous to the *lac* system (Figure 15–9).

To account for repression, they suggested the presence of a *normally inactive repressor* that alone cannot interact with the operator region of the operon. However, the repressor is an allosteric molecule that can bind to tryptophan. When this amino acid is present, the resultant complex of repressor and tryptophan attains a new conformation that binds to the operator, repressing transcription. Thus, when tryptophan, the end product of this anabolic pathway, is present, the system is repressed and enzymes are not made. Because the regulatory complex inhibits transcription of the operon, this repressible system is under negative control. And, as tryptophan participates in repression, it is referred to as a **co-repressor** in this regulatory scheme.

Genetic Evidence for the *trp* Operon

Support for the concept of a repressible operon was soon forthcoming, based primarily on the isolation of two distinct categories of constitutive mutations. The first class, *trpR⁻*, maps at a considerable distance from the structural genes. This locus represents the gene coding for the repressor. Presumably, the mutation either inhibits the interaction of the repressor with tryptophan or inhibits repressor formation entirely. Whichever the case, no repression ever occurs in cells with the *trpR⁻* mutation. As expected, if the *trpR* gene encodes a repressor molecule, the presence of an additional *trpR⁺* gene restores repressibility.

The second constitutive mutant is analogous to that of the operator of the lactose operon because it maps immediately adjacent to the structural genes. Furthermore, the addition of a wild-type operator gene into mutant cells (as a *trans*-acting element) does not restore enzyme repression. This is predictable if the mutant operator can no longer interact with the repressor–tryptophan complex.

The entire *trp* operon has now been well defined, as shown in Figure 15–9. Five contiguous structural genes (*trp E, D, C, B,* and *A*) are transcribed as a polycistronic message that directs translation of the enzymes that catalyze the biosynthesis of tryptophan. As in the *lac* operon, a promoter region (*trpP*) represents the binding site for RNA polymerase, and an operator region (*trpO*) binds the repressor. In the absence of binding, transcription is initiated within the overlapping *trpP–trpO* region and proceeds along a **leader sequence** 162 nucleotides prior to the first structural gene (*trpE*). Within this leader sequence, still another regulatory site has been demonstrated, called an attenuator, the subject of the next section of this chapter. As we shall see, this reg-

ulatory unit is an integral part of the control mechanism of this operon.

Attenuation

Charles Yanofsky, his co-worker Kevin Bertrand, and their colleagues observed that, even when tryptophan is present and the *trp* operon is repressed, initiation of transcription of the leader sequence of the operon usually still occurs. Thus, while the activated repressor binds to the operator region, it does not strongly inhibit the *initial expression* of the operon, suggesting that there must be a subsequent mechanism by which tryptophan inhibits enzyme synthesis. Yanofsky discovered that, in the presence of high concentrations of tryptophan, transcription of the leader sequence of the operon proceeds, but that transcription of mRNA is usually terminated prematurely. This process is called **attenuation**, indicative of the effect of severely diminishing genetic expression of the operon. However, when tryptophan is absent, or present in very low concentrations, the repressor is inactive and does not bind to the promoter. Transcription is initiated, but *not* subsequently terminated, instead proceeding through the leader sequence and into the structural genes. As a result, a polycistronic mRNA is transcribed and the enzymes essential to the biosynthesis of tryptophan are subsequently translated.

Identification of the site involved in attenuation was made possible by the isolation of deletion mutations in the region of the leader sequence. Such mutations abolish attenuation. Thus, the site is referred to as the **attenuator**. An explanation of how attenuation occurs and how it is overcome, put forward by Yanofsky and colleagues, involves folding of the RNA transcribed from the leader sequence. During attenuation (when tryptophan is abundant), the structure of the RNA transcribed *from the leader sequence* mimics that found at the end of mRNA molecules, forming a "hairpin loop." This configuration leads to the premature termination of transcription, reducing the amount of mRNA produced.

The phenomenon of attenuation appears to be a mechanism common to other bacterial operons that regulate the enzymes essential to the biosynthesis of amino acids. In addition to tryptophan, operons involved in threonine, histidine, leucine, and phenylalanine display attenuators in their leader sequences.

15.4 Genetic Regulation in Eukaryotes: An Overview

In 1960, the discovery of the operon in *Escherichia coli* by Jacob and Monod was the first step in unraveling the mechanisms that regulate gene expression. While some of the principles of regulation found in the *lac* operon are also present in eukaryotes, it is clear that eukaryotes have evolved a more complex system of gene regulation.

As we pointed out in the introduction of this chapter, in multicellular eukaryotes, differential regulation of gene expression is at the heart of cellular differentiation and

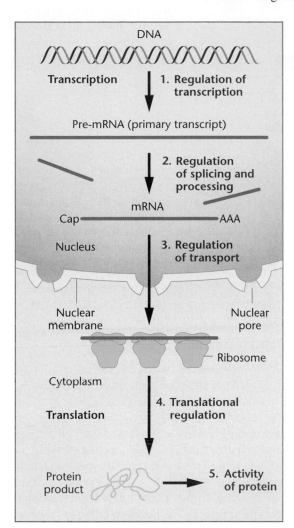

FIGURE 15–10 Various levels of regulation that are possible during the expression of the genetic material.

function. The central question is how an organism expresses a subset of genes in one cell type and a different subset of genes in another cell type. At the cellular level, this regulation is *not* accomplished by eliminating unused genetic information; instead, mechanisms have evolved to activate specific portions of the genome, and to repress the expression of other genes. The activation and repression of selected loci represent a delicate balancing act for an organism; the expression of a gene at the wrong time, in the wrong cell type, or in abnormal amounts can lead to a deleterious phenotype or death, even when the gene itself is normal.

Numerous factors help us to understand why gene regulation is more complex in eukaryotes than in prokaryotes:

1. Eukaryotic cells contain a much greater amount of genetic information than do prokaryotic cells, and this DNA is complexed with histones and other proteins to form chromatin. The structure of chromatin, either open (decondensed) and available to be transcribed, or closed (condensed) and not available, is an important factor during gene regulation.

2. In eukaryotes, transcription is spatially and temporally separated from translation—transcription occurs in the nucleus and translation occurs later in the cytoplasm. Thus, mRNA must be transported out of the nucleus in order to serve as the basis of protein synthesis.

3. The transcripts of eukaryotic genes are processed and reduced in size before their transport to the cytoplasm.

4. Eukaryotic mRNA has a much longer half-life ($t_{1/2}$) than does prokaryotic mRNA. If transcription is turned off in prokaryotes, the mRNA decays within minutes and translation ceases. This is not the case in eukaryotes.

5. Most eukaryotes are multicellular with differentiated cell types. These different cell types typically use different sets of overlapping genes to make different proteins even though each cell contains a complete set of genes.

As a result of the above factors, eukaryotic gene expression can potentially be controlled in many different ways. As shown in Figure 15–10, these include regulation during (1) transcription (as in prokaryotes); (2) processing and splicing of the pre-mRNA, referred to as posttranscription control; (3) transport to the cytoplasm; (4) translation (e.g., selecting which mRNAs are translated); and (5) posttranslational modification of the protein product.

15.5 Regulatory Elements, Transcription Factors, and Eukaryotic Genes

The internal structure of eukaryotic genes (see Chapter 12) includes several forms of regulatory elements that control transcription. These are of two main types: promoters and enhancers. These regulatory elements can be located next to, within, or at some distance from the gene. They interact with a wide variety of proteins that serve to modulate their activity. These are called transcription factors. In the following sections, we develop a general model that demonstrates how these elements and factors function in the regulation of eukaryotic genetic expression.

Promoters

Promoters, which have counterparts in bacteria, consist of *cis*-acting nucleotide sequences that serve as the recognition point for RNA polymerase binding. Therefore, they represent the region necessary to *initiate* transcription and are located immediately adjacent to the genes they regulate. Eukaryotic promoters recognized by RNA polymerase II (which transcribes genes into mRNA) consist of short modular DNA sequences usually located within 100 bp upstream (in the 5′-direction) of the gene (Figure 15–11, page 316). The core region of the promoter contains a sequence called the **TATA box**. Located about 25–30 bp upstream from the initial point of transcription (designated as −25 to −30), it consists of an 8-bp consensus sequence (a sequence conserved in most or all genes studied) composed only of T≡A pairs, often flanked on either side by G≡C-rich regions.

Transcription start site (+1)

GC Box   CAAT Box   TATA Box

GGGCGG   GGCCAATC   TATAAA

−110   −70   −30

FIGURE 15–11 The promoter at the 5'-end of eukaryotic genes consists of several modular elements, including the TATA box (−30), the CAAT box (−70), and the GC box (at about −110).

The critical importance of the TATA box is demonstrated by mutational studies (Figure 15–12). Mutations within the nucleotide sequence of this element severely reduce transcription, while those in adjacent bases have little or no effect on gene expression. Such a reduction is considered to be the result of losing the ability to bind to *trans*-acting transcription factors, which are responsible for stimulating transcription.

Farther upstream from the core region of promoter are several proximal elements that are also critical to the initiation of transcription. One of these is called the **CAAT box**. Its consensus sequence is CAAT or CCAAT, and it frequently appears in the region −70 to −80 bp from the start site of transcription. Like the TATA box, mutations in the sequence of the CAAT box severely diminish transcription (Figure 15–12). Mutations on either side of this element have little or no effect. Another proximal element is called the **GC box**, which has the consensus sequence GGGCGG and is usually found around position −110. Often present in multiple copies, its importance has also been documented by mutational analysis.

In different genes there are significant differences in the number and orientation of promoter elements and the distance between them. In addition to RNA polymerase II, other eukaryotic genes are transcribed by different polymerases, classified as type I (yielding ribosomal RNAs) and type III (yielding tRNAs, 5S rRNA, and several other small cellular RNAs). The promoters for both types of polymerase have a different sequence and bind different transcription factors.

Enhancers

In addition to the promoter regions, transcription of most, if not all, eukaryotic genes is affected by additional *cis*-acting DNA sequences called **enhancers**. Like promoters, these regions interact with *trans*-acting regulatory proteins. The im-

pact is to greatly increase the efficiency of the initiation of transcription, thus increasing the overall rate. Enhancers can be distinguished from promoters by several characteristics.

1. The position of the enhancer need not be fixed; it can be placed upstream, downstream, or within the gene it regulates.

2. Often, enhancers operate from quite a distance away from their target gene—as much as 50 kb.

3. The orientation of an enhancer can be inverted without significant effect on its action.

4. If an enhancer is moved to another location in the genome, or if an unrelated gene is placed near an enhancer, transcription of the adjacent gene is positively regulated.

Several examples serve to illustrate the varied nature of the location of enhancers. In the immunoglobulin heavy-chain genes, an enhancer is located *within* the gene it regulates, in an intron between two coding regions. Downstream enhancers are found in the human β-globin gene, and in chickens, an enhancer is located between the β-globin and the ε-globin genes.

The most intriguing question about enhancers is how they exert control over transcription when they are located at some distance from either promoters or the transcriptional start site. As we shall see, enhancers are bound by transcription factors that can alter the configuration of chromatin by bending or looping the DNA. This is thought to bring distant enhancers and promoters into close proximity in order to form activated transcription complexes. In the new configuration, transcription is stimulated above a basal level, increasing the overall rate of RNA synthesis.

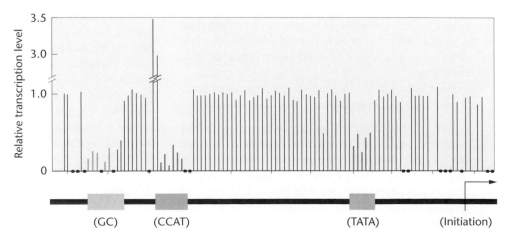

FIGURE 15–12 Summary of the effects of point mutations in the promoter region on transcription of the β-globin gene. Each line represents the level of transcription produced by a single nucleotide mutation (relative to wild type) in a separate experiment. Dots represent nucleotides where no mutation was obtained. Note that mutations within the specific elements of the promoter have the greatest effect on the level of transcription.

Transcription Factors

Overall, the picture of transcriptional regulation in eukaryotes is somewhat complex, but a number of generalizations can be drawn. Many proteins have been identified that are essential to the initiation of transcription, but not part of the RNA polymerase molecule that initiates and executes the process of transcription. These are called **transcription factors**. They control where, when, and at what rate genes are expressed. These proteins are modular and usually contain two functional domains (clusters of amino acids that carry out a specific function). One, the **DNA-binding domain**, binds to DNA sequences present in regulatory regions, including promoters and enhancers. The other, referred to as the ***trans-activating domain***, activates transcription via protein–protein interaction. For example, this domain of a transcription factor may bind to other transcription factors at the promoter or to RNA polymerase.

The regulation of transcription by protein factors is largely positive, although transcriptional repressors are now being recognized as important components of gene regulation. Before providing specific examples of such factors and discussing how they regulate gene expression, let's explore what is known about *how* they bind to nucleic acids. This represents a critical feature in genetic function.

Structural Motifs of Transcription Factors

The DNA-binding domains of eukaryotic transcription factors take on several forms. They have distinctive three-dimensional structural patterns or motifs. There are three major types of these structural motifs: helix–turn–helix, zinc finger, and basic leucine zippers. This classification is not exhaustive, and other new groups will undoubtedly be established as new factors are characterized.

The first DNA-binding domain to be discovered was the **helix–turn–helix (HTH) motif**. In prokaryotes, HTH motifs have been identified in the *lac* repressor, the *trp* repressor, and other proteins (Figure 15–13). Studies indicate that the HTH motif is present in many eukaryotic DNA-binding proteins. This motif is characterized by its geometric conformation rather than a distinctive amino acid sequence. Having two adjacent α-helices separated by a "turn" of several amino acids enables the protein to bind to DNA and provides the motif's name. Unlike several of the other DNA-binding mo-

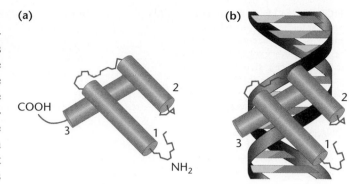

FIGURE 15–13 A *helix–turn–helix* or *homeodomain* where (a) three planes of the α helix of the protein are established. (b) These domains bind in the grooves of the DNA molecule.

tifs, the HTH pattern cannot fold or function alone, but is always part of a larger DNA-binding domain.

The potential for forming helix-turn-helix geometry has been recognized in distinct regions of a large number of eukaryotic genes known to regulate developmental processes. Present almost universally in eukaryotic organisms and called the **homeobox**, a stretch of 180 bp specifies a **homeodomain** sequence of 60 amino acids that can form a helix–turn–helix structure. Of the 60 amino acids, many are basic (arginine and lysine), and a conserved sequence is found among these many genes. We discuss homeobox-containing genes in Chapter 20 because of their significance to developmental processes.

Zinc fingers are one of the major structural families of eukaryotic transcription factors, and they are involved in many aspects of gene regulation. Originally discovered in a transcription factor in the frog, *Xenopus laevis*, this structural motif has now been identified in proto-oncogenes (Chapter 21), genes that regulate development in *Drosophila* (Chapter 20), and in proteins whose synthesis is induced by growth factors and differentiation signals, among others. There are several types of zinc finger proteins, each with a distinctive structural pattern.

One such zinc finger protein contains clusters of two cysteine and two histidine residues at repeating intervals (Figure 15–14). The interspersed cysteine and histidine residues covalently bind zinc atoms, folding the amino acids into loops (which are the zinc "fingers"). Each finger consists of approximately 23 amino

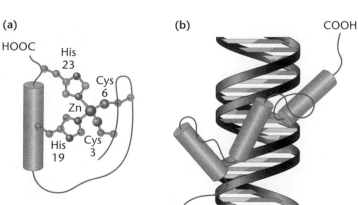

FIGURE 15–14 (a) A zinc finger where cysteine and histidine residues bind to a Zn^{++} atom. (b) This loops the amino acid chain out into a fingerlike configuration.

FIGURE 15–15 (a) A leucine zipper. Dimer formation occurs because of the presence of leucine residues at every other turn of the α helix in facing stretches of two polypeptide chains. (b) When the α-helical regions form a leucine zipper, the regions beyond the zipper form a Y-shaped region that grips the DNA in a scissors-like configuration.

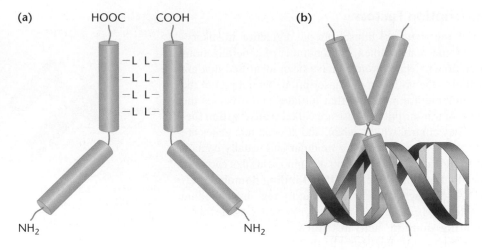

acids, with a loop of 12–14 amino acids between the Cys and His residues, and a linker between loops consisting of 7 or 8 amino acids. The amino acids in the loop interact with and bind to specific DNA sequences. Studies have shown that zinc fingers bind in the major groove of the DNA helix and wrap at least part way around the DNA. Within the major groove, the zinc finger makes contact with a set of DNA bases and may form hydrogen bonds with the bases, especially in G-rich strands. The number of fingers in a zinc finger transcription factor varies from 2 to 13. The length of the DNA-binding sequence also varies in size.

The third type of domain is represented by the **basic leucine zipper (bZIP)**. First seen as a stretch of 35 amino acids in a nuclear protein in rat liver, four leucine residues are spaced 7 amino acids apart, flanked by basic amino acids. The leucine-rich regions form a helix with leucine residues protruding at every other turn. When two such molecules dimerize (Figure 15–15), the leucine residues "zip" together. The dimer contains two alpha-helical regions adjacent to the zipper, which bind to phosphate residues and specific bases in DNA, making the dimer look like a pair of scissors.

In addition to domains that bind DNA, transcription factors contain domains that activate transcription. These regions can occupy from 30 to 100 amino acids and are distinct from the DNA-binding domains. Such stretches of amino acids interact with other transcription factors (such as those that bind to the TATA sequence) or directly with the RNA polymerase.

Assembly of the Transcription Complex

With the knowledge that transcription factors play a critical role in gene regulation and how they bind to nucleic acids, we can now examine how eukaryotic gene regulation specifically occurs. A number of insights into the transcriptional apparatus of type II genes in eukaryotes (those transcribed by RNA polymerase II) are now available, providing the basis for a model for the initiation of mRNA synthesis (Figure 15–16). Central to the model are the various *trans*-acting protein factors that facilitate template binding by RNA polymerase II. These transcription factors are well characterized and designated TFIIA, TFIIB, and so on. One of these, **TFIID**, is a multi-subunit protein com-

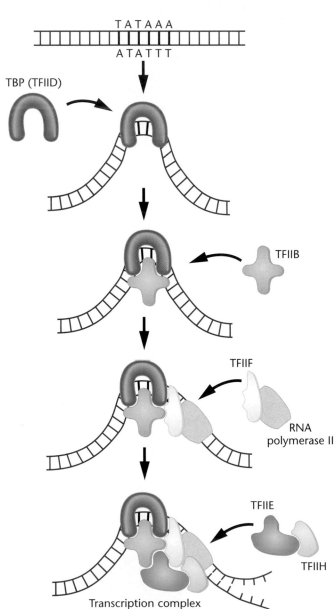

FIGURE 15–16 Assembly of the transcription complex in eukaryotes. The initial step involves binding of TBP to the TATA box, which bends DNA. A combination of transcription factors (TFIIB, etc.) and RNA polymerase II are added in a stepwise fashion prior to the initiation of transcription. TFIIH demonstrates helicase activity, which "opens" the helix during transcription.

plex that includes a protein that binds directly to the TATA-box DNA sequence of the promoter. Thus, it is called the **TATA-binding protein (TBP)**. TFIID includes at least 11 **TBP-associated factors**, called **TAFs**. In all genes studied, binding of the TBP at the TATA box of the promoter is essential to subsequent RNA polymerase binding at the start site. About 20 bp of DNA are involved in binding the TBP, and the other subunits then bind to the growing complex.

Other transcriptional factors are then assembled at the promoter region, in a specific order. TFIID (which includes TBP) responds to contact with activator proteins by undergoing conformational changes that expedite the binding of other transcription factors and RNA polymerase. As shown in Figure 15–16, TFIIB first binds, followed by TFIIF, which is complexed with RNA polymerase. Following the addition of TFIIE and TFIIH, the initiation complex is complete, and transcription is initiated at the start site just downstream from the TATA box. TFIIH is a large multi-subunit complex that demonstrates helicase activity, facilitating the separation of the DNA helix at the start site. As the transcription proceeds, TBF remains bound to the TATA sequence of the promoter, but the other transcription factors dissociate from the complex.

At this point, transcription of the DNA downstream will ensue at a low *basal* level. The final stage involves the achievement of the *induced* state, where transcription is stimulated at a much higher level. This state is not yet well defined, but involves other areas of the promoter region, enhancers, and numerous transcription factors, which control the assembly of the transcription complex and the rate at which RNA polymerase initiates transcription. Multiple transcription factors bind cooperatively to enhancers and then to the transcription complex. This loops out the DNA that separates the enhancer from the complex (Figure 15–17).

This general model is thought to constitute the mechanism for eukaryotic gene regulation. As we will see below, still other aspects of the genetic machinery also come into play.

Chromatin Conformation, DNA Methylation, and Gene Expression

We know that DNA in eukaryotes is complexed with proteins in the interphase nucleus and present in a form known as **chromatin**. As we will see in more detail in Chapter 17, chromatin is characterized by the presence of repeating structures called **nucleosomes**, each of which consists of an octamer of four different histone proteins complexed with about 150 bp of DNA. When complexed to histone proteins in this way, eukaryotic DNA is relatively inaccessible to transcription factors, and thus the formation of an active transcription complex cannot occur. Therefore, the normal structure of chromatin is thought to be sufficient to significantly repress gene activity. As a result, activation of genes requires what is called **chromatin remodeling**, whereby the conformation of chromatin is altered in a way that the proteins of the nucleosome are released from the DNA, allowing it to become accessible to transcription factors and RNA polymerase.

There are several ways in which this is thought to occur. One involves specialized proteins that actually disrupt the nucleosome particles and remove the histones, thus freeing up the DNA. For example, several specific proteins in yeast, called **SWI** and **SNF**, are part of a larger complex that functions to disrupt chromatin structure by displacing or removing the proteins making up the nucleosome, thereby facilitating transcription. Similar complexes have been found in other eukaryotes. In another mechanism, **acetylation** of histone proteins by specific enzymes lessens the attraction of histones to DNA, resulting in a remodeling of chromatin structure. Conversely, deacetylation reverses this process and is thought to repress gene activity. Specific enzymes are known to be involved.

Another type of change in chromatin that plays a role in gene regulation involves adding or removing methyl groups to the bases in DNA. The DNA of most eukaryotic organisms

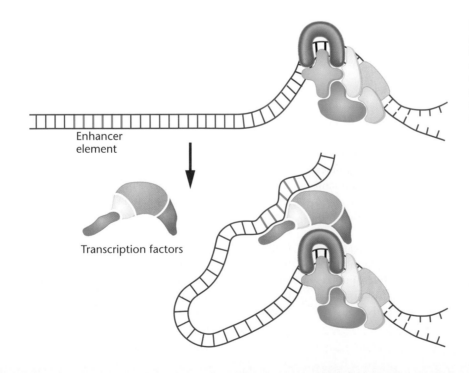

Enhancer element

Transcription factors

FIGURE 15–17 The influence of multiple transcription factors that bind cooperatively to an enhancer element, resulting in its interaction with the transcription complex at the promoter region. The effect is to increase transcription substantially above the basal level.

is modified after replication by the enzyme-mediated addition of methyl groups to bases and sugars. **DNA methylation** most often involves cytosine. Approximately 5 percent of the cytosine residues are methylated in the genome of any given eukaryotic species.

The ability of base methylation to alter gene expression is known from studies on the *lac* operon in *E. coli*. Methylation of DNA in the operator region, even at a single cytosine residue, can cause a marked change in the affinity of the repressor for the operator. Methylation of cytosine occurs at the number 5 carbon of the pyrimidine ring, causing the methyl group to protrude into the major groove of the DNA helix, where it can alter the binding of proteins to the DNA. This happens most often when the cytosine residue is part of a CG doublet in DNA, where the cytosines on both strands are affected:

$$5'\text{-mCpG-}3'$$
$$3'\text{- GpCm-}5'$$

Consistent with this information, analysis of the methylation of a given gene in different tissues shows that, in general, if a gene is expressed, its promoter region is not methylated, or has a low level of methylation. Further, in mammalian females, the X chromosome that forms a Barr body is almost totally inactive in gene expression. It has been shown to have a significantly higher level of methylation than its counterpart X chromosome, which is transcriptionally active. Within the chromosome condensed into the Barr body, small regions escape inactivation. These have much lower levels of methylation than those seen in adjacent, inactive regions.

While the absence of methyl groups in DNA is related to increased gene expression, methylation cannot, however, be regarded as a universal mechanism for gene regulation, because methylation is not a phenomenon characterizing all eukaryotes. In *Drosophila*, for example, little or no methylation of DNA has been observed. Thus, methylation may represent only one of a number of ways in which gene expression can be regulated by genomic changes.

15.6   Gene Regulation by Steroid Hormones

Our final consideration of eukaryotic gene regulation *at the level of transcription* involves a short discussion of how **steroid hormones** affect their target cells. Our knowledge of this system provides a nice recap of the various aspects of eukaryotic regulation discussed thus far. Steroid hormones are used to regulate growth and development and to maintain homeostasis. The major sex hormones are all steroids, as is vitamin D, and the homeostatic adrenal hormones that regulate glucose metabolism and mineral utilization.

The general scheme of hormone action is illustrated in Figure 15–18. Hormones enter the cell by passing through the plasma membrane and binding to a specific **hormone receptor protein** in the cytoplasm. The receptor–hormone complex is translocated to the nucleus and activates transcription of one or more specific genes.

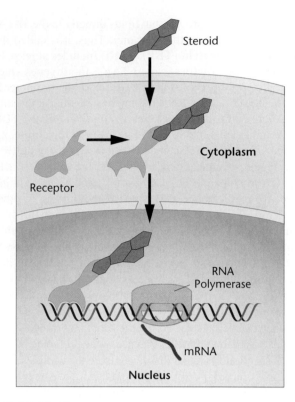

FIGURE 15–18 The effect of a steroid hormone on gene expression. The hormone enters the cell, binds to a specific receptor, leading to a conformational change in a DNA binding domain of the receptor. The complex enters the nucleus and behaves as a transcription factor, stimulating a target gene to initiate transcription.

Hormone receptors and the DNA sequences to which they bind have been studied intensively over the last decade. All receptors analyzed to date have three functional domains: a variable N-terminus domain, unique to each receptor; a short, highly conserved central domain that binds to DNA; and a C-terminus domain that binds to the hormone. The DNA-binding central domain contains two zinc fingers. The zinc fingers in hormone receptors bind to specific DNA sequences known as **hormone-responsive elements (HREs)**.

In general, HREs share some characteristics with promoters and enhancers. They are composed of short consensus sequences that are related but not always identical. HREs are often located several hundred bases upstream from the transcription start site and may be present in multiple copies. They are also often present in promoter or enhancer sequences.

While binding the receptor to the HRE is necessary for activating a specific gene, it may not be sufficient on its own to cause activation. Binding the receptor to the HRE may serve simply to facilitate the interaction of other transcription factors by altering the chromatin structure in the HRE and adjacent regions. Such an alteration may make other binding sites, including the promoter, available to bind transcription factors and RNA polymerase II, resulting in the initiation of transcription.

In some ways, then, steroid hormone regulation of gene transcription in eukaryotes is similar to the positive-control systems found in some of the operons in prokaryotes. An external effector binds to a cytoplasmic receptor, changes the configuration of the receptor, and moves the receptor to the DNA, where it acts as a transcription factor. In other ways, however, the regulation of gene expression is quite different in that the DNA-binding sequence is often at a great distance from the regulated gene, and that alterations in chromatin structure mediate the action of the receptor.

15.7 Posttranscriptional Regulation of Gene Expression

As we have seen, the regulation of genetic expression occurs at many points along the pathway from DNA to protein. Although transcriptional control is perhaps the most obvious and widely used mode of regulation in eukaryotes, **posttranscriptional modes** of regulation also occur in many organisms. Eukaryotic nuclear RNA transcripts are modified prior to translation, noncoding introns are removed, the remaining exons are precisely spliced together, and the mRNA is modified by the addition of a cap at the 5'-end and a poly-A tail at the 3'-end. The message is then complexed with proteins and exported to the cytoplasm. Each of these processing steps offers several possibilities for regulation. We shall examine one mechanism that is especially important in eukaryotes, alternative splicing of a single mRNA transcript to give multiple mRNAs.

Alternative Splicing of mRNA

Alternative splicing can generate different forms of a protein, so that expression of one gene can give rise to a family of related proteins. Figure 15–19 illustrates an example where the polypeptide products derived from a single type of pre-mRNA are distinct from one another. The initial bovine pre-mRNA transcript is processed into one or two **preprotachykinin mRNAs (PPT mRNAs)**. This precursor

mRNA molecule potentially includes the genetic information specifying two neuropeptides, called **P** and **K**. These two peptides, which are members of the family of sensory neurotransmitters referred to as **tachykinins**, are believed to play different physiological roles. While the P neuropeptide is largely restricted to tissues of the nervous system, the K neuropeptide is found more predominantly in the intestine and thyroid.

The RNA sequences for both neuropeptides are derived from the same gene. However, the processing of the initial RNA transcript can occur in two ways. In one case (Figure 15–19), exclusion of the K exon during processing results in the α-PPT mRNA, which upon translation yields neuropeptide P but not K. Conversely, processing that includes both the P and K exons yields β-PPT mRNA, which upon translation results in the synthesis of both the P and K neuropeptides. The analysis of the relative levels of the two types of RNA demonstrates striking differences between tissues. In nervous-system tissues, α-PPT mRNA predominates by as much as a threefold factor, while β-PPT mRNA is the predominant type in the thyroid and intestine.

Given the existence of alternative splicing, how many different polypeptides can be derived from the same pre-mRNA? Work on α-tropomyosin, an accessory protein that regulates muscle contraction, has provided a partial answer to this question. In rats, the α-tropomyosin gene contains a total of 14 exons, six of which make up three pairs that are alternatively spliced. Only one member of each pair ends up in the finished mRNA, never both. Alternative splicing of this pre-mRNA results in 10 different forms of α-tropomyosin, many of which are expressed in a tissue-specific manner. Another gene, for the muscle form of troponin T, a protein that regulates the calcium needed for contraction, produces 64 known forms of the protein from a single pre-mRNA by alternative splicing.

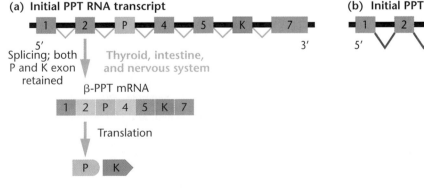

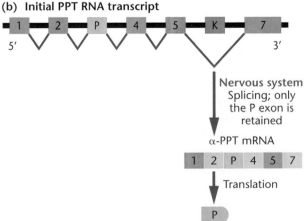

FIGURE 15–19 Alternative splicing of the initial RNA transcript of the preprotachykinin gene (PPT). Exons are either numbered or designated by the letters P or K. (a) the inclusion of P and K exons leads to β-PPT mRNA, which, upon translation, yields both the P and K tachykinin neuropeptides. (b) When the K exon is excluded, α-PPT mRNA is produced, and only the P neuropeptide is synthesized.

Chapter Summary

1. A system of genetic regulation must exist if the complete genome is not to be continuously active in transcription throughout the life of every cell of all species. Highly refined mechanisms have evolved that regulate transcription, maintaining genetic efficiency.

2. Both inducible and repressible operons, illustrated by the *lac* and *trp* gene complexes, respectively, have been documented and studied in bacteria. They involve genes of a regulatory nature in addition to structural genes that code for the enzymes of the system. Both operons are under the negative control of a repressor molecule.

3. The catabolite-activating protein (CAP) facilitates the binding of RNA polymerase to the promoter in the *lac* and other operons. Catabolite repression, a mechanism that represses the operon when glucose is present, has evolved presumably because glucose can be used more efficiently than lactose.

4. Transcription in eukaryotes is controlled by regulatory DNA sequences known as promoters and enhancers. Regulatory sequences, including the TATA box and the CCAAT box, are elements of promoters found near the transcriptional starting point. Enhancer elements, which appear to control the degree of transcription, can be located before, after, or within the gene expressed.

5. Transcription factors are proteins that bind to DNA recognition sequences within the promoters and enhancers and activate transcription through protein–protein interaction. Such factors display various types of motifs that are related to their potential binding to DNA.

6. Alteration of chromatin conformation is one way in which gene expression can be regulated. The degree of methylation of cytosine residues in DNA is believed to alter chromatin, with greater methylation ultimately reducing transcription.

7. Gene expression in eukaryotes can be regulated by steroid hormones originating outside the cell. Such signals are transduced by proteins that shuttle from the cytoplasm to the nucleus and regulate patterns of transcription.

8. Several types of posttranscriptional control of gene expression are possible in eukaryotes. One such mechanism is alternative splicing of a single class of pre-mRNAs to generate different mRNA species.

Key Terms

acetylation, 319
adenyl cyclase, 313
allosteric repressor, 309
alternative splicing, 321
attenuation, 314
attenuator, 314
β-galactosidase, 308
basic leucine zipper (bZIP), 318
CAAT box, 316
catabolite-activating protein (CAP), 312
catabolite repression, 312
chromatin remodeling, 319
cis-acting element, 307
constitutive mutant, 307
co-repressor, 314
DNA-binding domain, 317
DNA methylation, 320
enhancer, 316
equilibrium dialysis, 311
GC box, 316
gratuitous inducer, 308

helix–turn–helix (HTF) motif, 317
homeobox, 317
homeodomain, 317
hormone-receptor protein, 320
hormone-responsive element (HRE), 320
inducer, 307
inducible enzyme, 307
isopropylthiogalactoside (IPTG), 308
K neuropeptide, 321
lac operon, 308
lactose, 308
leader sequence, 314
merozygote, 310
negative control, 307
nucleosome, 319
operator region, 309
operon model, 309
P neuropeptide, 321
permease, 308
polycistronic mRNA, 308
positive control, 307

posttranscriptional mode of regulation, 321
preprotachykinin mRNAs (PPT mRNA), 321
promoter region, 312
repressible system, 307
repressor gene, 309
repressor molecule, 309
SNF protein, 319
SWI protein, 319
steroid hormone, 320
tachykinin, 321
TATA-binding protein (TBP), 319
TATA box, 315
TBP-associated factor (TAF), 319
TFIID transcription factor, 318
transacetylase, 308
trans-acting element, 307
trans-activating domain, 317
transcription factor, 317
tryptophan synthetase, 313
zinc finger, 317

Insights and Solutions

1. A theoretical operon (*theo*) in *E. coli* contains several structural genes encoding enzymes that are involved sequentially in the biosynthesis of an amino acid. Unlike the *lac* operon, where the repressor gene is separate from the operon, the gene encoding the regulator molecule is contained within the *theo* operon. When the end product (the amino acid) is present, it combines with the regulator molecule, and this complex binds to the operator, repressing the operon. In the absence of the amino acid, the regulatory molecule fails to bind to the operator, and transcription proceeds.

Characterize this operon; then consider the following mutations as well as the situation in which the wild-type gene is present along with the mutant gene in partially diploid cells (F'). In each case, will the operon be active or inactive in transcription, assuming that the mutation affects the regulation of the *theo* operon? Compare each response to the equivalent situation in the *lac* operon.

    (a) Mutation in the operator gene

    (b) Mutation in the promoter region

    (c) Mutation in the regulator gene

Solution: The operon is under negative control and is repressible. The regulatory molecule, when bound to the amino acid, binds to the operator region and inhibits gene expression.

(a) As in the *lac* operon, a mutation in the *theo* operator gene inhibits binding with the repressor complex, and transcription occurs constitutively. The presence of an F' plasmid bearing the wild-type allele would have no effect.

(b) A mutation in the *theo* promoter region would no doubt inhibit binding to RNA polymerase and therefore inhibit transcription. This would also happen in the lac operon. A wild-type allele present in an F' plasmid would have no effect.

(c) A mutation in the *theo* regulator gene, as in the *lac* system, may inhibit either its binding to the repressor or its binding to the operator gene. In both cases, transcription will be constitutive because the *theo* system is repressible. Both cases result in the failure of the regulator to bind to the operator, allowing transcription to proceed. In the *lac* system, failure to bind the co-repressor lactose would permanently repress the system. The addition of a wild-type allele would restore repressibility, provided that this gene was transcribed constitutively.

Problems and Discussion Questions

1. Contrast the need for the enzymes involved in the metabolism of lactose and tryptophan in bacteria in the presence and absence of lactose and tryptophan, respectively.
2. Contrast positive and negative control systems.
3. Contrast the role of the repressor in an inducible system and in a repressible system.

4. For the following *lac* genotypes, predict whether the structural genes of the operon are constitutive, permanently repressed, or inducible in the presence of lactose.

Genotype	Constitutive	Repressed	Inducible
$I^+O^+Z^+$			X
$I^-O^+Z^+$			
$I^+O^cZ^+$			
$I^-O^+Z^+/F'I^+$			
$I^+O^cZ^+/F'O^+$			
$I^SO^+Z^+$			
$I^SO^+Z^+/F'I^+$			

5. For the genotypes and condition (lactose present or absent) in the table below, predict whether functional enzymes are made, nonfunctional enzymes are made, or no enzymes are made.

Genotype	Condition	Functional Enzyme Made	Nonfunctional Enzyme Made	No Enzyme Made
$I^+O^+Z^+$	No lactose			X
$I^+O^cZ^+$	Lactose			
$I^-O^+Z^-$	No lactose			
$I^-O^+Z^-$	Lactose			
$I^-O^+Z^+/F'I^+$	No lactose			
$I^+O^cZ^+/F'O^+$	Lactose			
$I^+O^+Z^-/F'I^+O^+Z^+$	Lactose			
$I^-O^+Z^-/F'I^+O^+Z^+$	No lactose			
$I^SO^+Z^+/F'O^+$	No lactose			
$I^+O^cZ^+/F'O^+Z^+$	Lactose			

6. In a theoretical bacterial operon, genes *a*, *b*, *c*, and *d* represent the repressor gene, the promoter sequence, the operator gene, and the structural gene, but not necessarily in that order. This operon is concerned with the metabolism of a theoretical molecule (tm). From the data given below, first decide if the operon is inducible or repressible. Then assign *a*, *b*, *c*, and *d* to the four parts of the operon (AE = active enzyme, IE = inactive enzyme, NE = no enzyme).

Genotype	tm Present	tm Absent
$A^+B^+C^+D^+$	AE	NE
$A^-B^+C^+D^+$	AE	AE
$A^+B^-C^+D^+$	NE	NE
$A^+B^+C^-D^+$	IE	NE
$A^+B^+C^+D^-$	AE	AE
$A^-B^+C^+D^+/F'A^+B^+C^+D^+$	AE	AE
$A^+B^-C^+D^+/F'A^+B^+C^+D^+$	AE	NE
$A^+B^+C^-D^+/F'A^+B^+C^+D^+$	AE + IE	NE
$A^+B^+C^+D^-/F'A^+B^+C^+D^+$	AE	NE

7. Predict the level of genetic activity of the *lac* operon as well as the status of the *lac* repressor and the CAP protein under the cellular conditions listed in the accompanying table.

	Lactose	Glucose
(a)	−	−
(b)	+	−
(c)	−	+
(d)	+	+

8. Predict the effect on the inducibility of the *lac* operon of a mutation that disrupts the function of
   (a) the *crp* gene, which encodes the CAP protein
   (b) the CAP-binding site within the promoter

9. Why is gene regulation assumed to be more complex in a multicellular eukaryote than in a prokaryote? Why is the study of this phenomenon in eukaryotes more difficult?

10. List and define the potential levels of gene regulation in eukaryotes.

11. Distinguish between the regulatory elements referred to as promoters and enhancers.

12. Describe the role of transcription factors in the regulation of gene expression.

13. How are genes regulated by steroid hormones?

14. Contrast the modification of chromatin structure and posttranscriptional modes of gene regulation in eukaryotes.

15. Contrast and compare gene regulation in prokaryotes and eukaryotes.

16. A marine bacterium is isolated and shown to contain an inducible operon whose genetic products metabolize oil when it is encountered in the environment. Investigation demonstrates that the operon is under positive control and that there is a *reg* gene whose product interacts with an operator region (*o*) to regulate the structural genes designated *sg*.

    In the attempt to understand how the operon functions, a constitutive mutant strain and several partial diploid strains are isolated and tested with the results shown below. Draw all possible conclusions about the mutation as well as the nature of regulation of the operon. Is the constitutive mutation in the *trans*-acting *reg* element or in the *cis*-acting *o* (operator) element?

Host Chromosome	F' Factor	Phenotype
wild type	none	inducible
wild type	*reg* gene from mutant strain	inducible
wild type	operon from mutant strain	constitutive
mutant strain	*reg* gene from wild type	constitutive

Selected Readings

Aso, T., Shilatifard, A., Conaway, J. W., and Conaway, R. C. 1996. Transcription syndromes and the role of RNA polymerase II general transcription factors in human disease. *J. Clin. Investig.* 97:1561–69.

Beato, M. 1989. Gene regulation by steroid hormones. *Cell* 56:335–44.

Beckwith, J. R., and Zipser, D., eds. 1970. *The lactose operon.* Cold Spring Harbor, NY: Cold Spring Harbor Laboratory Press.

Berget, S. M. 1995. Exon recognition in vertebrate splicing. *J. Biol. Chem.* 270:2411–14.

Bertrand, K., et al. 1975. New features of the regulation of the tryptophan operon. *Science* 189:22–26.

Black, D. L. 2000. Protein diversity from alternative splicing: A challenge for bioinformatics and post-genomic biology. *Cell* 103:367–70.

Blumenthal, T. 1995. *Trans*-splicing and polycistronic transcription in *Caenorhabditis elegans*. *Trends Genet.* 11:132–36.

Buratowski, S. 1995. Mechanisms of gene activation. *Science* 270:1773–74.

Busch, S. J., and Sassone-Corsi, P. 1990. Dimers, leucine zippers and DNA-binding domains. *Trends Genet.* 6:36–40.

Cremer, T., and Cremer, C. 2001. Chromosome territories, nuclear architecture and gene regulation in mammalian cells. *Nature Reviews Genetics* 2:292–301.

Drapkin, R., Merino, A., and Reinberg, D. 1993. Regulation of RNA polymerase II transcription. *Curr. Opin. Cell Biol.* 5:469–76.

Dynan, W. S. 1988. Modularity in promoters and enhancers. *Cell* 58:1–4.

Edelson, E. 1990. Transcription factors: Governors for the genetic engine. *Mosaic* 21:2–9.

Gilbert, W., and Müller-Hill, B. 1966. Isolation of the *lac* repressor. *Proc. Natl. Acad. Sci. USA* 56:1891–98.

————. 1967. The *lac* operator in DNA. *Proc. Natl. Acad. Sci. USA* 58:2415–21.

Gilbert, W., and Ptashne, M. 1970. Genetic repressors, *Sci. Am.* (June) 222:36–44.

Hayes, J. J., and Wolffe, A. P. 1992. The interaction of transcription factors with nucleosomal DNA. *BioEssays* 14:597–603.

Jacob, F., and Monod, J. 1961. Genetic regulatory mechanisms in the synthesis of proteins. *J. Mol. Biol.* 3:318–56.

Jacobsen, A., and Peltz, S. 1996. Interrelationships of the pathway of mRNA decay and translation in eukaryotic cells. *Ann. Rev. Biochem.* 65:693–739.

Jacobson, R. H., and Tjian, R. 1996. Transcription factor IIA: A structure with multiple functions. *Science* 272:830–36.

Kass, S. U., Pruss, D., and Wolffe, A. P. 1997. How does DNA methylation repress transcription? *Trends in Genet.* 13:444–49.

Lewis, M., et al. 1996. Crystal structure of the lactose operon repressor and its complexes with DNA and inducer. *Science* 271:1247–54.

Littlewood, T. D., and Evans, G. I. 1994. Transcription factors 2: Helix-loop-helix. *Protein Profile* 1:639–709.

Lopez, A. J. 1995. Developmental role of transcription factor isoforms generated by alternate splicing. *Dev. Bio.* 172:396–411.

Lohr, D. 1997. Nucleosome transactions on the promoters of the yeast GAL and *PHO* genes. *J. Biol. Chem.* 272:26795–98.

Lugar, K. et.al. 1997. Crystal structure of the nucleosome core particle at 2.8 Å resolution. *Nature* 389: 251–56.

Maniatis, T., Goodbourn, S., and Fischer, J. A. 1987. Regulation of inducible and tissue-specific expression. *Science* 236:1237–45.

—————, and Ptashne, M. 1976. A DNA operator-repressor system. *Sci. Am.* (Jan.) 234:64–76.

McCarthy, J. E., and Brimacombe, R. 1994. Prokaryotic translation: The interactive pathway leading to initiation. *Trends Genet.* 10:402–7.

Mechanisms of eukaryotic transcription. *Genes Dev.* 10:367–381.

Meehan, R., et. al. 1992. Transcriptional repression by methylation of CpG. *J. Cell Sci.* Suppl. 16:9–14.

Miller, J. H., and Reznikoff, W. S. 1978. *The operon.* Cold Spring Harbor, NY: Cold Spring Harbor Laboratory.

Mitchell, P. J., and Tjian, R. 1989. Transcriptional regulation in mammalian cells by sequence-specific DNA binding proteins. *Science* 245:371–78.

Müller-Hill, B. 1996. *The lac operon: A short history of a genetic paradigm.* Hawthorne, NY: Walter de Gruyter.

Pieler, T., and Bellefroid, E. 1994. Perspectives on zinc finger protein function and evolution—an update. *Mol. Biol. Rep.* 20:1–8.

Ptashne, M., Johnson, A. D., and Pabo, C. O. 1982. A genetic switch in a bacterial virus. *Sci. Am.* (Nov.) 247:128–40.

Stringer, K. F., Ingles, C. J., and Greenblatt, J. 1990. Direct and selective binding of an acidic transcriptional activation domain to the TATA-box factor TFIID. *Nature* 345:783–86.

Struhl, K. 1993. Yeast transcription factors. *Curr. Opin. Cell Biol.* 5:513–20.

Tate, P., and Bird, A. 1993. Effects of DNA methylation on DNA-binding proteins and gene expression. *Curr. Opin. Genet. Dev.* 3:226–31.

Tjian, R. 1995. Molecular machines that control genes. *Sci. Am.* 272:54–61.

Workman, J. L., and Kingston, R. E. 1998. Alteration of nucleosome structure as a mechanism of transcriptional control. *Ann. Rev. Biochem.* 67:545–79.

Yankofsky, C. 1981. Attenuation in the control of expression of bacterial operons. *Nature* 289:751–58.

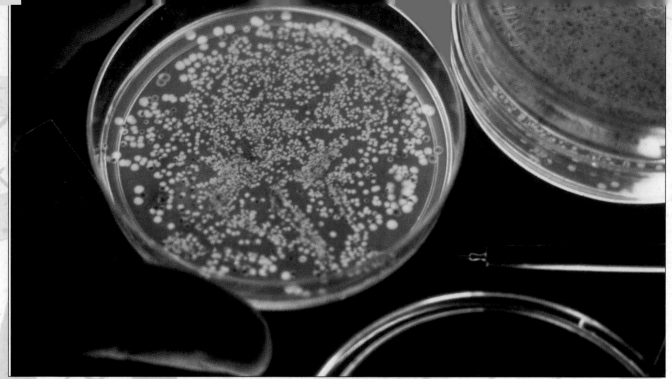

A Petri plate showing the growth of host cells after uptake of recombinant plasmids. (*Michael Gabridge/ Visuals Unlimited*)

16

Recombinant DNA Technology

CHAPTER CONCEPTS

Recombinant DNA technology depends in part on the ability to cleave and rejoin DNA molecules at specific nucleotide sequences. Once created, recombinant DNA molecules can be transferred to host cells and amplified, isolated and characterized. This technology has revolutionzed gene mapping, disease diagnosis, the commercial production of human gene products, and the transfer of genes between species of plants and animals.

The techniques used to create, replicate, and analyze recombinant DNA molecules were developed in the mid to late 1970s as a way for researchers to isolate and study specific DNA sequences. The first commercial application of recombinant DNA technology was approved in 1982, when the Food and Drug Administration permitted the marketing of human insulin produced by bacterial cells. This was a milestone in the development of the biotechnology industry. Today, recombinant DNA technology produces medicines, vaccines, industrial chemicals, genetically modified organisms and food products. The biotechnology industry is now a multi-billion dollar component of the national and world economy.

Recombinant DNA technology makes it possible to identify and isolate a single gene from the tens of thousands present in the human genome and to produce large quantities of this gene in the form of cloned DNA molecules. In addition, the gene can be transferred into cells that will synthesize its encoded gene product, which can then be recovered and purified for use in research, medicine, or industry.

Clones are identical organisms, cells, or molecules descended from a single ancestor. Cloning a gene involves producing many identical copies that can be used for numerous purposes, including research into the structure and organization of a gene and the commercial production of proteins such as insulin. In this chapter we review the basic methods of recombinant DNA technology used to isolate, replicate, and analyze genes. In Chapter 19, we discuss some applications of this technology to research, medicine, the legal system, agriculture, and industry.

16.1 Recombinant DNA Technology: An Overview

The term **recombinant DNA** refers to a new combination of DNA molecules that are not found together naturally. Although processes such as crossing over technically produce recombinant DNA, the term is generally reserved for DNA produced by joining molecules derived from different biological sources.

Recombinant DNA technology uses methods derived from nucleic acid biochemistry, coupled with genetic techniques originally developed for the study of bacteria and viruses. As described here, recombinant DNA technology is a powerful tool for isolating potentially unlimited quantities of a gene. Although several methods are available, the basic procedure involves these steps:

1. DNA is purified from cells or tissues.

2. Enzymes called **restriction endonucleases** or **restriction enzymes** generate specific DNA fragments. These enzymes recognize and cut DNA molecules at specific nucleotide sequences.

3. The fragments produced by treatment with restriction enzymes are joined to other DNA molecules that serve as **vectors**, or carrier molecules. A vector joined to a DNA fragment is a **recombinant DNA molecule**.

4. The recombinant DNA molecule, consisting of a vector carrying an inserted DNA fragment, is transferred to a host cell. Within the host cell, the recombinant molecule replicates, producing dozens of identical copies known as clones.

5. As host cells replicate, the recombinant DNA molecules within them are passed on to all their progeny, creating a population of host cells, each of which carries copies of the cloned DNA sequence.

6. The cloned DNA is recovered from host cells, purified, and analyzed.

7. The cloned DNA is transcribed, its mRNA translated, and the gene product isolated and used for research, or sold commercially.

16.2 Constructing Recombinant DNA Molecules

Recombinant DNA and gene-cloning technology provides scientists with methods for isolating large quantities of specific genes or other DNA sequences; this has facilitated studies of gene organization, structure, and expression.

Restriction Enzymes

The cornerstone of recombinant DNA technology is a class of enzymes called restriction endonucleases. These enzymes, isolated from bacteria, are so named because they restrict or prevent viral infection by degrading the invading viral DNA. Restriction enzymes recognize a specific nucleotide sequence and cut both strands of the DNA within that sequence. The 1978 Nobel Prize was awarded to Werner Arber, Hamilton Smith, and Daniel Nathans for their work on restriction enzymes. To date, over 200 restriction enzymes have been identified. Their usefulness in cloning derives from their ability to reproducibly cut DNA into fragments.

One of the first restriction enzymes identified was isolated from *E. coli* and is designated *Eco*RI (pronounced echo-r-one). Its recognition sequence and point of cleavage are shown in Figure 16–1. The DNA fragments produced by *Eco*RI

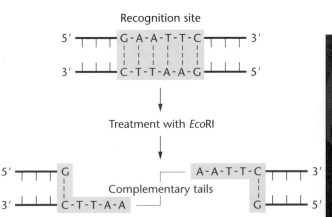

FIGURE 16–1 The restriction enzyme *Eco*RI recognizes and binds to the palindromic nucleotide sequence GAATTC. Cleavage of the DNA at this site produces complementary single-stranded tails. These single-stranded tails anneal with single-stranded tails from other DNA fragments to form recombinant DNA molecules.

digestion have overhanging single-stranded tails (called "sticky ends") that reanneal with complementary single-stranded tails on other DNA fragments. If they are mixed under the proper conditions, DNA fragments from two sources form recombinant molecules by hydrogen bonding of their sticky ends. The enzyme **DNA ligase** covalently links these fragments to form recombinant DNA molecules (Figure 16–2). Some common restriction enzymes and their recognition sequences are shown in Figure 16–3.

Vectors

Fragments of DNA produced by restriction enzyme digestion cannot directly enter bacterial cells for cloning; when a DNA fragment is joined to a vector, however, it can gain entry to a host cell, where it can be replicated or cloned into many copies. Vectors are, in essence, carrier DNA molecules. To serve as a vector, a DNA molecule must have several properties:

1. It must be able to independently replicate itself and the DNA segment it carries.

2. It should contain a number of restriction enzyme cleavage sites that are present only once in the vector. One site is cleaved with a restriction enzyme and used to insert a DNA segment cut with the same enzyme.

3. It should carry a selectable marker (usually in the form of antibiotic resistance genes or genes for enzymes missing in the host cell) to distinguish host cells that carry vectors from host cells that do not contain vectors.

4. It should be easy to recover from the host cell.

A number of vectors are currently in use, including those derived from **plasmids** and **bacteriophages**.

The first vectors were genetically modified plasmids, which are versatile and still widely used for cloning, but are unsuitable for cloning large fragments of DNA. Plasmids are naturally occurring, extrachromosomal, double-stranded DNA molecules that replicate autonomously within bacterial cells (Figure 16–4, page 330). The genetics of plasmids and their host bacterial cells has been discussed in Chapter 9. In this section we emphasize the role of plasmids as vectors. Although only a single plasmid may enter a host cell, many plasmids can increase their copy number in the host cell so that several hundred copies are present. When used as vectors, such plasmids allow more copies of cloned DNA to be produced. To use them as vectors, they are modified so that they contain a limited number of convenient restriction sites and marker genes that detect the presence of plasmids in host cells.

Many genetically engineered plasmid vectors are now available, and certain features make it easier to identify host cells carrying a plasmid with an inserted DNA fragment. One such plasmid is pUC18 (Figure 16–5, page 330). This plasmid has several useful features as a vector.

1. It is small (2686 bp), which allows it to carry relatively large DNA inserts.

2. In a host cell, it replicates 500 copies per cell, producing many copies of inserted DNA fragments.

3. A large number of restriction enzyme sites have been engineered into pUC18, and these are conveniently clustered in one region, called a **polylinker site**.

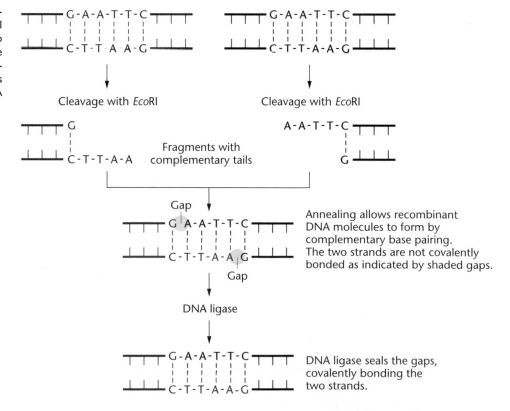

FIGURE 16–2 DNA from different sources is cleaved with *Eco*RI and mixed to allow annealing to form recombinant molecules. The enzyme DNA ligase then chemically bonds these annealed fragments into an intact recombinant DNA molecule.

Cleavage with *Eco*RI

Cleavage with *Eco*RI

Fragments with complementary tails

Gap

Gap

Annealing allows recombinant DNA molecules to form by complementary base pairing. The two strands are not covalently bonded as indicated by shaded gaps.

DNA ligase

DNA ligase seals the gaps, covalently bonding the two strands.

Enzyme	Recognition, cleavage sequence	Cleavage pattern	Source

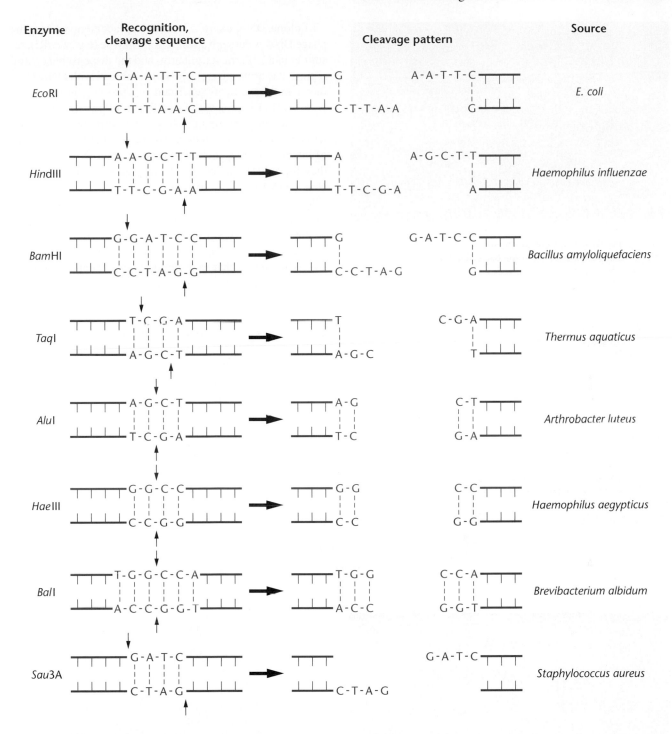

FIGURE 16–3 Some common restriction enzymes, with their recognition sequences, cutting sites, cleavage patterns, and sources.

4. It has a selection system that allows recombinant plasmids to be identified. It carries a fragment of the bacterial *lacZ* gene, and the polylinker is inserted into this fragment. Through the *lacZ* gene's action, bacterial host cells carrying pUC18 produce blue colonies when grown on medium containing a compound called Xgal. When a DNA fragment is inserted into the polylinker site, the *lacZ* gene is inactivated, and a bacterial cell carrying pUC18 with an inserted DNA fragment forms white colonies, making them easy to identify (Figure 16–6, page 330).

Plasmid vectors generally carry up to 10 kb of inserted DNA, but for many experiments, larger pieces of DNA are necessary. For these purposes, genetically modified strains of the bacterial virus lambda (λ), also called **lambda phage**, are used as vectors (Figure 16–7, page 330). Lambda genes have all been identified and mapped, and the nucleotide sequence of the entire genome is known. The central third of the lambda chromosome is dispensable, and this region can be replaced with foreign DNA without affecting the phage's ability to infect cells and form plaques (Figure 16–8, page 331).

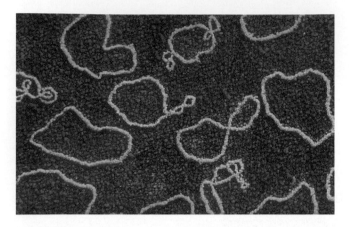

FIGURE 16–4 A color-enhanced electron micrograph of circular plasmid molecules isolated from the bacterium *E. coli.* Genetically engineered plasmids are used as vectors for cloning DNA. *(K.G. Murti/Visuals Unlimited)*

To clone DNA using a lambda-replacement vector, the phage DNA is cut with a restriction enzyme (e.g., *Eco*RI), resulting in a left arm, a right arm, and the dispensable central region. The arms are isolated and combined with DNA fragments that have been generated by cutting genomic DNA with *Eco*RI. Ligation with DNA ligase follows. Lambda vectors containing inserted DNA are then introduced into bacterial host cells, where they reproduce to form many particles of infective phage, each of which carries a DNA insert. As they reproduce, the phages form plaques from which the cloned DNA can be recovered.

Phage vectors can carry inserts of about 20 kb, more than twice as long as the DNA inserts in plasmid vectors. This is an important advantage when cloning entire genomes. In addition, some phage vectors only accept inserts of a minimum size. This means that in a cloning experiment where large inserts must be isolated, vectors do not end up carrying rel-

FIGURE 16–5 The plasmid pUC18 offers several advantages as a vector for cloning. Because of its small size, it accepts relatively large DNA fragments for cloning; it replicates to a high copy number, and has a large number of restriction sites in the polylinker, located within a *lacZ* gene. Bacteria carrying pUC18 produce blue colonies when grown on media containing Xgal. DNA inserted into the polylinker site disrupts the *lacZ* gene; this results in white colonies and allows direct identification of colonies carrying cloned DNA inserts.

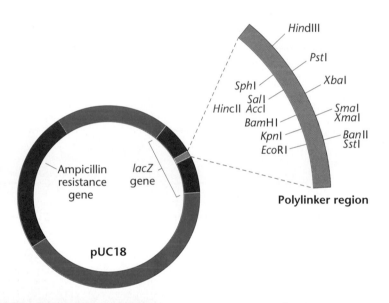

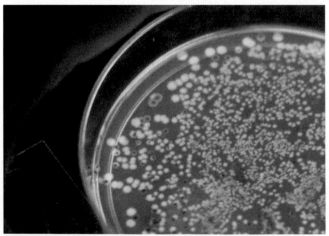

FIGURE 16–6 A Petri plate showing the growth of bacterial cells after uptake of recombinant plasmids. The medium on the plate contains a compound called Xgal. DNA inserts into the pUC18 vector disrupt the gene responsible for the formation of blue colonies. Cells in the blue colonies do not carry any cloned DNA inserts, whereas the white colonies contain vectors carrying DNA inserts. *(Michael Gabridge/Visuals Unlimited)*

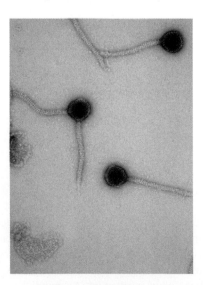

FIGURE 16–7 A colorized electron micrograph of a cluster of bacteriophage lambda, widely used as a vector in recombinant DNA work. *(M. Wurtz/Biozentrum, University of Basel/Science Photo Library/Photo Researchers, Inc.)*

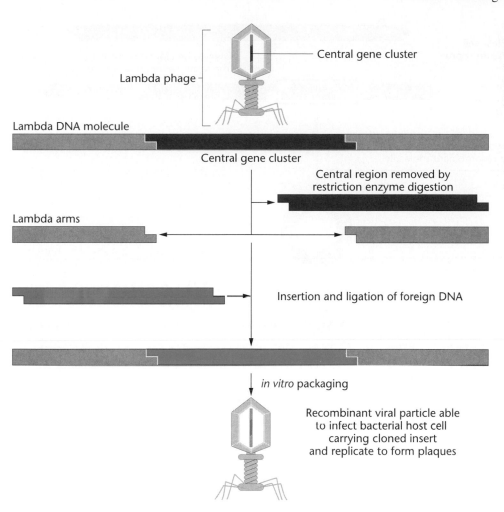

FIGURE 16–8 Phage lambda as a vector. DNA is extracted from a preparation of lambda phage, and the central gene cluster is removed by treatment with a restriction enzyme. The DNA to be cloned is cut with the same enzyme and ligated into the arms of the lambda chromosome. The recombinant chromosome is then packaged into phage proteins to form a recombinant virus. This virus infects bacterial cells and replicates its chromosome, including the DNA insert.

Central gene cluster

Lambda phage

Lambda DNA molecule

Central gene cluster

Central region removed by restriction enzyme digestion

Lambda arms

Insertion and ligation of foreign DNA

in vitro packaging

Recombinant viral particle able to infect bacterial host cell carrying cloned insert and replicate to form plaques

atively useless inserts of a few dozen or a few hundred nucleotides in length.

16.3 Cloning in *E. coli* Host Cells

As discussed earlier, scientists use biotechnology to construct *and* replicate recombinant DNA molecules, to produce many copies, or clones. This is accomplished by transferring recombinant molecules into host cells where replication takes place. The first cloning method, cell-based cloning, has been widely used to copy DNA fragments and used to study gene organization and function.

A variety of prokaryotic and eukaryotic cells serve as hosts for recombinant vector replication. One of the most commonly used hosts is a laboratory strain of the bacterium *E. coli* known as K12. *E. coli* strains such as K12 are genetically well characterized, and they serve as hosts for a wide range of vectors. Several steps are required to create recombinant DNA molecules (Figure 16–9, page 332) in an *E. coli* host cell.

1. The DNA to be cloned is isolated and treated with a restriction enzyme to create fragments that end in a specific sequence.

2. The fragments are ligated to plasmid molecules that have been cut with the same restriction enzyme, creating a recombinant vector.

3. The recombinant vector is transferred into an *E. coli* host cell, where it replicates to form dozens of copies.

4. The bacteria are grown on a nutrient plate where they form colonies and are screened to identify those that have taken up the recombinant plasmids.

Because the cells in each colony are derived from a single ancestral cell, all the cells in the colony and the plasmids they contain are genetically identical clones.

Similarly, phages containing foreign DNA are used to infect *E. coli* host cells, and, when plated on solid medium, each resulting plaque represents a cloned descendant of a single ancestral phage.

16.4 Cloning in Eukaryotic Host Cells

The yeast *Saccharomyces cerevisiae* is extensively used as a host cell for the growth and expression of cloned eukaryotic genes for five reasons. (1) Although yeast is a eukaryotic

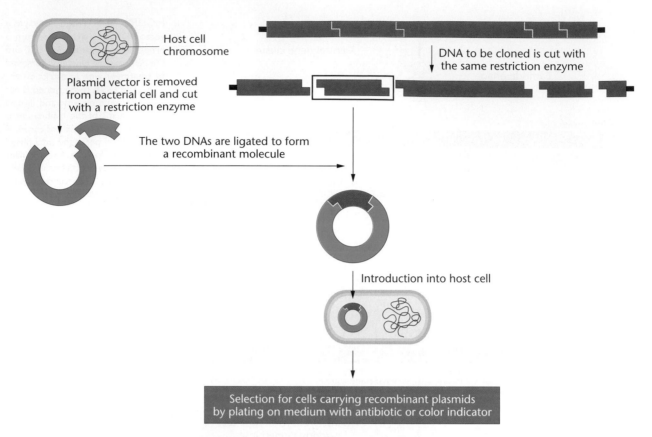

FIGURE 16–9 Cloning with a plasmid vector. Plasmid vectors are isolated and cut with a restriction enzyme. The DNA to be cloned is cut with the same restriction enzyme, producing a collection of fragments. These fragments are spliced into the vector and transferred to a bacterial host for replication. Bacterial cells carrying plasmids with DNA inserts are identified by growth on selective medium and isolated. The cloned DNA is then recovered from the bacterial host for further analysis.

organism, it can be grown and manipulated in much the same way as bacterial cells. (2) The genetics of yeast has been intensively studied, providing a large catalog of mutations and a highly developed genetic map. (3) The entire yeast genome has been sequenced, and most genes in the organism have been identified. (4) To study the function of some eukaryotic proteins, it is necessary to use a host cell that can posttranslationally modify the product of the cloned gene to convert it into a functional form. Bacterial host cells cannot carry out these modifications. (5) Because yeast has been used for centuries in the baking and brewing industries, it is considered a safe organism for producing proteins for vaccines and therapeutic agents (Table 16.1).

Several types of yeast cloning vectors have been developed, one of which is the **yeast artificial chromosome (YAC)** (Figure 16–10). Like natural chromosomes, a YAC contains telomeres at each end, an origin of replication (which initiates DNA synthesis), and a centromere. These components are joined to selectable markers (*TRP1* and *URA3*) and a cluster of restriction sites for DNA inserts. Yeast chromosomes range from 230 kb to over 1900 kb, making it possible to clone DNA inserts ranging from

TABLE 16.1	Recombinant Proteins Synthesized in Yeast Host Cells
Hepatitis B virus surface protein	
Malaria parasite protein	
Insulin	
Epidermal growth factor	
Platelet-derived growth factor	
α_1-antitrypsin	
Clotting factor XIIIA	

100 kb to 1000 kb in YACs. The ability to clone large pieces of DNA into these vectors makes them an important tool in genome projects, including the Human Genome Project (discussed in Chapters 18 and 19).

Although cloning in yeast vectors and host cells is currently the most advanced eukaryotic system used, other systems, including human artificial chromosome vectors with mammalian tissue culture cell hosts, are being developed.

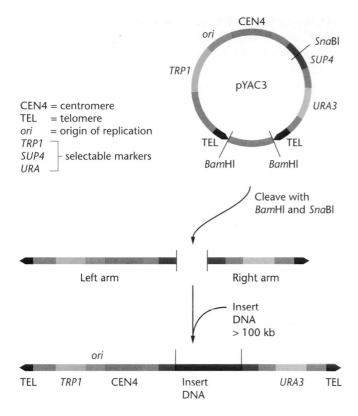

FIGURE 16–10 Cloning into a yeast artificial chromosome. The synthetic chromosome contains telomere sequences (TEL), a centromere (CEN4) derived from yeast chromosome 4, and an origin of replication (*ori*). These elements give the cloning vector the properties of a chromosome. *TRP1*, *SUP4*, and *URA3* are yeast genes that are selectable markers for the left and right arms of the chromosome, respectively. Within the *SUP4* gene is a restriction site for the enzyme *Sna*BI. Two *Bam*HI sites flank a spacer segment. Cleavage with *Sna*BI and *Bam*HI breaks the artificial chromosome into two arms. The DNA to be cloned is treated with the same enzyme, producing a collection of fragments. The arms and fragments are ligated together, and the artificial chromosome inserted into yeast host cells. Because yeast chromosomes are large, the artificial chromosome accepts inserts in the million base pair range (Mb = megabases).

16.5 Cloning Without Host Cells: The PCR Reaction

Developed in the early 1970s, recombinant DNA techniques revolutionized the way geneticists and molecular biologists conduct research and gave birth to the booming biotechnology industry. However, they are often labor-intensive and time-consuming. In 1986, another technique, called the **polymerase chain reaction (PCR)**, was developed. This advance again revolutionized recombinant DNA methodology, and further accelerated the pace of biological research. The significance of this method is underscored by the fact that the 1993 Nobel Prize in Chemistry was awarded to Kary Mullis for developing the PCR technique.

PCR is a rapid method of DNA cloning that has extended the power of recombinant DNA research and does away with the use of host cells as vehicles for cloning DNA fragments. Although host cell cloning is still widely used, PCR is now the method of choice in many applications, including molecular biology, human genetics, evolution, development, con-

servation, and forensics (the gathering of evidence in criminal cases). Some of these applications are discussed in Chapter 19.

PCR generates many copies of a specific DNA sequence through a series of reactions in a test tube (*in vitro*), and amplifies target DNA sequences present in infinitesimally small quantities in a population of other DNA molecules. To do this, some information about the nucleotide sequence of the target DNA is required. This information is used to synthesize two oligonucleotide primers that are added to a denatured (single-stranded) DNA sample. In the reaction, the primers hybridize to nucleotides that flank the sequence to be amplified. A heat-stable form of DNA polymerase (called *Taq* polymerase) synthesizes a second strand of the target DNA (Figure 16–11, page 334).

The PCR reaction involves three basic steps, and the amount of amplified DNA produced is limited theoretically only by the number of times these steps are repeated.

1. The DNA to be amplified is denatured into single strands. This DNA does not have to be purified and can come from any number of sources, including genomic DNA, forensic samples such as dried blood or semen, dried samples stored as part of medical records, single hairs, mummified remains, and fossils. Heating (at 90–95°C) denatures the double-stranded DNA, which dissociates into single strands (usually about 5 minutes).

2. The temperature of the reaction is lowered to somewhere between 50°C and 70°C, and at this temperature, the primers bind to the denatured DNA. These primers are synthetic oligonucleotides (15–30 nucleotides long) that bind specifically to sequences flanking the target segment. The primers serve as starting points for synthesizing new DNA strands complementary to the target DNA.

3. A heat-stable form of DNA polymerase (*Taq* polymerase, an enzyme from a bacterium, *Thermus aquaticus*, which lives in hot springs) is added to the reaction mixture. DNA synthesis is carried out at temperatures between 70°C and 75°C. The *Taq* polymerase extends the primers by adding nucleotides in the 5′–3′ direction, making a double-stranded copy of the target DNA.

Each set of three steps—**denaturation** of the double-stranded product, **annealing** of primers, and **extension** by polymerase—is referred to as a **cycle**. PCR is a chain reaction because the number of new DNA strands is doubled in each cycle, and the new strands and the old strands serve as templates in the next cycle. Each cycle, which takes 4–5 minutes, can be repeated, and 25–30 cycles (which can be completed in less than 3 hours) result in an over 1,000,000-fold increase in the amount of DNA. Machines called *thermocyclers* have automated the process. They can be programmed to carry out a predetermined number of cycles, yielding large amounts of the target DNA, which can be used in cloning, sequencing, clinical diagnosis, and genetic screening.

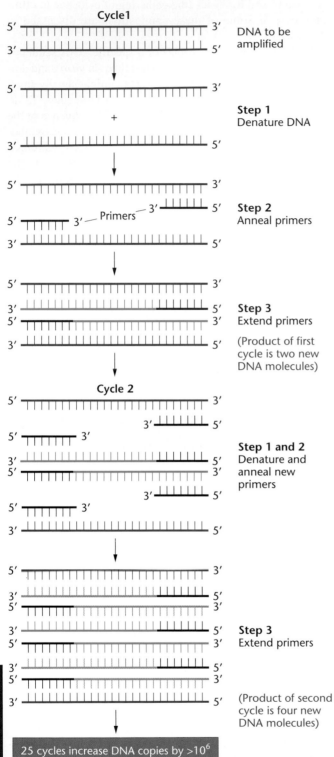

Cycle 1

Step 1
Denature DNA

Step 2
Anneal primers

Step 3
Extend primers

(Product of first
cycle is two new
DNA molecules)

Cycle 2

Step 1 and 2
Denature and
anneal new
primers

Step 3
Extend primers

(Product of second
cycle is four new
DNA molecules)

25 cycles increase DNA copies by >10⁶

FIGURE 16–11 PCR amplification. In the polymerase chain reaction, the target DNA is denatured into single strands; each strand is then annealed to a short, complementary primer. The primers are synthetic oligonucleotides that are complementary to sequences flanking the region to be amplified. DNA polymerase and nucleotides extend the primers in the 3′ direction, using the single-stranded DNA as a template. The result is a double-stranded DNA molecule with the primers incorporated into the newly synthesized strand. In a second PCR cycle, the products of the first cycle are denatured into single strands, primers are added, and DNA polymerase synthesizes new strands. Repeated cycles can amplify the original DNA sequence by more than a millionfold.

PCR-based DNA cloning has several advantages over host cell–based cloning. The PCR reaction is fast and can be carried out in a few hours, rather than the weeks required for host cell–based cloning. In addition, the design of PCR primers uses computer software, and the commercial synthesis of the oligonucleotides is also fast and economical. If desired, the products of PCR can be cloned into plasmid vectors for further use.

PCR is very sensitive and amplifies target DNA from vanishingly small DNA samples, including the DNA in a single cell. This feature of PCR is invaluable in several areas, including genetic testing, forensics, and molecular paleontology. DNA samples that have been partly degraded, mixed with other materials, or embedded in a medium (such as amber) can be used where conventional cloning would be difficult or impossible.

Although PCR is a valuable technique, it does have limitations. Some information about the nucleotide sequence of the target DNA must be known, and even minor contamination of the sample with DNA from other sources can cause difficulties. For example, cells shed from the skin of a laboratory worker while performing PCR can contaminate samples gathered from a crime scene or taken from fossils. Such contamination makes it difficult to obtain accurate results. PCR reactions must always be run with carefully designed and appropriate controls.

PCR is now one of the most widely used techniques in genetics and molecular biology. PCR and its variations have many other applications. It quickly identifies restriction-site variants and variations in tandemly repeated DNA sequences that can serve as genetic markers for gene mapping studies. Using gene-specific primers, PCR screens for mutations in genetic disorders, allow the location and nature of the mutation to be determined quickly. As we discuss in Chapter 19, primers can be designed to distinguish target sequences that differ by only a single nucleotide, making it possible to develop allele-specific probes for genetic testing. Random primers allow the indiscriminate amplification of DNA, which is particularly advantageous when studying samples from single cells, from fossils, or from crime scenes where a single hair or even a saliva-licked postage stamp is the source of the DNA. Researchers also explore uncharacterized DNA regions adjacent to known regions and even sequence DNA with PCR. PCR has been used to enforce the worldwide ban on the sale of certain whale products, and to settle arguments about the pedigree background of pure-bred dogs. In short, PCR is one of the most versatile techniques in modern genetics.

16.6 Libraries Are Collections of Cloned Sequences

Because each cloned DNA segment is relatively small, and may represent only a single gene or a portion of a gene, many separate clones are needed to cover even a small fraction of an organism's genome. A set of DNA clones derived from a single individual is called a cloned *library*.

These libraries can represent an entire genome, a single chromosome, or a set of genes that are expressed in a single cell type.

Genomic Libraries

Ideally, a **genomic library** contains at least one copy of all the sequences in an individual's genome. Genomic libraries are constructed using host cell cloning methods, since PCR-cloned DNA fragments are relatively small. In the process, DNA is extracted from cells or tissues, then cut with restriction enzymes and the fragments ligated into vectors. Since some vectors (such as plasmids) carry only a few thousand base pairs (kilobases) of inserted DNA, selecting the vector so that it contains a genome in the smallest number of clones is an important consideration in preparing a genomic library.

The number of clones required to carry all sequences in a genome depends on the average size of the cloned inserts and the size of the genome to be cloned. The number of clones in a library can be calculated as

$$N = \frac{\ln(1 - P)}{\ln(1 - f)}$$

where N is the number of required clones, P is the probability of recovering a given sequence, and f is the fraction of the genome in each clone.

Suppose we wish to prepare a library of the human genome using a lambda phage vector. The human genome contains approximately 3.0×10^6 kb of DNA, and the average size of a cloned insert that can be carried by the vector is 17 kb. Therefore, about 8.1×10^5 phages are required to establish a library containing all genomic sequences. If we select a plasmid vector with a capacity of 5 kb, several million clones will be needed. Thus, in this example, a phage vector is the best choice for constructing a library of the human genome.

cDNA Libraries

A library of the genes active in a specific cell type at a specific time can be constructed by taking advantage of the fact that almost all eukaryotic mRNA molecules contain a poly-A tail at their 3′-ends. This mRNA can be isolated and used to synthesize **complementary DNA (cDNA)** molecules that are subsequently cloned to form a **cDNA library**. After the poly-A–containing mRNA is isolated, a poly-dT primer is added to pair with the poly-A residues (Figure 16–12). The poly-dT primer serves as the starting point for synthesizing a complementary DNA (cDNA) strand with the enzyme **reverse transcriptase**. The result is an RNA–DNA double-stranded duplex molecule. The RNA strand is removed, and the remaining single-stranded DNA becomes the template for DNA synthesis with the enzyme **DNA polymerase I**. The 3′-end of the single-stranded cDNA often loops back on itself to form a hairpin loop. This loop serves as a primer for synthesizing the second strand by DNA polymerase I. The product is a double-stranded DNA molecule with the strands joined at one end. This loop can be opened using the enzyme S_1 nuclease, producing a double-stranded DNA mol-

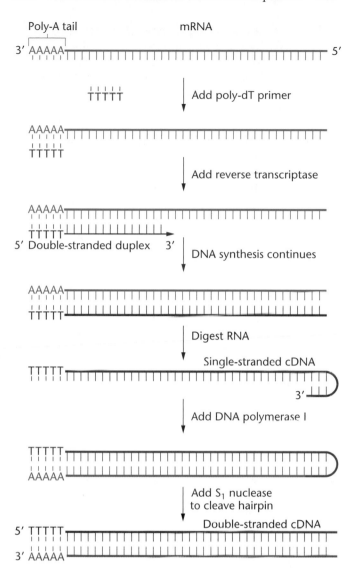

FIGURE 16–12 Producing cDNA from mRNA. Because many eukaryotic mRNAs have a polyadenylated tail (A) of variable length at their 3′-end, a short poly-dT oligonucleotide annealed to this tail serves as a primer for the enzyme reverse transcriptase. Reverse transcriptase uses the mRNA as a template to synthesize a complementary DNA strand (cDNA) and forms an mRNA/cDNA double-stranded duplex. The mRNA is digested with alkali treatment, or by the enzyme RNAseH. The 3′-end of the cDNA often folds back to form a hairpin loop. The loop serves as a primer for DNA polymerase, which is used to synthesize the second DNA strand. The S_1 nuclease opens the hairpin loop; the result is a double-stranded cDNA molecule that can be cloned into a suitable vector, or used as a probe for library screening.

ecule (called **complementary DNA** or **cDNA**), that can be cloned into a plasmid or phage vector.

PCR methods also generate cDNA from the 3′- or the 5′-end of mRNA molecules. This strategy, called RACE (rapid amplification of cDNA ends), depends on knowing a short nucleotide sequence contained in the coding region of the mRNA to be amplified. RACE identifies mRNAs that may be present in only 1–2 copies per cell, and the ends of cDNA molecules generated by RACE are used to search for cloned genes in genomic libraries.

16.7 Recovering Cloned Sequences from a Library

A genomic library can contain up to several hundred thousand clones. To select a gene of interest from this library, we need to identify and isolate only the clone or clones that contain that gene. We also must be able to determine whether a clone contains all or only part of that gene. Several methods allow us to sort through a library to recover the clones of interest; the choice often depends on the circumstances and available information about the gene being sought.

FIGURE 16–13 Screening a plasmid library to recover a cloned gene. The library, present in bacteria on Petri plates, is overlaid with a DNA binding filter, and colonies are transferred to the filter. Colonies on the filter are lysed, and the DNA denatured to single strands. The filter is placed in a hybridization bag along with buffer and a labeled single-stranded DNA probe. During incubation, the probe forms a double-stranded hybrid with complementary sequences on the filter. The filter is removed from the bag and washed to remove excess probe. Hybrids are detected by placing a piece of X-ray film over the filter and exposing it for a short time. The film is developed, and hybridization events are visualized as spots on the film. Colonies containing the insert that hybridized to the probe are identified from the orientation of the spots. Cells are picked from this colony for growth and further analysis.

Using Probes to Identify Specific Cloned Sequences

Many procedures employ probes to select specific clones from a library. A **probe** is any piece of DNA or RNA that has been labeled and is complementary to some part of a cloned sequence present in the library. The probe identifies complementary nucleic acid sequences present in one or more clones. Probes can be prepared as single- or double-stranded molecules, but are used in single-stranded form in hybridization reactions (hybridization reactions were described

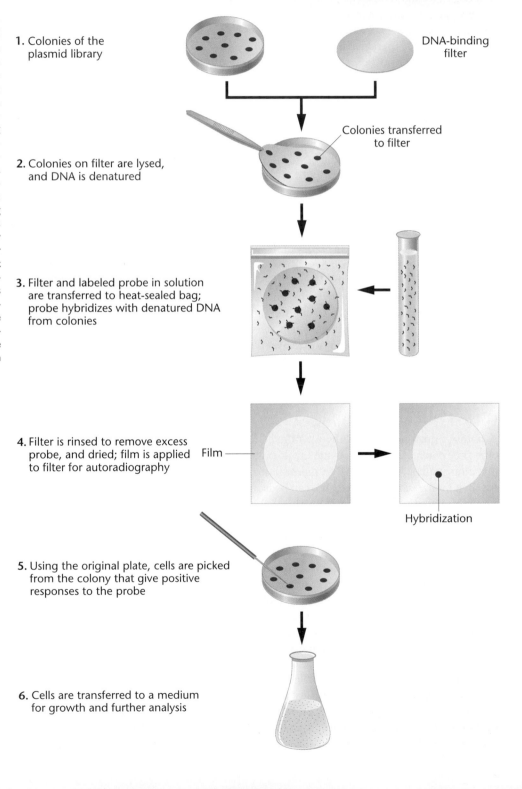

1. Colonies of the plasmid library

DNA-binding filter

Colonies transferred to filter

2. Colonies on filter are lysed, and DNA is denatured

3. Filter and labeled probe in solution are transferred to heat-sealed bag; probe hybridizes with denatured DNA from colonies

4. Filter is rinsed to remove excess probe, and dried; film is applied to filter for autoradiography

Film

Hybridization

5. Using the original plate, cells are picked from the colony that give positive responses to the probe

6. Cells are transferred to a medium for growth and further analysis

in Chapter 10). Often, probes are radioactive polynucleotides, but some probes depend on chemical or color reactions to indicate the location of a specific clone.

Probes are derived from a variety of sources; even related genes isolated from other species can be used if enough of the DNA sequence has been conserved. For example, extra-chromosomal copies of the ribosomal genes of the clawed frog *Xenopus laevis* can be isolated by centrifugation and cloned into plasmid vectors. Because ribosomal gene sequences have been highly conserved during evolution, clones carrying the human ribosomal genes can be recovered from a human genomic library using cloned *Xenopus* ribosomal DNA fragments as probes.

If the gene to be selected from a genomic library is expressed in certain cell types, a cDNA probe can be used. This technique is particularly helpful when purified or enriched mRNA for a gene product can be obtained. For example, β-globin mRNA is present in high concentrations in certain stages of red blood cell development. The mRNA purified from these cells can be copied by reverse transcriptase into a cDNA molecule, or amplified by PCR RACE. In fact, a cloned cDNA probe was originally used to recover the structural gene for human β-globin from a cloned genomic library.

Searching a Library for Specific Clones

To screen a plasmid library, clones from the library are grown on nutrient plates, forming hundreds or thousands of colonies (Figure 16–13). A replica of the colonies on the plate is made by gently pressing a nylon or nitrocellulose filter onto the plate's surface; this transfers the pattern of bacterial colonies from the plate to the filter. The filter is passed through solutions to lyse the bacteria, denature the double-stranded DNA into single strands, and bind these strands to the filter.

The DNA from the colonies is screened by incubation with a nucleic acid probe. If a radioactive DNA probe is used, it is denatured to form single strands and then added to a solution containing the filter for incubation. If the DNA sequence of any of the cloned DNA on the filter is complementary to the probe, a double-stranded DNA–DNA hybrid molecule will form between the probe and the cloned DNA. After incubation, unbound and/or excess probe is washed away, and the filter is assayed to detect hybrid molecules that may have formed. If a radioactive probe has been used, the filter is overlaid with a piece of X-ray film. Radioactive decay in the probe molecules hybridized to DNA on the filter will expose the film and produce dark spots on it; they represent colonies on the plate containing the cloned gene of interest (Figure 16–13). The corresponding colony is identified and recovered from the original nutrient plate, and the cloned DNA it contains can be used in further experiments. In nonradioactive probes, a chemical reaction emits photons of light (chemilumines-cence) to expose the photographic film and reveal the location of colonies carrying the gene of interest.

To screen a phage library, a slightly different method, called **plaque hybridization**, is used. A solution of phage carrying DNA inserts is spread over a lawn of bacteria growing on a plate. The phages infect the bacterial cells on the plate and form plaques as they replicate. Each plaque, which appears as a clear spot on the plate, represents the progeny of a single phage and is a clone. The plaques are transferred to a nylon or nitrocellulose membrane by pressing the membrane onto the plate's surface. The phage DNA is denatured into single strands and screened with a labeled probe. Phage plaques are much smaller than plasmid colonies, and many plaques can be screened on a single filter, making this method more efficient for screening large genomic libraries.

Identifying Nearby Genes: Chromosome Walking

In some cases, when the approximate location of a gene is known, it is possible to clone the gene by first cloning near-by sequences. Often these nearby sequences are identified by linkage analysis and serve as starting points for **chromosome walking**, in which adjacent clones are isolated from a library. In a chromosome walk, the end piece of a cloned DNA fragment is subcloned and used as a probe to recover another set of overlapping clones from the genomic library (Figure 16–14, page 338). These clones are analyzed to determine their degree of overlap. A subfragment from one end of the overlapping cloned DNA is used to recover another set of overlapping clones, and the analysis is repeated. In this way, it is possible to "walk" along the chromosome, clone by clone.

The gene in question can be identified by nucleotide sequencing of the recovered clones and searching for an **open reading frame (ORF)**. An ORF is a stretch of nucleotides that begins with a start codon followed by amino acid–encoding codons, and ends with one or more stop codons. Although laborious and time-consuming, chromosome walking has identified genes for several human genetic disorders, including those for cystic fibrosis and muscular dystrophy.

Chromosome walking has several limitations in complex eukaryotic genomes, including the human genome. If a probe contains a repetitive sequence, such as an *Alu* sequence (see Chapter 17 for a discussion of these and other repetitive sequences), it hybridizes to other clones in the genomic library containing that sequence. Most of these clones will not be adjacent to the clone from which the probe was derived, and the walk terminates. In some cases, a technique called **chromosome jumping** skips over the region containing the repetitive sequences and continues the walk.

16.8 Characterizing Cloned Sequences

The recovery and identification of genes and other specific DNA sequences by cloning or by PCR is a powerful tool for analyzing genomic structure and function. In fact, much of the Human Genome Project is based on such techniques. In the following sections, we consider some of these methods, which are answering questions about the organization and function of cloned sequences.

Initial adjacent sequence (B)

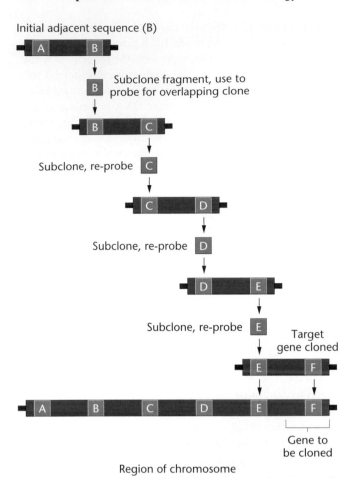

FIGURE 16–14 In chromosome walking, the approximate location of the gene to be cloned is known. A subcloned fragment of a linked adjacent sequence recovers overlapping clones from a genomic library. This process of subcloning and probing the genomic library is repeated to recover overlapping clones until the gene in question has been reached.

Restriction Mapping

One of the first steps in characterizing a DNA clone is the construction of a **restriction map**. A restriction map is a compilation of the number, order, and distance between restriction enzyme-cutting sites along a cloned segment of DNA. Map units are expressed in **base pairs (bp)** or, for longer distances, **kilobase pairs (kb)**. Restriction maps provide information about the length of a cloned insert and the location of restriction enzyme sites within the DNA. These data can be used to subclone fragments of a gene or compare its internal organization with that of other cloned sequences.

Figure 16–15 shows the construction of a restriction map from a cloned DNA segment. For this map, let's begin with a cloned DNA segment 7.0 kb in length. Three samples of the cloned DNA are digested with restriction enzymes—one with *Hind*III, one with *Sal*I, and one with both *Hind*III and *Sal*I. The generated fragments are separated by electrophoresis, stained with ethidium bromide, and photographed. They are then measured by comparing them to a set of standards run in adjacent lanes. The map is constructed by analyzing the fragments.

1. When the DNA is cut with *Hind*III, two fragments (0.8 and 6.2 kb) are produced; this confirms that the cloned insert is 7.0 kb in length and that there is only one cutting site for this enzyme (located 0.8 kb from one end).

2. When the DNA is cut with *Sal*I, two fragments (1.2 and 5.8 kb) result, indicating that this enzyme has one cutting site, located 1.2 kb from one end of the insert.

Taken together, the results show that each enzyme has one restriction site, but the relationship between the two sites is unknown. From the information available, two different maps are possible. In one map (model 1), the *Hind*III site is located 0.8 kb from one end and the *Sal*I site 1.2 kb from the same end. In the alternative map (model 2), the *Hind*III site is located 0.8 kb from one end and the *Sal*I site 1.2 kb from the other end.

The correct model is selected by considering the results from the sample digested with both *Hind*III and *Sal*I. Model 1 predicts that digestion with both enzymes will generate three fragments: 0.4, 0.8, and 5.8 kb in length; model 2 predicts that these fragments will be 0.8, 5.0, and 1.2 kb in length. The actual fragment pattern, observed after cutting with both enzymes, indicates that model 1 is correct (Figure 16–15).

Restriction maps are an important way to characterize a cloned DNA segment and can be constructed in the absence of information about the coding capacity or function of the mapped DNA. In conjunction with other techniques, restriction mapping can define the boundaries of a gene, dissect the molecular organization of a gene and its flanking regions, and locate mutational sites within genes.

Restriction maps can also refine genetic maps. To a large extent, the accuracy of genetic maps depends on the frequency of recombination between genetic markers and on the number of markers used to construct the map. In humans, for example, the genome size is large (3.1×10^9 bp) and the number of known genes is small, resulting in long distances between markers. Restriction enzyme-cutting sites are inherited and can be used as genetic markers, reducing the distance between markers, increasing their accuracy, and providing reference points for correlating genetic and physical maps.

Restriction sites play an important role in mapping genes to specific human chromosomes and to defined regions of individual chromosomes. In addition, if a restriction site maps close to a mutant gene, it can be used as a marker in a diagnostic test. These sites, known as **restriction fragment length polymorphisms**, or **RFLPs**, are described in Chapter 19. RFLPs have proven especially useful because the mutant genes underlying many human genetic diseases are poorly characterized at the molecular level, and restriction sites closely linked to mutant genes have successfully identified heterozygotes at risk for having affected children.

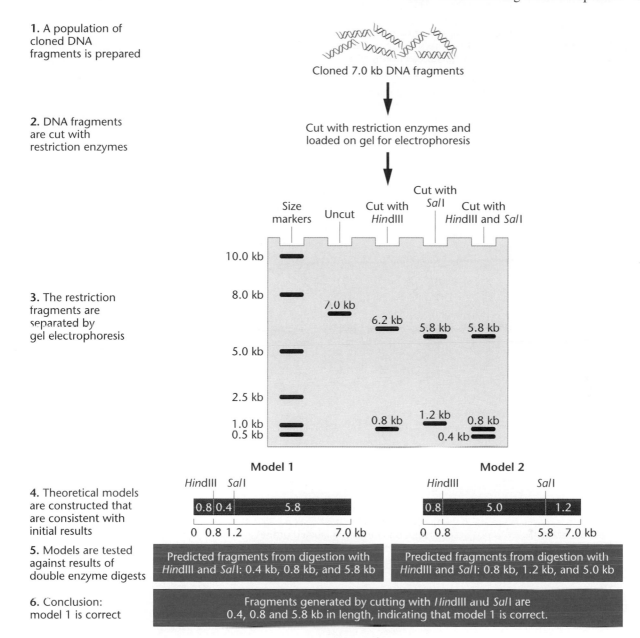

1. A population of cloned DNA fragments is prepared

Cloned 7.0 kb DNA fragments

2. DNA fragments are cut with restriction enzymes

Cut with restriction enzymes and loaded on gel for electrophoresis

3. The restriction fragments are separated by gel electrophoresis

Size markers | Uncut | Cut with *Hind*III | Cut with *Sal*I | Cut with *Hind*III and *Sal*I

10.0 kb
8.0 kb — 7.0 kb
6.2 kb — 5.8 kb — 5.8 kb
5.0 kb
2.5 kb
1.0 kb — 0.8 kb | 1.2 kb | 0.8 kb
0.5 kb — 0.4 kb

Model 1

*Hind*III *Sal*I

| 0.8 | 0.4 | 5.8 |

0   0.8 1.2   7.0 kb

Model 2

*Hind*III   *Sal*I

| 0.8 | 5.0 | 1.2 |

0   0.8   5.8 7.0 kb

4. Theoretical models are constructed that are consistent with initial results

5. Models are tested against results of double enzyme digests

Predicted fragments from digestion with *Hind*III and *Sal*I: 0.4 kb, 0.8 kb, and 5.8 kb

Predicted fragments from digestion with *Hind*III and *Sal*I: 0.8 kb, 1.2 kb, and 5.0 kb

6. Conclusion: model 1 is correct

Fragments generated by cutting with *Hind*III and *Sal*I are 0.4, 0.8 and 5.8 kb in length, indicating that model 1 is correct.

FIGURE 16–15 Constructing a restriction map. Samples of the 7.0-kb DNA fragment are digested with restriction enzymes: One sample is digested with *Hind*III, one with *Sal*I, and one with both *Hind*III and *Sal*I. The resulting fragments are separated by gel electrophoresis. The separated fragments are measured by comparing them with molecular-weight standards in an adjacent lane. Cutting the DNA with *Hind*III generates two fragments: 0.8 kb and 6.2 kb. Cutting with *Sal*I produces two fragments: 1.2 kb and 5.8 kb. Models are constructed to predict the fragment sizes generated by cutting with *Hind*III and with *Sal*I. Model 1 predicts that 0.4-, 0.8-, and 5.8-kb fragments will result from cutting with both enzymes, model 2 predicts that 0.8-, 1.2-, and 5.0-kb fragments will result. Comparing the predicted fragments with those observed on the gel indicates that model 1 is the correct restriction map.

MEDIA TUTORIAL Restriction Mapping

Nucleic Acid Blotting

Many of the techniques described in this chapter rely on hybridization between complementary nucleic acid (DNA or RNA) molecules. We now describe one of the most widely used methods for detecting such hybrids. This technique, called **Southern blotting** (after Edward Southern, who devised it), separates DNA fragments by gel electrophoresis, transfers them to filters, and screens the filters with labeled probes.

To make a Southern blot, cloned DNA is cut into fragments with one or more restriction enzymes, and the fragments are separated by gel electrophoresis (Figure 16–16, page 340). The DNA in the gel is denatured into single-stranded

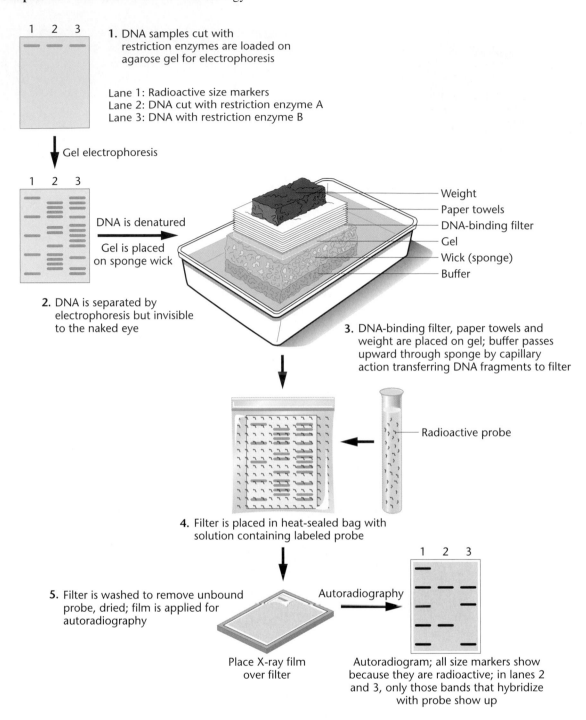

1. DNA samples cut with restriction enzymes are loaded on agarose gel for electrophoresis

Lane 1: Radioactive size markers
Lane 2: DNA cut with restriction enzyme A
Lane 3: DNA with restriction enzyme B

Gel electrophoresis

DNA is denatured

Gel is placed on sponge wick

2. DNA is separated by electrophoresis but invisible to the naked eye

Weight
Paper towels
DNA-binding filter
Gel
Wick (sponge)
Buffer

3. DNA-binding filter, paper towels and weight are placed on gel; buffer passes upward through sponge by capillary action transferring DNA fragments to filter

Radioactive probe

4. Filter is placed in heat-sealed bag with solution containing labeled probe

5. Filter is washed to remove unbound probe, dried; film is applied for autoradiography

Place X-ray film over filter

Autoradiography

Autoradiogram; all size markers show because they are radioactive; in lanes 2 and 3, only those bands that hybridize with probe show up

FIGURE 16–16 The Southern blotting technique. Samples of the DNA to be probed are cut with restriction enzymes and the fragments separated by gel electrophoresis. The pattern of fragments is visualized and photographed under ultraviolet illumination by staining the gel with ethidium bromide. The gel is then placed on a sponge wick in contact with a buffer solution and covered with a DNA-binding filter. Layers of paper towels or blotting paper are placed on top of the filter and held in place with a weight. Capillary action draws the buffer through the gel, transferring the pattern of DNA fragments from the gel to the filter. The DNA fragments on the filter are then denatured into single strands and hybridized with a labeled DNA probe. The filter is washed to remove excess probe and overlaid with a piece of X-ray film for autoradiography. The hybridized fragments show up as bands on the X-ray film.

fragments by treatment with an alkaline solution, and transferred to a sheet of DNA-binding material, usually nitrocellulose or a nylon derivative. To transfer the fragments, the membrane is placed on top of the gel, and a buffer solution flows through the gel and the membrane by capillary action (Figure 16–16). As the buffer solution flows through both, the DNA fragments move out of the gel and become immobilized on the membrane.

The DNA fragments on the membrane are hybridized with a labeled single-stranded DNA probe. Only DNA fragments complementary to the probe's nucleotide sequence will form double-stranded hybrids. Excess probe is washed away, and the hybridized fragments are visualized on a piece of film (Figure 16–17).

In addition to characterizing cloned DNAs, Southern blots map restriction sites within and near a gene, and identify DNA fragments carrying a single gene from a mixture of fragments and from related genes in different species. Southern blots also detect rearrangements and duplications in genes associated with human genetic disorders and cancers.

To determine whether a cloned gene is transcriptionally active in a given cell or tissue type, a related blotting technique probes for the presence of mRNA that is complementary to a cloned gene. This technique first extracts mRNA from one or several cell or tissue types, and fractionates the RNA by gel electrophoresis. The resulting pattern of RNA bands is transferred to a sheet of membrane, as in the Southern blot. The membrane is then hybridized to a single-stranded DNA probe, derived from the cloned gene. If RNA complementary to the DNA probe is present, it is detected as a band on the film. Because the original procedure (DNA bound to a filter) is known as a Southern blot, this procedure (RNA bound to a filter) is called a **northern blot**. (Following this somewhat perverse logic, another procedure involving proteins bound to a filter is known as a **western blot**.)

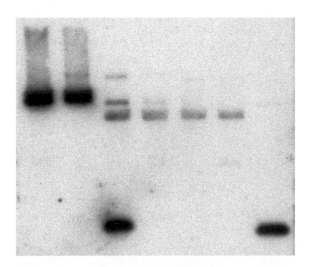

FIGURE 16–17 Southern blot after hybridization, exposure, and development. The bands show DNA fragments complementary to the nucleotide sequence of the probe. (*Dr. Suzanne McCutcheon*)

Northern blots provide information about the expression of specific genes and are used to study patterns of gene expression in embryonic and adult tissues. Northern blots also detect alternatively spliced mRNAs and multiple types of transcripts derived from a single gene, and are used to derive other information about transcribed mRNAs. If marker RNAs of known size are run in an adjacent lane, the size of a gene's mRNA can be measured. In addition, the amount of transcribed RNA present in the cell or tissue being studied is related to the density of the RNA band on the film. Measuring the density of the band gives the relative transcriptional activity. Thus, northern blots characterize and quantify the transcriptional activity of genes in different cells, tissues, and organisms.

16.9 DNA Sequencing: The Ultimate Way to Characterize a Clone

In a sense, a cloned DNA sequence is completely characterized when its nucleotide sequence is known. The ability to sequence cloned DNA has greatly enhanced our understanding of gene structure, gene function, and the mechanisms of regulation.

The most common method of **DNA sequencing** is based on **chain termination**. In this procedure, a single stranded DNA molecule whose sequence is to be determined serves as a template for synthesizing a series of complementary strands that terminate at specific nucleotides (Figure 16–18, page 342). In the first step, a short primer is annealed to the single-stranded template, and the primer-bound DNA is distributed into four tubes. In the second step, this primer is elongated in the 5′–3′ direction by adding the enzyme DNA polymerase and the four deoxyribonucleotide triphosphates (dATP, dCTP, dGTP and dTTP). In addition, each tube contains a small amount of one of the four base-specific analogs, called **dideoxynucleotides** (e.g., ddATP). As DNA synthesis takes place, the DNA polymerase occasionally inserts a dideoxynucleotide (instead of a deoxynucleotide) into a growing DNA strand. Since the analog lacks a 3′-hydroxyl group, it cannot form 3′ bond to add another nucleotide to the strand, and DNA synthesis terminates. For example, if ddATP is present, termination takes place at sites opposite to thymdine in the template strand (Figure 16–18). As the reaction proceeds, the tubes accumulate a series of DNA molecules that differ in length by one nucleotide at their 3′-ends. The fragments from each reaction tube are separated in four adjacent lanes (one for each tube) by gel electrophoresis. Electrophoresis separates DNA fragments in each lane that differ in size by a single nucleotide. The result is a series of bands forming a ladderlike pattern (Figure 16–19, page 342). The nucleotide sequence of the DNA can be read directly from bottom to top, corresponding to the 5′–3′ sequence of the DNA strand complementary to the template (Figure 16–19).

For genome sequencing, large-scale projects use automated machines that sequence several hundred thousand

1. Primer is bound to template strand

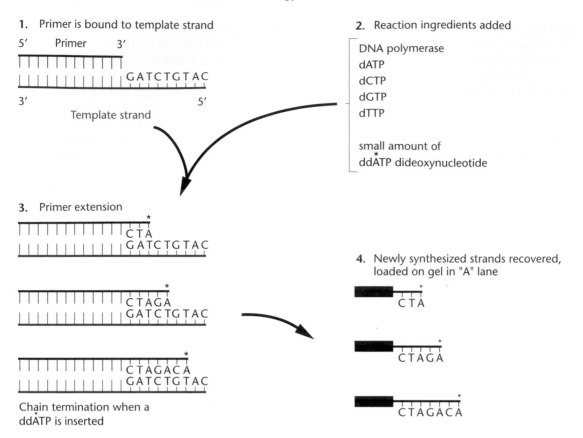

2. Reaction ingredients added

DNA polymerase
dATP
dCTP
dGTP
dTTP

small amount of
ddATP* dideoxynucleotide

3. Primer extension

Chain termination when a
ddATP* is inserted

4. Newly synthesized strands recovered, loaded on gel in "A" lane

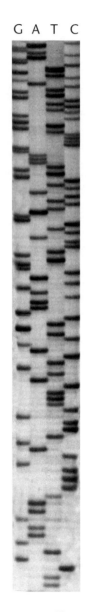

G A T C

FIGURE 16–18 DNA sequencing using the chain termination method. (1) A primer is annealed to a sequence adjacent to the DNA being sequenced (usually at the insertion site of a cloning vector). (2) A reaction mixture is added to the primer/template combination. This includes DNA polymerase, the four dNTPs (one of which is radioactively labeled) and a small amount of one dideoxynucleotide. Four tubes are used, each containing a different dideoxynucleotide (ddATP, ddCTP, etc.). (3) During primer extension, the polymerase occasionally inserts a ddNTP instead of a dNTP, terminating the synthesis of the chain, because the ddNTP does not have the 3′-OH group needed to attach the next nucleotide. In the figure, ddATP and the A inserted from this didexoynucleotide are indicated with an asterisk. Over the course of the reaction, all possible termination sites will have a ddNTP inserted. (4) The newly synthesized chains are removed from the template, and the mixture is placed on a gel. Chains terminating in A are loaded in the A lane, those ending in C are loaded In the C lane, and so forth.

FIGURE 16–19 DNA sequencing gel showing the separation of fragments in the four sequencing reactions (one per lane). To obtain the base sequence of the DNA fragment, the gel is read from the bottom, beginning with the lowest band in any lane, then the next lowest, then the next, and so on. For example, the sequence of the DNA on this gel begins with TTCGTGAAGAA. *(Dr. Suzanne McCutcheon)*

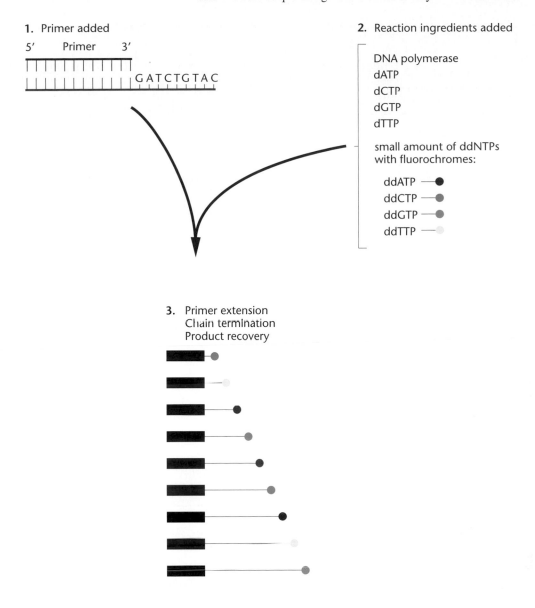

FIGURE 16–20 In DNA sequencing using dideoxynucleotides labeled with fluorescent dyes, all four ddNTPs are added to the same tube, and during primer extension, all combinations of chains are produced. The products of the reaction are added to a single lane on a gel, and the bands are read by a detector and imaging system. This process is now automated, and robotic machines, such as those used in the Human Genome Project, sequence several hundred thousand nucleotides in a 24-hour period, then store and analyze the data automatically.

nucleotides per day. In this procedure, the four dideoxynucleotide analogs are labeled with a fluorescent dye (Figure 16–20), so that chains terminating in A are labeled with one color, those ending in C with another color, and so forth. The reaction is carried out in a single tube, and loaded into one lane on the gel. The fluorescent detector in the sequencing machine reads the color of each band, and determines whether it represents an A, T, C, or G. The data are stored and analyzed using the appropriate software, or printed for reading (Figure 16–21, page 344).

DNA sequencing projects have now been completed on the genomes of over three dozen organisms. Sequencing provides information about the number, nature, and organization of genes in a genome, and elucidates the mutational events that alter both genes and gene products, confirming that genes and proteins are colinear molecules.

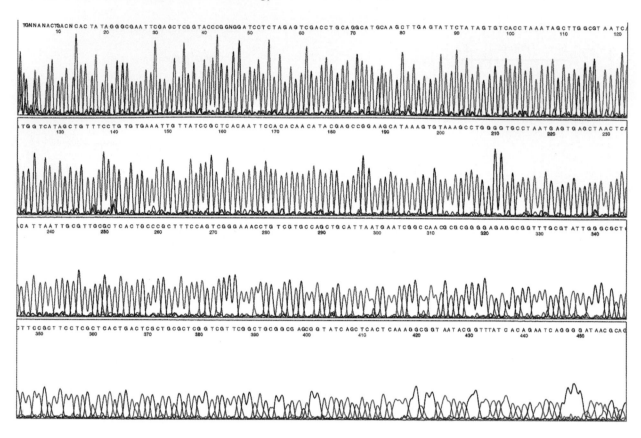

FIGURE 16–21 Automated DNA sequencing using fluorescent dyes, one for each base. The separated bases are read in order along the axis from left to right.

Chapter Summary

1. The cornerstone of recombinant DNA technology is a class of enzymes called restriction endonucleases, which cut DNA at specific recognition sites. The fragments produced are joined with DNA vectors to form recombinant DNA molecules.

2. Vectors replicate autonomously in host cells, and facilitate the manipulation of the newly created recombinant DNA molecules. Vectors are constructed from many sources, including bacterial plasmids and viruses.

3. Recombinant DNA molecules are transferred into a host, and cloned copies are produced during host cell replication. A variety of host cells may be used for replication, including bacteria, yeast, and mammalian cells. Cloned copies of foreign DNA sequences are recovered, purified, and analyzed.

4. The polymerase chain reaction (PCR) is a method for amplifying a specific DNA sequence that is present in a collection of DNA sequences, such as genomic DNA. The PCR method allows DNA to be cloned without host cells, and is a rapid, sensitive method with wide-ranging applications.

5. Once cloned, DNA sequences are analyzed through a variety of methods, including restriction mapping and DNA sequencing. Other methods such as Southern blotting hybridization identify genes and flanking regulatory regions within the cloned sequences.

Key Terms

annealing, 333

bacteriophage, 328

base pair (bp), 338

cDNA library, 335

chain termination, 341

chromosome jumping, 337

chromosome walking, 337

clone, 327

complementary DNA (cDNA), 335

denaturation, 333

dideoxynucleotide, 341

DNA ligase, 328

DNA polymerase I, 335

DNA sequencing, 341

genomic library, 335

Genetics, Technology, and Society

DNA Fingerprints in Forensics: The Case of the Telltale Palo Verde

The use of DNA analysis in forensics is making it harder and harder to get away with many crimes. It's now a simple matter to isolate DNA from tissue left at a crime scene, a splattering of blood, some skin left under a victim's fingernails, or even the cells clinging to the base of a hair shaft. A variety of techniques can be used to determine the likelihood that the sample came from a suspect in the case. In recent years, the PCR technique has been increasingly used in forensics, both because it is fast (taking only hours) and because it requires only a small amount of DNA (about a nanogram). One variation of the PCR method that may be especially valuable is called the random amplified polymorphic DNA (RAPD, pronounced "rapid") procedure. Under the right conditions, the RAPD procedure generates a DNA profile that is unique to each individual and that can thus be used for personal identification.

The RAPD procedure involves the amplification of a set of DNA fragments of unknown sequence. First, an amplification primer is mixed with a sample of DNA. The primer will anneal to all sites in the DNA sample that have a matching base sequence. For primers 10 nucleotides in length, binding will occur at a few thousand sites scattered randomly throughout the genome. When primers happen to bind to DNA sites that are not too far apart (within about 2000 nucleotides), DNA amplification can occur between these sites, eventually producing millions of copies of the DNA sequence lying between the binding sites. When the products of such a reaction are separated by electrophoresis on an agarose gel, each amplified DNA segment will appear as a distinct band. The locations of the primer binding sites in the genome of any two individuals in a population are likely to differ, due to minor DNA sequence differences (e.g., base changes, insertions, and deletions). Since these affect the initial binding to the primers, the number and locations of the DNA bands on the gel will vary from one individual to another. The resulting DNA profile is referred to as a DNA fingerprint. It represents a characteristic "snapshot" of the genome of an individual, allowing it to be distinguished from those of other individuals in the population.

When a DNA profile from tissue found at a crime scene matches that of a suspect, it does not prove that the tissue belongs to the suspect; instead, it excludes all those who have a different pattern. The strategy, therefore, is to generate five or more DNA profiles from the same sample, using different amplification primers. The more profiles that match between the sample and the suspect, the more unlikely it is that the sample at the crime scene came from someone other than the suspect. The likelihood that a complete match of all the banding patterns occurs simply by chance can be calculated, giving, for example, a 1-in-100,000 to a 1-in-1,000,000 probability of a "random match."

One of the most interesting uses of DNA fingerprinting in a criminal case did not involve the suspect's own DNA, but rather the DNA of plants growing at the crime scene. On the night of May 2, 1992, a Phoenix woman was strangled and her body dumped near an abandoned factory in the surrounding desert. Investigators discovered a pager near the body, making its owner a prime suspect. When questioned, this man admitted being with the woman the day of the murder, but claimed he had not been near the factory and suggested that the woman must have stolen his pager from his pickup truck. A search of the pickup provided the crucial clue, placing the suspect at the factory: In its bed were two seed pods from a palo verde tree.

The palo verde tree (*Cercidium floridum*), native to the desert of the Southwest, is a member of the bean family. In an adaptation to the hot, dry climate, it puts out very small leaves. To compensate for this, its stems become green and carry out photosynthesis. This gives the plant its name; palo verde is Spanish for "green stem." The tree may not resemble a bean plant in some respects, but, like a typical bean plant, it produces seeds in pods that drop to the ground when mature.

The homicide investigators assigned to the case wondered whether it could be proved that the seed pods found in the bed of the suspect's truck had fallen from a palo verde tree near the place where the body was found. If so, it would be a key piece of evidence placing the suspect at the scene. But how could this be demonstrated? The investigators turned to Dr. Timothy Helentjaris, then at the University of Arizona in nearby Tucson. Helentjaris proceeded to determine whether the DNA profile of the seed pods from the pickup truck matched that of any of the palo verde trees near the scene. He understood that for a match between a seed pod and a tree to have significance, he would first have to show that palo verdes differ genetically from one tree to the next. Otherwise, a pod could never be unambiguously assigned to one particular tree. Fortunately, he found considerable variation in DNA profiles among the palo verde trees.

Helentjaris was then given the two seed pods from the suspect's truck along with pods collected from 12 palo verde trees from the vicinity of the factory. The investigators knew which of the 12 trees was the key one at the crime scene, but they did not tell Helentjaris. He extracted DNA from seeds from each pod and performed RAPD reactions using an amplification primer that he had earlier found to reproducibly yield profiles of 10–15 distinct bands. The result of this controlled experiment was unmistakable: The DNA profile from one of the pods found in the truck exactly matched that of only 1 of the 12 trees, the one nearest to where the body was found. In an important additional test, Helentjaris showed that this DNA profile differed from that of pods collected from 18 other trees located at random sites around Phoenix. Collectively, Helentjaris's analysis led to the estimate of the likelihood of a random match at a little less than 1 in 1,000,000.

Helentjaris's analysis was admitted as evidence in the trial, placing the suspect at the crime scene. At the completion of the five-week trial, the suspect was found guilty of first-degree murder. His conviction was upheld upon appeal, and he is currently serving a life sentence without parole.

References

AYALA, F.J., and BLACK, B. 1993. Science and the courts. *Am. Sci.* 81:230–39.

KRAWCZAK, M., and SCHMIDTKE, J. 1994. *DNA fingerprinting*. Oxford: BIOS Scientific.

MARX, L. 1988. DNA fingerprinting takes the witness stand. *Science* 240: 1616–18.

Insights and Solutions

The recognition site for the restriction enzyme *Sau*3A is GATC (Figure 16–3). The recognition site for the enzyme *Bam*HI is GGATCC, where the four internal bases are identical to the *Sau*3A site. This means that the single-stranded ends produced by the two enzymes are identical. Suppose you have a cloning vector that contains a *Bam*HI site and foreign DNA that you have cut with *Sau*3A.

1. Can this DNA be ligated into the *Bam*HI site of the vector? Why?

Solution: DNA cut with *Sau*3A can be ligated into the vector's *Bam*HI site because the single-stranded ends generated by the two enzymes are identical.

2. Can the DNA segment cloned into this site be cut from the vector with *Sau*3A? With *Bam*HI? What potential problems do you see with the use of *Bam*HI?

Solution: The DNA can be cut from the vector with *Sau*3A because the recognition sequence for this enzyme (GATC) is maintained on each side of the insert. Recovering the cloned insert with *Bam*HI is more problematic. In the ligated vector, the conserved sequences are GGATC (left) and GATCC (right). Only about 25 percent of the time will the correct base follow the conserved sequence (to produce GGATCC on the left) and only about 25 percent of the time will the correct base precede the conserved sequence (and produce GGATCC on the right). Thus, only about 6 percent of the time (0.25 × 0.25) will *Bam*HI be able to cut the insert from the vector.

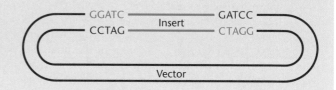

Problems and Discussion Questions

1. In recombinant DNA studies, what roles do restriction endonucleases, vectors, and host cells play?

2. Why is poly-dT an effective primer for reverse transcriptase?

3. The human insulin gene contains a number of introns. In spite of the fact that bacterial cells do not excise introns from mRNA, explain how can a gene like this be cloned into a bacterial cell and produce insulin.

4. Restriction enzymes recognize palindromic sequences in intact DNA molecules and cleave the double-stranded helix at these sites. Inasmuch as the bases are internal in a DNA double helix, how is this recognition accomplished?

5. Although the potential benefits of cloning in higher plants are obvious, the development of this field has lagged behind cloning in bacteria, yeast, and mammalian cells. Can you think of any reason for this?

6. Using DNA sequencing on a cloned DNA segment, you recover the following nucleotide sequence:

CAGTATCCTAGGCAT

Does this segment contain a palindromic recognition site for a restriction enzyme? What is the double-stranded sequence of the palindrome? What enzyme would cut at this site? (Consult Figure 16–3 for a list of restriction enzyme recognition sites.)

7. Figure 16–3 lists restriction enzymes that recognize sequences of four bases and six bases. How frequently should four- and six-base recognition sequences occur in a genome? If the recognition sequence consists of eight bases, how frequently will such sequences be encountered? Under what circumstances would you use such an enzyme?

8. You are given a cDNA library of human genes prepared in a bacterial plasmid vector. You are also given the cloned yeast gene that encodes EF-1a, a protein that is highly conserved in protein sequence among eukaryotes. Outline how you would use these resources to identify the human cDNA clone encoding EF-1a.

9. You recover a cloned DNA segment of interest, and determine that the insert is 1300 bp in length. To characterize this cloned segment, you isolate the insert and decide to construct a restriction map. Using enzyme I and enzyme II, followed by gel electrophoresis, you determine the number and size of the fragments produced by enzymes I and II alone and in combination as follows:

Enzymes	Restriction Fragment Sizes
I	350 bp, 950 bp
II	200 bp, 1100 bp
I and II	150 bp, 200 bp, 950 bp

Construct a restriction map from these data, showing the positions of the restriction sites relative to one another, and the distance between them in base pairs.

10. cDNA can be cloned into vectors to create a cDNA library. In analyzing cDNA clones, it is often difficult to find clones that are full length—that is, extend to the 5′-end of the mRNA. Why is this so?

11. Although the capture and trading of great apes has been banned in 112 countries since 1973, it is estimated that about 1000 chimpanzees are removed annually from Africa and smuggled into Europe, the United States, and Japan. This illegal trade is often disguised by private (such as zoo or circus) owners by simulating births in captivity. Until recently, genetic identity tests to uncover these illegal activities were not used because of the lack of highly polymorphic markers and the difficulties of obtaining chimp blood samples. Recently a study was reported in which DNA samples were extracted from freshly plucked chimp hair roots and used as templates for the polymerase chain reaction. The primers used in these studies flank highly polymorphic sites in human DNA that result from variable numbers of tandem nucleotide repeats. Several suspect chimp offspring and their putative parents were tested to determine whether the offspring are "legitimate" or the product of illegal trading. The data are shown at right. Examine these data carefully and choose the best conclusion:

   (a) None of the offspring are legitimate.

   (b) Offspring B and C are not the products of these parents and were probably purchased on the illegal market. The data are consistent with offspring A being legitimate.

   (c) Offspring A and B are products of the parents shown, but C is not and was therefore probably purchased on the illegal market.

   (d) There are not enough data to draw any conclusions. Additional polymorphic sites should be examined.

   (e) No conclusion can be drawn because "human" primers were used.

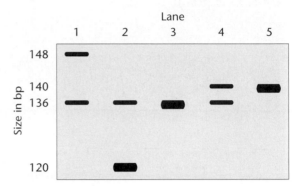

Lane 1: father chimp
Lane 2: mother chimp
Lanes 3–5: putative offspring A, B, C

12. Briefly describe the problem that a stretch of repeated sequences would cause in a chromosome walk and name the procedure used to overcome this problem.

Selected Readings

Alton, E. W., and Geddes, D. M. 1995. Gene therapy for cystic fibrosis: A clinical perspective. *Gene Ther.* 2:88–95.

Antonarakis, S. 1989. Diagnosis of genetic disorders at the DNA level. *N. Engl. J. Med.* 320:153–63.

Barker, D., et al. 1987. Gene for von Recklinghausen neurofibromatosis is in the pericentromeric region of chromosome 17. *Science* 236:1001–162.

Burger, S. L., and Kimmel, A. R. 1987. Guide to molecular cloning techniques. Methods in enzymology, vol. 152. San Diego, CA: Academic Press.

Carter, P. J. and Samulski, P. J. 2000. Adeno-associated viral vectors as gene delivery vehicles. *Int. J. Mol. Med.* 6:17–27.

Colosimo, A., Goncz, K. K., Holmes, A. R., et al. 2000. Transfer and expression of foreign genes in mammalian cells. *BioTechniques* 29:314–331.

Dube, I. D., and Cournoyer, D. 1995. Gene therapy: Here to stay. *Can. Med. Assoc. J.* 152:1605–13.

Eisensmith, R. C., and Woo, S. 1995. Molecular genetics of phenylketonuria: From molecular anthropology to gene therapy. *Adv. Genet.* 32:199–271.

Guyer, M. S., and Collins, F. S. 1993. The Human Genome Project and the future of medicine. *Am. J. Dis. Child.* 147:1145–52.

Knorr, D., and Sinskey, A. J. 1985. Biotechnology in food production and processing. *Science* 229:1224–29.

McKusick, V. A. 1988. The new genetics and clinical medicine. *Hosp. Pract.* 23:177–91.

Mullis, K. B. 1990. The unusual origin of the polymerase chain reaction. *Sci. Am.* (April) 262:56–65.

Old, R. W., and Primrose, S. B. 1994. Principles of genetic manipulation: An introduction to genetic engineering. 5th ed. Palo Alto, CA: Blackwell Scientific.

Oste, C. 1988. Polymerase chain reaction. *BioTechniques* 6:162–67.

Reiss, J., and Cooper, D. N. 1990. Application of the polymerase chain reaction to the diagnosis of human genetic disease. *Hum. Genet.* 85:1–8.

Sambrook, J., Fitch, E. F., and Maniatis, T. 1989. Molecular cloning: A laboratory manual, 2nd ed. Cold Spring Harbor, NY: Cold Spring Harbor Press.

Southern, E. 1975. Detection of specific sequences among DNA fragments separated by gel electrophoresis. *J. Mol. Biol.* 98: 503–507.

Tal, J. 2000. Adeno-associated virus-based vectors in gene therapy. *J. Biomed. Sci.* 7:279–91.

Thomas, T. L., and Hall, T. C. 1985. Gene transfer and expression in plants: Implications and potential. *BioEssays* 3:149–53.

Torrey, J. G. 1985. The development of plant biotechnology. *Am. Sci.* 73:354–63.

Welsh, J., and McClelland, M. 1990. Fingerprinting genomes using PCR with arbitrary primers. *Nucleic Acids Res.* 18:7213–18.

White, R. 1985. DNA sequence polymorphisms revitalize linkage approaches in human genetics. *Trends Genet.* 1:177–80.

Williams, J. G. K., et al. 1990. DNA polymorphisms amplified by arbitrary primers are useful as genetic markers. *Nucleic Acids Res.* 18:6531–35.

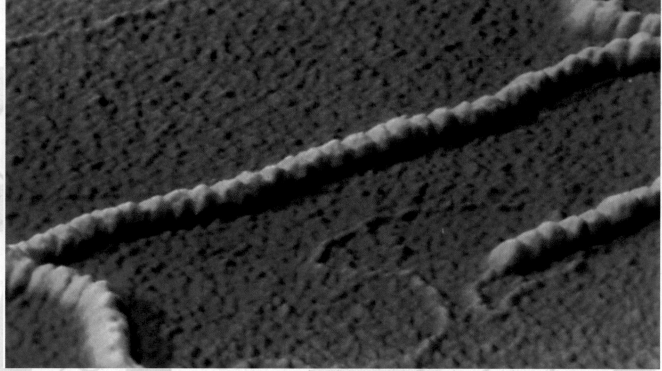

Chromatin fibers viewed using a scanning transmission electron microscope (STEM). *(Science VU/BMRL/Visuals Unlimited)*

17

Chromosome Structure and DNA Sequence Organization

CHAPTER CONCEPTS

Once it was understood that DNA houses genetic information, it was very important to determine how DNA is organized into genes and how these basic units of genetic function are organized into chromosomes. That is, how is the genetic material organized as it makes up the genome of organisms? Researchers are keenly interested in this question because the organization of the genetic material and associated molecules is pivotal to understanding many areas of genetics. For example, how the genetic information is stored, expressed, and regulated must be related to the molecular organization of the genetic molecule, DNA. Determining how genomic organization varies in different organisms—from viruses to bacteria to eukaryotes—will undoubtedly help us better understand the evolution of organisms.

17.1 Viral and Bacterial Chromosomes

17.2 Mitochondrial and Chloroplast DNA
Molecular Organization and Gene Products of Mitochondrial DNA
Molecular Organization and Gene Products of Chloroplast DNA

17.3 Specialized Chromosomes
Polytene Chromosomes
Lampbrush Chromosomes

17.4 Organization of Chromatin in Eukaryotes
Chromatin Structure and Nucleosomes
Heterochromatin

17.5 DNA Sequence Organization in Eukaryotes
Repetitive DNA and Satellite DNA
Centromeric and Telomeric DNA Sequences
Middle Repetitive Sequences: VNTRs and Dinucleotide Repeats
Repetitive Transposed Sequences: SINEs and LINEs
Middle Repetitive Multiple Copy Genes

17.6 The Eukaryotic Genome: What Proportion Encodes Genes?

In this chapter we focus on the various ways DNA is organized into chromosomes. Numerous approaches, including visualization by light microscopy, have been used to study the genetic material. These studies began with the nineteenth-century discovery of two types of unusually large chromosomes. Direct visualization has also been achieved by electron microscopy, and more recently molecular analysis has provided significant insights into chromosome organization.

We first survey what we know about chromosomes in viruses, bacteria, and two cellular organelles, the mitochondria, and chloroplasts. Then, we'll examine the large specialized structures called polytene and lampbrush chromosomes. In the second half of the chapter, we discuss how eukaryotic genomes are organized. For example, how is DNA complexed with proteins to form chromatin, and how are the chromatin fibers characteristic of interphase condensed into chromosome structures visible during mitosis and meiosis? We conclude the chapter by examining aspects of DNA sequence organization characteristic of eukaryotic genomes. Then, in Chapter 18, we turn to the more recent and very exciting studies that have elucidated how genes are organized in chromosomes.

17.1 Viral and Bacterial Chromosomes

Compared to eukaryotes, the chromosomes of viruses and bacteria are much less complicated. They usually consist of a single nucleic acid molecule, unlike the multiple chromosomes comprising the genome of higher forms. They are largely devoid of associated proteins and contain relatively little genetic information. These characteristics have greatly simplified analysis, and we now have a fairly comprehensive view of the structure of viral and bacterial chromosomes.

The chromosomes of viruses consist of a nucleic acid molecule—either DNA or RNA—that can be either single or double stranded. They can exist as circular structures (closed loops), or they can take the form of linear molecules. The single-stranded DNA of the **φX174 bacteriophage** and the double-stranded DNA of the **polyoma virus** are closed loops housed within the protein coat of the mature viruses. The **bacteriophage lambda (λ)**, on the other hand, possesses a linear double-stranded DNA molecule prior to infection, which closes to form a ring upon its infection of the host cell. Still other viruses, such as the T-even series of bacteriophages, have linear double-stranded chromosomes of DNA that do not form circles inside the bacterial host. Thus, circularity is not an absolute requirement for replication in some viruses.

Viral nucleic acid molecules have been visualized with the electron microscope. Figure 17–1 shows a mature bacteriophage λ with its double-stranded DNA molecule in the circular configuration. One constant feature shared by viruses, bacteria, and eukaryotic cells is the ability to package an exceedingly long DNA molecule into a relatively small volume. In λ, the DNA is 17 μm long and must fit into the phage head, which is less than 0.1 μm on any side.

Table 17.1 (page 350) compares the length of the chromosomes of several viruses to the size of their head structure. In each case, a similar packaging feat must be accomplished. We can compare the dimensions given for phage T2 with the micrograph of both the DNA and the viral particle shown in Figure 17–2 (page 350). Seldom does the space available in the head of a virus exceed the chromosome volume by more than a factor of 2. In many cases, almost all space is filled, indicating nearly perfect packing. Once packed within the head, the genetic material is functionally inert until it is released into a host cell.

Bacterial chromosomes are also relatively simple in form. They always consist of a double-stranded DNA molecule, compacted into a structure sometimes referred to as the

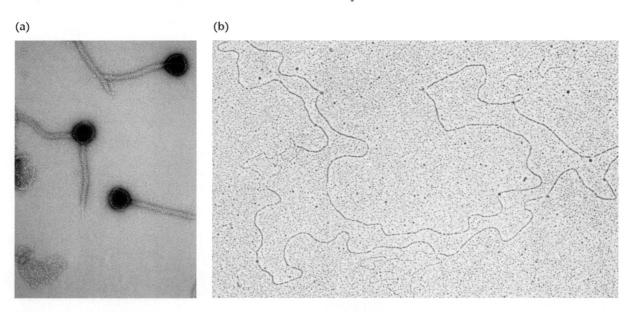

FIGURE 17–1 Electron micrographs of phage lambda (left) and the DNA isolated from it (right). The chromosome is 17 μm long. Note that the phages are magnified about five times more than the DNA.

TABLE 17.1   The Genetic Material of Representative Viruses and Bacteria

	Organism	Type	SS or DS*	Length (μm)	Overall Size of Viral Head or Bacteria (μm)
					Nucleic Acid
Viruses	φX174	DNA	SS	2.0	0.025 × 0.025
	Tobacco mosaic virus	RNA	SS	3.3	0.30 × 0.02
	Lambda phage	DNA	DS	17.0	0.07 × 0.07
	T2 phage	DNA	DS	52.0	0.07 × 0.10
Bacteria	*Haemophilus influenzae*	DNA	DS	832.0	1.00 × 0.30
	Escherichia coli	DNA	DS	1200.0	2.00 × 0.50

*SS = single-stranded; DS = double-stranded.

nucleoid. *Escherichia coli*, the most extensively studied bacterium, has a large, circular chromosome measuring approximately 1200 μm (1.2 mm) in length. When the cell is gently lysed and the chromosome released, it can be visualized under the electron microscope (Figure 17–3).

DNA in bacterial chromosomes is associated with several types of **DNA-binding proteins**. Two, called **HU** and **H**, are small but abundant in the cell and contain a high percentage of positively charged amino acids that can bond ionically to the negative charges of the phosphate groups in DNA. These proteins are structurally similar to molecules called histones that are associated with eukaryotic DNA. (We discuss histone organization later in this chapter.) Unlike the tightly packed chromosome present in the head of a virus, the bacterial chromosome is not functionally inert. Despite its somewhat compacted condition in the bacterial cell, the chromosome can be readily replicated and transcribed.

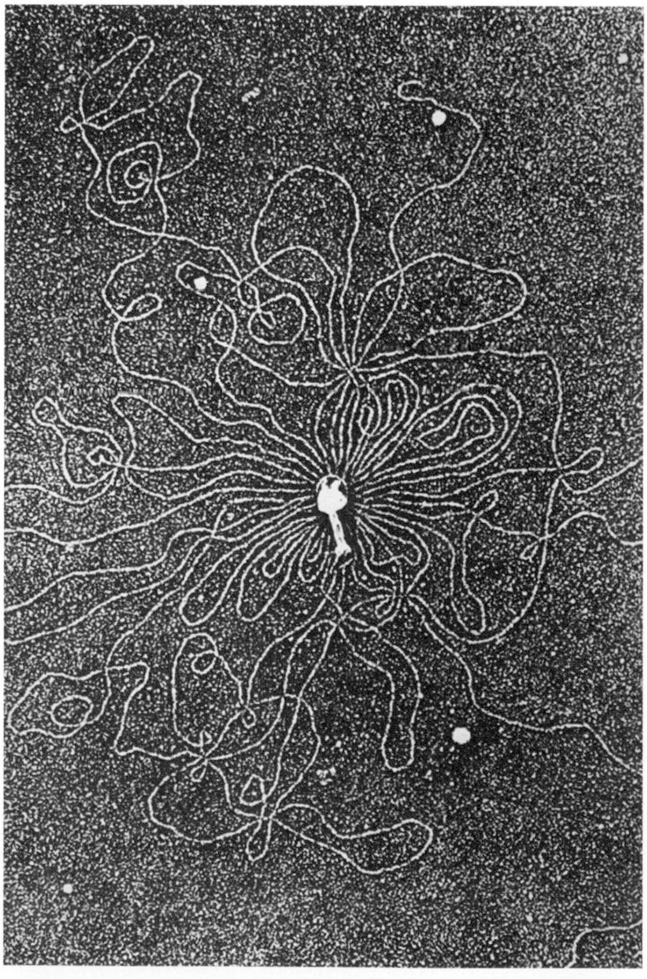

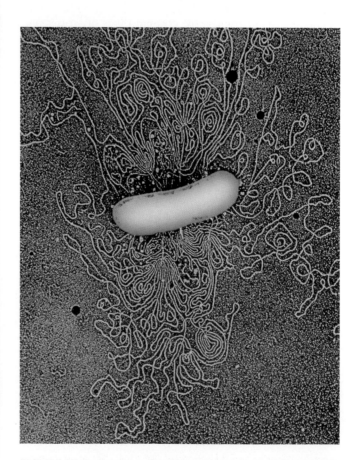

FIGURE 17–2 Electron micrograph of bacteriophage T2, which has had its DNA released by osmotic shock. The chromosome is 52 μm long.

FIGURE 17–3 Electron micrograph of the bacterium *Escherichia coli*, which has had its DNA released by osmotic shock. The chromosome is 1200 μm long.

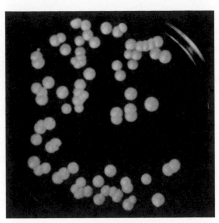

Normal colonies

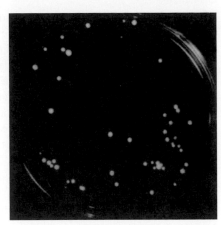

Petite colonies

FIGURE 17–4 Normal and mutant *petite* colonies in the yeast *Saccharomyces cerevisiae*.

17.2   Mitochondrial and Chloroplast DNA

Numerous studies demonstrate that both **mitochondria** and **chloroplasts** contain their own DNA and genetic system for expressing this information. This was first suggested by the discovery of mutations in yeast, other fungi, and plants that produce altered phenotypes that can be linked to these organelles. For example, several categories of mutations in yeast, called *petites*, exhibit abnormal mitochondria and are defective in cellular respiration, the primary function of mitochondria. As a result, the cells grow more slowly and the colony size is much smaller (Figure 17–4). Transmission of this trait occurs through the cytoplasm rather than through a mutant gene on a chromosome in the nucleus.

In other cases, the transmission of mutations demonstrates a strictly a *uniparental* mode of inheritance, with traits passed to offspring only through the egg. Because both mitochondria and chloroplasts are inherited through the maternal cytoplasm in most organisms, these observations suggested that the organelles house their own DNA, which, when mutated, may be responsible for the altered phenotypes.

Thus, geneticists set out to look for more direct evidence of DNA in these organelles. Electron microscopists not only documented the presence of DNA in both organelles, they also saw DNA in a form quite unlike that seen in the nucleus of the eukaryotic cells that house these organelles. This DNA looked remarkably similar to that seen in viruses and bacteria! This similarity, along with other observations, led to the idea that mitochondria and chloroplasts arose independently more than a billion years ago from free-living, prokaryotic-like organisms that possessed the abililty to undergo aerobic respiration or photosynthesis, respectively. This theory, championed by Lynn Margulis and others and called the **endosymbiotic hypothesis**, proposes that the prokaryotes were engulfed by larger primitive eukaryotic cells, which lacked these bioenergetic functions. A symbiotic relationship developed whereby the prokaryotic organisms eventually lost their ability to function independently, while the eukaryotic host cells gained the ability to either respire aerobically or undergo photosynthesis. Although many questions remain unanswered, the basic tenets of this theory are widely accepted.

Molecular Organization and Gene Products of Mitochondrial DNA

Extensive information is now available about the molecular aspects and gene products of **mitochondrial DNA (mtDNA)**. In most eukaryotes, mtDNA exists as a double-stranded closed circle (Figure 17–5) that replicates semiconservatively and is free of the chromosomal proteins characteristic of eukaryotic DNA. In size, mtDNA differs greatly among organisms, as demonstrated in Table 17.2 (page 352). In a variety of animals, including humans, mtDNA consists of about 16,000–18,000 bp (16–18 kb). However, yeast (*Saccharomyces*) contains 75 kb, while up to 367 kb may be present in plant mitochondria such as

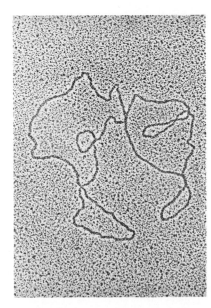

FIGURE 17–5 Electron micrograph of mitochondrial DNA (mtDNA) derived from *Xenopus laevis*.

TABLE 17.2 The Size of mtDNA in Different Organisms

Organism	Size (kb)
Human	16.6
Mouse	16.2
Xenopus (frog)	18.4
Drosophila (fruit fly)	18.4
Saccharomyces (yeast)	75.0
Pisum sativum (pea)	110.0
Arabidopsis (mustard plant)	367.0

TABLE 17.3 Sedimentation Coefficients of Mitochondrial Ribosomes

Kingdom	Examples	Sedimentation Coefficient (S)
Animals	Vertebrates	55–60
	Insects	60–71
Protists	*Euglena*	71
	Tetrahymena	80
Fungi	*Neurospora*	73–80
	Saccharomyces	72–80
Plants	Maize	77

Arabidopsis. Vertebrates have 5–10 such DNA molecules per organelle, while plants have 20–40 copies per organelle.

We can say several things about mtDNA. With only rare exceptions, no introns appear to be present in mitochondrial genes and gene repetitions are few or none. As noted in Chapter 12, expression of mitochondrial genes uses several modifications of the otherwise standard genetic code. Replication is dependent on enzymes encoded by nuclear DNA. In humans, mtDNA encodes 2 ribosomal RNAs (rRNAs), 22 transfer RNAs (tRNAs), as well as numerous

polypeptides essential to the cellular respiratory functions of the organelles. In most all cases, these polypeptides are part of multichain proteins, and the other polypeptides of each protein are encoded in the nucleus, synthesized in the cytoplasm, and transported into the organelle. Thus, the protein-synthesizing apparatus and the molecular components for cellular respiration are jointly derived from nuclear and mitochondrial genes.

As expected then, ribosomes found in the organelle are different from those present in the neighboring cytoplasm. Table 17.3 shows that mitochondrial ribosomes of different species vary considerably in their sedimentation coefficients, ranging from 55*S* to 80*S*. Since many are closer in their coefficient to bacterial counterparts than eukaryotes, this also supports the endosymbiont hypothesis.

Nuclear-coded gene products essential to biological activity in mitochondria include DNA and RNA polymerases, initiation and elongation factors essential for translation, ribosomal proteins, aminoacyl tRNA synthetases, and several tRNA species. These imported components are distinct from their cytoplasmic counterparts, even though both sets are coded by nuclear genes. For example, the synthetase enzymes essential for charging mitochondrial tRNA molecules (a process essential to translation) show a distinct affinity for the mitochondrial tRNA species as compared to the cytoplasmic tRNAs. Similar affinity has been shown for the initiation and elongation factors. Furthermore, while bacterial and nuclear RNA polymerases are known to be composed of numerous subunits, the mitochondrial variety consists of only one polypeptide chain. This polymerase is generally susceptible to antibiotics that inhibit bacterial RNA synthesis but not to eukaryotic inhibitors. The contributions of nuclear and mitochondrial gene products are contrasted in Figure 17–6.

Molecular Organization and Gene Products of Chloroplast DNA

Chloroplasts, like mitochondria, contain an autonomous genetic system distinct from that found in the nucleus and cytoplasm. This system includes DNA as a source of genetic

FIGURE 17–6 Gene products that are essential to mitochondrial function. Those shown entering the organelle are derived from the cytoplasm and encoded by the nucleus.

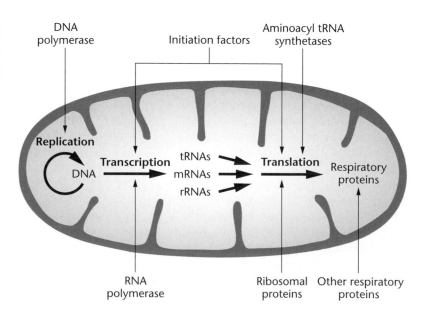

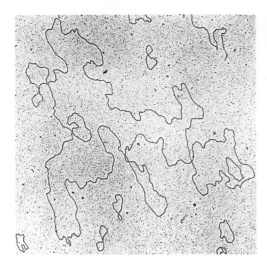

FIGURE 17–7 Electron micrograph of chloroplast DNA derived from lettuce.

information and a complete protein-synthesizing apparatus. And like the situation in mitochondria, the molecular components of the chloroplast translation apparatus are jointly derived from both nuclear and organelle genetic information. **Chloroplast DNA (cpDNA)**, shown in Figure 17–7, is much larger than mitochondrial DNA—usually 120–200 kb in length. Nevertheless, it also shares similarities to DNA found in prokaryotic cells. It is circular, double stranded, replicated semiconservatively, and free of the associated proteins characteristic of eukaryotic DNA. Compared with nuclear DNA of the same organism, it invariably shows a different buoyant density and base composition.

In the green alga *Chlamydomonas*, there are about 75 copies of the chloroplast DNA molecule per organelle. Each copy consists of a length of DNA that contains 195,000 bp (195 kb). In higher plants such as the sweet pea, multiple copies of the DNA molecule are present in each organelle, but the molecule (134 kb) is considerably smaller than that in *Chlamydomonas*. Interestingly, genetic recombination between the multiple copies of DNA within chloroplasts has been documented in *Chlamydomonas*.

Some chloroplast gene products function during translation. In a variety of higher plants (beans, lettuce, spinach, maize, and oats), two sets of the genes coding for the ribosomal RNAs—5*S*, 16*S*, and 23*S* rRNA—are present. Additionally, chloroplast DNA codes for at least 25 tRNA species and a number of ribosomal proteins specific to the chloroplast ribosomes. These ribosomes have a sedimentation coefficient slightly less than 70*S*, similar to that of bacteria. Even though chloroplast ribosomal proteins are encoded by both nuclear and chloroplast DNA, most, if not all, such proteins are distinct from their counterparts in cytoplasmic ribosomes.

Still other chloroplast genes specific to the photosynthetic function have been identified. Mutations in these genes may inactivate photosynthesis in chloroplasts with this mutation. One of the major photosynthetic enzymes is ribulose-1-5-bisphosphate carboxylase (RuBP). Interestingly, the small subunit of this enzyme is encoded by a nuclear gene, whereas the large subunit is encoded by cpDNA.

The great increase in size in cpDNA compared to mtDNA can to some extent be accounted for by an increased number of genes. However, the biggest difference appears to be due to the presence of long noncoding sequences of DNA as well as duplications of many DNA sequences. This observation is indicative of the independent evolution occurring in chloroplasts and mitochondria following their intial invasion of a primitive eukaryotic-like cell.

17.3 Specialized Chromosomes

We now consider two cases of genetic organization that demonstrate the specialized forms that chromosomes can take. Both types, polytene chromosomes and lampbrush chromosomes, are so large that their organization was discerned using light microscopy long before we understood how mitotic chromosomes form from interphase chromatin (see Chapter 2). The study of these chromosomes provided many of our initial insights into the arrangement and function of the genetic information.

Polytene Chromosomes

Giant **polytene chromosomes** are found in various tissues (salivary, midgut, rectal, and malpighian excretory tubules) in the larvae of some flies and in several species of protozoans and plants. Such structures were first observed by E. G. Balbiani in 1881. The vast amount of information obtained from studies of these genetic structures provided a model system for subsequent investigations of chromosomes. What is particularly intriguing about polytene chromosomes is that they can be seen in the nuclei of interphase cells.

Each polytene chromosome observed under the light microscope reveals a linear series of alternating bands and interbands (Figure 17–8, page 354). The banding pattern is distinctive for each chromosome in any given species. Individual bands are sometimes called **chromomeres**, a more generalized term describing lateral condensations of material along the axis of a chromosome. Each polytene chromosome is 200–600 μm long.

Extensive study using electron microscopy and radioactive tracers led to an explanation for the unusual appearance of these chromosomes. First, polytene chromosomes represent paired homologs. This is highly unusual in most organisms, since they are found in somatic cells, where chromosomal material is normally dispersed as chromatin and homologs are not paired. Second, their large size and distinctiveness result from the many DNA strands that compose them. The DNA of these paired homologs undergoes many rounds of replication, but without strand separation or cytoplasmic division. As replication proceeds, chromosomes with 1000–5000 DNA strands that remain in parallel register are created. Apparently, the parallel register of so many DNA strands gives rise to the distinctive band pattern along the axis of the chromosome.

The relationship between the structure of polytene chromosomes and the genes contained in them is intriguing. The presence of bands was initially interpreted as the visible manifestation of individual genes. The discovery that the strands

FIGURE 17–8 Polytene chromosomes derived from larval salivary gland cells of *Drosophila*.

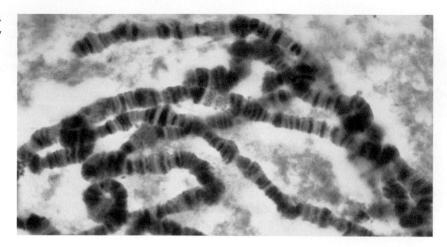

present in bands undergo localized uncoiling during genetic activity further strengthened this view. Each such uncoiling event results in what is called a **puff** because of its appearance (see Figure 17–9). That puffs are visible manifestations of gene activity (transcription that produces RNA) is evidenced by their high rate of incorporation of radioactively labeled RNA precursors, as assayed by autoradiography. Bands that are not extended into puffs incorporate fewer radioactive precursors or none at all.

The study of bands during development in insects, such as *Drosophila* and the midge fly *Chironomus*, reveals differential gene activity. A characteristic pattern of band formation, which is equated with gene activation, is observed as development proceeds. Despite attempts to resolve the issue, it is not yet clear how many genes are contained in each band. We do know that a band can contain up to 10^7 bp of DNA, certainly enough DNA to encode 50–100 average-size genes.

Lampbrush Chromosomes

Another specialized chromosome giving us insights into chromosomal structure is the **lampbrush chromosome**, so named because its resembles the brushes used to clean kerosene lamp chimneys in the nineteenth century. Lampbrush chromosomes were first discovered in 1892 in the oocytes of sharks and are now known to be characteristic of most vertebrate oocytes as well as the spermatocytes of some insects. Therefore, they are meiotic chromosomes.

Most of the experimental work has been done with material taken from amphibian oocytes.

These chromosomes are easily isolated from oocytes in the diplotene stage of the first prophase of meiosis, where they are active in directing the metabolic activities of the developing cell. The homologs are seen as synapsed pairs held together by chiasmata. However, instead of condensing, as most meiotic chromosomes do, lampbrush chromosomes often extend to lengths of 500–800 μm. Later in meiosis, they revert to their normal length of 15–20 μm. Based on these observations, lampbrush chromosomes are interpreted as extended, uncoiled versions of the normal meiotic chromosomes.

The two views of lampbrush chromosomes in Figure 17–10 provide significant insights into their morphology Part (a) shows the meiotic configuration under the light microscope. The linear axis of each structure contains a large number of condensed areas, and as with polytene chromosomes, these are referred to generally as chromomeres. Emanating from each chromomere is a pair of lateral loops, which give the chromosome its distinctive appearance. In part (b), the scanning electron micrograph reveals adjacent loops present along one of the two axes of the chromosome. As with bands in polytene chromosomes, much more DNA is present in each loop than is needed to encode a single gene. This SEM provides a clear view of the chromomeres and the chromosomal fibers emanating from them. Each chromosomal loop is thought to be composed of one DNA double helix, while the

FIGURE 17–9 Photograph of a puff within a polytene chromosome. The diagram depicts the uncoiling of strands within a band (B) region to produce a puff (P) in polytene chromosomes. Interband regions (IB) are also labeled.

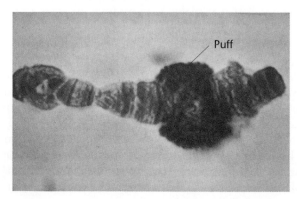

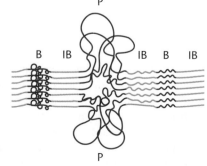

(b)

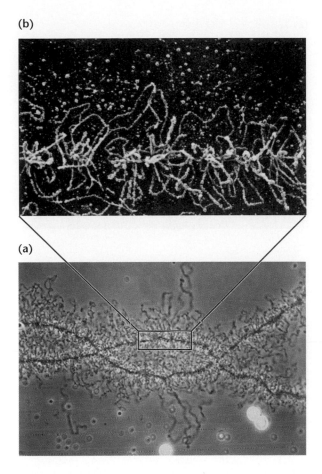

(a)

FIGURE 17–10 Lampbrush chromosomes derived from amphibian oocytes. Part (a) is a photomicrograph; part (b) is a scanning electron micrograph.

central axis is composed of two DNA helices. This hypothesis is consistent with the belief that each meiotic chromosome is composed of a pair of sister chromatids. Studies using radioactive RNA precursors reveal that the loops are active in the synthesis of RNA. The lampbrush loops, in a manner similar to puffs in polytene chromosomes, represent DNA that has been reeled out from the central chromomere axis during transcription.

17.4 Organization of Chromatin in Eukaryotes

The structure and organization of the genetic material in eukaryotic cells is much more intricate than in viruses or bacteria. This complexity is due to the greater amount of DNA per chromosome and the presence of a large number of proteins associated with eukaryotic DNA. For example, while DNA in the *E. coli* chromosome is 1200 μm long, the DNA in each human chromosome ranges from 19,000 to 73,000 μm in length. In a single human nucleus, all 46 chromosomes contain sufficient DNA to extend almost 2 meters. This genetic material, along with its associated proteins, is contained within a nucleus that usually measures about 5–10 μm in diameter.

Such intricacy parallels the structural and biochemical diversity of the many types of cells present in a multicellular eukaryotic organism. Different cells assume specific functions based on highly specific biochemical activity. While all cells carry a full genetic complement, different cells activate different sets of genes, so a highly ordered regulatory system governing the readout of this information must exist. Such a system must in some way be imposed on, or related to, the molecular structure of the genetic material.

Although bacteria can reproduce themselves, they never exhibit a complex process similar to mitosis. As described in Chapter 2, eukaryotic cells exhibit a highly organized cell cycle. During interphase, the genetic material and associated proteins are uncoiled and dispersed throughout the nucleus as **chromatin**. When mitosis begins, the chromatin condenses greatly, and during prophase it is compressed into recognizable chromosomes. This condensation represents a contraction in length of some 10,000 times for each chromatin fiber and is the basis of the **folded-fiber model** of chromosome structure (Figure 2–18). This highly regular cycle of condensation and uncoiling creates special organizational problems in eukaryotic genetic material.

Because of the limitation of light microscopy, early studies of the structure of eukaryotic genetic material concentrated on intact chromosomes, preferably large ones, such as the polytene and lampbrush chromosomes that we have discussed. Subsequently, new techniques for biochemical analysis, as well as the examination of relatively intact eukaryotic chromatin and mitotic chromosomes under the electron microscope, have greatly enhanced our understanding of chromosome structure.

Chromatin Structure and Nucleosomes

As we have seen, the genetic material of viruses and bacteria consists of strands of DNA or RNA nearly devoid of proteins. In eukaryotic chromatin, a substantial amount of protein is associated with the chromosomal DNA in all phases of the eukaryotic cell cycle. The associated proteins are divided into basic, positively charged **histones** and less positively charged **nonhistones**. Of the proteins associated with DNA, the histones clearly play the most essential structural role. Histones contain large amounts of the positively charged amino acids lysine and arginine, making it possible for them to bond electrostatically to the negatively charged phosphate groups of nucleotides. Recall that a similar interaction has been proposed for several bacterial proteins. There are five main types of histones (Table 17.4, page 356).

The general model for chromatin structure is based on the assumption that chromatin fibers, composed of DNA and protein, undergo extensive coiling and folding as they are condensed within the cell nucleus. X-ray diffraction studies confirm that histones play an important role in chromatin structure. Chromatin produces regularly spaced diffraction rings, suggesting that repeating structural units occur along the chromatin axis. If the histone molecules are chemically removed from chromatin, the regularity of this diffraction pattern is disrupted.

TABLE 17.4 Categories and Properties of Histone Proteins

Histone Type	Lysine-Arginine Content	Molecular Weight (Da)
H1	Lysine-rich	23,000
H2A	Slightly lysine-rich	14,000
H2B	Slightly lysine-rich	13,800
H3	Arginine-rich	15,300
H4	Arginine-rich	11,300

A basic model for chromatin structure was worked out in the mid-1970s. Several observations were particularly relevant to the development of this model.

1. Digestion of chromatin by certain endonucleases, such as micrococcal nuclease, yields DNA fragments that are approximately 200 bp in length, or multiples thereof. This demonstrates that enzymatic digestion is not random, for if it were, we would expect a wide range of fragment sizes. Thus, chromatin consists of some type of repeating unit, each of which is protected from enzymatic cleavage, except where any two units are joined. It is the area between units that is attacked and cleaved by the nuclease.

2. Electron microscopic observations of chromatin reveal that chromatin fibers are composed of linear arrays of spherical particles (Figure 17–11). Discovered by Ada and Donald Olins, the particles occur regularly along the axis of a chromatin strand and resemble beads on a string. These particles, initially referred to as *v*-bodies (*v* is the Greek letter nu), are now called **nucleosomes**. These findings conform nicely to the earlier observation, which suggests the existence of repeating units.

3. Studies of precise interactions of histone molecules and DNA in the nucleosomes constituting chromatin show that histones H2A, H2B, H3, and H4 occur as two types of tetramers, $(H2A)_2 \cdot (H2B)_2$ and $(H3)_2 \cdot (H4)_2$. Roger Kornberg predicted that each repeating nucleosome unit consists of one of each tetramer in association with about 200 bp of DNA. Such a structure is consistent with previous observations and provides the basis for a model that explains the interaction of histones and DNA in chromatin.

4. When nuclease digestion time is extended, some of the 200 bp of DNA are removed from the nucleosome, creating what is called a **nucleosome core particle** consisting of 146 bp. This number is consistent in all eukaryotes studied. The DNA lost in this prolonged digestion is responsible for linking nucleosomes. This linker DNA is associated with the fifth histone, H1.

5. On the basis of this information, as well as on X-ray and neutron-scattering analyses of crystallized core particles by John T. Finch, Aaron Klug, and others, a detailed model of the nucleosome was put forward in 1984.

(a)

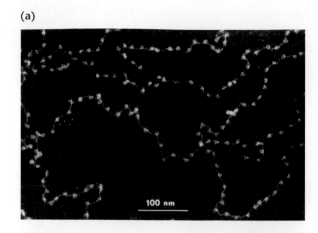

100 nm

(b)

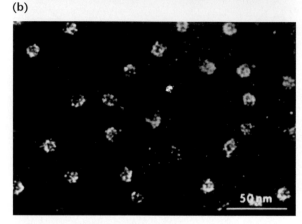

50 nm

FIGURE 17–11 (a) Dark-field electron micrograph of nucleosomes present in chromatin derived from a chicken erythrocyte nucleus. (b) Dark-field electron micrograph of nucleosomes produced by micrococcal nuclease digestion.

In this model, the 146-bp DNA core is coiled around an octamer of histones in a left-handed superhelix, which completes about 1.7 turns per nucleosome.

The extensive investigation of nucleosomes now provides the basis for predicting how the the chromatin fiber within the nucleus is formed, and how it coils up into a mitotic chromosome. This model is shown in Figure 17–12. The 2-nm DNA molecule is initially coiled into a nucleosome about 11 nm in diameter [Figure 17–12(a)], consistent with the longer dimension of the ellipsoidal nucleosome. Significantly, the formation of the nucleosome represents the first level of packing, whereby the DNA helix is reduced to about one-third of its original length.

In the nucleus, the chromatin fiber seldom, if ever, exists in the extended form described here. Instead, the 11-nm chromatin fiber is further packed into a thicker 30-nm fiber, which was initially called a **solenoid** [Figure 17–12(b)]. This larger fiber consists of numerous closely coiled nucleosomes, creating the second level of packing. The exact details of this structure are not completely clear, but 30-nm chromatin fibers are characteristically seen under the electron microscope.

In the transition to the mitotic chromosome, still another level of packing occurs. The 30-nm fiber forms a series of

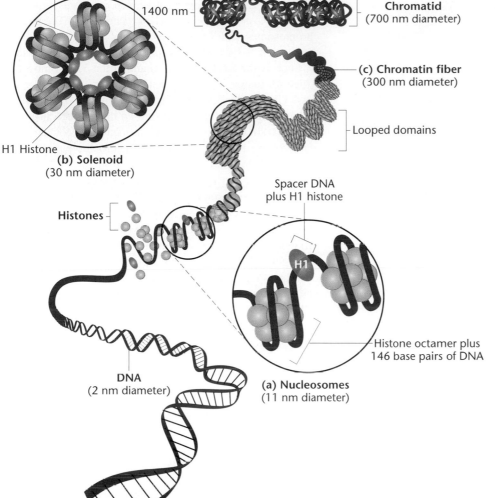

FIGURE 17–12 General model of the association of histones and DNA in the nucleosome, showing how the chromatin fiber can coil into a more condensed structure, ultimately producing a mitotic chromosome.

looped domains that further condense the structure into the chromatin fiber, which is 300 nm in diameter [Figure 17–12(c)]. The fibers are then coiled into the chromosome arms that constitute a chromatid, which is part of the metaphase chromosome [Figure 17–12(d)]. While we show the chromatid arms to be 700 nm in diameter, this value undoubtedly varies among different organisms. At a value of 700 nm, a pair of sister chromatids comprising a chromosome measures about 1900 nm.

More recent insights into the structure of the nucleosome occurred in 1997 when Timothy Richmond and members of his research team significantly improved the level of resolution in X-ray diffraction studies of nucleosome crystals (from 7 Å in the 1984 studies to 2.8 Å in the 1997 studies). At this resolution, most atoms are visible, thus revealing the subtle twists and turns of the superhelix of DNA that encircles the histones. One model based on their work is shown in Figure 17–13 (page 358). The double-helical ribbon represents 146 bp of DNA surrounding four pairs of histone proteins. This configuration is essentially repeated over and

over in the chromatin fiber and is the principal packaging unit of DNA in the eurkaryotic nucleus.

Further, details of the location of each histone entity have been revealed. Of particular interest is the observation that several unstructured histone tails are not packed into the folded histone domains within the core of the nucleosome. For example, tails devoid of any secondary structure extending from histones H3 and H2B protrude through the minor groove channels of the DNA helix. The significance of these and other histone tails is that they are seen to interact with neighboring structures, creating nucleosome to nucleosome interactions. These findings give us further insight into the structure of chromatin and will help unravel how various genetic functions such as transcription can occur on a DNA molecule that is complexed with histones.

The importance of the organization of DNA into chromatin and of chromatin into mitotic chromosomes is demonstrated in a human cell that stores its genetic material in a nucleus about 5–10 μm in diameter. The haploid genome contains 3.2×10^9 bp of DNA distributed among 23 chromosomes.

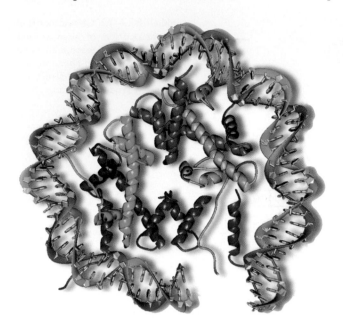

FIGURE 17–13 The nucleosome core particle derived from X-ray crystal analysis at 2.8-Å resolution. The double-helical DNA surrounds four pairs of histones.

The diploid cell contains twice that amount. At 0.34 nm per base pair, this amounts to a length of DNA of almost 2 m! One estimate suggests that about 25×10^6 nucleosomes per nucleus are complexed with this length of DNA.

In the transition from a fully extended DNA helix to the extremely condensed status of the mitotic chromosome, a packing ratio (the ratio of DNA length to the length of the structure containing it) of about 500 to 1 must be achieved. In fact, our model accounts for a ratio only of about 50 to 1. Obviously, the larger fiber can be further bent, coiled, and packed as even greater condensation occurs during the formation of a mitotic chromosome.

Heterochromatin

Evidence that the DNA of each eukaryotic chromosome consists of one continuous double-helical fiber along its entire length might suggest that the whole chromosome is structurally uniform. However, in the early part of this century, it was observed that some parts of the chromosome remain condensed and stain deeply during interphase, but most are uncoiled and do not stain. In 1928, the terms **heterochromatin** and **euchromatin** were coined to describe the parts of chromosomes that remain condensed and those that are uncoiled, respectively.

Subsequent investigation revealed a number of characteristics that distinguish heterochromatin from euchromatin. Heterochromatic areas are genetically inactive because they either lack genes or contain genes that are repressed. Also, heterochromatin replicates later during the S phase of the cell cycle than euchromatin does. The discovery of heterochromatin provided the first clues that parts of eukaryotic chromosomes do not always encode proteins. Instead, some chromosome regions are thought to be involved in maintenance of the chro-

mosome's structural integrity and in other functions, such as chromosome movement during cell division.

Heterochromatin is characteristic of the genetic material of eukaryotes. Early cytological studies showed that areas of the centromeres are composed of heterochromatin. The ends of chromosomes, called telomeres, are also heterochromatic. In some cases, in fact, whole chromosomes are heterochromatic. Such is the case with the mammalian Y chromosome, which for the most part is genetically inert. And, as we discussed in Chapter 5, the inactivated X chromosome in mammalian females is condensed into an inert heterochromatic Barr body. In some species, such as mealy bugs, all chromosomes of one entire haploid set are heterochromatic.

When certain heterochromatic areas from one chromosome are translocated to a new site on the same or another nonhomologous chromosome, genetically active areas sometimes become genetically inert if they now lie adjacent to the translocated heterochromatin. As we saw in Chapter 4, this influence on existing euchromatin is one example of what is more generally referred to as a **position effect**. That is, the position of a gene or groups of genes relative to all other genetic material may affect their expression.

17.5 DNA Sequence Organization in Eukaryotes

Thus far, we have looked at how DNA is organized into chromosomes in bacteriophages, bacteria, and eukaryotes. We now begin an examination of what we know about the organization of DNA sequences within the chromosomes making up an organism's genome, placing our emphasis on eukaryotes. Once we establish the pattern of genome organization, we'll focus on how genes themselves are organized within chromosomes in Chapter 18.

We learned in Chapter 10 that, in addition to single copies of unique DNA sequences that comprise genes, a great deal of the DNA sequences within chromosomes is repetitive in nature, and that various levels of repetition occur within the genome of organisms. Many studies have now provided insights into repetitive DNA, demonstrating various classes of these sequences and their organization within the genome. Figure 17–14 outlines the various categories of repetitive DNA. As you can see, some functional genes are present in more than one copy and are therefore repetitive in nature. However, the majority of repetitive sequences are nongenic and, in fact, most serve no known function. We explore three main categories: (1) heterochromatin found associated with centromeres and making up telomeres, (2) tandem repeats of both short and longer DNA sequences, and (3) transposable sequences that are interspersed throughout the genome of eukaryotes.

Repetitive DNA and Satellite DNA

The nucleotide composition of the DNA (e.g., the percentage of $G \equiv C$ versus $A = T$ pairs) of a particular species is reflected in its density, which can be measured with sedimen-

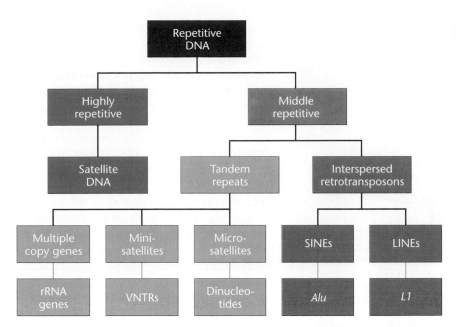

FIGURE 17–14 An overview of the various categories of repetitive DNA.

tation equilibrium centrifugation (introduced in Chapter 10). When eukaryotic DNA is analyzed in this way, the majority of it is present as a single main peak or band of fairly uniform density. However, one or more additional peaks represent DNA that differs slightly in density. This component, called **satellite DNA**, represents a variable proportion of the total DNA, depending on the species. For example, a profile of main-band and satellite DNA from the mouse is shown in Figure 17–15. By contrast, prokaryotes contain only main-band DNA.

The significance of satellite DNA remained an enigma until the mid-1960s, when Roy Britten and David Kohne developed a technique for measuring the reassociation kinetics of DNA that had previously been dissociated into single strands (see Chapter 10). They demonstrated that certain portions of DNA reannealed more rapidly than others. They concluded that rapid reannealing was characteristic of multiple DNA

fragments composed of identical or nearly identical nucleotide sequences—the basis for the descriptive term **repetitive DNA**.

When satellite DNA is subjected to analysis by reassociation kinetics, it falls into the category of *highly repetitive DNA*, which is known to consist of short sequences repeated a large number of times. Further evidence suggests that these sequences are present as tandem repeats clustered in very specific chromosomal areas known to be heterochromatic—the regions flanking *centromeres*. This was discovered in 1969 when several researchers, including Mary Lou Pardue and Joe Gall, applied the technique of *in situ* **molecular hybridization** to the study of satellite DNA. This technique (see Figure 10–15) involves the molecular hybridization between an isolated fraction of radioactively labeled DNA or RNA probes and the DNA contained in the chromosomes of a cytological preparation. Following the hybridization procedure, autoradiography is performed to locate the chromosome areas complementary to the fraction of DNA or RNA.

In their work, Pardue and Gall demonstrated that probes made from mouse satellite DNA hybridize with DNA of centromeric regions of mouse mitotic chromosomes (Figure 17–16, page 360). Several conclusions can be drawn. Satellite DNA differs from main-band DNA in its molecular composition, as established by buoyant density studies. It is also composed of short repetitive sequences. Finally, satellite DNA is found in the heterochromatic centromeric regions of chromosomes.

Centromeric and Telomeric DNA Sequences

Separation of chromatids is essential to the fidelity of chromosome distribution during mitosis and meiosis. Most estimates of infidelity during mitosis are exceedingly low: 1×10^{-5} to 1×10^{-6}, or 1 error per 100,000 to 1,000,000 cell divisions. As a result, it has been generally assumed that

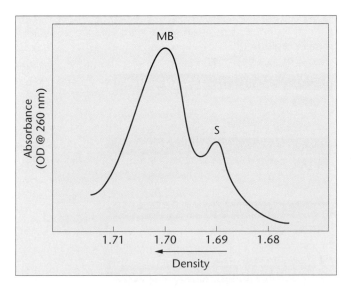

FIGURE 17–15 Separation of main-band (MB) and satellite (S) DNA from the mouse using ultracentrifugation in a CsCl gradient.

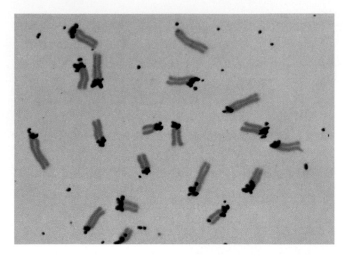

FIGURE 17–16 *In situ* hybridization between RNA transcribed from mouse satellite DNA and mitotic chromosomes. The grains in the autoradiograph localize the chromosome regions (the centromeres) containing satellite DNA sequences.

the analysis of the DNA sequence of centromeric regions will provide insights into the rather remarkable features of this chromosomal region. This DNA region is designated **CEN** and its function is now reasonably clear. Its structure within the heterochromatic region binds a platform of proteins forming the centromere, which includes the kinetechore that binds to the spindle fiber during division.

The analysis of the CEN regions of yeast *Saccharomyces cerevisiae* chromosomes provided the basis for a model system first described by John Carbon and Louis Clarke. Because each centromere serves an identical function, it is not surprising that all CENs were found to be remarkably similar in their organization. The CEN region of yeast chromosomes consists of about 225 bp, which can be divided into three regions (Figure 17–17). The first and third regions (I and III) are relatively short and highly conserved, consisting of only 8 and 26 bp, respectively. Region II, which is larger (80–85 bp) and extremely

A-T rich (up to 95%), varies in sequence among different chromosomes.

Mutational analysis suggests that regions I and II are less critical to centromere function than is region III. Mutations in the former regions are often tolerated, but mutations in region III can disrupt centromere function. While the DNA of this region appears to be essential to the eventual binding to the spindle fiber, DNA sequences are not unique to specific chromosomes. They can be exchanged between chromosomes experimentally without altering centromere function.

The amount of DNA associated with the centromeres of multicellular eukaryotes is much more extensive than in yeast. Recall from our discussion that highly repetitive "satellite" DNA is localized in the centromere regions of mice. Such sequences, absent from yeast but characteristic of most multicellular organisms, vary considerably in size. For example, the 10-bp sequence AATAACATAG is tandemly repeated many times in the centromeres of all four chromosomes of *Drosophila*. In humans, one of the most recognized satellite DNA sequences is the **alphoid family**. Found mainly in the centromere regions, alphoid sequences, each about 170 bp in length, are present in tandem arrays of up to 1 million base pairs. The role of this highly repetitive DNA in centromere function remains unclear.

The other prominent structural component of chromosomes is the **telomere**, found at the ends of linear chromosomes. Telomeres provide stability to the chromosome by rendering chromosome ends generally inert in interactions with other chromosome ends. It is thought that some aspect of the molecular structure of telomeres must be unique compared with most other chromosome regions.

As with centromeres, the analysis of telomeres was first approached by investigating the smaller chromosomes of simple eukaryotes, such as protozoans and yeast. The idea that all telomeres of all chromosomes in a given species might share a common nucleotide sequence has now been borne out.

FIGURE 17–17 Nucleotide sequence information derived from DNA of the three major centromere regions of chromosomes 3, 4, 6, and 11 of yeast.

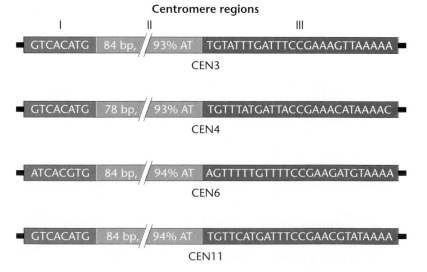

Two types of telomere sequences have been discovered. The first type, called simply **telomeric DNA sequences**, consists of short tandem repeats. It is this group that contributes to the stability and integrity of the chromosome. In the ciliate *Tetrahymena*, over 50 tandem repeats of the hexanucleotide sequence GGGGTT occur. In humans, the sequence GGGATT is repeated many times. The analysis of telomeric DNA sequences has shown them to be highly conserved throughout evolution, reflecting the critical role they play in maintaining the integrity of chromosomes.

The second type, **telomere-associated sequences**, is also repetitive and is found both adjacent to and within the telomere. These sequences vary among organisms, but their significance remains unknown.

Middle Repetitive Sequences: VNTRs and Dinucleotide Repeats

A brief review of still another prominent category of repetitive DNA sheds more light on our understanding of the organization of the eukaryotic genome. In addition to highly repetitive DNA, which constitutes about 5 percent of the human genome (and 10% of the mouse genome), a second category, **middle** (or **moderately**) **repetitive DNA**, recognized by C_0t analysis, is fairly well characterized. Because we are currently learning a great deal about the human genome, we will use our own species to illustrate this category of DNA in genome organization.

Middle repetitive DNA most prominently consists of either tandemly repeated or interspersed sequences. No function has been ascribed to these components of the genome. An example includes those called **variable number tandem repeats (VNTRs)**. The repeating DNA sequence of VNTRs can be 15–100 bp long and is found within and between genes. Many such clusters, often referred to as **minisatellites**, are dispersed throughout the genome.

The number of tandem copies of each specific sequence at each location varies in individuals, creating localized regions of 1000–5000 bp (1–5 kb) in length. As we shall see in Chapter 19, the variation in size (length) of these regions between individuals in humans is the basis for the forensic technique referred to as **DNA fingerprinting**.

Another group of tandemly repeated sequences is the dinucleotides, also referred to as **microsatellites**. Like VNTRs, they are dispersed throughout the genome and vary among individuals in the number of repeats present at any site. For example, in humans, the most common microsatellite is the dinucleotide $(CA)_n$, where *n* is the number of repeats. Most commonly, *n* is between 5 and 50. These clusters are also used forensically and serve as useful molecular markers during genome analysis.

Repetitive Transposed Sequences: SINEs and LINEs

Still another category of repetitive DNA consists of sequences that are interspersed throughout the genome, rather than being tandemly repeated. They can be either short or long and many have the added distinction of being **transposable sequences**, which are mobile and can move to different locations within the genome. A large portion of eukaryotic genomes are composed of such sequences.

For example, **short interspersed elements**, called **SINEs**, are less than 500 bp long and may be present 500,000 times or more in the human genome. The best characterized human SINE is a set of closely related sequences called the *Alu* **family** (the name is based on the presence of DNA sequences recognized by the restriction endonuclease *Alu*I). Members of this DNA family, also found in other mammals, are 200–300 bp long and are dispersed rather uniformly throughout the genome, both between and within genes. In humans, this family encompasses more than 5 percent of the entire genome.

Alu sequences are particularly interesting, although their function, if any, is yet undefined. Members of the *Alu* family are sometimes transcribed. The role of this RNA is not certain, but it may relate to their mobility in the genome. In fact, *Alu* sequences are thought to have arisen from an RNA element whose DNA complement was dispersed throughout the genome as a result of the activity of reverse transcriptase (an enzyme that synthesizes DNA on an RNA template).

The group of **long interspersed elements (LINEs)** represents still another category of repetitive transposable DNA sequences. In humans, the most prominent example is a family designated **L1**. Members of this sequence family are about 6400 bp long and are present up to 100,000 times, according to one estimate. Their 5′-end is highly variable, and their role has yet to be defined.

The basis for transposition of L1 elements is now clear. The L1 DNA sequence is first transcribed into an RNA molecule. The RNA then serves as the template for the synthesis of the DNA complement via the enzyme reverse transcriptase. This enzyme is encoded by a portion of the L1 sequence. The new L1 copy then integrates into the DNA of the chromosome at a new site. Because this mechanism of transposition resembles that used by retroviruses, LINEs are referred to as **retrotransposons**.

SINEs and LINEs represent a significant portion of human DNA. Both types of elements share the organizational feature of consisting of a mixture of about 70 percent unique and 30 percent repeating sequences within the DNA of each entity. Collectively, they constitute about 10 percent of the genome.

Middle Repetitive Multiple Copy Genes

In some cases, middle repetitive DNA includes functional genes present tandemly in multiple copies. For example, many copies exist of the genes encoding ribosomal RNA. *Drosophila* has 120 copies per haploid genome. Single genetic units encode a large precursor molecule that is processed into the 5.8*S*, 18*S*, and 28*S* rRNA components. In humans, multiple copies of this gene are clustered on the p arm of the acrocentric chromosomes 13, 14, 15, 21, and 22. Multiple copies of the genes encoding 5*S* rRNA are transcribed separately from multiple clusters found together on the terminal portion of the p arm of chromosome 1.

17.6   The Eukaryotic Genome: What Proportion Encodes Genes?

Given the information above involving various forms of repetitive DNA in eukaryotes, it is of interest to pose an important question: *What proportion of the eukaryotic genome actually encodes functional genes?* Taken together, the various forms of highly repetitive and moderately repetitive DNA comprise up to 40 percent of the human genome. Such an observation is not uncommon in eukaryotes. In addition to repetitive DNA, a large amount of single-copy DNA sequences as defined by C_0t analysis appears to be noncoding. A small portion of them are called **pseudogenes**, which represent evolutionary vestiges of duplicated copies of genes that have undergone sufficient mutations to render them untranscrible. In some cases, multiple copies of such genes exist.

While the proportion of the genome consisting of repetitive DNA varies among organisms, one feature seems to be shared: Only a very small part of the genome actually codes for proteins. For example, the 20,000–30,000 genes encoding proteins in sea urchin occupy less than 10 percent of the genome. In *Drosophila*, only 5–10 percent of the genome is occupied by genes coding for proteins. In humans, it appears that the estimated 30,000 functional genes occupies less than 5 percent of the genome.

The study of the various forms of repetitive DNA has significantly enhanced our understanding of genome organization. In the next chapter we explore the organization of genes within chromosomes.

Chapter Summary

1. The organization of the molecular components that form chromosomes is essential to understanding the function of the genetic material. Largely devoid of associated proteins, bacteriophage and bacterial chromosomes contain DNA molecules in a form equivalent to the Watson–Crick model.

2. Mitochondria and chloroplasts contain DNA that encodes products essential to their biological function. This DNA is remarkably similar in form and appearance to some bacterial and phage DNA, lending support to the endosymbiotic hypothesis, which suggests that these organelles were once free-living prokaryotic-like organisms.

3. Polytene and lampbrush chromosomes are examples of specialized structures that have extended our knowledge of genetic organization and function.

4. The eukaryotic chromatin fiber is a nucleoprotein organized into repeating units called nucleosomes. Composed of 146 bp of DNA and an octamer of four types of histones, the nucleosome facilitates the conversion of the extensive chromatin fiber characteristic of interphase into the highly condensed chromosome seen in mitosis.

5. The structural heterogeneity of the chromosome axis has been established as a result of both biochemical and cytological investigation. Heterochromatin, prematurely condensed in interphase, is genetically inert. The centromeric and telomeric regions, the Y chromosome, and the Barr body are examples.

6. DNA analysis reveals unique nucleotide sequences in both the centromere and telomere regions of chromosomes, which no doubt impart the characteristic they share of being heterochromatic.

7. Eukaryotic genomes demonstrate complex sequence organization characterized by numerous categories of repetitive DNA.

8. Repetitive DNA consists of either tandem repeats clustered in various regions of the genome or single sequences interspersed uniformly throughout the genome. In the former group, the size of each cluster varies among individuals, providing one form of biochemical identity. The latter group of sequences may be short or long, such as *Alu* and L1, respectively and are transposable elements.

9. The vast majority of a eukaryotic genome does not encode functional genes. In humans, for example, less than 5 percent of the genome is used to encode the 30,000 genes found in our genome.

Key Terms

φX174 bacteriophage, 349
alphoid family, 360
Alu family, 361
bacteriophage lambda, 349
CEN region, 360
chloroplast DNA (cpDNA), 353
chromatin, 355
chromomeres, 353
DNA fingerprinting, 361

DNA-binding proteins, 350
endosymbiotic hypothesis, 351
euchromatin, 358
folded-fiber model, 355
H and HU proteins, 350
heterochromatin, 358
histones, 355
in situ molecular hybridization, 359
L1, 361

lampbrush chromosomes, 354
lateral loops, 354
long interspersed elements (LINEs), 361
microsatellites, 361
middle repetitive DNA, 361
minisatellites, 361
mitochondrial DNA (mtDNA), 351
moderately repetitive DNA, 361
nucleoid, 350

Insights and Solutions

A previously undiscovered single-celled organism was found living at a great depth on the ocean floor. Its nucleus contained only a single linear chromosome containing 7×10^6 nucleotide pairs of DNA coalesced with three types of histonelike proteins. Consider the following questions:

1. A short micrococcal nuclease digestion yielded DNA fractions consisting of 700, 1400, and 2100 bp. Predict what these fractions represent. What conclusions can be drawn?

Solution: The chromatin fiber may consist of a variation of nucleosomes containing 700 bp of DNA. The 1400- and 2100-bp fractions respectively represent two and three linked nucleosomes. Enzymatic digestion may have been incomplete, leading to the latter two fractions.

2. The analysis of individual nucleosomes reveals that each unit contained one copy of each protein, and that the short linker DNA contained no protein bound to it. If the entire chromosome consists of nucleosomes (discounting any linker DNA), how many are there, and how many total proteins are needed to form them?

Solution: Since the chromosome contains 7×10^6 bp of DNA, the number of nucleosomes, each containing 7×10^2 bp, is equal to

$$7 \times 10^6 / 7 \times 10^2 = 10^4 \text{ nucleosomes}$$

The chromosome thus contains 10^4 copies of each of the three proteins, for a total of 3×10^4 molecules.

3. Analysis then revealed the organism's DNA to be a double helix similar to the Watson–Crick model, but containing 20 bp per complete turn of the right-handed helix. The physical size of the nucleosome was exactly double the volume occupied by that found in all other known eukaryotes, by virtue of increasing the distance along the fiber axis by a factor of 2. Compare the degree of compaction of this organism's nucleosome to that found in other eukaryotes.

Solution: The unique organism compacts a length of DNA consisting of 35 complete turns of the helix (700 bp per nucleosome/20 bp per turn) into each nucleosome. The normal eukaryote compacts a length of DNA consisting of 20 complete turns of the helix (200 bp per nucleosome/10 bp per turn) into a nucleosome one-half the volume of that in the unique organism. The degree of compaction is therefore less in the unique organism.

4. No further coiling or compaction of this unique chromosome occurs in the unique organism. Compare this to a eukaryotic chromosome. Do you think an interphase human chromosome 7×10^6 base pairs in length would be a shorter or longer chromatin fiber?

Solution: The eukaryotic chromosome contains still another level of condensation in the form of "solenoids," which are dependent on the H1 histone molecule associated with linker DNA. Solenoids condense the eukaryotic fiber by still another factor of 5.

The length of the unique chromosome is compacted into 10^4 nucleosomes, each with an axis length twice that of the eukaryotic fiber. The eukaryotic fiber consists of $7 \times 10^6 / 2 \times 10^2 = 3.5 \times 10^4$ nucleosomes, 3.5 more than the unique organism. However, they are compacted by the factor of 5 in each solenoid. Therefore, the chromosome of the unique organism is a longer chromatin fiber.

Problems and Discussion Questions

1. Compare and contrast the chemical nature, size, and form assumed by the genetic material of viruses and bacteria.
2. Contrast the DNA associated with mitochondria and chloroplasts.
3. How are giant polytene chromosomes formed?
4. What genetic process is occurring in a "puff" of a polytene chromosome? How do we know?
5. During what genetic process are lampbrush chromosomes present in vertebrates?
6. Why might we predict that the organization of eukaryotic genetic material will be more complex than that of viruses or bacteria?
7. Describe the sequence of research findings leading to the model of chromatin structure. What is the molecular composition and arrangement of the nucleosome? Describe the transitions that occur as nucleosomes are coiled and folded, ultimately forming a chromatid.
8. When chloroplasts and mitochondria are isolated, they are found to contain ribosomes that are similar to prokaryotic cells. How does this observation relate to the endosymbiotic hypothesis? What other evidence supports this hypothesis?
9. Provide a comprehensive definition of heterochromatin and list as many examples as you can.

10. Mammals contain a diploid genome consisting of at least 10^9 bp. If this amount of DNA is present as chromatin fibers where each group of 200 bp of DNA is combined with nine histones into a nucleosome, and each group of six nucleosomes is combined into a solenoid, achieving a final packing ratio of 50, determine the following:
    (a) The total number of nucleosomes in all fibers.
    (b) The total number of solenoids in all fibers.
    (c) The total number of histone molecules combined with DNA in the diploid genome.
    (d) The combined length of all fibers.

11. Assume that a viral DNA molecule is a 50 μm-long circular strand of a uniform 20 Å diameter. If this molecule is contained in a viral head that is a 0.08 μm-diameter sphere, will the DNA molecule fit into the viral head, assuming complete flexibility of the molecule? Justify your answer mathematically.

12. How many base pairs are in a molecule of phage T2 DNA, which is 52 μm long?

13. If a human nucleus is 10 μm in diameter, and it must hold as much as 2 m of DNA, which is complexed into nucleosomes that during full extension are 11 nm in diameter, what percentage of the volume of the nucleus is occupied by the genetic material?

Selected Readings

Angelier, N., et al. 1984. Scanning electron microscopy of amphibian lampbrush chromosomes. *Chromosoma* 89:243–53.

Beerman, W., and Clever, U. 1964. Chromosome puffs. *Sci. Am.* (Apr.) 210:50–58.

Callan, H. G. 1986. *Lampbrush chromosomes*. New York: Springer-Verlag.

Carbon, J. 1984. Yeast centromeres: Structure and function. *Cell* 37:352-53.

Corneo, G., et al. 1968. Isolation and characterization of mouse and guinea pig satellite DNA. *Biochemistry* 7:4373–79.

DuPraw, E. J. 1970. *DNA and chromosomes*. New York: Holt, Rinehart & Winston.

Gall, J. G. 1963. Kinetics of deoxyribonuclease on chromosomes. *Nature* 198:36–38.

————— 1981. Chromosome structure and the C-value paradox. *J. Cell Biol.* 91:3s–14s.

Green, B. R., and Burton, H. 1970. *Acetabularia* chloroplast DNA: Electron microscopic visualization. *Science* 168:981–82.

Hewish, D. R., and Burgoyne, L. 1973. Chromatin sub-structure. The digestion of chromatin DNA at regularly spaced sites by a nuclear deoxyribonuclease. *Biochem. Biophys. Res. Comm.* 52:504–10.

Hill, R. J., and Rudkin, G. T. 1987. Polytene chromosomes: The status of the band–interband question. *BioEssays* 7:35–40.

Jeffreys, A. J., Wilson, V., and Thein, S. L. 1985. Hypervariable minisatellite regions in human DNA. *Nature* 314:66-73.

Korenberg, J. R. and Rykowski, M. C. 1988. Human genome organization: *Alu*, LINES, and the molecular organization of metaphase chromosome bands. *Cell* 53:391–400.

Kornberg, R. D. 1975. Chromatin structure: A repeating unit of histones and DNA. *Science* 184:868–71.

Kornberg, R. D., and Klug, A. 1981. The nucleosome. *Sci. Am.* (Feb.) 244:52–64.

Lorch, Y., Zhang, N, and Kornberg, R. D. 1999. Histone octamer transfer by a chromatin-remodeling complex. *Cell.* 96:389–92.

Luger, K., et. al 1997. Crystal structure of the nucleosome core particle at 2.8 A resolution. *Nature* 389:251–56.

Moyzis, R. K. 1991. The human telomere. *Sci. Am.* (Aug) 265:48–55.

Olins, A. L., and Olins, D. E. 1974. Spheroid chromatin units (v bodies). *Science* 183:330–32.

————— 1978. Nucleosomes: The structural quantum in chromosomes. *Am. Sci.* 66: 704–11.

Singer, M. F. 1982. SINEs and LINEs: Highly repeated short and long interspersed sequences in mammalian genomes. *Cell* 28:433–34.

Van Holde, K. E. 1989. *Chromatin*. New York: Springer-Verlag.

Verma, R. S., ed. 1988. *Heterochromatin: Molecular and structural aspects.* Cambridge: Cambridge University Press.

Arabidopsis thaliana serves as a model system for molecular and developmental genetic investigations. *(J. Mylne & J. Botella, Department of Botany, University of Queensland, Australia)*

18
Genomics and Proteomics

CHAPTER CONCEPTS

Analysis of genomes has moved from the discovery and mapping of mutations to nucleotide sequencing of entire genomes. This direct approach has generated an array of computer-based and automated techniques called genomics that is changing the way geneticists study organisms. Several dozen genomes have been sequenced, and the study of this information is providing new insights into gene function and evolution. Accompanying this revolution in genetics is an emphasis on studying the set of proteins expressed in a cell under different conditions. This set of proteins is called the proteome, and proteomics provides a way of studying the functions and interactions of gene products in the inner workings of the cell.

ince the early years of the twentieth century, geneticists have worked to identify and map genes in organisms of interest. A wide range of organisms continue to be studied; however, emphasis has gradually been placed on a few organisms, such as *Drosophila*, maize, mice, bacteria and yeast. For mapping studies, geneticists developed a set of methods using two main approaches. The first approach looked for spontaneous mutations or mutations created using chemical or physical means. Once mutations were found, the second approach was the generation of genetic maps by linkage analysis. In some organisms, such as *Drosophila*, physical maps of genes' locations on chromosomes were also created. These approaches, developed 70–90 years ago, were efficient and widely used in genetics. The drawback of such approaches is that at least one mutation for each gene in the genome is required, and obtaining mutations is a difficult, labor-intensive endeavor. In addition, mutations often have a lethal phenotype, making it difficult or impossible to map the mutated gene.

Beginning in the mid-1980s, geneticists moved away from the classical approaches of mutagenesis and mapping, and began using recombinant DNA methods for genetic analysis. In this approach, a collection of clones called a genomic library is established. The clones are pieced together into overlapping sets, and assembled into genetic and physical maps that encompass the entire genome. In the final step, the clones are sequenced, with all genes in the genome identified by their nucleotide sequence. Collectively, these methods are called **genomics**. A number of genomic methods were discussed in Chapter 16, and more will be described in Chapter 19. **Proteomics** is the study of gene products encoded by a

genome, including a list of which genes are expressed, their time of expression, the type and extent of any posttranslational modification of the gene product, the function of the encoded protein, and its location in various cellular compartments.

In many cases, genomic analysis is a large-scale, labor-intensive endeavor, often requiring the coordinated effort of many laboratories. For this reason, geneticists have formed genome projects. One of the largest and best known of these efforts is the Human Genome Project (HGP). The **Human Genome Project** is a coordinated, international effort whose ultimate goal is to determine the sequence of the 3.2 billion base pairs in the haploid human genome. In the United States, a project to map and sequence the human genome was proposed in 1986, and in 1988 the National Institutes of Health and the U.S. Department of Energy created a joint committee to develop a plan for the project. The HGP got under way in 1990. Other countries, notably France, Britain, and Japan, began similar projects, all of which are now coordinated by an international organization, the Human Genome Organization (HUGO). Under the umbrella of the HGP, scientists are working on the human genome as well as on genomes of a number of organisms, including bacteria (*E. coli*), yeast (*S. cerevisiciae*), nematode (*C. elegans*), fruit fly (*D. melanogaster*), and the mouse (*M. musculus*). The timeline for these projects is shown in Figure 18–1.

Because sequencing our own genome has widespread implications, the HGP is also studying the ethical, legal, and social implications of genomic research. Issues under study include uses of genetic testing, privacy, and the fair use of genetic information in insurance, employment, and health care. The Ethical, Legal, and Social Implications (ELSI) Program

FIGURE 18–1 A timeline of genome projects funded by the National Institutes of Health (NIH), including the Human Genome Project. *(International Human Genome Sequencing Consortium. 2001. Initial sequencing and analysis of the human genome. Figure 1, p. 862. Nature 409:860–921.)*

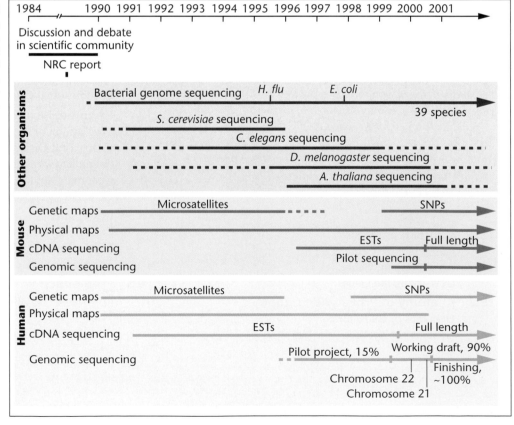

of the Human Genome Project addresses these issues and involves scientists, health professionals, policy makers, and the public in formulating policy recommendations and legislation. This program and its goals are discussed in Chapter 19.

In this chapter we outline the technology of genome analysis and review the significant findings from a variety of genome projects using prokaryotic and eukaryotic organisms. This review includes insights into the organization of genes in chromosomes, genome evolution, findings from comparative genomics, and the minimum number of genes necessary for life, as well as the evolution of gene families. We also discuss the emerging field of proteomics, and how geneticists are studying patterns of gene expression and the applications of this knowledge in various fields.

18.1 Genomic Analysis

Geneticists use two main methods for sequencing genomes (Figure 18–2, page 368). The **clone-by-clone method** was the first to be developed. It begins with the construction of genomic libraries of fragments covering all the genomic DNA (genomic clones) of an organism. Using genetic markers, overlapping clones are assembled to establish genetic and physical maps that encompass the entire genome. The nucleotide sequence is determined on a clone-by-clone basis until the entire genome is sequenced.

In the second method, called **shotgun cloning**, genomic libraries are prepared and randomly selected clones are sequenced until all clones in the library are analyzed. Assembler software organizes the nucleotide sequence information into a genome sequence. This method, developed by Craig Venter and his colleagues at The Institute for Genome Research (TIGR) was used to sequence the genome of *Haemophilus influenzae* in 1995, the first organism to have its genome completely sequenced. After refining the method and using it to sequence the genomes of other prokaryotes, Venter formed a new company, Celera, to sequence eukaryotic genomes, including *Drosophila* and humans. Using the shotgun method, Celera began a privately funded human genome project in September 1999 and completed the sequence in June 2000.

Compiling the Sequence

To ensure that the nucleotide sequence of a genome is complete and error-free, the genome is sequenced more than once. For example, using the shotgun method on the genome of the bacterium *Pseudomonas aeruginosa*, researchers sequenced the 6.3 million nucleotides seven times to ensure that the sequence was accurate and free from errors. Even with this level of redundancy, the assembler software recognized 1604 regions that required further clarification. These regions were reanalyzed and resequenced to improve accuracy. Finally, the accuracy of the shotgun method's sequence was compared to the sequence obtained by conventional cloning of two widely separated genomic regions using a high-capacity vector. The sequence of the 81,843 nucleotides cloned in the two

vectors was in perfect agreement with the sequence obtained by the shotgun method. This level of care is not unusual. Similar precautions are used in all genome projects.

The HGP plans to sequence the 3.2 billion base pairs of the human genome a total of 12 times. The privately based shotgun cloning project at Celera used a strategy of sequencing from both ends of DNA fragments, and covered the genome 35.6 times. Although a draft of the human genome sequence is finished, several other tasks are yet to be completed. These include obtaining the remaining sequence and correcting errors (proofreading the genome); filling sequence gaps (which amounted to about 150 Mb in mid-2001); and then sequencing the 15 percent of the genome that contains heterochromatin. Heterochromatic regions of the genome were excluded by design, as they contain long stretches of repetitive DNA sequences, and were initially thought to contain no genes. However, in sequencing the genome of *Drosophila*, researchers discovered that heterochromatic regions do contain a small number of genes (about 50 in *Drosophila*). As a result of this discovery, the heterochromatic regions of the human genome must now be sequenced to ensure that all genes are identified. Once the human genome or any other genome is sequenced, compiled, and proofread, the next stage—annotation—begins.

Annnotating the Sequence

After a genome sequence has been obtained, organized, and checked for accuracy, the next task is to find all the genes that encode proteins. This is the first step in **annotation**, a process that identifies genes, their regulatory sequences, and their function(s). Annotation also identifies the nonprotein coding genes (such as those for ribosomal RNA, transfer RNA, and small nuclear RNAs), finds and characterizes mobile genetic elements, and repetitive sequence families that may be present in the genome.

Finding the protein-coding genes is done by inspecting the sequence by eye or by computer. Several features are specific to genes. Protein-coding genes are composed of **open reading frames (ORFs)**, a series of nucleotides that specify an amino acid sequence. ORFs begin with an initiation sequence (usually ATG) and end with a termination sequence (TAA, TAG, or TGA). Scanning a DNA sequence for ORFs bordered by an ATG and followed by a termination codon is one strategy for finding genes. However, the task is complicated by the fact that a DNA sequence has six different ORFs: three on one strand, and three in the opposite direction on the complementary strand (coding is discussed in Chapter 12).

ORF scanning, usually by computer, is an effective method for annotating bacterial genomes (Figure 18–3, page 369). Eukaryotic genomes, including the human genome, have several properties that make this method less useful. First, many genes in eukaryotes are organized as exons, interrupted by introns. This means that many genes are not organized as continuous ORFs. As a result, scanning software often interprets exons as individual genes because termination codons are common features of introns. Second, as we shall see in a

FIGURE 18–2 The two genome sequencing strategies. (a) In the clone-by-clone method, a genomic library is prepared, and clones are organized into genetic and physical maps by observing the inheritance pattern of genetic markers in heterozygous families. After the clones are arranged into physical maps, they are broken into smaller, overlapping clones that cover each chromosome. Each smaller clone is sequenced, and the genomic sequence is assembled by stringing together the nucleotide sequence of the clones. (b). In the shotgun method, a genomic library is constructed from fragments of genomic DNA. Clones are selected from the library at random, and sequenced. The sequence is assembled by looking for sequence overlaps between clones from different libraries. This is usually done by computer, using assembler software designed for genomic analysis.

(a) CLONE-BY-CLONE METHOD

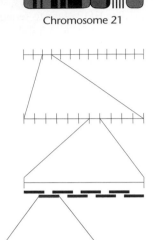

Chromosome 21

Genetic map of markers, such as RFLPs, STSs spaced about 1 Mb apart. This map is derived from recombination studies

Physical map with RFLPs, STSs showing order, physical distance of markers. Markers spaced about 100,000 base pairs apart

Set of overlapping ordered clones each covering 0.5–1.0 Mb

Each overlapping clone will be sequenced, sequences assembled into genomic sequence of 3.2×10^9 nucleotides

ATGCCCGATTGCAT

(b) SHOTGUN METHOD

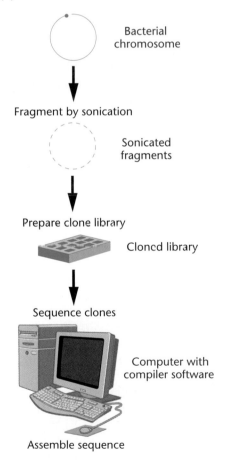

Bacterial chromosome

Fragment by sonication

Sonicated fragments

Prepare clone library

Cloned library

Sequence clones

Computer with compiler software

Assemble sequence

TAATAACCGGGCAGGCCATGTCTGCCCGTATTTCGCGTAAGGAAATCCATTATGTACTATTTAAAAAACACAAACTTTTG

GATGTTCGGTTTATTCTTTTTCTTTTACTTTTTTATCATGGGAGCCTACTTCCCGTTTTTCCCGATTTGGCTACATGACATCAACCATATCAGCAAAAGTGATACGGGTATTATTTTTGCCGC

TATTTCTCTGTTCTCGCTATTATTCCAACCGCTGTTTGGTCTGCTTTCTGACAAACTCGGGCTGCGCAAATACCTGCTGTGGATTATTACCGGCATGTTAGTGATGTTTGCGCCGTTC

TTTATTTTTATCTTCGGGCCACTGTTACAATACAACATTTTAGTAGGATCGATTGTTGGTGGTATTTATCTAGGCTTTTGTTTTAACGCCGGTGCGCCAGCAGTAGAGGCATTTATTGA

GAAAGTCAGCCGTCGCAGTAATTTCGAATTTGGTCGCGCGCGGATGTTTGGCTGTGTTGGCTGGGCGCTGTGTGCCTCGATTGTCGGCATCATGTTCACCATCAATAATCAGTTTG

TTTTCTGGCTGGGCTCTGGCTGTGCACTCATCCTCGCCGTTTTACTCTTTTTCGCCAAAACGGATGCGCCCTCTTCTGCCACGGTTGCCAATGCGGTAGGTGCCAACCATTCGGCAT

TTAGCCTTAAGCTGGCACTGGAACTGTTCAGACAGCCAAAACTGTGGTTTTTGTCACTGTATGTTATTGGCGTTTCCTGCACCTACGATGTTTTTGACCAACAGTTTGCTAATTTCTTTA

CTTCGTTCTTTGCTACCGGTGAACAGGGTACGCGGGTATTTGGCTACGTAACGACAATGGGCGAATTACTTAACGCCTCGATTATGTTCTTTGCGCCACTGATCATTAATCGCATCGGT

GGGAAAAACGCCCTGCTGCTGGCTGGCACTATTATGTCTGTACGTATTATTGGCTCATCGTTCGCCACCTCAGCGCTGGAAGTGGTTATTCTGAAAACGCTGCATATGTTTGAAGTACC

GTTCCTGCTGGTGGGCTGCTTTAAATATATTACCAGCCAGTTTGAAGTGCGTTTTTCAGCGACGATTTATCTGGTCTGTTTCTGCTTCTTTAAGCAACTGGCGATGATTTTTATATGTCG

TACTGGCGGGCAATATGTATGAAAGCATCGGTTTCCAGGGCGCTTATCTGGTGCTGGGTCTGGTGGCGCTGGGCTTCACCTTAATTTCCGTGTTCACGCTTAGCGGCCCCGGCCCG

CTTTCCCTGCTGCGTCAGGTGAATGAAGTCGCTTAAGCAATCAATGTCGGATGCGG

FIGURE 18–3 ORF analysis of the *lacY* gene of *E. coli*. The gene's ORF is underlined in green. Other ORFs longer than 50 codons are marked in red. Most genes are much longer than 50 codons, and ORF length is one indication of an actual gene.

subsequent section, genes in humans and other eukaryotes are often widely spaced, increasing the chances of finding false ORFs (over 70% of the human genome is composed of intergenic spacer DNA).

Recent versions of ORF scanning software for eukaryotic genomes make scanning more efficient. These programs search for such features as codon bias, intron/exon junctions, upstream regulatory sequences, the 3′ AATAAA poly A signal, and in some genomes, CpG islands. CpG islands are regions of DNA containing many copies of this nucleotide doublet. CpG islands are found in gene-rich regions of the mammalian genome, often adjacent to genes. Codon bias is the selective use of one or two codons to encode amino acids that can be encoded by a number of different codons. For example, alanine can be encoded by GCA, GCT, GCC, and GCG, yet in the human genome, GCC is used 41 percent of the time, and GCG only 11 percent of the time. Codon bias is present in exons, but should not be present in introns or intergenic spacers. However, searching for codon bias and other features of eukaryotic genes is not foolproof, and scanning by eye is a common backup technique.

Classifying the Genes

After a genomic sequence is annotated, the predicted ORFs are examined one at a time in a homology search. This analysis has several parts: searching databases such as GenBank to find similar genes isolated from other organisms; comparing the predicted ORFs with those from well-characterized bacterial genes; and lastly, looking for functional motifs, regions of DNA that encode protein domains such as ion channels, DNA binding regions, or secretion/export signals. In addition, researchers examine the ORFs to identify genes with specialized functions that may not be present in other organisms. Table 18.1 (page 370) shows the results of an ORF analysis in the *Pseudomonas aeruginosa* genome; the ORFs are classified on a scale according to confidence level. As in other bacterial genomes sequenced to date, about half the ORFs have no known or proposed function (confidence level 4). The rest of the ORFs are assigned to functional classes 1–3 (Table 18.2, page 370). The remaining task in the analysis of the *P. aeruginosa* genome is to identify functions for the rest of the genes, and to understand how these genes and the known genes interact in the biology of *P. aeruginosa*.

18.2 Anatomy of Prokaryotic Genomes

The sequencing of several dozen genome projects is now complete for both prokaryotes and eukaryotes. The results provide insights into genome anatomy, and indicate that

TABLE 18.1   Analysis of the *Pseudomonas aeruginosa* Genome

General Features

Genome size (bp)	6,264,403
G+C content	66.6%
Coding regions	89.4%

Coding Sequences

Confidence level	ORFs	(%)	Definition
1	372	(6.7)	*P. aeruginosa* genes with demonstrated function
2	1,059	(19.0)	Strong homologs of genes with demonstrated function from other organisms
3	1,590	(28.5)	Genes with proposed function based on motif searches or limited homology
4	769	(13.8)	Homologs of reported genes of unknown function
4	1,780	(32.0)	No homology to any reported sequences
Total	5,570	(100)	

Source: Stover, et al., 2000. Complete genome sequence of *Pseudomonas aeruginosa* PA01, an opportunistic pathogen. *Nature* 406:959–964. Table is derived from Table 1, p. 961.

TABLE 18.2   Functional Classes of Predicted Genes in *Pseudomonas*

	ORFs	
Functional Class	*Number*	*%*
Adaptation, protection (for example cold shock proteins)	60	1.1
Amino acid biosynthesis and metabolism	150	2.7
Antibiotic resistance and susceptibility	19	0.3
Biosynthesis of cofactors, prosthetic groups and carriers	119	2.1
Carbon compound catabolism	130	2.3
Cell division	26	0.5
Cell wall	83	1.5
Central intermediary metabolism	64	1.1
Chaperones & heat shock proteins	52	0.9
Chemotaxis	43	0.8
DNA replication, recombination, modification and repair	81	1.5
Energy metabolism	166	3.0
Fatty acid and phospholipid metabolism	56	1.0
Membrane proteins	7	0.1
Motility & attachment	65	1.2
Nucleotide biosynthesis and metabolism	60	1.1
Protein secretion/export apparatus	83	1.5
Putative enzymes	409	7.3
Related to phage, transposon or plasmid	38	0.7
Secreted factors (toxins, enzymes, alginate)	58	1.0
Transcription, RNA processing and degradation	45	0.8
Transcriptional regulators	403	7.2
Translation, post-translational modification, degradation	149	2.7
Transport of small molecules	555	10.0
Two-component regulatory systems	118	2.1
Hypothetical	1,774	31.8
Unknown (conserved hypothetical)	757	13.6
Total	5,570	100

Source: Stover, et al., 2000. Complete genome sequence of *Pseudomonas aeruginosa* PA01, an opportunistic pathogen. *Nature* 406:959–964.

TABLE 18.3   Genome Size and Gene Number in Selected Prokaryotes

	Genome Size (Mb)	Number of Genes
Archaea		
Archaeoglobus fulgidis	2.17	2,493
Methanococcus jannaschii	1.66	1,813
Thermoplasma acidophilum	1.56	1,509
Eubacteria		
Escherichia coli	4.64	4,397
Bacillus subtilis	4.21	4,212
Haemophilus influenzae	1.83	1,791
Aquifex aeolicus	1.55	1,552
Rickettsia prowazekii	1.11	834
Mycoplasma pneumoniae	0.82	710
Mycoplasma genitalium	0.58	503

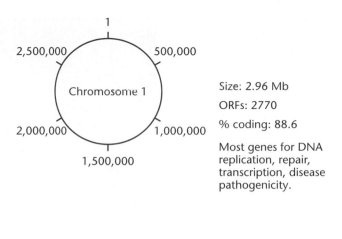

Size: 2.96 Mb

ORFs: 2770

% coding: 88.6

Most genes for DNA replication, repair, transcription, disease pathogenicity.

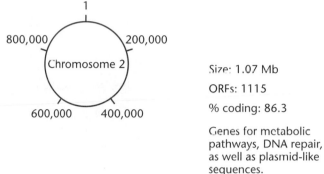

Size: 1.07 Mb

ORFs: 1115

% coding: 86.3

Genes for metabolic pathways, DNA repair, as well as plasmid-like sequences.

FIGURE 18–4 The *Vibrio cholerae* genome is contained in two chromosomes. The larger chromosome (chromosome 1) contains most of the genes for essential cellular functions and infectivity. Most of the genes on chromosome 2 (52% of 1115) are of unknown function. The bias in gene content and the presence of plasmid-like sequences on chromosome 2 suggest that this chromosome was a mega-plasmid captured by an ancestral *Vibrio* species.

genome organization in prokaryotes and eukaryotes differs somewhat. Consequently, we consider the anatomy of these genomes separately. Because prokaryotes have a small genome amenable to shotgun cloning and sequencing, more than 50 genome projects have been completed for two types of prokaryotes, eubacteria and archaea (Table 18.3), with more than 200 additional projects underway. The prokaryotic genomes sequenced include those that cause human diseases such as cholera, tuberculosis, and genital herpes.

Genomes of Eubacteria

Based on genome project results, we can make a number of generalizations concerning the size and organization of bacterial genomes (eubacteria, or true bacteria). The bacterial genome is usually thought of as relatively small (less than 5 Mb), and organized as a single circular DNA molecule. This traditional viewpoint is now being modified by the new data. Although most prokaryotic genomes are small, sizes in prokaryotic genomes vary widely with some overlap between the larger bacterial genomes (30 Mb in *Bacillus megaterium*) and the smaller eukaryotic genomes (12.1 Mb in yeast).

The majority of prokaryotic genomes examined to date are in fact circular DNA molecules, but an increasing number of genomes organized as linear DNA molecules are being identified, including *Borrelia burgdorferi*, the organism that causes Lyme disease. In addition, some species of *Streptomyces* have linear genomes.

Perhaps more importantly, new findings on plasmids are redefining our view of the bacterial genome as a single, circular DNA molecule. Most plasmids carry nonessential genes, can be transferred from one cell to another, and the same plasmid is often present in bacteria from different species, all of which suggests that plasmid genes should not be included as part of a bacterial genome. However, *B. burgsdorferi* carries approximately 17 plasmids, which contain at least 430 genes, and some appear to be essential, including

genes for purine biosynthesis and membrane proteins. Recently, sequencing of the genome of *Vibrio cholerae*, the organism responsible for cholera, revealed the presence of two circular chromosomes (Figure 18–4). Chromosome 1 (2.96 Mb) contains 2770 ORFs, and chromosome 2 (1.07 Mb) contains 1115 ORFs, some of which encode essential genes, such as ribosomal proteins. Chromosome 2 apparently derives from a plasmid captured by an ancestral species, since two chromosomes are present in other species of *Vibrio*. The existence of multichromosomal prokaryotic genomes, where one chromosome is plasmid-derived, raises several important questions. For example, when should a plasmid be considered a chromosome, and what regulatory mechanisms control gene expression and metabolism in a multichromosome prokaryotic system? Answers to these questions as well as the findings presented above may redefine many of our ideas about prokaryotic genomes, and provide clues about the evolution of multichromosome eukaryotic genomes.

We can make a number of generalizations concerning the organization of protein-coding genes in bacterial chromosomes. First, gene density is very high, averaging about one gene per kilobase pair of DNA. *E. coli* has a large genome (4.6 Mb), with 4288 protein-coding genes, a density of

FIGURE 18–5 The *E. coli* genome. The origin and terminus of replication is shown. The outer circle of bars represents genes transcribed in a clockwise direction, and the inner circle represents genes transcribed in a counterclockwise direction. (*Dr. Fred Blattner, Laboratory of Genetics, University of Wisconsin–Madison*)

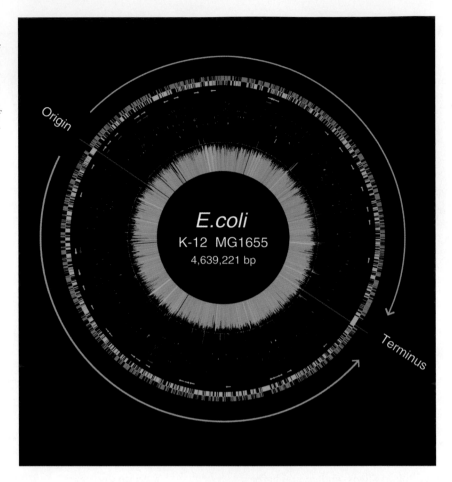

almost one gene per kilobase pair (Figure 18–5). *M. genitalium* is an example of a bacterium with a small genome (0.6 Mb). It contains 503 genes, also almost one gene per kilobase pair. This close packing of genes results in a very high proportion of the DNA (approximately 85–90%) serving as coding DNA. In this organism, the average amount of DNA between genes is only 110–125 bp. Typically, less than 1 percent of bacterial DNA is noncoding DNA, usually in the form of transposable elements, that can move from one place to another in the genome (discussed in Chapter 17). Introns are extremely rare in bacteria.

A second generalization we can make is that bacterial genomes are characterized by the presence of operons (operons are discussed in Chapter 15). In *E. coli*, 27 percent of the predicted transcription units are in operons (almost 600 operons).

In *Aquifex aeolicus*, most of the genes are in polycistronic transcription units we would usually call operons. However, the definition that operons contain genes for syn-

thesizing enzymes that control a single biochemical pathway does not hold true in this organism. In *A. aeolicus* (Figure 18–6), one operon contains six genes: two for DNA recombination, one for lipid synthesis, one for nucleic acid synthesis, and one for protein synthesis, and a protein for cell motility. This finding and results of related genome projects provide new insights into the nature of operons and their role in biochemical regulation in bacterial cells, and may redefine our concept of the operon.

Figure 18–7 shows a small portion of the *E. coli* gene map. The lines above the genes show how those genes are organized into transcription units with the promoters that locate the start of transcription for each unit. Some of these units are organized as operons. Note that there are three cases of overlapping genes, where one gene is nested inside another on the same DNA strand (shown in green). Overlapping genes are common in viruses (where space is at a premium) but are uncommon in bacteria.

FIGURE 18–6 An operon from the *A. aeolicus* genome. This operon contains genes for protein synthesis (*gatC*), for DNA recombination (*recA* and *recJ*), for a motility protein (*pilU*), for nucleotide biosynthesis (*cmk*), and for lipid biosynthesis (*pgsA1*). This organization challenges the conventional idea that genes in an operon encode products that control a common biochemical pathways.

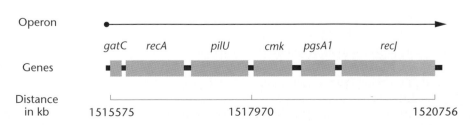

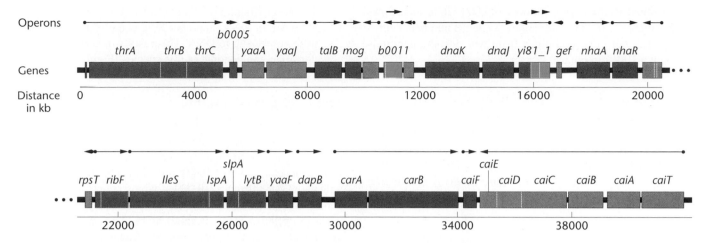

FIGURE 18–7 A portion of the *E. coli* chromosome showing genes and operons. A dot indicates the promoter for each gene or operon. Arrows and color indicate the direction of transcription: dark blue genes are transcribed left to right, light blue are transcribed right to left. Overlapping genes are shown in green.

Genomes of Archaea

Archaea (formerly known as archaebacteria) are one of the three major divisions of living organisms. The other two are the eubacteria (true bacteria, without a membrane-bound nucleus, such as *Haemophilus* and *Mycoplasma*) and eukarya (containing a true membrane-bound nucleus). Archaea, like the eubacteria, are prokaryotes; that is, they have no nucleus. They have only recently been recognized as a separate kingdom. Carl Woese proposed the original classification of Archaea as a separate kingdom in 1977, based on nucleotide sequence of the DNA encoding ribosomal RNA (rDNA). This classification was later confirmed by protein sequences and metabolic pathway analysis.

Archaea are typically extremophiles; they live in extreme conditions of very high temperature, high salt, high pressure, and/or extreme pH. Even though Archaea have an overall cellular resemblance to eubacteria, it was recognized early on that in many aspects of metabolism they more closely resemble eukaryotes. With the completion of several genome projects from the Archaea, we can now examine this relationship more closely.

The sequence of the archaeon, *Methanococcus jannaschii*, was completed in 1996. It has a circular, double-stranded DNA genome of 1.7 Mb, that contains 1738 protein-coding genes. This organism was originally isolated from a high-temperature deep-sea vent 2600 meters below sea level. It has an optimum growth temperature of 85°C and can survive up to 94°C (water boils at 100°C). The *M. jannaschii* genome consists of three chromosomes: a large, circular chromosome of 1.66 Mb and two small, circular chromosomes of 58.4 and 16.5 kb. Most genes in this organism, 58 percent, do not match any other known genes. The majority of genes involved in energy production, cell division, and general metabolism closely resemble those of eubacteria. Gene organization also resembles eubacteria; they are densely packed, have operons, and do not have introns.

On the other hand, archaeal genes also have significant similarities to eukaryotic genes. The genes involved in RNA syn-

thesis, protein synthesis, and DNA synthesis more closely resemble those in eukaryotes. Most surprising is the presence of histone chromosomal proteins and evidence that chromosomal DNA is organized into chromatin. Although they have no introns in their protein-coding genes, archaea do have introns in their tRNA genes, as do eukaryotes.

18.3 Anatomy of Eukaryotic Genomes

Eukaryotic nuclear genomes are usually divided into a series of linear DNA molecules, each contained in an individual chromosome. In addition, eukaryotes have a second genome, the mitochondrial genome, present as a circular DNA molecule. Plants and other photosynthetic organisms also have a third genome, present in the chloroplast as a circular DNA molecule. Our focus in this chapter is the nuclear genome. Although the basic features of the eukaryotic genome are similar in different species, genome size is highly variable in eukaryotes (Table 18.4). Genome sizes range

TABLE 18.4 Genome Size and Gene Number in Selected Eukaryotes

Organism	Genome Size (Mb)	Number of Genes
S. cerevisiae (yeast)	12	6,548
P. falciparum (malaria)	30	~6,500
C. elegans (nematode)	97	>20,000
A. thalania (mustard plant)	120	~20,000
D. melanogaster (fruit fly)	170	~16,000
O. sativa (rice)	415	~20,000
Z. mays (maize)	2,500	~20,000
H. sapiens (human)	3,300	~35,000
H. vulgare (barley)	5,300	~20,000

from about 10 Mb in fungi to over 100,000 Mb in some flowering plants (a 10,000-fold range), while the number of chromosomes per genome ranges from two into the hundreds (about a 100-fold range).

General Features of the Eukaryotic Genome

Compared with prokaryotes, eukaryotes have a relatively low gene density as shown in Figure 18–8. In general, more complex eukaryotes have less compact genomes, with lower levels of gene density. If we compare a region of the yeast genome from chromosome III [Figure 18–8(b)] and a region of the human genome from chromosome 7 [Figure 18–8(c)], several features are apparent.

- **Gene density.** This 50-kb region of yeast chromosome III contains over 20 genes, while the 50-kb segment of human chromosome 7 contains only 6 genes.

- **Introns.** While not shown in Figure 18–8, none of the yeast genes contain introns, while each human gene contains introns. In fact, the entire yeast genome has only 239 introns, while some single genes in the human genome contain over 100 introns.

- **Repetitive sequences.** The presence of introns and repetitive sequences are two major reasons for the wide range of genome sizes in eukaryotes. In some plants, such as maize, repetitive sequences are the dominant feature of the genome (repetitive sequences are dis

cussed in Chapter 17). The maize genome has about 5000 Mb of DNA, and over 80 percent of this is composed of repetitive DNA. This results in very low gene density in the genome of this species [Figure 18–8(d)].

The *C. elegans* Genome: A Surprising Organization

Although eukaryotic genomes share many features, specific genomic features vary widely. The genome of the nematode *C. elegans* contains 97 Mb organized into 6 chromosomes, with about 20,000 genes, 19,099 of which are protein-coding genes. The gene density is much lower than in yeast—three times more genes and eight times more DNA, with an average density of about 1 gene per 5 kb. About half the *C. elegans* genome is composed of intergenic spacer DNA, and a significant fraction of the genome is middle-repetitive DNA and highly repetitive simple-sequence DNA (sequences such as ATTAT repeated tens of thousands of times).

The arrangement of *C. elegans* genes differs dramatically from most eukatyotes. Approximately 25 percent of *C. elegans* genes are organized into polycistronic transcription units, or operons, like those in bacteria (Figure 18–9). Compared with yeast, introns are far more prevalent in *C. elegans* genes; the average gene has five introns. In fact, 26 percent of the genome is introns. In addition, *C. elegans* genes are commonly found within introns of other genes (Figure 18–10).

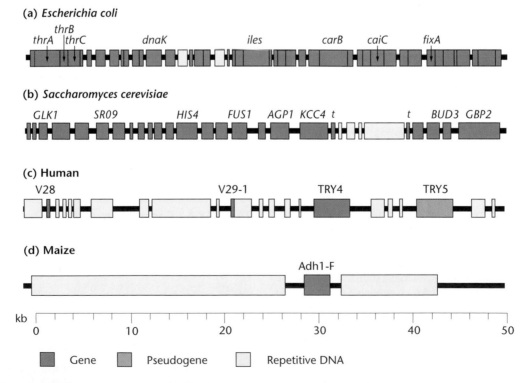

FIGURE 18–8 Gene density in four organisms. (a) A 50-kb region at the beginning of the *E. coli* genome. (b) A region from yeast chromosome III. (c) A 50-kb region from human chromosome 7 encoding a cell surface receptor. (d) The region of the maize genome surrounding the *AdhI-F* gene.

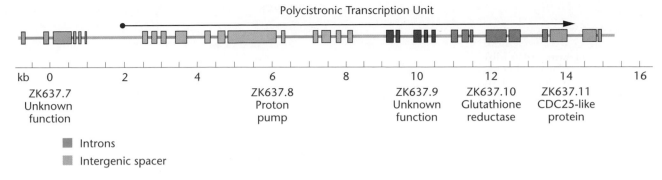

FIGURE 18–9 A region on chromosome III of *C. elegans* showing the location and organization of five genes. The central genes, *ZK637.8*, *ZK637.9*, and *ZK637.10* are part of an operon, and are transcribed as a set to form a polycistronic mRNA. Operons are common in bacterial genomes, but rare in most eukaryotic genomes. In *C. elegans*, however, about 25 percent of all genes are part of operons.

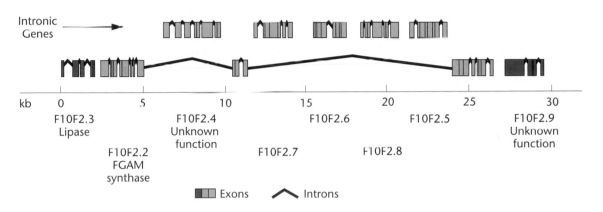

FIGURE 18–10 A 30-kb section of the *C. elegans* genome showing the location and organization of eight genes. One gene, *F10F2.3*, which encodes a lipase, spans just over 25 kb and contains two introns. Five other genes (*F10F2.4*, *F10F2.7*, *F10F2.6*, *F10F2.8*, and *F10F2.5*) are located in the introns of the lipase gene.

Genome Organization in Higher Plants

To geneticists, the small flowering plant *Arabidopsis thaliana* is the *Drosophila* of the plant world. Its very small genome, 120 Mb distributed in five chromosomes, is comparable to *C. elegans* and *Drosophila* in size and gene number. It contains an estimated 20,000 genes with a gene density of 1 gene per 5 kb, which also resembles *C. elegans* and *Drosophila*. In fact, at least half its genes are closely related to genes found in bacteria and humans. Because of its compact genome, *Arabidopsis* is a model organism for studying other plants with larger genomes, such as the economically important grasses, rice, maize and barley.

Although the sequencing of the *Arabidopsis* genome is not yet complete, analysis of the nucleotide sequence of chromosomes 2 and 4 of *Arabidopsis* reveals a dynamic genome. Both chromosomes have many tandem gene duplications (239 duplications on chromosome 2, involving 539 genes) and larger duplications, involving four blocks of DNA sequences, spanning 2.5 Mb. There is some interchromosomal duplication, with 37 genes on chromosome 4 also present on chromosome 5. This emphasizes the role of genomic and gene duplications in evolution, which we discuss in a later section.

Other plants such as maize and barley have genomes that are more than an order of magnitude larger than *Arabidopsis* but have about the same number of genes (Table 18.4). These large-genome plants are characterized by genes clustered in long stretches of DNA that are separated by even longer stretches of intergenic DNA (Figure 18–11, page 376). The gene clusters are collectively known as the gene space, and occupy only 12–24 percent of the genomic DNA. In the case of maize, the intergenic sequences are composed mainly of transposons (see Chapter 17). Because plants closely related to *Arabidopsis* have large genomes, the small genome of *Arabidopsis* is thought result from a genomic contraction, in which almost all of the gene-empty regions disappeared, along with most of the transposon sequences. In addition, there may have been a reduction in the size and number of introns, accounting for the compact genome size in this plant.

Organization of the Human Genome: The Human Genome Project

The public and private genome projects announced jointly in June 2000 that they had completed a draft sequence of the human genome, and in February 2001, they each published an analysis, which covers about 96 percent of the

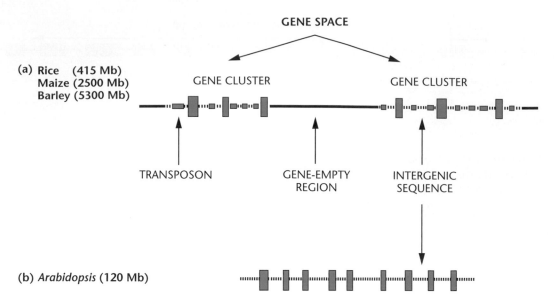

FIGURE 18–11 Genome organization in (a) several large-genome plants and (b) the compact genome of *Arabidopsis*. In (a), genes (vertical boxes) are located in clusters, separated by long, gene-empty spaces of repetitive DNA sequences. Within the gene clusters, the intergenic spaces contain many transposons (small horizontal boxes). In the genome of *Arabidopsis* (b), there is a loss of gene-empty regions, and a reduction or loss of transposable elements. The result is a much smaller genome with genes at a much higher density throughout the genome. (*Barakat, et al. 1998. Figure 4, p. 10048. Proc. Nat. Acad. Sci. USA 95:10044–10049.*)

euchromatic region of the genome. Much work remains to be done in completing the sequence, filling in gaps, and in the analysis of the vast amount of data gathered in this project. The human genome is about 25 times larger than the genome of any organism sequenced to date, and is the first vertebrate genome to be completed. In this section, we will present a summary of the major features of the genome analysis completed to date.

Although the human genome contains over 3 billion nucleotides, the DNA sequences that encode proteins make up only about 5 percent of the genome. At least 50 percent of the genome is derived from transposable elements, such as LINE and Alu sequences. The genes are distributed over 24 chromosomes in clusters of gene-rich regions, separated by gene-poor regions, often referred to as gene deserts. The gene-poor regions correlate with the G bands seen in karyotypes. Chromosome 19 has the highest gene density, and chromosomes 13 and the Y chromosome have the lowest gene density. The chromosomal organization of genes in the human genome is discussed in the following section.

Not all the genes in the genome have been identified, but it appears that humans have 30,000 to 40,000 genes, a number much lower than the previous estimate of 80,000 to 100,000. The average gene size, including introns and exons, is about 27 kb. The number of introns in human genes ranges from zero (histone genes) to 234 (titin, a gene that encodes a muscle protein). An unexpected finding is that hundreds of genes have been transferred directly from bacteria into vertebrate genomes, including our own, by a mechanism that remains unknown. Functions have been assigned to about 60 percent of the identified genes (Figure 18–12).

Chromosomal Organization of the Human Genome

Although sequencing is not complete, the human genome appears to be a typical eukaryotic genome. About 3000 Mb in length, it is organized into 24 chromosomes, which range from about 55 Mb to 250 Mb. It was originally estimated that the human genome might contain between 80,000 and 100,000 genes. More recently, results (discussed below) from chromosome sequencing indicate that there may only be between 35,000 and 40,000 genes in the genome. Human genes tend both to be larger and to contain more and larger introns than the genes and introns in invertebrate genomes such as *Drosophila*. The largest human gene known so far is the gene encoding dystrophin. This gene, associated in mutant form with muscular dystrophy, is 2.5 Mb in length and is larger than many bacterial chromosomes. Most of the transcription unit in this gene is composed of introns. It is not uncommon to find 30, 40 or even 50 introns in some human genes.

Sequencing of the long arms of human chromosomes 21 and 22 is now complete, and the genetic landscape of these two chromosomes reveals some interesting features (Figure 18–13). One of the most interesting findings is the difference in gene density on these two chromosomes. The long arm of chromosome 21 is about 33.65 Mb, and contains 225 genes (about 1 gene per 150 kb of DNA), while the long arm of chromosome 22 is about 34.65 Mb, and has 541 genes (about 1 gene per 64 kb of DNA).

Closer examination of gene distribution reveals that genes on these chromosomes are not evenly spaced. The

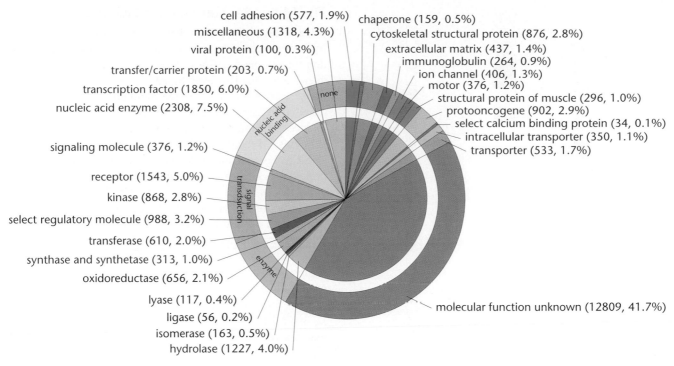

FIGURE 18–12 A preliminary list of assigned functions for 26,588 genes in the human genome. These are based on similarity to proteins of known function. Among the most common genes are those involved in nucleic acid metabolism (7.5% of all genes identified), receptors (5%), protein kinases (2.8%) and cytoskeletal structural proteins (2.8%). A total of 12,809 predicted proteins (41%) have unknown functions, reflecting the work needed to fully decipher our genome. *(Venter, et al. 2001. The sequence of the human genome. Figure 15, p. 1335. Science 291:1304–1351.)*

proximal half of the long arm of chromosome 21 (corresponding to the large G band) is gene-poor, and averages 1 gene per 304 kb of DNA (Figure 18–14). The distal (telomeric) half of the long arm has a much higher gene density, with a gene every 95 kb of DNA. In addition, two regions of chromosome 21 can be regarded as almost empty of genes. One region spanning 7 Mb contains only one gene, and another three regions of 1 Mb each contain no genes. Together, these gene-poor regions add up to 10 Mb, or about one-third the length of the long arm. Chromosome 22 has a 2.5-Mb region near the telomere, and two smaller regions (about 1 Mb each) located elsewhere that lack genes.

Chromosomes 21 and 22 each contain duplicated regions. On chromosome 21, a 220-kb region is duplicated near both ends of the long arm, and another 10-kb region

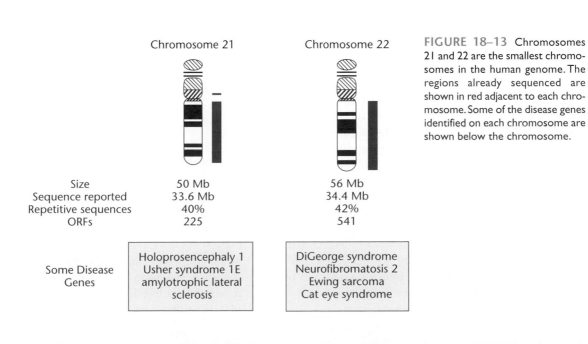

FIGURE 18–13 Chromosomes 21 and 22 are the smallest chromosomes in the human genome. The regions already sequenced are shown in red adjacent to each chromosome. Some of the disease genes identified on each chromosome are shown below the chromosome.

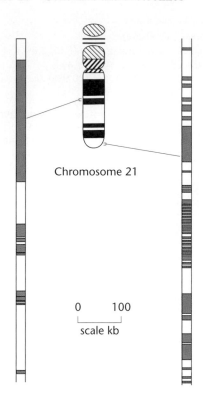

Chromosome 21

0   100

scale kb

FIGURE 18–14 The organization, number, and arrangement of genes in two regions of chromosome 21. The region in the dark-staining band is relatively gene-poor, and the region in the light-staining band near the distal tip is gene-rich. This organization parallels that for the whole chromosome.

is duplicated in the region around the centromere. On chromosome 22, a 60-kb segment is duplicated. These duplicated regions are separated only by a 12-Mb segment. A study of chromosome breakpoints associated with translocations and deletions on chromosome 21 indicates that breakpoints cluster within and near duplicated regions, suggesting that these regions may mediate events involved in chromosomal rearrangement. Future analysis of the duplicated regions and their associated repetitive sequences may provide a molecular explanation for the events in chromosome breakage.

18.4 Genome Evolution

The fossil record indicates that about 3.5 billion years ago, cells similar to bacteria were present on our planet. We can only assume that the genomes of these organisms, or organisms that appeared shortly thereafter, were composed of a double-stranded DNA molecule. The first eukaryotic fossils (resembling single-celled algae) date from about 1.4 billion years ago. As both prokaryotes and eukaryotes underwent changes in size, shape, and complexity, their genomes also changed dynamically. These changes were driven by genetic mechanisms that include mutation, recombination, transposition, gene transfer, as well as gene deletion and duplication. By examining and comparing genomes of organisms that exist today, we gain insights into how these mechanisms have shaped genomes.

The Minimum Genome for Living Cells

What is the minimum number of genes essential for viability? We cannot fully answer this question yet because we don't know the functions of each gene in the genomes sequenced to date. However, based on genes with known functions, we can speculate on the minimum number of genes required to carry out cell functions. To do this, we use sequence information from two of the smallest bacterial genomes, *Mycoplasma genitalium* and *M. pneumoniae*. These two organisms are among the simplest self-replicating prokaryotes known. *M. genitalium* has a genome of 0.6 Mb while that of *M. pneumoniae* is 0.8 Mb. They are members of a group of bacteria that lack a cell wall and invade and often cause disease in a wide range of hosts, including insects, plants, and humans (genital and respiratory infections).

The *M. genitalium* genome contains 467 ORFs, while *M. pneumoniae* has 677. All 467 genes from *M. genitalium* are found in the 677 genes of *M. pneumoniae*. In contrast, *E. coli* has a 4.6-Mb genome with 4288 ORFs, and *H. influenzae* has a 1.8-Mb genome with 1727 ORFs. Table 18.5 summarizes the functions of some genes in these bacteria. Obviously, cells must contain genes that encode products required for DNA replication and repair; transcription and translation; transport proteins; general cellular processes, including cell division and secretion; and for the myriad biochemical pathways involving metabolism.

It is interesting to compare the genes required for a given function among species. For example, *E. coli* has 131 genes for amino acid metabolism, *H. influenzae* has 68, and *M. genitalium* has only 1. Despite having nearly four times as many genes, the physiological capabilities of *Haemophilus* are very similar to *Mycoplasma*. Apparently many more genes are involved in each function category. The greatest advance in *Haemophilus* involves its marked increase in biosynthetic capability. This more complex bacterium has 68 genes involved in amino acid biosynthesis, whereas *Mycoplasma* has only one such gene. Without this capability, *Mycoplasma* must rely on numerous metabolic products from its host.

These comparisons enable us to speculate on the minimal number of genes necessary for organisms to exist as independent, self-reproducing organisms. These estimates are based on comparisons of *M. genitalium* and *M. pneumoniae* genomes, and on mutation experiments in *M. genitalium*. It is now estimated that living organisms require a minimum of 250–350 genes.

Genes That Distinguish Organisms

Using genomic sequencing data, scientists can also speculate about which genes distinguish a species. A comparison of the 467 genes in *M. genitalium* showed that 350 of them are found in a distantly related cousin, *B. subtilis*. This suggests that about 115–120 genes encode the biochemical and functional properties that make *Mycoplasma* different from *Bacillus*. However, *M. genitalium* is a parasite, while *B. subtilis* is a free-living soil bacterium. These adaptations may be a result of the genetic differences of these two

TABLE 18.5 Functional Classes of Genes in Three Bacterial Species

Functional Class	E. coli	H. influenzae	M. genitalium
Protein-coding genes	4,288	1,727	470
DNA replication, repair	115	87	32
Transcription	55	27	12
Translation	182	141	101
Regulatory proteins	178	64	7
Amino acid biosynthesis	131	68	1
Nucleic acid biosynthesis	58	53	19
Lipid metabolism	48	25	6
Energy metabolism	243	112	31
Uptake, transport proteins	427	123	34

species. The genome of another parasite, *B. burgdorferi*, also lacks many of the *M. genitalium* genes, even though they too are very distantly related.

Origin and Evolution of the Eukaryotic Genome

Eukaryotes are traditionally distinguished from prokaryotes by several distinctive features, including a membrane-bound nucleus, cytoplasmic membrane systems such as the endoplasmic reticulum, and a cytoskeleton. As genome projects give us more information, comparisons between prokaryotic and eukaryotic genomes are providing clues about the origins of the eukaryotic genome and how eukaryotes arose from prokaryotes. Several lines of evidence, including amino acid analysis of proteins, gene sequences, and metabolic pathways, indicate that the eukaryotic genome is a mosaic that has received major contributions from both the Archaea and the eubacteria. The eukaryotic nuclear genome has few if any operons, and its genes contain introns, both of which are features of the Archae. The eukaryotic mitochondrial genome strongly resembles that of alpha-proteobacteria. To explain these observations, it has been proposed that eukaryotes arose as a consequence of a symbiotic association or fusion between an anerobic archaebacterial host and an alpha-proteobacterium (like *Rickettsia*), which evolved into the mitochondrion. The monophyletic nature of all eukaryotes suggests that this event occurred successfully only once in the history of the earth.

Genomic Duplications

Such events may account for the origin of the eukaryotic genome, but not for the large, complex genomes that distinguish eukaryotes from prokaryotes. As discussed earlier in this chapter, gene duplication played an important role in eukaryotic genome evolution. Most attention has focused on duplications of single genes, but Susumo Ohno has proposed that whole genome duplications are an important evolutionary mechanism. Analysis of nucleotide sequence data from genome projects supports the idea that much of the difference

in gene number that separates prokaryotes from eukaryotes resulted from genome expansions. We now recognize that a major expansion in eukaryote genomic size resulted from a genome duplication event associated with the appearance of the vertebrates, but duplications have occurred throughout eukaryotic evolution.

A recent analysis of the yeast genome shows traces of an ancient expansion by genomic duplication. Kenneth Wolfe and Denis Shields conducted a search through the yeast genome in which they compared the sequence of each yeast gene to every other yeast gene. Their search uncovered 55 duplicated regions that contain 376 genes; this covers 50 percent of the genome. Of the 55 regions, 50 are in the same relative position on different chromosomes (Figure 18–15, page 380). For example, chromosomes XI and XII contain a duplicated block of genes (block 43) in which the gene order and orientation with respect to the centromere has been conserved. Other chromosomes contain internal duplications, such as block 53, located on either side of the centromere on chromosome XII. Phylogenetic analysis (see Chapter 23) indicates that this genome duplication event took place about 100 million years ago.

Analysis of the human genome also shows evidence of an ancient large-scale duplication followed by rearrangements and gene loss in some duplicated regions. These duplications involve small segments covering only a few genes up to large stretches that almost cover a chromosome. The larger duplications appear to date to the origin of the vertebrates, about 500 million years ago. Altogether, there are 1077 blocks of duplicated regions in the human genome, containing just over 10,000 genes. One such duplication between chromosomes 18 and 20 is shown in Figure 18–16.

Gene Duplications

The almost uninterrupted flow of nucleotide sequence data from genome projects is providing evidence that multigene families are present in many if not all genomes. In addition to genomewide duplications, small blocks of genes and single genes can be duplicated by several mechanisms, including

- **unequal crossing over**—a recombination event between members of a homologous pair of chromosomes in which a DNA segment is duplicated in one of the recombination products

- **unequal sister chromatid exchange**—a recombination event similar to unequal crossing over that occurs between the two chromatids of a single chromosome

- **replication errors**—during replication of a template molecule, slippage can cause the insertion of a short segment into the newly synthesized strand

Once generated, members of multigene families may remain linked on a single chromosome, or may disperse to other parts of the genome. Several mechanisms drive this process, including inversions, translocations, and transposition by mobile elements.

Using the techniques of molecular phylogenetics (Chapter 23), we can trace the ancestry of individual members of gene families. One of the best studied examples is that of the globin gene superfamily (Figure 18–17). In this case, duplication of an ancestral gene encoding an oxygen transport protein occurred some 800 million years ago. This split produced two sister genes, one of which evolved into the modern day myoglobin gene. Myoglobin is an oxygen-carrying protein found in muscle. The other gene became the ancestral globin gene. About 500 million years ago, the ancestral globin gene duplicated to

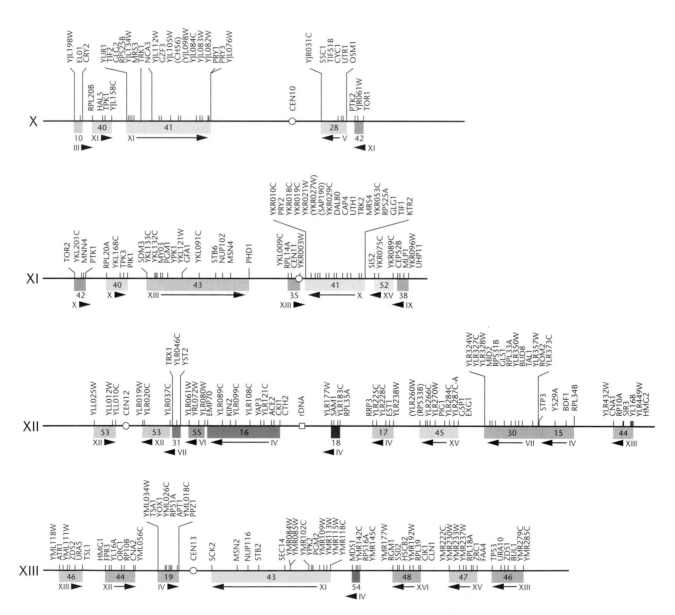

FIGURE 18–15 Location of duplicated regions on yeast chromosomes X, XI, XII, and XIII. The duplicated blocks are numbered, and roman numerals below the blocks give the location of the duplicate block. Arrows show the relative orientation of the blocks. Genomewide orientation relative to the centromere is conserved except for five blocks. (*Wolfe and Shields. 1997. Molecular evidence for an ancient duplication of the entire yeast genome. Figure 2, p. 710. Nature 387:708–713.*)

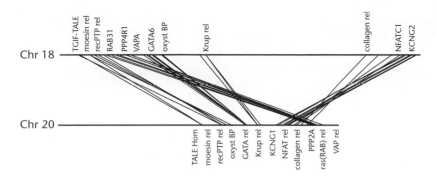

FIGURE 18–16 A segmental duplication of genes on chromosomes 18 and 20. This duplication involves a total of 64 genes. For clarity, only 12 duplicated pairs are shown here. In the genome, there are 1077 duplicated blocks of genes, containing a total of 10,310 genes. (*Venter, et al. 2001. The sequence of the human genome. Figure 13, p. 1332. Science 291:1304–1351.*)

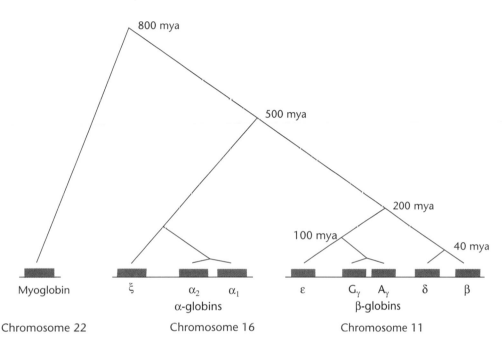

FIGURE 18–17 Evolutionary history of the globin gene superfamily. About 800 million years ago (mya), a duplication event in an ancestral gene gave rise to two lineages. One led to the myoglobin gene, which in humans is located on chromosome 22. The other lineage underwent a second duplication event about 500 mya, giving rise to the ancestors of the alpha and beta subfamilies. Duplications about 200 mya produced the alpha and beta globin subfamilies. In humans, the α-globin genes are located on chromosome 16 and the β-globin genes are on chromosome 11.

form the prototypes of the α- and β-globin subfamilies. The alpha and beta globin genes encode the proteins found in hemoglobin; the oxygen-carrying molecule in red blood cells (see Chapter 13). Additional duplications within the α- and β-globin genes occurred in the last 200 million years. Subsequent chromosomal events dispersed the members of this superfamily, and each is now on separate chromosomes.

Similar patterns of evolution are observed in other gene families, including the trypsin–chymotrypsin family of proteases, the homeotic selector genes of animals, and the rhodopsin family of visual pigments.

18.5 Comparative Genomics: Multigene Families

As outlined earlier, members of **multigene families** share DNA-sequence homology, descend from a single ancestral gene, and their gene products frequently have similar functions. Members of multigene families are often, but not always, found together in a single location along a chromosome. To see how analysis of multigene families provides insight into eukaryotic genome organization, we first examine cases where groups of genes encode very similar, but not identical, polypeptide chains that become part of proteins with closely related functions. Multiple proteins that arise from single-gene duplications are known as **paralogs**. The globin gene family responsible for encoding the various polypeptides that are part of hemoglobin molecules, exemplifies a multigene family that arose by duplication and dispersal to different chromosomal sites.

The Globin Gene Family

The human **α-globin** and **β-globin genes** are subfamilies of the globin gene superfamily and are two of the most intensively studied regions of the human genome. There are three α-globin genes on the short arm of chromosome 16, and five β-globin genes on the short arm of chromosome 11. Members of both subfamilies show homology to one another, but members of the same subfamilies show more homology to each other.. The genes of both subfamilies encode globins

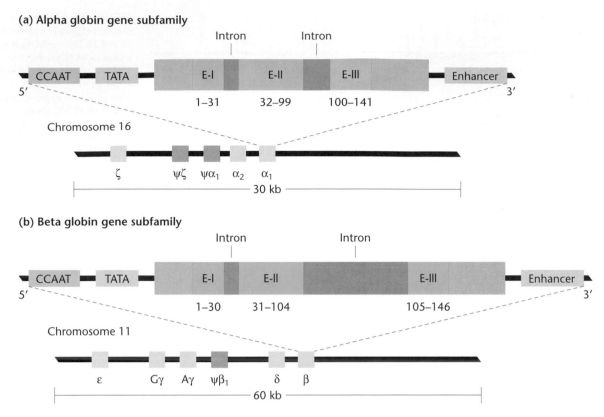

FIGURE 18–18 Organization of the alpha globin gene subfamily (a) on chromosome 16 and the beta globin gene subfamily on chromosome 11 (b). Also shown is the internal organization of the α_1 gene and the β gene. Each gene contains three exons (E-I, E-II, E-III) and two introns. The numbers below the exons indicate the amino acids in the gene product encoded by each exon.

that combine into a tetrameric molecule containing two α- and two β-polypeptides. The tetramer incorporates four heme groups that reversibly bind oxygen, forming functional hemoglobin molecules. Within each subfamily, genes are coordinately turned on and off during embryonic, fetal, and adult stages of development. For both the alpha and beta subfamilies, this expression occurs in the same order in which the genes are arranged on the chromosome.

The alpha subfamily [Figure 18–18(a)], spans more than 30 kb and contains three genes: the ζ (zeta) gene, expressed only in the early embryonic stage, and two copies of the α gene, expressed during the fetal (α_1) and adult stages (α_2). In addition, two nonfunctional **pseudogenes** ($\psi\zeta$ and $\psi\alpha_1$) are present in the cluster. Pseudogenes are designated by the prefix ψ (psi), followed by the symbol of the gene they most resemble. Thus, the designation $\psi\alpha_1$ indicates a pseudogene of the adult α_1 gene. Pseudogenes are nonfunctional versions of genes that resemble other gene sequences but contain significant nucleotide substitutions, deletions, and duplications that prevent their expression.

The organization of the alpha subfamily members and the location of their introns and exons reveals several interesting features. First, as is common in eukaryotes, the DNA containing the three functional α genes occupies only a small portion of the chromosomal region that houses the subfamily. Most of the DNA in this region consists of intergenic spacer DNA. Second, each functional gene in this subfamily contains two introns at precisely the same positions. Third, the nucleotide sequences within corresponding exons are nearly identical in the ζ and α genes. Both genes encode polypeptide chains that are 141 amino acids long. However, the intron sequences are extremely divergent, even though they are about the same size. Significantly, much of the nucleotide sequence within each gene is contained in these noncoding introns.

In humans, the β-globin gene cluster is larger than the α-globin gene cluster, and contains five genes spaced over 60 kb of DNA [Figure 18–18(b)]. As with the alpha subfamily, the sequence of genes along the chromosome parallels the order of expression during development. Of the five genes, three are expressed prior to birth: The ε (epsilon) gene is expressed only during embryogenesis, while the two nearly identical γ genes (G_γ and A_γ) are expressed only during fetal development. The polypeptide products of the two γ genes differ only by a single amino acid. The two remaining genes, δ and β, are expressed following birth. Finally, a single pseudogene $\psi\beta_1$ is present within the subfamily. All five functional genes encode products that are 146 amino acids long and contain two similarly sized introns at exactly the same positions. The second intron in the β genes is significantly larger than its counterpart in the functional α genes. These similarities reflect the evolutionary history of each subfamily and the events such as gene duplication, nucleotide substitution, and chromosome translocations that produced the present-day globin superfamily.

The Immunoglobulin Gene Family

The immune system is an effective barrier against the successful invasion of potentially harmful foreign substances, which are recognized as nonself and subsequently destroyed by the immune system. We now know that genomic alterations are a hallmark of *normal* events in the maturation and function of the immune system. These alterations include rounds of recombination-associated deletion of DNA sequences in specific somatic cells. From a set of less than 300 genes, recombination events, deletions, and mismatching of rejoined DNA segments in immature cells of the immune system create a vast potential for immune response.

One component of the immune response produces antibodies against foreign substances. Agents that elicit antibody production are termed **antigens**. Many different molecules can act as antigens, including proteins, polysaccharides, and, rarely, molecules such as nucleic acids. Usually, a distinctive structural feature of an antigen, called an **epitope**, stimulates antibody production. Antigens can be free molecules, or they can be part of the surface of a cell, microorganism, or virus. **Antibodies** are proteins produced and secreted by the **B cells** of the immune system. Organisms with an immune system can make antibodies against any antigen they encounter. Among vertebrates, each individual can produce millions of different types of antibodies, each responding to a different antigen. In the following sections we examine the molecular basis of antibody diversity.

In humans, antibodies are produced by plasma B cells, a type of lymphocyte (white blood cell). Five classes of antibodies or **immunoglobulins (Ig)** are recognized: IgM, IgD, IgG, IgA, and IgE (Table 18.6). IgM antibodies are the first ones secreted by a plasma cell in response to an antigen, and they are part of the early stages of the immune response. The IgD antibodies are associated with the surface of B cells and may regulate their action, but little is known about their mechanism of action. The IgG class represents about 80 percent of the antibodies found in the blood and is the most extensively characterized group of antibodies. IgG is also involved in immunological memory. Immunoglobulins of the IgA class (found in breast milk) can be secreted across plasma membranes and help resist infections of the respiratory and digestive tracts. The IgE class of antibodies fights parasitic infections and is also associated with allergic responses.

A typical antibody molecule (Figure 18–19) consists of two different polypeptide chains, each present in two copies

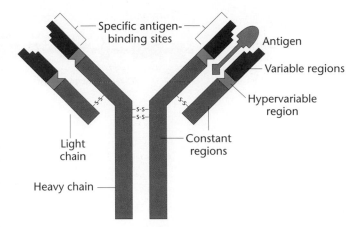

FIGURE 18–19 A typical antibody (immunoglobulin) molecule. The molecule is Y-shaped, and contains four polypeptide chains. The longer arms are H chains and the shorter arms are L chains. The chains are joined by disulfide bonds. Each chain contains a variable region and a constant region. The variable and hypervariable regions of a pair of L and H chains form a combining site that interacts with a specific antigen.

and held together by disulfide bonds. Each larger or **heavy chain (H)** contains approximately 440 amino acids. The sequence of the first 110 amino acids at the N-terminus differs among heavy chains and is known as the **variable region** (V_H). The remaining C-terminal amino acids are the same and make up the **constant region** (C_H) of the heavy chain. In humans, genes on the long arm of chromosome 14 encode the H chains.

Each **light chain (L)** consists of 220 amino acids, with the first 110 amino acids making up the variable region (V_L). The remaining amino acids at the C-terminus make up the constant region (C_L) of the light chain. Two types of L chains exist: **kappa chains**, encoded by genes on human chromosome 2, and **lambda chains**, encoded by genes on the long arm of chromosome 22. The variable regions of the heavy and light chains together form the **antibody-combining site**. Each combining site has a unique structural conformation that allows it to bind to a specific antigen, like a key in a lock.

One characteristic feature of the immune system is its ability to generate new cells that contain different antibodies. Because there are billions of such combinations, it is impossible for separate genes to encode each combination. There is simply not enough DNA in the human genome to encode tens or hundreds of millions of antibodies. One of the long-standing question in immunogenetics is how this vast molecular variability in antibodies and antigen receptors is encoded in the genome.

Antibody diversity results from a number of special features of the genome. The key to understanding how tens of millions of antibodies can be produced lies in understanding the immunoglobulin gene structure. In humans, the kappa-chain gene (Figure 18–20, page 384) consists of several elements: the **L-V (leader-variable) region**, the **J (joining) region**, and a **C (constant) region**. There are 70–100 segments in the L-V region, each with a different nucleotide sequence and a

TABLE 18.6   Classes and Components of Immunoglobulins

Ig Class	Light Chain	Heavy Chain	Tetramers	
IgM	κ or λ	μ	$\kappa_2\mu_2$	$\lambda_2\mu_2$
IgD	κ or λ	δ	$\kappa_2\delta_2$	$\lambda_2\delta_2$
IgG	κ or λ	γ	$\kappa_2\gamma_2$	$\lambda_2\gamma_2$
IgE	κ or λ	ε	$\kappa_2\epsilon_2$	$\lambda_2\epsilon_2$
IgA	κ or λ	α	$\kappa_2\alpha_2$	$\lambda_2\alpha_2$

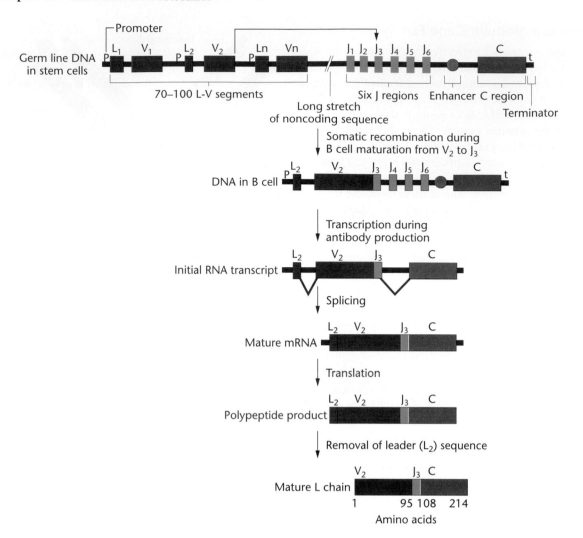

FIGURE 18–20 Formation of the DNA segments encoding a human kappa chain and the subsequent transcription, mRNA splicing, and translation leading to the final polypeptide chain. In germ-line DNA, 70–100 different L-V (leader-variable) segments are present. These are separated from the J regions by a long noncoding sequence. The J regions are separated from a single C segment by an intron that must be spliced out of the initial mRNA transcript. Following translation, the amino acid sequence derived from the leader RNA is cleaved off as the mature polypeptide chain passes across the cell membrane.

promoter, but no regulator. A significant number of the L-V segments are pseudogenes. The J region contains six different segments, and the C region contains only one C segment. A regulator, but no promoter, is located between the last J segment and the C segment. This feature ensures that individual segments cannot be transcribed by themselves.

During B-cell maturation, one of the 100 L-V regions (L_2-V_2) and its promoter are *randomly* joined by a recombination event to one of the six J regions (J_3) and the C region (and its regulator). This recombination event forms a functional L-chain gene [Figure 18–20 (line 2)]. The DNA between the L-V segment selected and the J_3 segment is excised and destroyed. This form of recombination differs from the reciprocal recombination in meiosis and involves different enzymes. The joining event between V_2 and J_3 is *imprecise* and can occur over a span of about 6 bp. In addition, during this recombination event, the ends of the DNA are subjected to a **break-nibble-add mechanism**, in which a few bases

are randomly removed and a few bases are randomly added. The newly created gene contains three exons—L, V-J, and C—and two introns, one between L and V-J and another between J_3 and C. Segments J_4–J_6, not selected, become part of the second intron. The L-chain gene is transcribed, processed, and translated to form kappa-chain proteins that become part of an antibody molecule. The rearranged gene is stable and is passed on to all progeny of the B cell.

Antibody diversity in kappa-chain production results from combining any of the 100 segments in the L-V regions with any of the six segments in the J regions, and generates about 600 different κ genes. Since each joining reaction can occur over several base pairs, the number of possible genes increases to several thousand. Additional diversity comes from the break-nibble-add mechanism. The organization of the lambda-chain genes is somewhat different from that of the κ genes, but recombination events also generate a similar number of different λ genes.

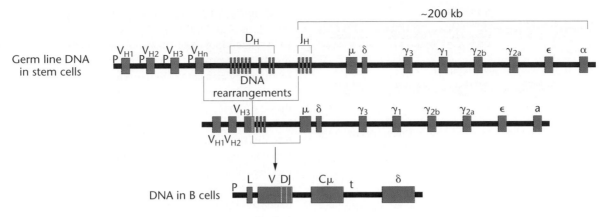

FIGURE 18–21 The variable region of the H chain gene is assembled by joining three different DNA segments together: L-V, D, and J. In embryonic cells, these segments occur in clusters separated by long intervals of other DNA sequences adjacent to the μ gene region. In a maturing B cell, a random combination of one L-V region with one of the 20 D regions and one of the four J regions produces the μ-chain gene. In a mature B cell, this gene is transcribed from the promoter (P) to the terminator (t) and translated to form an H chain. In other B cells, different combinations of H-chain segments are joined, producing a large number of different μ chains.

The heavy-chain genes in humans extend over a large region of DNA and include four types of regions: L-V (leader-variable), D (diversity), J (joining), and C (constant). There are approximately 300 different V segments, 10–30 different D segments, and 6 different J segments. In addition, there are 9 C segments for the five classes of immunoglobulins (Figure 18–21). The μ heavy-chain protein is coded by the μ region, the δ chain, by the δ region, the γ chains by γ regions, the ϵ chain by an ϵ region, and the α chain by an α region. During B cell maturation, recombination randomly joins an L-V segment with a D segment and then with a J segment. The L-V-D-J composite assembled by recombination lies adjacent to the C segment of the μ gene. This joining resembles that described earlier for the L chain, and includes a break-nibble-add step.

The random recombination of H-chain components generates a large number of H-chain genes. If we assume there are 25 D segments, combining any of these with any of the 300 V segments and 6 J segments generates 45,000 H-chain genes. Add in imprecise joining and break-nibble-add for both J to D and V to D-J, and the total climbs markedly. The potential for overall antibody diversity is estimated by multiplying the combination of all H-chain genes with all L-chain genes, resulting in hundreds of millions of possible antibody genes from a few hundred coding sequences.

To summarize, antibody diversity is due to four special features found in the immunoglobulin gene system: (1) multiple numbers of variable segments, (2) multiple numbers of diversity and joining segments, (3) multiple splice locations with break-nibble-add joining, and (4) multiple combination of L chains with H chains. As antibody-forming B cells mature, DNA recombination rearranges these genes so that each mature B lymphocyte comes to encode, synthesize, and secrete only one specific type of antibody. Each mature B cell can make one type of light chain (kappa or lambda) and one type of heavy chain. When an antigen is present, it stimulates the B cell, which encodes an antibody against that antigen to divide and differentiate. As a result, populations of differentiated plasma cells are produced, all of which synthesize one type of antibody that interacts with the antigen.

The Histone Gene Family

The histone genes are a variation on the organizational theme established in the globin and immunoglobulin gene families. Here a cluster of five related but nonidentical genes, separated from each other by highly divergent intergenic spacer regions, is tandemly repeated many times. Recall from Chapter 17 that histones are positively charged (basic) proteins that interact with the negatively charged phosphate groups of DNA to form nucleosomes. In rapidly dividing cells of some organisms, histone synthesis must keep pace with DNA replication, and the necessary quantity of histones must appear each time, as one cell cycle runs into the next. The multiple copies of the histone cluster allow the coordinate synthesis of DNA and histones during the S phase of the cell cycle.

Many interesting correlations support this idea. Yeast cells, with much less DNA than other eukaryotes, have only two copies of each of four histone genes (they lack histone H1). In contrast, the entire cluster of five histone genes is tandemly repeated from 10 to 800 times in many complex organisms (Table 18.7). Sea urchins are noted for their extremely

TABLE 18.7 Number of Histone Gene Clusters in Selected Eukaryotes

Organism	Repeats
Yeast	2
Chickens	10
Mammals	20
Xenopus	40
Drosophila	100
Sea urchin	300–600
Newt	600–800

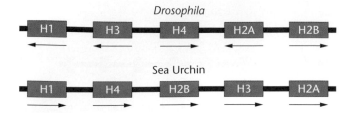

FIGURE 18–22 The histone gene cluster in *Drosophila* and in the sea urchin. Arrows indicate the direction of transcription.

rapid rate of cell division during development, and also have one of the largest sets of histone gene clusters.

Beyond its repeated organization, the histone gene family shows several other differences from other families. First, almost all histone genes lack introns. Second, individual genes within the cluster of a given species often orient in opposite directions with respect to transcription. As shown in Figure 18–22, the arrangement of genes and their polarity of transcription, shown by the direction of the arrow, vary in the sea urchin and *Drosophila*, two organisms where this cluster has been well studied. Studies also show polarity differences other organisms. Despite polarity differences, the amino acid sequences of the various histones (reflecting the DNA encoding each one) are remarkably similar in highly divergent organisms. The homology of histone genes is one of the best examples of sequence conservation through evolution.

In mice and humans, members of the histone gene family appear to be clustered, but not present as a tandem array of repeat units. For example, a region of the distal tip of the long arm of the human chromosome 7 contains all members of the histone gene family, but they can be interspersed with other nonhistone genes. Therefore, no single pattern of histone gene organization applies to all organisms.

18.6 Proteomics

Proteome is a relatively new term, coined to define the complete set of proteins expressed and modified during a cell's entire lifetime. In a narrower sense, it also describes the set of proteins expressed in a cell at a given time. **Proteomics**, the study of the proteome, uses technologies ranging from genetic analysis to mass spectrometry.

As genome projects provide information about the number and kinds of genes present in prokaryotic and eukaryotic genomes, attention is turning to the question of protein function. In genomes ranging from bacteria to complex eukaryotes, it is clear that many newly discovered genes have no known function and others have only presumed functions assigned by analogy with known genes. For example, in two intensively studied organisms, *E. coli* and the yeast *S. cerevisiae*, more than half the genes encoded by their genomes have no known function. About 41% of the predicted proteins identified in the Human Genome Project have

no assigned functions. Some of these may be derived from retroviruses and not be real genes.

Understanding gene function involves more than identifying the gene products. Once made, many gene products are modified by cleavage of end groups such as signal sequences, pro-peptides, or initiator methionine residues, or by the addition of chemical groups (methyl-, acetyl-, phosphoryl) or linkage to sugars and lipids. Proteins are internally and externally cross-linked, and even processed by removing internal amino acid sequences (called **inteins**). Over a hundred mechanisms of posttranslational modification are known, in addition to the high level of diversity produced by the alternate splicing of mRNA. Thus, the human genome, which may have 35,000–40,000 protein-coding genes, may produce almost a half million different gene products. The goal of proteomics is to provide for each protein encoded in a genome a set of information that includes function, structure, posttranslational modifications, cellular localization, variants, and relationships (shared domains, evolutionary history) to other proteins.

Proteomics Technology

Genome projects have a highly developed array of techniques, including vectors, PCR, automated DNA sequencing, and bioinformatics software. Proteomics requires a different range of techniques, many of which are still in various stages of development. The basic techniques in proteomics involve separating and identifying proteins isolated from cells. Currently, proteins are separated by two-dimensional gel electrophoresis (Figure 18–23). In this technique, proteins extracted from a cell are placed on a polyacrylamide gel, and an electric charge is applied across the gel. This separates the proteins according to their molecular weight. The gel is rotated 90 degrees, and during another round of elec-

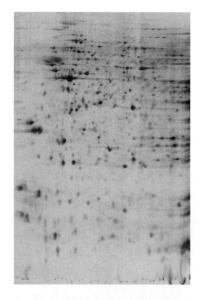

FIGURE 18–23 A two-dimensional protein gel, showing the separated proteins as spots. Several thousand proteins can be displayed on such gels. (*Laboratory of Biochemical Genetics/National Institute of Mental Health, NIH*)

trophoresis, the proteins are separated in a second dimension according to their molecular mass. When the gels are stained, the proteins are revealed as spots; typical gels show 200–10,000 spots (Figure 18–23). To identify individual proteins, spots are cut out of the gel, and the proteins are digested with enzymes such as trypsin, producing a set of characteristic fragments. The fragments are analyzed by mass spectrometry in a method called peptide mass fingerprinting (Figure 18–24). To identify the proteins, the peptide mass is compared with the predicted masses using genetic or protein databases. To increase the speed and accuracy of proteome analysis, researchers are working to adapt DNA chip technology (see Chapter 19 for a description of DNA chips).

In the following sections, we discuss two applications of proteomics, an analysis of a bacterial proteome, and a description of the architecture of a subcellular structure, the nuclear pore.

The Bacterial Proteome

As outlined earlier, *M. genitalium*, with a reduced genome of 480 genes, represents one of the simplest, independently living organisms known. Several groups are analyzing the proteome of *M. genitalium* in an attempt to define the minimum set of biochemical and metabolic reactions needed for a living system. In a recent study, Valerie Wasinger and her colleagues used proteomics to provide a snapshot of gene expression during the exponential and postexponential growth phases in *M. genitalium*.

Using two-dimensional electrophoresis, they identified a total of 427 protein spots in exponentially growing cells. Of these, 201 proteins were analyzed and identified by peptide digestion, mass spectrometry, and comparison to known proteins. The analysis uncovered 158 known proteins (33% of the proteome) and 17 unknown proteins. The remaining spots include fragments derived from larger proteins, different forms of the same protein (isoforms), and posttranslationally modified products. The identified proteins include enzymes involved in energy metabolism, DNA replication, transcription, translation, and transport of materials across the cell membrane.

The transition from exponential growth to stationary phase is associated with a 42 percent reduction in the number of proteins synthesized. During this transition, some new proteins appeared, and other proteins underwent dramatic changes in abundance. These changes are apparently a consequence of nutrient depletion, increased acidity of the growth medium, and other adaptations to environmental changes.

Wasinger's analysis helps establish the minimum number of expressed genes required for independent existence, and the changes in gene expression that accompany the transition to postexponential growth. This study also shows that it is unlikely that conditions can be found under which a cell will express its entire proteome. In *M. genitalium*, only 33 percent of the proteome was expressed under conditions optimized for maximal growth. By virtue of technical limitations, these represent the most abundantly expressed proteins. The remaining 67 percent of the proteome are proteins likely to be expressed under different environmental conditions, those present in very low abundance (too low to be detected on gels), or proteins that cannot be solubilized and recovered by the extraction methods used in Wasinger's study. In spite of these limitations, proteomic analysis provides a wide range of information that cannot be obtained by genome sequencing.

Architecture of the Nuclear Pore Complex

Nuclear pores are complex structures on the nuclear membrane (Figure 18–25). They connect the cell's cytoplasmic and nuclear compartments and allow the exchange of materials between the nucleus and the cytoplasm. The structural and functional complexity of such structures has been a barrier to understanding their molecular organization.

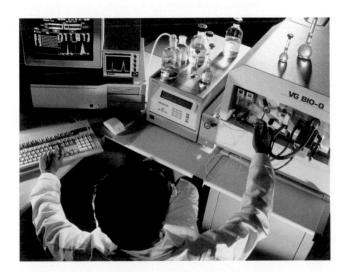

FIGURE 18–24 Proteins are digested and analyzed by mass spectrometry to identify the gene products present in a cell at a given time. *(Geoff Tompkinson/Science Photo Library/Photo Researchers, Inc.)*

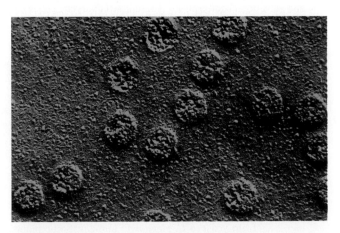

FIGURE 18–25 A transmission electron micrograph of the cytoplasmic side of the nuclear pore complex. *(Richard Kessel/Visuals Unlimited, Inc.)*

Recently, Michael Rout and his colleagues at the Rockefeller Institute used proteomics to define the molecular architecture of this subcellular structure, and to provide insight into its functions. Using mass spectrometry on peptide digests of proteins purified from nuclear pore complexes, they identified about 30 different proteins, organized into subunits. Each nuclear pore contains 16 subunits, 8 on the pore's nuclear side, and 8 on its cytoplasmic side (Figure 18–26). In addition, there are filament-base structures on both sides of the pore, organized into a basket on the nuclear side. Similar approaches studying protein–protein interactions should lead to the macromolecular characterization of other cellular organelles, and dramatically change the direction of cell biology.

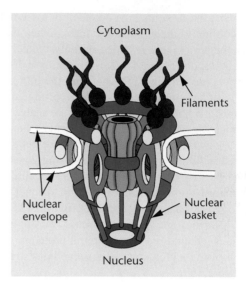

FIGURE 18–26 The nuclear pore complex reconstructed from proteomic analysis. The pore is made up of subunits, each of which contains about 30 different proteins. Each pore has 16 subunits arranged in an octagonal structure extending across the nuclear envelope. Additional gene products that function as transport-factor binding proteins form filamentous structures on the cytoplasmic side and to the nuclear basket on the nuclear side of the pore. *(Blobel, G., and Wozniak, R. W. 2000. Proteomics for the pore. Figure 1, p. 835. Nature 403:835–836.)*

Chapter Summary

1. Once obtained, genome sequences are analyzed in several steps to ensure that the sequence is accurate, to identify all encoded genes, and to classify known genes into functional categories.

2. Bacterial chromosomes are usually small, circular, double-stranded DNA. Gene density is very high, averaging one gene per kilobase pair of DNA. Typically, as much as 90 percent of the chromosome is gene coding. Many genes are organized into polycistronic transcription units that do not contain introns. Archaea are a major prokaryotic kingdom. Their chromosome and gene organization strongly resemble eukaryotic genes and genomes.

3. Eukaryotic genomes are organized into two or more chromosomes, each containing a linear double-stranded DNA molecule. The gene density is much lower than that in bacteria. Genes typically are not organized into operons, rather each is a separate transcription unit. However, the nematode *C. elegans* has many genes organized into operons. Eukaryotic genes are often interrupted with introns.

4. Complex multicellular eukaryotes differ from the less complex yeast in a number of ways. They have more genes and much more DNA. This results in gene densities falling to 1 gene per 5 kb and even 1 gene per 10–20 kb or more. A higher proportion of genes have introns, the number of introns per gene increases, and the size of introns increases as complexity increases from yeast to *C. elegans* to humans. Some plants, such as *Arabidopsis*, have a gene structure and organization that is indistinguishable from animals. Other plants, such as maize, have a different organization, with vast blocks of transposable elements separating islands of genes.

5. In the human genome, large differences exist in gene density on different chromosomes, with gene-rich regions alternating with gene-poor regions. Duplicated segments are a common feature of the chromosomes sequenced to date.

6. Many eukaryotic genes have undergone duplication followed by sequence divergence, leading to multigene families. The globin, immunoglobulin, and histone gene clusters are prime examples of this phenomenon.

7. Proteomics is used to study the expression of genes in bacterial cells under different growth conditions and is providing insight into the gene sets cells use during growth. These techniques have also been used to study the architecture of subcellular elements, such as the nuclear pore complex.

Genetics, Technology, and Society

Completion of the Human Genome Project: The Hype and the Hope

Not since Neil Armstrong walked on the surface of the moon have politicians and the media expressed such exuberance about a technical achievement. On June 26, 2000, President Bill Clinton, joined via satellite by Prime Minister Tony Blair in London, announced the "completion of the first survey of the entire human genome." Present at the news conference were U.S. senators, ambassadors from the United Kingdom, Japan, Germany, and France, Nobel prizewinner James Watson, and scientists, including Dr. Francis Collins (Director of the International Human Genome Project) and Dr. Craig Venter (President of Celera Genomics). Clinton praised the revelation of "nearly all the 3 billion letters of our miraculous genetic code" and stated that "With this profound new knowledge, humankind is on the verge of gaining immense new power to heal. It will revolutionize the diagnosis, prevention, and treatment of most, if not all, human diseases." He compared the human genome sequencing to Galileo's discoveries and the mapping of the American west by Lewis and Clark. Others were less restrained. They described the sequence of the human genome as more important than the invention of the wheel, as the first glimpse of God's instruction book, as one of the most outstanding achievements in human history, on a par with the discovery of penicillin, the Manhattan Project, the *Apollo* Project, the Holy Grail—keeping poets and philosophers inspired for millennia, providing us with the blueprints to a human being's biological functions and susceptibility to illness, soon changing all our lives.

Media reports were peppered with predictions: Doctors will soon be able to diagnose your risk of developing genetic diseases, repair your genetic defects with gene therapy, and treat diseases with side-effect-free drugs that are designed specifically for you, and so on. Some scientists stated that it was hard to exaggerate the importance of the Human Genome Project's announcement.

Although the sequencing of the entire human genome represents a laudable technical tour de force, can it live up to such high expectations?

The first sobering caveat to the Human Genome Project announcement is that the project is, in fact, not finished. It will take several years for the sequencing to be completed, checked, and double-checked, and gaps filled in. It will take much longer to identify all the protein-coding regions within the 3.2 billion base pairs of DNA. As yet, it is unclear how to do this with any accuracy. Even if the protein-coding regions can be reliably identified, it will take a monumental research effort (much greater than the Human Genome Project itself), over many decades, to assign functions to all gene products. With this in mind, the announcement of the working draft of the human genome sequence could more accurately be described as a milestone marking the end of the beginning of deciphering the human genetic blueprint.

The second caveat is that having access to the human genome sequence will remove only one of the numerous obstacles on the road to effective gene therapies, designer drugs and genetic tests. In addition, the hyperbole and high expectations that accompany gene research may do more to slow progress than to speed it.

An illustration is provided by a decade of gene therapy. Since 1990, scientists have conducted over 400 gene therapy clinical trials—for conditions as diverse as cancers, hemophilias, and immunodeficiencies. Although several promising initial reports have appeared, there have been no unequivocal cures. As an example, there was considerable enthusiasm for cystic fibrosis gene therapy when the CFTR gene was first identified in 1989. Because cystic fibrosis is a single-gene defect and the gene had been cloned, it was thought to be straightforward to introduce the normal gene into cystic fibrosis patients and correct the defect. In reality, it has proven to be extremely difficult. Several biotechnology companies quickly initiated gene therapy trials, but none of these trials has brought benefit to cystic fibrosis patients. These companies have now pulled back from clinical trials until they can resolve fundamental problems such as designing efficient vectors with few side effects, understanding how to find and target the appropriate lung cells, and how to measure whether the therapy is working.

Another serious blow was dealt in September, 1999, with the death of a gene therapy clinical trial patient, Jesse Gelsinger. Mr. Gelsinger was involved in a University of Pennsylvania Phase I clinical trial for ornithine decarboxylase gene deficiency. He died from a massive immune response to the adenovirus vector following its injection. Although the reasons for his severe response to the virus are unclear, his death resulted in a shutdown of a number of clinical trials, tighter federal guidelines over gene therapy clinical trials, and a profound chill over the gene therapy field. It also resulted in calls for more cautious, balanced reporting of gene therapy research. Some observers worry that the excitement and unrealistic expectations surrounding gene therapy put pressure on researchers and biotechnology companies to get the therapies working as quickly as possible, at the expense of methodical approaches. These observers emphasize that we resign ourselves to, and indeed applaud and encourage, the painstaking, slow, incremental progress that characterizes most scientific research and medical developments.

The availability of the entire sequence of the human genome will undoubtedly contribute to faster and more efficient isolation and characterization of disease genes. It will also facilitate whole-genome studies and comparative genomics. However, the practical medical benefits of the Human Genome Project will likely be realized only gradually, over the next century. In the meantime, it may be responsible and prudent to promise little and deliver much.

References

Anderson, W. F. 2000. The best of times, the worst of times. *Science* 288:627–29.

Lemonick, M. D. 2000. The genome is mapped. Now what? *Time* 156:24–9 (July 3).

Marshall, E. 2000. Gene therapy on trial. *Science* 288:951–957.

Rosenberg, L. E., and Schechter, A. N. 2000. Gene therapist, heal thyself. *Science* 287:1751.

Websites

Garber, K. 1999. High noon for gene therapy. In *Signals*, an online magazine of analysis for biotechnology executives. **http://www.signalsmag.com/signalsmag.nsf**

The Human Genome Program of the U.S. Department of Energy. **http://www.ornl.gov/hgmis/**

Key Terms

α-globin gene, 381
β-globin gene, 381
annotation, 367
antibody, 383
antibody-combining site, 383
antigen, 383
B cell, 383
break-nibble-add mechanism, 383
C (constant) region, 383
clone-by-clone method, 367
epitope, 383
gene density, 374

genomics, 366
heavy chain (H), 383
Human Genome Project, 366
immunoglobulin (Ig), 383
intein, 386
intron, 374
J (joining), 383
kappa chain, 383
lambda chain, 383
light chain (L), 383
L-V (leader-variable) region, 383
multigene families, 381

open reading frames (ORFs), 367
paralog, 381
proteome, 386
proteomics, 366
pseudogene, 382
repetitive sequence, 374
replication error, 381
shotgun method, 367
unequal crossing over, 381
unequal sister chromatid exchange, 381
V (variable) region, 383

Insights and Solutions

1. An antibody molecule contains two identical H chains and two identical L chains. This results in antibody specificity, with the antibody binding to a specific antigen. Recall that there are five classes of H-chain genes and two classes of L-chain genes. Because of the high degree of variability in the genes that encode the H and L chains, it is possible (even likely) that an antibody-producing cell contains two different alleles of the H-chain gene and two different alleles of the L-chain gene. Yet the antibodies produced by the plasma cell contain only a single type of H chain and a single type of L chain. How can you account for this, based on the number of classes of H- and L-chain genes and the possibility of heterozygosity?

Solution: Although an antibody-producing cell contains different classes of H- and L-chain genes, and although it is entirely possible and even likely that a given antibody-producing cell will contain different alleles for the H-chain gene and for the L-chain gene, a phenomenon known as allelic exclusion allows the expression of only one H-chain allele and one L-chain allele in any given plasma cell at a given time. That is not to say that the same antibody is produced over the life span of the antibody-producing cell, however. At first, many plasma cells produce antibodies of the IgM class, yet at later times they may produce antibodies of a different class (e.g., IgG). Even in switching between different H-chain genes, allelic exclusion is maintained, with only one allele of an H-chain gene (or L-chain gene) being expressed at a given time.

Problems and Discussion Questions

1. Compare and contrast the chemical nature, size, and form assumed by the genetic material of bacteria and yeast.
2. Why might we predict that the organization of eukaryotic genetic material is more complex than that of viruses or bacteria?
3. Compare the gene organization of bacterial genes to that of eukaryotic genes.
4. The β-globin gene family consists of 60 kb of DNA, yet only 5 percent of the DNA encodes β-globin gene products. Account for as much of the remaining 95 percent of the DNA as you can.
5. What do the symbols V_L, C_H, IgG, J, and D represent in immunoglobulin structure?
6. If germ-line DNA contains 10 V, 30 D, 50 J, and 3 C segments, how many unique DNA sequences can be formed by recombination?
7. If there are 5 V, 10 D, and 20 J regions available to form a heavy-chain gene, and 10 V and 100 J regions available to form a L-chain gene, how many unique antibodies can be formed?

Selected Readings

Baltimore, D. 2001. Our genome unveiled. *Nature* 409:814–816.

Blackstock, W. P., et al. 1999. Proteomics: quantitative and physical mapping of cellular proteins. *Trends Biotechnol.* 17:121–127.

Blattner, F. R., et al. 1997. The complete genome sequence of *Escherichia coli* K-12. *Science* 277:1453–74.

Brown, P. O., and Botstein, D. 1999. Exploring the world of the genome with DNA microarrays. *Nature Genetics* (Suppl). 21:33–37.

Bult, C. J., et al. 1996. Complete genome sequence of the methanogenic Archaeon, *Methanococcus jannaschii*. *Science* 273:1058–72.

Coller, H., et al. 2000. Expression analysis with oligonucleotide arrays reveals that MYC regulates genes involved in growth, cell cycle, signaling, and adhesion. *Proc. Nat. Acad. Sci.* 97:3260–65.

Fraser, C. M., et al. 1995. The minimal gene complement of *Mycoplasma genitalium*. *Science* 270:397–403.

Freeman, W. M., et al. 2000. Fundamentals of DNA hybridization arrays for gene expression analysis. *BioTechniques* 29:1042–1055.

Gall, J. G. 1981. Chromosome structure and the *C*-value paradox. *J. Cell Biol.* 91:3s–14s.

Gardner, M. J., et al. 1998. Chromosome 2 sequence of the human malaria parasite *Plasmodium falciparum*. *Science* 282:1126–32.

Gellert, M. 1996. A new view of V-D-J recombination. *Genes to Cells* 1:269–75.

Goffeau, A., et al. 1996. Life with 6000 genes. *Science* 274:546–67.

Golub, T., et al. 2000. Molecular classification of cancer: class discovery and class prediction by gene expression monitoring. *Science* 286:531–537.

Hamedeh, H., and Afshari, C. 2000. Gene chips and functional genomics. *Am. Scientist* 88:508–515.

Heintz, N. 1991. The regulation of histone gene expression during the cell cycle. *Biochim. Biophys. Acta* 1088:327–39.

International Human Genome Sequencing Consortium. 2001. Initial sequencing and analysis of the human genome. *Nature* 409:860–921.

Iyer, V. R., et al. 1999. The transcriptional program in the response of human fibroblasts to serum. *Science* 283:83–87.

Meinke, D. W., et al. 1998. *Arabidopsis thaliana*: A model plant for genome analysis. *Science* 282:662–82.

Mewes, H. W., et al. 1997. Overview of the yeast genome. *Nature* 387:5–105.

Myers, E. W., et al. 2000. A whole-genome assembly of *Drosophila*. *Science* 287:2196–2204.

Orkin, S. H. 1995. Regulation of globin gene expression in erythroid cells. *Eur. J. Biochem.* 231:271–81.

San Miguel, P., et al. 1998. The paleontology of intergene retrotransposons of maize. *Nature Genet.* 20:43–47

Stover, C. K., et al., 2000. Complete genome sequence of *Pseudomonas aeruginosa* PA01, an opportunistic pathogen. *Nature* 406:959–964.

The *C. elegans* Sequencing Consortium 1998. Genome sequence of the nematode *C. elegans*: A platform for investigating biology. *Science* 282:2012–18.

Venter, J. C., et al. 1998. Shotgun sequencing of the human genome. *Science* 280:1540–42.

————. 2001. The sequence of the human genome. *Science* 291:1304–1351.

Williams, S. C., et al. 1996. Sequence and evolution of the human germline V lambda repertoire. *J. Mol. Bio.* 264:200–210.

Wolfe, K. H., and Shields, D. C., 1997. Molecular evidence for an ancient duplication of the entire yeast genome. *Nature* 387:708–713.

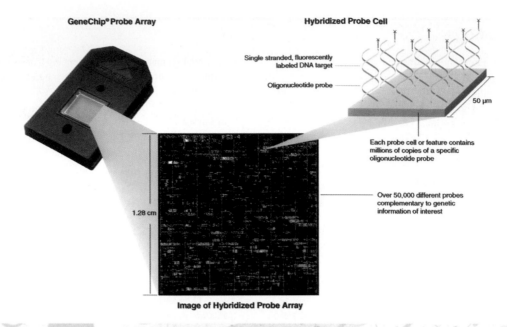

GeneChip® Probe Array

Hybridized Probe Cell

Single stranded, fluorescently labeled DNA target

Oligonucleotide probe

50 μm

Each probe cell or feature contains millions of copies of a specific oligonucleotide probe

Over 50,000 different probes complementary to genetic information of interest

1.28 cm

Image of Hybridized Probe Array

Affymetrix GeneChip® probe arrays are used to detect DNA mutations and monitor gene expression. *(Affymetrix, Inc.)*

19

Biotechnology and Its Implications for Society

CHAPTER CONCEPTS

Techniques used to create, replicate, and analyze recombinant DNA molecules build on discoveries made in molecular biology, biochemistry and bacterial genetics in the mid to late 1970s. The commercial application of recombinant DNA technology began in 1982 when the Food and Drug Administration approved the sale of human insulin produced in bacterial cells. Today, recombinant DNA technology is used to produce medicine, vaccines, industrial chemicals, genetically modified food products and organisms. Biotechnology is now a multibillion-dollar component of the national and world economy, but its use raises social, ethical and policy problems that are still being debated and resolved.

In Chapter 16, we discussed methods for creating and analyzing recombinant DNA molecules. In this chapter we describe how these tools are used in research, medicine, forensics, and commercial applications, and discuss some of the ethical dilemmas posed by this technology. We consider a cross section of applications that demonstrate the power of recombinant DNA technology in mapping and identifying human genes, diagnosing and treating disease, and generating new plants and animals. We begin by considering how recombinant DNA has changed basic methods of genetic mapping and generated projects that are sequencing the genomes of many species of bacteria, plants, and animals. Next, we examine the impact of recombinant DNA on human genetics and medicine, from the prenatal analysis of genotypes to the treatment of genetic disorders and the mapping of the human genome. Finally, we discuss some of the products of the multibillion-dollar biotechnology industry.

19.1 Mapping Human Genes

When the first human genetic disorders were mapped, the phenotype was usually accompanied by the presence of a mutant gene product that was defective in affected individuals. This information established the pattern of inheritance, and in a few cases, was used to determine the chromosomal locus of the gene. However, in the majority of human genetic disorders, the function of the normal gene product is unknown and conventional methods of mapping cannot be used. Although the genetic and molecular basis for a handful of diseases was known before the recombinant DNA revolution, real progress in the mapping of genes has been made only in the last two decades. Now, with our increasingly detailed knowledge of the human genome, we can map a gene without information about its product.

The first step in this process identifies a genetic marker linked to the disease phenotype. With this linkage in hand, researchers either look in the chromosomal region where the marker resides for potential gene sequences or, with an increasing number of human genes localized to chromosomal sites, check the identified chromosomal area for genes that have already been mapped.

RFLPs as Genetic Markers

Now that the Human Genome Project has sequenced essentially all human chromosomes, genes can be directly identified by looking for coding regions in the nucleotide sequence. Variations in nucleotide sequence occur throughout the human genome (mostly in noncoding regions) with a frequency of about 1 in every 200 nucleotides. These nucleotide changes occur at specific sites and are created by substitutions, deletions, or insertions of one or more nucleotide pairs. Such variations can create or destroy restriction enzyme cutting sites. If a restriction site created by nucleotide variation is present on one chromosome but absent on its homolog, the two chromosomes can be distinguished by their pattern of restriction fragments on a DNA blot (Figure 19–1). The region of chromosome A shown in the figure contains three

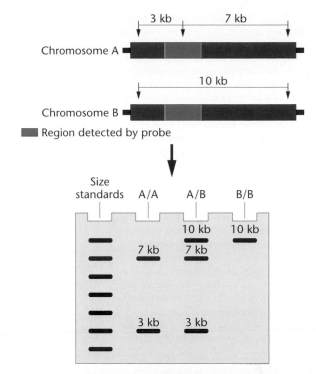

Genotypes	Fragment sizes
Homozygous for chromosome A (A/A)	3 kb, 7 kb
Heterozygous (A/B)	3 kb, 7 kb, 10 kb
Homozygous for chromosome B (B/B)	10 kb

FIGURE 19–1 Restriction fragment length polymorphisms (RFLPs). The alleles on chromosome A and chromosome B represent DNA segments from homologous chromosomes. The region that hybridizes to a probe is shown. Small arrows indicate the location of restriction enzyme cutting sites that define the alleles. On chromosome A, three cutting sites generate fragments of 7 kb and 3 kb. On chromosome B, only two cutting sites are present, generating a 10-kb fragment. The absence of the cutting site in B may be the result of a single-base mutation within the site. Because these differences are inherited in a codominant fashion, there are three possible genotypes: AA, AB, and BB. The allele combination carried by any individual can be detected by restriction digestion of genomic DNA (obtained from a blood sample or skin fibroblasts), followed by gel electrophoresis, transfer to a DNA-binding filter, and hybridization to the appropriate probe. The fragment patterns for the three possible genotypes are shown as they would appear on a Southern blot.

BamHI sites; its homolog, chromosome B, contains two such sites. When chromosome A is cut with BamHI, 3-kb and 7-kb fragments are generated, whereas only a single 10-kb fragment is generated when chromosome B is cut. Using a probe from this chromosomal region, these fragments can be visualized on a Southern blot (Figure 19–1).

Variations in DNA fragment length generated by cutting with a restriction enzyme are called **restriction fragment length polymorphisms (RFLPs)**. RFLPs are quite common; thousands have been identified in the human genome, and most have been assigned to specific chromosomes. These variations are inherited as codominant alleles. They can be mapped to regions on individual chromosomes and

FIGURE 19–2 RFLP analysis and the inheritance of a dominant trait. The family shown in the pedigree has members affected by a dominant trait, as indicated by the filled symbols. Members of this family also carry two alleles for a restriction site: The *A* allele is represented as a 5.0-kb fragment, and the *B* allele is composed of two fragments, one of 2 kb and the other of 3 kb. Only the 2-kb fragment can be detected by the probe. The Southern blot pattern of RFLP alleles for each family member is shown below the pedigree. Note that individual II-5 is from outside the family and is homozygous for the normal trait alleles, even though she carries the RFLP *B* allele. Her daughter, III-1, who is unaffected, probably received the *A* allele from her father and the *B* allele from her mother. Her oldest son (III-2) is affected, and probably received the *A* allele from his mother and the *B* allele from his father. The youngest son (III-3), also affected, received a *B* allele from each parent. The pedigree and the Southern blot indicate that the mutant allele responsible for the disorder is carried on a chromosome with the *B* (2.0-kb) allele. Establishing linkage to a chromosome is the first step in mapping a gene using RFLP analysis.

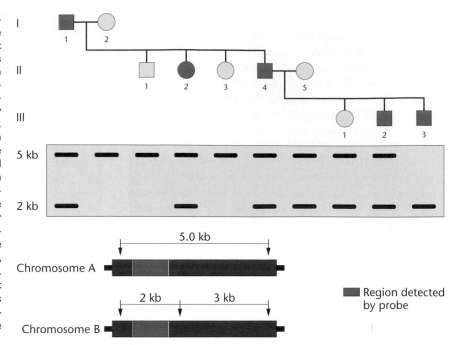

used as markers to follow the inheritance of genetic disorders from generation to generation in an affected family.

Using RFLPs to map the chromosomal locus of a genetic disorder involves identifying an RFLP marker that cosegregates with the genetic disorder in a multigenerational family (three generations or more). Selecting which RFLP to use in the family being studied is a matter of trial and error. Many different loci are tested in order to find one for which most members of the family are heterozygous. This allows each member of the chromosome pair being tested to be identified as it is passed from generation to generation, and provides a way to establish linkage between a specific chromosome (identified by the RFLP marker) and the disease phenotype.

Linkage Analysis Using RFLPs

To map a genetic disorder, the inheritance of a given RFLP and the disorder are traced through a family (Figure 19–2). If loci for the RFLP and the disorder are near each other on the same chromosome, they will show linkage. In such studies, most RFLPs will not show linkage with the disease phenotype, either because the RFLP and the disease loci are on different chromosomes, or there is frequent recombination between the RFLP and the disease locus. However, these data are not useless because they help identify chromosomes that do *not* carry the disease locus.

To analyze the results of such experiments, researchers use probability to determine whether an RFLP marker and a genetic disorder are linked. The probabilities are ultimately expressed as the **logarithm of the odds** or **lod score**. A lod

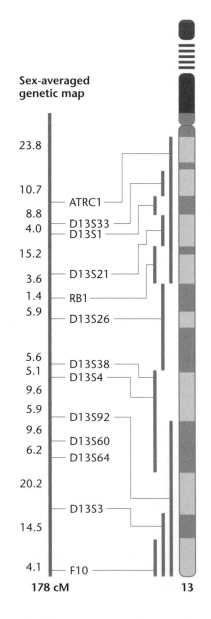

FIGURE 19–3 A genetic and a physical map of human chromosome 13. The genetic map is 178 cM in length, as shown at the left. The locations of markers on the physical map are indicated by the vertical bars adjacent to the chromosome.

score of 3 means that the odds are 1000 times more likely that the RFLP and disease locus are linked than not linked; a lod score of 4 means that the odds are 10,000 times greater in favor of linkage than non-linkage. Lod scores of 3–4 are usually taken as evidence that the gene and the marker are linked.

In mapping human chromosomes, the unit of linkage is the **centimorgan (cM)**, named after the geneticist T. H. Morgan. One centimorgan is equal to a recombination frequency of 1 percent between two loci. The genetic distance between two genes expressed in centimorgans is not directly correlated with the physical distance expressed in nucleotides, but in human chromosomes a distance of 1 cM corresponds roughly to 1–3 million nucleotides of DNA. Many genes have been mapped in humans using RFLP analysis, and by compiling studies from many families, locations of numerous markers can be determined and genetic maps for human chromosomes constructed (Figure 19–3).

Positional Cloning: The Gene for Neurofibromatosis

The search for the chromosomal locus of the gene for **type 1 neurofibromatosis (NF1)** is an excellent example of RFLP analysis in gene mapping. NF1 is inherited as an autosomal dominant condition, with an incidence of about 1 in 3000. It is associated with a range of nervous system defects, including benign tumors and an increased incidence of learning disorders. The spontaneous mutation rate at this locus is very high, and up to 50% of all cases may represent new mutations. To map this gene by conventional methods would require the identification of large, multigenerational families with a long history of NF1. To be useful, each family must also carry one or more genetic disorders already mapped to an autosome. In practice, it would be almost impossible to assemble such family groups, and the gene was mapped using recombinant DNA techniques.

The *NF1* gene was mapped by RFLP analysis in several steps. First, laboratories around the world compared the inheritance of *NF1* and dozens of RFLP markers in multigenerational families. Each RFLP marker represents a specific human chromosome or chromosome region. The resulting **exclusion map** indicated which RFLPs are not linked to the disease and, therefore, which chromosomes do *not* carry the *NF1* locus. This work also produced evidence of linkage and pointed to chromosomes 5, 10, and 17 as sites where the *NF1* gene might reside. The next stage focused on these chromosomes and produced conclusive evidence that the disorder is closely linked to an RFLP that maps to a site near the centromere of chromosome 17 (Figure 19–4). Then, more than 30 RFLP markers from this region of chromosome 17 were used to analyze 13,000 individuals from NF1 families, and the gene was mapped to region 17q11.2. Finally, using a collection of genomic clones that spanned a small region of 17q11.2, the locus for *NF1* was identified in 1990 by chromosome walking (see Chapter 16) and DNA sequencing. The gene's identity was confirmed when mutant versions of the gene were found in individuals with NF1.

Once the gene was identified, the amino acid sequence of the gene product was reconstructed from the DNA sequence and found to consist of 2485 amino acids. Databases with protein-sequence information show that the NF1 gene product is similar to proteins that play a role in signal transduction. Further analysis confirmed that the NF1 protein, **neurofibromin**, is involved in the transduction of intracellular signals and the down-regulation of a gene that controls cell growth. Mutation in the *NF1* gene leads to a loss of control of cell growth, causing the production of the small tumors that are characteristic of this disorder.

The mapping, cloning, and sequencing of this gene and the identification of the gene product took a little over 3 years. Researchers began with no direct knowledge of the nature of the gene product or the mutational events that produce the NF1 phenotype. The mapping and cloning

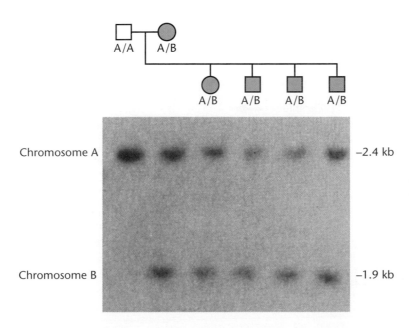

FIGURE 19–4 The segregation of a 2.4-kb RFLP allele with type 1 neurofibromatosis (NF1) in each of four affected offspring and their mother. This RFLP is detected by probe pA10-41, which is known to be a DNA segment near the centromere of human chromosome 17. On the basis of this and results from other probes, the locus for NF1 was assigned to chromosome 17. (*Photo from Barker, D., et al. 1987. "Gene for von Recklinghausen Neurofibromatosis is in the Pericentromeric Region of Chromosome 17." Science 236:1100–02, fig. 1. © 1987 by the American Association for the Advancement of Science.*)

of the *NF1* gene described here is an example of **positional cloning**. This recombinant DNA-based method is a departure from previous methods of mapping, which worked from an identified gene product to the gene locus. In positional cloning, researchers map, isolate, and clone the gene with no knowledge of the gene product. Using this strategy, a large number of human genes have been mapped and isolated.

19.2   Genetic Disorders: Diagnosis and Screening

In many cases, the diagnosis for a genetic disease is made prenatally. The most widely used methods for prenatal diagnosis of genetic disorders are **amniocentesis** and **chorionic villus sampling (CVS)**. In amniocentesis, a needle is used to withdraw amniotic fluid (Figure 19–5), and the fluid and cells it contains are analyzed for chromosomal or single-gene disorders. In CVS, a catheter is inserted into the uterus and a small tissue sample of the fetal chorion retrieved. Cytogenetic, biochemical, and recombinant DNA-based testing is then performed on the tissue.

Coupled with these sample recovery methods, recombinant DNA technology has proven to be a highly sensitive and accurate tool for prenatal detection of genetic disorders. Cloned DNA sequences have expanded the range of prenatal testing because the fetal genotype is examined directly, rather than relying on the few tests available for normal or mutant gene products. This is particularly important because frequently the gene product cannot be detected before birth, even in cases where tests are available. For example, defects in the adult β-globin protein cannot be detected prenatally because the β-globin gene is not expressed until a few days after birth.

Sickle Cell Anemia and Prenatal Genotyping

Sickle-cell anemia is an autosomal recessive condition common in people with family origins in areas of West Africa, the Mediterranean basin, parts of the Middle East and India (see discussion in Chapter 13). Sickle-cell anemia is caused by a single amino acid substitution in the β-globin protein. This change is caused by a single nucleotide substitution, which, coincidentally eliminates a cutting site for the restriction enzymes *Mst*II and *Cvn*I. As a result, this mutation alters the pattern of restriction fragments seen on Southern blots. These differences in RFLP patterns are used to diagnose sickle-cell anemia prenatally and to determine the genotypes of parents and other family members who may be heterozygous carriers of this condition.

For prenatal diagnosis, fetal cells are obtained by amniocentesis or CVS. DNA is extracted from these cells and digested with a restriction enzyme, such as *Mst*II. This enzyme cuts three times in the region of the normal β-globin gene, producing two small DNA fragments. In the mutant allele, the middle *Mst*II site has been destroyed by the mutation, and one large restriction fragment is produced by digestion with *Mst*II (Figure 19–6). The restriction-digested DNA fragments are separated by gel electrophoresis, transferred to a nylon membrane, and visualized by Southern blot hybridization.

In Figure 19–6, the parents (I-1 and I-2) are both heterozygous carriers of the mutation. Digestion of DNA from each parent produces a large band (the mutant allele) and two smaller bands (the normal allele). Their first child (II-1) is homozygous normal because she has only the two smaller bands. The second child (II-2) has sickle-cell anemia; he has only one large band and is homozygous for the mutant allele. The fetus (II-3) has a large band and two small bands

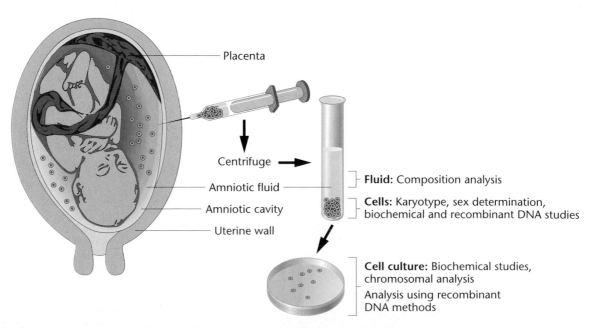

Placenta

Centrifuge →

Amniotic fluid

Amniotic cavity

Uterine wall

Fluid: Composition analysis

Cells: Karyotype, sex determination, biochemical and recombinant DNA studies

Cell culture: Biochemical studies, chromosomal analysis

Analysis using recombinant DNA methods

FIGURE 19–5 The technique of amniocentesis. The position of the fetus is first determined by ultrasound, and then a needle is inserted through the abdominal and uterine walls to recover fluid and fetal cells for cytogenetic and/or biochemical analysis.

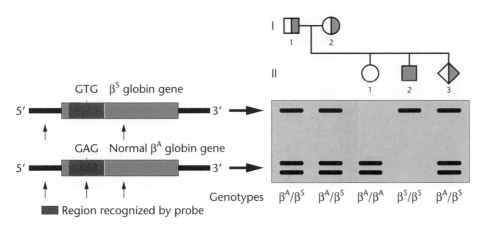

GTG β^S globin gene

5′ ━━━ 3′

GAG Normal β^A globin gene

5′ ━━━ 3′

■ Region recognized by probe

Genotypes   β^A/β^S   β^A/β^S   β^A/β^A   β^S/β^S   β^A/β^S

FIGURE 19–6 Southern blot diagnosis of sickle-cell anemia. Small arrows represent the location of restriction enzyme cutting sites. In the mutant (β^S) globin gene, a single nucleotide mutational change has destroyed a restriction enzyme cutting site, resulting in a single large fragment on the Southern blot. The normal (β^A) globin gene produces two small fragments with MstII. In the pedigree, the family has one unaffected homozygous normal daughter (II-1), an affected son (II-2), and an unaffected fetus (II-3). The genotypes of each family member can be read directly from the blot, as shown below it.

and is therefore heterozygous for sickle-cell anemia. He or she will be unaffected, but will be a carrier of this condition.

Only about 5–10 percent of all nucleotide substitutions can be detected by restriction enzyme analysis. However, if a mutant gene has been well characterized and the mutated region sequenced, synthetic oligonucleotides can be used as probes to detect mutant alleles as described in the next section.

Allele-Specific Nucleotides in Genetic Screening

Alleles that differ by as little as a single nucleotide can be distinguished by synthetic probes known as **allele-specific oligonucleotides (ASO)**. In contrast to restriction enzyme analysis, which is limited to cases where a mutation changes a restriction site, ASOs detect single nucleotide changes of all types, including changes that do not affect restriction enzyme cutting sites. As a result, this method offers increased resolution and wider application. Under proper conditions, an ASO hybridizes only with its complementary sequence and not with other sequences, which might vary by as little as a single nucleotide.

A method using ASOs and PCR is now available to screen for many genetic disorders, including sickle-cell anemia. In this procedure, DNA from white blood cells is extracted and denatured into single strands and a region of the β-globin gene is amplified by PCR. A small amount of the amplified DNA is spotted onto filters, and each filter is hybridized to an ASO (Figure 19–7). After visualization, the genotype is read directly from the filters. Using an ASO for the normal sequence [Figure 19–7(a)], the homozygous normal (AA) genotype produces a dark spot (two copies of the normal allele), and the heterozygous genotype (AS) produces a light spot (one copy of the normal allele). The homozygous recessive sickle-cell genotype does not bind the probe, and no spot is visible. Using a probe for the mutant allele [Figure 19–7(b)], the pattern is reversed. This rapid, inexpensive, and highly accurate technique is becoming the method of choice for diagnosing a wide range of genetic disorders caused by point mutations.

In cases where the nucleotide sequence of the normal allele is known and the molecular nature of the mutant gene

has been identified, ASOs are synthesized directly from normal and mutant copies of the gene to screen for heterozygous carriers of the genetic disorder. For example, in cystic fibrosis a three-nucleotide deletion (called Δ508) is found in 70 percent of all mutant copies of the gene. Cystic fibrosis is an autosomal recessive disorder associated with a defect in a protein called the **cystic fibrosis transmembrane conductance regulator (CFTR)**, which regulates chloride ion transport across the plasma membrane. To

DNA extracted from white blood cells

Codon 6

5′ ━━━ 3′

▨ Region covered by ASO probes

DNA is spotted onto binding filters, hybridized with ASO probe

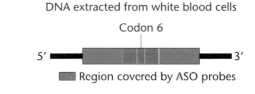

(a) Genotypes   AA   AS   SS

Normal (β^A) ASO: 5′ – CTCCTGAGGAGAAGTCTGC – 3′

(b) Genotypes   AA   AS   SS

Mutant (β^S) ASO: 5′ – CTCCTGTGGAGAAGTCTGC – 3′

FIGURE 19–7 Genotype determinations using allele-specific oligonucleotides (ASOs). In this technique, the β-globin gene is amplified by PCR using DNA extracted from blood cells. The amplified DNA is denatured and spotted onto strips of DNA-binding filters. Each strip is hybridized to a specific ASO and visualized on X-ray film after hybridization and exposure. If all three genotypes are hybridized to an ASO from the normal β-globin gene, the pattern in (a) will be observed. If two copies of the normal β-globin gene are present (AA), the spot will show heavy hybridization. If one normal β-globin gene and one mutant gene are present (AS), the spot will show weaker hybridization. If no normal copy of the β-globin gene is present (SS), there will be no hybridization to the ASO probe. (b) The same genotypes hybridized to the probe for the sickle-cell β-globin gene will show the reverse pattern: no hybridization by the AA genotype, weak hybridization by the heterozygote (AS), and strong hybridization by the homozygous sickle-cell genotype (SS).

detect heterozygous carriers for the *Δ508* mutation, allele-specific oligonucleotides are made by PCR from cloned samples of the normal allele and the mutant allele. DNA extracted from the white blood cells of those to be tested is spotted on a nylon filter and hybridized to each ASO (Figure 19–8). In affected individuals, only the ASO made from the mutant allele will hybridize; in heterozygotes, both ASOs hybridize; and in normal homozygotes, only the ASO from the normal allele hybridizes.

Because cystic fibrosis (CF) affects approximately 1 in 2000 individuals of northern European descent, screening for CF can be used in these populations to detect heterozygous carriers and counsel them about their status for CF. However, not all of the known mutations for this gene (over 500 mutations have been identified) can be screened, so a negative result does not eliminate someone as a heterozygous carrier, and it is likely that more CF mutations remain to be identified. Consequently, CF screening is not widespread, but will no doubt become commonplace once tests can cover 98–99 percent of all possible CF mutations.

DNA Microarrays and Genetic Screening

DNA probes similar to allele-specific nucleotides coupled with semiconductor technology produce **DNA microarrays** (also called DNA chips). The microarrays are made of glass and are divided into fields (small squares); each one can be as small as half the width of a human hair. A field contains a specific DNA probe about 20 nucleotides in length attached to the glass (Figure 19–9). Along a row, the sequence of the probe differs by one nucleotide from field to field. Thus, a set of four fields (one for each nucleotide) is needed to test a given position for its nucleotide content. The current generation of chips holds between 280,000 and 560,000 fields, but chips with several million fields are being developed. For genetic testing, DNA is extracted from cells and cut with one or more restriction enzymes. The resulting fragments are tagged with a fluorescent dye, melted into single strands, and pumped into the microarray. Fragments with a nucleotide se-

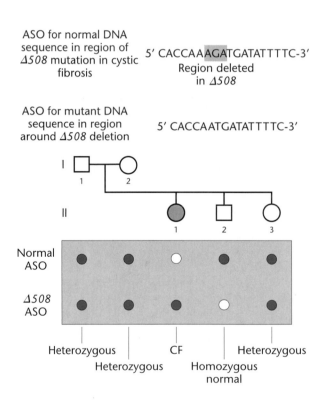

ASO for normal DNA sequence in region of *Δ508* mutation in cystic fibrosis

5′ CACCAAGATGATATTTTC-3′
Region deleted in *Δ508*

ASO for mutant DNA sequence in region around *Δ508* deletion

5′ CACCAATGATATTTTC-3′

FIGURE 19–8 Screening for cystic fibrosis (CF) by allele-specific oligonucleotides (ASOs). ASOs for the region spanning the most common mutation in CF, a three-nucleotide deletion (*Δ508*), are prepared from normal CF genes and *Δ508* CF genes. In screening, the CF genes are amplified by PCR using DNA extracted from blood samples and spotted on a DNA-binding membrane. The membrane is hybridized to a mixture of the two ASOs. The genotypes of each family member are read directly from the filter. DNA from I-1 and I-2 hybridizes to both ASOs, indicating that they carry a normal allele and a mutant allele and are therefore heterozygous. The DNA from II-1 hybridizes only to the *Δ508* ASO, indicating that she is homozygous for the mutation and has cystic fibrosis. The DNA from II-2 hybridizes only to the normal ASO, indicating that he carries two normal alleles. II-3 has two hybridization spots, and is heterozygous.

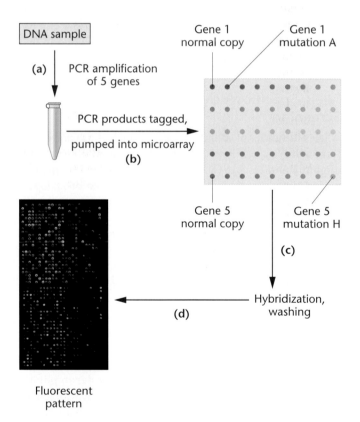

Fluorescent pattern

FIGURE 19–9 Gene screening using a DNA microarray. (a) DNA extracted from a blood sample is amplified by PCR. In this example, primers for five genes are used, but in practice, many more are used. (b) The microarray contains single-stranded probes for the normal (column 1) allele and seven mutant alleles for each of the five genes (1 row = 1 gene). Microarrays containing several hundred thousand probes are now used. The probes are attached to the glass substrate, and each occupies a different field on the array. The single-stranded PCR products are tagged with fluorescent probes and pumped into the microarray. (c) The probes and the tagged PCR products are allowed to hybridize, and the unbound DNA is washed away. (d) The resulting hybridization is revealed by the pattern and color of the fluorescent spots on the microarray. These can be automatically scanned, and the data presented in several forms. *(Courtesy Cancer Genetic Branch/National Human Genome Research Institute)*

quence that exactly matches the probe sequence will bind, and those with a sequence that doesn't match are washed off. A laser scanner reads the pattern of fluorescence [Figure 19–9(d)]. A linked software program analyzes the fluorescent pattern of hybridization, and the data can be presented in several forms.

DNA microarrays are already used to scan for mutations in the *p53* gene, which is mutated in 60 percent of all cancers, and to screen for mutations in the *BRCA1* gene, which predisposes women to breast cancer.

19.3 Genetic Testing and Ethical Dilemmas

In an earlier section, we described two types of genetic testing, prenatal diagnosis and screening for heterozygous carriers of recessive disorders. Using current technology, genetic testing can also predict risk of disease, thereby identifying those who are presently healthy but at high risk of contracting a genetic disease in the future. In the next few years, DNA microarrays may be used to test for 50–100 diseases at a time, including many that may not develop for years or decades. These advances will affect our health, reproductive patterns, and medical care in fundamental ways. Likewise, this technology also raises profound legal, social, and ethical issues that will not be easy to resolve. For example, what should people know before deciding to have a genetic test? How can we protect the information revealed by a genetic test? How can we define and prevent genetic discrimination? We know that sickle-cell anemia is a condition that makes the carrier resistant to malaria. This may be true of other genetic diseases. How do we protect beneficial aspects of mutations as we strive to eliminate their destructive aspects? Some mutations have horrific consequences. Some mutations gave rise to our very existence as humans. We know a great deal and yet we know very little. Thoughtful and wide-ranging public debate is imperative as we negotiate these uncharted waters.

Many of the potential risks and benefits of genetic testing are still unknown. We have the ability to test for many genetic diseases but do not as yet have effective treatments to cure these disorders or mitigate their effects. With present technology, the fact that a genetic test gives a negative result does not rule out future development of the disease, nor does a positive result always mean that one will get the disease.

Public policy and laws on genetic testing are lagging the development and application of the technology. The **Ethical, Legal, and Social Implications (ELSI) Program** of the Human Genome Project (described later in this chapter) has set up task forces to identify issues related to genetic testing. This program is charged with developing recommendations for policy makers and lawmakers that will retain the benefits of genetic testing, but reduce or eliminate the potential harm. In addition, other groups made up of scientists, health-care professionals, lawmakers, ethicists, and consumers are debating these issues and formulating policy options.

19.4 Gene Therapy

Gene *products* such as insulin have been used for decades in therapeutic treatment of genetic disorders. Methods for introducing specific *genes* into mammalian cells, originally developed as a research tool, are now used to treat genetic disorders, a process known as **gene therapy**. In theory, gene therapy transfers a normal allele into a somatic cell that carries one or more mutant alleles. The delivery of these structural genes and their regulatory sequences is accomplished via a vector or gene transfer system.

Several methods transfer genes into human cells. These include using viruses as vectors, chemically assisting the transfer of genes across cell membranes, and fusing cells with artificial vesicles that contain cloned DNA sequences.

In the first generation of gene therapy trials, the most common method of gene transfer used retroviruses as vectors. The first retroviral vectors were based on a mouse virus called Moloney murine leukemia virus (Figure 19–10). The procedure creates a recombinant virus carrying a human gene. First, a cluster of three genes is removed from the virus and the cloned human gene inserted. After packaging into a viral protein coat, the recombinant vector can infect cells, but cannot replicate itself because of the missing viral genes. Once inside the cell, the recombinant virus with the inserted human gene moves to the nucleus and integrates into a chromosome, where they become a part of the cell's genome. In the initial stages of gene therapy, several heritable disorders, including **severe combined immunodeficiency (SCID)**, **familial hypercholesterolemia**, and **cystic fibrosis** were treated.

Gene Therapy for Severe Combined Immunodeficiency (SCID)

The first attempts at gene therapy began in 1990, with the treatment of a young girl, Ashanti DeSilva, who has a genetic disorder called severe combined immunodeficiency

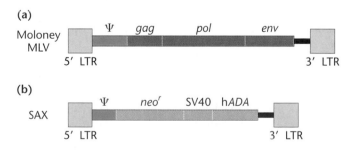

FIGURE 19–10 A retroviral vector constructed from the Moloney murine leukemia virus (Moloney MLV). (a) The native MLV genome contains a psi sequence (Ψ) required for encapsulation, and genes that encode viral coat proteins (*gag*), an RNA-dependent DNA polymerase (*pol*), and surface glycoproteins (*env*). At each end, the genome is flanked by long terminal repeat (LTR) sequences that control transcription and integration into the host genome. (b) The SAX vector, engineered from MLV retains the LTR and Ψ sequences, and includes a bacterial neomycin resistance (*neo^r*) gene that serves as a selective marker. As shown, the vector carries a cloned human adenosine deaminase (*hADA*) gene, fused to an SV40 early-region promoter/enhancer. The SAX construct is typical of retroviral vectors now being used in human gene therapy.

FIGURE 19–11 Gene therapy for treatment of severe combined immunodeficiency (SCID), a fatal disorder of the immune system caused by absence of the enzyme adenosine deaminase (ADA). The cloned human *ADA* gene is transferred into a viral vector, which is then allowed to infect white blood cells removed from the patient. The transferred *ADA* gene is incorporated into a chromosome and becomes active. After growth to enhance their numbers, the cells are reimplanted in the patient, where they produce ADA, enabling an immune response to develop.

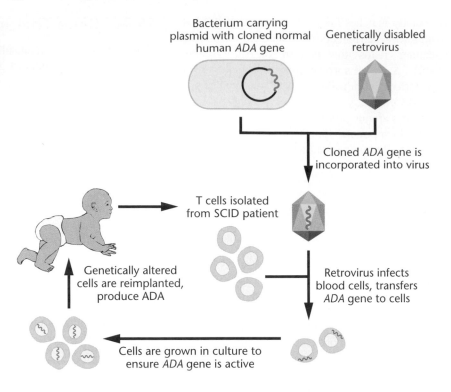

Bacterium carrying plasmid with cloned normal human *ADA* gene

Genetically disabled retrovirus

Cloned *ADA* gene is incorporated into virus

T cells isolated from SCID patient

Retrovirus infects blood cells, transfers *ADA* gene to cells

Cells are grown in culture to ensure *ADA* gene is active

Genetically altered cells are reimplanted, produce ADA

(SCID). In this disorder, affected individuals have no functional immune system and usually die from what would otherwise be minor infections. Mutations in several different genes can cause SCID, one of which is a mutation in the gene encoding the enzyme **adenosine deaminase (ADA)**.

The first step in the gene therapy isolates white blood cells called T cells from the patient (Figure 19–11). These cells, which are part of the immune system, are mixed with a genetically modified retrovirus carrying a normal copy of the *ADA* gene (Figure 19–11). The virus infects the T cells, and inserts a functional copy of *ADA* into the cell's genome. The genetically modified T cells are grown in the laboratory to ensure that the transferred gene is expressed, and the patient is treated by injecting a billion or so of the altered T cells into the bloodstream.

After treatment, Ashanti has maintained normal ADA protein levels in 25–30 percent of her T cells, and she now leads a normal life (Figure 19–12). Unfortunately, a second child treated a short time later had the normal gene in only 0.1–1 percent of her white blood cells after treatment, a level not high enough to be effective. In later trials, attempts were made to transfer the *ADA* gene into bone marrow cells that give rise to T cells, but were mostly unsuccessful. Other gene therapy trials, including those for cystic fibrosis and familial hypercholesterolemia, have also had poor results, most of which have been attributed to inefficient vectors.

Gene Therapy: A New Beginning

Disappointing results from first-generation gene therapy trials led to a widespread crisis of confidence in the late 1990s in both the medical and scientific communities. However, with some promising results in several animal models, the tide may finally have turned. This is thanks primarily to improvements in existing vectors and the use of new vectors; both circumvent several of the problems encountered with earlier vectors.

In 2000, gene therapy trials for an X-linked form of SCID (*SCID-X1*) reported some initial success. In these trials, a modified Moloney vector was used, and a more efficient transfer of the normal gene into blood cells was achieved.

FIGURE 19–12 Ashanti DiSilva, the first person to be treated by gene therapy. *(Courtesy of Van de Silva)*

TABLE 19.1   Vectors for Gene Therapy

Vector	Cell Targets	Cloning Capacity	Advantages	Disadvantages
Adenovirus	Lung, respiratory tract	7.5 kb	Efficient transfection	Strong immune response
Adeno-associated virus	Fibroblasts, T cells, others	4.5 kb	Transfects many cell types	Small insert size
Retroviruses	Proliferating cells	8 kb	Prolonged expression	Low transfection efficiency
Lentiviruses	Stem cells, proliferating cells	8 kb	Efficient transfection	Related to HIV

The two patients were treated at ages 8 and 11 months. Some 10 months after treatment, the disease phenotype was fully corrected, and they had functional immune systems.

Other researchers have focused on developing new vectors. First-generation vectors such as Moloney virus and adenovirus, have several drawbacks: (1) Integrating the retroviral genome (including the cloned human gene) into the host cell genome occurs only if the host cells replicate their DNA. In many highly differentiated human tissue types there are few suitable target cells. (2) Most of these viruses eventually elicit an immune response in the host. (3) Insertion of viral genomes into the host chromosome can inactivate or mutate an indispensable gene. (4) Retroviruses have a low cloning capacity and cannot carry inserted sequences much larger than 8 kb; many human genes, even without introns, exceed this size. (5) Finally, an infectious virus may be produced if a recombination event takes place between the vector and retroviral genomes already present in the host cell. For these reasons, new viral vectors and new strategies for targeting cells are being developed. Some of the second-generation vectors are listed in Table 19.1. It is hoped that these vectors will overcome the problems associated with the early vectors, and will help fulfill the promise of gene therapy.

Using an AAV vector, recent gene therapy studies have been successful in obtaining gene transfer and expression in individuals with hemophilia B, an X-linked blood-clotting disorder. Further trials will use higher doses of the gene, and in the near future, it may be possible to completely correct the phenotype of hemophilia with gene therapy. In addition to treating heritable genetic disorders, gene therapy is now in clinical trials for treating several forms of cancer, infectious diseases, and cardiovascular diseases (Table 19.2). In the last few years, more than a dozen biotechnology companies have been founded to develop products for gene therapy, and in the coming decade, gene therapy may be a commonplace treatment for a large number of disorders.

TABLE 19.2   Some Current Gene Therapy Trials (1999)

Disease/Disorder	Number of Trials
Cancer	216
Single-gene disorders	49
Infectious diseases	24
Cardiovascular diseases	8

Ethical Issues in Gene Therapy

Gene therapy obviously raises many ethical concerns, and is the source of intense debate. As a result, gene therapy trials currently underway or in the planning stages are restricted to the use of somatic cells as targets for gene transfer. Ethical guidelines for undertaking gene therapy in its present form are well established. This experimental treatment is initiated only after careful review at several levels, and trials are monitored to protect the patient's interests. New guidelines strengthen these protections, and coordinate the oversight of government agencies that regulate gene therapy trials. The transfer of genes into somatic cells is called **somatic gene therapy**. In this therapy, only one individual is affected, and the therapy is done with the permission and informed consent of the patient.

Two other potential approaches in gene therapy have yet to be approved, primarily because of unresolved ethical issues. The first is **germ-line therapy**, where germ cells (cells that give rise to gametes—i.e., sperm and eggs) or mature gametes are the targets. In this approach, the corrective DNA is incorporated into all the cells of the individual produced from the genetically altered gamete, including his or her own germ cells. This means that individuals in future generations may also be affected, without their consent. Is this ethical? Do we have the right to make this decision for future generations? Thus far, concerns have outweighed potential benefits, and such research is prohibited.

The second type of gene therapy, **enhancement gene therapy**, raises an even greater ethical dilemma. In this therapy, human potential can be enhanced for some desired trait. This potential use is extremely controversial and strongly opposed in many quarters. Here is the key question: If such genes are identified and cloned, should we use them to increase height, enhance athletic ability, or extend intellectual potential? Presently, the consensus is that enhancement and germ-line therapies are unacceptable uses. However, the debate continues and will no doubt grow more heated as our facility with these technologies expands. The outcome of these debates may affect not only the fate of individuals, but that of our species as well.

19.5   DNA Fingerprints

As discussed earlier, the presence or absence of restriction sites in the human genome can be used as genetic markers. Another type of nucleotide sequence variation, discovered

FIGURE 19–13 VNTR loci and DNA fingerprints. VNTR alleles at two loci (A and B) are shown for each individual. Small arrows mark restriction enzyme cutting sites flanking the VNTRs. Restriction digestion produces a series of fragments that can be detected as bands on a Southern blot (below). Because of differences in the number of repeats at each locus differ, the overall pattern of bands is distinct for each individual, even though one band is shared (the *B2* allele band). Such a pattern is known as a DNA fingerprint.

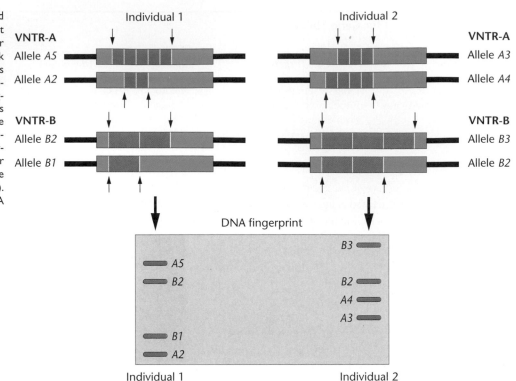

Minisatellites and VNTRs

One of the most useful forms of RFLPs arises from variations in the number of tandemly repeated DNA sequences present between two restriction enzyme sites. These sequences are **minisatellites**, or clusters of 2–100 nucleotides. For example, the sequence

GGAAGGGAAGGGAAGGGAAG

is composed of four tandem repeats of the sequence GGAAG. Clusters of such sequences are widely dispersed in the human genome. Typically, each repeat contains between 14 and 100 nucleotides, and the number of repeats at each site ranges from 2 to more than 100. These loci, known as **variable-number tandem repeats (VNTRs)**, were introduced in Chapter 17 as examples of middle repetitive DNA. What varies is the number of repeats at a given locus; each variation constitutes a VNTR allele. Many loci have dozens of alleles; as a result, heterozygosity is common.

A pattern of bands is produced when VNTR sequences are cut with restriction enzymes and visualized by Southern blotting. This pattern is one example of a DNA fingerprint (Figure 19–13). These patterns are the equivalent of actual fingerprints because the band pattern is always the same for

in the mid-1980s, depends on variation in the length of repetitive DNA sequence clusters. These polymorphisms in human DNA serve as the basis for a technique called **DNA fingerprinting**. DNA fingerprinting is now used for a wide variety of applications, from identifying individuals for paternity and forensics cases to enforcing bans on the sale of whale products.

a given individual (no matter what tissue is used as the source of the DNA), and it varies from individual to individual. In fact, there is so much variation in the band patterns that, theoretically, each person's pattern is unique. DNA fingerprint analysis can be performed on very small samples of material (less than 60 μL of blood) and on samples that are quite old (VNTR analysis has been performed on Egyptian mummies over 2400 years old), thus increasing its usefulness in a variety of circumstances.

Forensic Applications of DNA Fingerprints

DNA fingerprints have been accepted as evidence in criminal trials in the United States since 1988. The method has also been used in a wide range of other applications, including immigration cases, disputes involving purebred dogs, paternity cases, and in animal conservation studies. At present, standard forensic testing employs just over a dozen different VNTR probes (Figure 19–14). Most of these are on different chromosomes, and each probe visualizes the VNTR pattern at a particular locus. Standard forensic tests with 4–10 VNTR probes are used to develop a detailed DNA fingerprint profile. The results of DNA fingerprints are analyzed and interpreted with statistics, probability, and population genetics.

The frequency for each of the alleles in the standard set of VNTR probes has been measured in many population groups across the U.S. Using this information, the probability of having any combination of these alleles can be calculated. For example, if an allele of locus 1 is carried by 1 in 333 individuals, and an allele of locus 2 is carried by 1 in 83 individuals, then the probability of having both these alleles is

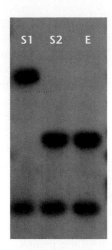

FIGURE 19–14 DNA fingerprinting in a forensic case. The DNA profile of suspect 2 (S2) matches that of the blood sample obtained as evidence (E). *(William S. Klug)*

1 in 28,000 (the product of their individual frequencies in the population). This overall probability is not especially convincing, but if an allele at a third locus (carried by 1 in 25 individuals) and an allele of a fourth locus (carried by 1 in 100 individuals) are included, the probability of that allele combination becomes 1 in 70 million. That is, the chance that anyone has this combination of alleles is 1 in 70 million, or about 4 individuals in the U.S. population.

If suspects' DNA fingerprints do not match that of the evidence, they can be excluded as the criminal (exclusions occur in about 30% of all cases). When a match is made between DNA obtained at a crime scene and the DNA of a suspect, there are two possible interpretations: The DNA fingerprint came from the suspect, or it is from someone else with the same pattern of bands. In this case, the DNA fingerprint evidence does not prove the suspect is guilty, but is

one piece of evidence that must be considered along with other facts in the case.

19.6 Genome Projects

Over the past 90 years, geneticists have developed efficient methods to generate and map mutations in order to identify and locate genes. One drawback to these methods is that genes can be characterized only if mutants are isolated. To characterize a genome exhaustively, at least one mutation for each gene in the genome is required, and because of lethal alleles and other problems, this is probably an impossible goal.

Geneticists now use recombinant DNA techniques to identify all the genes in an organism's genome directly. Instead of screening for mutants and constructing genetic maps, a genomic library is established, and the nucleotide sequence of the clones in the library is determined. The result is a genomic map, with all genes in the genome identified by location and nucleotide sequence. Analysis of derived amino acid sequences and the generation of gene-specific mutants helps establish the gene product's function. In addition, sequencing provides information about the nongene-containing parts of a genome, regions that are beyond the reach of current mutation mapping strategies.

Genomes of over 60 organisms are completely sequenced, with 203 prokaryotic and 132 eukaryotic genomes (including the human genome) partially completed (Table 19.3). Several of these projects are part of the publicly funded Human Genome Project, while others have been sequenced by organizations such as The Institute for Genome Research (TIGR) and companies (Celera). Nonhuman organisms included in the Human Genome Project were selected for several reasons, including genome size, the availability of detailed genetic maps, and the value of comparisons on gene number, location, and function across species ranging from bacteria to humans.

TABLE 19.3 Some Sequenced Genomes

Kingdom: Archaea	Genomes sequenced: 9		
Organism	*Number of Genes*	*Number of Chromosomes*	*Reference*
Methanococcus jannaschii	1,738	1	*Science* (1996) 273:1058–1073
Archaeoglobus fulgidis	2,493	1	*Nature* (1997) 390:364–370
Kingom: Bacteria	**Genomes Sequenced: 36**		
Organism	*Number of Genes*	*Number of Chromosomes*	*Reference*
Haemophilus influenzae	1,850	1	*Science* (1995) 269:496–512
Mycoplasma genitalium	482	1	*Science* (1995) 270:397–403
Escherichia coli	4,288	1	*Science* (1997) 277:1453–1474
Kingdom: Eukarya	**Genomes Sequenced: 10**		
Organism	*Number of Genes*	*Number of Chromosomes*	*Reference*
Saccharomyces cerevisiae	6,241	16	*Nature* (1997) 387:5–105
Caenorhabditis elegans	18,425	6	*Science* (1998) 282:2012–2018
Drosophila melanogaster	13,601	4	*Science* (2000) 287:2185–2195

19.7   The Human Genome Project: An Overview

The **Human Genome Project (HGP)** began as a coordinated, international effort to sequence the more than 3 billion nucleotide pairs in the haploid human genome. In the United States, the project started in 1990 as a joint program of the National Institutes of Health (NIH) and the U.S. Department of Energy. Other countries, notably France, Britain, and Japan, began similar projects, which are now coordinated by an international organization, the Human Genome Organization (HUGO).

The task of sequencing the human genome was planned as a series of well-defined phases. However, work on all project phases has progressed more or less simultaneously in laboratories around the world. The HGP originally used a clone-by-clone method (see Chapter 18 for review), and includes these phases:

1. Construction of high-resolution genetic maps for each human chromosome, with a 2– to 5–cM average distance between markers. As you may recall, genetic maps are based on recombination frequencies and give the order and distance between different loci along the length of a chromosome. For this phase, researchers used identified genes, RFLPs, and other markers to assemble the maps, which have an average distance of 2 million base pairs of DNA between markers.

2. Physical maps of each chromosome. Physical maps show the order and physical distance between markers expressed in base pairs of DNA. Several methods have been used to make these maps; they consist of at least 30,000 markers, spaced at intervals of about 100 kb, for each chromosome.

3. An overlapping set of clones for each chromosome. This phase uses YACs (yeast artificial chromosomes) and other large-capacity cloning vectors to generate sets of overlapping clones that completely cover each chromosome. The original strategy divided chromosomes into a number of segments, then isolated and mapped the clones that cover each segment.

4. Genome sequencing. The project's ultimate goal is to determine the nucleotide sequence in the human genome. This phase was started in six U.S. centers in 1998 to improve sequencing technologies and to begin large-scale sequencing. Similar centers were established in other countries.

A Progress Report

In addition to publicly funded efforts led by NIH in the United States, a private effort was launched in 1998. Craig Venter, the scientist who developed the shotgun method of genome sequencing, and used it to sequence several prokaryotic genomes, formed a company called Celera to sequence the human genome. In partnership with PE Biosystems, which manufactures automated DNA-sequencing machines, Venter announced he could sequence the human genome in 3 years at a cost of $200–300 million (compared to 10 years and about $3 billion for the publicly funded HGP).

The NIH-funded project and the Celera corporate project announced in June 2000 that they had each finished a draft sequence of the human genome and were beginning to annotate and analyze the sequence data. Each project published an analysis of the data in a draft sequence of the genome in February 2001. The analysis is still incomplete because hundreds of small gaps in the sequence remain to be filled. Each team estimates that it will take 18–20 months to completely fill in the gaps. Some of the features revealed by the analysis of the draft sequence were discussed in Chapter 18.

The genomes of four of the five model organisms in the HGP, *E. coli* (bacteria), *S. cerevisiae* (yeast,), and *C. elegans* (a metazoan roundworm), and the fruit fly (*D. melanogaster*) have been completed. Both genome teams are sequencing the genome of the mouse, and expect to complete this project in the next two years.

It is apparent that sequencing the human genome will profoundly advance our knowledge of human genetics, and will have a great impact on biomedical research and health care. Applications of this knowledge raise unique ethical, social, and legal issues.

The Ethical, Legal, and Social Implications Program

Scientists and the public have raised numerous concerns about how the HGP genome information will be used, and how the interests of both individuals and society can be protected. To address these concerns, the Ethical, Legal, and Social Implications (ELSI) Program was established. Part of the HGP, this program is considering a number of issues, including impacts on individuals; privacy and confidentiality; and implications for medical practice, genetic counseling, and reproductive decision making. Through research grants, workshops, and public forums, ELSI is formulating policy options to address these issues.

Currently, ELSI is focusing on four areas: (1) privacy and fairness in the use and interpretation of genetic information, (2) how genetic knowledge should be transferred from the research laboratory to clinical practice, (3) informed consent for participants in genetic research, and (4) public and professional education. Hopefully, as the HGP moves from generating information about the genetic bases of disease to improved treatments, prevention, and cures, ethical issues will be identified and an international consensus will be developed on appropriate policies and laws.

After the Genome Projects

Although analysis of the human genome is still in its early stages, important insights into human genetic disorders are already emerging. A new mechanism of mutation, the expansion of trinucleotide repeats (see Chapter 14), has been discovered. This mechanism is responsible for several neurodegenerative disorders, including Huntington disease, myotonic dystrophy, and spinal/bulbar muscular atrophy. Another unexpected finding is that gene dosage can be increased by duplicating small regions of chromosomes; this leads to disorders such as Charcot–Marie–Tooth syndrome. Most intriguing, we now know that mutations in different parts of a single gene can give rise to different genetic disorders (e.g., multiple endocrine neoplasia, thyroid carcinoma, and Hirschprung disease).

Analysis on the sequence data on the long arms of chromosomes 21 and 22 has provided new and unexpected insight into the actual number of genes carried in the human genome. On the long arm of chromosome 21, 225 genes have been identified over a distance of 33.55 Mb (million base pairs). This region represents about 1 percent of the genome, and should contain 800–1000 genes if the total number of human genes is about 100,000 (as predicted). Coupled with the relatively low gene density on the long arm of chromosome 22 (545 genes over 33.46 Mb), and the sequence data from the HGP, the number of genes in the human genome may turn out to be much lower than predicted, perhaps on the order of 30,000–35,000.

The shift in research focus on the model organisms such as yeast and *Drosophila* gives us an idea of what to expect once a genome has been sequenced. In these organisms, most of the genes identified from sequence data are newly discovered. Attention is now shifting to gene expression. What genes are active in a cell? When they are active? What protein functions do they encode? As outlined in Chapter 18, researchers are analyzing the **transcriptome**, the set of mRNA molecules produced by a cell at any given time, and the **proteome**, the expressed set of proteins present in the cell. As these fields develop, we can expect to see a new generation of drugs, derived from functional analyses of the expressed genome and the proteome.

19.8   Biotechnology

Although recombinant DNA techniques were originally developed to facilitate basic research on gene organization and regulation of expression, scientists have not been blind to the commercial possibilities of this technology. As a result, researchers have helped form biotechnology companies that use recombinant DNA technology to develop products such as hormones, clotting factors, herbicide-resistant plants, enzymes for food production, and vaccines. In the last decade, the biotechnology industry has grown into a multibillion-dollar segment of the economy.

Insulin Production by Bacteria

The first human gene product manufactured using recombinant DNA and licensed for therapeutic use was human insulin, which became available in 1982. Insulin is a protein hormone that regulates sugar metabolism. An inability to produce insulin results in diabetes, a disease that in its more severe form affects more than 2 million individuals in the United States.

Clusters of cells embedded in the pancreas synthesize a precursor peptide known as preproinsulin. As preproinsulin is secreted from the cell, amino acids are cleaved from the end and middle of the chain. This produces the mature insulin molecule, which consists of two polypeptide chains (A and B chains), joined by disulfide bonds.

Although human insulin is now produced by another process, a look at the original process is instructive, as it shows both the promises and difficulties of recombinant DNA technology. The functional insulin protein consists of two polypeptide chains, A and B. The A subunit has 21 amino acids; the B subunit has 30. In the original process, synthetic genes for the A and B subunits were constructed by oligonucleotide synthesis (63 nucleotides for A polypeptide and 90 nucleotides for B polypeptide). Each synthetic oligonucleotide was inserted into a vector at a position adjacent to a gene encoding the bacterial form of the enzyme, β-galactosidase. When transferred to a bacterial host, the *β-gal* gene and the adjacent synthetic oligonucleotide were transcribed and translated as a unit. The product, a **fusion polypeptide**, included the amino acid sequence for β-galactosidase attached to the amino acid sequence for one of the insulin subunits (Figure 19–15, page 406). The fusion proteins were purified from bacterial extracts and treated with cyanogen bromide to cleave the insulin subunits from the β-galactosidase. The insulin subunits were then separated and purified.

Each insulin subunit was produced separately by this process. When mixed, the two subunits spontaneously united, forming an intact, active insulin molecule. The purified insulin was then packaged for use by diabetics who must take insulin injections.

A number of genetically engineered proteins for therapeutic use have been produced by similar processes or are in clinical trials (Table 19.4, page 406).

Transgenic Animal Hosts and Pharmaceutical Products

Bacterial host cells were used to produce the first generation of recombinant proteins, even though there are some disadvantages in using prokaryotic hosts to synthesize eukaryotic proteins. For example, bacterial cells are unable to process and modify eukaryotic proteins. Thus, they cannot add sugars and phosphate groups that are often needed for full biological activity. In addition, eukaryotic proteins produced in prokaryotic cells often don't fold into the proper three-dimensional configuration and, as a result, are inactive. To overcome these difficulties and to increase yields, second-generation methods use eukaryotic hosts. Rather than being produced by host cells grown in tissue culture, human proteins such as α-1-antitrypsin are produced in the milk of livestock.

A deficiency of the enzyme α-1-antitrypsin is associated with the heritable form of emphysema, a progressive and fatal respiratory disorder common among those of European

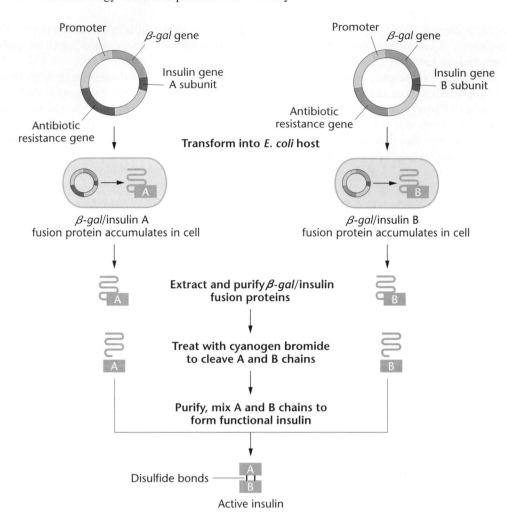

FIGURE 19–15 The method used to synthesize recombinant human insulin. Synthetic oligonucleotides encoding the insulin A and B chains are inserted at the tail end of a cloned *E. coli* β-galactosidase (*β-gal*) gene. These recombinant plasmids are transferred to *E. coli* hosts, where the *β-gal*/insulin fusion protein is synthesized and accumulates in the host cells. Fusion proteins are extracted from the host cells and purified. Insulin chains are released from the β-galactosidase by treatment with cyanogen bromide, and the insulin subunits purified and mixed to produce a functional insulin molecule.

TABLE 19.4   Genetically Engineered Pharmaceutical Products Now Available or in Clinical Testing

Gene Product	*Condition Treated*
Atrial natriuretic factor	Heart failure, hypertension
Epidermal growth factor	Burns, skin transplants
Erythropoietin	Anemia
Factor VIII	Hemophilia
Gamma interferon	Cancer
Granulocyte colony-stimulating factor	Cancer
Hepatitis B vaccine	Hepatitis
Human growth hormone	Dwarfism
Insulin	Diabetes
Interleukin-2	Cancer
Superoxide dismutase	Transplants
Tissue plasminogen activator	Heart attacks

ancestry. To produce α-1-antitrypsin by genetic engineering, the human gene was cloned into a vector at a site adjacent to a sheep promoter sequence that regulates the expression of milk-associated proteins. Genes placed next to this promoter are expressed only in mammary tissue. This fusion gene was then microinjected into sheep zygotes fertilized *in vitro* (Figure 19–16), which in turn were implanted into foster mothers. The resulting transgenic sheep developed normally and, after mating, produced milk that contains high concentrations of functional human α-1-antitrypsin. This human protein is present in concentrations up to 35 g/L of milk. It is easy to envision that a small herd of lactating sheep could provide an adequate supply of this protein and that herds of other transgenic animals, acting as biofactories, might become a routine part of the pharmaceutical repertoire. In fact, the cloning of Dolly the sheep (see essay) was done to establish a flock of sheep that will produce high levels of human proteins.

Transgenic Crop Plants and Herbicide Resistance

Damage from weed infestation destroys about 10 percent of crops worldwide. To combat this problem, more than 100 different herbicides are used, at an annual cost of more than 10 billion dollars. Some herbicides also kill crop plants, while others remain in the environment, or are present in runoff and contaminate water supplies. Creating herbicide-resistant plants using recombinant DNA technology is one way to improve crop yields. Vectors transfer the traits for herbicide resistance to crop plants. For example, the herbicide **glyphosate** is effective at very low concentrations, is not toxic to humans, and is rapidly degraded by soil microorganisms. However, glyphosate works by inhibiting the action of a chloroplast enzyme called EPSP synthase. This enzyme is important in amino acid biosynthesis in both bacteria and plants. Without the ability to synthesize these vital amino acids, plants, including crop plants, wither and die.

The development of glyphosate-resistant crop plants began with the isolation and cloning of an EPSP synthase gene from a glyphosate-resistant strain of *E. coli*. This gene was cloned into a vector adjacent to plant promoter sequences and upstream from plant transcription-termination sequences. The recombinant vector was transferred into the bacterium *Agrobacterium tumifaciens* (Figure 19–17, page 408). Plasmid-carrying bacteria were in turn used to infect cells in discs cut from plant leaves. Calluses formed from these discs were then selected for their ability to grow on glyphosate. Transgenic plants generated from glyphosate-resistant calluses were grown and sprayed with glyphosate at concentrations four times higher than needed to kill wild-type plants. The transgenic plants overproducing the EPSP synthase thrived, while the control plants withered and died. Glyphosate-resistant corn and soybeans developed in this same manner are now on the market.

While use of genetically engineered crops should reduce herbicide costs to the farmer, they present an ethical dilemma. Is it appropriate to generate crops that encourage the use of herbicides? Another ethically debatable application of genetic engineering to agriculture is the introduction of genes into crop plants that make the plants resistant to viruses, bacteria, fungi, and insect infestation. Other work seeks to improve drought resistance and the nutritional value of crops such as soybeans and corn. Many of these projects are in the developmental stage, and their transgenic products will begin to reach the marketplace over the next few years.

Edible Vaccines

One of the most beneficial applications of recombinant DNA technology may be in the production of vaccines. Vaccines stimulate the immune system to produce antibodies against a disease-causing agent and thereby confer immunity against the disease. Two types of vaccines are commonly used: inactivated vaccines, prepared from killed samples of the infectious virus or bacteria, and attenuated vaccines, which are live viruses or bacteria that can no longer reproduce and cause disease when present in the body.

Using recombinant DNA technology, a new type of vaccine called a subunit vaccine is being produced. These vaccines consist of one or more surface proteins of the virus or bacterium. The protein acts as an antigen that stimulates the immune system to make antibodies against the virus or bacterium. One of the first licensed subunit vaccines was for a surface protein of hepatitis B, a virus that causes liver damage and cancer. The gene for the hepatitis B surface protein is cloned into a yeast-expression vector and produced using yeast as the host. The protein is extracted and purified from the host cells and then packaged for use as a vaccine.

Recombinant DNA technology is currently used to produce the hepatitis B surface protein in edible plants. The idea is to have plant leaves or fruits serve as the source of an oral vaccine. Plant-produced vaccines offer several advantages. They are inexpensive, as industrial investment for production and extensive purification is unnecessary. In addition, such

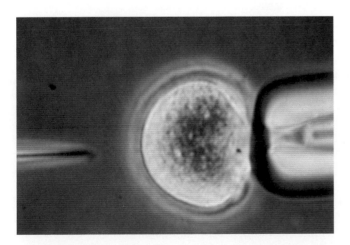

FIGURE 19–16 Injection of cloned DNA. A micropipette transfers cloned genes into the nucleus of a mammalian zygote. The injected zygote will then be transferred to the uterus of a surrogate mother for development. (*Hank Morgan/Photo Researchers, Inc.*)

FIGURE 19–17 Gene transfer of glyphosate resistance. The EPSP gene is fused to a promoter from cauliflower mosaic virus. This chimeric or fusion gene is then transferred to a Ti plasmid vector, and the recombinant vector is inserted into an *Agrobacterium* host. *Agrobacterium* infection of cultured plant cells transfers the EPSP fusion gene into a plant cell chromosome. Cells that acquire the gene are able to synthesize large quantities of EPSP synthase, making them resistant to the herbicide glyphosate. Resistant cells are selected by growth in herbicide-containing medium. Plants regenerated from these cells are herbicide resistant.

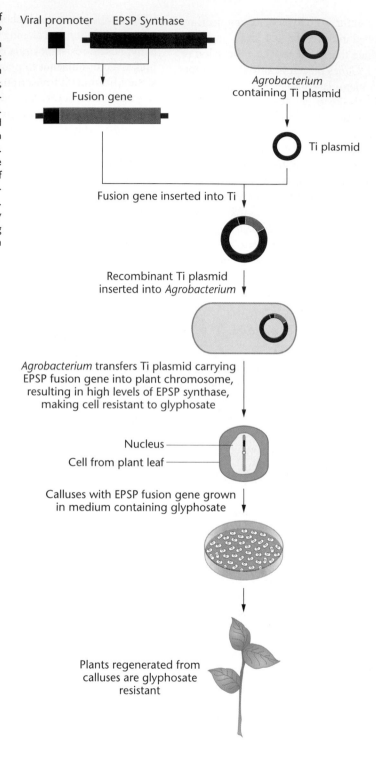

vaccines do not require injection and will be especially useful in developing countries where medical services and health-care delivery are not widely available.

As a model, the antigenic subunit of hepatitis B vaccine has been transferred to tobacco plants and expressed in the leaves (Figure 19–18). To use it as a source of vaccine, the gene will be inserted into food plants such as grains or vegetables.

Recent clinical tests have used genetically engineered potatoes that carry recombinant bacterial antigens. Human volunteers ate small quantities of these potatoes (50–100 g) and developed antibodies to the antigen, establishing that edible vaccines may be feasible. Similar tests using genetically engineered bananas are underway. If such tests are successful, genetically engineered edible plants will soon be used to vaccinate people against many diseases.

Plants and animals were domesticated some 8000–10,000 years ago, and we have been modifying these organisms by selective breeding ever since, producing the diversity of domesticated plants and animals present today. Recombinant DNA technology has changed the rate at which new plants and animals can be developed and, by transferring genes between species, has altered the types of changes that can be made. Unlike selective breeding programs, biotechnology has generated concerns about releasing genetically modified organisms into the environment and about the safety of eating such products. If biotechnology is to achieve a new green revolution, these concerns need to be addressed through prudent research and education.

FIGURE 19–18 Tobacco leaves expressing the hepatitis B antigen. Transgenic tobacco plants carrying an antigenic subunit of hepatitis B virus were first generated, then leaves from transgenic plants were treated with antibodies against hepatitis B antigen. The results showed that the plants produced the antigen. The central leaf is from a normal tobacco plant, and is unstained. *(Charles J. Arntzen, President & CEO, Boyce Thompson Institute, Ithaca, NY)*

Genetics, Technology, and Society

Beyond Dolly: The Cloning of Humans

The birth of Dolly, a healthy white-faced lamb, took the world by surprise. Until February 1997, the idea that an animal could be cloned from the cells of another adult animal had heretofore been the stuff of science fiction—something from *Brave New World* or *The Boys from Brazil*. For many people, the notion that animals could be cloned conjured up scenarios with multiple Adolf Hitlers or creations gone astray, as in Shelley's Frankenstein. For decades, scientists had comforted the public, asserting that it would be impossible to clone a mammal from cells of an adult. It was thought that cells from adult animals could not be reprogrammed in such a way as to form an entire new organism. Then came Dolly.

Dolly—the first mammal cloned from adult cells—was brought into being by a group of embryologists led by Ian Wilmut and Keith Campbell at the Roslin Institute in Scotland. Their goal was to clone identical transgenic animals that secrete pharmaceutical products such as blood-clotting factors or insulin into their

milk. The cloning method that Wilmut and Campbell used to create Dolly—a procedure called nuclear transfer—involves replacing the nucleus of an egg with the nucleus from an adult cell, thereby creating a reconstructed zygote containing the nucleus from one animal and the egg cytoplasm from another. In theory, the genetic information in the donor nucleus should direct all subsequent embryonic development, and the new organism should be a genetic replica of the adult that donated the nucleus. Although this procedure is simple in theory, it proved to be extremely difficult in practice. Even when technical procedures such as collecting eggs, removing egg nuclei, and fusing donor cell and recipient egg were perfected, the cloning procedure would work only if the donor nucleus came from an embryonic cell. Because adult nuclei come from specialized cells such as skin, liver and kidney cells, they are "differentiated" and express only a small subset of all the genes in the nuclear genome. As many genes in the genome must be actively expressed in an embryo, adult nuclei proved to be inadequate for cloning.

Wilmut and Campbell accomplished the impossible by reprogramming the adult donor cells before transferring their nuclei into recipient eggs. They removed donor cells from the udder of an adult ewe and starved the cells, so that they became quiescent and entered the G0 phase of the cell cycle. For reasons unknown, this starvation procedure allows the silent genes within the udder cell nucleus to be turned on after the nucleus is transferred to the cytoplasm of a sheep's egg. In addition, the researchers passed an electric current through the egg to facilitate nuclear transfer, and to stimulate the new reconstructed zygote to begin dividing. To create Dolly, over 200 nuclear transfers were done using udder cell nuclei. Of these, only 29 developed into embryos and 13 were implanted into surrogate mother ewes. One pregnancy resulted, which culminated in the birth of Dolly.

Since 1997, cloning of farm animals has become common. Nuclear transfer technology has been used to clone cows, mice, goats, and pigs, as well as transgenic sheep. For example, a cloned transgenic sheep ("Polly") bears the human gene for blood-clotting factor IX, and secretes factor IX into her milk, paving the way for production of pharmaceuticals from cloned farm animals. Besides benefits for drug production, cloning may be used to rescue endangered species, cure genetic diseases in farm animals, produce pig organs for human transplantation, and provide animal models of human diseases for which there are presently no research models.

Beyond these agricultural and medical benefits lies the controversial topic of *human cloning*. The idea that humans could be cloned has been termed "immoral," "repugnant," and "ethically wrong." Within days of Dolly's announcement, bills were introduced into the U.S. Congress, prohibiting research into human cloning, and worldwide bans were called for. Frightening scenarios were proposed: rich and powerful people cloning themselves out of vanity, people with serious illnesses cloning replicas to act as organ donors, and legions of human clones suffering loss of autonomy, individuality, and kinship ties. But is it really possible to clone humans? And if so, should human cloning ever be done?

The answers to these questions are not straightforward. First, it may not be possible to clone humans using nuclear transfer techniques. The only success in cloning primates (two rhesus monkeys, cloned from embryonic cells) has not been duplicated, suggesting that nuclear transfer may be inadequate for primate cloning. Second, the prospect of cloning whole human beings (human reproductive cloning) presents serious safety and ethical problems. Cloning of animals is still a very inefficient process, with less than 2 percent of nuclear transfers resulting in live births. Many cloned animals exhibit developmental defects, and it is still uncertain whether cloned humans would undergo accelerated aging or be subject to higher cancer rates. Many feel it is unacceptable to subject human clones to such risks. If we did clone humans, would this seriously alter our concepts of uniqueness? Would cloning be misused for unscrupulous social or political goals? Despite these serious reservations, it is possible to foresee positive aspects to human reproductive cloning. Infertile couples, or those who suffer genetic disease on one side of the family, could choose to clone one of the partners in order to raise a child who is biologically related. However, this, too, raises profound moral issues.

The prospect of cloning human cells for therapeutic purposes has also been greeted with mixed reviews. Human therapeutic cloning is a method of producing cloned cells from a patient, with the goal of transplanting these cells back to the patient to replace damaged tissues. The patient's nuclei would be introduced into oocytes, the reconstructed zygotes induced to divide to the blastocyst stage, and the embryonic stem cells removed and grown in culture. The stem cells would then be differentiated into various tissues such as cardiomyocytes (to replace damaged heart tissue), insulin-producing β-cells (for diabetes transplant), or dopaminergic neurons (to treat Parkinson's disease). Many technical hurdles still exist before human therapeutic cloning can be used; however, many people believe that the benefits of therapeutic cloning outweigh the ethical concerns. Ethical concerns include worries that therapeutic cloning opens the door to human reproductive cloning, as well as reservations about creating and then destroying potential human embryos.

Thanks to Dolly, we are entering an era in which human cloning is no longer a fantasy. We must now consider all aspects of human reproductive and therapeutic cloning, to ensure responsible and beneficial outcomes of this new technology.

References

Cibelli, J. B., et al. 1998. Cloned transgenic calves produced from nonquiescent fetal fibroblasts. *Science* 280:1256–58.

Pennisi, E. 1998. After Dolly, a pharming frenzy. *Science* 279:646–48.

Robertson, J. A. 1998. Human cloning and the challenge of regulation. *N. Eng. J. Med.* 339:118–25.

Solter, D. 1998. Dolly is a clone—and no longer alone. *Nature* 394: 315–16.

Wilmut, I. Cloning, for medicine. *Scientific American*, pp. 58–63. December, 1998.

Wilmut, I., et al. 1997. Viable offspring derived from fetal and adult mammalian cells. *Nature* 385:810–13.

Additional References

Lanza, R. P., et al. 1999. Prospects for the use of nuclear transfer in human transplantation. *Nature Biotechnology* 17:1171–74.

Pennisi, E., and Vogel, G. 2000. Clones: A hard act to follow. *Science* 288:1722–27.

Chapter Summary

1. Recombinant DNA technology offers a new approach to genetic analysis. Instead of relying on isolation and mapping of mutant genes, large cloned segments of the genome are manipulated to make genetic and physical maps that use molecular markers rather than phenotypes visible at the level of the organism.

2. Genome projects are complete or underway for several organisms, including *E. coli*, yeast, *Drosophila*, the mouse, and humans. The goals of these genome projects are to identify and map all genes in the genome and to determine the complete nucleotide sequence of the genome.

3. Cloned DNA is finding a wide variety of applications, including gene mapping and the identification and isolation of genes responsible for genetic disorders. The method known as positional cloning, based on recombinant DNA methodology, allows a gene to be mapped and identified with no knowledge of the nature or function of the gene product.

4. Recombinant DNA techniques are used in the prenatal diagnosis of human genetic disorders. This method allows direct examination of the genotype, whereas previous methods relied on gene expression and the identification of the gene product. These methods can also identify carriers of genetic disorders, and they are the basis of proposals to screen the population for a number of genetic disorders, including sickle-cell anemia and cystic fibrosis.

5. The availability of cloned human genes has led to their use in gene therapy. In somatic gene therapy, a cloned normal copy of a gene is transferred into a vector; the vector then transfers the gene to a target tissue that takes up and expresses the cloned copy of the gene, thereby altering the mutant phenotype. The development of new and more effective vector systems probably means that gene therapy will become a standard method for treating of genetic disorders in the near future.

6. Recombinant DNA techniques that detect allelic variants of variable tandem nucleotide repeats (DNA fingerprints) have found applications in forensics, paternity testing, and a wide range of fields including archeology, conservation biology, and public health.

7. The biotechnology industry uses recombinant DNA methods to produce human gene products in a variety of hosts, ranging from bacteria to farm animals. In addition, gene transfer techniques are improving crop plants by transferring herbicide resistance. In the near future, it may be possible to vaccinate against infectious disease using food plants, making resistance to infectious agents almost universal.

Key Terms

adenosine deaminase (ADA), 400
allele-specific oligonucleotide (ASO), 397
amniocentesis, 396
centimorgan (cM), 395
chorionic villus sampling (CVS), 396
cystic fibrosis, 399
cystic fibrosis transmembrane conductance regulator (CFTR), 397
DNA fingerprinting, 402
DNA microarrays, 398
enhancement gene therapy, 401
Ethical, Legal, and Social Implications (ELSI) Program, 399

exclusion map, 395
familial hypercholesterolemia, 399
fusion polypeptide, 405
gene therapy, 399
germ line therapy, 401
glyphosate, 407
Human Genome Project, 404
lod score, 395
logarithm of the odds (lod) score, 395
minisatellite, 402
neurofibromin, 395
positional cloning, 396

proteome, 405
restriction fragment length polymorphism (RFLP), 393
severe combined immunodeficiency (SCID), 399
somatic gene therapy, 401
transcriptome, 405
type 1 neurofibromatosis (NF1), 395
variable-number tandem repeat (VNTR), 402

Problems and Discussion Questions

1. In attempting to vaccinate against diseases by eating antigens, the antigen (such as the cholera toxin) must be presented to the cells of the small intestine. What are some potential problems with this method? Why don't absorbed food molecules stimulate the immune system and make you allergic to the food you eat?

2. Outline the steps involved in transferring glyphosate resistance to a crop plant. Do you envision that this trait could escape from the crop plant and make weeds glyphosate-resistant? Why or why not?

3. Gene therapy for human genetic disorders involves transferring a copy of the normal human gene into a vector and using the vector to transfer the cloned human gene into target tissues. Presumably, the gene enters the target tissue, becomes active, and the gene product relieves the symptoms. Although gene therapy is now used to treat several disorders, there are some unresolved problems with this method.
   (a) Why are disorders such as muscular dystrophy difficult to treat by gene therapy?
   (b) What are the potential problems in using retroviruses as vectors?
   (c) In the long run, should gene therapy involve germ tissue instead of somatic tissue?

   (d) What are some ethical dilemmas associated with this approach?

4. In producing physical maps of markers and cloned sequences, what advantage does *Drosophila* offer that other organisms including humans do not?

5. Outline the steps in identifying a gene by positional cloning. What steps can cause difficulty in this process? Once a region on a chromosome has been identified as containing a given gene, what kinds of mutations would speed the process of identifying the locus?

6. The phenotype of many behavior traits such as bipolar disorder or schizophrenia may be controlled by several genes, each at a different locus. Can positional cloning be used to map and isolate such genes? What if a trait is controlled by six genes, each contributing equally in an additive way to the phenotype? Can positional cloning be used in this case? Why or why not?

7. Suppose you develop a screening method for cystic fibrosis that allows you to identify the predominant mutation (*Δ508*) and the next six most prevalent mutations. What do you need to consider before using this method to screen the population at large for this disorder?

Insights and Solutions

1. DNA fingerprints are used to test forensic specimens, identify criminal suspects, and settle immigration cases and paternity disputes. Probes for DNA fingerprinting can be derived from a single locus or multiple loci. Two multiple-loci probes have been widely employed in both criminal and civil cases and derive from minisatellite loci on chromosome 1 (1cen-q24) and chromosome 7 (7q31.3). These probes, widely used because they produce a highly individual fingerprint, have determined paternity in thousands of cases over the last few years.

(a) Shown here are the results of DNA fingerprinting of a mother (M), putative father (F), and child (C) using these probes.

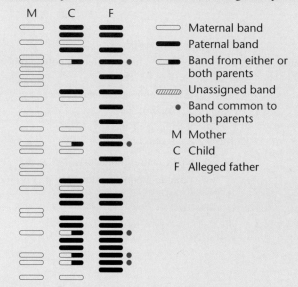

As shown, the child has 6 maternal bands, 11 paternal bands, and 5 bands shared between the mother and the alleged father. Based on this fingerprint, can you conclude that the man tested is the father of the child?

Solution: All bands present in the child can be assigned as coming from either the mother or the father. In other words, all the bands in the child's DNA fingerprint that are not maternal are present in the father. Because the father and the child have 11 bands in common and the child has no unassigned bands, paternity can be assigned with confidence. In fact, the chance that the man tested is *not* the father is on the order of 10^{13}.

(b) The fingerprints of a second mother, alleged father, and child are as follows:

In this case, the child has 8 maternal bands, 15 paternal bands, 6 bands that are common to both the mother and the alleged father, and 1 band not present in either the mother or the alleged father. What are the possible explanations for the presence of this band? Based on your analysis of the band pattern, which explanation is most likely?

Solution: In this case, one band in the child cannot be assigned to either parent. Two possible explanations for this finding are that the child is mutant for one band, or that the man tested is not the father. In estimating probability of paternity, the mean number of resolved bands (*n*) is determined, and the probability (*p*) that a band in individual A matches that in a second unrelated individual,

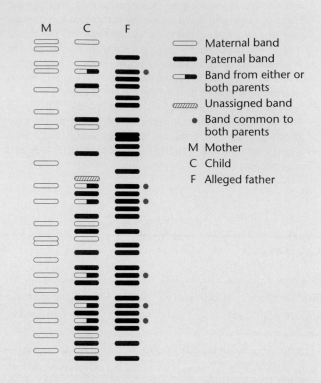

B, is calculated. In this case, because the child and the father share 15 bands in common, the probability that the man tested is *not* the father is very low (10^7 or lower). As a result, the most likely explanation is that the child is a mutant for a single band. In fact, in 1419 cases of genuine paternity resolved by the minisatellite probes on chromosomes 1 and 7, single-mutant bands in the children were recorded in 399 cases, accounting for 28 percent of all cases.

2. Infection by HIV (human immunodeficiency virus) destroys cells in the immune system and eventually results in the symptoms of AIDS (acquired immunodeficiency syndrome). HIV infects and kills cells of the immune system that carry a cell-surface receptor known as CD4. An HIV surface protein known as gp120 binds to the CD4 receptor and allows the virus entry into the cell. Now that the gene that encodes the CD4 protein has been cloned, how can this clone and recombinant DNA techniques be used to combat HIV infection?

Solution: Several methods that use the *CD4* gene to combat HIV infection are being explored. First, because HIV infection depends on an interaction between the viral gp120 protein and the CD4 protein, the cloned *CD4* gene has been modified to produce a soluble form of the protein (sCD4). The idea is that HIV can be prevented from infecting cells if the gp120 protein of the virus is bound with sCD4, and thus made unable to bind to CD4 proteins. Studies in cell culture systems indicate that the presence of sCD4 effectively prevents HIV infection of tissue culture cells. However, studies in HIV-positive humans have been somewhat disappointing, mainly because the strains of HIV used in the laboratory are different from those found in infected individuals.

HIV-infected cells carry the viral gp120 protein on their surface. To kill such cells, a second approach fuses the *CD4* gene with those that encode bacterial toxins. The resulting fusion protein contains the CD4 regions that bind to gp120 and the toxin regions that kill the infected cell. In tissue culture experiments, the fusion protein kills HIV-infected cells, whereas uninfected cells survive. It is hoped that the targeted delivery of drugs and toxins can be used in therapeutic applications to treat HIV infection.

8. The DNA sequence for normal and mutant genes surrounding the site of the sickle-cell mutation in β-globin is shown here:

5′ G A C T C C T G A G G A G A A G T 3′

3′ C T G A G G A C T C C T C T T C A 5′

Normal DNA

5′ G A C T C C T G T G G A G A A G T - 3′

3′ C T G A G G A C A C C T C T T C A - 5′

Sickle-cell DNA

Each type of DNA is denatured into single strands and applied to a filter. The paper containing the two spots is hybridized to an ASO of the following sequence:

5′-GACTCCTGAGGAGAAGT-3′

Which pattern of hybridization will be observed? Why?

9. Dominant mutations can be categorized according to whether they increase or decrease overall activity of a gene or gene product. Although a "loss-of-function" mutation (i.e., a mutation that inactivates the gene product) is usually recessive, for some genes one dose of the gene product is not sufficient to produce a normal phenotype. In this case, a loss-of-function mutation in the gene will be dominant and the gene is said to be *haploinsufficient*. A second category of dominant mutations, "gain-of-function" mutations, results in increased activity or expression of the gene or gene product. The phenotype of these mutations results from too much gene product. The gene therapy technique currently used in clinical trials involves adding a normal copy of a gene to somatic cells. In other words, a normal copy of the gene is inserted into the genome of the mutant somatic cell, but the mutated copy of the gene is not removed or replaced. Will this strategy work for either of these two types of dominant mutations?

10. One form of hemophilia, an X-linked disorder of blood clotting, is caused by mutation in clotting factor VIII. Many single-nucleotide mutations of this gene have been described, making detection of mutant genes by Southern blots inefficient. However, an RFLP for the enzyme *Hind*III contained in an intron of the factor VIII gene can often be used in screening, as shown here.

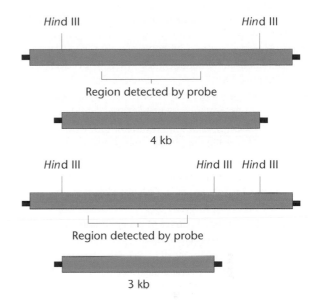

A female whose brother has hemophilia is at 50 percent risk of being a carrier for this disorder. To test her status, DNA is obtained from white blood cells from family members, cut with *Hind*III, and the fragments probed and visualized by Southern blotting. The results are shown below.

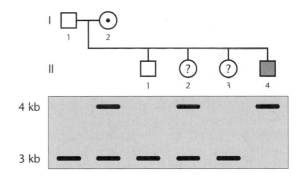

Determine whether any females in generation II are carriers for hemophilia.

11. The human insulin gene contains a number of introns. Despite the fact that bacterial cells will not excise introns from mRNA, how can a gene like this be cloned into a bacterial cell to produce insulin?

12. In mice transfected with the rabbit β-globin gene, the rabbit gene is active in a number of tissues including spleen, brain, and kidney. In addition, some mice suffer from thalassemia, caused by an imbalance in the coordinate production of α- and β-globins. What problems associated with gene therapy are demonstrated by these findings?

Selected Readings

Anderson, W. F. 2000. The best of times, the worst of times. *Science* 288:627–28.

Amos, B., and Pemberton, J. 1993. DNA fingerprinting in nonhuman populations. *Curr. Opin. Genet. Dev.* 2:857–60.

Barker, D., et al. 1987. Gene for von Recklinghausen neurofibromatosis is in the pericentromeric region of chromosome 17. *Science* 236:1001–02.

Cavazzana-Calvo, M., et al. 2000. Gene therapy of human severe combined immunodeficiency (SCID)-X1 disease. *Science* 288:669–672.

Chang, J. C., and Kan, Y. W. 1981. Antenatal diagnosis of sickle-cell anemia by direct analysis of the sickle mutation. *Lancet* 2:1127–29.

Cipriano, F. and Palumbi, S. R. 1999. Genetic tracking of a protected whale. *Nature* 397:307–08.

Collins, F. S. 1995. Ahead of schedule and under budget: The Genome Project passes its fifth birthday. *Proc. Natl. Acad. Sci. USA* 92:10821–23.

Crystal, R. G. 1995. Transfer of genes to humans: Early lessons and obstacles to success. *Science* 270:404–10.

Danna, K., and Nathans, D. 1971. Specific cleavage of simian virus 40 DNA by restriction endonuclease of *Hemophilus influenzae*. *Proc. Natl. Acad. Sci. USA* 68:2913–17.

Dietrich, W. F., et al. 1995. Mapping the mouse genome: current status and future prospects. *Proc. Natl. Acad. Sci. USA* 92:10849–53.

Dujon, B. 1996. The yeast genome project: What did we learn? *Trends Genet.* 12:263–70.

Fleischmann, R. D., et al. 1995. Whole genome random sequencing and assembly of *Haemophilus influenzae* Rd. *Science* 269:496–512.

Fraser, C. M., et al. 1995. The minimal gene complement of *Mycoplasma genitalium*. *Science* 270:397–403.

Goodman, H. M., Ecker, J. R., and Dean, C. 1995. The genome of *Arabadopsis thaliana*. *Proc. Natl. Acad. Sci. USA* 92:10831–35.

Graves, D. J. 1999. Powerful tools for genetic analysis come of age. *Trends Biotech.* 17:127–134.

Guyer, M. and Collins, F. S. 1995. How is the Human Genome Project doing, and what have we learned so far? *Proc. Natl. Acad. Sci. USA* 92:10841–48.

Hacia, J. G., Collins, F. S. 1999. Mutational analysis using oligonucleotide microarrays. *J. Med. Genet.* 36:730–36.

Hermeren, G. 1999. Neonatal screening; ethical aspects. *Acta Paediatr.* Suppl. 432:99–103.

Hudson, T. J., et al. 1995. An STS-based map of the human genome. *Science* 270:1945–54.

Kay, M. A., et al. 2000. Evidence for gene transfer and expression of factor IX in haemophilia B patients treated with an AAV vector. *Nat. Genet.* 24:257–61.

Krumlauff, R., Jeanpierre, M., and Young, B. D. 1982. Construction and characterization of genomic libraries from specific human chromosomes. *Proc. Natl. Acad. Sci. USA* 79:2971–75.

Maniatis, T., Fritsch, E. F., and Sambrook, J. 1988. *Molecular cloning: A laboratory manual*, 2nd ed. Cold Spring Harbor, NY: Cold Spring Harbor Laboratory Press.

Marshall, E. 1995. Gene therapy's growing pains. *Science* 269:1050–55.

Marshall, E., and Pennisi, E. 1996. NIH launches the final push to sequence the genome. *Science* 272:188–89.

—————. 1998. Hubris and the human genome. *Science* 280:994–995.

Mason, H., Lam, D., and Arntzen, C. 1992. Expression of hepatitis B surface antigen in transgenic plants. *Proc. Natl. Acad. Sci. USA* 89:11745–49.

Mountain, A. 2000. Gene therapy: The first decade. *Trends Biotech.* 18:119–128.

Mullis, K. B. 1990. The unusual origin of the polymerase chain reaction. *Sci. Am.* (Apr.) 262:56–65.

Richter, L., Kipp, P. B. 1999. Transgenic plants as edible vaccines. *Curr. Top. Microbiol. Immunol.* 240:159–176.

Rudolph, N. 1999. Biopharmaceutical production in transgenic livestock. *Trends Biotech.* 17:367–374

Saiki, R. K., et al. 1986. Analysis of enzymatically amplified beta-globin and HLA-DQ alpha DNA with allele-specific oligonucleotide probes. *Nature* 324:163–66.

Tacket, C. O., et al. 1998. Immunogenicity in humans of a recombinant bacterial antigen delivered in a transgenic potato. *Nature Medicine* 4:607–609.

Travis, J. 1998. Another human genome project. Science News 153:334–335.

Venter, C., et al. 1998. Shotgun sequencing of the human genome. *Science* 280:1540–42.

Villa-Komaroff, L., et al. 1978. A bacterial clone synthesizing proinsulin. *Proc. Natl. Acad. Sci. USA* 75:3727–31.

Waterson, R., and Sulston, J. 1995. The genome of *Caenorhabditis elegans*. *Proc. Natl. Acad. Sci.* 92:10836–40.

This unusual four-winged *Drosophila* has developed an extra set of wings as a result of a homeotic mutation. *(Courtesy of E. B. Lewis, California Institute of Technology Archives)*

20

Genes and Development

CHAPTER CONCEPTS

The genetic basis of development is differential gene action. Gene expression varies over time in the same tissue and between different tissues present at the same developmental stage. Differential gene expression is required to bring about and maintain adult structures. Processes such as determination, differentiation and cell–cell interaction regulate the embryonic program of gene expression.

20.1 Basic Concepts in Developmental Genetics

20.2 Genetics of Embryonic Development in *Drosophila*
Overview of *Drosophila* Development
Genetic Analysis of Embryogenesis
Zygotic Genes and Segment Formation

20.3 Flower Development in *Arabidopsis*: Role of Homeotic Selector Genes

20.4 Cell–Cell Interactions in *C. elegans* Development
Overview of *C. elegans* Development
Genetic Analysis of Vulva Formation

In multicellular plants and animals, a fertilized egg, without further stimulus, begins a cycle of developmental events that ultimately give rise to an adult member of the species from which the egg and sperm were derived. Thousands, millions, or even billions of cells are organized into a cohesive and coordinated unit that we perceive as a living organism. Developmental biologists study the series of events through which organisms attain their final adult forms. This area of study is perhaps the most intriguing in biology. Not only does the comprehension of developmental processes require knowledge of many different biological disciplines—such as molecular, cellular, and organismal biology—it also requires an understanding of how biological processes change over time and ultimately transform the organism.

In the past hundred years, investigations in embryology, genetics, biochemistry, molecular biology, cell physiology, biophysics, and evolution have all contributed to our knowledge of development. The findings emphasize the tremendous complexity of developmental processes. Unfortunately, knowing what happens does not answer the "why" and "how" of development. Over the last two decades, however, genetic analysis and molecular biology have identified genes that regulate developmental processes, and we are beginning to understand how the action and interaction of these genes control development in a wide range of organisms.

In this chapter, emphasis will be placed on the role of gene action in regulating development. The field of developmental genetics is making tremendous strides in our understanding of developmental processes because genetic information is required both for the molecular and cellular functions mediating developmental events and for determining the phenotype of the newly formed organism.

20.1 Basic Concepts in Developmental Genetics

In higher eukaryotes, development begins with the formation of the **zygote**, the cell generated by the fusion of the sperm and oocyte. The oocyte is a cell with a cytoplasm that is heterogeneous and nonuniform in distribution. Following fertilization and early cell divisions, the nuclei of progeny cells find themselves in different environments as the maternal cytoplasm is distributed into the new cells. Evidence suggests that the cytoplasm in these cells exerts influence on the genetic material, causing differential transcription at specific points during development. Therefore, **cytoplasmic localization**, or the inheritance of particular cytoplasmic components by individual embryonic cells plays a major role in development.

Gene products synthesized soon after zygote formation further alter the cytoplasm of each cell, producing yet another cellular environment that in turn activates other genes, and so on. In addition, as cells increase in number, they influence one another through **cell–cell interaction**. The environment acting on the genetic material therefore now includes the cytoplasm as well as signals sent from other cells. Although most cells early in embryogenesis show no evidence of structural or functional specialization, the combination of their localized cytoplasmic components and their position in the developing embryo appears to determine their ultimate form. It is as if their fate has been programmed prior to the actual events leading to specialization.

As cells respond to their external and internal environments, which are continually changing, they embark on a pathway comprising a series of developmental stages. The most important stage on this pathway is **determination**, when, in response to many external and internal cues, the cell's specific developmental fate becomes fixed. We now know that determination precedes the actual events of **differentiation**, the process by which the cell achieves its final form and function.

In the following sections we examine how patterns of gene expression relate to the processes of determination, differentiation, and cell–cell interaction during development. We begin by examining the role of differential gene expression in the progressive restriction of developmental options.

20.2 Genetics of Embryonic Development in *Drosophila*

Why certain cells turn specific genes on or off during development is a central issue in developmental biology. And we do not have a simple answer. However, information derived from the study of many different organisms gives us a starting point. One example, embryonic development in *Drosophila*, reflects the scope of our knowledge in this area, and highlights the key role played by molecular components placed in the oocyte cytoplasm during oogenesis. Let's start with an overview of *Drosophila* development.

Overview of *Drosophila* Development

Beginning with the fertilized egg, *Drosophila* passes through a preadult period of development with five distinct phases: the embryo, three larval stages, and the pupal stage. The adult fly emerges from the pupal case about 10 days after fertilization (Figure 20–1). Externally, the *Drosophila* egg has a number of structures that delineate the anterior, posterior, dorsal, and ventral regions (Figure 20–2). The anterior end of the egg contains the micropyle, a specialized conical structure that helps sperm enter the egg, while the posterior end is rounded and marked by a series of aeropyles (openings that allow gas exchange during development). The dorsal side of the egg is flattened and contains the chorionic appendages, while the ventral side is curved. Internally, the egg cytoplasm is organized into a series of maternally derived molecular gradients. These gradients play a key role in establishing the developmental fates of nuclei that migrate into specific regions of the embryo.

Immediately after fertilization, the zygote nucleus undergoes a series of divisions. After nine rounds of division, most of the approximately 512 nuclei migrate to the egg's outer surface, or cortex, where further divisions take place [Figure 20–3 (c) and (d)]. This is the **syncytial blastoderm** stage of embryonic development (a syncytium is any cell

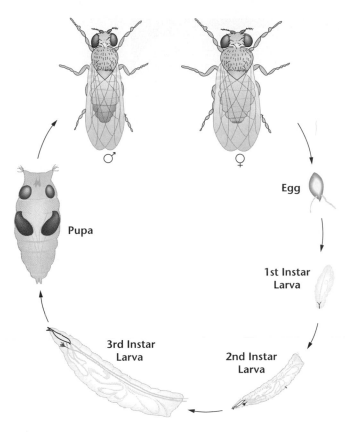

FIGURE 20–1 *Drosophila* life cycle.

with more than one nucleus). The nuclei are now surrounded by cytoplasm that contains maternally derived transcripts and proteins localized in these regions. Under the influence of these cytoplasmic components, a program of gene expression is initiated in the nuclei, leading to various developmental programs.

The formation of the germ cells at the posterior pole of the embryo demonstrates the regulatory role of these localized

FIGURE 20–2 Scanning electron micrograph of a newly oviposited *Drosophila* egg. The micropyle is the small white projection at the anterior tip (at the bottom). The dorsal surface (with the two chorionic appendages) faces up. (*William S. Klug, The College of New Jersey*)

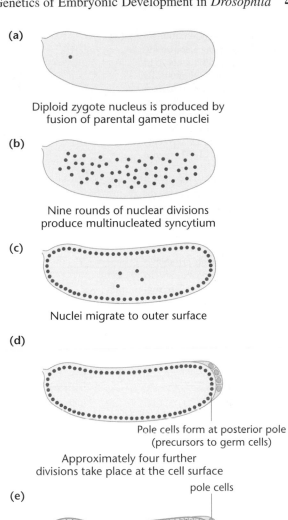

(a) Diploid zygote nucleus is produced by fusion of parental gamete nuclei

(b) Nine rounds of nuclear divisions produce multinucleated syncytium

(c) Nuclei migrate to outer surface

(d) Pole cells form at posterior pole (precursors to germ cells)
Approximately four further divisions take place at the cell surface

(e) pole cells
Nuclei become enclosed in membranes, forming a single layer of cells over embryo surface

FIGURE 20–3 Early stages of embryonic development in *Drosophila*. (a) Fertilized egg with zygotic nucleus, about 30 minutes after fertilization. (b) Nuclear divisions occur about every 10 minutes, producing a multinucleate cell, the syncytial blastoderm. (c) After approximately nine divisions (512 nuclei), the nuclei migrate to the outer surface or cortex of the egg. (d) At the surface, four additional rounds of nuclear division occur. A small cluster of cells, the pole cells, form at the posterior pole about 2.5 hours after fertilization. These cells will form the germ cells of the adult. (e) About 3 hours after fertilization, the nuclei become enclosed in plasma membranes and form a single layer of cells over the embryo surface creating, the cellular blastoderm.

cytoplasmic components [Figure 20–3(d) and (e)]. Experiments have shown that any nucleus placed in the posterior cytoplasm (the pole plasm) will generate cells that differentiate into germ cells. Hence the pole plasm contains maternal components that direct the posterior nuclei to give rise to germ cells. The transcriptional programs triggered in migrating nuclei also result in a body plan organized along anterior–posterior and dorsal–ventral axes of symmetry.

FIGURE 20–4 Segmentation in *Drosophila*. (a) Scanning electron micrograph of a *Drosophila* embryo at about 10 hours after fertilization. By this stage, the segmentation pattern of the body is clearly established. (b) The segments, compartments, and parasegments of the *Drosophila* embryo. The Ma, Mx, and Lb segments form head structures. T1–T3 are thoracic segments, and A1–A9 are abdominal segments. Each segment is divided into anterior (A) and posterior (P) compartments. Parasegments represent an early pattern specification that is later refined into the segmental plan of the body. Note that all the parasegments are shifted forward by one compartment to form the segments. (c) The segmented embryo and the adult structures that will form each segment. *(Photo: F. R. Turner/Department of Biology/Indiana University)*

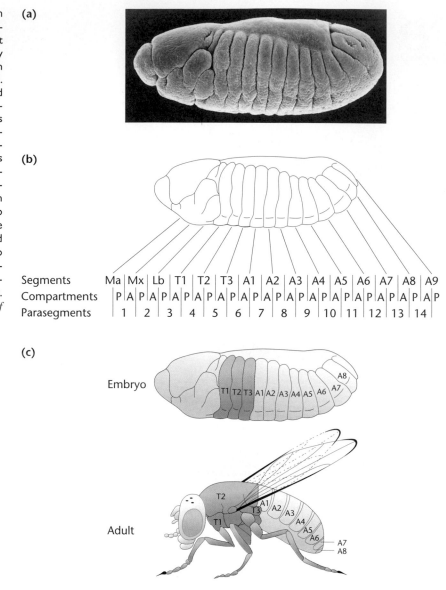

Following the organization of the syncytial blastoderm, plasma membranes form around individual nuclei, creating the **cellular blastoderm** [Figure 20–3(e)]. Later, the embryo organizes into a series of segments, called parasegments, defined by the expression patterns of various embryonic genes (Figure 20–4). These later give rise to the adult segments, though they are shifted by one-half segment. Cells within the segments first become determined to form either an anterior or a posterior portion (called a **compartment**) of the segment. In the following stages, further restrictions of developmental options occur so that cells in each compartment are eventually restricted to forming a single structure in the adult body.

The adult body plan of *Drosophila* derives its overall organization from the larval body plan and is composed of head, thoracic, and abdominal segments. In addition, many of the adult body structures form from small groups of cells, or **imaginal discs**, which developed during the larval stages. There are 12 bilaterally paired discs—the eye-antennal discs, leg discs, wing discs, and so forth—and one genital disc.

Figure 20–5 indicates which imaginal discs will give rise to which adult structures.

Genetic Analysis of Embryogenesis

Although our knowledge of anatomical development in *Drosophila* has been useful, genetic analysis provides the most significant information about events in embryogenesis. Genes that control embryonic development are either **maternal-effect genes** or **zygotic genes**. Maternal-effect gene products (mRNA and proteins) are deposited in the developing egg during oogenesis, often distributed in a gradient or concentrated in specific regions of the egg. The phenotype of flies with mutations in maternal-effect genes is expected to be female sterility, since none of the embryos of females homozygous for a recessive mutation would receive wild-type gene products from their mother and would therefore develop abnormally. In *Drosophila*, these maternal-effect genes encode transcription factors, receptors, and proteins that regulate translation. During embryonic development,

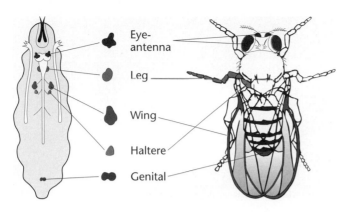

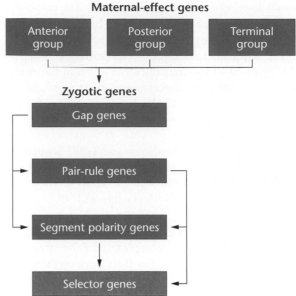

FIGURE 20–5 Imaginal discs of the *Drosophila* larva and the adult structures that derive from them.

these gene products activate or repress the expression of zygotic genes in a temporal and spatial sequence.

Zygotic genes are those expressed in the developing embryo, and are therefore transcribed after fertilization. The phenotype of mutations in this class show embryonic lethality. In a cross between two flies heterozygous for a recessive zygotic mutation, one-fourth of the embryos (the homozygotes) fail to develop. In *Drosophila*, many of the zygotic genes are transcribed in patterns that depend on the distribution of maternal-effect proteins.

Much of what we know about the zygotic genes that regulate development comes from a mutagenic analysis performed by Christiane Nüsslein-Volhard and Eric Wieschaus, who systematically screened for mutations that affect embryonic development. They examined thousands of dead F_2 offspring of mutagenized flies for recessive embryonic lethal mutations with defects in external structures. The parents were thus identified as carriers of these mutations, which fall into three classes, gap, pair-rule, and segment polarity genes. In a paper published in 1980, Nüsslein-Volhard and Wieschaus proposed a model in which embryonic development is initiated by gradients of maternal-effect gene products. The positional information laid down by these molecular gradients along the anterior–posterior axis of the embryo is interpreted by two sets of zygotic genes: (1) those identified in their screen (gap, pair-rule, and segment polarity genes), collectively called the **segmentation genes** (which divide the embryo into a series of stripes or segments and define the number, size, and polarity of each segment), and (2) **selector genes** (which specify the identity or fate of each segment) (Figure 20–6).

Figure 20–7 shows the model developed by Nüsslein-Volhard and Wieschaus. The process begins when maternal-effect gene products responsible for forming the anterior–posterior axis, placed in the egg during oogenesis, are activated immediately after fertilization [Figure 20–7(a)]. Their activity restricts cells to forming either anterior or posterior structures. Gene products from this gradient activate the transcription of the gap genes that divide the embryo into a limited number of broad regions [Figure 20–7(b)]. Gap proteins are transcriptional factors that activate pair-rule genes whose products divide the embryo into regions

FIGURE 20–6 The hierarchy of genes involved in establishing the segmented body plan in *Drosophila*. Gene products from the maternal genes regulate the expression of the first three groups of zygotic genes (gap, pair-rule, and segment polarity; collectively called the segmentation genes), which in turn control expression of the selector genes.

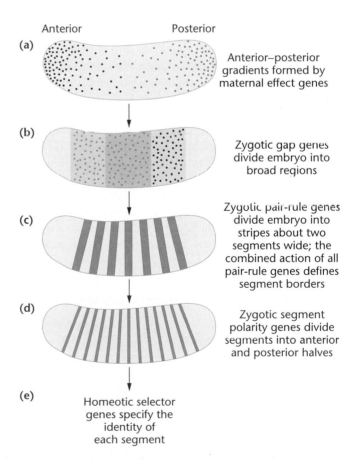

FIGURE 20–7 Progressive restriction of cell fate during development in *Drosophila*. Gradients of maternal proteins are established along the anterior–posterior axis of the embryo (a). The three groups of segmentation genes progressively define the body segments (b), (c), and (d). Individual segments are given identity by the selector genes (e).

about two segments wide [Figure 20–7(c)]. The combined action of all gap genes defines segment borders. The pair-rule genes in turn, activate the segment polarity genes that divide the segments into anterior and posterior compartments [Figure 20–7(d)]. The collective action of the maternal genes that form the anterior–posterior axis and the segmentation genes defines the fields of action for the homeotic selector genes, which specify the identity of each segment [Figure 20–7(e)].

In a second screening, Wieschaus and Trudi Schupbach screened thousands of flies for maternal-effect mutations that affected the external structures of the embryo. They estimated that about 40 maternal-effect genes and 50–60 zygotic genes regulate embryonic development. This means that about 100 genes control normal embryogenesis, a surprisingly low number given the complexity of the structures formed from the fertilized egg. For their work on the genetic control of development in *Drosophila*, Nüsslein-Volhard and Wieschaus, along with E. B. Lewis—the geneticist who initially identified and studied many of these genes in the 1970s—were awarded the 1995 Nobel Prize for Physiology or Medicine.

Zygotic Genes and Segment Formation

Zygotic genes are activated or repressed in a positional gradient by the maternal-effect gene products. Three subsets of these genes, and the segmentation genes divide the embryo into a series of segments along the anterior–posterior axis. The segmentation genes are normally transcribed in the developing embryo, and their mutations have embryonic lethal phenotypes.

Over 20 segmentation loci have been identified (Table 20.1). They are classified on the basis of their mutant phenotypes: (1) Gap genes delete a group of adjacent segments. (2) Pair-rule genes affect every other segment and eliminate a specific part of each affected segment. (3) Segment polarity genes cause defects in homologous portions of each segment.

Gap Genes The transcription of **gap genes** is activated or inactivated by gene products previously expressed along the anterior–posterior axis and by other genes of the maternal

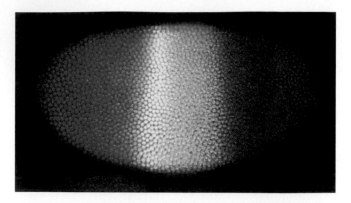

FIGURE 20–8 Expression of several gap genes in a *Drosophila* embryo. The hunchback protein is shown in orange and Krüppel in green. The yellow stripe is created when cells contain both hunchback and Krüppel proteins. *(Stephen Paddock)*

gradient systems. When mutated, these genes produce large gaps in the embryo's segmentation pattern. *Hunchback* mutants lose head and thorax structures, *Krüppel* mutants lose thoracic and abdominal structures, and *knirps* mutants lose most abdominal structures. Gap gene transcription divides the embryo into a series of broad regions (roughly, the head, thorax, and abdomen). Within these regions, different combinations of gene expression eventually specify both the type of segment that forms and the proper order of segments in the body of the larva, pupa, and adult. To date, all gap genes that have been cloned encode transcription factors with zinc finger DNA-binding motifs. The expression of gap genes correlates roughly with their mutant phenotypes: *hunchback* at

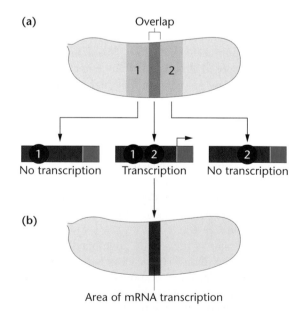

FIGURE 20–9 New patterns of gene expression can be generated by overlapping regions containing gene products. (a) Transcription factors 1 and 2 are present in an overlapping region of expression. If both transcription factors must bind sites in a promoter to trigger the expression of a target gene, the gene will be active only in cells containing both factors (most likely in the zone of overlap). (b) The expression of the target gene in the restricted region of the embryo.

TABLE 20.1   Segmentation Genes in *Drosophila*

Gap Genes	Pair-Rule Genes	Segment Polarity Genes
Krüppel	hairy	engrailed
knirps	even-skipped	wingless
hunchback	runt	cubitis interruptus[D]
giant	fushi-tarazu	hedgehog
tailless	odd-paired	fused
huckebein	odd-skipped	armadillo
	sloppy-paired	patched
		gooseberry
		paired
		naked
		dishelved

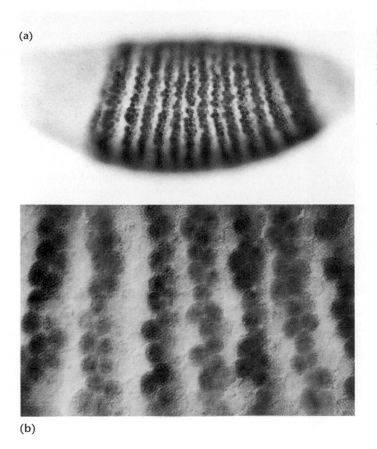

(a)

(b)

FIGURE 20–10 Stripe pattern of pair-rule gene expression in *Drosophila* embryo. This embryo is stained to show expression patterns of the genes *even-skipped* and *fushi-tarazu*. Low-power view (a), and high-power view (b) of the same embryo. *(Peter A. Lawrence and P. Johnson, Development 105:761–67 [1989].)*

the anterior, *Krüppel* in the middle, and *knirps* at the posterior (Figure 20–8). Gap genes also control the transcription of pair-rule genes.

Pair-Rule Genes **Pair-rule genes** divide the broad regions established by gap genes into sections about one segment wide. Mutations in pair-rule genes eliminate segment-size sections at every other segment. The pair-rule genes are expressed in narrow bands or stripes of nuclei that extend around the circumference of the embryo. The expression of this gene set first establishes the boundaries of segments and then establishes the developmental fate of the cells within each segment by controlling the segment polarity genes. At least eight pair-rule genes act to divide the embryo into a series of stripes. However, the boundaries of these stripes overlap, meaning that cells in the stripes express different combinations of pair-rule genes in an overlapping fashion (Figure 20–9).

Many pair-rule genes encode transcription factors containing helix-turn-helix homeodomains. The transcription of pair-rule genes is mediated by the action of gap gene products, but their pattern is resolved into highly delineated stripes by interaction among the gene products of the pair-rule genes themselves (Figure 20–10).

Segment Polarity Genes Expression of **segment polarity genes** is controlled by transcription factors encoded by pair-rule genes. Each segment polarity gene becomes active in a single band of cells within each segment that extends around the embryo (Figure 20–11). These expression patterns divide the embryo into 14 segments and the gene products con-

trol cellular identity within each segment. Some segment polarity genes, including *engrailed*, encode transcription factors. Rather than activating transcription, however, the engrailed protein competitively inhibits activation by other homeodomain proteins, resulting in the establishment of a segment border.

Selector Genes As segment boundaries are established by the action of the segmentation genes, selector genes are activated. These genes determine the structures to be formed by each body segment, including the antennae,

FIGURE 20–11 The 14 stripes of expression of the segment polarity gene *engrailed* in a *Drosophila* embryo. *(Jim Langeland, Stephen Paddock, and Sean Carroll)*

(a)

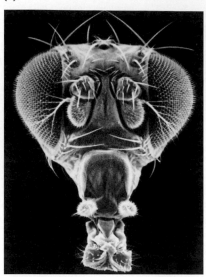

(b)

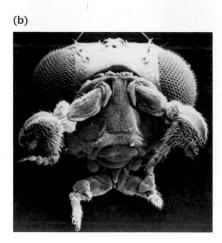

FIGURE 20–12 *Antennapedia* (*Antp*) mutation in *Drosophila*. (a) Head from wild-type *Drosophila* showing the structure of the antenna and other head parts. (b) Head from an *Antp* mutant, showing the replacement of normal antenna structures with legs. This is caused by activation of the *Antp* gene in the head region. *(From T. Kaufmann, et al. 1990. Advances in Genetics 27:309–62. Department of Biology, Indiana University/F.R. Turner/F. Rudolph Turner)*

mouth parts, legs, wings, thorax, and abdomen. Mutants of these genes are known as **homeotic mutants** (from the Greek word for "same") because the identity of one segment is transformed so that it is the same as that of a neighboring segment. For example, the wild-type allele of the *Antennapedia* (*Antp*) gene specifies structures in the sec-

ond thoracic segment (which carries a leg). Dominant gain-of-function *Antp* mutations lead to the expression of the gene in the head and thorax, and the antennae of mutant flies acquire the identity of structures on the thorax and thereby form legs (Figure 20–12).

The *Drosophila* homeotic selector genes are found in two clusters on chromosome 3 (Table 20.2). The **Antennapedia (ANT-C) complex** contains five genes required for specifying structures in the head and first two thoracic segments [Figure 20–13(a)]. The **bithorax (BX-C) complex** contains three genes required for specifying structures formed by the posterior portion of the second thoracic segment, the entire third thoracic segment, and the abdominal segments [Figure 20–13(b).

Gap genes and pair-rule genes activate the selector genes. Although selector gene expression is confined to certain do-

(a)

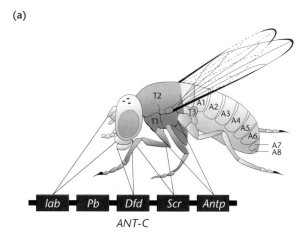

(b)

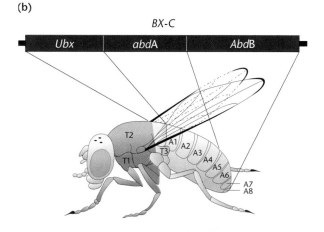

FIGURE 20–13 (a) Genes of the *Antennapedia* complex (*ANT-C*) and the adult structures they specify. The *labial* (*lab*) and *Deformed* (*Dfd*) genes control the formation of head segments. The *Sex comb reduced* (*Scr*) and *Antennapedia* (*Ant*) genes specify the identity of the first two thoracic segments. The remaining gene in the complex, *Proboscipedia* (*Pb*), may not act during embryogenesis, but may be required to maintain the differentiated state in adults. In mutants, the labial palps are transformed into legs. (b) Genes of the *bithorax* (*BX-C*) complex and the adult structures they specify. *Ultrabithorax* (*Ubx*) controls structures in the posterior compartment of T2 and structures in T3. The two other genes, *abdominal A* (*abdA*) and *Abdominal B* (*AbdB*), specify the segmental identities of the eight abdominal segments (A1–A8).

TABLE 20.2   Homeotic Selector Genes of *Drosophila*

ANT-C	BX-C
labial	*Ultrabithorax*
Antennapedia	*Abdominal a*
Sex comb reduced	*Abdominal B*
Deformed	
proboscipedia	

mains in the embryo, the timing and patterns of expression are rather complex and involve interactions between segmentation and selector genes as well as interactions among the various selector genes.

A number of genes that control selector gene expression have been identified, including *extra sex combs* (*esc*), *Polycomb* (*Pc*), *super sex combs** (*sxc*), and *trithorax* (*trx*). In the mutant *extra sex combs* (*esc*), some of the head and all of the thoracic and abdominal segments develop as posterior abdominal segments (Figure 20–14), indicating that this gene normally controls *BX-C* gene expression in all body segments. The mutation does not affect either the number or the polarity of the segments. It does affect their developmental fate, indicating that the *esc*+ gene product that is synthesized and stored in the egg by the maternal genome may be required to correctly interpret the information gradient in the egg cortex.

Several genes that regulate selector gene expression, including the *Polycomb* genes, operate by binding to many sites along chromosomes, rather than just to the promoters and enhancers of individual genes. This may explain an intriguing observation about the expression pattern of selector genes. When embryos are stained to reveal mRNA for each homeotic selector gene, the anterior borders for each gene's expression parallel its position on the chromosome. Genes at the 3′-end of the two gene clusters are expressed at the embryo's anterior end, while in genes located toward the 5′-end of the clusters, the pattern of expression moves toward the embryo's posterior end (Figure 20–15). In addition, expression patterns overlap in the more posterior regions such that several genes are expressed in the same segments. These overlapping patterns of expression suggest two things: (1) The regulation of selector gene expression may be coordinated during development and, (2) the formation of each segment requires the activity of a particular combination of homeotic genes.

E. B. Lewis has proposed that genes in the bithorax complex, and others involved in segmentation arose from a common ancestral gene by tandem duplication and subsequent divergence of structure and function. In fact, each homeotic selector gene listed in Table 20.2 encodes a transcription factor that includes a DNA-binding domain encoded by a 180-bp sequence known as a **homeobox**. The homeobox encodes a sequence of 60 amino acids known as a **homeodomain** (see Chapter 16). Similar sequences have been found in the genomes of other eukaryotes with segmented body plans, including *Xenopus*, chicken, mice, and humans. In mammals, the complexes of selector genes are called ***Hox*** **gene clusters** (Figure 20–16, page 424). Homeodomains from all organisms examined to date are very similar in amino acid sequence and encode a protein associated with the transcriptional regulation of a specific gene set. This suggests that the segmented body plan may have evolved only once.

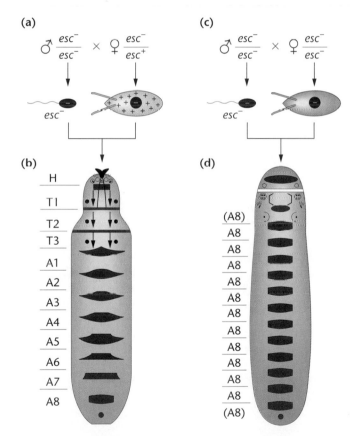

(a) Expression domains of homeotic genes

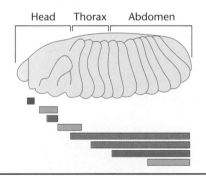

(b) Chromosomal locations of homeotic genes

FIGURE 20–14 Action of the *extra sex combs* (*esc*) mutation in *Drosophila*. (a) Heterozygous females form wild-type *esc* gene product and store it in the oocyte. (b) When the *esc*− egg formed at meiosis by the heterozygous female is fertilized by *esc*− sperm, it produces wild-type larva with a normal segmentation pattern (maternal rescue). Borderlines between the head and thorax and between the thorax and abdomen are marked with small arrows. (c) Homozygous *esc*− females produce defective eggs, which, when fertilized by *esc*− sperm, produce a larva (d) in which most of the segments of the head, thorax, and abdomen are transformed into the eighth abdominal segment. Maternal rescue demonstrates that the *esc*− gene product is produced by the maternal genome and is stored in the oocyte for use in the embryo. (H, head; T1–T3, thoracic segments; A1–A8, abdominal segments.)

FIGURE 20–15 Colinear relationship between the spatial pattern of expression and chromosomal locations of homeotic genes in *Drosophila*. (a) *Drosophila* embryo and the domains of homeotic gene expression in the embryonic epidermis and central nervous system. (b) Chromosomal location of homeotic genes. Note that the order of genes on the chromosome correlates with the sequential pattern of their expression.

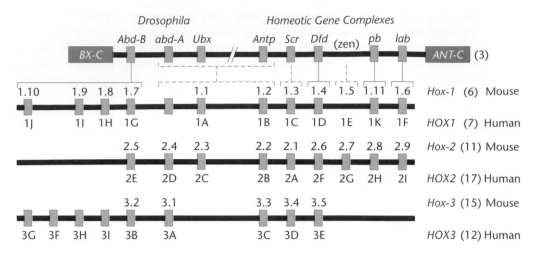

FIGURE 20–16   The *Hox* genes of mammals (human and mouse) and their organizational alignment with the *BX-C* and *ANT-C* complexes of *Drosophila*.

To summarize, genes that control development in *Drosophila* act in a temporally and spatially ordered cascade, beginning with genes that establish the anterior–posterior and dorsal–ventral axes of the egg and early embryo. Gradients of maternal mRNAs and proteins along the anterior–posterior axis activate the gap genes, which subdivide the embryo into broad bands. Gap genes in turn activate the pair-rule genes, which divide the embryo into segments. The final group of segmentation genes, the segment polarity genes, divide each segment into anterior and posterior regions arranged linearly along the anterior–posterior axis. These segments are then given identity by the selector genes. Therefore, this progressive restriction of the developmental potential of the *Drosophila* embryo's cells, (all of which occurs during the first one-third of embryogenesis) involves a cascade of gene action, with regulatory proteins acting at both the transcriptional and translational levels.

20.3   Flower Development in *Arabidopsis*: Role of Homeotic Selector Genes

Flower development in *Arabidopsis thalania* (Figure 20–17), a small plant in the mustard family, has been used to study pattern formation in plants. A cluster of undifferentiated cells called the floral meristem gives rise to flowers (Figure 20–18). Each flower consists of four organs: sepals, petals, stamens and carpels that develop from concentric whorls of cells within the meristem (Figure 20–19). Each organ develops from a different whorl. Three classes of floral homeotic genes control the development of these organs: A class genes give rise to sepals, A and B class genes specify petals, B and C class

FIGURE 20–17   The flowering plant, *Arabidopsis thalania*, used as a model organism in plant genetics. (*Elliot M. Meyerowitz/California Institute of Technology, Division of Biology*)

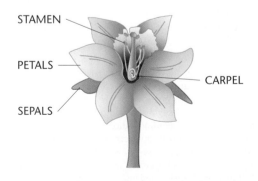

FIGURE 20–18   Parts of a flower. The floral organs are arranged concentrically. The sepals form the outermost ring, followed by petals and stamens, with carpels on the inside.

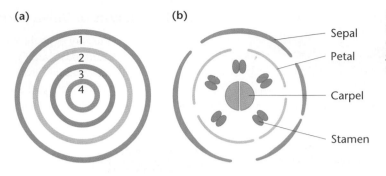

FIGURE 20–19 Cell arrangement in the floral meristem. The four concentric rings or whorls, labeled 1–4 give rise to the sepals, petals, stamens and carpels, respectively.

FIGURE 20–20 (a) Wild-type flowers of *Arabidopsis* have four petals, six stamens, a pistil fused from the two carpels, and sepals (not visible in this photo.) (b) Homeotic *agamous* mutants have petals and sepals at places where stamens and carpels should form. (*Left photo: Elliot M. Meyerowitz/ California Institute of Technology, Division of Biology); right photo: J. Bowman/Elliot M. Meyerowitz/ California Institute of Technology, Division of Biology*)

genes control stamen formation, and C class genes give rise to carpels [Figure 20–20(a)]. The genes in each class are listed in Table 20.3. The A class genes are active in the first two whorls (sepals and petals), B class genes act in the second and third whorls (petals and stamens) and C class genes act in the third and fourth whorls (stamens and carpals). The organ formed depends on the expression pattern of the three class of genes. If only class A is expressed in a region, sepals will form. In regions where classes A and B are expressed, petals form. Activity of classes B and C leads to stamen formation, and if only class C is expressed, carpels form.

As in *Drosophila*, mutations in homeotic genes cause normal organs to form in abnormal locations. For example, in *AG* mutants (lacking activity of a C class gene), the order of organs is sepal, petal, petal, sepal instead of sepal, petal, stamen carpel [Figure 20–20(b)]. If both A and B class activity are eliminated by mutation, then the whorls will have only

C class activity (which will spread to the first and second whorls because class A is absent), and only carpels will form.

The evolutionary conservation of developmental processes and the homology between distantly related species is emphasized by the fact that the ABC gene classes of *Arabidopsis* are transcription factors, expressed in overlapping patterns, as are the homeotic genes in *Drosophila*. In *Arabidopsis*, however, the floral homeotic genes are members of a different family of transcription factors, called the **MADS-box proteins**. Each member of this family contains a common sequence of 58 amino acids. The action of the floral homeotic genes is controlled by a gene called *CURLY LEAF*,* that shares significant homology with members of the *Drosophila Polycomb* gene family. Thus, both plants and animals use the same mechanism, the activation of unique gene sets in each neighboring region of a particular structure, to determine its developmental pathway.

20.4 Cell–Cell Interactions in *C. elegans* Development

During development in multicellular organisms, cells influence the transcriptional patterns and developmental fate of other cells; these interactions involve the generation and reception of signal molecules. Cell–cell interaction is an important process in the development of most eukaryotic organisms, including *Drosophila*, as well as vertebrates such as *Xenopus*, mice, and humans.

TABLE 20.3 Homeotic Selector Genes in *Arabidopsis*

Class A	*APETALA1 (AP1)**
	APETALA2 (AP2)
Class B	*APETALA3 (AP3)*
	PISTILLATA (P1)
Class C	*AGAMOUS (AG)*

*By convention, wild-type genes in *Arabidopsis* use capital letters.

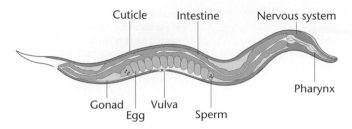

FIGURE 20–21 An adult *C. elegans*. This nematode, about 1 mm in length, consists of 959 cells and has been used to study many aspects of the genetic control of development.

Overview of *C. elegans* Development

The nematode *Caenorhabditis elegans*, which we first encountered in our discussion of sex determination in Chapter 5, is widely used to study the genetic control of development. This organism has two advantages for such studies: Its genetics are well known, and adults have a small number of cells that follow a developmental program that does not change from individual to individual. The adult is about 1 mm long and matures from a fertilized egg in about 2 days (Figure 20–21). The life cycle consists of an embryonic stage (about 16 hours), four larval stages (L1–L4), and the adult stage. Adults are of two sexes—XX self-fertilizing hermaphrodites that can make both eggs and sperm, and XO males.

The adult hermaphrodite consists of 959 somatic cells (and about 2000 germ cells). The exact cell lineage from fertilized egg to adult has been mapped (Figure 20–22) and is invariant from individual to individual. Knowing the lineage of each cell, we can easily follow the events that result either from mutational alterations in cell fate or from the killing of cells by laser microbeams or ultraviolet irradiation.

In *C. elegans* hermaphrodites, the fate of cells in the developing reproductive system is determined by cell–cell interactions and gives us insight into how gene expression and cell–cell interactions work together to specify developmental pathways.

Genetic Analysis of Vulva Formation

Adult hermaphrodites lay eggs through the **vulva**, an opening located about midbody (see Figure 20–21). The vulva forms in stages during larval development; the process involves several rounds of cell–cell interactions.

During development in *C. elegans*, two neighboring cells, Z1.ppp and Z4.aaa, interact so that one becomes the gonadal anchor cell and the other a precursor to the ventral uterus. The determination of which is which occurs during the second larval stage (L2) and is controlled by the *lin-12* gene, which encodes a cell-surface receptor protein. In recessive *lin-12(0)* mutants (a loss-of-function mutant) both cells become anchor cells. The dominant mutant *lin-12(d)* (a gain-of-function mutation) causes both to become uterine precursors. Thus, it appears that the *lin-12* gene causes the uterine pathway to be selected, since in the absence of the LIN-12 protein, both cells become anchor cells. As shown in Figure 20–23, both cells normally synthesize and secrete a chemical signal (as yet unknown) for uterine differentiation and also synthesize the LIN-12 protein, which is a receptor for the signal. At a critical time in L2, the cell that by chance secretes more of the signal molecule causes its neighbor to increase transcription of the *lin-12* gene, thus increasing production of the receptor. The cell with more LIN-12 receptors becomes the ventral uterine precursor, and the other cell becomes the anchor cell. The critical factor in this first round of cell–cell interaction and determination is the LIN-12 gene product.

A second round of cell–cell interactions in L3 involves the anchor cell, located in the gonad, and six precursor cells located in the skin (hypodermis) adjacent to the gonad. The precursor cells are named P3.p, P4.p, P5.p, P6.p, P7.p, and P8.p, and collectively are called Pn.p cells. The fate of each Pn.p cell is specified by its position relative to the anchor cell. Figure 20–24 shows the determination pathway, which is described in the following paragraphs.

Sometime in L3, a gene called *lin-3* is activated in the anchor cell and secretes a protein related to the vertebrate epidermal growth factor (EGF). All six Pn.p cells express a cell

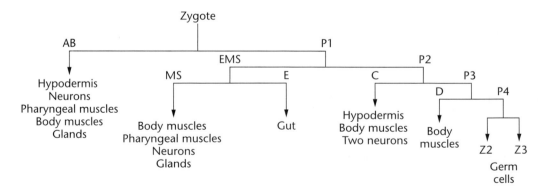

FIGURE 20–22 A truncated lineage chart for *C. elegans* that shows early cell divisions and the tissues and organs formed. Each vertical line represents a cell division, and horizontal lines connect the two cells produced. For example, the first cell division creates two new cells from the zygote, AB and P1. The cells in this chart refer to those present in the first-stage larva L1. During subsequent larval stages, further cell divisions will produce the 959 somatic cells of the adult hermaphrodite worm.

(a)

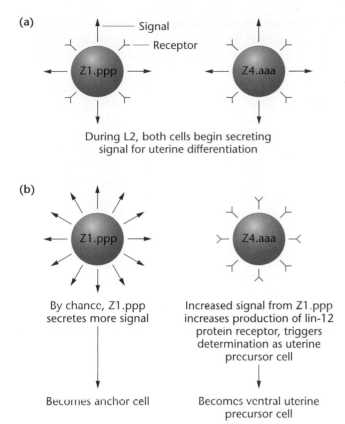

During L2, both cells begin secreting
signal for uterine differentiation

(b)

By chance, Z1.ppp
secretes more signal

↓

Becomes anchor cell

Increased signal from Z1.ppp
increases production of lin-12
protein receptor, triggers
determination as uterine
precursor cell

↓

Becomes ventral uterine
precursor cell

FIGURE 20–23 Cell–cell interaction in anchor cell determination. (a) During L2, two neighboring cells secrete chemical signals that induce uterine differentiation. (b) By chance, cell Z1.ppp secretes more of these signals, causing cell Z4.aaa to increase production of the receptor. The action of increased signals causes Z4.aaa to become the ventral uterine precursor cell and allows Z1.ppp to become the anchor cell.

surface receptor encoded by the *let-23* gene, similar to the vertebrate EGF receptor. Binding of the LIN-3 protein to the LET-23 receptor triggers an intracellular cascade of events that determines whether the Pn.p cells form vulval precursor cells or secondary vulvar cells. The *let-23* gene establishes the primary and secondary fates of Pn.p cells. Recessive loss-of-function mutations in *let-23* cause the Pn.p cells to develop as hypodermis (a tertiary fate). In *let-23* mutants, Pn.p cells act as though they have not received a signal from the anchor cell, and no vulva is formed.

The signal transduction cascade from anchor cells to Pn.p cells also involves the *let-60* gene. Recessive mutations in *let-60* cause the Pn.p cells to develop as though they have not received a signal from the anchor cell. Dominant *let-60* alleles have the opposite phenotype: They cause all Pn.p cells to respond to form vulval cells, forming multiple vulvas. The *let 60* gene has been cloned, and it is the *C. elegans* homolog of the *ras* gene, which in mammals acts

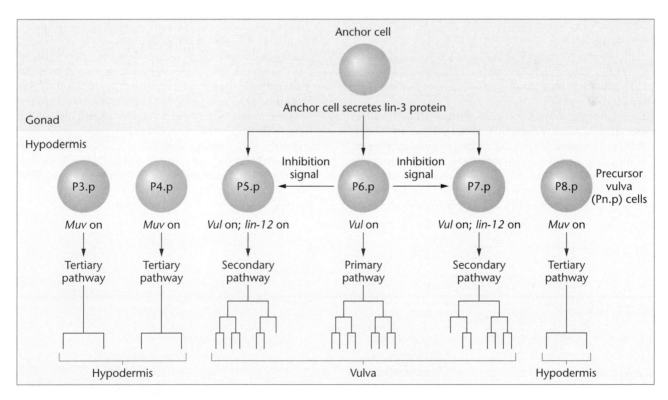

FIGURE 20–24 Cell lineage determination in *C. elegans* vulva formation. A signal from the anchor cell in the form of LIN3 protein is received by three precursor vulval cells (Pn.p cells). The cells closest to the anchor cell become primary vulval precursor cells, and adjacent cells become secondary precursor cells. Primary cells secrete a signal that activates the *lin-12* gene in secondary cells, preventing them from becoming primary cells. Flanking Pn.p cells, which receive no signal from the anchor cell *Muv* gene, become hypodermis cells instead of vulval cells.

downstream of receptor proteins in signal transduction. The *let-60* dominant gain-of-function mutant that causes multiple vulva formation has a Gly-Glu mutation at amino acid 13, the same mutation that converts *c-ras* to an oncogene (see Chapter 21).

Data on the signal transduction cascade initiated by binding of LIN-3 to LET-23 suggest that the cell (P6.p) closest to the anchor cell receives the strongest signal. In P6.p, expression of the *Vulvaless* (*Vul*) gene (named for its mutant phenotype) is activated; the primary fate of this cell is to divide three times to produce vulva cells. The two neighboring cells receive a lower amount of signal; this specifies a secondary fate—asymmetric division to form more vulva cells. To reinforce this determination, the primary vulval precursor activates the *lin-12* gene in the two neighboring cells. This lateral inhibition signal prevents the neighboring secondary cells from adopting the division pattern of the primary cell. In other words, cells in which both *Vul* and *lin-12* are active cannot become primary vulvar cells. The three remaining Pn.p cells receive no signal from the anchor cell. In these cells (P3.p, P4.p, and P8.p), the *Multivulva* (*Muv*) gene is expressed, *Muv* represses *Vul*, and the three cells develop as hypodermal (skin) cells.

Thus, three levels of cell–cell interaction are required to specify the developmental pathway that leads to vulva formation in *C. elegans*. First, two neighboring cells interact during L2 to establish the identity of the anchor cell. Second, in L3, the anchor cell interacts with three Pn.p cells to establish the primary vulvar precursor cell and two secondary cells. Third, the primary vulvar cell interacts with the secondary cells to suppress their choice of the primary vulvar pathway. Each interaction is accompanied by the secretion of molecular signals and by the reception and processing of these signals by neighboring cells.

This theme of cell–cell interaction acting in a spatial and temporal cascade to specify the developmental fates of individual cells is repeated over and over in organisms from prokaryotes to higher vertebrates, including mice and humans.

Chapter Summary

1. The role of genetic information during development and differentiation is a major research topic in biology and has been studied extensively. Geneticists are isolating developmental mutations and identifying the genes that control developmental processes.

2. Determination is the regulatory event whereby cell fate becomes fixed during early development. Determination precedes the actual differentiation or specialization of distinctive cell types.

3. During embryogenesis, the internal environment of the cell (localized cytoplasmic components) appears to affect specific gene activity. The regulation of early events is mediated by the maternal cytoplasm, which then influences zygotic gene expression. As development proceeds, both the cell's internal and external environments become further altered by the presence of early gene products and by communication with other cells.

4. In *Drosophila*, both genetic and molecular studies confirm that the egg contains information that specifies the body plan of the larva and adult, and that interactions of embryonic nuclei with the maternal cytoplasm initiate transcriptional programs characteristic of specific developmental pathways.

5. Extensive genetic analysis of embryonic development in *Drosophila* has identified maternal-effect genes that lay down the anterior–posterior and dorsal–ventral axes of the embryo. In addition, these maternal-effect genes activate sets of zygotic segmentation genes, initiating a cascade of gene regulation that ends with the selector genes determining segment identity.

6. The stereotyped lineage of all cells in *C. elegans* allows developmental biologists to study the cell–cell signaling required for organogenesis and to determine which genes are required for the normal process of programmed cell death.

Key Terms

Antennapedia (*ANT-C*) complex, 422
bithorax (*BX-C*) complex, 422
cell–cell interaction, 416
cellular blastoderm, 418
compartment, 418
cytoplasmic localization, 416
determination, 416
differentiation, 416

gap gene, 420
homeobox, 423
homeodomain, 423
homeotic mutant, 422
Hox gene clusters, 423
imaginal disc, 418
MADS-box proteins, 425
maternal-effect genes, 418

pair-rule gene, 421
segment polarity gene, 421
segmentation gene, 419
selector gene, 419
syncytial blastoderm, 416
vulva, 426
zygote, 416
zygotic gene, 418

Insights and Solutions

1. The timing of differential gene action during development is key to normal developmental programs. If a gene has been cloned, the time of action and range of cell types in which the gene is active can be determined using a number of molecular techniques. However, when a cloned gene or its transcripts are not available, genetic techniques can establish a comparative order of gene action among two or more genes. By extending this analysis to include several genes, a relative order of gene action can be established. This order serves as a starting point for investigations at the molecular level by providing developmental time scales for gene action. The following example shows how a relative time scale for the action of two genes is constructed.

In *Drosophila*, the autosomal recessive gene *lozenge-clawless* (*lzcl*) produces multiple abnormalities of the female genitals, eyes, and tarsal regions of the legs. Homozygous females have abnormal genitals, no sperm storage organs, no ovarian glands, and are sterile. Homozygous mutant males, on the other hand, have normal genitals and are fully fertile. A second autosomal recessive gene, *transformer* (*tra*), converts XX females into phenotypic males (recall that in *Drosophila*, XX flies are female, XY flies are male). Thus, XX flies homozygous for both *lzcl* and *tra* are phenotypically male with normal genitals. What do these results say about the normal sequence of action of these two genes? Which one acts first during development?

Solution: The flies in question are genetically female, since they have two X chromosomes. Females that are homozygous for the *lzcl* gene are expected to have abnormal genitalia. In this case, the action of the *tra* gene in changing the female phenotype into the male phenotype (with the development of male genitalia) must take place before the action of the *lzcl* gene, since the XX flies have a normal male phenotype.

Problems and Discussion Questions

1. Carefully distinguish between the terms *differentiation* and *determination*. Which phenomenon occurs initially during development?

2. The *Drosophila* mutant *spineless aristapedia* (*ss*ᵃ) results in the formation of a miniature tarsal structure (normally part of the leg) on the end of the antenna. This is a homeotic mutation. From your knowledge of imaginal discs, what insight is provided by *ss*ᵃ concerning the role of genes during determination?

3. In the sea urchin, early development up to gastrulation can occur even in the presence of actinomycin D, which inhibits RNA synthesis. However, if actinomycin D is present early on but removed at the end of blastula formation, gastrulation does not proceed. In fact, if actinomycin D is present only between the 6th and 11th hours of development, gastrulation (normally occurring at the 15th hour) is arrested. What conclusions can you draw concerning the role of gene transcription between hours 6 and 15?

4. How can you determine whether a particular gene is being transcribed in different cell types?

5. A particular gene is transcribed during development. How can you tell whether the expression of this gene is under transcriptional or translational control?

6. Both the *ftz* gene and the *engrailed* gene encode homeobox transcription factors and are capable of eliciting the expression of other genes. Both genes work at about the same time during development and in the same region to specify cell fate in body segments. The question is, Does *ftz* regulate the expression of *engrailed*, or does *engrailed* regulate *ftz*? Or are they both regulated by another gene? To answer these questions, mutant analysis is performed. In *ftz⁻* embryos (*ftz/ftz*), engrailed protein is absent; in *engrailed⁻* embryos (*eng/eng*), *ftz* expression is normal. What does this tell you about the regulation of these two genes? Does the *engrailed* gene regulate *ftz*? Does the *ftz* gene regulate *engrailed*?

7. Define what is meant by a homeotic mutant. If it were possible to introduce a homeotic gene from *Drosophila* into an *Arabidopsis* embryo that is homozygous for a homeotic flowering gene, would you expect any of the *Drosophila* genes to be able to prevent expression of the *Arabidopsis* mutation? Why or why not?

8. Nuclei from almost any source can be injected into *Xenopus* oocytes. Studies have shown that these nuclei remain active in transcription and translation. How can such an experimental system be useful in developmental genetic studies?

9. The concept of epigenesis indicates that an organism develops by forming cells that acquire new structures and functions, which become greater in number and complexity as development proceeds. This theory contradicts the preformationist doctrine that miniature adult entities are contained in the egg that must merely unfold and grow to give rise to a mature organism. What sorts of evidence presented in this chapter might have led to the preformation doctrine? Why is the epigenetic theory held as correct today?

Selected Readings

Beachy, P., Helfand, S., and Hogness, D. 1985. Segmental distribution of bithorax complex proteins during *Drosophila* development. *Nature* 313:545–51.

Brenner, S. 1974. The genetics of *Caenorhabditis elegans*. *Genetics* 77:71–94.

DeRobertis, E., Oliver, G., and Wright, C. 1990. Homeobox genes and the vertebrate body plan. *Sci. Am.* (July) 262:46–52.

Duboule, D., and Morata, G. 1994. Colinearity and functional hierarchy among genes of the homeotic complexes. *Trends Genet.* 10:358–64.

Fay, D. S. and Han, W. 2000. The synthetic multivulval genes of *C. elegans*: functional redundancy, Ras-antagonism, and cell fate determination. *Genesis* 26:279–84.

Grant, K., et al. 2000. *sem-4* promotes vulval cell-fate determination in *Caenorhabditis elegans* through regulation of *lin-39 Hox*. *Dev. Biol.* 2244:496–506.

Gurdon, J. 1968. Transplanted nuclei and cell differentiation. *Sci. Am.* (Dec.) 219:24–35.

Honma, T., and Goto, K. 2000. The *Arabidopsis* floral homeotic gene PISTILLA is regulated by discrete cis-elements responsive to induction and maintenance signals. *Development* 127:2021–30.

Jenik, P. D., and Irish, V. F. 2000. Regulation of cell proliferation patterns by homeotic genes during *Arabidopsis* floral development. *Development* 127:1267–76.

Koornneef, M., et al. 1998. Genetic control of flowering time in *Arabidopsis*. *Ann Rev. Plant Mol. Biol.* 49:345–370.

Lawrence, P. A., and Morata, G. 1994. Homeobox genes: Their function in *Drosophila* segmentation and pattern formation. *Cell* 78:181–89.

Manseau, L., and Schupbach, T. 1989. The egg came first, of course! Anterior–posterior pattern formation in *Drosophila* embryogenesis and oogenesis. *Trends Genet.* 5:400–5.

Meyerowitz, E. 1994. The genetics of flower development. *Sci. Am.* 271:(Nov) 56–65.

McGinnis, W., and Krumlauf, R. 1992. Homeobox genes and axial patterning. *Cell* 68:283–302.

Parcy, F., et al. 1998. A genetic framework for floral patterning. *Nature* 395:561–566.

Schultz, C., and Tantz, D. 1995. Zygotic caudal regulation by *hunchback* and its role in abdominal segment formation of the *Drosophila* embryo. *Development* 121:1023–28.

Slack, J., and Tannahill, D. 1992. Mechanism of anteroposterior axis specification in vertebrates. Lessons from the amphibians. *Development* 114:285–302.

Tantz, D., and Sommer, R. J. 1995. Evolution of segmentation genes in insects. *Trends Genet.* 11:23–27.

Wilson, D. S., and Desplan, C. 1995. Homeodomain proteins. Cooperating to be different. *Curr. Biol.* 5:32–34.

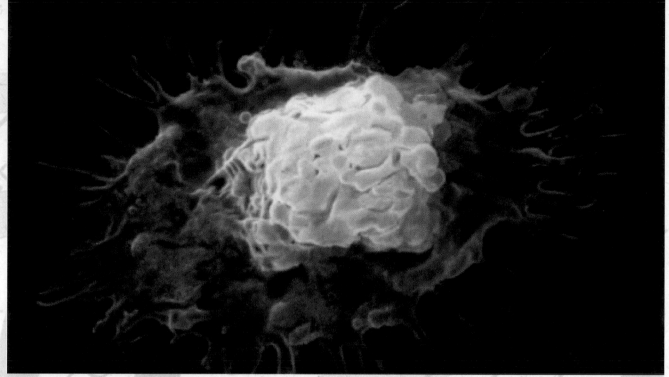

A human breast cancer cell. *(AMC/Albany Medical College/Custom Medical Stock Photo, Inc.)*

21

The Genetic Basis of Cancer

Although often viewed as a single disease, cancer is actually a complex group of diseases affecting a wide range of cells and tissues. It is also a serious health problem. In the United States, the lifetime risk of developing cancer is 1 in 2 for males and 1 in 3 for females (Table 21.1). A genetic link to cancer was first proposed early in the twentieth century, and this idea has served as a foundation for cancer research. Mutations that alter gene expression are now regarded as a common feature of all cancers. In most cancer cases, mutations arise in somatic cells and are not passed on to future generations through the germ cells. However, in about 1 percent of cases, germ-line mutations of various genes are transmitted to offspring and cause susceptibility to cancer. Although the frequency of these cases is low, studies of these mutations provide many insights into the origins of cancer. Sometimes for cancer to occur, the inherited mutation is itself insufficient and must be accompanied by an additional somatic mutation at the homologous locus, creating homozygosity. Whichever the case, cancer is now considered a genetic disorder at the cellular level.

Genomic alterations associated with cancer can involve small-scale changes, such as single nucleotide substitutions, or large-scale events, such as chromosome rearrangements, chromosome gain or loss, or even the integration of viral genomes into chromosomal sites. Large-scale genomic alterations are a common feature of cancer; the majority of human tumors are characterized by visible chromosomal changes. Some chromosomal changes, particularly in leukemia, are so characteristic they are used to diagnose the disorder and to make accurate predictions about the severity and course of the disease.

That cancer runs in families has been known for over 200 years. Most often, no clear-cut pattern of inheritance is discernable for these familial forms, primarily because these patients inherit only one mutant allele of a cancer-causing gene. The likelihood that an individual will ultimately develop cancer depends on the particular mutant allele, mutations in other genes, and environmental factors. These variables may influence the age of onset and the severity of the disease. Thus, we can identify a class of genes called **cancer susceptibility genes** that increase the risk of cancer. Variant alleles of these cancer susceptibility genes may have

an important role in sporadic cancers as well as familial forms of cancer.

Because of the background rate of spontaneous mutation, there will always be a baseline rate of cancer. Over and above this baseline rate, environmental agents that promote mutation are also active in the development of cancer. Almost all known environmental **carcinogens** (cancer-causing agents), such as ionizing radiation, chemicals, and viruses, act by generating mutations. Given that mutations play such a central role in cancer, in this chapter we explore questions about how mutations convert normal cells into malignant tumors, which mutant genes are most likely to result in cancer, and how many mutations are required to cause cancer.

To answer these questions, let's consider the properties of cancer cells that distinguish them from normal cells and ask what genes control these properties. Cancer cells have two properties in common: (1) uncontrolled growth, and (2) the ability to **metastasize**, or spread, from their original site to other locations in the body. Cell division is the result of cells traversing the cell cycle; in cancer cells, control over the cell cycle is lost, and cells proliferate rapidly. Investigations into the genetic control of the cell cycle are now providing insights into the origins of cancer.

The metastasis of cancer cells is controlled by gene products that become localized on the cell surface, and the genetics of metastasis relates to how cells interact with the extracellular matrix and with other cells through cell surface molecules. Although this field is less well developed than that of the cell cycle, it is beginning to provide insights into the secondary events in tumor progression.

We also consider the relationship between genes and cancer, with emphasis on the relationship between the cell cycle and genetic disorders. Finally, we examine how mutations, chromosomal changes, and environmental agents play a role in the development of cancer.

21.1 The Cell Cycle and Cancer

The cell cycle represents the sequence of events occurring between mitotic divisions in a eukaryotic cell. Because this cycle is closely related to the genetics of cancer, we first dis-

TABLE 21.1 Cancer Probabilities in the United States

		Birth to 39	*40–59*	*60–79*	*Birth to death*
All sites	Male	1 in 62	1 in 12	1 in 3	1 in 2
	Female	1 in 52	1 in 11	1 in 4	1 in 3
Breast	Female	1 in 235	1 in 25	1 in 15	1 in 8
Prostate	Male	<1 in 10,000	1 in 53	1 in 7	1 in 6
Colon-rectum	Male	1 in 1500	1 in 124	1 in 29	1 in 18
	Female	1 in 1900	1 in 149	1 in 33	1 in 18
Lung-bronchus	Male	1 in 2500	1 in 78	1 in 16	1 in 12
	Female	1 in 2900	1 in 106	1 in 25	1 in 18

Source: American Cancer Society

cuss what we currently know about events in the cell cycle and the genes that regulate progression through the cycle.

As outlined in Chapter 2, the cell cycle progresses from a period of chromosomal DNA replication (S phase) to the segregation of chromosomes into two nuclei during mitosis (M phase). Interspersed between these phases are two gaps, called G1 and G2. Together, G1, S, and G2 make up interphase (Figure 21–1). The G1 phase begins after mitosis; the synthesis of many cytoplasmic elements including ribosomes, enzymes, and membrane-derived organelles occurs at this time. In S, DNA replication produces a duplicate copy of each chromosome. Then the second period of growth and synthesis—G2—occurs as a prelude to mitosis.

While the cell cycles of some cells, such as dermal cells in human skin, is continuous, other cell types, including many nerve cells, withdraw from G1 and permanently enter a nondividing state known as G0. Still other cell types such as white blood cells can be recruited from G0 and reenter the cell cycle. Taken together, these observations suggest that the cell cycle is tightly regulated and is dependent on a cell's life history and its differentiated state. We next look at what we currently know about the genetic regulation of the cell cycle.

Control of the Cell Cycle

Much of the basic research on the control of the cell cycle has been conducted by two groups—geneticists working with yeasts, especially *Saccharomyces cerevisiae* and *Schizosaccharomyces pombe*, and developmental biologists studying newly fertilized eggs of organisms such as frogs, sea urchins, and newts. Both groups have succeeded in identifying and characterizing genes involved in the cell cycle, and their work is now converging and overlapping with important areas of cancer biology, particularly studies on growth factors and the genes that suppress or promote tumor formation.

Cell Cycle Checkpoints

The emerging picture indicates that the cell cycle is regulated at two main checkpoints: the G1/S transition and the G2/M transition (see Figure 21–1). At both points, a decision is made to proceed or halt progression through the cell cycle. This decision is controlled through the interaction of two classes of proteins. One is a class of enzymes called **protein kinases** that when activated, selectively phosphorylate target proteins. Although a large number of different protein kinases exist in the cell, only a few are involved in cell-cycle regulation; some of these are **cyclin-dependent kinases (CDKs)**. The second class of proteins are called **cyclins**, which control progression through the cell cycle. First identified in the embryos of developing invertebrates, cyclins are synthesized and degraded in a synchronous pattern related to the different stages of the cell cycle (Figure 21–2). Altogether, almost a dozen different cyclins have been identified, and a growing number of cyclin-dependent kinases are being described, indicating that multiple checkpoints in the cell cycle exist or that these kinases and cyclins have multiple functions.

Physical interaction between kinases and cyclins produces a regulatory molecule that controls the cell's movement through the cycle. Different CDKs control the initiation of the S and M phases. At the G1/S control point, a kinase known as CDK4 binds to a cyclin D molecule and activates the transcription of a set of genes required for S phase. The onset of mitosis (M phase) in most eukaryotic cells is controlled by a kinase called CDK1 (cyclin-dependent kinase), which was biochemically characterized in maturing amphibian eggs and genetically identified in yeast as the product of the *cdc2* gene.

Several events mark the entry from G2 into mitosis, including the condensation of chromatin into chromosomes, breakdown of the nuclear membrane, and reorganization of the cytoskeleton. Major events in this transition are regulated by the formation of an active CDK1/cyclin B complex. When bound

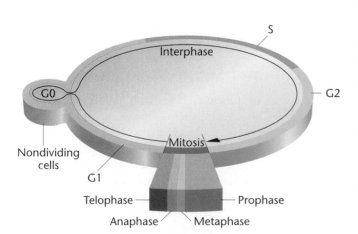

FIGURE 21–1 The cell cycle is controlled at several checkpoints, including one at the G2/M transition, and another in late G1 before entry into S phase. These checkpoints involve interactions between transitory proteins (called cyclins) and kinases that add phosphate groups to proteins. Phosphorylation of target proteins triggers a cascade of events allowing progress through the cell cycle.

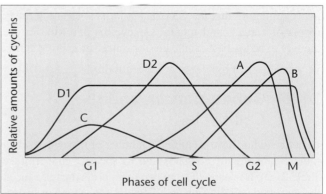

FIGURE 21–2 Relative expression times and amounts of cyclins during the cell cycle. D1 accumulates early in G1 and is expressed at a constant level through most of the cycle. Cyclin C accumulates in G1, reaches a peak, and declines by mid-S phase. Cyclin D2 begins accumulating in the last half of G1, reaches a peak just after the beginning of S, and then declines by early G2. Cyclin A appears in late G1, accumulates through S, reaches a peak in G2, and is degraded rapidly as M phase begins. Cyclin B appears in mid-S phase, peaks at the G2/M transition, and is rapidly degraded.

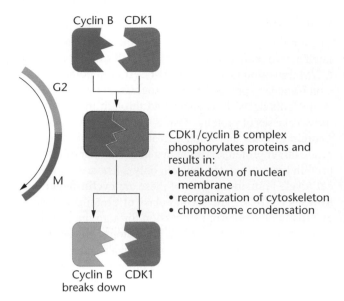

FIGURE 21–3 Transition from G2 to M is controlled by CDK1 and cyclin B. These molecules interact to form a complex that adds phosphate groups to cellular components that break down the nuclear membrane (lamins A, B, and C), reorganize the cytoskeleton (caldesmon), and initiate chromosome condensation (histone H1). Cyclin B may specify cellular localization of target molecules. Other cyclins (especially cyclin A) are thought to be involved during this transition, but their functions are not yet known.

to cyclin B, CDK1 catalyzes phosphorylation of a cytoplasmic protein, caldesmon, which brings about nuclear membrane breakdown and rearrangement of the cytoskeleton. CDK1 also phosphorylates histone H1, which may play a role in chromatin condensation (Figure 21–3). Although several experiments suggest that cyclin A is also involved in the progression from G2 to M, its functions are not clearly understood.

Mutations that disrupt any step in cell cycle regulation are candidates for the study of cancer-causing genes. For example, mutations in genes that encode the kinases and cyclins, or their target genes, are candidates for such genes. Evidence is accumulating that the G1 checkpoint is aberrant in many forms of cancer, and mutant G1 cyclins and kinases are thought to be the best candidates for cancer-promoting genes.

21.2 Genes, Cancer, and the Cell Cycle

Genetic studies have identified a number of genes that when mutated, confer a predisposition to specific cancers (see Table 21.2). It is clear that the two main properties of cancer, uncontrolled cell division and the ability to spread or metastasize, are the result of genetic alterations. As we mentioned earlier, these alterations can involve large-scale genomic instability that results in chromosome loss, chromosome rearrangement, or the insertion of foreign (often viral) DNA sequences into loci on human chromosomes. Smaller-scale alterations, such as changes in nucleotide sequence, or more subtle modifications that alter only the amount of a gene product or the time over which it is active may also be involved.

TABLE 21.2   Inherited Predispositions to Cancer

Tumor Predisposition Gene	Chromosome
Early-onset familial breast cancer	17q
Familial adenomatous polyposis	5q
Familial melanoma	9p
Gorlin syndrome	9q
Hereditary nonpolyposis colon cancer	2p
Li-Fraumeni syndrome	17p
Multiple endocrine neoplasia, type 1	11q
Multiple endocrine neoplasia, type 2	22q
Neurofibromatous, type 1	17q
Neurofibromatous, type 2	22q
Retinoblastoma	13q
Von Hippel-Lindau syndrome	3p
Wilms tumor	11p

In general, the control of cell division is regulated in two ways: (1) by genes that normally function to suppress cell division, and (2) by genes that normally function to promote cell division. The first group of regulatory genes is called **tumor suppressor genes**. When expressed, these genes halt passage through the cell cycle and prevent mitotic division. For cell division to take place, these genes, and/or their gene products, must be inactive or absent. If tumor suppressor genes become permanently inactivated or deleted through mutation, control over cell division is lost, and the mutant cell begins to proliferate in an uncontrolled fashion.

Genes that normally promote cell division are called **proto-oncogenes**. These genes can be "on" or "off," and when they are "on," they promote cell division. To halt cell division, these genes and/or their gene products must be inactivated. If these genes become permanently switched on, then uncontrolled cell division occurs, leading to tumor formation. Mutant forms of proto-oncogenes are known as **oncogenes**.

In the following sections we examine how mutations in tumor suppressor genes can lead to a loss of control over cell division and the development of cancer. Then we consider the role of proto-oncogenes and oncogenes.

21.3 Tumor Suppressor Genes

Many studies document families with high frequencies of certain types of cancers, such as breast, colon, or kidney cancers. In most cases, identifying a clear, simple pattern of inheritance has proved difficult. However, genetic studies indicate that in such cases, mutations in single genes do indeed predispose cells to becoming malignant (Table 21.2). One example is the inheritance of a predisposition to **retinoblastoma (RB)**, a cancer of the retinal cells of the eye.

Retinoblastoma

Retinoblastoma occurs with a frequency that ranges from 1 in 14,000 to 1 in 20,000, and most often appears between the ages of 1 and 3 years. Two forms of retinoblastoma are known. In the familial form (about 40% of all cases), individuals who carry one mutant *RB* allele are far more susceptible to developing retinoblastoma than those with two normal wild-type *RB* alleles. Thus, predisposition to this form of cancer is inherited as an autosomal dominant trait although the mutation itself is actually recessive, as we will explain below. In fact, 90 percent of individuals who inherit a mutant *RB* allele will develop retinal tumors, usually in both eyes. In addition, these individuals are also predisposed to developing other forms of cancer, such as osteosarcoma, a bone cancer, even if they don't develop retinoblastoma.

The second form of retinoblastoma (the remaining 60%) is not familial, and tumors develop spontaneously. This sporadic form is characterized by the appearance of tumors only in one eye, and onset occurs at a much later age than in the familial form.

By studying the two types of retinoblastoma, Alfred Knudson and his colleagues developed a model that requires the presence of two mutated copies of the *RB* gene in the same retinal cell for tumor formation; in other words, tumor development is a recessive trait. In the familial form, one mutant *RB* allele is inherited and carried by all cells of the body, including cells of the retina [(Figure 21–4(a)]. If the second allele of the *RB* gene mutates in a retinal cell, a retinal tumor results. Therefore, individuals carrying an inherited mutation of the *RB* gene are predisposed to develop retinoblastoma, as only one additional mutational event is required to cause tumor formation. This does not happen in all cases. About 10 percent of those inheriting a mutant *RB* allele do not develop cancer. In these cases, the wild-type *RB* allele does not mutate in any retinal cells.

In the nonfamilial sporadic cases [Figure 21–4(b)], independent mutations in both normal *RB* alleles must occur in the same retinal cell for a tumor to develop. As might be expected, these events are far less frequent and occur at a later age. As predicted by Knudson's model, such sporadic forms of retinoblastoma are more likely to occur in a single eye.

Similar studies on predisposition to other cancers have led researchers to conclude that the number of mutations in a single cell necessary to bring about the development of cancer ranges from 2 to perhaps as many as 20 (Table 21.3).

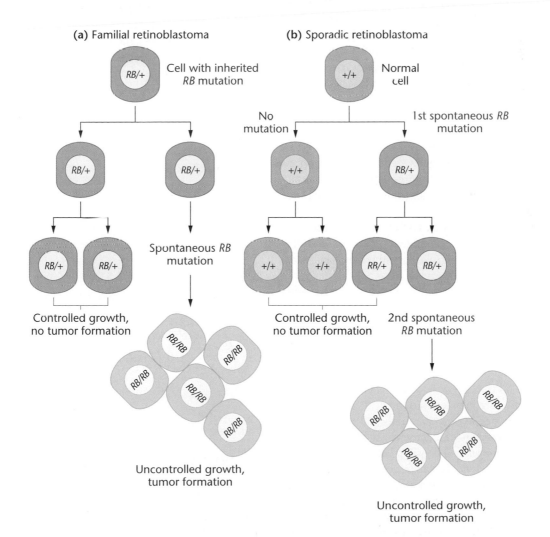

FIGURE 21–4 (a) In familial retinoblastoma, one mutation is inherited and present in all cells. A second mutation at the retinoblastoma locus in any retinal cell results in uncontrolled cell growth and tumor formation. (b) In spontaneous retinoblastoma, two mutations in the retinoblastoma gene in a single cell are acquired sequentially, causing uncontrolled cell growth and division, and resulting in tumor formation.

TABLE 21.3 Number of Mutations Associated with Some Cancers

Cancer	Chromosome Sites	Minimum Number of Mutations Required
Retinoblastoma	13q	2
Wilms tumor	11p	2
Colon cancer	5p, 12p, 17p, 18q	4–5
Small-cell lung cancer	3p, 11p, 13q, 17p	10–15

The retinoblastoma gene (*RB1*) located on chromosome 13 encodes a protein designated **pRB**. The pRB protein is present in the nuclei of retinal cells and all cell types examined so far. Found in both resting (G0) cells and actively dividing cells, pRB is present at all stages of the cell cycle. The pRB protein is part of a regulatory pathway that is important in cell cycle progression. pRB acts as a molecular switch that controls the passage of cells from G1 into S phase. Because the normal gene is expressed ubiquitously in the body, cells can progress through the G1/S transition only when pRB is inactivated by phosphorylation. pRB is regulated by cyclin-dependent kinases (CDK). When CDK4 binds to cyclin D, the kinase phosphorylates pRB. pRB is phosphorylated in the S and the G2/M phases, but is not phosphorylated in the G0 and G1 phases. When pRB is un-phosphorylated (its active state), it binds to members of the E2F family of transcription factors. Recall from Chapter 12 that transcription factors are proteins that bind to the promoter region of genes and regulate transcription. E2F transcription factors control the expression of some 20–30 genes required to move the cell from G1 into S phase. In its un-phosphorylated form, pRB binds to E2F, blocking transcription, and keeping the cell in G1 (Figure 21–5).

The phosphorylation of pRB by CDK4 takes place in late G1. When phosphate groups are added, pRB releases E2F, allowing transcription of the genes that move the cell from G1 into S phase.

Direct evidence for the role of pRB in regulating the cell cycle comes from several experiments. Cultured osteosarcoma (bone cancer) cells carry two *RB* mutations, and these cells do not produce pRB. When osteosarcoma cells are injected into a cancer-prone strain of mice, tumors are formed. If a normal *RB* gene is transferred to the cancer cells, pRB is produced and no tumors are formed when these genetically modified cells are injected into mice.

In a separate two-step experiment, a normal *RB* gene was transferred into osteosarcoma cells grown in culture. These cells produced pRB and stopped cell division. When D or E cyclins were added to these pRB-blocked cells, cell division resumed. The analysis of pRB indicates that adding cyclins causes pRB to become phosphorylated, presumably by activating CDK4. These results demonstrate that active pRB stops cell division, that pRB is a target of a G1 CDK/cyclin complex, and that inactivating pRB allows passage through the cell cycle and subsequent division, confirming pRB's role in G1 control.

In normal retinal cells, pRB is activated by CDK4/cyclin D phosphorylation; it then prevents passage into S phase by in-teracting with the transcription factor E2F. In retinoblastoma cells, both copies of the *RB1* allele are defective, and no pRB is present. As a result, E2F activates genes required for passage through the G1/S checkpoint. This important checkpoint is continually overridden in these cells, resulting in uncontrolled cell growth and tumor formation.

p53: **Guardian of the Genome**

Another tumor suppressor gene, *p53*, has been implicated in human cancer. The *p53* gene encodes a nuclear protein that acts as a transcription factor. Mutations of *p53* are found in a wide range of cancers, including breast, lung, bladder, and colon cancers. It is estimated that 50 to 60 percent of all cancers are associated with mutations in the *p53* gene, suggesting that *p53* controls one or more key events in the proliferation of all cells, and is not involved in a cell- or tissue-specific form of regulation. Inherited mutations in the *p53* gene are associated with the Li-Fraumeni syndrome, an autosomal dominant condition with a predisposition to develop cancers in several tissues at a high frequency.

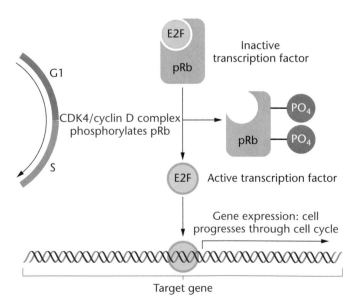

FIGURE 21–5 In the nucleus during G1, pRB, the product of the retinoblastoma gene, interacts with and inactivates transcription factor E2F. As the cell moves from G1 to S, a CDK/cyclin D complex forms and adds phosphate groups to pRB. As pRB becomes hyperphosphorylated, E2F is released and becomes transcriptionally active, allowing the cell to pass through S phase. Phosphorylation of pRB is transitory; as cyclin is degraded, phosphorylation declines.

Normally the p53 protein is present in cells at a low level, in a rapidly degraded, inactive form. Several types of signals can cause a shutdown of p53 degradation, leading to a rapid increase in the cellular concentration of the p53 protein. These signals include chemical damage to DNA, double-stranded breaks in DNA induced by ionizing radiation, or the presence of DNA repair intermediates generated by exposure to ultraviolet light. Activation of the p53 protein can lead to several responses, including (1) DNA repair; (2) cell cycle arrest; and (3) **apoptosis**, a genetically programmed pathway of cell death. These tasks are accomplished by the activation of target genes by p53 acting as a transcription factor. The expression of more than 20 target genes are mediated by the action of p53.

Cell cycle arrest by p53 can occur at several phases, including G1. To arrest the cell cycle at this phase, p53 stimulates the transcription of a gene encoding a CDK inhibitor called p21. The p21 protein targets a number of CDK/cyclin complexes, including the CDK4/cyclin D complex (discussed above).

This keeps pRB in its active configuration, repressing the transcription of genes needed to move the cell from G1 into S phase. Cells lacking functional p53 protein are unable to arrest in G1 following irradiation, and they move immediately from G1 into S. These cells therefore do not repair the DNA damage and, as a result, have a high rate of mutation. Thus the *p53* gene is often referred to as the "guardian of the genome."

As mentioned above, more than 50 percent of all human cancers carry a mutation in the *p53* gene. Because the activated p53 protein is organized as a tetramer, mutation of one *p53* allele usually abolishes all p53 activity, since almost all tetramers contain at least one defective subunit. This means that mutations in the *p53* gene act as dominant negative mutations, and that individuals heterozygous for a *p53* mutation will develop cancer with a frequency of 90–95 percent. The central role of the *p53* gene in controlling the cell cycle emphasizes the relationships between cancer and the cell cycle and between genes that regulate cell growth and cancer.

Breast Cancer Genes

Mutations in *BRCA1*, a gene that maps to the long arm of chromosome 17, are associated with a predisposition to breast cancer; this predisposition is inherited as an autosomal dominant trait. About 85 percent of women carrying one mutant *BRCA1* gene will have a mutation in the second *BRCA1* allele, and develop breast cancer. These women also have an increased risk of ovarian cancer.

A second breast cancer gene, *BRCA2*, located on the long arm of chromosome 13, is also inherited as an autosomal dominant predisposition to breast cancer, but is not associated with an increased risk of ovarian cancer. Together, these genes account for a large majority of breast cancer associated with a genetic predisposition (about 10% of all cases of breast cancer). The mutant forms of these genes play no role in sporadic cases of breast cancer, which make up 90 percent of all cases.

Both the *BRCA1* and the *BRCA2* genes encode large proteins that are confined to the nucleus, and expressed in many tissues. Expression is at its highest during S phase of the cell cycle. Circumstantial evidence, including their common pattern of expression, the similar phenotype of the mutant alleles (breast cancer), cellular localization, and experimental evidence from studies in mice, suggests that both genes have similar functions, and are involved in DNA repair. More direct evidence for the role of the BRCA1 protein in DNA repair also exists. In cells exposed to ionizing radiation (which produces double-stranded DNA breaks), BRCA1 is phosphorylated (a sign of molecular activation). Current efforts center on identifying the kinase proteins that carry out this phosphorylation, in an attempt to place the *BRCA1* and the *BRCA2* gene products in a pathway associated with DNA repair.

Recent evidence suggests that the *BRCA1* gene product can be phosphorylated by two different kinases, and that either of these kinases are activated by double-stranded DNA breaks (Figure 21–6). In this model, DNA damage activates

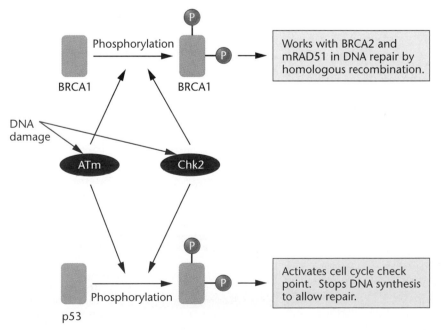

FIGURE 21–6 DNA damage activates the kinase ATM, which in turn activates the kinase Chk2. Once activated, these kinases stimulate the phosphorylation of the nuclear proteins p53 and BRCA1. Phosphorylation stabilizes p53, leading to increased levels of the protein. As the concentration of p53 increases, it activates a cell cycle control point, halting DNA replication. The phosphorylated BRCA1 protein interacts with a number of other proteins, including BRCA2 and mRAD51 to bring about repair of double-stranded DNA breaks by homologous recombination.

a kinase called ATM and/or a kinase called Chk2. In turn, these kinases phosphorylate BRCA1 and p53. The activated p53 protein arrests replication during S phase to allow DNA repair. The activated BRCA1 protein participates in DNA repair with the BRCA2 protein, the mRAD51 protein, and other nuclear proteins involved in DNA repair.

The role of BRCA1 and BRCA2 proteins in DNA repair may also explain why they are not associated with sporadic cases of breast cancer. In sporadic cases, at least three mutations would have to accumulate in a breast cell for it to become cancerous: Both copies of the *BRCA1* or *BRCA2* genes would have to be mutant, and at least one other mutation in a cell-cycle regulation gene would have to occur. On the other hand, those who inherit a mutant copy of either the *BRCA1* or *BRCA2* gene already have one mutation and need only two more to trigger breast cancer.

21.4 Oncogenes

We now discuss the second category of genes involved in cell cycle regulation: those that normally function to promote cell division, called **proto-oncogenes**. When they are expressed, these genes promote cell division. To stop cell division, these genes and/or their gene products must be inactivated. In mutant form, as **oncogenes**, they induce or maintain uncontrolled cellular proliferation associated with cancer. The protein products of proto-oncogenes are found throughout the cell, including the plasma membrane, cytoplasm, and nucleus (Table 21.4). In spite of their wide-ranging locations within the cell, which suggest varying functions, all products characterized to date alter gene expression either directly or indirectly.

Unlike most tumor suppressor genes, where mutations in both alleles of a gene are necessary to promote the development of cancer, only one of the two copies of a proto-oncogene needs to mutate to induce malignancy, resulting in a dominant phenotype.

Rous Sarcoma Virus and Oncogenes

In 1910, Francis Peyton Rous first inferred the existence of specific genes associated with transforming normal cells into cancerous cells. Rous studied a connective tissue tumor in chickens known as a **sarcoma**. He injected cell-free extracts from such tumors into healthy chickens and induced the formation of sarcomas. He postulated the existence of an agent that transmitted the disease, which decades later was shown by other investigators to be a virus. It is now known as the **Rous sarcoma virus (RSV)**. Rous received the Nobel Prize in 1966 for his pioneering work in establishing the relationship between viruses and cancer.

RSV is able to infect cells and reproduce within them. Once inside a cell, the single-stranded RNA genome of RSV is first transcribed by the enzyme **reverse transcriptase**; this converts the RNA genome into a single-stranded DNA molecule. The single-stranded DNA is used as a template to synthesize the complementary strand, creating a double-stranded DNA molecule that integrates into the genome of the infected cell, forming a **provirus**. At a later time, the DNA is transcribed into RNA, which is translated into viral proteins. Packaging RNA molecules into these proteins forms new virus particles. Because the replication cycle of viruses like RSV "reverses" the flow of genetic information, they are called **retroviruses**.

In RSV, the tumor-forming ability results from a single gene, the *src* gene, present in the viral genome. This gene, re-

TABLE 21.4   Cellular Location of c-onc and v-onc Proteins

Gene	Location of c-onc Protein	Location of v-onc Protein
src	Membranes	Membranes
ras	Membranes	Membranes
myc	Nucleus	Nucleus
fps	Cytoplasm	Cytoplasm and membranes
abl	Nucleus	Cytoplasm
erbB	Plasma membrane	Plasma membrane and Golgi

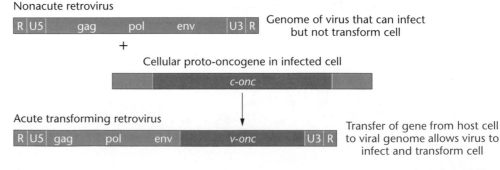

FIGURE 21–7 A transforming retrovirus has acquired a copy of a gene from the host genome, converting it from a *c-onc*, or proto-oncogene, into an oncogene that confers on the virus the ability to transform a specific type of host cell into a cancerous cell.

TABLE 21.5 Origin and Function of Representative Oncogenes

Oncogene	Origin	Species	Cellular Function
src	Rous sarcoma virus	Chicken	Tyrosine kinase signal protein
fos	FBJ osteosarcoma virus	Mouse	Transcription factor
sis	Simian sarcoma virus	Monkey	Platelet-derived growth factor
abl	Abelson murine leukemia virus	Mouse	Tyrosine kinase
myc	Avian myelocytomatosis virus	Chicken	Transcription factor
erbB	Avian erythroblastosis virus	Chicken	EGF receptor
N-ras	Neuroblastoma, leukemia	Human	GTP-binding protein

sponsible for inducing tumor formation in chicken cells, is an oncogene. Retroviruses which carry oncogenes are known as **acute transforming viruses**. Other retroviruses that do not carry oncogenes but that induce the activity of cellular genes that bring about tumor formation are known as **nonacute viruses**.

Origin of Oncogenes

Oncogenes (*onc*) carried by acute transforming viruses are acquired from the host's genome during infection, when a portion of the viral genome is exchanged for a cellular proto-oncogene (Figure 21–7). The oncogenes carried by a retroviruses are called *v-onc*; the normal, cellular version of the gene is *c-onc* (also called a proto-oncogene). Retroviruses that carry a *v-onc* gene are able to infect and transform a specific type of host cell into a tumor cell. For RSV, the oncogene captured from the chicken genome, *v-src*, confers the ability to transform chicken cells into sarcomas. The cellular version of the same gene, found in the chicken genome, is *c-src*. More than 20 oncogenes have been identified by their presence in retroviral genomes, with over 60 oncogenes identified to date. Some of these are listed in Table 21.5. It is important to note that although oncogenes were first identified in retroviruses, not all oncogenes are mutant versions of cellular genes carried by retroviruses. Some oncogenes arise by mutagenic events that occur spontaneously. We must, therefore, consider a broad range of mechanisms in oncogenetic events.

Formation of Oncogenes

At least three mechanisms can explain how proto-oncogenes are converted into oncogenes: **point mutations**, **translocations**, and **overexpression** (see Table 21.6). Some of these events are

TABLE 21.6 Conversion of the Proto-Oncogene *c-onc* to Oncogenes

Mechanism	Oncogene
Point mutation	*ras*
Translocation	*abl*
Overexpression of gene product	
New promoter by viral insertion	*mos, myb*
New enhancer by viral insertion	*myc*
Amplification of proto-oncogene	*myc*

mediated by viruses, and others by intracellular events that occur in the absence of retroviruses.

Mutations in Ras proteins demonstrate how a point mutation converts a proto-oncogene into an oncogene. The ***ras* gene family** encodes signal transduction proteins that play a major role in regulating cell growth and division. Ras proteins act as molecular switches that transmit signals from the extracellular environment to the cytoplasm. More than 30 percent of all human cancers carry a mutant *ras* oncogene. Ras proteins are part of the plasma membrane, and cycle between an inactive (switch "off") state and an active (switch "on") state. Activated Ras proteins interact with cytoplasmic proteins, starting a cascade of events that transmits a signal from the cytoplasm to the nucleus. In the nucleus, the signal activates the transcription of genes that initiate cell division.

Comparing the amino acid sequences of Ras proteins from a number of different human carcinomas (tumors of epithelial tissue) reveals that *ras* mutations involve single amino acid substitutions at either position 12 or 61 (Figure 21–8, page 440) in the 189 amino acid Ras protein. Each of these changes can be created by a single nucleotide substitution in the *ras* gene. Because of this nucleotide substitution, mutant Ras proteins cannot cycle from the active to the inactive form, and are locked into the "on" position, continually signaling for cell division (Figure 21–9, page 440). This change represents one step in the transformation from a normal cell to a malignant one.

One well-characterized example of oncogene activation by translocation is that of *c-abl*, an oncogene associated with chronic myelogenous leukemia (CML). In this case, described in detail in a later section, the translocation results in altered gene activity that causes tumor formation.

At least three separate mechanisms of proto-oncogene activation are associated with overexpression. First, the proto-oncogene may acquire a new promoter, which causes the level of transcript production to increase or a silent locus to activate. This is the case in avian leukosis, where strong viral promoters are inserted adjacent to a proto-oncogene, causing an increase in both mRNA production and the amount of the gene product. A second mechanism of overexpression involves the acquisition of new upstream regulatory sequences, including enhancers. The third mechanism involves the amplification of the proto-oncogene. In human tumors, members of the *myc* family of oncogenes are frequently amplified. The *c-myc* proto-oncogene is found amplified up to several hundred copies in some human tumors.

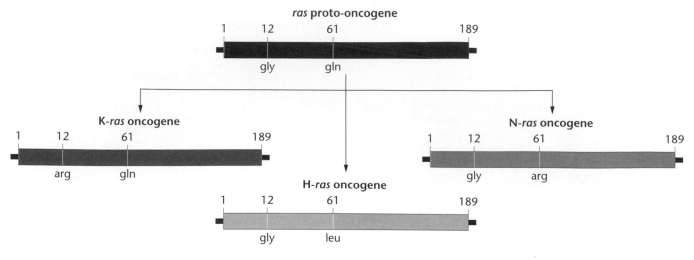

FIGURE 21–8 The *ras* proto-oncogene encodes a protein of 189 amino acids. In the normal protein, glycine is encoded at position 12, and glutamine at position 61. Analysis of *ras* oncogene proteins from several tumors shows a single amino acid substitution at one of these positions, which converts a proto-oncogene into a tumor-promoting oncogene.

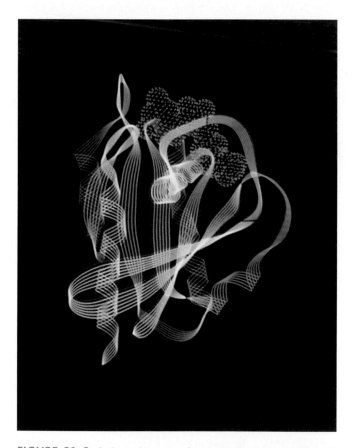

FIGURE 21–9 A three-dimensional computer-generated image of Ras proteins in two different conformations. Normal Ras proteins act as molecular switches controlling cell growth and differentiation. The switch is "on" (conformation shown in blue) when GTP binds to the protein, and "off" (conformation shown in yellow) when the GTP is hydrolyzed to GDP. Switching the protein between states alters the conformation of the protein in the two regions (blue and yellow). Oncogenic mutations of *ras* are stuck in the "on" state, continuously signaling for cell growth. *(Sung-Hou Kim, University of California, Dept. of Chemistry and Lawrence Berkeley National Laboratory, Berkeley, CA)*

21.5 A Genetic Model of Cancer: Colon Cancer

Cancer is clearly a multistep process that results from a number of specific genetic alterations. Studies of tumors such as retinoblastoma have established that in some cases, only a few steps are required to transform a normal cell into a malignant one. However, most cancers develop in multiple steps, with intermediate levels of genetic and cellular transformation.

For several reasons, the study of colorectal cancer has been used to obtain detailed information about the nature and order of the genetic and cellular events that result in a malignant growth. First, the malignant form of colorectal tumors develops from preexisting benign tumors. Second, several discrete precancerous stages occur, and these stages can be isolated and studied. Furthermore, both hereditary and sporadic forms of colorectal cancer exist. As a result, colorectal cancer serves as a useful model for studying the interaction of genetic and environmental factors in tumor formation.

Two forms of genetic predisposition to colon cancer are known: (1) an autosomal dominant trait, known as **familial adenomatous polyposis (FAP)**, and (2) a genetically complex trait, known as **hereditary nonpolyposis colorectal cancer (HNPCC)**. FAP is associated with about 1 percent of all cases, and HNPCC is responsible for 2–4 percent of all colon cancers. The remaining 95 percent of all cases are sporadic. Researchers have developed genetic models of both forms of colon cancer, which provide insight into the events that cause normal cells to become cancerous.

FAP and Colon Cancer

In this model, spontaneous cases of FAP-associated colon cancer begin with a mutation in the *APC* gene, which maps to the long arm of chromosome 5. This mutation takes place

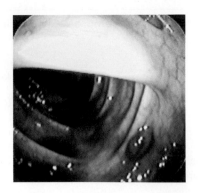

FIGURE 21–10   Polyps in the colon. *(Albert Paglialunga/Phototake)*

in a normal epithelial cell in the lining of the colon. Those who inherit a mutant copy of the *APC* gene carry the mutation in all cells of their body. In spontaneous cases, the presence of an *APC* mutation causes the epithelial cell to partially escape cell cycle control, and the cell divides to form a small cluster of cells called a **polyp** (Figure 21–10). Individuals heterozygous for a mutant *APC* gene form hundreds or thousands of polyps. Each polyp consists of a clone of cells, all of which carry an *APC* mutation. The loss of the corresponding allele on the homologous copy of chromosome 5 is *not* necessary for polyp formation. However, in the majority of cases, the second *APC* allele becomes mutant in a later stage, and the *APC* gene is regarded as a tumor suppressor gene. The first step and the relative order of subsequent mutations are shown in Figure 21–11.

Subsequent mutations of genes within the polyp cause intermediate stages of tumor formation. Mutations in the *ras* oncogene on chromosome 12 in polyp cells (with a preexisting mutation of the *APC* gene) cause the polyp to grow larger and to develop a number of fingerlike (villous) outgrowths. This is the intermediate adenoma stage. To progress further, a polyp cell carrying an *APC* mutation must acquire mutations in a gene on chromosome 18, called *DCC* (deleted in colon cancer). Mutations in both *DCC* alleles result in the formation of late-stage adenomas. Finally, a mutation on

17p, involving the loss or inactivation of *p53*, causes the transition to a cancerous cell. As discussed earlier, mutations in the *p53* gene are pivotal to the development of a number of cancers, including lung, brain, and breast cancers as well as colon cancer. Recall that the normal *p53* allele is a tumor suppressor gene that regulates the passage of cells from late G1 to S phase. Metastasis occurs after the formation of colon cancer, and it involves an unknown number of mutational steps.

HNPCC and Colon Cancer

The HNPCC form of colon cancer develops after a small number of polyps are formed, rather than the hundreds or thousands associated with FAP. Genes associated with HNPCC have been mapped by linkage analysis to loci at 2p16 and 3p21. Mutations at these two loci generate a cascade of mutations in short, tandemly repeated microsatellite sequences located throughout the genome.

The gene on chromosome 2 associated with HNPCC is a DNA repair gene called *MSH2*. Inactivation of this gene causes a rapid accumulation of mutations and also causes the development of colorectal cancer. A second DNA repair gene, *MLH1*, associated with HNPCC has been mapped to chromosome 3. At least two other DNA repair genes related to *MLH1* and *MSH2* have been identified, and remain to be mapped. These genes are called mismatch repair genes (*MMR*) and generate a condition called microsatellite instability (microsatellites are discussed in Chapter 17). Mutations in any of the *MMR* genes promote genomewide genetic instability, accelerating the rate at which mutations accumulate, with colon cancer as one outcome. It may turn out that mutations in any one of these genes is enough to cause HNPCC.

The mutations leading to HNPCC are similar to those in FAP-associated colon cancer (Figure 21–11) but have several important differences. FAP begins with a mutation in *APC*, and the formation of hundreds or thousands of polyps. Each polyp progresses slowly toward cancer by accumulating mutations in other genes (*ras*, *DCC*, etc.). However, because

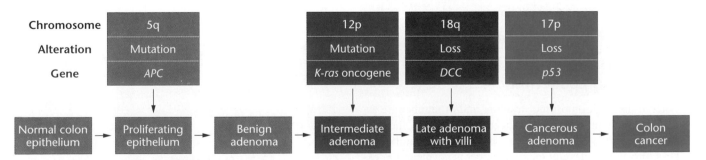

FIGURE 21–11   A model for the multistep production of colon cancer. The first step is the loss or inactivation of both alleles of the *APC* gene on chromosome 5. In familial cases, one mutation of the *APC* gene is inherited. Loss of both alleles leads to the formation of benign adenomas. Subsequent mutations involving genes on chromosomes 12, 17, and 18 in cells of the benign adenomas can lead to a malignant transformation that results in colon cancer. Although the mutations in chromosomes 12, 17, and 18 usually occur at a later stage than those involving chromosome 5, the sum of the changes is more important than the order in which they occur.

there are so many polyps, there is a high probability that at least one polyp will accumulate all the mutations necessary to cause colon cancer. HNPCC begins with mutation in DNA repair genes in an epithelial cell of the colon, leading to genomewide mutations, one of which can include *APC*. Mutation of *APC* is a relatively rare event, so only a few polyps will form. However, even though the polyps are relatively few in number, the continued high mutation rate ensures that at least one polyp will acquire the mutations necessary to become cancerous. FAP-associated colon cancer is therefore a disease of rapid tumor initiation, with slow progression to cancer, and HNPCC is a disease with slow initiation of tumors, but rapid progression to cancer. Both forms of colon cancer are mediated by mutations in the *APC* gene.

21.6 Gatekeeper and Caretaker Genes

The two pathways to colon cancer in FAP and HNPCC offer an insight into the nature of genes that control cancer predisposition. These differences have led to the idea that mutations in two types of genes cause predisposition to cancer: **gatekeeper genes** and **caretaker genes**. The *FAP* gene is a gatekeeper; it normally inhibits cell growth. In general, tumor suppressor genes are gatekeepers. In different cell types, only one or a few genes serve as gatekeepers, and if both copies of the gatekeeper mutate, a specific cancer, such as retinoblastoma or colon cancer, develops. Individuals predisposed to a specific form of cancer inherit one mutant copy of such a gatekeeper, and need only one additional mutation to initiate tumor formation. In spontaneous cases, both copies of the gatekeeper must mutate for cancer to develop.

Caretaker genes maintain the integrity of the genome, including DNA repair genes such as *MSH2* and *MLH1*. These genes normally function to repair damage to DNA caused by environmental agents such as ultraviolet light, or mismatches that occur during DNA replication. Mutation of a caretaker gene does not promote tumor formation directly, but leads to genetic instability that increases the mutation rate of all genes, including gatekeeper genes. If a gatekeeper gene mutates and begins tumor formation, the process is accelerated by the high rate of mutation in the cell.

21.7 Chromosomal Translocations and Leukemia

Alterations in chromosome structure and/or number are associated with many forms of cancer. In most cases, the relationship between changes in chromosome number or structure and the development of cancer is not clear. For example, individuals with Down syndrome carry an extra copy of chromosome 21. This quantitative alteration in genetic content is also associated with a 20-fold increased risk of leukemia as compared to the general population. In a limited number of cases in which specific alterations in chromosome structure are associated with cancer, more direct

TABLE 21.7 Specific Chromosome Aberration and Cancer

Cancer	Chromosome Alteration*
Chronic myelogenous leukemia	t(9;22)
Acute promyelocytic leukemia	t(15;17)
Acute lymphocytic leukemia	t(4;11)
Acute myelogenous leukemia	t(8;21)
Prostate cancer	del(10q)
Synovial sarcoma	t(X;18)
Testicular cancer	inv(12p)
Retinoblastoma	del(13q)
Wilms tumor	del(11p)

*t = translocation, del = deletion, inv = inversion.

information indicates how structural rearrangements are associated with the development and/or maintenance of the cancerous state.

Changes in chromosome structure or number are often associated with various forms of cancer. In selected cases, the relationship between the chromosome aberration and the development and/or maintenance of the cancerous state is clear. Such a connection is most clearly seen in leukemias, where the presence of a specific translocation of a chromosome is well defined (Table 21.7). One of the best-studied examples is the translocation between chromosomes 9 and 22 that is associated with **chronic myelogenous leukemia (CML)** (Figure 21–12). Originally, this translocation was described as an abnormal chromosome 21 and called the **Philadelphia chromosome** (because it was discovered in that city). Later, Janet Rowley showed that the Philadelphia chromosome re-

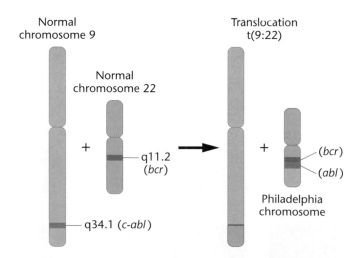

FIGURE 21–12 The Philadelphia chromosome. A reciprocal translocation involving the long arms of chromosomes 9 and 22 results in the production of a characteristic chromosome, the Philadelphia chromosome, which is associated with cases of chronic myelogenous leukemia (CML). The t(9;22) translocation results in the fusion of the *c-abl* oncogene on chromosome 9 with the *bcr* gene on chromosome 22. The fusion protein is a powerful hybrid molecule that allows cells to escape control of the cell cycle, resulting in leukemia.

sults from the exchange of genetic material between chromosomes 9 and 22. This translocation is never seen in a normal cell and is found only in the white blood cells involved in CML. These observations indicate that the translocation is a primary and causal event in the generation of CML, and that this form of cancer may originate from a single cell bearing this translocated chromosome.

By examining a large number of cases involving the Philadelphia chromosome, researchers were able to determine the exact location of the breakpoints on chromosomes 9 and 22. Genetic mapping studies using recombinant DNA techniques established that a proto-oncogene, *c-abl* maps to the breakpoint region on chromosome 9, and that the gene *bcr* maps near the breakpoint on chromosome 22. The normal c-Abl protein is a kinase and the normal Bcr protein activates a phosphorylation reaction. In the translocation event, all or most of the *c-abl* gene is translocated to a region within the *bcr* gene, generating a hybrid *bcr/c-abl* oncogene that is transcriptionally active. The abnormal hybrid 200-kDa protein product of the fused genes has been implicated in the generation of CML.

Both cytogenetic and molecular approaches have been applied to the cytogenetic events in translocations involving chromosome 8 [these include t(8;14), t(8;22), and t(2;8)] in a disease related to leukemia, known as **lymphoma**. The breakpoint on chromosome 8 in all these translocations is the same, and the proto-oncogene *c-myc* has been mapped to this locus. The loci at the breakpoint on the other chromosomes involved in this series of translocations all have immunoglobulin genes at the breakpoints. The movement of the *c-myc* gene to a position near these immunoglobulin genes leads to the overexpression of the *c-myc* gene, resulting in the transformation of the lymphoid cells.

Other genes at translocation breakpoints in leukemia have been isolated and characterized. In these cases, as in CML, the translocation results in an abnormal gene product that causes the cell to undergo a malignant transformation even though a second and presumably normal copy of the gene is present and active in synthesizing a normal gene product. It is hoped that, if the abnormal gene products at other translocation loci associated with leukemia can be identified, therapeutic strategies will be developed to administer high levels of the normal gene product, or to inactivate the abnormal gene product.

21.8 Environmental Factors and Cancer

The relationship between environmental agents and the genesis of cancer is often elusive, and in the early stages of an investigation, this relationship is based on indirect evidence. Such studies often begin with an epidemiological survey, comparing the cancer death rates among different geographic populations. When differences are found in death rates for a type of cancer, an age group, or a cluster of related occupations such as chemical workers, further work is necessary to identify one or more environmental factors that correlate with these cancer deaths. These correlations are not conclusive, but they identify factors that, upon additional investigation, may directly relate to the development of cancer. Finally, extensive laboratory investigations may establish the mechanism by which an environmental agent generates cancer. In other cases, such as for certain viruses, the relationship between cancer and the environment is more straightforward, as we see next.

Hepatitis B and Cancer

Epidemiological surveys show that individuals who develop a form of liver cancer known as **hepatocellular carcinoma (HCC)** are infected with the **hepatitis B virus (HBV)** (Figure 21–13). In fact, for those carrying HBV, the risk of cancer increases by a factor of 100. Aside from the risk of cancer, HBV infection is a public health risk affecting some 300 million people worldwide; it is most prevalent in Asia and tropical regions of Africa. The infection produces a wide range of responses, from a chronic, self-limiting infection with few symptoms, to active hepatitis and cirrhosis, to fatal liver disorders and hepatocellular carcinoma. In the last few years, efforts have centered on understanding the mechanism by which HBV replicates and its role in causing liver cancer.

The genome of HBV is a mostly double-stranded DNA molecule of 3200 nucleotides. After cellular infection, the DNA moves to the nucleus, where it inserts into a chromosome. The viral DNA is transcribed into an RNA molecule and packaged into a viral capsid. The capsid containing the RNA pregenome moves to the cytoplasm, where it is reverse-transcribed into a DNA strand, which in turn is made double-stranded. The copied DNA genome is repackaged into a new capsid for release from the cell or returns to the nucleus for another round of replication.

The key to the role of HBV in carcinogenesis apparently resides in its entry into the nucleus. While in the nucleus, the HBV genome can insert itself at many different sites into

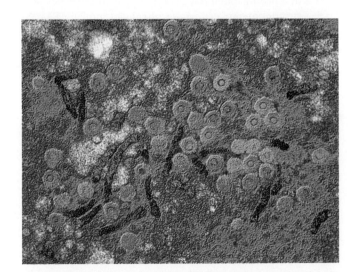

FIGURE 21–13 A false-color transmission electron micrograph of a hepatitis B virus. Infection with this virus often results in cancer of the liver. Worldwide, viral infections of all kinds are thought to be responsible for about 15 percent of all cancers. (*Meckes/Ottawa/Photo Researchers, Inc.*)

FIGURE 21–14 The possible outcomes following cellular infection with hepatitis B virus (HBV). Following the integration of the viral chromosome into the human genome, oncogenes can be activated, which results in tumor formation. Integration of the HBV genome can also cause chromosome instability, including the production of translocations and deletions. Insertions can also activate a cyclin A gene, causing abnormal regulation of the cell cycle and cellular proliferation, and resulting in cancer formation.

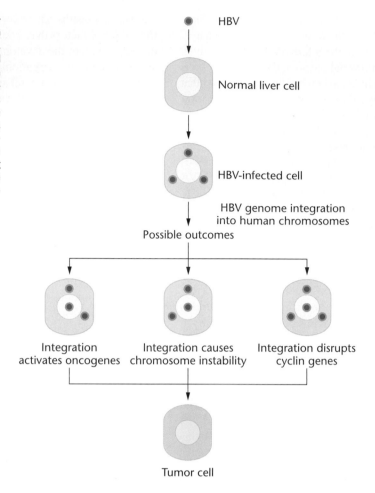

human chromosomes (Figure 21–14). Insertion often results in cytogenetic alterations of the host genome that include translocations, deletions, or amplifications of adjacent regions. As outlined earlier, most forms of cancer are associated with chromosomal rearrangements of these types, and such rearrangements may trigger the development of a cancerous transformation. The HBV genome has been shown to integrate into the cyclin A gene (*CCNA2*), and abnormal expression of this gene can disrupt the cell cycle and normal control of proliferation. Similarly, HBV integration results in activation of proto-oncogene loci, and altered expression of such loci is clearly implicated in the development of malignant cell growth. Thus, a strong link exists between the HBV virus as an environmental agent and the development of cancer in infected individuals.

Environmental Agents

Many surveys of cancer incidence (Table 21.8) point to the role of the physical environment and personal behavior as factors contributing to the development of cancer. The precise role of the environment in the genesis of cancer can be difficult to ascertain. Obviously, the induction of cancer involves an interaction between the genotype and environmental agents. The differential expression of a specific genotype in various environments is often neglected in addressing the role of environment in cancer, but researchers estimate that at least 50 percent of all cancers are environmentally induced. Environmental agents responsible for cancer include background levels of radiation, occupational exposure to physical and chemical agents, exposure to sunlight, and personal behavior such as diet and the use of to-

TABLE 21.8   Epidemiology of Cancer in Various Countries

	Death Rate per 100,000	
Country	*Male*	*Female*
Australia	212	125
Canada	214	136
Dominican Republic	54	48
Egypt	39	18
England	248	156
Greece	188	103
Israel	174	145
Japan	190	109
Nicaragua	22	35
Portugal	180	108
Singapore	249	130
United States	216	137
Venezuela	135	128

bacco. To show how environmental factors are identified, let's consider a recent study on the risk of colon cancer.

Epidemiological surveys reveal that cancer of the colon occurs with a much higher frequency in North America and Western Europe than in Asia, Africa, or other parts of the world. When people migrate from low-risk areas to high-risk areas, their risk of colon cancer increases to match that of the native residents. This finding suggests that environmental factors, including diet, may play a role in the incidence of colon cancer. In a long-range health study, the dietary habits of more than 88,000 registered nurses across the United States have been monitored since 1980. In that population, 150 cases of colon cancer were observed through 1986. Detailed analysis indicated that nurses who consumed daily meals of pork, beef, or lamb had a 2.5-fold higher risk of colon cancer than those who consumed such meals less than once a month. A more detailed analysis of diets strongly suggests that animal fat in the diet is the environmental risk factor for colon cancer. At the same time,

a negative correlation exists between eating skinless chicken meat and the incidence of colon cancer. Laboratory studies have been undertaken to identify the mechanisms by which fat brings about a cancerous transformation of the intestinal epithelium. However, even in the absence of information about the mechanism of action, it would seem prudent to reduce one's intake of animal fat in order to reduce the risk of colon cancer.

Research on the role of external factors as a cause of cancer indicate that neither the environment in general nor pollution in particular is responsible for a large fraction of cancer cases. Rather, diet, tobacco, and other agents, including medical and dental X-rays, and drugs, play significant roles in cancer development. It is estimated that these agents cause 50 percent of all cancer cases; thus, personal choices (e.g., the composition of food in the diet, tobacco use, and suntanning) figure prominently in the cancer equation. Education and judicious changes in lifestyles could prevent a high percentage of all human cancers.

Genetics, Technology, and Society

The Double-Edged Sword of Genetic Testing: The Case of Breast Cancer

The world is increasingly amazed—and perhaps bewildered—by the remarkable achievements in genetics and molecular biology over the last two decades. Both scientists and the media speculate that we will soon be able to test a person's gene sequences, and accurately predict that person's risk of developing diseases such as heart disease and cancers. Of course, the prospect of using molecular biology to prevent and cure a whole range of diseases that have genetic components is exciting. In our enthusiasm for the new genetic technologies, however, we often forget that these achievements have significant limitations and profound ethical concerns. The story of breast cancer genetic testing illustrates how we must temper our high expectations with respect for uncertainty.

Breast cancer is the most common type of cancer among women and the second leading cause of cancer deaths (after lung cancer). Each year, more than 180,000 new cases are diagnosed in the United States. Breast cancer is not limited to women; about 1400 men are diagnosed with breast cancer each year. The likelihood of developing breast cancer is low before age 35, but the risk increases after that. A woman's lifetime risk of developing breast cancer is about 10 percent.

Approximately 5–10 percent of breast cancers are familial, defined by the appearance of several cases of breast or ovarian cancer amongst near blood relatives and by an early onset of the disease. The clustering of breast and ovarian cancers in certain families suggests that genetic mutations are involved in the development of these cancers. In 1994, two genes were identified that appear to be linked to familial breast cancers. Germline mutations in these genes (BRCA1 and BRCA2) are associated with the majority of familial breast cancers; however, mutations in several other genes may also contribute. The molecular functions of BRCA1 and BRCA2 are still uncertain, although they may involve repair of damaged DNA or transcription regulation of hormone-responsive genes. Mutations in these genes act as autosomal dominants with variable penetrance. Women who bear mutations in BRCA1 have a 50–80 percent lifetime risk of developing breast cancer and a 20–40 percent risk of developing ovarian cancer. Men with germline mutations in BRCA2 have a 6 percent lifetime breast cancer risk—a 100-fold increase over the general male population.

Recently, genetic testing for mutations in the BRCA1 and BRCA2 genes has become available to families at high risk of familial breast cancer. These tests are extremely accurate, detecting any of the hundreds of mutations that can occur within the coding region of these genes. But the tests have limitations. They do not detect mutations in regulatory regions outside of the coding region, mutations that could result

in aberrant expression of these genes. Also, little is known about how any particular mutation manifests itself in terms of cancer risk, as penetrance can be affected by environmental factors and interactions with other susceptibility genes.

Some familial breast cancer patients and their families opt to take the BRCA1 and BRCA2 genetic tests. These patients feel that test results will help them to prevent recurrence of the disease, will guide them in childbearing decisions, and allow them to inform family members at risk. But none of these options has clear-cut positive benefits.

Women whose test results are negative are relieved that "they are not subject to familial breast cancer." However, their risk of developing breast cancer is still 10 percent (the population risk), and they should continue to monitor for the disease. Also, negative BRCA1 and BRCA2 tests do not eliminate the possibility that the patient bears a familial breast cancer mutation in another, unknown gene or that BRCA1 or BRCA2 mutations do exist, but occur outside of the gene coding regions, inaccessible to current genetic tests.

Those whose tests are positive face difficult choices. Treatment options are poor, consisting of close monitoring, prophylactic mastectomy and oophorectomy (removal of breasts and ovaries), and taking birth control pills or tamoxifen. Prophylactic surgery can reduce the risks but cannot eliminate them, as cancers still occur in tissues that remain after surgery. Drugs such as tamoxifen can also reduce the risk but have serious side effects. Genetic tests do not only affect the patient, they also affect the patient's entire family. People often experience fear, anxiety, and guilt on learning that they are carriers of a genetic disease. Studies suggest that even people who refuse genetic test results suffer from increased anxiety. Confidentiality is a major concern. Patients fear that their genetic information may be leaked to insurance companies or employers, jeopardizing their prospects for jobs and affordable health insurance.

Genetic testing is so new that the health system is lagging behind the science. Because genetic test results have both psychological side effects and medical ambiguities, genetic counseling is imperative for patients and their families. However, there are insufficient numbers of qualified genetic counselors with experience in genetic testing, and the issues in the most qualified of hands are still complex and difficult. Physicians often have limited knowledge of human clinical genetics and feel inadequate to advise their patients, and insurance companies have yet to develop policies concerning genetic tests and genetic information. It is incumbent on governments to develop regulations and standards for the safety and efficacy of genetic tests and for the uses of genetic information, to limit discrimination and ensure confidentiality. Given the unclear interpretation of *BRCA1/2* genetic tests, the reasonably ineffective treatment options, and the potential psychological and societal side effects, it is not surprising that only about 60 percent of familial breast cancer patients and their families decide to take the genetic tests.

The unanswered questions about *BRCA1/2* genetic testing are many and important. What risks are associated with which mutations? What types of cancers result from which mutations? Should all people be allowed access to *BRCA1/2* tests? Can we develop effective treatments for these diseases? How can we ensure that the high costs of genetic tests and counseling do not limit this new technology to only a portion of the population? Our struggle with these and similar questions is just beginning as we develop genetic tests for more and more diseases over the next few decades.

References

Kahn, P. 1996. Coming to grips with genes and risk. *Science* 274:496–98.

Martin, A-M., and Weber, B. L. 2000. Genetic and hormonal risk factors in breast cancer. *Journal of the National Cancer Institute* 92:1126–1135.

Miki, Y., et al. 1994. A strong candidate for the breast and ovarian cancer susceptibility gene BRCA1. *Science* 266:66–71.

Wopster, R., et al. 1995. Identification of the breast cancer susceptibility gene BRCA2. *Nature* 378:789–92.

Website

Health Technology Advisory Committee, Minnesota Department of Health. 1998. Genetic testing for susceptibility to breast cancer.
http://www.health.state.mn.us/htac/gt.htm

Chapter Summary

1. Cancer is a genetic disorder at the cellular level that can result from the mutation of a subset of genes or from alterations in the timing and amount of gene expression. Although some forms of cancer show familial patterns of inheritance, few show clear evidence for Mendelian inheritance.

2. Cancer results from the uncontrolled proliferation of cells and from the ability of such cells to metastasize or migrate to other sites and form secondary tumors. Mutant forms of genes involved in regulating the cell cycle are obvious candidates for cancer-causing genes. The cell cycle is regulated at several checkpoints. Two types of gene products, cyclins and kinases, are involved in regulating some checkpoints. The actions of these genes and the genes they control are important in regulating cell division, and links between these genes and the process of tumor formation are under intense scrutiny.

3. Although most cancers do not show clear-cut patterns of Mendelian inheritance, certain genes predispose individuals to cancer. Studies of retinoblastoma provide insight into the action of mutations that result in the development of tumors, and confirm that cell cycle regulation and cancer are linked.

4. Tumor suppressor genes normally act to suppress cell division. When these genes mutate or alter expression, control over cell division is lost. Molecular analysis of the retinoblastoma (*RB*) gene indicates that it controls gene expression by acting as a transcription factor. Mutations in both alleles of a tumor suppressor gene are necessary to promote the development of cancer.

5. Oncogenes normally function to initiate or maintain cell division; these genes must be mutated or inactivated to halt cell division. If these genes escape control and become permanently switched on, cell division occurs in an uncontrolled fashion. In contrast to tumor suppressor genes, a mutation in only one copy of an oncogene can promote the development of cancer.

6. In most cases of cancer, a series of mutations is necessary to cause the malignant state. Colon cancer is a useful model for demonstrating the multistep nature of cancer.

7. The cells of most tumors have visible chromosomal alterations, and the study of these aberrations provides insight into the steps involved in the development of cancer. This relationship between cancer and chromosome alterations has been best studied in leukemias, where the formation of hybrid genes or substitution of regulatory sequences is associated with the transformation of normal cells into malignant tumors.

8. Although cancer results from genetic alterations, the environment appears to be an important factor in cancer induction. Environmental agents include occupational exposure to physical or chemical substances, viruses, diet, and other personal choices such as the use of tobacco and suntanning.

Key Terms

acute transforming virus, 439

apoptosis, 437

carcinogen, 432

caretaker gene, 442

cancer susceptibility genes, 432

chronic myelogenous leukemia (CML), 442

cyclin-dependent kinase (CDK), 433

cyclin, 433

familial adenomatous polyposis (FAP), 440

gatekeeper gene, 442

hepatitis B virus (HBV), 443

hepatocellular carcinoma (HCC), 443

hereditary nonpolyposis colorectal cancer
    (HNPCC), 440

lymphoma, 443

metastasis, 432

nonacute virus, 439

oncogene, 434

overexpression, 439

p53 gene, 436

Philadelphia chromosome, 442

point mutation, 439

pRB protein, 436

protein kinase, 433

proto-oncogene, 434

provirus, 438

ras gene family, 439

retinoblastoma (RB), 434

retrovirus, 438

reverse transcriptase, 438

Rous sarcoma virus (RSV), 438

sarcoma, 438

translocation, 439

tumor suppressor gene, 434

Insights and Solutions

1. In disorders such as retinoblastoma, a mutation in one allele of the retinoblastoma (*RB*) gene can be inherited from the germ line, causing an autosomal dominant predisposition to the development of eye tumors. To develop tumors, a somatic mutation in the second copy of the *RB* gene is necessary, indicating that the mutation itself acts as a recessive trait. In sporadic cases, two independent mutational events, involving both *RB* alleles, are necessary for tumor formation. Given that the first mutation can be inherited, in what ways can a second mutational event occur?

Solution: In considering how this second mutation arises, we must look at several levels of mutational events, including changes in nucleotide sequence and events that involve whole chromosomes or chromosome parts. Retinoblastoma results when both copies of the *RB* locus are lost or inactivated. With this in mind, perhaps the best way to proceed is to list the phenomena that can result in a mutational loss or inactivation of a gene.

One way the second *RB* mutation can occur is by a nucleotide alteration that converts the remaining normal *RB* allele to a mutant form. This alteration can occur through a nucleotide substitution or by a frameshift mutation caused by the insertion or deletion of nucleotides during replication. A second mechanism involves the loss of the chromosome carrying the normal allele. This event would take place during mitosis, resulting in chromosome 13 monosomy, leaving the mutant copy of the gene as the only *RB* allele. This mechanism does not necessarily involve loss of the entire chromosome; deletion of the long arm (*RB* is on 13q) or an interstitial deletion involving the *RB* locus and some surrounding material would have the same result.

Alternatively, a chromosome aberration involving loss of the normal copy of the *RB* gene might be followed by a duplication of the chromosome carrying the mutant allele. Two copies of chromosome 13 would be restored to the cell, but no normal *RB* allele would be present. Finally, a recombination event followed by chromosome segregation could produce a homozygous combination of mutant *RB* alleles.

More can be discovered about the mechanisms involved in RB by analyzing cells tumors using a combination of cytogenetic and molecular techniques (such as RFLP analysis and hybridizations to look for deletions) to see which mechanisms are actually found in tumors and to what extent they play a role in generating the second mutation. Although such analysis is still in the preliminary stages, all the mechanisms proposed have been found in tumors, indicating that a variety of spontaneous events brings about the second mutation that triggers retinoblastoma.

Problems and Discussion Questions

1. As a genetic counselor, you are asked to assess the risk for a couple thinking about having children, who have a family history of retinoblastoma. In this case, both the husband and wife are phenotypically normal, but the husband has a sister with familial retinoblastoma in both eyes. What is the probability that this couple will have a child with retinoblastoma? Are there any tests that you could recommend to help in this assessment?

2. Review the phases of the cell cycle. What events occur in each phase? Which phase is most variable in length?

3. Where are the major regulatory points in the cell cycle?

4. Progression through the cell cycle depends on the interaction of two types of regulatory proteins, kinases and cyclins. List the functions of each and describe how they interact to cause cells to move through the cell cycle.

5. What is the difference between saying that cancer is inherited and saying that predisposition to cancer is inherited?

6. Define tumor suppressor genes. Why are most tumor suppressor genes expected to have a recessive phenotype?

7. Distinguish between oncogenes and proto-oncogenes. In what ways can proto-oncogenes be converted to oncogenes?

8. How do translocations such as the Philadelphia chromosome lead to oncogenesis?

9. Given that 50 percent of all cancers are environmentally induced, and most of these are caused by lifestyle choices such as smoking, suntanning, and diet, how much of the money spent on cancer research do you think should be devoted to research and education on preventing cancer rather than on finding a cure?

10. The compound benzo[*a*]pyrene is found in cigarette smoke. This compound chemically modifies guanine bases in DNA. Such abnormal bases are typically removed by an enzyme, which hydrolyzes the base, leaving an apurinic site. If such a site is left unrepaired, an adenine is preferentially inserted *across from* the apurinic site. In a study of lung cancer patients, tumor cells from 15 out of 25 patients had a G to T transversion in the *p53* gene, which has a known role in cancer formation. You are testifying as an expert witness in the following court case. A widow of a man who died of lung cancer is suing R. J. Reynolds for selling tobacco products that killed her husband (who was a life-long smoker). What do you tell the jury? (Science only; no personal expositions on lawyers or the legal system.) [Reference: *Nature* 350:377–78 (1991).]

Selected Readings

Aaltonen, L. A. 1993. Clues to the pathogenesis of familial colorectal cancer. *Science* 260:812–16.

Ames, B., Magraw, R., and Gold, L. 1990. Ranking possible cancer hazards. *Science* 236:71–80.

Ames, B., Profet, M., and Gold, L. 1990. Dietary pesticides (99.99% all natural). *Proc. Natl. Acad. Sci. USA* 87:7777–81.

Anwar, S., et al. 2000. Hereditary non-polyposis colorectal cancer: an updated review. *Eur. J. Surg. Oncol.* 26:635–45.

Beijersbergen, R. L., and Bernards, R. 1996. Cell cycle regulation by the retinoblastoma family of growth inhibitory proteins. *Biochim. Biophys. Acta* 1287:103–20.

Benedict, W., et al. 1990. Role of the retinoblastoma gene in the initiation and progression of a human cancer. *J. Clin. Invest.* 85:988–93.

Bieche, I., and Lidereau, R. 1995. Genetic alterations in breast cancer. *Genes Chromosomes Cancer* 14:227–51.

Birrer, M., and Minna, J. 1989. Genetic changes in the pathogenesis of lung cancer. *Annu. Rev. Med.* 40:305–17.

Brown, M. A. 1997. Tumor suppressor genes and human cancer. *Adv. Genet.* 36:45–135.

Cavenee, W. K., and White, R. L. 1995. The genetic basis of cancer. *Sci. Am.* (Mar.) 272:72–79.

Chen, J., et al. 1998. Stable interaction between the products of the *BRCA1* and *BRCA2* tumor suppressor genes in mitotic and meiotic cells. *Molec. Cell* 2:317–28.

Compagni, A., and Christofori, G. 2000. Recent advances in research on multistage tumorigenesis. *Brit. J. Cancer* 83:1–5.

Cornelis, J. F., et al. 1998. Metastasis. *Am. Scient.* 86:130–41.

Croce, C., and Klein, G. 1985. Chromosome translocations and human cancer. *Sci. Am.* (Mar.) 252:54–60.

Damm, K. 1993. *ErbA*: Tumor suppressor turned oncogene? *FASEB J.* 7:904–09.

Easton, D., et al. 1993. Genetic linkage analysis in familial breast and ovarian cancer: Results from 214 families. *Am. J. Hum. Genet.* 52:678–701.

Elledge, S. J. 1996. Cell cycle checkpoints: Preventing an identity crisis. *Science* 274:1664–72.

Evan, G., and Littlewood, T. 1998. A matter of life and cell death. *Science* 281:1317–22.

Fearon, E. R. 1997. Human cancer syndromes: Clues to the origin and nature of cancer. *Science* 278:1043–50.

Feunteun, J., and Lenoir, G. M. 1996. *BRCA1*, a gene involved in inherited predisposition to breast cancer. *Biochim. Biophys. Acta* 1242:177–180.

Giaccone, G. 1996. Oncogenes and antioncogenes in lung tumorigenesis. *Chest 109 (Suppl. 5):* 130S–35S.

Hatkeyama, M., et al. 1994. The cancer cell and the cell cycle clock. *Cold Spring Harbor Symp. Quant. Biol.* 59:1–10.

Huan, B., and Siddique, A. 1993. Regulation of hepatitis B virus gene expression. *J. Hepatol. 17(Suppl. 3):* 520–S23.

Kinzler, K. W., and Vogelstein, B. 1996. Lessons from hereditary colorectal cancer. *Cell* 87:156–70.

—————. 1997. Gatekeepers and caretakers. *Nature* 386: 761–63.

Lengauer, C., Kinzler, K. W., and Vogelstein, B. 1997. Genetic instability in colorectal cancer. *Nature* 386: 623–27.

Nurse, P. 1997. Checkpoint pathways come of age. *Cell* 91:865–67.

Olsson, H., and Borg, A. 1996. Genetic predisposition to breast cancer. *Acta Oncol.* 35:1–8.

Peltomaki, P., et al. 1993. Genetic mapping of a locus predisposing to human colorectal cancer. *Science* 260:810–12.

Pines, L. 1996. Cyclin from sea urchin to HeLas: Making the human cell cycle. *Biochem. Soc. Trans.* 24:15–33.

Raff, M. 1998. Cell suicide for beginners. *Nature* 396:119–22.

Sherr, C. J. 1996. Cancer cell cycles. *Science* 274:1672–77.

Shibata, D., and Aaltonen, L. A. 2001. Genetic predisposition and somatic diversification in tumor development and progression. *Adv. Cancer Res.* 80:83–114.

Shito, K., et al. 2000. Pathogenesis of non-familial colorectal carcinomas with high microsatellite instability. *J. Clin. Pathol.* 53:841–45.

Smith, M. L., and Fornace, A. J., Jr. 1996. The two faces of tumor suppressor p53. *Am. J. Pathol.* 148:1019–22.

Solomon, E., et al. 1987. Chromosome 5 allele loss in human colorectal carcinomas. *Nature* 328:616–29.

Soussi, T. 2000. The p53 tumor suppressor gene: from molecular biology to clinical investigation. *Ann. N.Y. Acad. Sci.* 910:121–37.

Stahl, A., et al. 1994. The genetics of retinoblastoma. *Ann. Genet.* 37:172–178.

Timar, J., et al. 1995. Interaction of tumor cells with elastin and the metastatic phenotype. *Ciba Found. Symp.* 192:321–35.

von Lindern, M., et al. 1992. Translocation t(6;9) in acute non-lymphocytic leukaemia results in the formation of a *DEK-CAN* fusion gene. *Baillieres Clin. Haematol.* 5:857–79.

Weinberg, R. A. 1995. The molecular basis of oncogenes and tumor suppressor genes. *Ann. NY Acad. Sci.* 758:331–338.

————. 1995. The retinoblastoma protein and cell cycle control. *Cell* 81:323–30.

————. 1996. How cancer arises. *Sci. Am.* (Sept.) 275:62–70.

————. 1996. E2F and cell proliferation: A world turned upside down. *Cell* 85:457–59.

Zang, H., Tombline, G., and Weber, B. L. 1998. *BRCA1, BRCA2,* and DNA damage response: Collision or collusion? *Cell* 92:433–36.

These lady-bird beetles, from the Chiricahua Mountains in Arizona, show considerable pheno-typic variation. *(Edward S. Ross/California Academy of Sciences)*

22

Population Genetics

CHAPTER CONCEPTS

Individuals can carry only two different alleles of a given gene. A group of individuals can carry a larger number of different alleles, giving rise to a reservoir of genetic diversity. The diversity contained in the population can be measured by the Hardy–Weinberg law. Mutation is the ultimate source of genetic variation, and other factors such as drift, migration, and selection can alter the amount of genetic variation in populations.

Alfred Russel Wallace and Charles Darwin first identified natural selection as the mechanism of adaptive evolution in the mid-nineteenth century, based on a series of observations of populations of organisms: (1) Phenotypic variations exist among individuals within populations; (2) these differences are passed from parents to offspring; (3) more offspring are born than will survive and reproduce; and (4) some variants are more successful at surviving and/or reproducing than others. In populations where all four factors operate, the relative abundance of the population's different phenotypes changes across generations. In other words, the population evolves.

Although Wallace and Darwin described how organisms evolve by natural selection, there was no accurate model of the mechanisms responsible for variation and inheritance. Gregor Mendel published his work on the inheritance of traits in 1865, but it received little notice at the time. When Mendel's work was rediscovered in 1900, biologists immediately recognized its importance and began a 30-year effort to reconcile Mendel's concept of genes and alleles with the theory of evolution. In a key insight, biologists realized that changes in the relative abundance of different phenotypic traits in a population are tied to changes in the relative abundance of the alleles that influence the traits. The discipline within evolutionary biology that studies changes in allele frequencies is known as **population genetics**.

Early in the twentieth century a number of workers, including G. Udny Yule, William Castle, Godfrey Hardy, and Wilhelm Weinberg, formulated the basic principles of population genetics. For many years, theorists focused on developing mathematical models that would describe the genetic structure of populations. Prominent among the theoreticians who developed these models were Sewall Wright, Ronald Fisher, and J. B. S. Haldane. Following their work, experimentalists and field workers tested the models using biochemical and molecular techniques that measure variation directly at the protein and DNA levels. These experiments examined allele frequencies and the forces that alter the frequencies, such as selection, mutation, migration, and random genetic drift. In this chapter we consider some general aspects of population genetics and also discuss other areas of genetics that relate to evolution.

22.1 Populations and Gene Pools

Members of a species often range over a wide geographic area. A **population** is a group of individuals from the same species that lives in the same geographic area, and that actually or potentially interbreeds. If we consider a single genetic locus in this population, we may find that individuals within the population have different genotypes. To study population genetics, we compute frequencies at which various alleles and genotypes occur, and how these frequencies change from one generation to the next.

When we consider generational changes in alleles and genotypes, we look at gene pools. A **gene pool** consists of all gametes made by all the breeding members of a population

in a single generation. These gametes will combine to form the zygotes that become the next generation. The eggs and sperm that constitute the gene pool are haploid and thus contain only a single allele for each genetic locus. When we consider a single locus, we may find that different gametes carry different alleles. The proportion of gametes in a gene pool that carry a particular allele represents the frequency with which this allele occurs in the population.

Populations are dynamic; they grow and expand and diminish and contract through changes in birth and death rates, through migration, or through contact with other populations. The dynamic nature of populations has important consequences and can, over time, lead to changes in the population's gene pool.

22.2 Calculating Allele Frequencies

Most genetic population researchers first measure the frequencies at which alleles occur at a particular locus. To do this, the genotypes of a large number of individuals in the population must be determined. In some cases, researchers can infer genotypes directly from phenotypes. In other cases, proteins or DNA sequences are analyzed to determine genotypes. To understand how allele frequencies are calculated, we consider an example that involves HIV infection rates.

Our example comes from research into the genetic factors that influence an individual's susceptibility to infection by HIV-1, the virus responsible for the bulk of AIDS cases worldwide. Of particular interest to AIDS researchers is the small number of individuals who make high-risk choices (such as unprotected sex with HIV-positive partners), yet remain uninfected. In 1996, Rong Liu and colleagues discovered that two exposed-but-uninfected individuals were homozygous for a mutant allele of a gene called *CC-CKR-5*.

The *CC-CKR-5* gene, located on chromosome 3, encodes a protein called the C-C chemokine receptor-5, often abbreviated CCR5. Chemokines are cell-surface signaling molecules found on cells in the immune system. When cells with CCR5 receptor proteins bind to chemokine signals, the cells move into inflamed tissues to fight infection.

The CCR5 protein is a co-receptor for strains of HIV-1. To gain entry into cells, a protein called Env (short for envelope protein) on the surface of HIV-1 binds to the CD4 protein on the surface of the host cell. Binding to CD4 causes Env to change shape and form a second binding site. This second site binds to CCR5, which in turn initiates the fusion of the viral protein coat with the host cell membrane. The merging of viral envelope with cell membrane transports the HIV viral core into the host cell's cytoplasm.

The mutant allele of the *CCR5* gene contains a 32-bp deletion in one of its coding regions. As a result, the protein encoded by the mutant allele is shortened and made nonfunctional. The protein never makes it to the cell membrane, so HIV-1 cannot enter these cells. The gene's normal allele is called *CCR51* (or *1*) and its allele with the 32-bp deletion is called *CCR5-Δ32* (or *Δ32*). The two uninfected individuals described by Liu both had the genotype *Δ32/Δ32*.

TABLE 22.1   *CCR5* Genotypes and Phenotypes

Genotype	Phenotype
1/1	Susceptible to sexually transmitted strains of HIV-1
1/Δ32	Susceptible, but may progress to AIDS slowly
Δ32/Δ32	Resistant to most sexually transmitted strains of HIV-1

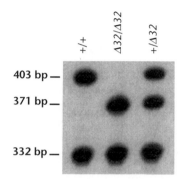

FIGURE 22–1 Allelic variation in the *CCR5* gene. Michel Samson and colleagues used PCR to amplify a part of the *CCR5* gene containing the site of the 32-bp deletion, cut the resulting DNA fragments with a restriction enzyme, and ran the fragments on an electrophoresis gel. Each lane reveals the genotype of a single individual. The *1* allele produces a 332-bp fragment and a 403-bp fragment; the *Δ32* allele produces a 332-bp fragment and a 371-bp fragment. Heterozygotes produce three bands. (From Samson, et al. 1996.) *(Michel Samson, Frederick Libert, et al., "Resistance to HIV-1 infection in Caucasian individuals bearing alleles of the CCR-5 chemokine receptor gene." Reprinted with permission from Nature 382, Aug. 22, 1996, p. 725, Fig. 3. © 1996 Macmillan Magazines Ltd.)*

As a result, they had no CCR5 on the surface of their cells, and were resistant to infection by strains of HIV-1 that require CCR5 as a co-receptor.

At least three *Δ32/Δ32* individuals who are infected with HIV-1 have been found. Curiously, researchers have not discovered any adverse effects associated with this genotype. Heterozygotes with genotype *1/Δ32* are susceptible to HIV-1 infection, but evidence suggests that they progress more slowly to AIDS. Table 22.1 summarizes the genotypes possible at the *CCR5* locus, and the phenotypes associated with each.

The discovery of the *CCR5-Δ32* allele, and the fact that it provides some protection against AIDS, generates two important questions: Which human populations harbor the *Δ32* allele, and how common is it? To address these questions, several teams of researchers surveyed a large number of people from a variety of populations. Genotypes were determined by direct analysis of DNA (Figure 22–1). We'll look at one sample population of 100 French individuals from a survey in Brittany in which 79 individuals have genotype *1/1*, 20 have genotype *1/Δ32*, and 1 has genotype *Δ32/Δ32*. This sample has 158 *1* alleles (carried by the *1/1* individuals) plus 20 *1* alleles carried by *1/Δ32* individuals, for a total of 178. The frequency of the *CCR51* allele in this population is thus 178/200 = 0.89 = 89 percent. The *CCR5-Δ32* allele count shows 20 carried by *1/Δ32* individuals, plus 2 carried by the *Δ32/Δ32* individual, for a total of 22. The frequency of the *CCR5-Δ32* allele is thus 22/200 = 0.11 = 11 percent. Notice that 0.89 + 0.11 = 1.00, which confirms that we have accounted for the entire gene pool. Table 22.2 shows two methods for computing the frequencies of the *1* and *Δ32* alleles in the Brittany sample.

Figure 22–2 shows the frequency of the *CCR5-Δ32* allele in the 18 European populations surveyed. The studies show that populations in Northern Europe have the highest frequencies of the *Δ32* allele. The frequency of the allele declines to the south and to the west. In populations without European an-

---

TABLE 22.2   Methods of Determining Allele Frequencies from Data on Genotypes

A. Counting Alleles

Genotype	*1/1*	*1/Δ32*	*Δ32/Δ32*	Total
Number of individuals	79	20	1	100
Number of *1* alleles	158	20	0	178
Number of *Δ32* alleles	0	20	2	22
Total number of alleles	158	40	2	200

Frequency of *CCR51* in sample: 178/200 = 0.89 = 89%
Frequency of *CCR5-Δ32* in sample: 22/200 = 0.11 = 11%

B. From Genotype Frequencies

Genotype	*1/1*	*1/Δ32*	*Δ32/Δ32*	Total
Number of individuals	79	20	1	100
Genotype frequency	79/100 = 0.79	20/100 = 0.20	1/100 = 0.01	1.00

Frequency of *CCR1* in sample: 0.79 + (0.5) 0.20 = 0.89
Frequency of *CCRT-Δ32* in sample: (0.5) 0.20 + 0.01 = 0.11

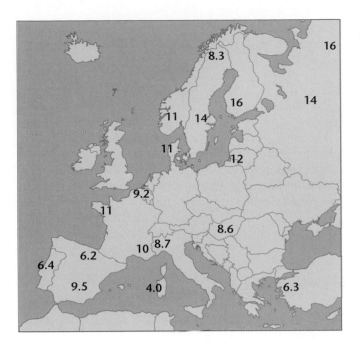

FIGURE 22–2 The frequency (percentage) of the *CCR5-Δ32* allele in 18 European populations. (From F. Leibert, et al. The *CCR5* mutation conferring protection against HIV-1 in Caucasian populations has a single and recent origin in Northeastern Europe. *Hum. Molec. Genet.* 7:399–406 by permission of Oxford University Press.)

cestry the *Δ32* allele is essentially absent. The highly patterned global distribution of the *Δ32* allele presents an evolutionary puzzle that we'll return to later in the chapter.

22.3 The Hardy–Weinberg Law

The large variation in the frequency of the *CCR5-Δ32* allele among populations raises a number of questions. For example, can we expect the allele to increase in populations in which it is currently rare? Population genetics explores such questions using a mathematical model developed independently by the British mathematician Godfrey H. Hardy and the German physician Wilhelm Weinberg. This model, called the **Hardy–Weinberg law**, shows what happens to allele and genotype in an "ideal" population (free of many of the complications that affect real populations) using a set of simple assumptions.

1. Individuals of all genotypes have equal rates of survival and equal reproductive success; that is, there is no selection.

2. No new alleles are created or converted from one allele into another by mutation.

3. Individuals do not migrate into or out of the population.

4. The population is infinitely large, which in practical terms means that the population is large enough that sampling errors and other random effects are negligible.

5. Individuals in the population mate randomly.

The Hardy–Weinberg law demonstrates that an "ideal" population has these properties:

1. The frequency of alleles does not change from generation to generation; in other words, the population does not evolve.

2. After one generation of random mating, offspring genotype frequencies can be predicted from the parent allele frequencies.

A model in which allele frequencies do not change from one generation to the next may not seem like a promising tool for studying evolution. What makes the Hardy–Weinberg law useful, however, is its assumptions. By specifying the assumptions under which the population cannot evolve, the Hardy–Weinberg law identifies the real-world forces that cause allele frequencies to change. In other words, by holding certain conditions constant, the Hardy–Weinberg law isolates the forces of evolution.

A Demonstration of the Law

To demonstrate how the Hardy–Weinberg law works, we begin with a specific case, and then consider the general case. In both examples, we focus on a single locus with two alleles, *A* and *a*.

Imagine a population in which the frequency of allele *A*, in both eggs and sperm, is 0.7, and the frequency of allele *a* is 0.3. Note that $0.7 + 0.3 = 1$, indicating that all the alleles for gene A present in the gene pool are accounted for. We assume, per Hardy–Weinberg requirements, that individuals mate randomly, which we visualize as follows. We place all the gametes in the gene pool, in a barrel and stir. We then randomly draw eggs and sperm from the barrel and pair them to make zygotes. What genotype frequencies does this give us? For any one zygote, the probability or chance that the egg will contain *A* is 0.7, and the probability that the sperm will contain *A* is 0.7. The probability that both egg *and* sperm will contain *A* is $0.7 \times 0.7 = 0.49$. In other words, we predict that genotype *AA* will occur 49 percent of the time. The probability that a zygote will be formed from an egg carrying *A* and a sperm carrying *a* is $0.7 \times 0.3 = 0.21$, and the probability that a zygote will be formed from an egg carrying *a* and a sperm carrying *A* is $0.3 \times 0.7 = 0.21$. Thus, the frequency of genotype *Aa* is $0.21 + 0.21 = 0.42 = 42$ percent. Finally, the probability that a zygote will be formed from an egg carrying *a* and a sperm carrying *a* is $0.3 \times 0.3 = 0.09$, meaning that the predicted frequency of genotype *aa* is 9 percent. As a check on our calculations, note that $0.49 + 0.42 + 0.09 = 1.0$, which confirms that we have accounted for all of the zygotes. These calculations are summarized in Figure 22–3 (page 454).

We started with the frequency of a particular allele in a specific gene pool and calculated the probability that certain genotypes would be produced from this pool. When the zygotes develop into adults and reproduce, what will be the frequency of distribution of alleles in the new gene pool? Recall that under the Hardy–Weinberg law, we assume that

	Sperm fr(A) = 0.7	fr(a) = 0.3
Eggs fr(A) = 0.7	fr(AA) = 0.7 × 0.7 = 0.49	fr(Aa) = 0.7 × 0.3 = 0.21
fr(a) = 0.3	fr(aA) = 0.3 × 0.7 = 0.21	fr(aa) = 0.3 × 0.3 = 0.09

FIGURE 22–3 Calculating genotype frequencies from allele frequencies. Gametes represent withdrawals from the gene pool to form the genotypes of the next generation. In this population, the frequency of the *A* allele is 0.7, and the frequency of the *a* allele is 0.3. The frequencies of the genotypes in the next generation are calculated as 0.49 for *AA*, 0.42 for *Aa*, and 0.09 for *aa*. Under the Hardy–Weinberg law, the frequencies of *A* and *a* remain constant from generation to generation.

all genotypes have equal rates of survival and reproduction. This means that in the next generation, all genotypes contribute equally to the new gene pool. The *AA* individuals constitute 49 percent of the population, and we can predict that the gametes they produce will constitute 49 percent of the gene pool. These gametes all carry allele *A*. Likewise, *Aa* individuals constitute 42 percent of the population, so we predict that their gametes will constitute 42 percent of the new gene pool. Half (0.5) of these gametes will carry allele *A*. Thus, the frequency of allele *A* in the gene pool is $0.49 + (0.5)\,0.42 = 0.7$. The other half of the gametes produced by *Aa* individuals will carry allele *a*. The *aa* individuals constitute 9 percent of the population, so their gametes will constitute 9 percent of the new gene pool. These gametes all carry allele *a*. Thus, we can predict that the allele *a* in the new gene pool is $(0.5)\,0.42 + 0.09 = 0.3$. As a check on our calculation, note that $0.7 + 0.3 = 1.0$, accounting for all of the gametes in the gene pool of the new generation.

We have arrived back where we began, with a gene pool in which the frequency of allele *A* is 0.7 and the frequency of allele *a* is 0.3. These calculations demonstrate the Hardy–Weinberg law: Allele frequencies in our population do not change from one generation to the next, and after just one generation of random mating the genotype frequencies can be predicted from the allele frequencies. In other words, this population does not evolve.

For the general case of the Hardy–Weinberg law, we use variables instead of numerical values for the allele frequencies. Imagine a gene pool in which the frequency of allele *A* is p and the frequency of allele *a* is q, such that $p + q = 1$. If we randomly draw an egg and sperm from the gene pool and pair them to make a zygote, the probability that both egg and sperm will carry allele *A* is $p \times p$. Thus, the frequency of genotype *AA* among the zygotes is p^2. The probabilities that the egg carries *A* and the sperm carries *a* is $p \times q$, and that the egg carries *a* and the sperm carries *A* is $q \times p$. Thus the frequency of genotype *Aa* among the zygotes is $2pq$. Finally, the probability that the egg and sperm will both carry *a* is $q \times q$, making the frequency of genotype *aa* among the zygotes q^2. Thus, the distribution of genotypes among the zygotes is:

$$p^2 + 2pq + q^2 = 1$$

Figure 22–4 shows these calculations.

For our general case we ask what the allele frequencies in the new gene pool will be when these zygotes develop into adults and reproduce. All gametes from *AA* individuals carry allele *A*, as do half of the gametes from *Aa* individuals. Thus we predict that the frequency of allele *A* in the new gene pool will be

$$p^2 + (1/2)\,2pq = p^2 + pq$$

Recall that $p + q = 1$. We can therefore substitute $(1 - p)$ for q in the above equation, which gives

$$p^2 + p(1 - p) = p^2 + p - p^2 = p$$

Likewise, half the gametes from *Aa* individuals carry allele *a*, as do all gametes from *aa* individuals. Thus, we predict that the frequency of allele *a* in the new gene pool will be

$$(1/2)2pq + q^2 = pq + q^2$$

We substitute $(1 - q)$ for p in this equation and get

$$(1 - q)q + q^2 = q - q^2 + q^2 = q$$

Once again, we arrive back where we began, with a gene pool in which the predicted proportion of allele *A* is p and that of allele *a* is q. The Hardy–Weinberg law holds true for any frequencies of *A* and *a*, so long as the frequencies add to 1 and the five assumptions are invoked. A population in which the allele frequencies remain constant from generation to generation and in which the genotype frequencies can be predicted from the allele frequencies is said to be in a state of **Hardy–Weinberg equilibrium**.

	Sperm fr(A) = p	fr(a) = q
Eggs fr(A) = p	fr(AA) = p^2	fr(Aa) = pq
fr(a) = q	fr(aA) = qp	fr(aa) = q^2

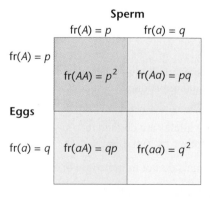

FIGURE 22–4 The general case of allele and genotype frequencies under Hardy–Weinberg assumptions. The frequency of allele *A* is p and the frequency of allele *a* is q. After mating, the three genotypes *AA*, *Aa*, and *aa* have the frequencies p^2, $2pq$, and q^2, respectively.

MEDIA TUTORIAL Populations: Allele Frequencies

Consequences of the Law

The Hardy–Weinberg law has several important consequences. First, it shows that dominant traits do not necessarily increase from one generation to the next. Second, it demonstrates that **genetic variability** can be maintained in a population since, once established in an ideal population, allele frequencies remain unchanged. Third, if we invoke Hardy–Weinberg assumptions, then knowing the frequency of just one genotype enables us to calculate the frequencies of all other genotypes. This relationship is particularly useful in human genetics because we can now calculate the frequency of heterozygous carriers for recessive genetic disorders even when all we know is the frequency of affected individuals (see later in the chapter).

We began this discussion by asking whether we can expect the *CCR5-Δ32* allele to increase in populations in which it is currently rare. From what we now know about the Hardy–Weinberg law, we can say for the general case that if (1) individuals of all genotypes have equal rates of survival and reproduction, (2) there is no mutation, (3) no one migrates into or out of the population, (4) the population is extremely large, (5) individuals in the population choose their mates randomly, then the frequency of the *Δ32* allele will not change. Of course, for real populations, few if any of these assumptions are likely to hold.

The general case demonstrates the most important role of the Hardy–Weinberg law: It is the foundation upon which population genetics is built. By showing that "ideal" populations do not evolve, we can use the Hardy–Weinberg law to identify forces that do cause populations to evolve. As we shall see later in the chapter, when the assumptions of the Hardy–Weinberg law are broken—because of natural selection, mutation, migration, and random sampling errors (also known as genetic drift)—the allele frequencies in a population may change from one generation to the next. Nonrandom mating does not, by itself, alter *allele* frequencies, but by altering *genotype* frequencies it indirectly affects the course of evolution. The Hardy–Weinberg law tells geneticists where to look to find the causes of evolution in real populations. We return to the *CCR5-Δ32* allele later in the chapter to see how a population genetics perspective has generated fruitful hypotheses for further research.

Testing for Equilibrium

One way we establish whether one or more of the Hardy–Weinberg assumptions do not hold in a given population is by determining whether the population's genotypes are in equilibrium. To do this, we first determine the frequencies of the genotypes, either directly from the phenotypes (if heterozygotes are recognizable) or by analyzing proteins or DNA sequences. We then calculate the allele frequencies from the genotype frequencies, as demonstrated earlier. Finally, we use the parents' allele frequencies to predict the offspring's genotype frequencies. According to the Hardy–Weinberg law, the genotype frequencies are predicted to fit the $p^2 + 2pq + q^2 = 1$ relationship. If they do not, then one or more of the assumptions are invalid for the population in question.

We will use the *CCR5* genotypes of a population in Britain to demonstrate the Hardy–Weinberg law. The population includes 283 individuals, of which 223 have genotype *1/1*, 57 have genotype *1/Δ32*, and 3 have genotype *Δ32/Δ32*. These numbers represent genotype frequencies of 223/283 = 0.788, 57/283 = 0.201, and 3/283 = 0.011, respectively. From the genotype frequencies we compute the *CCR51* allele frequency as 0.89 and the frequency of the *CCR5-Δ32* allele as 0.11. From these allele frequencies, we can use the Hardy–Weinberg law to determine whether this population is in equilibrium. The allele frequencies predict the genotype frequencies as follows:

Expected frequency of genotype *1/1* = $p^2 = (0.89)^2 = 0.792$
Expected frequency of genotype *1/Δ32* = $2pq = 2(0.89)(0.11) = 0.196$
Expected frequency of genotype *Δ32/Δ32* = $q^2 = (0.11)^2 = 0.012$

These expected frequencies are nearly identical to the observed frequencies. Our test of this population has failed to provide evidence that Hardy–Weinberg assumptions are being violated. This conclusion is confirmed by a χ^2 analysis (see Chapter 3). The χ^2 value in this case is tiny: 0.00023. To be statistically significant at even the most generous, accepted level, $p = 0.05$, the χ^2 value would have to be 3.84. (In a test for Hardy–Weinberg equilibrium, the degrees of freedom are given by $k - 1 - m$, where k is the number of genotypes and m is the number of independent allele frequencies estimated from the data. Here, $k = 3$ and $m = 1$, since calculating only one allele frequency allows us to determine the other by subtraction. Thus, we have $3 - 1 - 1 = 1$ degree of freedom.)

On the other hand, if the Hardy–Weinberg test had demonstrated that the population is not in equilibrium, it would indicate that one or more assumptions are not being met. To illustrate this, imagine two hypothetical populations, one living on East Island, the other living on West Island. The East Island population all have genotype *1/1*, so the *CCR51* allele constitutes 100 percent of this population. The West Island population all have genotype *Δ32/Δ32*, so the frequency of the *CCR5-Δ32* allele is 100 percent. Both island populations are in Hardy–Weinberg equilibrium.

Now imagine that 500 people from each island move to the previously uninhabited Central Island. The allele and genotype frequencies for the source populations and the new population on Central Island appear in Table 22.3 (page 456). Both the *1* allele and the *Δ32* allele occur at frequencies of 0.5 in the new population. Given these allele frequencies, the Hardy–Weinberg law predicts that the genotype frequencies will be 0.25 + 0.50 + 0.25 = 1. These expected frequencies obviously differ from the observed frequencies; there are no heterozygotes at all on Central Island. Therefore, population is not in Hardy–Weinberg equilibrium (and a χ^2 test would confirm this).

Two assumptions are violated in the new population. First, everyone living on Central Island is a migrant. Second, Central Island's population are not the products of random mating. Everyone on Central Island moved there, and has either two parents from East Island or two parents from West Island.

TABLE 22.3   *Frequencies of CCR51 and CCR5-D32 Alleles in a Hypothetical Population Composed of 500 1/1 Individuals and 500 Δ32/Δ32 Individuals*

	Allele Frequencies		Genotype Frequencies		
Population	*CCR51*	*CCR5-Δ32*	*1/1*	*1/Δ32*	*Δ32/Δ32*
Source populations					
East Island	1.0	0	1.0	0	0
West Island	0	1.0	0	0	1.0
New population on Central Island					
Observed	$p = 0.5$	$q = 0.5$	0.5	0	0.5
Expected			$p^2 = 0.25$	$2pq = 0.5$	$q^2 = 0.25$

However, it would take only one generation of random mating on Central Island to bring the offspring to the expected allele frequencies, as shown in Figure 22–5. Therese Markow and colleagues documented a real human population that is not in Hardy–Weinberg equilibrium. These researchers studied 122 Havasupai, a population of Native Americans in Arizona. They determined the genotype of each Havasupai individual at two loci in the major histocompatibility complex (MHC). These genes, *HLA-A* and *HLA-B*, encode proteins that are involved in the immune system's discrimination between self and nonself. The immune systems of individuals heterozygous at MHC loci appear to recognize a greater diversity of foreign invaders and thus may be better able to fight disease. Markow and colleagues observed significantly more individuals who are heterozygous at both loci, and significantly fewer homozygous individuals than would be expected under the Hardy–Weinberg law. Violation of either or both of two Hardy–Weinberg assumptions could explain the excess of heterozygotes among the Havasupai. First, Havasupai fetuses, children, and adults who are heterozygous for *HLA-A* and *HLA-B* may have higher rates of survival than individuals who are homozygous. Second, rather than choosing their mates randomly, Havasupai people may somehow prefer mates whose MHC genotypes differ from their own.

22.4   Extensions of the Hardy–Weinberg Law

We commonly find several alleles of a single locus in a population. The ABO blood group in humans (discussed in Chapter 4) is such an example. The locus *I* (isoagglutinin) has three alleles (I^A, I^B, and I^O), yielding six possible genotypic combinations ($I^A I^A$, $I^B I^B$, $I^O I^O$, $I^A I^B$, $I^A I^O$, $I^B I^O$). Recall that in this case *A* and *B* are codominant alleles, and both of these are dominant to *O*. The result is that homozygous *AA* and heterozygous *AO* individuals are phenotypically identical, as are *BB* and *BO* individuals, so we can distinguish only four phenotypic combinations.

By adding another variable to the Hardy–Weinberg equation, we can calculate both the genotype and allele frequencies for the situation involving three alleles. Let *p*, *q*, and *r* represent the frequencies of alleles *A*, *B*, and *O*, respectively. Note that because there are three alleles

$$p + q + r = 1$$

Under Hardy–Weinberg assumptions, the frequencies of the genotypes are given by

$$(p + q + r)^2 = p^2 + q^2 + r^2 + 2pq + 2pr + 2qr = 1$$

If we know the frequencies of blood types for a population, we can then estimate the frequencies for the three alleles of the ABO system. For example, in one population sampled, the following blood-type frequencies are observed: A = 0.53, B = 0.13, AB = 0.08, and O = 0.26. Because the *O* allele is recessive, the population's frequency of type O blood equals the proportion of the recessive genotype r^2. Thus,

$$r^2 = 0.26$$
$$r = \sqrt{0.26}$$
$$r = 0.51$$

Using *r*, we can estimate the allele frequencies for the *A* and *B* alleles. The *A* allele is present in two genotypes, *AA* and *AO*. The frequency of the *AA* genotype is represented by p^2, and the *AO* genotype by $2pr$. Therefore, the combined frequency of type A blood and type O blood is given by

$$p^2 + 2pr + r^2 = 0.53 + 0.26$$

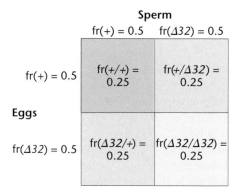

Sperm

	fr(+) = 0.5	fr(Δ32) = 0.5
fr(+) = 0.5	fr(+/+) = 0.25	fr(+/Δ32) = 0.25
fr(Δ32) = 0.5	fr(Δ32/+) = 0.25	fr(Δ32/Δ32) = 0.25

Eggs

FIGURE 22–5   Hardy–Weinberg equilibrium. If mating is random in a population that also meets the other assumptions of the Hardy–Weinberg law, equilibrium will be reached in one generation. Here the genotype frequencies (fr) after one generation of random mating are 0.25, 0.5, and 0.25; compare these with the genotype frequencies shown in Table 22.3.

TABLE 22.4   Calculating Genotype Frequencies for Multiple Alleles Where the Frequency of Allele A = 0.38, Allele B = 0.11, and Allele O = 0.51

Genotype	Genotype Frequency	Phenotype	Phenotype Frequency
AA	$p^2 = (0.38)^2 = 0.14$	A	0.53
AO	$2pr = 2(0.38)(0.51) = 0.39$		
BB	$q^2 = (0.11)^2 = 0.01$	B	0.12
BO	$2qr = 2(0.11)(0.51) = 0.11$		
AB	$2pq = 2(0.38)(0.11) = 0.084$	AB	0.08
OO	$r^2 = (0.51)^2 = 0.26$	O	0.26

If we factor the left side of the equation and take the sum of the terms on the right, we get

$$(p + r)^2 = 0.79$$
$$p + r = \sqrt{0.79}$$
$$p = 0.89 - r$$
$$p = 0.89 - 0.51 = 0.38$$

Having estimated p and r, the frequencies of allele A and allele O, we can now estimate the frequency for the B allele:

$$p + q + r = 1$$
$$q = 1 - p - r$$
$$= 1 - 0.38 - 0.51$$
$$= 0.11$$

The phenotypic frequencies and genotypic frequencies for this population are summarized in Table 22.4.

22.5 Using the Hardy–Weinberg Law: Calculating Heterozygote Frequency

In one application, the Hardy–Weinberg law allows us to estimate the frequency of heterozygotes in a population. The frequency of a recessive trait can usually be determined by counting such individuals in a sample of the population. With this information and the Hardy–Weinberg law, we can then calculate the allele and genotype frequencies.

Cystic fibrosis, an autosomal recessive trait, has an incidence of about $1/2500 = 0.0004$ in people of northern European ancestry. Individuals with cystic fibrosis are easily distinguished from the population at large by such symptoms as salty sweat, excess amounts of thick mucus in the lungs, and susceptibility to bacterial infections. Because this is a recessive trait, individuals with cystic fibrosis must be homozygous. Their frequency in a population is represented by q^2, provided that mating has been random in the previous generation.

The frequency of the recessive allele therefore is

$$q = \sqrt{q^2} = \sqrt{0.0004} = 0.02$$

Since $p + q = 1$, then the frequency of p is

$$p = 1 - q = 1 - 0.02 = 0.98$$

In the Hardy–Weinberg equation, the frequency of heterozygotes is $2pq$.

$$2pq = 2(0.98)(0.02)$$
$$= 0.04, \text{ or } 4\%, \text{ or } 1/25$$

Thus, heterozygotes for cystic fibrosis are rather common in the population (4%), even though the incidence of homozygous recessives is only 1/2500, or 0.04 percent.

In general, the frequencies of all three genotypes can be estimated once the frequency of either allele is known and Hardy–Weinberg assumptions are invoked. The relationship between genotype and allele frequency is shown in Figure 22–6. It is important to note that heterozygotes increase rapidly in a population as the values of p and q move from 0 or 1. This observation confirms our conclusion that when a recessive trait such as cystic fibrosis is rare, the majority of those carrying the allele are heterozygotes. In populations in which the frequencies of p and q are between 0.33 and 0.67, heterozygotes occur at higher frequency than either homozygote.

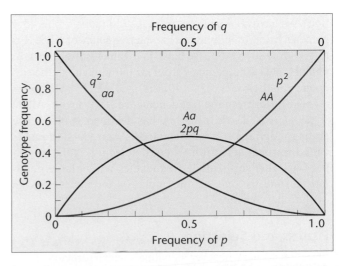

FIGURE 22–6 The relationship between genotype and allele frequencies derived from the Hardy–Weinberg equation.

22.6   Factors That Alter Allele Frequencies in Populations

We have noted that the Hardy–Weinberg law establishes an ideal population that allows us to estimate allele and genotype frequencies in populations in which the assumptions of random mating, absence of selection and mutation, and equal viability and fertility hold. Obviously, it is difficult to find natural populations in which all these assumptions hold. In nature, populations are dynamic, and changes in size and gene pool are common. The Hardy–Weinberg law allows us to investigate populations that vary from the ideal. In this and following sections, we discuss factors that prevent populations from reaching Hardy–Weinberg equilibrium, or that drive populations toward a different equilibrium, and the relative contribution of these factors to evolutionary change.

22.7   Natural Selection

The first assumption of the Hardy–Weinberg law is that individuals of all genotypes have equal rates of survival and equal reproductive success. If this assumption does not hold, allele frequencies may change from one generation to the next. To see why, let's imagine a population of 100 individuals in which the frequency of allele A is 0.5 and that of allele a is 0.5. Assuming the previous generation mated randomly, the genotype frequencies in the present generation are $(0.5)^2 = 0.25$ for AA, $2(0.5)(0.5) = 0.5$ for Aa, and $(0.5)^2 = 0.25$ for aa. Because our population contains 100 individuals, we have 25 AA individuals, 50 Aa individuals, and 25 aa individuals. Now suppose that individuals with different genotypes have different rates of survival: All 25 AA individuals survive to reproduce, 90 percent or 45 of the Aa individuals survive to reproduce, and 80 percent or 20 of the aa individuals survive to reproduce. When the survivors reproduce, each contributes two gametes to the new gene pool, giving us $2(25) + 2(45) + 2(20) = 180$ gametes. What are the frequencies of the two alleles in the surviving population? We have 50 A gametes from AA individuals, plus 45 A gametes from Aa individuals, so the frequency of allele A is $(50 + 45)/180 = 0.53$. We have 45 a gametes from Aa individuals, plus 40 a gametes from aa individuals, so the frequency of allele a is $(45 + 40)/180 = 0.47$.

These differ from the frequencies we started with. Allele A has increased, while allele a has declined. A difference among individuals in survival and/or reproduction rate is called **natural selection**. Natural selection is the principal force that shifts allele frequencies within large populations and is one of the most important factors in evolutionary change.

Fitness and Selection

Selection occurs whenever individuals with a particular genotype enjoy an advantage in survival or reproduction over other genotypes. However, selection may be weak or strong, depending on the magnitude of the advantage. In the example above, selection was strong. Weak selection might involve just a fraction of a percent difference in the survival rates of different genotypes. Advantages in survival and reproduction ultimately translate into increased genetic contribution to future generations. An individual's genetic contribution to future generations is called **fitness**. Thus, genotypes associated with high rates of survival and/or high reproductive success are said to have high fitness, whereas genotypes associated with low rates of survival and/or low reproductive success are said to have low fitness.

Hardy–Weinberg analysis also allows us to examine fitness. By convention, population geneticists use the letter w to represent fitness. Thus, w_{AA} represents the relative fitness of genotype AA, w_{Aa} the relative fitness of genotype Aa, and w_{aa} the relative fitness of genotype aa. Assigning the values $w_{AA} = 1$, $w_{Aa} = 0.9$, and $w_{aa} = 0.8$ could mean, for example, that all AA individuals survive, 90 percent of the Aa individuals survive, and 80 percent of the aa individuals survive, as in the example above.

Let's consider selection against deleterious alleles. Fitness values $w_{AA} = 1$, $w_{Aa} = 1$, and $w_{aa} = 0$ describe a situation in which allele a is a lethal recessive. As homozygous reces-

Generation	p	q	p^2	$2pq$	q^2
0	0.50	0.50	0.25	0.50	0.25
1	0.67	0.33	0.44	0.44	0.12
2	0.75	0.25	0.56	0.38	0.06
3	0.80	0.20	0.64	0.32	0.04
4	0.83	0.17	0.69	0.28	0.03
5	0.86	0.14	0.73	0.25	0.02
6	0.88	0.12	0.77	0.21	0.01
10	0.91	0.09	0.84	0.15	0.01
20	0.95	0.05	0.91	0.09	<0.01
40	0.98	0.02	0.95	0.05	<0.01
70	0.99	0.01	0.98	0.02	<0.01
100	0.99	0.01	0.98	0.02	<0.01

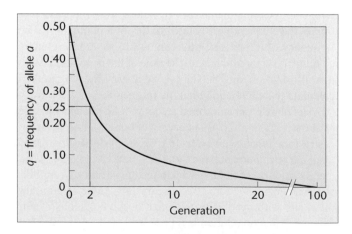

FIGURE 22–7 Change in the frequency of a lethal recessive allele, a. The frequency of a is halved in two generations, and halved again by the sixth generation. Subsequent reductions occur slowly because the majority of a alleles are carried by heterozygotes.

sive individuals die without leaving offspring, the frequency of allele *a* will decline. The decline in the frequency of allele *a* is described by the equation

$$q_g = \frac{q_0}{1 + gq_0}$$

where q_g is the frequency of allele *a* in generation *g*, q_0 is the starting frequency of *a* (i.e., the frequency of *a* in generation zero), and *g* is the number of generations that have passed.

Figure 22–7 shows what happens to a lethal recessive allele with an initial frequency of 0.5. At first, because of the high percentage of *aa* genotypes, the frequency of allele *a* declines rapidly. The frequency of *a* is halved in only two generations. By the sixth generation, the frequency is halved again. By now, however, the majority of *a* alleles are carried by heterozygotes. Because *a* is recessive, these heterozygotes are not selected against. This means that as time continues to pass, the frequency of allele *a* declines ever more slowly. As long as heterozygotes continue to mate, it is difficult for selection to completely eliminate a recessive allele from a population.

Of course, a deleterious allele may not be completely recessive; many other scenarios are possible. A deleterious allele may be codominant, so that heterozygotes have intermediate fitness. Or an allele may be deleterious in the homozygous state, but beneficial in the heterozygous state, like the allele for sickle-cell anemia.

Figure 22–8 shows examples in which a deleterious allele is codominant, but the intensity of selection varies from strong to weak. In each case, the frequency of the deleterious allele, *a*, starts at 0.99 and declines over time. However,

the rate of decline depends heavily on the strength of selection. When only 90 percent of the heterozygotes and 80 percent of the *aa* homozygotes survive (red curve), the frequency of allele *a* drops from 0.99 to less than 0.01 in 85 generations. However, when 99.8 percent of the heterozygotes and 99.6 percent of the *aa* homozygotes survive (blue curve), it takes 1000 generations for the frequency of allele *a* to drop from 0.99 to 0.93. Two important conclusions can be drawn from this graph. First, given thousands of generations, even weak selection can cause substantial changes in allele frequencies; because evolution occurs over vast spans of time, selection is a powerful force of evolution. Second, for selection to produce rapid changes in allele frequencies, changes measurable within a human life span, the differences in fitness among genotypes must be large.

The manner in which selection affects allele frequencies allows us to make some inferences about the *CCR5-Δ32* allele that we discussed earlier. Because individuals with genotype *Δ32/Δ32* are resistant to most sexually transmitted strains of HIV-1, while individuals with genotypes *1/1* and *1/Δ32* are susceptible, we might expect AIDS to act as a selective force causing the frequency of the *Δ32* allele to increase over time. Indeed it probably will, but the increase in frequency is likely to be slow in human terms.

We can do a rough calculation as follows. Imagine a population in which the current frequency of the *Δ32* allele is 0.10. Under Hardy–Weinberg assumptions, the genotype frequencies in this population are 0.81 for 1/1, 0.18 for *1/Δ32*, and 0.01 for *Δ32/Δ32*. Imagine also that 1 percent of the *1/1* and *1/Δ32* individuals in this population will contract HIV and die of AIDS.

Based on our assumptions, we can assign fitness levels to the genotypes as follows: $w_{1/1} = 0.99$; $w_{1/\Delta32} = 0.99$; $w_{\Delta32/\Delta32} = 1.0$. Given the assigned fitness, we can predict that the frequency of the *CCR5-Δ32* allele in the next generation will be 0.100091. In fact, it will take about 100 generations (about 2000 years) for the frequency of the *Δ32* allele to reach just 0.11 (Figure 22–9). In other words, the frequency of the *Δ32* allele will probably not change much over the next few generations in most populations that currently harbor it.

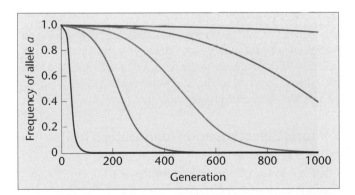

Selection against allele *a*				
Strong ⟵			⟶	Weak
—	—	—	—	—
w_{AA} 1.0	1.0	1.0	1.0	1.0
w_{Aa} 0.90	0.98	0.99	0.995	0.998
w_{aa} 0.80	0.96	0.98	0.99	0.996

FIGURE 22–8 The effect of selection on allele frequency. The rate at which a deleterious allele is removed from a population depends heavily on the strength of selection.

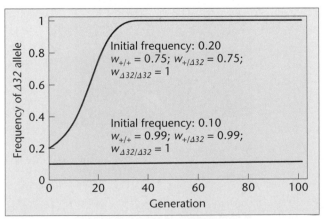

FIGURE 22–9 The rate at which the frequency of the *CCR5-Δ32* allele changes in hypothetical populations with different initial frequencies and different fitnesses.

A population genetic perspective sheds light on the *CCR5-Δ32* story in other ways as well. Two research groups have analyzed genetic variation at marker loci closely linked to the *CCR5* gene. Both groups concluded that most, if not all present-day copies of the *Δ32* allele are descended from a single ancestral copy that appeared in northeastern Europe at most a few thousand years ago. In fact, one group estimates that the common ancestor of all *Δ32* alleles existed just 700 years ago. How could a new allele rise from a frequency of virtually zero to as high as 20 percent in roughly 30 generations?

It seems that there must have been strong selection in favor of the *Δ32* allele. The agent of selection cannot have been HIV-1, because HIV-1 moved from chimpanzees to humans too recently. Because selection occurred about 700 years ago, J. C. Stephens suggests that the agent of selection was bubonic plague. During the Black Death of 1346–1352, between a quarter and a third of all Europeans died from plague. Bubonic plague is caused by the bacterium *Yersinia pestis*. This bacterium manufactures a protein that kills some kinds of white blood cells. Stephens hypothesizes that the process through which the bacterial protein kills white cells involves the *CCR5* gene product. If true, some mechanism makes individuals homozygous for the *Δ32* allele more likely to survive plague epidemics.

If past epidemics of bubonic plague are responsible for the high frequency of the *CCR5-Δ32* allele in European populations, then the virtual absence of the allele in non-European populations is at first somewhat puzzling. Perhaps plague has been more common in Europe than elsewhere. Alternatively, perhaps other populations have different alleles of the *CCR5* gene that also confer protection against the plague. Teams of researchers looking for other alleles of the *CCR5* gene in various populations have found a total of 20 mutant alleles, including *Δ32*. Sixteen of these alleles encode proteins different in structure from that encoded by the *CCR51* allele. Some, like *Δ32*, are loss-of-function alleles.

FIGURE 22–10 Pesticides used to control insects act as selective agents, changing allele frequencies of resistance genes in the populations exposed to the pesticide. *(Larsh K. Bristol/Visuals Unlimited, Inc.)*

Some of the alleles appear confined to Asian populations, others to African populations. Their frequencies are as high as 3–4 percent. Together, these discoveries are consistent with the hypothesis that alteration or loss of the CCR5 protein protects against an as-yet-unidentified infectious disease or diseases.

Selection in Natural Populations

Geneticists have done a great deal of research on the effect of natural selection on allele frequencies in both laboratory and natural populations. Among the most detailed studies of natural populations are those that involve insects exposed to pesticides. Christine Chevillon and colleagues, for example, studied the effect of the insecticide chlorpyrifos on allele frequencies in populations of house mosquitoes (Figure 22–10). Chlorpyrifos kills mosquitoes by interfering with the function of the enzyme acetylcholinesterase (ACE), which under normal circumstances breaks down the neurotransmitter acetylcholine. An allele of the gene encoding ACE called *Ace^R* encodes a slightly altered version of ACE that is immune to interference by chlorpyrifos.

Chevillon measured the frequency of the *Ace^R* allele in nine populations. In the first four locations, chlorpyrifos had been used to control mosquitoes for 22 years; in the last five locations, chlorpyrifos had never been used. Chevillon predicted that the frequency of *Ace^R* would be higher in the exposed populations. The researchers also predicted that the frequencies of alleles for enzymes unrelated to the physiological effects of chlorpyrifos would show no such pattern. Among the control enzymes studied was aspartate amino transferase 1.

The results appear in Figure 22–11. As the researchers predicted, the frequency of the *Ace^R* allele was significantly higher in the exposed populations. Also as predicted, the frequencies of the most common alleles of the control enzyme genes showed no such trends. The explanation is that during the 22 years of exposure to the pesticide, mosquitoes had higher rates of survival if they carried the *Ace^R* allele. In other words, the *Ace^R* allele had been favored by natural selection.

Natural Selection and Quantitative Traits

Most phenotypic traits are controlled not by alleles at a single locus but by the combined influence of the individual's genotype at many different loci and the environment. Because selection is a consequence of the organism's genotypic/phenotypic combination, polygenic or quantitative traits (those controlled by a number of genes that might be susceptible to environmental influences) also respond to selection. Such quantitative traits, including adult body height and weight in humans, often demonstrate a continuously varying distribution resembling a bell-shaped curve. Selection for such traits can be classified as (1) directional, (2) stabilizing, or (3) disruptive.

In **directional selection**, important to plant and animal breeders, desirable traits, often representing phenotypic extremes, are selected. If the trait is polygenic, the most extreme phenotypes that the genotype can express will appear

House mosquito, *Culex pipiens*

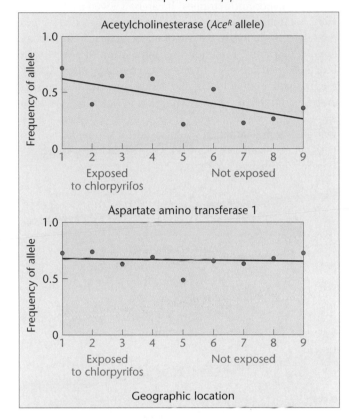

Acetylcholinesterase (*Ace^R* allele)

Aspartate amino transferase 1

Geographic location

FIGURE 22–11 The effect of selection on allele frequencies in natural populations. (a) The frequency of the *Ace^R* allele, which confers resistance to the insecticide chlorpyrifos, is higher in house mosquito populations exposed to chlorpyrifos. (b) The frequency of an allele for an enzyme unrelated to chlorpyrifos metabolism (aspartate amino transferase I) shows no such pattern. *(Photo: Hans Pfletschinger/Peter Arnold, Inc.)*

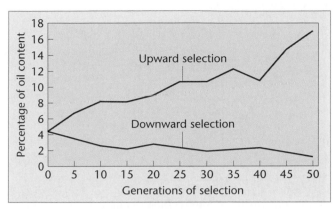

FIGURE 22–12 Long-term selection to alter the oil content of corn kernels.

in the population only after prolonged selection. An example of directional selection is the long-running experiment at the State Agricultural Laboratory in Illinois selecting for high and low oil content in corn kernels. At the start of the experiment in 1896, a population of 163 corn ears was surveyed. The 24 ears highest in oil content were used to produce the next generation. In each succeeding generation, the ears with the highest oil content were used to breed. In this line of upward directional selection, the oil content has been raised after 50 generations from about 4 percent to just over 16 percent (Figure 22–12), and there is no sign that a plateau has been reached. A second line was founded with 12 ears with the lowest oil content, and downward directional selection from this founding population has lowered the oil content from 4 percent to less than 1 percent in

the 50-generation period (Figure 22–12). In this case, the high and low graphs have diverged to the point where the distribution of oil content in each graph does not overlap with that of the other one.

In the upwardly selected line, the alleles for high oil content increase in frequency and replace the alleles for lower oil content. At some point, all individuals in the line will have the genotype for the highest oil content, and all will express the most extreme phenotype for high oil content. In the downwardly selected line, the opposite situation will occur, eventually producing a population with a genotype and phenotype for the lowest oil content. When the lines fail to respond to further selection, no genetic variance for oil content will remain, and each line will have a homozygous genotype for oil content.

In nature, directional selection can occur when one of the phenotypic extremes becomes selected for or against, usually as a result of changes in the environment. A carefully documented example comes from research by Peter and Rosemary Grant and their colleagues using the medium ground finches (*Geospiza fortis*) of Daphne Major island in the Galapagos Islands. The beak size of these birds varies enormously. In 1976, for example, some birds in the population had beaks less than 7 mm deep, while others had beaks more than 12 mm deep. Beak size is heritable, which means that large-beaked parents tend to have large-beaked offspring, and small-beaked parents tend to have small-beaked offspring. In 1977, a severe drought on Daphne killed some 80 percent of the finches. Big-beaked birds survived at higher rates than small-beaked birds because when food became scarce, the big-beaked birds were able to eat a greater variety of seeds. When the drought ended in 1978 and the survivors paired off and bred, the offspring inherited their parents' big beaks. Between 1976 and 1978, the beak depth of the average finch in the Daphne Major population increased by just over 0.5 mm, shifting the average beak size toward one phenotypic extreme.

Stabilizing selection, in contrast, tends to favor intermediate types, with both extreme phenotypes being selected against. One of the clearest demonstrations of stabilizing selection is provided by the data of Mary Karn and Sheldon Penrose on human birth weight and survival for 13,730

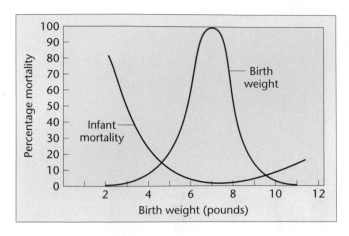

FIGURE 22–13 Relationship between birth weight and mortality in humans.

children born over an 11-year period. Figure 22–13 shows the distribution of birth weight and the percentage of mortality at 4 weeks of age. Infant mortality increases on either side of the optimal birth weight of 7.5 pounds, and quite dramatically so at the low end. At the genetic level, stabilizing selection acts to keep a population well adapted to its environment. In this situation, individuals closer to the average for a given trait will have higher fitness.

Disruptive selection is selection against intermediates and for both phenotypic extremes. It can be viewed as the opposite of stabilizing selection because the intermediate types are selected against. In one set of experiments, John Thoday applied disruptive selection to a population of *Drosophila* based on bristle number. In every generation he allowed only the flies with high or low bristle numbers to breed. After several generations, most of the flies could be easily placed in a low or high bristle category (Figure 22–14). In natural populations, such a situation might exist for a population in a heterogeneous environment. The types and effects of selection are summarized in Figure 22–15.

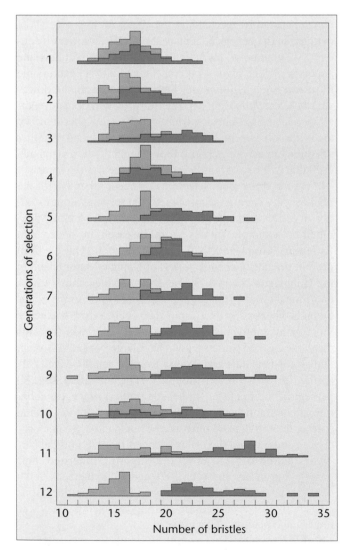

FIGURE 22–14 The effect of disruptive selection on bristle number in *Drosophila*. When individuals with the highest and lowest bristle number were selected, the population showed a nonoverlapping divergence in only 12 generations.

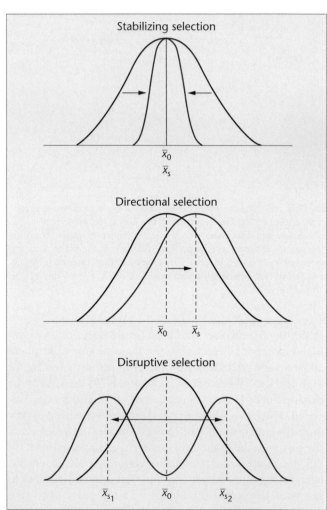

FIGURE 22–15 The impact of stabilizing, directional, and disruptive selection. In each case, $\bar{x}_0$ is the mean of an original population (blue), and $\bar{x}_s$ is the mean of the population following selection (red).

22.8 Mutation

Within a population, the gene pool is reshuffled each generation to produce new genotypes in the offspring. Because the number of possible genotypic combinations is so large, population members alive at any given time represent only a fraction of all possible genotypes. The enormous genetic reserve present in the gene pool allows Mendelian assortment and recombination to produce new genotypic combinations continuously. But assortment and recombination do not produce new alleles. **Mutation** alone acts to create new alleles. It is important to keep in mind that mutational events occur at random—that is, without regard for any possible benefit or disadvantage to the organism. In this section we consider whether mutation is, by itself, a significant factor in causing allele frequencies to change.

To determine whether mutation is a significant force in changing allele frequencies, we measure the rate at which mutations are produced. As most mutations are recessive, it is difficult to observe mutation rates directly in diploid organisms. Indirect methods using probability and statistics or large-scale screening programs are employed. For certain dominant mutations, however, a direct method of measurement can be used. To ensure accuracy, several conditions must be met:

1. The allele must produce a distinctive phenotype that can be distinguished from similar phenotypes produced by recessive alleles.

2. The trait must be fully expressed or completely penetrant so that mutant individuals can be identified.

3. An identical phenotype must never be produced by nongenetic agents such as drugs or chemicals.

Mutation rates can be stated as the number of new mutant alleles per given number of gametes. Suppose that for a given gene that undergoes mutation to a dominant allele, 2 out of 100,000 births exhibit a mutant phenotype. In these two cases, the parents are phenotypically normal. Because the zygotes that produced these births each carry two copies of the gene, we have actually surveyed 200,000 copies of the gene (or 200,000 gametes). If we assume that the affected births are each heterozygous, we have uncovered 2 mutant alleles out of 200,000. Thus, the mutation rate is 2/200,000 or 1/100,000, which in scientific notation is written as 1×10^{-5}.

In humans, a dominant form of dwarfism known as **achondroplasia** fulfills the requirements for measuring mutation rates. Individuals with this skeletal disorder have an enlarged skull, short arms and legs, and can be diagnosed by X-ray examination at birth. In a survey of almost 250,000 births, the mutation rate (μ) for achondroplasia has been calculated as

$$\mu = 1.4 \times 10^{-5} \pm 0.5 \times 10^{-5}$$

Knowing the rate of mutation, we can estimate the extent to which mutation can cause allele frequencies to change from one generation to the next. We represent the normal allele as d, and the allele for achondroplasia as D.

Imagine a population of 500,000 individuals in which everyone has genotype dd. The initial frequency of d is 1.0, and the initial frequency of D is 0. If each individual contributes 2 gametes to the gene pool, the gene pool will contain 1,000,000 gametes, all carrying allele d. While the gametes are in the gene pool, 1.4 of every 100,000 d alleles mutates into a D allele. The frequency of allele d is now $(1,000,000 - 14)/1,000,000 = 0.999986$, and the frequency of D is $14/1,000,000 = 0.000014$. From these numbers, it will clearly be a long time before mutation, by itself, causes any appreciable change in the allele frequencies in this population.

More generally, if we have two alleles, A with frequency p and a with frequency q, and if μ represents the rate of mutations converting A into a, then the frequencies of the alleles in the next generation are given by

$$p_{g+1} = p_g - \mu p_g \quad \text{and} \quad q_{g+1} = q_g + \mu p_g$$

where p_{g+1} and q_{g+1} represent the allele frequencies in the next generation, and p_g and q_g represent the allele frequencies in the present generation.

Figure 22–16 shows the replacement rate (change over time) in allele A for a population in which the initial frequency of A is 1.0, and the rate of mutations (μ) converting A into a is 1.0×10^{-5}. At this mutation rate, it will take about 70,000 generations to reduce the frequency of A to 0.5. Even if the rate of mutation increases through exposure to higher levels of radioactivity or chemical mutagens, the impact of mutation on allele frequencies will be extremely weak. The ultimate source of the genetic variability, mutation provides the raw material for evolution, but *by itself* plays a relatively insignificant role in changing allele frequencies. Instead, the fate of alleles created by mutation is more likely to be determined by natural selection (discussed previously) and genetic drift (discussed later).

The frequency for the mutant alleles causing cystic fibrosis an autosomal recessive disorder, which we discussed previously in relation to calculating heterozygote frequencies, is about 2 percent in European populations. Until recently, most individuals with two mutant alleles died before reproducing, meaning that selection against homozygous recessives was

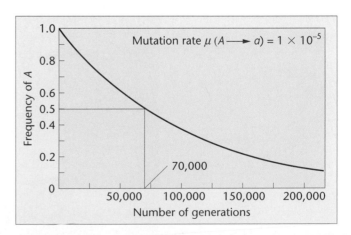

FIGURE 22–16 Replacement rate of an allele by mutation alone, assuming an average mutation rate of 1.0×10^{-5}.

rather strong. This creates a puzzle: In the face of selection against them, what has maintained the mutant alleles at an overall frequency of 2 percent? Although several hypotheses have been put forward, many evolutionary geneticists prefer the heterozygote superiority hypothesis. According to the **heterozygote superiority hypothesis**, selection against homozygous mutant individuals is counterbalanced by selection in favor of heterozygotes.

Recent work suggests that cystic fibrosis heterozygotes may have enhanced resistance to typhoid fever. Typhoid fever is caused by the bacterium *Salmonella typhi*, which infiltrates cells of the intestinal lining. In laboratory studies, mouse intestinal cells that were heterozygous for *CFTR-ΔF508*, the analog of the most common cystic fibrosis mutation in humans, acquired 86 percent fewer bacteria than did cells homozygous for the wild-type allele. Whether humans heterozygous for *CFTR-Δ508* also enjoy resistance to typhoid fever remains to be established. If they do, then cystic fibrosis will join sickle-cell anemia as an example of heterozygote superiority.

22.9   Migration

Occasionally, a species divides into populations that to some extent are separated geographically. Various evolutionary forces, including selection can establish different allele frequencies in such populations. **Migration** occurs when individuals move between the populations. Imagine a species in which a single locus has two alleles, *A* and *a*. There are two populations of this species, one on a mainland, and one on an island. The frequency of *A* on the mainland is represented by p_m and the frequency of *A* on the island is p_i. Under the influence of migration from the mainland to the island, the frequency of *A* in the next generation on the island ($p_{i'}$) is given by

$$p_{i'} = (1-m)p_i + mp_m$$

where *m* represents migrants from the mainland to the island.

Under these conditions, the frequency of *A* in the next generation on the island ($p_{i'}$) will be affected by migration. For example, assume that $p_i = 0.4$ and $p_m = 0.6$, and that 10 percent of the parents of the next generation are migrants from the mainland, so that $m = 0.1$. In the next generation, the frequency of allele *A* on the island will be

$$\begin{aligned} p_{i'} &= [(1 - 0.1) \times 0.4] + (0.1 \times 0.6) \\ &= 0.36 + 0.06 \\ &= 0.42 \end{aligned}$$

In this case, migration from the mainland has changed the frequency of *A* on the island from 0.40 to 0.42 in a single generation.

If either *m* is large or if p_m is very different from p_i, then a rather large change in the frequency of *A* can occur in a single generation. If migration is the only force acting to change the allele frequency on the island, then an equilibrium will be attained only when $p_i = p_m$. These calculations reveal that the change in allele frequency attributable to migration is proportional to the differences in allele frequency between the donor and recipient populations and to the rate of migration. As *m* can have a wide range of values, the effect of

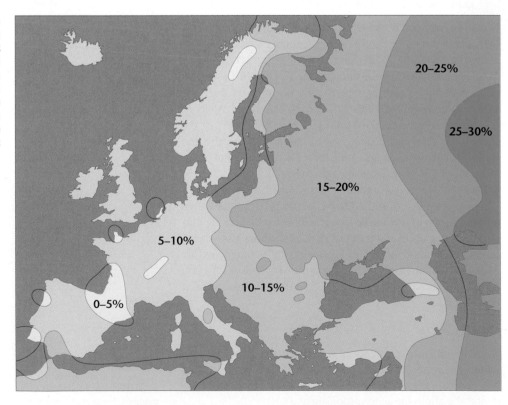

FIGURE 22–17 Migration as a force in evolution. The *B* allele of the *ABO* locus is present in a gradient from east to west. This allele shows highest frequency in Central Asia; and the lowest is in northeastern Spain. The gradient parallels the waves of Mongol migration into Europe following the fall of the Roman Empire and is a genetic relic of human history.

20–25%

25–30%

15–20%

5–10%

10–15%

0–5%

migration can substantially alter allele frequencies in populations, as shown for the *B* allele of the ABO blood group in Figure 22–17. Although migration can be difficult to quantify, it can often be estimated.

Migration can also be regarded as the flow of genes between populations that were once, but are no longer, geographically isolated. Esteban Parra and colleagues measured allele frequencies for several different DNA sequence polymorphisms in African-American and European-American populations, and in African and European populations representative of the ancestral populations from which the two American populations are descended. One locus they studied, a restriction site polymorphism called *FY-NULL*, has two alleles, *FY-NULL*1* and *FY-NULL*2*. Figure 22–18 shows the frequency of *FY-NULL*1* in each population. The frequency of this allele is 0 in the three African populations, and 1.0 in the three European populations, but lies between these extremes in all African-American and European-American populations. The simplest explanation for these data is that genes have mixed between American populations with predominantly African ancestry and American populations with predominantly European ancestry. Based on *FY-NULL* and several other loci, researchers estimate that African-American populations derive between 11.6 and 22.5 percent of their ancestry from Europeans, and that European-American populations derive between 0.5 and 1.2 percent of their ancestry from Africans.

22.10 Genetic Drift

In laboratory crosses, one condition essential to realizing theoretical genetic ratios (e.g., 3:1, 1:2:1, 9:3:3:1) is a fairly large sample size. A large sample size is also important to the study of population genetics as allelic and genotype frequencies are examined or predicted. For example, if a population consists of 1000 randomly mating heterozygotes (*Aa*), the next generation will consist of approximately 25 percent *AA*, 50 percent *Aa*, and 25 percent *aa* genotypes. Provided that the initial and subsequent populations are large, only minor deviations from this mathematical ratio will occur, and the frequencies of *A* and *a* will remain about equal (at 0.50).

However, if a population consists of only one set of heterozygous parents and they produce only two offspring, the allele frequency can change drastically. We can also predict genotypes and allele frequencies in the offspring in this case. As Table 22.5 shows, in 10 of 16 times such a cross is made, the allele frequencies would be altered. In 2 of 16 times, either the *A* or *a* allele would be eliminated in a single generation.

Founding a population with only one set of heterozygous parents that produce two offspring is an extreme example, but it illustrates the point that large interbreeding populations are essential to maintain Hardy–Weinberg equilibrium. In small populations, significant random fluctuations in allele frequencies are possible by chance deviation. The degree of fluctuation increases as the population size decreases, a situation known as **genetic drift**. In the extreme case, genetic drift can lead to the chance fixation of one allele to the exclusion of another allele.

To study genetic drift in laboratory populations of *Drosophila melanogaster*, Warwick Kerr and Sewall Wright set up over 100 lines, with four males and four females as the parents for each line. Within each line, the frequency of the sex-linked bristle mutant *forked* (*f*) and its wild-type allele (*f*+) was 0.5. In each generation, four males and four females were chosen at random to parent the next generation. After 16 generations, the complete loss of one allele and the fixation of the other had occurred in 70 lines—29 in which only the *forked* allele was present and 41 in which the wild-type

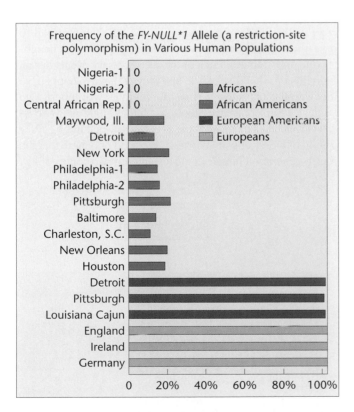

FIGURE 22–18 Frequency histogram of a restriction site polymorphism allele in several populations. These and other data demonstrate that African-American and European-American populations are of mixed ancestry.

TABLE 22.5 All Possible Pairs of Offspring Produced by Two Heterozygous Parents (*Aa* × *Aa*)

Possible Genotype of Two Offspring	Probability	Allele Frequency A	a
AA and *AA*	(1/4)(1/4) = 1/16	1.00	0.00
aa and *aa*	(1/4)(1/4) = 1/16	0.00	1.00
AA and *aa*	2(1/4)(1/4) = 2/16	0.50	0.50
Aa and *Aa*	(2/4)(2/4) = 4/16	0.50	0.50
AA and *Aa*	2(1/4)(2/4) = 4/16	0.75	0.25
aa and *Aa*	2(1/4)(2/4) = 4/16	0.25	0.75

allele had become fixed. The remaining lines were still segregating the two alleles or had gone extinct. If fixation had occurred randomly, then an equal number of lines should have become fixed for each allele. In fact, the experimental results do not differ statistically from the expected ratio of 35:35, demonstrating that alleles can spread through a population and eliminate other alleles by chance alone.

How are small populations created in nature? In one scenario, a large population may be split by some event (like war), creating a small isolated subpopulation. A disaster such as an epidemic might occur, leaving a small number of survivors to breed. Or, a small group might emigrate from the larger population and become founders in a new environment, such as a volcanically created island.

Allele frequencies in certain isolated human populations demonstrate the role of drift as an evolutionary force in natural populations. The Pingelap atoll in the western Pacific Ocean (lat. 6° N, long. 160° E) has in the past been devastated by typhoons and famine, and around 1780 there were only about nine surviving males. Today there are fewer than 2000 inhabitants, all of whose ancestry can be traced to the typhoon survivors. About 4–10 percent of the current population is blind from infancy. These people are affected by an autosomal recessive disorder, **achromatopsia**, which causes ocular disturbances, a form of color blindness, and cataract formation. The disorder is extremely rare in the human population as a whole. However, the mutant allele is present at a relatively high frequency in the Pingelap population. Genealogical reconstruction shows that one of the original survivors (about 30 individuals) was heterozygous for the condition. If we assume that he was the only carrier in the founding population, the initial gene frequency was 1/60, or 0.016. On average, about 7 percent of the current population is affected (a homozygous recessive genotype), so the frequency of the allele in the present population has increased to 0.26. The story of the inhabitants of Pingelap is told by Oliver Sacks, in his book, *The Island of the Colorblind.*

22.11 Nonrandom Mating

We have explored how violations of the first four assumptions of the Hardy–Weinberg law, in the form of selection, mutation, migration, and genetic drift, can cause allele frequencies to change. The fifth assumption is that the members of a population mate at random. Nonrandom mating itself does not directly alter the frequencies of alleles. It can, however, alter the frequencies of genotypes in a population and thereby indirectly affect the course of evolution.

The most important form of nonrandom mating, and the form we focus on here, is **inbreeding**, mating between relatives. For a given allele, inbreeding increases the proportion of homozygotes in the population. Over time, with complete inbreeding, only homozygotes will remain in the population. To demonstrate this concept, we consider the most extreme form of inbreeding, **self-fertilization**.

Inbreeding

Figure 22–19 shows the results of four generations of self-fertilization starting with a single individual heterozygous for one pair of alleles. By the fourth generation, only about 6 percent of the individuals are still heterozygous, and 94 percent of the population is homozygous. Note, however, that alleles *A* and *a* still remain at 50 percent.

In humans, inbreeding (called **consanguineous marriage**) is related to population size, mobility, and social customs governing marriages among relatives. To describe the amount of inbreeding in a population, Sewall Wright devised the coefficient of inbreeding. Expressed as *F*, the **coefficient of inbreeding** is defined as the probability that the two alleles of a given gene in an individual are identical *because they are descended from the same single copy of the allele in an ancestor.* If $F = 1$, all individuals are homozygous, and both alleles in every individual are derived from the same ancestral copy. If $F = 0$, no individual has two alleles derived from a common ancestral copy.

Figure 22–20 shows a pedigree of a first-cousin marriage. The fourth-generation female (shaded pink) is the daughter of first cousins (purple). Suppose her great-grandmother (green) was a carrier of a recessive lethal allele, *a*. What is the probability that the fourth-generation female will inherit two copies of her great-grandmother's lethal allele? For this to happen, (1) the great-grandmother had to pass a copy of the allele to her son, (2) her son had to pass it to his daughter, and (3) his daughter has to pass it to her daughter (the pink female). Also, (4) the great-grandmother had to pass a copy of the allele to her daughter, (5) her daughter had to pass it to her son, and (6) her son has to pass it to his daughter (the pink female). Each of the six necessary events has an individual probability of 1/2, and they *all* have to happen, so the overall probability that the pink female will inherit two copies of her great-grandmother's lethal allele is $(1/2)^6 = 1/64$. This takes us most of the way toward calculating *F* for a child of a first-cousin marriage. We need only note that the fourth-generation female could also inherit two copies of any

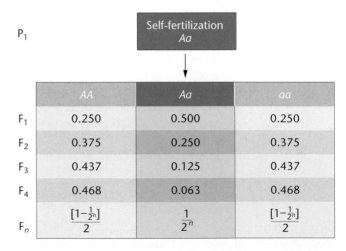

	AA	*Aa*	*aa*
P_1		Self-fertilization *Aa*	
F_1	0.250	0.500	0.250
F_2	0.375	0.250	0.375
F_3	0.437	0.125	0.437
F_4	0.468	0.063	0.468
F_n	$\dfrac{[1-\frac{1}{2^n}]}{2}$	$\dfrac{1}{2^n}$	$\dfrac{[1-\frac{1}{2^n}]}{2}$

FIGURE 22–19 Reduction in heterozygote frequency brought about by self-fertilization. After *n* generations, the frequencies of the genotypes can be calculated according to the formulas in the bottom row.

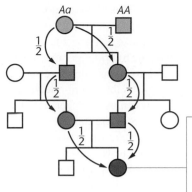

FIGURE 22–20 Calculating the coefficient of inbreeding (*F*) for the offspring of a first-cousin marriage.

The chance that this female will inherit 2 copies of her great-grandmother's *a* allele is:

$$F = \frac{1}{2} \times \frac{1}{2} \times \frac{1}{2} \times \frac{1}{2} \times \frac{1}{2} \times \frac{1}{2} = \frac{1}{64}$$

Because the female's 2 alleles could be identical by descent from any of four different alleles,

$$F = 4 \times \frac{1}{64} = \frac{1}{16}$$

of the other three alleles present in her great-grandparents. Because any of four possibilities would give the pink female two alleles identical by descent from an ancestral copy, $F = 4 \times (1/64) = 1/16$.

Genetic Effects of Inbreeding

Inbreeding results in the production of individuals homozygous for recessive alleles that were previously concealed in heterozygotes. Because many recessive alleles are deleterious when homozygous, one consequence of inbreeding is an increased chance that an individual will be homozygous for a recessive deleterious allele. Inbred populations often have a lowered mean fitness. **Inbreeding depression** is a measure of the loss of fitness caused by inbreeding. In domesticated plants and animals, inbreeding and selection have been used for thousands of years, and these organisms already have a high degree of homozygosity at many loci. Further inbreeding will usually produce only a small loss of fitness. However, inbreeding among individuals from large, randomly mating populations can produce high levels of inbreeding depression. This effect can be seen by examining the mortality rates in offspring of inbred animals in zoo populations (Table 22.6). To counteract inbreeding depression, many zoos use DNA fingerprinting to estimate the relatedness of animals used in breeding programs and choose the least related animals as parents.

As natural populations of endangered species decrease, the concern about inbreeding is a factor in designing programs to restore these species. One example is a project in Scandanavian zoos to preserve the Fennoscandic wolf. Conservationists established a captive-bred wolf population with four founding individuals. Later, two Russian wolves were added to the population to reduce the level of inbreeding, a strategy some conservationists later came to regret because it introduced foreign alleles into the Fennoscandic gene pool. Because the captive wolf line was founded by such a small number of individuals, it is severely inbred, and individuals show inbreeding depression in the form of smaller body weight, reduced reproductive success, and reduced longevity. Furthermore, a number of wolves in the line are blind.

Researchers concluded that blindness in the captive wolf line is caused by an autosomal recessive allele at a single locus. The bad news is that all the remaining pure-bred Fennoscandic wolves (i.e., individuals with no genes from the two Russian wolves) have at least a 6 percent chance of being

TABLE 22.6 Mortality in Offspring of Inbred Zoo Animals

Species	N		Non-inbred	Inbred	Inbreeding Coefficient
Zebra	32	Lived:	20	3	0.250
		Died:	7	2	
Eld's deer	24	Lived:	13	0	0.250
		Died:	4	7	
Giraffe	19	Lived:	11	2	0.250
		Died:	3	3	
Oryx	42	Lived:	35	0	0.250
		Died:	2	5	
Dorcas gazelle	92	Lived:	36	17	0.269
		Died:	14	25	

carriers for blindness. Some pure-bred individuals have a 67 percent chance of being carriers. The good news is that individuals with a greater than 30 percent chance of being carriers can be removed from the breeding population without reducing the remaining genetic variation in the population by more than 10 percent. Removing the likely carriers would reduce the frequency of the blindness allele from 14 percent to 7 percent, and would improve the long-term prospects of maintaining a viable Fennoscandic wolf population.

In humans, inbreeding increases the risk of spontaneous abortions, neonatal deaths, congenital deformities and recessive genetic disorders. Although less common than in the past, inbreeding occurs in many regions of the world where social customs favor marriage between first cousins. Alan Bittles and James Neel analyzed data from numerous studies on different cultures. They found that the rate of child mortality (i.e., death in the first several years of life) varies dramatically from culture to culture. No matter what the baseline mortality rate for children of unrelated parents, however, children of first cousins virtually always have a higher death rate—typically by about 4.5 percentage points. Over many generations, inbreeding should eventually reduce the frequency of deleterious recessive alleles (if homozygotes die before reproducing). However, some studies indicate that parents who are first cousins tend to have more children to compensate for those lost to genetic disorders. On average, two-thirds of the surviving offspring are heterozygous carriers of the deleterious allele.

It is important to note that inbreeding is not always harmful. Indeed, inbreeding has long been recognized as a useful tool for breeders of domesticated plants and animals. When an inbreeding program is initiated, homozygosity increases, and some breeding stocks become fixed for favorable alleles and others for unfavorable alleles. By selecting the more viable and vigorous plants or animals, the proportion of individuals carrying desirable traits can be increased.

If members of two favorable inbred lines are mated, hybrid offspring are often more vigorous in desirable traits than either of the parental lines. This phenomenon is called **hybrid vigor**. When such an approach was used in breeding programs established for maize, crop yields increased tremendously. Unfortunately, the hybrid vigor extends only through the first generation. Many hybrid lines are sterile, and those that are fertile show subsequent declines in yield. Consequently, the hybrids must be regenerated each time by crossing the original inbred parental lines.

Hybrid vigor has been explained in two ways. The first theory, the **dominance hypothesis**, incorporates the obvious reversal of inbreeding depression, which inevitably must occur in out-crossing. Consider a cross between two strains of maize with the following genotypes:

$$\text{Strain A} \qquad \text{Strain B} \qquad F_1$$
$$aaBBCCddee \times AAbbccDDEE \longrightarrow AaBbCcDdEe$$

The F_1 hybrids are heterozygotes at all loci shown. The deleterious recessive alleles present in the homozygous form in the parents is masked by the more favorable dominant alleles in the hybrids. Such masking is thought to cause hybrid vigor.

The second theory, **overdominance**, holds that in many cases the heterozygote is superior to either homozygote. This may relate to the fact that in the heterozygote two forms of a gene product may be present, providing a form of biochemical diversity. Thus, the cumulative effect of heterozygosity at many loci accounts for the hybrid vigor. Most likely, hybrid vigor results from a combination of phenomena explained by both hypotheses.

We have seen that nonrandom mating can drive the genotype frequencies in a population away from their expected values under the Hardy–Weinberg law. This can indirectly affect the course of evolution. As in stocks purposely inbred by animal and plant breeders, inbreeding in a natural population may increase the frequency of homozygotes for a deleterious recessive allele. As in domestic stocks, this increases the efficiency with which selection removes the deleterious allele from the population.

Chapter Summary

1. The Hardy–Weinberg law specifies what will happen to allele frequencies in a population under a set of simple assumptions. If there is no selection, no mutation, and no migration, if the population is large, and if individuals mate randomly, then allele frequencies will not change from one generation to the next. The population will not evolve.

2. The Hardy–Weinberg law can be used to investigate whether or not a population is in evolutionary equilibrium, and to estimate the frequency of heterozygotes in a population from the frequency of homozygous recessives.

3. By specifying the conditions under which allele frequencies will not change, the Hardy–Weinberg law identifies the forces that can cause evolution. Selection, mutation, migration, and genetic drift can cause allele frequencies to change and are thus forces of evolution. Nonrandom mating does not alter the frequencies of alleles, but by altering the frequencies of genotypes, nonrandom mating can indirectly affect the course of evolution.

4. Natural selection is the most powerful of the forces that can alter a population's allele frequencies. The rate of evolution under natural selection depends on the initial allele and on the relative fitness of different genotypes. In small populations, random genetic drift can also be an important force altering allele frequencies.

5. Mutation and migration introduce new alleles into a population, but by themselves usually have little effect on allele frequencies. Instead, the retention or loss of introduced alleles depends on the fitness they confer and the action of selection.

6. Genetic drift is change in allele frequencies as a result of chance events; it is an important force of evolution in small populations.

7. The most important form of nonrandom mating is inbreeding, or mating between relatives. Inbreeding increases the frequency of homozygotes in a population, and decreases the frequency of heterozygotes.

Key Terms

Insights and Solutions

1. Tay–Sachs disease is caused by mutations in a gene on chromosome 15 that encodes a lysosomal enzyme. Tay–Sachs is inherited as an autosomal recessive condition. Among Jews of central European ancestry, about 1 in 3600 children is born with the disease. What proportion of the individuals in this population are carriers?

Solution: If we let p represent the frequency of the wild-type enzyme allele, and q the frequency of recessive loss-of-function alleles, and if we assume that the population is in Hardy–Weinberg equilibrium, then the frequencies of the genotypes are given by p^2 for homozygous wild-type, $2pq$ for carriers, and q^2 for individuals with Tay–Sachs. The frequency of Tay-Sachs alleles is thus

$$q = \sqrt{q^2} = \sqrt{\frac{1}{3600}} = 0.017$$

Since $p + q = 1$, we have

$$p = 1 - q = 1 - 0.017 = 0.983$$

Therefore, we can estimate that the frequency of carriers is

$$2pq = 2(0.983)(0.017) = 0.033 \quad \text{or} \quad \text{1 in 30}$$

2. Eugenics is the term employed for the selective breeding of humans to bring about improvements in populations. As a eugenic measure, it has been suggested that individuals suffering from serious genetic disorders should be prevented (sometimes by force of law) from reproducing (by sterilization, if necessary) in order to reduce the frequency of the disorder in future generations.

Suppose that such a recessive trait is present in the population at a frequency of 1 in 40,000, and that affected individuals do not reproduce. In 10 generations, or about 250 years, what would be the frequency of the condition? Are the eugenic measures effective in this case?

Solution: Let q represent the frequency of the recessive allele responsible for the disorder. Because the disorder is recessive, we can estimate that

$$q = \sqrt{\frac{1}{40,000}} = 0.005$$

If all affected individuals are prevented from reproducing, then in an evolutionary sense the disorder is lethal: Affected individuals have zero fitness. That means we can predict the frequency of the recessive allele 10 generations in the future using the equation:

$$q_g = \frac{q_0}{1 + gq_0}$$

In this case, $q_0 = 0.005$, and $g = 10$, so we have

$$q_{10} = \frac{0.005}{1 + (10 \times 0.005)}$$

$$= 0.0048$$

If $q_{10} = 0.0048$ then the frequency of homozygous recessive individuals will be roughly

$$(q_{10})^2 = (0.0048)^2 = 0.000023 = \frac{1}{43,500}$$

The frequency of the genetic disorder has been reduced from 1 in 40,000 to 1 in 43,500 in 10 generations, indicating that this eugenic measure has limited effectiveness.

Problems and Discussion Questions

1. The ability to taste the compound PTC is controlled by a dominant allele T, while individuals homozygous for the recessive allele t are unable to taste PTC. In a genetics class of 125 students, 88 can taste PTC, and 37 cannot. Calculate the frequency of the T and t alleles in this population, and the frequency of the genotypes.

2. Calculate the frequencies of the AA, Aa, and aa genotypes after one generation if the initial population consists of 0.2 AA, 0.6 Aa, and 0.2 aa genotypes. What genotype frequencies will occur after a second generation?

3. Consider rare disorders in a population caused by an autosomal recessive mutation. From the frequencies of the disorder in the population given, calculate the percentage of heterozygous carriers:
   (a) 0.0064   (b) 0.000081   (c) 0.09
   (d) 0.01   (e) 0.10

4. What must be assumed in order to validate the answers in Problem 3?

5. In a population where only the total number of individuals with the dominant phenotype is known, how can you calculate the percentage of carriers and homozygous recessives?

6. Determine whether the following two sets of data represent populations that are in Hardy–Weinberg equilibrium. Use χ^2 analysis if necessary.
   (a) CCR5 genotypes: $1/1$, 60 percent; $1/\Delta32$, 35.1 percent; $\Delta32/\Delta32$, 4.9 percent
   (b) Sickle-cell hemoglobin: AA, 75.6 percent; AS, 24.2 percent; SS, 0.2 percent

7. If 4 percent of a population in equilibrium expresses a recessive trait, what is the probability that the offspring of two individuals who do not express the trait will express it?

8. Consider a population in which the frequency of allele A is $p = 0.7$ and the frequency of allele a is $q = 0.3$, and where the alleles are codominant. What will be the allele frequencies after one generation if the following occurs?
   (a) $w_{AA} = 1$, $w_{Aa} = 0.9$, and $w_{aa} = 0.8$
   (b) $w_{AA} = 1$, $w_{Aa} = 0.95$, and $w_{aa} = 0.9$
   (c) $w_{AA} = 1$, $w_{Aa} = 0.99$, and $w_{aa} = 0.98$
   (d) $w_{AA} = 0.8$, $w_{Aa} = 1$, and $w_{aa} = 0.8$

9. If the initial allele frequencies are $p = 0.5$ and $q = 0.5$, and allele a is a lethal recessive, what will be the frequencies after 1, 5, 10, 25, 100, and 1000 generations?

10. Determine the frequency of allele A in an island population after one generation of migration from the mainland under the following conditions:
    (a) $p_i = 0.6$; $p_m = 0.1$; $m = 0.2$
    (b) $p_i = 0.2$; $p_m = 0.7$; $m = 0.3$
    (c) $p_i = 0.1$; $p_m = 0.2$; $m = 0.1$

11. Assume that a recessive autosomal disorder occurs in 1 of 10,000 individuals (0.0001) in the general population. Assume that in this population about 2 percent (0.02) of the individuals are carriers for the disorder. Estimate the probability of this disorder occurring in the offspring of a marriage between first cousins. Compare this probability to the population at large.

12. What is the basis of inbreeding depression?

13. Describe how inbreeding can be used in the domestication of plants and animals. Discuss the theories underlying these techniques.

14. Evaluate the following statement: Inbreeding increases the frequency of recessive alleles in a population.

15. In a breeding program to improve crop plants, which of the following mating systems should be employed to produce a homozygous line in the shortest possible time?
    (a) self-fertilization
    (b) brother–sister matings
    (c) first-cousin matings
    (d) random matings

    Illustrate your choice with pedigree diagrams.

16. If the recessive trait albinism (a) is present in 1/10,000 individuals in a population at equilibrium, calculate the frequency of
    (a) the recessive mutant allele
    (b) the normal dominant allele
    (c) heterozygotes in the population
    (d) matings between heterozygotes

17. One of the first Mendelian traits identified in humans was a dominant condition known as *brachydactyly*. This gene causes an abnormal shortening of the fingers and/or toes. At the time, it was thought by some that this dominant trait would spread until 75 percent of the population would be affected (since the phenotypic ratio of dominant to recessive is 3:1). Show that this reasoning is incorrect.

18. Achondroplasia is a dominant trait that causes a characteristic form of dwarfism. In a survey of 50,000 births, five infants with achondroplasia were identified. Three of the affected infants had affected parents, while two had normal parents. Calculate the mutation rate for achondroplasia and express the rate as the number of mutant genes per given number of gametes.

19. A prospective groom, who is normal, has a sister with cystic fibrosis, an autosomal recessive disease state. Their parents are normal. The brother plans to marry a woman who has no history of cystic fibrosis (CF) in her family. What is the probability that they will produce a CF child? They are both Caucasian and the overall frequency of CF in the Caucasian population is 1/2500; that is, 1 affected child per 2500. (Assume the population meets the Hardy–Weinberg assumptions.)

Selected Readings

Ansari-Lari, M. A., et al. 1997. The extent of genetic variation in the CCR5 gene. *Nature Genetics* 16:221–22.

Ballou, J., and Ralls, K. 1982. Inbreeding and juvenile mortality in small populations of ungulates: A detailed analysis. *Biol. Conserv.* 24:239–72.

Barton, N. H., and Turelli, M. 1989. Evolutionary quantitative genetics: How little do we know? *Annu. Rev. Genet.* 23:337–70.

Bittles, A. H., and Neel, J. V. 1994. The costs of human inbreeding and their implications for variations at the DNA level. *Nature Genet.* 8:117–21.

Carrington, M., and Kissner, T., et al. 1997. Novel alleles of the chemokine-receptor gene CCR5. *Am. J. Hum. Genet.* 61:1261–67.

Chevillon, C., et al. 1995. Population structure and dynamics of selected genes in the mosquito *Culex pipiens. Evolution* 49:997–1007.

Crow, J. F. 1986. *Basic concepts in population, quantitative and evolutionary genetics.* New York: W. H. Freeman.

Dudley, J. W. 1977. 76 generations of selection for oil and protein percentage in maize. Ed. E. Pollack, et al. In *Proceedings of the International Conference on Quantitative Genetics*, pp. 459–73. Ames, IA: Iowa State University Press.

Fisher, R. A. 1930. *The genetical theory of natural selection.* Oxford: Clarendon Press. (Reprinted Dover Press, 1958.)

Freeman, S., and Herron, J. C. 2001. *Evolutionary analysis.* Upper Saddle River, NJ: Prentice Hall.

Freire-Maia, N. 1990. Five landmarks in inbreeding studies. *Am. J. Med. Genet.* 35:118–20.

Gayle, J. S. 1990. *Theoretical population genetics.* Boston: Unwin-Hyman.

Grant, P. R. 1986. *Ecology and evolution of Darwin's finches.* Princeton, NJ: Princeton University Press.

Hartl, D. L. and Clark, A. G. 1997. *Principles of population genetics*, 3rd ed. Sunderland, MA: Sinauer Associates.

Jones, J. S. 1981. How different are human races? *Nature* 293:188–90.

Karn, M. N., and Penrose, L. S. 1951. Birth weight and gestation time in relation to maternal age, parity and infant survival. *Ann. Eugen.* 16:147–64.

Kerr, W. E., and Wright, S. 1954. Experimental studies of the distribution of gene frequencies in very small populations of *Drosophila melanogaster.* I. *Forked. Evolution* 8:172–77.

Khoury, M. J., and Flanders, W. D. 1989. On the measurement of susceptibility to genetic factors. *Genet. Epidemiol.* 6:699–701.

Laikre, L., Ryman, N., and Thompson, E. A. 1993. Hereditary blindness in a captive wolf (*Canis lupus*) population: Frequency reduction of a deleterious allele in relation to gene conservation. *Conservation Biol.* 7:592–601.

Leibert, F., et al. 1998. The DCCR5 mutation conferring protection against HIV-1 in Caucasian populations has a single and recent origin in northeastern Europe. *Hum. Molec. Genet.* 7:399–406.

Liu, R., et al. 1996. Homozygous defect in HIV-1 coreceptor accounts for resistance in some multiply-exposed individuals to HIV-1 infection. *Cell* 86:367–77.

Lucotte, G., and Mercier, G. 1998. Distribution of the CCR5 gene 32-bp deletion in Europe. *J. Acquired Immune Deficiency Syndrome and Human Retrovirology* 19:174–77.

Markow, T., et al. 1993. HLA polymorphism in the Havasupai: Evidence for balancing selection. *Am. J. Hum. Genet.* 53:943–52.

Martinson, J. J., et al. 1997. Global distribution of the *CCR5* gene 32-bp deletion. *Nature Genet.* 16:100–103.

Mettler, L. E., Gregg, T., and Schaffer, H. E. 1988. *Population genetics and evolution*, 2nd ed. Englewood Cliffs, NJ: Prentice-Hall.

Parra, E. J., et al. 1998. Estimating African American admixture proportions by use of population-specific alleles. *Am. J. Hum. Genet.* 63:1839–51.

Pier, G. B., et al. 1998. *Salmonella typhi* uses CFTR to enter intestinal epithelial cells. *Nature* 393:79–82.

Quillent, C., et al. 1998. HIV-1-resistance phenotype conferred by combination of two separate inherited mutations of *CCR5* gene. *The Lancet* 351:14–18.

Renfrew, C. 1994. World linguistic diversity. *Sci. Am.* (Jan.) 270:116–23.

Roberts, D. F. 1988. Migration and genetic change. *Hum. Biol.* 60:521–39.

Samson, M., et al. 1996. Resistance to HIV-1 infection in Caucasian individuals bearing mutant alleles of the *CCR-5* chemokine receptor gene. *Nature* 382:722–25.

Spiess, E. B. 1989. *Genes in populations*, 2nd ed. New York: John Wiley.

Stephens, J. C., et al. 1998. Dating the origin of the *CCR5-Δ32* AIDS-resistance allele by the coalescence of haplotypes. *Am. J. Hum. Genet.* 62:1507–15.

Wallace, B. 1989. One selectionist's perspective. *Quart. Rev. Biol.* 64:127–45.

Woodworth, C. M., Leng, E. R., and Jugenheimer, R. W. 1952. Fifty generations of selection for protein and oil in corn. *Agron. J.* 44:60–66.

Yudin, N. S., et al. 1998. Distribution of *CCR5-Δ32* gene deletion across the Russian part of Eurasia. *Hum. Genet.* 102:695–98.

Light and dark forms of the peppered moth *Biston betularia* on light tree bark.
(*Breck P. Kent/Animals Animals*)

23

Genetics and Evolution

CHAPTER CONCEPTS

Natural selection acting on the gene pool of a population brings about speciation. Selection leads to changes in genetic diversity and the ability to undergo evolutionary divergence. The process of speciation divides one gene pool into two or more reproductively separate pools. This division may be accompanied by changes in morphology, physiology, and adaptation to the environment. The genetic differences between populations or species can be used to reconstruct evolutionary history and relationships.

Evolutionary history involves two processes: the transformation and the splitting of lineages to produce a diversity of populations and species. In Chapter 22, we described how populations evolve by changes in allele frequencies. We also outlined the forces that cause allele frequencies to change. Mutation, migration, selection, and drift, individually and collectively, bring about evolutionary divergence and species formation. This process depends not only on genetic divergence but also on environmental or ecological diversity. If a population is spread over a geographic range encompassing a number of subenvironments or **niches**, the populations occupying them will adapt to the conditions in these niches, and become genetically differentiated. Differentiated populations are dynamic: They may continue to exist, become extinct, merge with the parental population, or diverge from the parental population until they become reproductively isolated and form new species.

It is difficult to define the exact moment when a new species forms. A **species** is a group of interbreeding or potentially interbreeding populations that is reproductively isolated in nature from all other such groups. In sexually reproducing organisms, speciation divides a single gene pool into two or more separate gene pools. Changes in morphology, physiology, and adaptation to an ecological niche may also occur, but are not necessary components of the speciation event. Speciation can take place gradually or within a few generations.

Because population and species divergence is accompanied by genetic differentiation, we can use patterns of genetic differences to reconstruct evolutionary history. After exploring the genetic structure of populations, their divergence across space and time, and the process of speciation, we'll discuss how genetic data can be used to answer questions that have an evolutionary context.

23.1 Evolutionary History: Models of Speciation

Figure 23–1 shows an evolutionary tree, or **phylogeny**, that describes the history of several hypothetical lizard species. The passage of time is plotted horizontally, and the change in the form, or phenotype, of the lizards is plotted vertically. The graph traces the phenotype of the average lizard in each species over time.

The history of these lizards begins with species 1. For some time, species 1 experiences evolutionary **stasis**; that is, it does not change. Species 1 then undergoes a period of steady transformation, also called **phyletic evolution**, or **anagenesis**, and becomes species 2. Species 2 then undergoes **cladogenesis**; it gives rise to two distinct and independent daughter species. Further transformation of these daughter species produces the species we see today, species 3 and species 4.

In his 1859 book, *On the Origin of Species*, Charles Darwin amassed evidence that all species derive from a single common ancestor by transformation and speciation:

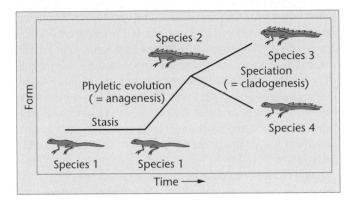

FIGURE 23–1 In phyletic evolution, or anagenesis, one species is transformed over time into another species. At all times, only one species exists. In cladogenesis, one species splits into two or more species.

"…all living things have much in common, in their chemical composition, their germinal vesicles, their cellular structure, and their laws of growth and reproduction…. Therefore I should infer… that probably all the organic beings which have ever lived on this earth have descended from some one primordial form…."

Everything biologists have since learned supports Darwin's conclusion that only one tree of life exists. To understand evolution, we must understand the mechanisms responsible for transforming one species into another, and for splitting one species into two or more new species.

Chapter 22 examined the mechanisms responsible for the transformation of species. Chief among them is natural selection, discovered independently by Alfred Russel Wallace and Darwin. The Wallace–Darwin concept of natural selection can be summarized as follows:

1. Among individuals of a species, variations in phenotype exist. These can include differences in size, agility, coloration, or ability to obtain food.

2. Many of these variations, even small and seemingly insignificant ones, are heritable and passed on to offspring.

3. Organisms tend to reproduce in an exponential fashion. More offspring are produced than can survive. This causes members of a species to engage in a struggle for survival, competing with other members of the species for scarce resources.

4. In the struggle for survival, individuals with particular phenotypes will be more successful than individuals with others, allowing them to survive and reproduce at higher rates.

As a consequence of natural selection, species change over time. The phenotypes that confer improved ability to survive and reproduce become more common, and the phenotypes that confer poor prospects for survival and reproduction disappear.

Although Wallace and Darwin proposed that natural selection explains how evolution occurs, they could not explain how the raw material for evolution, the variations themselves, arise nor how such variations are passed from parents to offspring. In the twentieth century, as biologists applied the principles of Mendelian genetics to populations, the source of variation (mutation) and the mechanism of inheritance (segregation of alleles) became apparent. We now view evolution as changes in allele frequencies in populations over time. This union of population genetics with natural selection generated a new view of the evolutionary process, called neo-Darwinism.

In this chapter we consider the mechanisms responsible for speciation. As with natural selection, our understanding of speciation builds on key insights about genetic variation.

23.2   Estimating Genetic Variation

We might assume that the members of a well-adapted population are genetically homogeneous because the most favorable allele at each locus has become fixed. Certainly, examinations of most populations of plants and animals reveal many phenotypic similarities among individuals. However, considerable evidence indicates that most populations contain a high degree of heterozygosity. This built-in genetic diversity is concealed, so to speak, because it is not necessarily apparent in the phenotype. Detecting this concealed genetic variation is not an easy task. Nevertheless, such investigations have been successful using the techniques we discuss next.

Protein Polymorphisms

Gel electrophoresis separates protein molecules on the basis of differences in size and electrical charge. If a nucleotide variation in a structural gene results in the substitution of a charged amino acid (such as glutamic acid) for an uncharged amino acid (such as glycine), the net electrical charge on the protein will be altered. This difference in charge can be detected as a change in the rate at which proteins migrate through an electrical field. In the mid-1960s, John Hubby

and Richard Lewontin used gel electrophoresis to measure protein variation in natural populations of *Drosophila*. Now researchers routinely use gel electrophoresis to study genetic variation in a wide range of organisms (Table 23.1).

The electrophoretically distinct forms of a protein produced by different alleles are called allozymes. As shown in Table 23.1, a surprisingly large percentage of loci examined from diverse species produce distinct allozymes. Of the populations shown in Table 23.1, approximately 30 loci per species were examined, and about 30 percent of the loci are polymorphic, with an approximate average of 10 percent allozyme heterozygosity per diploid genome.

These values apply only to genetic variation detectable by altered protein migration in an electric field. Electrophoresis probably detects only about 30 percent of the actual variation due to amino acid substitutions because many substitutions do not change the net electric charge on the molecule. Richard Lewontin has estimated that about two-thirds of all loci are polymorphic in a population and that, in any individual within that group, about one-third of the loci exhibit genetic variation in the form of heterozygosity.

The significance of genetic variation as detected by electrophoresis is controversial. Some argue that allozymes are biochemically equivalent and therefore do not play a role in adaptive evolution. We address this argument later in this section.

Variations in Nucleotide Sequence

The most direct way to estimate genetic variation is to compare the nucleotide sequences of genes carried by individuals in a population. With the development of techniques for cloning and sequencing DNA, nucleotide sequence variations are being cataloged for an increasing number of gene systems. Using restriction endonucleases to detect polymorphisms, Alec Jeffreys surveyed 60 unrelated individuals in order to estimate the total number of DNA-sequence variants in humans. His results show that within the genes of the β-globin cluster, 1 in 100 bp shows polymorphic variation. If this region is representative of the genome, this in-

---

TABLE 23.1   Allozyme Heterozygosity at the Molecular Level

Species	Populations Studied	Loci Examined	Polymorphic Loci per Population (%)	Heterozygotes per Locus (%)
Homo sapiens (humans)	1	71	28	6.7
Mus musculus (mouse)	4	41	29	9.1
Drosophila pseudoobscura (fruit fly)	10	24	43	12.8
Limulus polyphemus (horseshoe crab)	4	25	25	6.1

Note: A polymorphic locus is one for which a population harbors more than one allele.
Source: From Lewontin, (1974, p. 117).

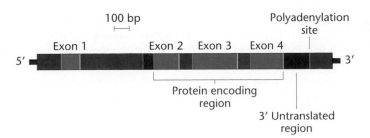

FIGURE 23–2 Organization of the *Adh* locus of *Drosophila melanogaster*.

dicates that at least 3×10^7 nucleotide variants per genome are possible.

In another study, Martin Kreitman investigated the *alcohol dehydrogenase* locus (*Adh*) in *Drosophila melanogaster* (Figure 23–2). This locus encodes two allozymic variants, the *Adh-f* and the *Adh-s* alleles. These differ by a single amino acid (Thr versus Lys at codon 192). To determine whether the genetic variation at the protein level (one amino acid difference) corresponds to that at the nucleotide level, Kreitman cloned and sequenced *Adh* loci from five natural populations of *Drosophila*.

The 11 cloned loci contain 43 nucleotide variations from the consensus *Adh* sequence of 2721 bp. These variations are distributed throughout the locus: 14 variations in the exon coding regions, 18 in the introns, and 11 in the nontranslated and flanking regions. Of the 14 in the coding regions, only one leads to an amino acid replacement, accounting for the two observed electrophoretic variants. The other 13 nucleotide substitutions do not lead to amino acid replacements. Are the differences in the number of allozyme and nucleotide variants the result of natural selection? If so, of what significance is this? We discuss these issues next.

Among the most intensively studied loci to date is the locus encoding the cystic fibrosis transmembrane conductance regulator (CFTR). Recessive loss-of-function mutations in the *CFTR* locus cause cystic fibrosis, a disease with such symptoms as salty skin, excessive amounts of thick mucus in the lungs, and susceptibility to bacterial infections. Geneticists have analyzed the *CFTR* locus in some 30,000 chromosomes from individuals with cystic fibrosis and have found over 500 different mutations that can cause the disease. These include missense mutations, amino acid deletions, nonsense mutations, frameshifts, and splice defects.

Figure 23–3 shows a map of the 27 exons in the *CFTR* locus, with most exons identified by function. The histogram above the map shows the locations of the known disease-causing mutations and the number of copies that have been found of each. A single mutation, a 3-bp deletion in exon 10 called *ΔF508*, accounts for 67 percent of the cystic fibrosis allele copies found, but several other mutations have been found in at least 100 chromosomes. In populations of European ancestry, between 1 in 44 and 1 in 20 individuals are heterozygous carriers of cystic fibrosis alleles. Note that Figure 23–3 includes only the sequence variants that alter the function of the CFTR protein. There are undoubtedly many more *CFTR* alleles with "silent sequence" variants that neither change the structure of the protein nor affect its function. The *CFTR* locus, by itself, shows considerable genetic variation.

Studies of other organisms, including the rat and the mouse, have produced similar estimates of nucleotide diversity. Thus, an enormous reservoir of genetic variability exists within most populations, and at the DNA level, most and perhaps all genes exhibit diversity from individual to individual.

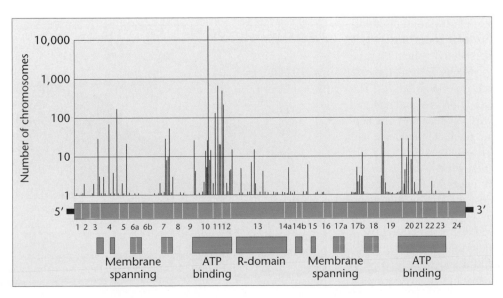

FIGURE 23–3 The locations of disease-causing mutations in the cystic fibrosis gene. The histogram shows the number of copies of each mutation geneticists have found (the vertical axis is on a logarithmic scale). The genetic map below the histogram shows the locations and relative sizes of the 27 exons of the *CFTR* locus. The boxes at the bottom indicate the functions of different domains of the CFTR protein. (*Reprinted from* Trends Genet. 8:392–98, L. Tsui, The spectrum of cystic fibrosis mutations, 1992, with permission from Elsevier Science.)

23.3   Evolution and Genetic Variation

As we mentioned earlier, the finding that populations harbor considerable genetic diversity at the level of amino acid and nucleotide sequences came as a surprise to many evolutionary biologists. The early consensus was that selection would favor a single optimal allele at each locus, and as a result, populations would be genetically homogenous. This expectation was obviously wrong, and considerable research and argument ensued concerning the forces that maintain genetic variation.

One view on why so much genetic variation exists in species argues that it primarily reflects the action of mutation and genetic drift. The **neutral theory** of molecular evolution, proposed by Motoo Kimura, argues that mutations leading to amino acid substitutions are rarely favorable. They are sometimes detrimental, but most often are neutral or genetically equivalent to the allele that is replaced. Those few polymorphisms that are favorable or detrimental are either preserved or removed from the population, respectively, by natural selection. However, the vast majority of mutations are neutral genetic changes that are not affected by selection. The frequency of these alleles in a population will be determined by mutation rates and random genetic drift. Some neutral mutations will drift to fixation in the population; other neutral mutations will be lost. At any given time, the population will contain several neutral alleles at any particular locus. The diversity of alleles at most polymorphic loci does not, however, reflect the action of natural selection.

Opposed to the neutral theory are the **selectionists**. These geneticists point to examples where enzyme or protein polymorphisms are associated with adaptation to certain environmental conditions. The well-known advantage of sickle-cell anemia heterozygotes in infection by malarial parasites exemplifies this type of adaptation.

Selectionists also stress that enzyme polymorphisms often appear to offer no advantage, but exist in such frequency that it cannot be explained as random occurrences. Thus, even though no currently available analytical technique can detect any physiological difference, some slight advantage associated with amino acid substitution may exist. Perhaps having two forms of a given protein allows optimum performance under a wider range of cellular conditions.

The neutralist and selectionist perspectives should not be viewed as mutually incompatible positions. Instead, they are best seen as the endpoints of a spectrum of possibilities. The neutralists do not discount natural selection as a guiding force in evolution. Rather, they suggest that some features of an organism's genotype are nonadaptive, fluctuate randomly, and may have been fixed by genetic drift. On the other hand, the selectionists certainly don't deny that genetic drift is an important factor in establishing differences in allele frequencies. It is difficult to argue against the notion that some genetic variation must be neutral. The difference between the two theories is in their proposed degree of neutrality.

Current data are insufficient to determine what fraction of molecular genetic variation is neutral and what fraction is naturally selected. The neutral theory nonetheless serves a crucial function. By pointing out that some genetic variation is expected simply as a result of mutation and drift, the neutral theory provides a working hypothesis for studies of molecular evolution. In other words, biologists must find positive evidence that selection is acting on the allele frequencies at a particular locus before they can reject the simpler assumption that only mutation and drift are at work.

23.4   Changes in the Genetic Structure of Populations

As geneticists discovered that most populations harbor considerable genetic diversity, they also found that the genetic structure of populations varies across space and time. To illustrate, we consider studies on *Drosophila pseudoobscura* conducted by Theodosius Dobzhansky and his colleagues. This species is found over a wide range of environmental habitats, including the western and southwestern United States. Although the flies throughout this range are morphologically similar, Dobzhansky's team discovered that populations from different locations vary in the arrangement of genes on chromosome 3. They found several different inversions in this chromosome that can be detected by loop formations in larval polytene chromosomes. Each inversion sequence is named after the locale in which it was first discovered (e.g., AR Arrowhead, British Columbia; and CH Chiricahua Mountains). The inversion sequences were compared with one standard sequence, designated ST.

Figure 23–4 compares three arrangements found in populations at three elevations in the Sierra Nevada chain in California. The ST arrangement is most common at low elevations; at 8000 feet, AR is the most common and ST least common. In these populations, the frequency of the CH arrangement gradually increases with elevation. The gradual change in inversion frequencies is probably the result of natural selection and parallels the gradual environmental changes occurring at ascending elevations. As the populations along this gradient show a continuous and gradual

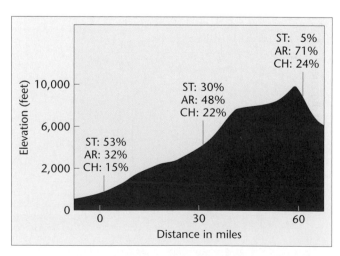

FIGURE 23–4 Inversions in chromosome 3 of *D. pseudoobscura* at different elevations in the Sierra Nevada mountain range near Yosemite National Park.

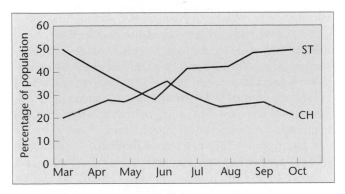

FIGURE 23–5 Changes in the ST and CH arrangements in *D. pseudoobscura* throughout the year.

change in inversion frequencies, it is difficult to classify a fly as a member of one racial group.

Dobzhansky's team also found that if populations are collected at a single site throughout the year, the inversion frequencies also change. That is, cyclic variation in chromosome arrangements occurs through the seasons, as shown in Figure 23–5. Such variation was consistently observed over a period of several years. The frequency of ST always declined during the spring, with a concomitant increase in CH during the same period.

To test the hypothesis that this cyclic change is a response to natural selection, Dobzhansky's team devised the following laboratory experiment. They constructed a large population cage from which samples of *D. pseudoobscura* could be periodically removed and studied, and began with a population of a known inversion frequency, 88 percent CH and 12 percent ST. They reared it at 25°C and sampled it over a 1-year period. As shown in Figure 23–6, ST increased gradually until it was present at a level of 70 percent. At this point, an equilibrium between ST and CH was reached. When the same experiment was performed at 16°C, no change in inversion frequency occurred. They concluded that the equilibrium reached at 25°C is a response to the elevated temperature, the only variable in the experiment.

The results indicate that a balance in the frequency of the two inversions and their respective gene arrangements in a population is superior to either inversion by itself. The equilibrium attained presumably represents the highest mean fitness in the population under controlled laboratory conditions. This interpretation of Dobzhansky's experiment suggests that natural selection is the driving force maintaining the diversity in chromosome 3 inversions.

In a more extensive study, Dobzhansky and his colleagues sampled *D. pseudoobscura* populations over a broader geographic range. They found 22 different chromosome arrangements in populations from 12 locations. Figure 23–7 shows the relative frequencies of five of these inversions according to geographic location. The differences are largely quantitative, with most populations differing only in the relative percentages of inversions.

Collectively, Dobzhansky's data show that the genetic structure of *D. pseudoobscura* populations changes from place to place and from one time to another. At least some of this variation in population genetic structure is the result of natural selection.

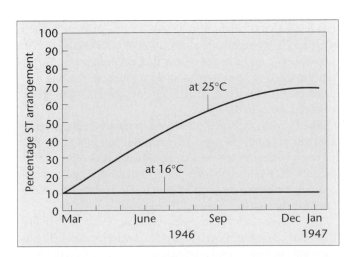

FIGURE 23–6 Increase in the ST arrangement of *D. pseudoobscura* in population cages under laboratory conditions.

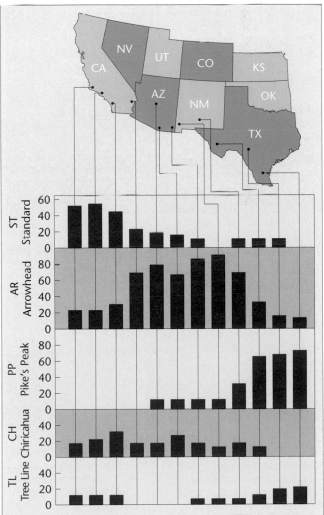

FIGURE 23–7 Relative frequencies (percentages) of five chromosomal inversions in *D. pseudoobscura* in different geographic regions.

Another example of how the genetic structure of a species varies among populations is provided by the work of Dennis Powers and Patricia Schulte on the mummichog (*Fundulus heteroclitus*). The mummichog is a small fish (5–10 cm long) that lives in inlets, bays, and estuaries along the Atlantic coast of North America from Florida to Newfoundland. These workers studied allele frequencies at the locus encoding the enzyme lactate dehydrogenase-B (LDH-B). LDH-B is made in the liver, heart, and red skeletal muscle. It converts lactate to pyruvate and is thus pivotal in both the synthesis of glucose and in aerobic metabolism. Two allozymes of LDH-B are electrophoretically distinguishable; they differ at two amino acid positions. The alleles encoding the allozymes are *Ldh-B^a* and *Ldh-B^b*.

The frequencies of the *Ldh-B* alleles vary dramatically among mummichog populations [Figure 23–8(a)]. In northern populations, where the mean water temperature is about 6°C, *Ldh-B^b* predominates. In southern populations, where the mean water temperature is about 21°C, *Ldh-B^a* predominates. Between the geographic extremes, allele frequencies are intermediate.

(a)

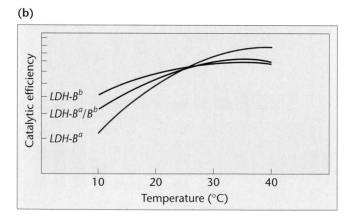

(b)

FIGURE 23–8 Variation in genetic structure among mummichog populations. (a) Frequencies of the *Ldh-B^b* allele in populations along the Atlantic coast of North America. (b) Catalytic efficiency of LDH-B allozymes as a function of temperature. (*From D.A. Powers and P.M. Schulte. Evolutionary adaptations of gene structure and expression in natural populations in relation to a changing environment, The Journal of Experimental Zoology, ©1998. Reprinted by permission of Wiley-Liss, Inc., a subsidiary of John Wiley & Sons, Inc. and with permission from Annu. Rev. Genet. 25, © 1991 by Annual Reviews.*)

To determine whether the geographic variation in *Ldh-B* allele frequencies is due to natural selection, Powers and Schulte studied the biochemical properties of the LDH-B allozymes. They found that the enzyme encoded by *Ldh-B^b* shows higher catalytic efficiency at low temperatures, whereas the gene product of *Ldh-B^a* is more efficient at high temperatures [Figure 23–8(b)]. A mixture of the two forms has intermediate efficiency at all temperatures.

In addition to the functional differences between the LDH-B allozymes, the *Ldh-B* alleles have sequence differences in their regulatory regions. The *Ldh-B^b* allele's relative rate of transcription is more than twice that of the *Ldh-B^a* allele. As a result, northern fish have higher concentrations of the LDH-B enzyme in their cells.

The differences in catalytic efficiency and relative transcription rate associated with the *Ldh-B* allele are consistent with the hypothesis that mummichog populations are adapted to the temperatures at which they live. The fish are ectotherms; their body temperature is determined by their environment. In general, low body temperatures slow an ectotherm's metabolic rate. The higher transcription rate of the *Ldh-B^b* allele and the superior low-temperature catalytic efficiency of its gene product appear to help northern fish compensate for the tendency of their cold environment to reduce their metabolic rate. The superior high-temperature catalytic efficiency and lower transcription rate associated with the *Ldh-B^a* allele appear to allow southern fish to economize on the resources devoted to the machinery for glucose production and aerobic metabolism. Based on this and other evidence, Powers suggests that the differences among mummichog populations in *Ldh-B* allele frequencies are the result of natural selection.

23.5   Formation of Species

We have seen that most populations harbor considerable genetic variation, and that different populations within a species may have different alleles and/or allele frequencies at a variety of loci. The divergence of populations within a species can be caused by natural selection, genetic drift, or both. In Chapter 22, we saw that the migration of individuals between populations, and the gene flow that accompanies this migration tends to homogenize allele frequencies among populations. In other words, migration counteracts the tendency of populations to diverge.

When gene flow between two populations is reduced or absent, the populations may diverge to the point that members of one population are no longer able to successfully interbreed with members of the other. When populations reach the point where they are reproductively isolated from one another, they have become different species.

The barriers that prevent interbreeding between populations can be physiological, behavioral, or mechanical. The biological and behavioral properties of organisms that prevent or reduce interbreeding are called **reproductive isolating mechanisms**. These mechanisms are classified in Table 23.2.

TABLE 23.2 Reproductive Isolating Mechanisms

Prezygotic Mechanisms (prevent fertilization and zygote formation)

1. Geographic or ecological: The populations live in the same regions but occupy different habitats
2. Seasonal or temporal: The populations live in the same regions but are sexually mature at different times
3. Behavioral (only in animals): The populations are isolated by different and incompatible behavior before mating
4. Mechanical: Cross-fertilization is prevented or restricted by differences in reproductive structures (genitalia in animals, flowers in plants)
5. Physiological: Gametes fail to survive in alien reproductive tracts

Postzygotic Mechanisms (fertilization takes place and hybrid zygotes are formed, but these are nonviable or give rise to weak or sterile hybrids)

1. Hybrid nonviability or weakness
2. Developmental hybrid sterility: Hybrids are sterile because gonads develop abnormally or meiosis breaks down before completion
3. Segregational hybrid sterility: Hybrids are sterile because of abnormal segregation into gametes of whole chromosomes, chromosome segments, or combinations of genes
4. F_2 breakdown: F_1 hybrids are normal, vigorous, and fertile, but the F_2 contains many weak or sterile individuals

Source: From G. Ledyard Stebbins, *Processes of organic evolution*, 3rd ed., copyright 1977, p. 143. Reprinted by permission of Prentice-Hall, Upper Saddle River, NJ.

Prezygotic isolating mechanisms prevent individuals from mating in the first place. Individuals from different populations do not find each other at the right time, do not recognize each other as suitable mates, or try to mate but find that they are unable to do so.

Postzygotic isolating mechanisms create reproductive isolation even when the members of two populations are willing and able to mate with each other. Perhaps genetic divergence has reached the stage where the viability or fertility of hybrids is reduced. Hybrid zygotes can be formed, but all or most will be inviable. Alternatively, the hybrids are viable but sterile, or have reduced fertility. In another possibility, the hybrids themselves are fertile, but their progeny has lowered viability or fertility. These postzygotic mechanisms act at or beyond the level of the zygote and are generated by genetic divergence.

In some models of speciation, postzygotic isolation develops first, and is followed by prezygotic isolation. Postzygotic isolating mechanisms waste gametes and zygotes and lower the reproductive fitness of hybrid survivors. Selection therefore favors the spread of alleles that reduce the formation of hybrids and lead to the development of prezygotic isolating mechanisms. These mechanisms prevent interbreeding and the formation of hybrid zygotes and offspring. In animal evolution, the most effective prezygotic mechanism is behavioral isolation, involving courtship behavior.

Observing Speciation

We consider two examples of speciation, one from a laboratory study, the other from a field study. Diane Dodd and her colleagues were interested in the evolution of digestive physiology in *Drosophila pseudoobscura*. They collected flies from a wild population and used them to establish several separate laboratory populations. Some of the laboratory populations were raised on starch-based medium, and others on maltose-based medium. Both food sources were stressful for the flies. It was only after several months of evolution by natural selection that the populations adapted to their artificial diets and began to thrive.

Dodd's team wanted to know whether the starch-adapted populations and the maltose-adapted populations, which had been diverging under strong selection and in the absence of gene flow, had become different species. Roughly a year after the populations were established, a series of mating trials were performed. For each trial they placed 48 flies in a bottle: 12 males and 12 females from a starch-adapted population, and 12 males and 12 females from a maltose-adapted population. Then they simply noted which flies mated.

Dodd's team predicted that if the populations adapted to different media had speciated, then the flies would prefer to mate with members of their own population. If the populations had not speciated, then the flies would mate at random. The results appear in Table 23.3. Roughly 600 of the 900 matings they observed in the mating trials occurred between males and females from the same population. In other words, the differently adapted fly populations showed partial premating isolation. They concluded that the populations had begun to speciate, but had not yet completed the process.

The Isthmus of Panama, which created a land bridge connecting North and South America and simultaneously separated the Caribbean Sea from the Pacific Ocean, formed roughly 3 million years ago. Diane Knowlton and her colleagues took advantage of a natural experiment that the formation of the Isthmus of Panama had "performed" on sev-

TABLE 23.3 Incipient Speciation in Laboratory Populations of *Drosophila pseudoobscura* (Number of Matings Involving Each Kind of Male–Female Pair)

		Female	
		Starch-adapted	*Maltose-adapted*
Male	Starch-adapted	290	153
	Maltose-adapted	149	312

Source: Compiled from Dodd, D. M. B. 1989. Reproductive isolation as a consequence of adaptive divergence in *Drosophila pseudoobscura. Evolution* 43:1308–11.

eral species of snapping shrimps (Figure 23–9). After identifying seven Caribbean species, they matched each species with a Pacific species to form a natural pair. The members of each species pair were more similar to each other in structure and appearance than either was to any other species in its own ocean. Analysis of allozyme allele frequencies and mitochondrial DNA sequences confirmed that the members of each pair were one another's closest genetic relatives.

The Knowlton team's interpretation of these data is that prior to the formation of the Isthmus, the ancestors of each species pair were a single species. When the Isthmus closed, the seven ancestral species divided into two separate populations, one in the Caribbean, the other in the Pacific.

Meeting in a dish in Knowlton's lab for the first time in 3 million years, would Caribbean and Pacific members of a species pair recognize each other as suitable mates? Knowlton placed males and females together and noted their behavior toward each other. She then calculated the relative inclination of Caribbean/Pacific couples to mate versus that of Caribbean/Caribbean or Pacific/Pacific couples. For three of the seven species pairs, transoceanic couples refused to mate altogether. For the other four species pairs, transoceanic couples were 33, 45, 67, and 86 percent as likely to mate with each other as were same-ocean pairs. Of the same-ocean couples that mated, 60 percent produced viable clutches of eggs. Of the transoceanic couples that mated, only 1 percent produced viable clutches. We can conclude from these data that 3 million years of separation has resulted in complete

or nearly complete speciation, involving strong pre- and postzygotic isolating mechanisms, for all seven species pairs.

The Minimum Genetic Divergence Required for Speciation

How much genetic divergence is required between two populations before they become different species? We consider two examples, an insect and a plant, which demonstrate that in some cases the answer is not very much.

Researchers estimate that *Drosophila heteroneura* and *D. silvestris*, found only on the island of Hawaii, diverged from a common ancestral species only about 300,000 years ago. They are thought to be descended from *D. planitibia* colonists from the older island of Maui (Figure 23–10). The two species are clearly separated from each other by different and incompatible courtship and mating behaviors (a prezygotic isolating mechanism), by morphology, and by body and wing pigmentation (Figure 23–11).

In spite of their morphological and behavioral divergence, demonstrating significant differences between these species in chromosomal inversion patterns or protein polymorphisms is difficult. When DNA-hybridization studies are carried out on these species, the sequence diversity between the two is

FIGURE 23–9 A snapping shrimp (genus *Alpheus*). *(Carl C. Hansen/Nancy Knowlton/Smithsonian Institution Photo Services)*

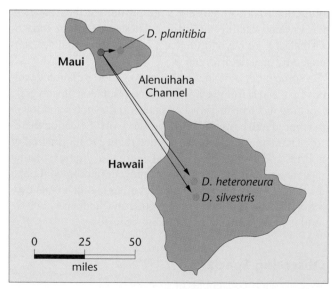

FIGURE 23–10 Proposed pathway for Hawaii's colonization by members of the *D. planitibia* species. The open circle represents a population ancestral to the three present-day species.

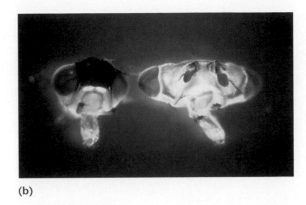

(a)   (b)

FIGURE 23–11 (a) Differences in pigmentation patterns in *D. sylvestris* (left) and *D. heteroneura* (right). (b) Head morphology in *D. sylvestris* (left) and *D. heteroneura* (right). *(Kenneth Kaneshiro/University of Hawaii/CCRT)*

only about 0.55 percent (Figure 23–12). Thus, nucleotide sequence diversity may precede the development of protein or chromosomal polymorphisms.

Genetic evidence suggests that the differences between *D. heteroneura* and *D. silvestris* are controlled by a relatively small number of genes. For example, as few as 15–19 major loci may be responsible for the morphological differences between the species, demonstrating that the process of speciation need only involve a small number of genes.

Studies using two closely related species of monkey flowers, a plant that grows in the Sierra Nevada, Rocky Mountain, and Cascade mountain ranges of the western United States, confirm that species can be separated by only a few genetic differences. One species, *Mimulus cardinalis* is fertilized by hummingbirds and does not interbreed with *M. lewisii*, which is fertilized by bumblebees. H. D. Bradshaw and his colleagues studied genetic differences related to reproduction in the two species: flower shape, size, and color, and nectar production. For each trait, a difference in a single gene provided at least 25 percent of the variation observed among laboratory-created *M. cardinalis* × *lewisii* hybrids. *M. cardinalis* makes 80 times more nectar than *M. lewisii*, and a single gene is responsible for at least half the difference. A single gene also controls a large part of the differences in flower color between the two species (Figure 23–13). In this

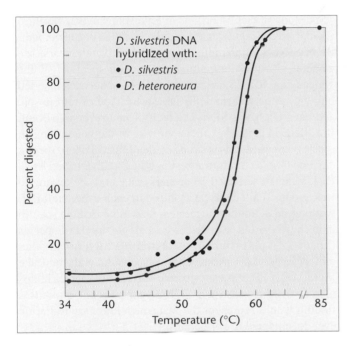

FIGURE 23–12 Nucleotide sequence diversity in the *Drosophila planitibia* species complex. The shift to the left by the heterologous hybrids indicates the degree of nucleotide sequence divergence.

(a)   (b)

FIGURE 23–13 Flowers of two closely related species of monkey flowers, (a) *Mimulus cardinalis* and (b) *Mimulus lewisii*. *(Courtesy of Toby Bradshaw and Doug W. Schemske, University of Washington. Photo by Jordan Rehm.)*

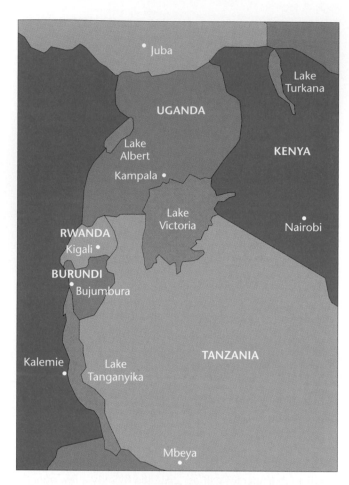

FIGURE 23–14 Lake Victoria in the Rift Valley of East Africa is home to more than 400 species of cichlids.

case, as in the Hawaiian *Drosophila*, species differences can be traced to a relatively small number of genes.

The Rate of Speciation

How much time is required for speciation? In many cases, speciation takes place slowly over a long period of time. In other cases, however, speciation is surprisingly rapid. The Rift Valley lakes of East Africa support hundreds of species

of cichlid fish (Figure 23–14). Lake Victoria, for example, contains over 400 species. Cichlids have small amounts of morphological diversity, but, more dramatically, they are highly specialized for different niches (Figure 23–15). Some eat algae floating on the water's surface, others are bottom feeders, insect feeders, mollusk eaters, and predators on other fish species. Lake Tanganyika and Lake Malawi have a similar array of species. Genetic analyses indicate that the species in a given lake are all more closely related to each other than to species from other lakes. The implication is that most or all of the species in, say, Lake Tanganyika and Lake Victoria are descended from a single common ancestor, and that they evolved within their home lake.

Lake Victoria is between 250,000 and 750,000 years old, and evidence suggests that the lake dried out nearly completely less than 14,000 years ago. Is it possible that the 400 or so species present in the lake today all evolved in less than 14,000 years from a single ancestral species?

In a study of cichlid origins in Lake Tanganyika, Norihiro Okada and colleagues examined the insertion of a novel family of SINES (short, interspersed repetitive elements; see Chapter 17) into the genomes of cichlid species in Lake Tanganyika. SINES are repetitive DNA sequences inserted into the genome at random. They are a type of retroposon, and integration of a SINE at a locus is an irreversible event. If a SINE is present at the same loci in the genome of all species examined, this is strong evidence that all these species descend from a common ancestor. Using a SINE called AFC, Okada's team screened 33 species of cichlids belonging to four groups (called species tribes). In each tribe, the SINE was present at all the sites tested, indicating that the species in each tribe are descended from a single ancestral species (Figure 23–16). If further research using SINES and other molecular markers produces similar results with the Lake Victoria species, it means that the 400 cichlid species in this lake evolved in less than 14,000 years. If confirmed, this finding would represent the fastest evolutionary radiation ever documented in vertebrates.

Even faster speciation is possible through the special mechanism of polyploidy. The formation of animal species by polyploidy is rare, but polyploidy is an important factor in the

FIGURE 23–15 Cichlids occupy a diverse array of niches, and each species is specialized for a distinct food source. *(Dr. Paul V. Loiselle)*

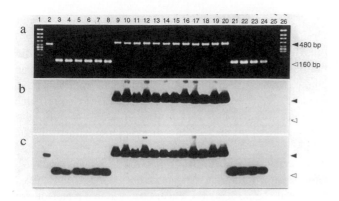

FIGURE 23–16 Tracing species relationships in cichlids from Lake Tanganyika using repetitive DNA elements. (a) Photograph of an agarose gel showing PCR fragments generated from primers that flank the AFC family of SINES. Large fragments containing the SINE are present in all samples of genomic DNA from members of the Lamprologini tribe of cichlids (lanes 9–20). DNA from the species in lane 2 has a similar, but shorter fragment, which may represent another repetitive sequence. DNA from other species (lanes 3–8 and 21–24) produce short, non-SINE containing fragments. (b) A Southern blot of the gel from (a) probed with DNA from the AFC family of SINES. This SINE is present in DNA from all species in the Lamprologini tribe (lanes 9–20) but not in DNA from other species (lanes 3–8 and 21–24). (c) A second Southern blot of the gel in (a) probed with the genomic sequence at which the SINE inserts. All species examined (lanes 2–24) contain the insertion sequence. The larger fragments in Lamprologini DNA (lanes 9–20) correspond to those in the previous Southern, showing the SINE is inserted at the same site in all tribal species. In summary, these data show that the AFC SINE is present only in members of this tribe, is present in all member species, and is inserted at the same site in all cases. These results are interpreted as showing a common origin for the species of this tribe. *(From Takahashi, K., et al. 1998. A novel family of short interspersed repetitive elements from cichlids: The pattern of insertion of SINES at orthologous loci support the supposed monophyly of four major groups of cichlid fishes in Lake Tanganyika. Molec. Biol. Evol. 15(4):391–407. Taken from Figure 5, part A, p. 400.)*

FIGURE 23–17 The cultivated tobacco plant *Nicotiana tabacum* is the result of hybridization between two other species. *(C. B. & D. W. Frith/Bruce Coleman, Inc.)*

evolution of plants. It is estimated that one-half of all flowering plants have evolved by this mechanism. One form of polyploidy is allopolyploidy (see Chapter 7), produced by doubling the chromosome number in an interspecific hybrid.

If two species of related plants have the genetic constitution SS and TT (where S and T represent the haploid set of chromosomes in each species), then the F_1 hybrid would have the chromosome constitution ST. Normally such a plant is sterile because few or no homologous chromosome pairs exist and aberrations would arise during meiosis. However, if the hybrid undergoes a spontaneous doubling of chromosome number, a tetraploid SSTT plant is produced. This chromosomal aberration might occur during mitosis in somatic tissue, giving rise to a partially tetraploid plant that produces some tetraploid flowers. Alternatively, aberrant meiotic events may produce ST gametes, which when fertilized, yield SSTT zygotes. The SSTT plants are fertile because they possess homologous chromosomes producing viable ST gametes. This new, true-breeding tetraploid would have a combination of characters derived from the parental species and would be reproductively isolated from them because F_1 hybrids are triploids and consequently sterile.

The tobacco plant *Nicotiana tabacum* ($2n = 48$) is the result of the doubling of the chromosome number in the hybrid between *N. otophora* ($2n = 24$) and *N. silvestris* ($2n = 24$) (Figure 23–17). The origin of *N. tabacum* is an example of virtually instantaneous speciation, occurring in one generation.

23.6 Reconstructing Evolutionary History

Early in this chapter we noted that evolution involves the transformation of one species into another, and the splitting of species into two or more new species. Our examples have demonstrated that speciation is associated with changes in genetic structure of populations and genetic divergence. If this is true, then we should be able to use genetic differences among species to reconstruct their evolutionary histories.

In an important early example of phylogeny reconstruction, W. M. Fitch and E. Margoliash assembled data in the 1960s on the amino acid sequence for cytochrome c in a variety of organisms. **Cytochrome c** is a respiratory pigment found in the mitochondria of eukaryotes, and its amino acid sequence has evolved very slowly. For example, the amino acid sequence in humans and chimpanzees is identical; humans and rhesus monkeys show only one amino acid difference. This is remarkable considering that the fossil record indicates that the lines leading to humans and monkeys diverged from a common ancestral species approximately 20 million years ago.

TABLE 23.4 **Amino Acid Differences and the Minimal Mutational Distances Between Cytochrome c in Humans and Other Organisms**

Organism	(a) Amino Acid Differences	(b) Minimal Mutational Distance
Human	0	0
Chimpanzee	0	0
Rhesus monkey	1	1
Rabbit	9	12
Pig	10	13
Dog	10	13
Horse	12	17
Penguin	11	18
Moth	24	36
Yeast	38	56

Source: From W. M. Fitch and E. Margoliash, Construction of phylogenetic trees, *Science* 155:279–84, 20 January 1967. Copyright 1967 by the American Association for the Advancement of Science.

Column (a) of Table 23.4 lists the number of amino acid differences between cytochrome c in humans and a variety of other organisms. The table is broadly consistent with our intuitions about our relatedness to these other species. For example, we are more closely related to other mammals than to insects, and we are more closely related to insects than we are to fungi. Likewise, our cytochrome c differs in 13 amino acids from that of dogs, in 36 amino acids from that of moths, and in 56 amino acids from that of yeast.

However, more than one nucleotide change may be required to affect a given amino acid. When the nucleotide changes necessary for all amino acid differences observed in a protein are totaled, the **minimal mutational distance** between the genes of any two species is established. Column (b) in Table 23.4 shows such an analysis of the genes encoding cytochrome c. As expected, these values are larger than the corresponding number of amino acids separating humans from the other nine organisms.

Fitch used data on the minimal mutational distances between the cytochrome c genes of 19 organisms to reconstruct their evolutionary history. The result is an estimate of the evolutionary tree, or **phylogeny**, that unites the species (Figure 23–18). The blue dots on the tips of the branches represent species in existence today. These are connected to the inferred common ancestors, represented by red dots. The ancestral species evolved and diverged to produce modern organisms. The common ancestors are connected to still ear-

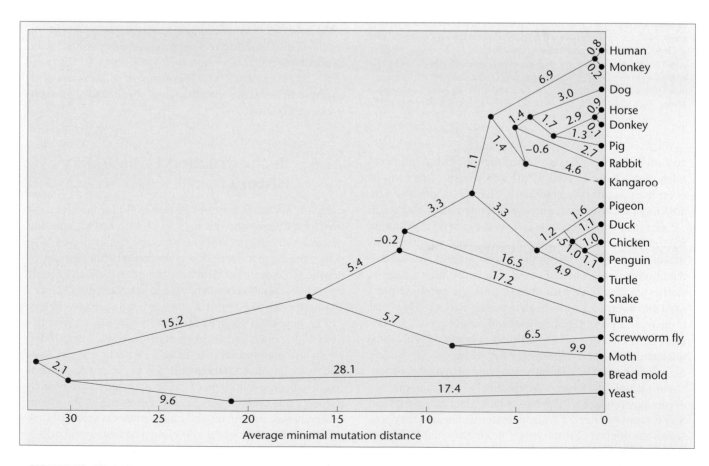

FIGURE 23–18 Phylogeny constructed by comparing homologies in cytochrome c amino acid sequences. *(From W. M. Fitch and E. Margoliash, Construction of phylogenetic trees,* Science *155:279–84, 20 January 1967. Copyright 1967 by the American Association for the Advancement of Science.)*

lier common ancestors, culminating in a single common ancestor for all the species on the tree, represented by the red dot on the extreme left.

Constructing Evolutionary Trees

Many methods use data on genetic differences to estimate phylogenies. It is beyond the scope of this chapter to review all of them. Instead we present just one method, called the *un*weighted *p*air *g*roup *m*ethod using *a*rithmetic averages, or **UPGMA**. UPGMA is not the most powerful method for estimating phylogenies from genetic data, but it is, in spite of its name, intuitively straightforward. Furthermore, UPGMA works reasonably well under many circumstances. The underlying assumption in this method is that the more genetically related two species are, the more likely they are to share a common ancestor.

The starting point for UPGMA is a table of genetic distances among a group of species [Figure 23–19(a)]. To demonstrate UPGMA, we'll use data from DNA-hybridization studies by Charles Sibley and Jon Alquist. They computed genetic distances of humans and four species of apes: the common chimpanzee, the gorilla, the siamang, and the common gibbon.

The method uses the following steps:

1. Search for the smallest genetic distance between any pair of species. In Figure 23–19(a), this value is 1.628, the distance between human and chimpanzee. Once identified, this species pair is placed on neighboring branches of an evolutionary tree [Figure 23–19(b)]. The length of each branch is half the genetic distance between the species (1.628/2), so the branches connecting human and chimpanzee to their common ancestor are about 0.81.

2. Recalculate the table of genetic distances between this species pair and the other species [Figure 23–19(c)]. The genetic distance between the human-chimpanzee (Hu-Ch) species cluster and the other species is the average of the distance between each cluster member and the other species. For example, the genetic distance between the human–chimp cluster and gorilla is the average of the human–gorilla distance and the chimp–gorilla distance. This is $(2.267 + 2.21)/2 = 2.2385$.

3. Repeat steps 1 and 2 until all the species have been added to the tree.

(a)

	Human	Chimp	Gorilla	Siamang	Gibbon
Human	–				
Chimp	**1.628**	–			
Gorilla	2.267	2.21	–		
Siamang	4.7	5.133	4.543	–	
Gibbon	4.779	4.76	4.753	1.95	–

(c)

	Hu-Ch	Gorilla	Siamang	Gibbon
Hu-Ch	–			
Gorilla	2.2385	–		
Siamang	4.9165	4.543	–	
Gibbon	4.7695	4.753	**1.95**	–

(e)

	Hu-Ch	Gorilla	Si-Gi
Hu-Ch	–		
Gorilla	**2.239**	–	
Si-Gi	4.843	4.648	–

(g)

	Hu-Ch-Go	Si-Gi
Hu-Ch-Go	–	
Si-Gi	**4.778**	–

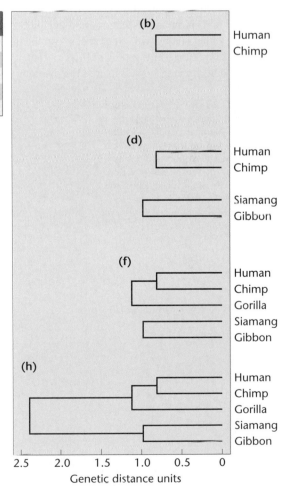

FIGURE 23–19 Phylogeny reconstruction by UPGMA. *(From Table 1, p. 126 from Journal of Molecular Evolution 26:99–122.)*

The smallest distance in the recalculated table is 1.95, the distance between siamang and gibbon [Figure 23–19(c)]. To build the next section of the tree, siamang and gibbon are placed on neighboring branches, with branch lengths equal to 1/2 of 1.95, or 0.98 [Figure 23–19(d)].

Recalculating the table again, there are two clusters plus gorilla [Figure 23–19(e)]. The genetic distance between the human–chimp cluster and the siamang–gibbon cluster is the average of four distances: human–siamang, human–gibbon, chimp–siamang, and chimp–gibbon.

Now the smallest genetic distance is 2.239, the distance between the human–chimp cluster and gorilla. We therefore add a gorilla branch to the tree and connect it to the common ancestor of the human–chimp cluster [Figure 23–19(f)]. The branches are drawn so that the distance between the tips of any two branches in the human–chimp–gorilla cluster is 2.239.

Redoing the table for the final time [Figure 23–19(g)], we calculate the genetic distance between the human–chimp–gorilla cluster and the siamang–gibbon cluster as the average of six genetic distances: human–siamang, human–gibbon, chimp–siamang, chimp–gibbon, gorilla–siamang, and gorilla–gibbon. This distance is 4.778, which allows us to complete our evolutionary tree [Figure 23–19(h)].

Our evolutionary tree indicates that humans and chimpanzees are one another's closest relatives. That is not to say that humans evolved from chimpanzees. Rather, this analysis indicates that humans and chimpanzees share a more recent common ancestor than either shares with other species on the tree.

The chief shortcoming of UPGMA is that it provides no way to determine how well its tree fits the data, compared to other possible trees. For example, how much better does the tree we just created fit our data than a tree that shows chimpanzees and gorillas as closest relatives?

Evolutionary geneticists have developed a variety of techniques for searching the set of all possible trees connecting a group of species, and calculating the relative performance of each. One set of techniques, called **parsimony** methods, compares trees using the minimum number of evolutionary changes each requires, then selects the simplest possible tree. Another set of analytical tools, called **maximum likelihood** methods, starts with a model of the evolutionary process, calculates how likely it is that evolution will produce each possible tree under the model, and selects the most likely tree. It is important to keep in mind that a phylogeny produced from data on genetic distances is not necessarily the way species actually evolved, but is instead a reasonable estimate of its evolution, based on the method used. Phylogenies based on combined data sets from several loci are usually more reliable than those based on a single locus or protein.

Molecular Clocks

In many cases, we would like to estimate not only which of a set of species are most closely related, but also when their common ancestors lived. Sometimes we can do so, thanks to molecular clocks. **Molecular clocks** are amino acid sequences or nucleotide sequences in which evolutionary changes accumulate at a constant rate over time.

Research by Walter M. Fitch and colleagues on the influenza A virus shows how molecular clocks are used. They sequenced

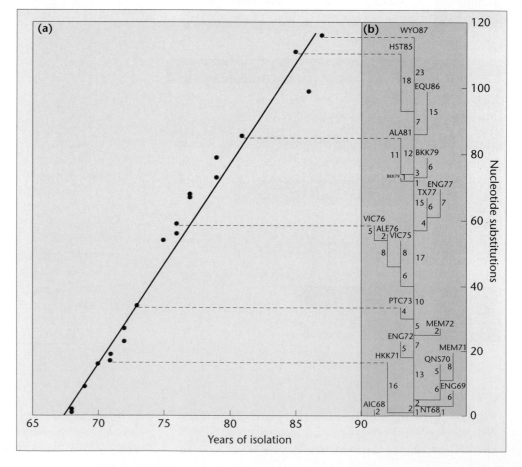

FIGURE 23–20 Molecular clock in the influenza A hemagglutinin gene. (a) Number of nucleotide differences between the first isolate and each subsequent isolate as a function of year of isolation. (b) Estimate of the phylogeny of the isolates. *(From Fitch, et al. 1991. Positive Darwinian evolution in human influenza A viruses. PNAS 88:4270–73.)*

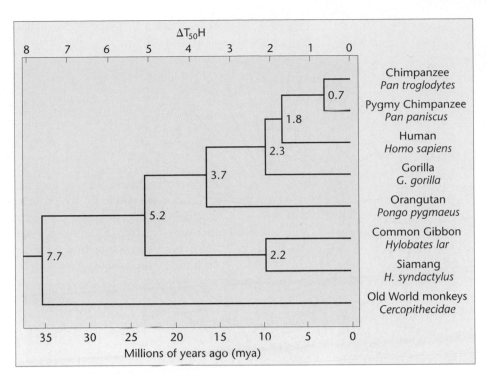

FIGURE 23–21 Phylogenetic sequence of hominoid primates and Old World monkeys as estimated by DNA hybridization. The numbers at the branch points are $DT_{50}H$ measurements. The evolutionary branch points are dated using the fossil record and nucleotide divergence.

part of the hemagglutinin gene from flu viruses that had been isolated at different times over a 20-year period. They calculated the number of nucleotide differences among the different viruses and constructed an evolutionary tree [Figure 23–20(b)]. Most strains have gone extinct; they have no descendents among the more recently isolated viruses. Fitch's team then plotted the number of nucleotide substitutions between the first isolate and each subsequent isolate against the year in which the strain was isolated [Figure 23–20(a)]. The points all fall very close to a straight line, indicating that nucleotide substitutions in this gene have accumulated at a steady rate. The hemagglutinin gene thus serves as a molecular clock. Molecular clocks are used to compare the sequences of new flu viruses as they appear each year, and estimate the time that has passed since each diverged from their common ancestor.

Molecular clocks must be carefully calibrated and used with caution. For example, Fitch's data indicate that strains of influenza A that jump from birds to humans have evolved much more rapidly than strains that remain in birds. Thus, a molecular clock calibrated from human strains of the virus would be highly misleading if applied to bird strains.

Figure 23–21 shows a phylogeny based on Sibley's data for genetic differences among humans and apes. Using a molecular clock based on the fossil record, this information is placed on a time scale that estimates ages for the common ancestors of each species cluster. Sibley's tree suggests that the most recent common ancestor of humans and chimps lived between 5 and 10 mya.

23.7 Using Evolutionary History

Evolutionary analysis addresses a wide range of questions, some of which relate to evolution directly, while others deal with contemporary issues and even criminal activity. In the

following sections, we consider several examples of how evolutionary trees can be used.

Transmission of HIV

In late 1986, a Florida dentist tested positive for HIV. Several months later he was diagnosed with AIDS. He continued to practice dentistry for two more years, until one of his patients, a young woman with no obvious risk factors, discovered that she was infected with HIV. When the dentist publicly urged his other patients to have themselves tested, several more were found to be HIV-positive. Did this dentist transmit HIV to his patients, or did the patients become infected by some other means?

At first glance, this situation appears to be a long way from a discussion of evolution; however, the movement of a virus from one individual to another is not unlike the founding of a new island population by a small number of migrants. Gene flow between the ancestral population and the new population is nonexistent. The populations are free to diverge, as a result of genetic drift, adaptation to different environments, or both.

If the dentist passed his HIV infection to his patients, then the HIV strains isolated from the patients should be more closely related to each other and to the dentist's strain than to strains from other individuals living in the same area. Chin-Yih Ou and colleagues sequenced portions of the gene for the HIV envelope protein from viruses collected from the dentist, 10 of his patients, and several other HIV-infected individuals from the area who served as local controls.

An evolutionary tree for the HIV isolates, produced from this sequence data, appears in Figure 23–22 on page 488 (this tree is circular to fit it more easily on the page). The viruses isolated from the study participants are at the tips of the branches around the outside of the tree; the inferred common ancestors are toward the center. The HIV strains taken from

FIGURE 23–22 A phylogeny of HIV strains taken from a dentist, his patients, and several local controls (LC). The group of viral strains in the shaded portion of the phylogeny are all derived from a common ancestral strain. *(From Hillis (1998), Current Biology 7:R129–R131.)*

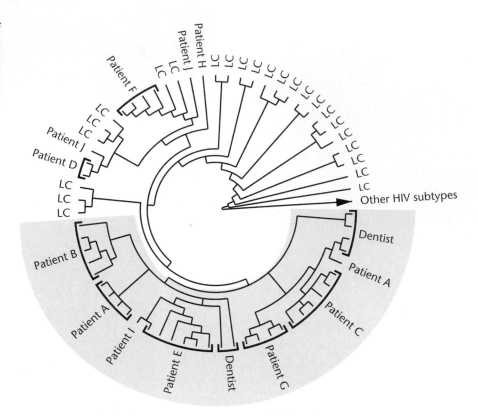

patients A, B, C, E, G, and I all share a more recent common ancestor with each other and the dentist's strain than with HIV from other local individuals. The evolutionary relationship among these HIV strains indicates that the dentist did, indeed, transmit his infection to these patients. In contrast, the HIV strains from patients D, F, H, and J are all more closely related to strains from local controls (LC) than they are to the dentist's strain. These patients appear to have gotten their infection from someone other than the dentist. Patient J, in fact, seems to have acquired his infection from two different sources.

Neanderthals and Modern Humans

Paleontological evidence indicates that the Neanderthals, *Homo neanderthalensis*, lived in Europe and Western Asia from some 300,000–30,000 years ago. For at least 30,000 years, the Neanderthals coexisted with anatomically modern humans (*H. sapiens*) in several areas. Several questions about Neanderthals and modern humans remain unresolved. (1) Can Neanderthals be regarded as direct ancestors to modern humans? (2) Did Neanderthals and *H. sapiens* interbreed, so that the descendants of Neanderthals are alive today? Or (3), did the Neanderthals die off and become extinct?

To resolve these questions, several groups have focused their attention on the recovery and analysis of DNA extracted from Neanderthal bones. In 1997, Svante Pääbo, Matthias Krings, and their colleagues extracted fragments of mitochondrial DNA from a Neanderthal skeleton found in Feldhofer cave near Düsseldorf, Germany. After confirming that they had indeed isolated Neanderthal gene fragments, the researchers placed the Neanderthal sequences on a phylogeny with over 2000 modern humans [Figure 23–23 (a)]. Based on this analy-

sis, Neanderthals appear to be a distant relative of modern humans. Using a molecular clock calibrated with chimpanzees and humans, the researchers calculated that the last common ancestor between Neanderthals and modern humans lived roughly 600,000 years ago, four times as long ago as the last common ancestor of all modern humans. However, considerable caution is required when drawing conclusions based on a single Neanderthal specimen.

More recently, Igor Ovchinnikov and his colleagues analyzed mitochondrial DNA recovered from Neanderthal remains discovered in Mezmaiskaya cave in the Caucasus Mountains east of the Black Sea. Although the two Neanderthal sequences are from individuals from different geographic regions more than 1000 miles apart, they vary only by about 3.5 percent, indicating they derive from a single gene pool. Further, the amount of variation between the two Neanderthal sequences is comparable to that seen among modern humans. Phylogenetic analysis places the two Neanderthals in a group that is distinct from modern humans [Figure 23–23 (b)]; the conclusion here is that although Neanderthals and humans have a common ancestor, the Neanderthals were a separate hominid line, and did not contribute mitochondrial genes to *H. sapiens*. Taken together, these two studies suggest that when the Neanderthals disappeared, their lineage went extinct with them.

The Origin of Mitochondria

Mitochondria are cellular organelles that contain their own genome in the form of a circular DNA molecule. In humans, the mitochondrial genome encodes 13 proteins required for oxidative phosphorylation and ATP production, 22 tRNA mol-

(a)

(b)

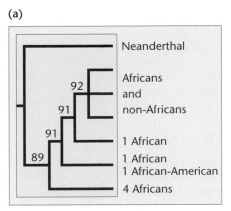

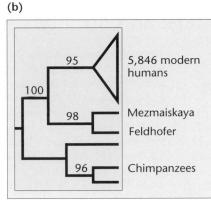

FIGURE 23–23 (a) A phylogeny estimated from mitochondrial DNA sequences of one Neanderthal and over 2000 modern humans. *(From M. Krings, A. Stone, R.W. Schmitz, H. Krainitzki, M. Stoneking, and S. Pääbo. 1997. ©1997 Cell Press. Cell 90:19–30. Fig. 7A, p. 26.)* (b) A phylogeny of the Feldhofer and Mezmaiskaya Neanderthal specimens compared with over 5,000 modern humans. *(From Ovchinnikov, I. V., et al., 2000. Molecular analysis of Neanderthal DNA from the northern Caucasus. Nature 404:490–493.)*

ecules and 2 rDNA genes. The organization and function of the mitochondrial genome bears many similarities to that of bacterial genomes. Therefore, mitochondria are thought to derive from bacteria, that is, they are descended from free-living prokaryotic organisms that became intracellular symbionts in nucleated host cells. Once incorporated into a nucleated cell, the genome of the mitochondrion became smaller over time, and the reduced genome we now see in mitochondria is thought to be the product of gene loss and gene transfer to the nucleus of the host cell. However, many questions about the origin and evolution of mitochondria remain. Among them is one relating to the bacterial origins of mitochondria: Which group of bacteria are they descended from?

Researchers investigating this question have used information derived from nucleotide sequencing of the mitochondrial genes for small-subunit ribosomal RNAs (SSU rRNA). These sequences are under strong functional constraint, and so are highly conserved. Phylogenetic analysis of these sequences identified a group of bacteria known as the α-proteobacteria (purple bacteria) as the closest living relatives of mitochondria. More recent work has divided the α-proteobacteria into two subdivisions, one of which is called the rickettsial subdivision.

Based on the SSU phylogeny, the rickettsia have been identified as the bacteria most closely related to mitochondria. Interestingly, rickettsia, like mitochondria, live only inside eukaryotic cells.

Unlike mitochondria, however, rickettsia are disease-causing parasites, responsible for typhus, one of the most serious diseases in the history of our species. Genomic sequencing of one rickettsial species, *Rickettsia prowazekii*, provides additional insight into the relationship between these bacteria and mitochondria. Using the genome sequence information for genes involved in ATP synthesis derived from *R. prowazekii*, other bacteria, and mitochondria from several sources, Siv Andersson and colleagues constructed a phylogenetic tree (Figure 23–24). Their tree indicates a close evolutionary relationship between *R. prowazekii* and mitochondria. Evidence from SSU analysis results in a similar tree. Thus, two lines of evidence, from SSUs and proteins involved in ATP synthesis, provide strong evidence that the mitochondria we carry in our cells derive from an ancient ancestor of the rickettsia.

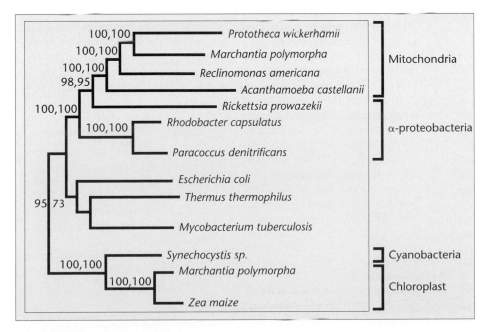

FIGURE 23–24 The evolutionary relationships of mitochondria, α-proteobacteria, cyanobacteria, and chloroplasts. Note that although *R. prowazekii* is an α-proteobacterium, it is more closely related to mitochondria than to other α-proteobacteria. *(From Anderssen, S. G. E., et al., 1998. The genome sequence of* Rickettsia prowazekii *and the origin of mitochondria. Nature 396:133–143.)*

Genetics, Technology, and Society

What Can We Learn from the Failure of the Eugenics Movement?

The eugenics movement had its origins in the ideas of the English scientist Francis Galton, who became convinced from his study of the appearance of geniuses within families (including his own) that intelligence is inherited. Galton concluded in his 1869 book *Hereditary Genius* that it would be "quite practicable to produce a highly-gifted race of men by judicious marriages during several consecutive generations." The term eugenics, coined by Galton in 1883, refers to the improvement of the human species by such selective mating. Once Mendel's principles were rediscovered in 1900, the *eugenics* movement flourished.

The eugenicists believed that a wide range of human attributes were inherited as Mendelian traits, including many aspects of behavior, intelligence, and moral character. Their overriding concern was that the presumed genetically "feeble-minded" and immoral in the population were reproducing faster than the genetically superior, and that this differential birthrate would result in the progressive deterioration of the intellectual capacity and moral fiber of the human race. Several remedies were proposed. *Positive eugenics* called for the encouragement of especially "fit" parents to have more children. More central to the goals of the eugenicists, however, was the *negative eugenics* approach aimed at discouraging the reproduction of the genetically inferior or, better yet, eliminating it altogether.

In the United States, the eugenics movement enjoyed wide popular support for a time and had a significant impact on public policy. Partially at the urging of prominent eugenicists, 30 states passed laws compelling the sterilization of criminals, epileptics, and inmates in mental institutions; most states enacted laws invalidating marriages between the "feebleminded" and others considered eugenically unfit. The crowning legislative achievement of the eugenics movement, however, was the passage of the Immigration Restriction Act of 1924, which severely limited the entry of immigrants from eastern and southern Europe due to their perceived mental inferiority.

Throughout the first two decades of the century, most geneticists passively accepted the views of eugenicists, but by the 1930s critics recognized that the goals of the eugenics movement were determined more by racism, class prejudice, and anti-immigrant sentiment than by sound genetics. Increasingly, prominent geneticists began to speak out against the eugenics movement, among them William Castle, Thomas Hunt Morgan, and Hermann Muller. When the horrific extremes to which the Nazis took eugenics became known, a strong reaction developed that all but ended the eugenics movement.

Paradoxically, the eugenics movement arose at the same time basic Mendelian principles were being developed, principles that eventually undermined the theoretical foundation of eugenics. Today, any student who has completed an introductory course in genetics should be able to identify several fundamental mistakes the eugenicists made:

1. They assumed that complex human traits such as intelligence and personality were strictly inherited, completely disregarding any environmental contribution to the phenotype. Their reasoning was that because certain traits ran in families, they must be genetically determined.

2. They assumed that these complex traits were determined by single genes with dominant and recessive alleles. This belief persisted despite research showing that multiple gene pairs contribute to many phenotypes.

3. They assumed that a single ideal genotype existed in humans. Presumably, such a genotype would be highly homozygous in order to be sustained. This precept runs counter to current evidence suggesting that a high level of homozygosity is often deleterious, supporting the superiority of the heterozygote.

4. They assumed that the frequency of recessively inherited defects in the population could be significantly lowered by preventing homozygotes from reproducing. In fact, for recessive traits that are relatively rare, most of the recessive alleles in the population are carried by asymptomatic heterozygotes who are spared from such selection. Negative eugenic practices, no matter how harsh, are relatively ineffective at eliminating such traits.

5. They assumed that those deemed genetically unfit in the population were out-reproducing those thought to be genetically fit. This is the exact reverse of the Darwinian concept of fitness, which equates reproductive success with fitness. (Galton should have understood this, being Darwin's first cousin!)

More than seven decades have passed since the eugenics movement was in full bloom. We now have a much more sophisticated understanding of genetics, as well as a greater awareness of its potential misuses. But the application of current genetic technologies makes possible a "new eugenics" of a scope and power that Francis Galton could not have imagined. In particular, prenatal genetic screening and *in vitro* fertilization enable the selection of children according to their genotype, a power that will dramatically increase as more and more genes are associated with inherited diseases and perhaps even behaviors.

As we move into this new genetic age, we must not forget the mistakes made by the early eugenicists. We must remember that phenotype is a complex interaction between the genotype and the environment, and not lapse into a new hereditarianism that treats a person as only a collection of genes. We must keep in mind that many genes may contribute to a particular phenotype, whether a disease or a behavior, and that the alleles of these genes may interact in unpredictable ways. We must not fall prey to the assumption that there is an ideal genotype. The success of all populations in nature is correlated with genetic diversity. And most of all, we must not use genetic information to advance ideological goals. We may find that there is a fine line between the legitimate uses of genetic technologies, such as having healthy children, and other eugenic practices. It will be up to us to decide exactly where the line falls.

References

Allen, G. E., Jacoby, R., and Glauberman, N. (eds.). 1995. Eugenics and American social history, 1880-1950. *Genome* 31:885-889.

Hartl, D. L. 1988. A primer of population genetics. 2nd ed. Sunderland, MA: Sinauer.

Kevles, D. J. 1985. In the name of eugenics: Genetics and the uses of human heredity. Berkeley: Univ. of California Press.

Chapter Summary

1. Today's organisms are the products of an evolutionary history that includes the transformation, splitting, and divergence of lineages. Alfred Russel Wallace and Charles Darwin formulated the theory of natural selection, which provides a mechanism for the transformation of lineages. The genetic basis of evolution and the role of natural selection in changing allele frequencies were discovered in the twentieth century.

2. When geneticists began studying the genetic structure of populations, they discovered that most populations harbor considerable genetic diversity. This diversity is apparent in electrophoretic studies of proteins and at the level of DNA sequences. Whether the genetic diversity of populations is maintained primarily by mutation plus genetic drift or by natural selection is a matter of some debate.

3. The geographic ranges of most species encompass a degree of environmental diversity. As a result of both adaptation to different environments and genetic drift, different populations within a species may have different alleles and/or allele frequencies at many loci.

4. Gene flow among populations tends to homogenize their genetic composition. When gene flow is reduced, genetic drift and adaptation to different environments can cause populations to diverge. Eventually populations may become so different that the individuals in one population either will not or cannot mate with the individuals in the other. At this point, the divergent populations have become different species.

5. Because speciation is associated with genetic divergence, we can use the genetic differences among species to infer their evolutionary history. By comparing amino acid or nucleotide sequences, we can determine the genetic distances among species. The genetic distances are used to reconstruct evolutionary trees. The simplest methods for reconstructing phylogenies assume that the least divergent species are one another's closest relatives.

6. The reconstruction of evolutionary trees is a key technique in answering a surprising diversity of interesting questions. Examples include tracing the transmission route in disease transmission, deciphering recent events in human evolution, and determining evolutionary relationships among all organisms.

Key Terms

anagenesis, 473
cladogenesis, 473
cytochrome c, 483
maximum likelihood, 486
minimal mutational distance, 484
molecular clock, 486

neutral theory, 476
niche, 473
parsimony, 486
phyletic evolution, 473
phylogeny, 473
postzygotic isolating mechanisms, 479

prezygotic isolating mechanisms, 479
reproductive isolating mechanism, 478
selectionist, 476
species, 473
stasis, 473
UPGMA, 485

Insights and Solutions

1. Sequence analysis of DNA can be accomplished by a number of techniques. Protein sequencing, on the other hand, is made more complex by the fact that 20 different subunits need to be unambiguously identified and enumerated, rather than the four nucleotides of DNA. Because of their unique properties, the N-terminal and C-terminal amino acids in a protein are easy to identify, but the array in between offer a difficult challenge, since many proteins contain hundreds of amino acids. So how is protein sequencing accomplished?

Solution: The strategy for protein sequencing is the same as for DNA sequencing—divide and conquer. To accomplish this, specific enzymes are used that reproducibly cleave proteins between certain amino acids. These different enzymes produce overlapping fragments. Each fragment is isolated and its amino acid sequence is determined by chemical means. Sequences from overlapping fragments are then assembled to give the sequence for the entire protein. Alternatively, researchers first sequence the gene that encodes the protein, then infer the protein's amino acid sequence from the genetic code.

2. A single plant twice the size of others in the same population suddenly appears. Normally, plants of this species reproduce by self-fertilization and by cross-fertilization. Is this new giant plant simply a variant, or is it a new species? How would you determine this?

Solution: One of the most widespread mechanisms of speciation in higher plants is polyploidy, the multiplication of entire chromosome sets. The result of polyploidy is usually a larger plant with larger flowers and seeds. There are two ways to test this new variant to determine whether it is a new species. First, the giant plant is crossed with a normal-size plant to see whether it produces viable, fertile offspring. If it does not, the two different types of plants are probably reproductively isolated. Second, the giant plant is cytogenetically screened to examine its chromosome complement. If it has twice the number of its normal-size neighbors, it is a tetraploid that may have arisen spontaneously. If the chromosome number differs by a factor of 2 and the new plant is reproductively isolated from its normal-size neighbors, it is a new species.

Problems and Discussion Questions

1. Discuss the rationale behind the statement that inversions in chromosome 3 of *Drosophila pseudoobscura* represent genetic variation.

2. Describe how populations with substantial genetic differences can form. What is the role of natural selection?

3. What types of nucleotide substitutions will not be detected by electrophoretic studies of a gene's protein product?

4. Shown here are two homologous lengths of the alpha and beta chains of human hemoglobin. Consult the genetic code dictionary (Figure 12–7) and determine how many amino acid substitutions may have occurred as a result of a single nucleotide substitution. For any that cannot occur as the result of a single change, determine the minimal mutational distance.

Alpha:	Ala	Val	Ala	His	Val	Asp	Asp	Met	Pro
Beta:	Gly	Leu	Ala	His	Leu	Asp	Asn	Leu	Lys

5. Determine the minimal mutational distances between these amino acid sequences of cytochrome c from various organisms. Compare the distance between humans and each organism.

Human:	Lys	Glu	Glu	Arg	Ala	Asp
Horse:	Lys	Thr	Glu	Arg	Glu	Asp
Pig:	Lys	Gly	Glu	Arg	Glu	Asp
Dog:	Thr	Gly	Glu	Arg	Glu	Asp
Chicken:	Lys	Ser	Glu	Arg	Val	Asp
Bullfrog:	Lys	Gly	Glu	Arg	Glu	Asp
Fungus:	Ala	Lys	Asp	Arg	Asn	Asp

6. The data below are taken from a paper by Heui-Soo Kim and Osamu Takenaka. They are short DNA sequences from five species: human, chimpanzee, gorilla, orangutan, and baboon. The sequences are each 50 bp long. They represent a short piece of a gene for testis-specific protein Y, which is located on the Y chromosome. The complete human sequence is given. The sequences for the other four species are shown only in the spots where they differ from the human sequence. Calculate the genetic difference between each pair of species. For example, chimps and gorillas differ in 2 out of 50 bases, or 4 percent. Thus, the genetic difference between chimps and gorillas is 0.04. Then use UPGMA to reconstruct the phylogeny for these five species. Is your phylogeny consistent with those shown in Figures 23–19 and 23–21?

7. The genetic difference between two species of *Drosophila*, *D. heteroneura* and *D. sylvestris*, as measured by nucleotide diversity, is about 1.8 percent. The difference between chimpanzees (*Pan troglodytes*) and humans (*Homo sapiens*) is about the same, yet the latter species are classified in different genera. In your opinion, is this valid? If so, why; if not, why not?

8. As an extension of Problem 7, consider the following: In sorting out the complex taxonomic relationships among birds, species with DT$_{50}$H values of 4.0 are placed in the same genus, even by traditional taxonomy based on morphology. Using the data in Figure 23–21, construct a classification that obeys this rule, using appropriate genus names (*Pongo*, *Pan*, *Homo*), or creating new ones.

9. The use of nucleotide-sequence data to measure genetic variability is complicated by the fact that the genes of higher eukaryotes are complex in organization and contain 5′ and 3′ flanking regions as well as introns. Slightom and colleagues have compared the nucleotide sequence of two cloned alleles of the γ-globin gene from a single individual and found a variation of 1 percent. Those differences include 13 substitutions of one nucleotide for another, and three short DNA segments that have been inserted in one allele or deleted in the other. None of the changes take place in the gene's exons (coding regions). Why do you think this is so, and should it change our concept of genetic variation?

10. Discuss the arguments supporting the neutralist hypothesis. What counterarguments are proposed by the selectionists?

11. Of what value to our understanding of genetic variation and evolution is the debate concerning the neutralist hypothesis? What genetic changes take place during speciation?

Human	A G A G G T T T T T C A G T G A A T G A A G C T A T T T T T A A G G G A G T G T G A T T G C T G C C
Chimpanzee	C
Gorilla	T C C
Orangutan	T G T C C C C
Baboon	C C G G T C G G C C C G

Selected Readings

Andersson, S. G. E., et al. 1998. The genome sequence of *Rickettsia prowazekii*. *Nature* 396:133–43.

Anderson, W., et al. 1975. Genetics of natural populations: XLII. Three decades of genetic change in *Drosophila pseudoobscura*. *Evolution* 29:24–36.

Armour, J. A., et al. 1996. Minisatellite diversity supports a recent African origin for modern humans. *Nat. Genet.* 13:154–60.

Avise, J. C. 1990. Flocks of African fishes. *Nature* 347:512–13.

Ayala, F. J., 1984. Molecular polymorphism: How much is there, and why is there so much? *Dev. Genet.* 4:379–91.

Bradshaw, H. D., Jr. et al. 1998. Quantitative trait loci affecting differences in floral morphology between two species of monkeyflower (*Mimulus*). *Genetics* 149:367–82.

Bowcock, A. M., et al. 1994. High resolution of human evolutionary trees with polymorphic microsatellites. *Nature* 368:455–57.

Carson, H. 1970. Chromosome tracers of the origin of species. *Science* 168:1414–18.

————. 1975. The genetics of speciation at the diploid level. *Am. Natural.* 109:83–92.

Collard, M., and Wood, B. 2000. How reliable are human phylogenetic hypotheses? *Proc. Nat. Acad. Sci.* 97:5003–06.

Coyne, J. A. 1992. Genetics and speciation. *Nature* 355:511–15.

Diamond, J. 1992. *The third chimpanzee: The evolution and future of the human animal.* New York: HarperCollins.

Dobzhansky, T. 1947. Adaptive changes induced by natural selection in wild populations of *Drosophila*. *Evolution* 1:1–16.

————. 1948. Genetics of natural populations, XVI. Altitudinal and seasonal changes produced by natural selection in certain populations of *Drosophila pseudoobscura* and *Drosophila persimilis*. *Genetics* 33:158–76.

————. 1955. *Genetics of the evolutionary process.* New York: Columbia University Press.

Dobzhansky, T. et al. 1966. Genetics of natural populations: XXXVIII. Continuity and change in populations of *Drosophila pseudoobscura* in western United States. *Evolution* 20:418–27.

Eldredge, N. 1985. *Time frames: The evolution of punctuated equilibria.* Princeton, NJ: Princeton University Press.

Fitch, W. M. 1973. Aspects of molecular evolution. *Annu. Rev. Genet.* 7:343–80.

Fitch, W. M., and Margoliash, E. 1967. Construction of phylogenetic trees. *Science* 155:279–84.

————. 1970. The usefulness of amino acid and nucleotide sequences in evolutionary studies. *Evol. Biol.* 4:67–109.

Freeman, S., and Herron, J. C. 2001. *Evolutionary analysis.* Upper Saddle River, NJ: Prentice Hall.

Fryer, G. 1997. Biological implications of a suggested Late Pleistocene desiccation of Lake Victoria. *Hydrobiologia* 354:177–82.

Gillespie, J. H. 1992. *The causes of molecular evolution.* New York: Oxford University Press.

Gray, M. W., et al. 1999. Mitochondrial evolution. *Science* 283:1476–81.

Gould, S. J. 1982. Darwinism and the expansion of evolutionary theory. *Science* 216:380–87.

Hardison, R. 1999. The evolution of hemoglobin. *Amer. Scient.* 87:126–37.

Hillis, D. M. 1998. Phylogenetic analysis. *Curr. Biol.* 7:R129–31.

Hillis, D. M., et al. 1992. Experimental phylogenetics: Generation of a known phylogeny. *Science* 255:589–92.

Hillis, D. M., Huelsenbeck, J. P., and Cunningham, C. W. 1994. Application and accuracy of molecular phylogenies. *Science* 264:671–77.

Höss, M. 2000. Neanderthal population genetics. *Nature* 404:453–54.

Hunt, J., et al. 1981. Evolution distance in Hawaiian *Drosophila*. *J. Mol. Evol.* 17:361–67.

Johnson, T. C., et al. 1996. Late pleistocene desiccation of Lake Victoria and rapid evolution of cichlid fishes. *Science* 273: 1091–93.

Kim, H.-S., and Takenaka, O. 1996. A comparison of TSPY genes from Y-chromosomal DNA of the great apes and humans: Sequence, evolution, and phylogeny. *Am. J. Phys. Anthropol.* 100:301–9.

Kimura, M. 1979a. Model of effectively neutral mutations in which selective constraint is incorporated. *Proc. Natl. Acad. Sci. USA* 76:3440–44.

————. 1979b. The neutral theory of molecular evolution. *Sci. Am.* (Nov.) 241:98–126.

————. 1989. The neutral theory of molecular evolution and the world view of the neutralists. *Genome* 31:24–31.

King, M. C., and Wilson, A. C. 1975. Evolution at two levels: Molecular similarities and biological differences between humans and chimpanzees. *Science* 188:107–16.

Knowlton, N., et al. 1993. Divergence in proteins, mitochondrial DNA, and reproductive compatibility across the Isthmus of Panama. *Science* 260: 1629–32.

Kreitman, M. 1983. Nucleotide polymorphism at the alcohol dehydrogenase locus of *Drosophila melanogaster*. *Nature* 304:412–17.

Krings, M., et al. 1997. Neandertal DNA sequences and the origin of modern humans. *Cell* 90:19–30.

Lewontin, R. C., and Hubby, J. L. 1966. A molecular approach to the study of genic heterozygosity in natural populations: II. Amount of variation and degree of heterozygosity in natural populations of *Drosophila pseudoobscura*. *Genetics* 54:595–609.

Meyer, A., Kocher, T. D., Basasibwaki, P., and Wilson, A. C. 1990. Monophyletic origin of Lake Victoria cichlid fishes suggested by mitochondrial DNA sequences. *Nature* 347:550–53.

Müller, M., and Martin, W. 1999. The genome of *Rickettsia prowazekii* and some thoughts on the origin of mitochondria and hydrogenosomes. *Bioessays* 21:377–81.

Ou, C.-Y., et al. 1992. Molecular epidemiology of HIV transmission in a dental practice. *Science* 256:1165–71.

Ovchinnikov, I. V., et al. 2000. Molecular analysis of Neanderthal DNA from the northern Caucasus. *Nature* 404:490–493.

Pääbo, S. 1993. Ancient DNA. *Sci. Am.* (Nov.) 269:86–92.

Powers, D. A., and Schulte, P. M. 1998. Evolutionary adaptations of gene structure and expression in natural populations in relation to a changing environment: A multidisciplinary approach to address the million-year saga of a small fish. *J. Exper. Zool.* 282:71–94.

Schemske, D. W., and Bradshaw, H. D., Jr. 1999. Pollinator preference and the evolution of floral traits in monkeyflowers. *Proc. Nat. Acad. Sci.* 96:11910–15.

Scholz, M., et al. 2000. Genomic differentiation of Neanderthals and anatomically modern man allows a fossil-DNA-based classification of morphologically indistinguishable hominid bones. *Am. J. Hum. Genet.* 66:1927–32.

Sibley, C., and Ahlquist, J. 1984. The phylogeny of the hominoid primates, as indicated by DNA–DNA hybridization. *J. Mol. Evol.* 20:2–15.

Sibley, C. G., Comstock, J. A., and Ahlquist, J. E. 1990. DNA evidence of hominoid phylogeny: A reanalysis of the data. *J. Mol. Evol.* 30:202–36.

Smithies, O., et al. 1981. Co-evolution and control of globin genes. In *Levels of genetic control in development*, ed. S. Subtelny and U. Abbot, pp. 185–200. New York: Alan R. Liss.

Stiassny, M. L. J., and Meyer, A. 1999. Cichlids of the Rift Lakes. *Sci. Am.* (Feb.) 280:64–69.

Stoneking, M. 1995. Ancient DNA: How do you know when you have it and what can you do with it? *Am. J. Hum. Genet.* 57:1259–62.

Takahashi, K., et al. 1998. A novel family of short interspersed repetitive elements (SINEs) from cichlids: The pattern of insertion of SINEs at orthologous loci support the proposed monophyly of four major groups of cichlid fishes in Lake Tanganyika. *Mol. Biol. Evol.* 15:391–407.

Templeton, A. R. 1985. Phylogeny of the hominoid primates: A statistical analysis of the DNA–RNA hybridization data. *Mol. Biol. Evol.* 2:420–33.

Thorne, A. G., and Wolpoff, M. H. 1992. The multiregional evolution of humans. *Sci. Am.* (Apr.) 266:76–83.

Tishkoff, S. A., et al. 1996. Global patterns of linkage disequilibrium at the CD4 locus and modern human origins. *Science* 271:1380–87.

Tsui, L.-C. 1992. The spectrum of cystic fibrosis mutations. *Trends in Genet.* 8:392–98.

Val, F. C. 1977. Genetic analysis of the morphological differences between two interfertile species of Hawaiian *Drosophila*. *Evolution* 31:611–29.

Wilson, A. C., and Cann R. L. 1992. The recent African genesis of humans. *Sci. Am.* (Apr.) 266:68–73.

Yang, D., et al. 1985. Mitochondrial origins. *Proc. Nat. Acad. Sci. USA* 82: 4443–47.

Humpback whales, the "poster children" of endangered species. *(James Watt/Animals Animals)*

24

Conservation Genetics

CHAPTER CONCEPTS

Accelerating human population growth has had a tremendous impact on nonhuman species on our planet. Direct or indirect influences have resulted in the loss of substantial genetic diversity. Diminished population size in threatened or endangered species also limits genetic diversity, which is eroded through drift, inbreeding, or reduced gene flow. This affects the capacity of a population to adapt to a changing environment and may reduce fitness through increasing homozygosity for deleterious alleles. Conservation geneticists work to assess genetic diversity and to maintain population numbers for long-term species survival.

As we enter the twenty-first century, the diversity of life on the earth is under increasing pressure from the direct and indirect effects of explosive human population growth. Approximately 10 million *Homo sapiens* lived on the planet 10,000 years ago. This number grew to 100 million 2,000 years ago and to 2.5 billion by 1950. Within the span of a single lifetime, the world's human population more than doubled to 5.5 billion in 1993, and is projected to reach as high as 19 billion by 2100 (Figure 24–1). The effect of accelerating human population growth on other species has been dramatic: Data from the World Conservation Union (IUCN) show that globally 25 percent of all mammal species, 11 percent of birds, 20 percent of reptiles, 25 percent of amphibians and 34 percent of fish species are vulnerable or endangered. The situation for plants is no better: based on IUCN surveys 12 percent of all vascular plant species worldwide— some 34,000 species—are threatened. Not just wild species are at risk. Genetic diversity in domesticated plants and animals is also being lost as many traditional crop varieties and livestock breeds disappear. The Food and Agriculture Organization (FAO) estimates that since 1900 75 percent of the genetic diversity in agricultural crops has been lost, and that of approximately 5,000 different breeds of domesticated farm animals worldwide, one-third are at risk of being lost.

Why should we be concerned about losing **biodiversity**, the biological variation represented by these different plants and animals? As the fossil record shows, unrelated to the human influences, many different plants and animals that once inhabited the planet have become extinct over millions of years, and others have taken their place. Biologists are concerned, however, at the accelerated rate of species extinctions we are witnessing today, all of which can be ascribed to direct or indirect human impacts. Deliberate hunting or harvesting of plants and animals by humans, habitat destruction through human development activities, and the indirect effects of global climate change are making it increasingly difficult for many species to survive. Some biologists fear that ecosystems, the complex webs of interdependence of different plants and animals found together in the same environment, may collapse if key sustaining species are lost. This portends possible consequences for our own long-term survival. Other scientists have pointed out that we may be losing the unknown economic potential or other benefits of unexploited plants and animals if we allow them to go extinct. A much-publicized recent example is the Pacific yew (*Taxus brevifolia*), a rare tree found in coastal forests of the western United States. This slow-growing species was ignored due to its small size and poor timber qualities and was frequently destroyed during logging operations as a "trash tree." In 1991, it was found to be a source of taxol, a compound now widely used as a powerful drug for treating cancer. Quite apart from practical benefits, scientists and nonscientists have made the case that human life will be diminished both spiritually and aesthetically if we do not strive to maintain the fascinating and often beautiful variety of living organisms with which we share the planet.

Conservation biologists work to understand and maintain biodiversity, studying the factors that lead to species decline and the ways in which species can be preserved. The new field of **conservation genetics** has emerged in the last 20 years as scientists began to recognize that genetics will be an important tool in maintaining and restoring population viability. The applications of genetics to conservation biology are multifaceted; in this chapter we explore only a few of them. Underlying the increasingly important role of genetics in conservation biology, however, is the recognition that biodiversity depends on **genetic diversity**, and that maintaining biodiversity in the long term is unlikely if genetic diversity is lost.

24.1 Genetic Diversity

While biodiversity encompasses the variation represented by all existing species of plants and animals on this planet at any given time, genetic diversity is not as easy to define and study.

Genetic diversity can be considered on two levels: **interspecific diversity** and **intraspecific diversity**. Diversity between species (interspecific diversity) is reflected in the number of different plant and animal species present in an ecosystem. Some ecosystems have a very high level of species diversity, such as a tropical rainforest in which hundreds of different plant and animal species may be found within a few square meters [Figure 24–2(a)]). Other ecosystems, especially where plants and animals must adapt to a harsh environment, may have much lower levels of interspecific diversity [Figure 24–2(b)]. Lists of the different plant and animal species found in a particular environment are used to compile species inventories. Such inventories are used to identify diversity hotspots: geographic areas with especially high levels of interspecific diversity, where conservation efforts can be focused. Concerns of conservation biologists working at the ecosystem level are to prevent species being lost, or to restore species that were once part of the system but that are no longer present. The reestablishment of the grey wolf (*Canis lupus*) in Yellowstone

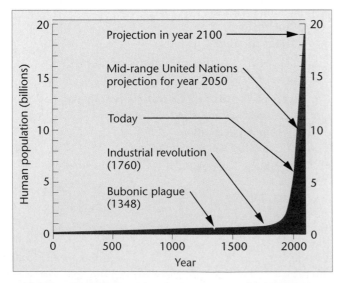

FIGURE 24–1 Growth in human population over the past 2000 years.

FIGURE 24–2 (a) Tropical rainforest: an ecosystem with high interspecific diversity. (b) coastal marsh in North Carolina: an ecosystem with low interspecific diversity. *[Photos: (a) David Austen/ Stone; (b) Sarah M. Ward]*

National Park, and the release of captive-bred California condors (*Gymnogyps californianus*) into the mountain ranges they previously occupied in the southwest United States are examples of current attempts by conservation biologists to restore missing species to their ecosystems.

Intraspecific diversity (diversity within a species) is reflected in the level of genetic variation occurring between individuals within a single population of a given species (intrapopulation diversity) or between different populations of the same species (interpopulation diversity). Genetic variation within populations can be measured as the frequency of individuals in the population that are heterozygous at a given locus, or as the number of different alleles at a locus that are present in the population gene pool. When DNA profiling techniques are used, the percentage of polymorphic loci—those that are represented by varying bands on the DNA profile in different individuals—can be calculated to indicate the extent of genetic diversity in a population. In outbreeding species, most intraspecific genetic diversity is found at the intrapopulation level. Significant interpopulation diversity can occur if populations are separated geographically and there is no migration or exchange of gametes between them. Predominantly inbreeding species such as self-fertilizing plants, on the other hand, tend to have greater levels of interpopulation than intrapopulation diversity, with a limited number of genotypes dominating an individual population but greater variation between different populations. Understanding the breeding system and the distribution of genetic variation in an endangered species is important to its conservation, not only to ensure continued production of offspring but also to determine the best strategy for maintaining intraspecific diversity. For example, would it be more effective to preserve a few large populations or many small distinct ones? This information can then be used to guide conservation or restoration efforts.

Loss of Genetic Diversity

Loss of genetic diversity in nondomesticated species is usually associated with a reduction in population size. This may be due to excessive hunting or harvesting; for example, bi-

ologists blame commercial overfishing for the collapse in the early 1990s of the deep-sea cod populations off the Newfoundland coast, once one of the most productive fisheries in the world. Habitat loss is also a major cause of population decline. As the global human population increases, more land is developed for housing and transport systems or is put into agricultural production, reducing or eliminating areas that were once home to wild plants and animals. The shrinking available habitat not only reduces populations of wild species, but often also isolates them from each other as individual populations become trapped in pockets of undeveloped land surrounded by areas taken over for agriculture, urban development, or other human uses. This process is known as **population fragmentation**. Where populations are no longer in contact with each other, gene flow via migration or gamete exchange between them ceases, and an important mechanism for maintaining genetic variation is lost. We discuss this in more detail later in this chapter.

Loss of genetic diversity in domesticated species is not usually the result of habitat loss or collapsing population numbers; there is little risk that cows or corn as species will become extinct any time soon. Reduction in diversity within domesticated species can be traced to changes in agricultural practice and consumer demand. Modern farming techniques have greatly increased production levels, but they have also led to greater genetic uniformity. As farmers switch to new crop varieties or improved livestock strains on a large scale, they abandon cultivation of many older local types, which may then disappear if efforts are not made to preserve them. For example, in 1900 more than 100 different kinds of potato were available in the United States. Today three-quarters of commercial potato production in this country depends on just nine varieties, with a single type, Russet Burbank, comprising 43 percent of the total acreage planted. Modern crop varieties or livestock strains may be better adapted to meet the demands of modern agricultural production, but older types often contain useful genes that can still play a vital role in survival functions, such as resistance to disease, cold, or drought. When the Russian wheat aphid

(*Diuraphis noxia*), an insect that causes serious damage to cereal crops, invaded the United States in 1986, plant breeders eventually found genes conferring resistance to this pest, not in modern American wheat lines, but in old traditional varieties from the former Soviet Union, which fortunately had been collected and preserved. The old Russian wheats were crossed with U.S. wheat plants to transfer the resistance genes, creating new commercial wheat varieties that resist Russian wheat aphid attack and saving U.S. wheat growers from crop losses in the millions of dollars.

Identifying Genetic Diversity

For many years population geneticists based estimates of intraspecific diversity on phenotypic differences between individuals, such as different colors of seeds or flowers, or variation in markings (Figure 24–3). More recently, techniques have been developed for analyzing intraspecific diversity with greater precision at the molecular level. The more traditional method used is **isozyme analysis**. Isozymes are multiple versions of a single enzyme occurring within a single species. Isozymes catalyze the same reaction, but differences in the polypeptide chains forming the bulk of the molecule result in various forms of the enzyme that differ in size or net electrical charge and can be separated by electrophoresis. The presence of isozyme variation in a population can be used as an indicator of genetic variation, as the different versions of the molecule are encoded by different alleles at a locus.

The use of **DNA profiles** has now become the dominant tool for detecting and assessing intrapopulation and interpopulation genetic variation between individuals drawn from a population. Nuclear, mitochondrial, and chloroplast DNA can all be analyzed to determine levels of genetic variation. We described earlier applications in human genetics of DNA fingerprinting techniques using either restriction enzymes or PCR (see Chapter 19). These techniques have been used with DNA from many different species that are of concern to conservation biologists. **AFLPs, amplified fragment length polymorphisms**, are detected by a recently developed DNA

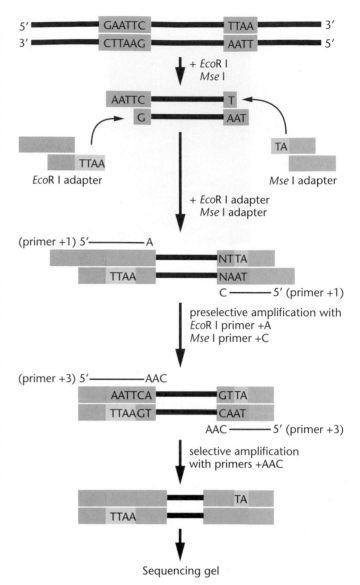

FIGURE 24–4 The procedure used in preparing AFLP profiling samples.

profiling technique, using both restriction enzymes and PCR (Figure 24–4). This method is especially powerful at detecting very small amounts of genetic diversity in a population and is therefore valuable for work with threatened and endangered species. AFLPs are generated by first cutting genomic DNA into fragments using two restriction enzymes, then attaching short single-stranded pieces of DNA to the cut ends of the fragments. Because the sequence of the attached DNA oligonucleotide is known, a subset of the genomic fragments can then be amplified via PCR by using primers that match the attached piece but are 1–3 bases longer. Only those fragments that have bases at each end complementary to the oligonucleotide extension will be amplified. AFLPs can thus produce different fingerprint patterns for individuals whose DNA sequence differs by as little as a single base pair.

AFLP profiling (Figure 24–5) has been used extensively to analyze genetic variation in crop species and some livestock,

FIGURE 24–3 Phenotypic variation in seed color and markings in common bean (*Phaseolus vulgaris*) reveals high levels of intraspecific diversity. (*Sarah M. Ward*)

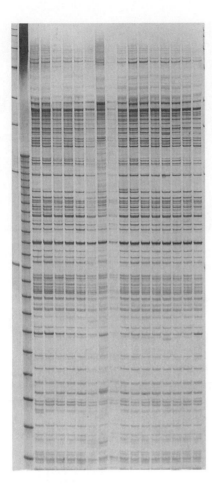

FIGURE 24–5 A representative AFLP gel used in DNA profiling, derived from analysis of jointed goat grass. *(Gel image courtesy of Todd A. Pester and Sarah M. Ward, Colorado State University)*

and is now being used to examine variation and guide conservation efforts in nondomesticated species. In a recent study, AFLP analysis was performed on the only remaining population of a critically endangered plant species (*Limonium cavanillesii*) growing on the Mediterranean coast of Spain. Researchers found very low levels of diversity, but did detect genetically distinct individual plants within the population from which seed could be collected. In another recent study, AFLP analysis of the southwestern willow flycatcher, an endangered migratory bird species that nests in the southwestern United States, found significant levels of genetic diversity within populations at different breeding sites but little interpopulation diversity. This indicated to scientists studying the flycatcher that migration of individual birds between breeding populations is important in maintaining diversity in the species, and conservation efforts should be directed to preserving nesting sites that allow this to continue.

We read in the Palo Verde story in Chapter 16 (see the Genetics, Technology, and Society essay on page 345) about the use of DNA fingerprinting of plants to help solve a murder. Forensic applications of DNA fingerprinting are also used in conservation biology to uncover illegal trade in endangered species that are protected by law. Since 1986, an international moratorium on commercial whaling has been in place, with only limited hunting of a few species

permitted. In 1994, scientists based in New Zealand and Hawaii used PCR to amplify mitochondrial DNA sequences extracted from meat on sale in markets in Japan, where whale meat is prized as a delicacy. The DNA fingerprints revealed several meat samples from humpback and fin whales, which are protected species. Two meat samples turned out not to be whale at all, but dolphin! This information assisted Japanese customs officials in tracking down sources of illegal whale meat imports. Unlawful trade in animal products also occurs in the United States. In a recent study by scientists at the University of Florida, amplification of mitochondrial DNA sequences from turtle meat sold in Louisiana and Florida revealed that one-quarter of the samples were actually alligator, not turtle at all. Most of the rest were from small freshwater turtles, indicating that populations of the larger freshwater turtle species were declining through overharvesting and in need of protection.

24.2 Population Size and Species Survival

Some species have never been numerous, especially those that are adapted to survive in unusual habitats; biologists refer to such species as *naturally rare*. "Newly rare" species, on the other hand, are those whose numbers are in decline due to pressures such as habitat loss. Populations of such species may not only be small in number, but also fragmented and isolated from other populations. Both decreased population size and increased population isolation have important genetic consequences, leading to an increased risk of loss of diversity.

How small must a population be before it is considered endangered? This varies somewhat with species, but populations can quickly become vulnerable to genetic problems that increase the risk of extinction. In general, a population of less than 100 individuals is considered extremely sensitive to these problems, including genetic drift, inbreeding, and reduction in gene flow. The effects of these problems on species survival are substantial, and we discuss them in more detail below. Studies of bighorn sheep, for example, have shown that populations of less than 50 are highly likely to go extinct within 50 years. Projections based on computer models show that for all species, populations of fewer than 10,000 are likely to be limited in adaptive genetic variation, and at least 100,000 individuals must be present if a population is to show long-term sustainability.

Determining the number of individuals a population must contain for it to have long-term sustainability is complicated by the fact that not all members of a population are equally likely to produce offspring: Some will be infertile, too young, or too old. The **effective population size** (N_e) is defined as the number of individuals in a population having an equal probability of contributing gametes to the next generation, and N_e is almost always smaller than the **absolute population size** (*N*). The effective population size can be calculated in different ways, depending on the factors that are preventing all individuals in a population from contributing equally to the next generation. In a sexually reproducing population that contains different numbers of males and females, for example, the effective population size is calculated as

$$N_e = \frac{4\,(N_m N_f)}{N_m + N_f}$$

where N_m is the number of males and N_f the number of females in the population.

A population of 100 males and 100 females would have an effective size of 4(100 × 100)/(100 + 100) = 200. In contrast, if there were 180 males and only 20 females, the effective population size would be 4(180 × 20)/(180 + 20) = 72.

Effective population size is also affected by fluctuations in absolute population size from one generation to the next. Here the effective population size is the harmonic mean of the numbers in each generation, so

$$N_e = \frac{1}{\dfrac{1}{t}\left(\dfrac{1}{N_1} + \dfrac{1}{N_2 + \cdots + N_t}\right)}$$

where t = the total number of generations being considered. For example, if a population went through a temporary reduction in size in generation 2, so that $N_1 = 100$, $N_2 = 10$, and $N_3 = 100$, then

$$N_e = \frac{1}{\dfrac{1}{3}\left(\dfrac{1}{100} + \dfrac{1}{10} + \dfrac{1}{100}\right)} = \frac{1}{0.04} = 25$$

In this case, although the mean actual number of individuals in the population over three generations was 70, the effective population size during that time was only 25. A severe temporary reduction in size such as this is known as a **population bottleneck**. Bottlenecks occur when a population or species is reduced to a few reproducing individuals whose offspring then increase in numbers over subsequent generations to reestablish the population. Although the number of individuals may be restored to healthier levels, genetic diversity in the newly expanded population is often severely reduced, as gametes from the handful of surviving individuals functioning as parents do not represent all the different allelic frequencies present in the original gene pool. Captive breeding programs, where a few surviving individuals from an endangered species are removed from the wild and their offspring raised in a protected environment to rebuild the population, inevitably create **population bottlenecks**. Bottlenecks also occur naturally when a small number of individuals from one population migrate to establish a new population elsewhere. When a new population derived from a small subset of individuals has significantly less genetic diversity than the original population, it is said to be exhibiting a **founder effect**. Reduced levels of genetic diversity due to founder effect can persist for many generations, as shown by studies of two species of Antarctic fur seal (*Arctocephalus gazella* and *Arctocephalus tropicalis*). Seal hunters in the eighteenth and nineteenth century severely reduced populations of these species, eliminating them from parts of their natural range in the Southern Ocean. Although Antarctic fur seal numbers have now rebounded, and the two species have recolonized much of their original habitat, mtDNA fingerprinting reveals

FIGURE 24–6 The cheetah (*Acinonyx jubatus*), a species with reduced genetic variation following population bottlenecks. (*Johnny Johnson/DRK Photo*)

that a founder effect can still be detected in *A. gazella*, with reduced genetic variation in current populations that are descended from a handful of surviving individuals.

The cheetah (*Acinonyx jubatus*) (Figure 24–6) is a another well-studied example of a species with reduced genetic variation as the result of at least one severe population bottleneck having occurred in its recent history. Isozyme studies of South African cheetah populations have shown levels of genetic variation less than 10 percent of those found in other mammals. The abnormal spermatozoa and poor reproductive rates commonly observed in cheetahs are thought to be linked to this lack of genetic diversity, although when and how the population bottleneck occurred in this species is still unclear.

24.3 Genetic Effects of Decreased Population Size

Small isolated populations such as those found in threatened and endangered species are especially vulnerable to genetic drift, inbreeding, and reduction in gene flow. These phenomena act on the gene pool in different ways, but ultimately have similar effects in that they can all further reduce genetic diversity and long-term species viability.

Genetic Drift

If the number of breeding individuals in a population is small, fewer gametes will form the next generation. The alleles carried by those gametes may not be a representative sample of all those present in the population; purely by chance, some alleles may be under-represented or not present at all. This will result in changes in allele frequency over time. This phenomenon, known as **genetic drift**, has been previously described in Chapter 22. A serious result of genetic drift in populations with a small effective population size is the loss of genetic variation. Genetic drift is a random process, so both deleterious and advantageous alleles can become fixed within a small population. This means a useful allele can be lost even if it has the potential to increase fitness or long-term

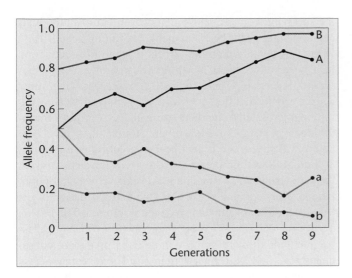

FIGURE 24–7 Change in frequencies over 10 generations for two sets of alleles, A/a and B/b in a theoretical population subject to genetic drift.

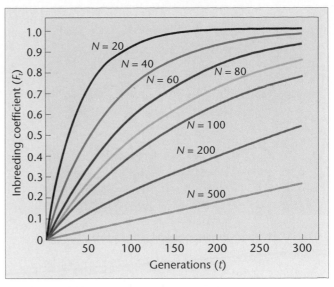

FIGURE 24–8 Increase in inbreeding coefficient (F) in theoretical populations with different effective population size N_e.

adaptability. If we consider a single locus with two alleles, A and a, genetic drift may result in one of the alleles eventually disappearing while the other becomes fixed—in other words, it becomes the only version of that gene present in the gene pool of the population. A simple correlation describes the likelihood of the fixation or loss of an allele. The probability that an allele will be fixed through drift is the same as its initial frequency. So if $p(A) = 0.8$, the probability of A being fixed is 0.8 or 80 percent and the probability of A being lost through drift is $(1 - 0.8) = 0.2$ or 20 percent . The theoretical effects of genetic drift are examined in Figure 24–7.

Inbreeding

In small populations, the chance of **inbreeding** (matings between closely related individuals) is increased. As described in Chapter 22, inbreeding increases the proportion of homozygotes in a population, thus increasing the possibility that an individual may be homozygous for a deleterious allele. The **inbreeding coefficient (F)** (also described in Chapter 21) measures the extent of inbreeding occurring in a population; F measures the probability that two alleles of a given gene in an individual are derived from a common ancestral allele. The inbreeding coefficient is inversely related to the frequency of heterozygotes in the population, and can be calculated as

$$F = \frac{2pq - H}{2pq}$$

where $2pq$ is the expected frequency of heterozygotes based on the Hardy–Weinberg principle and H is the actual frequency of heterozygotes in the population. In a declining population that has become small enough for drift to occur, heterozygosity (H) will decrease with each generation. The smaller the effective population size, the more rapid the decrease in H and the resulting increase in F, as can be seen from the following equation:

$$H_t = (1 - \tfrac{1}{2} Ne)^t H_0$$

where H_0 = initial frequency of heterozygotes, H_t = frequency of heterozygotes after t generations, and N_e = effective population size. Figure 24–8 compares rates of increase in F for different effective population sizes.

What effect does inbreeding have on the long-term survival of a population? In some cases, inbreeding may not immediately reduce the amount of genetic variation present in the overall gene pool of a species where numbers of individuals remain high. Self-pollinating plants, for example, often show high levels of homozygosity and relatively little genetic variation within a single population, but considerable variation between different populations, each of which has adapted to slightly different local environmental conditions. On the other hand, in most outcrossing species, including all mammals, inbreeding is associated with reduced fitness and lower survival rates among the offspring. This is referred to as **inbreeding depression**, which can result from increased homozyosity for deleterious alleles. The number of deleterious alleles present in the gene pool of a population is referred to as the **genetic load**. In some species, inbreeding accompanied by selection against less fit individuals homozygous for deleterious alleles has resulted in the elimination of these alleles from the gene pool, a process known as "purging the genetic load." Species that have successfully purged their genetic load do not show continued reduction in fitness, even with many generations of inbreeding. This is true of many domesticated species, especially self-pollinating plants such as wheat. However, computer simulation experiments have shown that it may take 50 generations or more to complete the purging process, during which time inbreeding depression will still occur.

Alternatively, inbreeding depression can result from heterozygous individuals having a higher level of fitness than either of the corresponding homozygotes. In this case, the long-term survival of the population requires that inbreeding be avoided and that levels of all alleles in the gene pool be maintained—both of which can be difficult in a species that has already suffered a significant reduction in population size.

FIGURE 24–9 Isle Royale gray wolf (*Canis lupus*). *(Art Wolfe/Stone)*

The effects of inbreeding depression and loss of genetic variation in a small isolated population have been documented in the case of the Isle Royale gray wolves (Figure 24–9). Around 1950, a pair of gray wolves apparently crossed an ice bridge from the Canadian mainland to Isle Royale in Lake Superior. The island had no other wolves and an abundance of moose, which became the wolves' main food source. By 1980, the Isle Royale wolf population had increased to over 50 individuals. Over the next decade, however, wolf numbers declined to fewer than a dozen with no new litters being born, despite plentiful food and no apparent sign of disease. The genetic variation of the remaining wolves was examined by mtDNA analysis and nuclear DNA fingerprinting, and it was found that the Isle Royale wolves had levels of homozygosity twice as high as wolves in an adjacent mainland population. Furthermore, the wolves all possessed the same mtDNA genotype, consistent with descent from the same female, so the degree of relatedness between individual Isle Royale wolves was equivalent to that of full siblings. This suggests that the wolves' reproductive failure was due to inbreeding depression, a phenomenon that has also been seen in captive wolf populations.

Reduction in Gene Flow

Gene flow, the gradual exchange of alleles between two populations, is brought about by the dispersal of gametes or the migration of individuals. It is an important mechanism for introducing new alleles into a gene pool and increasing genetic variation. Migration is the main route for gene flow to occur in animals. We have previously examined some of the genetic effects of migration in Chapter 22. In plants, gene flow occurs not through movement of individuals, but as a result of cross-pollination between different populations and, to some extent, through seed dispersal. Isolation and fragmentation of populations in rare and declining species significantly reduces gene flow and the potential for maintaining genetic diversity. As we have already seen, habitat loss is a major threat to species survival. It is not unusual for a threatened or endangered species to be restricted to small separate pockets of the remaining habitat. This isolates and fragments

the surviving populations so that movement of individuals can no longer occur between them, preventing gene flow.

One species in which the effects of gene flow reduction have been studied is the North American brown bear (*Ursus arctos*). A team of Canadian and U.S. researchers measured heterozygosity in different brown bear populations using amplified microsatellite markers to create DNA profiles for individual bears. The researchers found that levels of heterozygosity in the brown bear population living in and around Yellowstone Park were only two-thirds as high as those in Canadian and mainland Alaskan brown bear populations. They concluded this was due to the isolation of the Yellowstone bears and their reduced migration. The habitat of the Canadian and Alaskan bear populations was much less fragmented, allowing migration of individuals and consequent gene flow between populations. The researchers also examined an island population of brown bears on the Kodiak Archipelago off the Alaskan coast. Here heterozygosity was even lower—less than one-half that of the mainland populations—providing further evidence for the effect of restricted migration and gene flow on reduced genetic diversity.

24.4 Genetic Erosion: The Loss of Genetic Diversity

The loss of previously existing genetic diversity from a population or a species is referred to as **genetic erosion**. Why does it matter if a population loses genetic diversity, especially if the numbers of individuals remain high?

Genetic erosion has two important effects on a population. First, it can result in the loss of potentially useful alleles from the gene pool, reducing the ability of the population to adapt to changing environmental conditions and increasing its risk of extinction. Several decades before the dawn of modern genetics, Charles Darwin recognized the importance of diversity to long-term species survival and evolutionary success. In *The Origin of Species* (1859), Darwin wrote:

> The more diversified the descendants of any one species...by so much will they be better enabled to seize on many widely diversified places in the polity of nature, and so enabled to increase in numbers.

The story of the peppered moth *Biston betularia* in the nineteenth century, (see the opening photograph in Chapter 23) illustrates the importance of maintaining allelic diversity in the gene pool. The allele producing the darker melanic phenotype did not confer any obvious advantage to the species until the moth's environment in parts of Great Britain was significantly altered as a result of the increasing industrialization and the accompanying pollution. The continued presence of the melanic allele in the moth's gene pool, however, enabled it to adapt to the new environmental conditions and survive, just as the persistence of the nonmelanic allele in the population now allows the moth to readapt as its environment changes once again. What might have happened to the peppered moth if the melanic allele had been lost from its gene pool before the Industrial Revolution?

The second important effect of genetic erosion is a reduction in levels of heterozygosity. At the population level, reduced heterozygosity will be seen as an increase in the number of individuals homozygous at a given locus. At the individual level, a decrease in the number of heterozygous loci within the genotype of a particular plant or animal will occur. As we have seen, loss of heterozygosity is a common consequence of reduced population size. Alleles may be lost through genetic drift, or because individuals carrying them die without reproducing. Smaller populations also increase the likelihood of inbreeding, which inevitably increases homozygosity. Obviously, once an allele is lost from a gene pool, the potential for heterozygosity is greatly reduced, or completely eliminated if there are only two alleles at the locus in question and one is now fixed. The level of homozygosity that can be tolerated varies with species, but studies of populations showing higher than normal levels of homozygosity have documented a range of deleterious effects, including reduced sperm viability and reproductive abnormalities in African lions, increased offspring mortality in elephant seals, and reduced nesting success in spotted woodpeckers.

As yet, no evidence has emerged from field studies conclusively linking the extinction of a wild population to genetic erosion. However, laboratory studies using the fruit fly, *Drosophila melanogaster*, to model evolutionary events show that loss of genetic variation does reduce the ability of a population to adapt to changing environmental conditions. Fruit flies, with their small size and a rapid generation time of only 10–14 days, are a useful model organism for studying evolutionary events, since large populations can be readily developed and maintained through multiple generations. In one set of experiments carried out by scientists at Macquarie University in Sydney, fruit fly populations were reduced to a single pair for up to three generations to create a population bottleneck, and then allowed to rebuild in number. The capacity of the bottlenecked populations to tolerate increasing levels of sodium chloride was compared with that of normal outbred populations, and researchers found that the bottlenecked populations went extinct at lower salt concentrations (Figure 24–10). Experiments comparing inbred and outbred fruit fly populations exposed to environmental stresses such as high temperatures and ethanol have also found that populations with even a low level of inbreeding have a much greater probability of extinction at lower levels of environmental stress than the outbred populations. Experimental results such as these indicate that genetic erosion does indeed reduce the long-term viability of a population by reducing its capacity to adapt to changing environmental conditions.

24.5 Conservation of Genetic Diversity

Scientists working to maintain biological diversity face several dilemmas. Should they focus on preserving individual populations, or should they take a broader approach by trying to conserve not just one species but all the interdependent plants and animals in an ecosystem? How can genetic diversity be maintained in a species whose numbers are declining? Can genetic diversity lost from a population be restored? Early conservation efforts often focused simply on the population size of an endangered species. Biologists now recognize that a complex interplay of different factors must be considered during conservation efforts, including the need to examine the habitat and role of a species within an ecosystem as well as the importance of genetic variation for long-term survival.

Ex situ Conservation

***Ex situ* conservation** (literally "off-site") involves the removal of plants or animals from their original habitat to an artificially maintained location such as a zoo or botanic garden. Living collections such as these have played a major role in the conservation and restoration of several species close to extinction. One example is the recovery of the black-footed ferret (*Mustela nigripes*) (Figure 24–11), once wide-

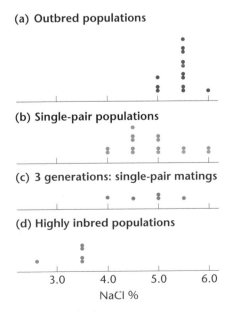

(a) Outbred populations

(b) Single-pair populations

(c) 3 generations: single-pair matings

(d) Highly inbred populations

3.0   4.0   5.0   6.0
NaCl %

FIGURE 24–10 Effects of bottlenecks in various populations on evolutionary potential in fruit flies, as shown by distributions of NaCl concentrations at extinction.

FIGURE 24–11 The black-footed ferret (*Mustela nigripes*). (*Jim Brandenburg/Minden Pictures*)

spread throughout the plains of the western United States, but by the 1970s thought to be extinct after decades of trapping and poisoning of both the ferret itself and its main prey, the prairie dog. A small colony of black-footed ferrets was discovered on a ranch in Wyoming in 1981 and transferred to a captive breeding facility. Starting with the 18 animals originally captured, the breeding program has produced over 3000 ferrets, and this species is now being reintroduced to parts of its original range. With such a small founder group, the population has obviously passed through a severe bottleneck, and the risk of inbreeding and further genetic erosion in the captive breeding program is very high. Genetic management strategies, such as using DNA markers to identify the most genetically varied individuals and maintaining careful pedigree records to avoid mating closely related animals, have helped maintain the genetic diversity that remained in the species. Conservation geneticists working on the recovery project estimate that all existing black-footed ferrets today share about 12 percent of their genome, roughly the equivalent of being first cousins. The long-term effects of this degree of genetic similarity in the expanding ferret population remain to be seen, but so far there is no evidence of reduced fitness through inbreeding.

Another form of *ex situ* conservation is provided by **gene banks**. In contrast to housing entire animals or plants, these collections instead provide long-term storage and preservation for reproductive components such as sperm, ova, and frozen embryos in the case of animals, or seed, pollen, and cultured tissue in the case of plants. Many more individual genotypes can be preserved for longer periods in a gene bank than in a living collection. Deep-frozen gametes or seeds can be used to reconstitute lost or endangered animals or plants after many years in storage. Because they are expensive to construct and maintain, most gene banks are used to conserve the above components of domesticated species of economic value. Gene banks have now been established in many countries to help preserve genetic material of agricultural importance, such as traditional crop varieties that are no longer grown, or old livestock breeds that are becoming rare. One of the most important *ex situ* collections in the United States is the National Seed Storage Laboratory, a U.S. Department of Agriculture facility in Fort Collins, Colorado, which maintains more than 275,000 different accessions of crop varieties and related wild species. Some of the accessions are stored as seeds, and others as cryogenically preserved tissue from which whole plants can be regenerated (Figure 24–12).

Ex situ conservation, while often vital, has a number of disadvantages. A major problem with gene banks is that even large collections cannot contain all the genetic variation present in a species. Conservation geneticists attempt to address this problem by identifying a **core collection** for a species. The core collection is a subset of individual genotypes that represents as much as possible of the genetic variation within a species; preserving the core collection takes priority over collecting and preserving large numbers of genotypes at random. Another disadvantage of *ex situ* conservation is that the artificial conditions under which a species is preserved in a living collection or gene bank often

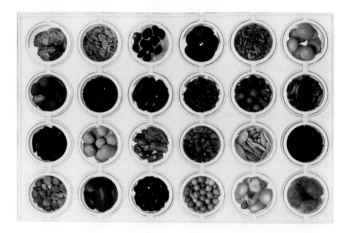

FIGURE 24–12 Cryogenically preserved seeds of rare crop varieties at the U.S. Department of Agriculture National Seed Storage Laboratory. *(Photo by Scott Bauer/ARS Photo Unit/USDA)*

create their own selection pressures. When seeds of a rare plant species are maintained in cold storage, for example, selection may occur for those genotypes better adapted to withstand the lower temperatures, and genotypes may be lost that would actually have greater fitness in the natural environment. Still an additional problem posed by *ex-situ* conservation is that, while the greatest biological diversity in both domesticated and nondomesticated species is frequently found in underdeveloped countries, most *ex situ* collections are in developed countries that have the resources to establish and maintain them. This leads to conflict over who owns and has access to the potentially valuable genetic resources maintained in such collections.

In Situ Conservation

In situ **conservation** (literally "on-site") attempts to preserve the population size and biological diversity of a species while maintaining it in its original habitat. The use of species inventories to identify diversity hotspots, as we learned earlier, is an important tool for determining the best places to establish parks and reserves where plants and animals are protected from hunting or collecting, and where their habitat can be maintained. For domesticated species, there is increasing interest in "on-farm" preservation, where farmers are encouraged through provision of additional resources and financial incentives to maintain traditional crop varieties and livestock breeds. Nondomesticated species with economic potential have also been targeted for *in situ* preservation. In 1998, the U.S. Department of Agriculture established its first *in situ* conservation sites for a wild plant, protecting populations of the native rock grape (*Vitis rupestris*) in several eastern states. The rock grape is prized by winegrowers not for its fruit but for its roots; grape vines grafted onto wild rock grape rootstock are resistant to phylloxera, a serious pest of wine grapes. The advantage of *in situ* conservation for the rock grape, as with other species, is that larger populations with greater genetic diversity can be maintained. Species conserved *in situ* will also continue to live and reproduce in the environments to which they are adapted, reducing the likelihood that novel selection pressures will produce undesirable changes in allele frequency. However, as the global human population continues to rise, setting aside suitable areas for *in situ* conservation becomes a greater challenge. As we have already seen in the case of the North American brown bear, even in large preserves such as Yellowstone National Park, a species' migration and gene flow may be eliminated with the consequent loss of genetic diversity. This problem is even more acute in smaller and more fragmented areas of protected habitat.

Population Augmentation

What genetic considerations should accompany efforts to restore populations or species that are in decline? As we have already seen, populations that go through bottlenecks continue to suffer from low levels of genetic diversity, even after their numbers have recovered. We have also seen that inbreeding and drift contribute to genetic erosion in small populations, and fragmentation interrupts migration and gene flow, further reducing diversity. Captive breeding programs to restore a critically endangered species from a few surviving individuals risk genetic erosion in the renewed population from founder effect and inbreeding, as in the case of the black-footed ferret.

An alternative strategy used by conservation biologists is **population augmentation**—boosting the numbers of a declining population by transplanting and releasing individuals of the same species captured or collected from more numerous populations elsewhere. Population augmentation projects in the United States have involved bighorn sheep and grizzly bears in the Rocky Mountains. It has also been employed with the Florida panther (*Felis coryi*), an isolated population of less than 50 animals confined to the area around the Big Cypress Swamp and Everglades National Park in south Florida. As discussed in the essay at the end of this chapter, DNA profiling patterns show high levels of in-

breeding in the Florida panther, with reduced fitness due to severe reproductive abnormalities and increased susceptibility to parasite infections. Seven unrelated animals from a captive population of South and North American panthers were released into the Everglades in the 1960s and have interbred with the Florida population, producing more genetically diverse family groups. Further population augmentation in this species has been controversial, however, as some biologists argue that the unique features that allow the Florida panther to be classified as a separate subspecies will be lost if mating with other panthers takes place.

Despite such controversies, population augmentation would appear to be a valuable restoration tool. This strategy cannot only increase population numbers, but also genetic diversity, if the transplanted individuals are unrelated to those in the population to which they are introduced. A potential problem with population augmentation, however, is the recently identified phenomenon of **outbreeding depression**, where reduced fitness occurs in the progeny from matings between genetically diverse individuals. Outbreeding depression occurring in the F_1 generation is thought to be due to the offspring being less well-adapted to local environmental conditions than the parents. This phenomenon has been documented in some plant species where seed of the same species, but from a different location, was used to revegetate a damaged area. Outbreeding depression that occurs in the F_2 and later generations is thought to be due to the disruption of coadapted gene complexes—groups of alleles that have evolved to work together to produce the best level of fitness in an individual. This type of outbreeding depression has been documented in F_2 hybrid offspring from matings between fish from different salmon populations in Alaska. Studies such as this suggest that restoring the most beneficial type and amount of genetic diversity in a population is more complicated than previously thought, reinforcing the argument that the best long-term strategy for species survival is to prevent the loss of diversity in the first place.

Chapter Summary

1. Biodiversity is being lost as an increasing number of plant and animal species are threatened with extinction. Conservation genetics applies principles of population genetics to the preservation and restoration of threatened species. A major concern of conservation geneticists is the maintenance of genetic diversity.

2. Genetic diversity includes interspecific differences, reflected by the number of different species present in an ecosystem, and intraspecific differences, reflected by genetic variation within a population or between different populations of the same species. Genetic diversity can be measured by examining different phenotypes in the population, or at the molecular level by using isozyme analysis or DNA profiling techniques.

3. Major declines in a species' population numbers reduce genetic diversity and contribute to their risk of extinction as a result of genetic drift, inbreeding, or loss of gene flow. Populations that suffer severe reductions in effective size and then recover

are said to have passed through a population bottleneck and often show reduced genetic diversity.

4. Loss of genetic diversity reduces the capacity of a population to adapt to changing environmental conditions as useful alleles may disappear from the gene pool. Reduced genetic diversity also results in greater levels of homozygosity in a population, often leading to an accumulation of deleterious alleles and inbreeding depression.

5. Conservation of genetic diversity depends on *ex situ* methods, such as living collections, captive breeding programs, and gene banks, and *in situ* approaches such as the establishment of parks and preserves.

6. Population augmentation, where individuals are transplanted into a declining population from a more numerous population of the same species located elsewhere, can be used to increase numbers and genetic diversity. However, the risk of outbreeding depression accompanies this process.

Genetics, Technology, and Society

Gene Pools and Endangered Species: The Plight of the Florida Panther

In the last 400 years, more than 700 of Earth's animal and plant species have become extinct. In the U.S. alone, at least 30 species have suffered extinction in the last decade. In addition, hundreds of genetically distinct plant and animal species are now endangered. The rates of species and ecosystem losses are accelerating rapidly and are predicted to continue unabated throughout this century. This dramatic loss of biological diversity is a direct consequence of human activity. As we humans increase our numbers and spread over Earth's surface, we harness more of Earth's resources for our use—clearing the forests, polluting the water, and inexorably and permanently altering Earth's natural balance.

Although the destruction continues, we are beginning to understand the implications of our actions and to make efforts to save some of Earth's more endangered life forms. However, despite our best efforts to save threatened species, we often intercede after population numbers have suffered severe declines. This means that much of the genetic diversity that existed in the species is lost, and the population must be rebuilt from a reduced gene pool. The resulting genetic uniformity can reduce fitness, as well as exposing genetic diseases. This reduced fitness may further deplete the numbers of the threatened plant or animal, and the population spirals downward to extinction.

The story of the Florida panther provides a poignant example of how an animal can be brought to the verge of extinction, and how the efforts of scientists and the general public might restore this creature to a healthy place in the ecosystem. The Florida panther, *Felis concolor coryi*, is one of 30 subspecies of cougar (puma or mountain lion), and is one of the most endangered mammals in the world. Florida panthers once roamed the southeastern corner of North America, from South Carolina and Arkansas to the southern tip of Florida. As people settled the southern and eastern states, panthers were killed as potential threats to livestock and humans. In the 19th and 20th centuries, panthers were killed by

hunting, highway collisions, poisoning, and loss of habitat. Today, only 30–50 Florida panthers remain, isolated in southern Florida in the Big Cypress Swamp and Everglades National Park regions. Population projections indicate that, unless drastic measures are taken, there is an 85 percent likelihood that the Florida panther will be extinct in 25 years.

The geographical isolation and inbreeding of the few remaining Florida panthers have resulted in the loss of genetic variability and declining health. The panther has been separated from other cougar subspecies for 15 to 25 generations. As a result, Florida panthers have the lowest levels of genetic heterozygosity of any subspecies of cougar. This loss of genetic diversity has manifested itself in the appearance of several severe genetic defects. Almost 80 percent of panther males born after 1989 in the Big Cypress region show a rare heritable (autosomal dominant or sex-linked recessive) condition known as cryptorchidism, manifested as the failure of one or both testicles to descend. This defect is associated with lower testosterone levels and reduced sperm count. Florida panther males in general have the poorest seminal quality of any cat species or any cougar subspecies, with approximately 93 percent abnormal sperm. Life-threatening congenital heart defects are also appearing in the Florida panther population, possibly due to an autosomal dominant gene defect. In addition, some immune deficiencies are appearing, and these may have a genetic component. Reduced immunity could leave the small panther population susceptible to diseases that could wipe out the population. Other less serious genetic features have appeared, such as a kink in the tail and a whorl of fur on the back.

Over the last two decades, a faint glimmer of hope has appeared for the Florida panther. Federal and state agencies, as well as private individuals, are implementing a "Florida Panther Recovery Program." The goal of the plan is to exceed 130 breeding animals (wild and captive) soon after the year 2000, and to approach 500 by the year 2010. If successful, the plan would grant the panther a 95 percent probability of survival while retaining 90 percent of its ge-

netic diversity. The plan is multifaceted and includes a captive breeding program, strict protection, increasing and improving the panther habitat, and educating the public and private landowners. Wildlife underpasses have been constructed on highways in panther territory, and these have significantly reduced panther highway fatalities (which account for half of panther deaths). In 1995, a genetic restoration program began. In order to introduce genetic diversity into the Florida panther population and to retard the detrimental effects of inbreeding, eight female wild Texas panthers (a related subspecies from western Texas) were released into Florida panther territory. Two of the females have been killed—one from an automobile collision and one from gunshot wounds. However, the remaining Texas females have given birth to 12 healthy kittens (as of the summer of 1998), and one of these F_1 offspring has produced three kittens of her own. None of the kittens appear to have the "kinked" tail of the inbred Florida panther.

The survival of the Florida panther is far from certain. Its recovery will require years of monitoring and frequent intervention. In addition, people must be willing to share their land with wild creatures that do not directly further their self-interests. However, as public support for the return of the Florida panther has been strong, there may be hope for this unique, impressive animal.

References

Fergus, C. 1991. The Florida panther verges on extinction. *Science* 251: 1178–80.

Hedrick, P.W. 1995. Gene flow and genetic restoration: The Florida panther as a case study. *Conservation Biology* 9: 996–1007.

O'Brien, S.J. 1994. A role for molecular genetics in biological conservation. *Proc. Natl. Acad. Sci. USA* 91: 5748–55.

Website:

The Florida Panther Society
http://members.atlantic.net/~oldfla/panther

Key Terms

Insights and Solutions

1. Is a rare species found as several fragmented subpopulations more vulnerable to extinction than an equally rare species found as one larger population? What factors should be considered when managing fragmented populations of a rare species?

Solution: A rare species where the remaining individuals are divided among smaller isolated subpopulations can appear to be less vulnerable. If one subpopulation becomes extinct through local causes such as disease or habitat loss, then the remaining sub populations may still survive. However, genetic drift will cause smaller populations to experience more rapidly increasing homozygosity over time compared to larger populations. Even with random mating, the change in heterozygosity from one generation to the next due to drift can be calculated as

$$H_1 = H_0(1 - \tfrac{1}{2}N)$$

where H_0 is the frequency of heterozygotes in the present generation, H_1 is the frequency of heterozygotes in the next generation, and N is the number of individuals in the population. Thus, in a small population of 50 individuals with an initial heterozygote frequency of 0.5, in just one generation heterozygosity will decline to $0.5(1 - 1/100) = 0.495$, a loss of 0.5%. In a larger population of 500 individuals and the same initial heterozygote frequency, after one generation heterozygosity will be $0.5(1 - 1/1000) = 0.4995$, a loss of only 0.05%.

This means that smaller populations are likely to show the effects of homozygosity for deleterious alleles sooner than larger populations, even with random mating. If populations are fragmented so that movement of individuals or gametes between them is prevented, management options could include transplanting individuals from one subpopulation to another to enable gene flow to occur. Establishment of "wildlife corridors" of undisturbed habitat that will connect fragmented populations could be considered. In captive populations, exchange of breeding adults (or their gametes through shipment of preserved semen or pollen) can be undertaken. Management for increased population numbers, however, is vital to prevent further genetic erosion through drift.

Problems and Discussion Questions

1. A wildlife biologist studied four generations of a population of rare Ethiopian jackals. When the study began, there were 47 jackals in the population, and analysis of microsatellite loci from these animals showed a heterozygote frequency of 0.55. In the second generation, an outbreak of distemper occurred in the population, and only 17 animals survived to adulthood. These produced 20 surviving offspring, which in turn gave rise to 35 progeny in the fourth generation.

   (a) What was the effective population size for the four generations of this study?
   (b) Based on this effective population size, what is the heterozygote frequency of the jackal population in generation 4?
   (c) Assuming an inbreeding coefficient of zero at the beginning of the study, no change in microsatellite allele frequencies in the gene pool, and random mating in all generations, what is the inbreeding coefficient in generation 4?

2. Chondrodystrophy, a lethal form of dwarfism, has recently been reported in captive populations of the California condor where it has killed embryos in 5 out of 169 fertile eggs. Chondrodystrophy in condors appears to be caused by an autosomal recessive allele with an estimated frequency of 0.09 in the gene pool of this species. How do you think California condor populations should be managed in the future to minimize the effect of this lethal allele? What are the advantages and disadvantages of attempting to eliminate it from the gene pool?

3. A geneticist is studying three loci, each with one dominant and one recessive allele, in a small population of rare plants. She estimates the frequencies of the alleles at each of these loci as follows:

$A = 0.75$	$a = 0.25$
$B = 0.80$	$b = 0.20$
$C = 0.95$	$c = 0.05$

What is the probability that all the recessive alleles will be lost from the population through genetic drift?

4. How are genetic drift and inbreeding similar in their effects on a population? How are they different?

5. You are manager of a game park in Africa with a native herd of just 16 black rhinos, an endangered species worldwide. Describe how you would manage this herd to establish a viable population of black rhinos in the park. What genetic factors would you take into consideration in your management plan?

6. Compare the causes and effects of inbreeding depression and outbreeding depression.

7. Cloning, using the techniques similar to those pioneered by the Scottish scientists who produced Dolly the sheep, has been proposed as a way to increase the numbers of some highly endangered mammal species. Discuss the advantages and disadvantages of using such an approach to aid long-term species survival.

Selected Readings

Avise, J. C., and Hamrick, J. L. (eds.) 1996. *Conservation genetics: Case histories from nature.* New York: Chapman & Hall.

Baker, C. S., Palumbi, S. R. 1994. Which whales are hunted? A molecular genetic approach to monitoring whaling. *Science* 265:1538–1539.

Beatiie, A., and Erlich, P. 2001. *Wild solutions: How biodiversity is money in the bank.* New Haven: Yale University Press.

Bongaarts, J., and Bulatao, R. A. (eds.) 2000. *Beyond six billion: Forecasting the world's population.* Washington DC: National Academy Press.

Bonnell, M. L., Selander, R. K. 1974. Elephant seals: genetic variation and near extinction. *Science* 184:908–90.

Frankham, R. 1995. Conservation genetics. *Ann. Rev. Genet.* 29:305–327.

——————. 1999. Quantitative genetics in conservation biology. *Genet. Res.* 74:237–44.

Gerber, L. R., DeMaster, D. P., and Roberts, S. P. 2000. Measuring success in conservation. *Amer. Scient.* 88:316–21.

Gharrett, A. J., and Smoker, W. W. 1991. Two generations of hybrids between even-year and odd-year pink salmon (*Oncorhynchus gorbuscha*): a test for outbreeding depression? *Canadian J. Fish. Aquat. Sci.* 48:426–238.

Lacy, R. C. 1997. Importance of genetic variation to the viability of mammalian populations. *J. Mammalogy* 78:320–335.

O'Neill, B. C., MacKellar, F. L., and Lutz, W. 2001. *Population and climate change.* Cambridge: Cambridge University Press.

Paetkau, D., Waits, L. P., Clarkson, P. L., Craighead, L., Vyse, E., Ward, R., and Strobeck, C. 1998. Variation in genetic diversity across the range of North American brown bears. *Conserv. Biol.* 12:418–429.

Palacios, C., and Gonzalez-Candelas, F. 1999. AFLP analysis of the critically endangered *Limonium cavanillesii*. *J. Heredity* 90:485–489.

Ralls, K., Ballou, J. D., Rideout B. A., and Frankham, R. 2000. Genetic management of chondrodystrophy in California condors. *Animal Conservation* 3:145–153.

Roman, J., and Bowen, B. W. 2000. The mock turtle syndrome: genetic identification of turtle meat purchased in the southeastern United States of America. *Animal Conservation* 3:61–65.

Storfer, A. 1996. Quantitative genetics: A promising approach for the assessment of genetic variation in endangered species. *Trends Ecol. Evol.* 11:343–47.

Wayne, R. K., et. al. 1991. Conservation genetics of the endangered Isle Royale gray wolf. *Conserv. Biol.* 5:41–51.

Wilson, E. O. (ed.) 1988. *Biodiversity*. Washington, D.C.: National Academy of Sciences.

Wynen, L. P., et al. 2000. Postsealing genetic variation and population structure of two species of fur seal. *Molecular Ecology* 9:299–314.

Appendix
Answers to Selected Problems

Chapter 1

2. *Pangenesis* refers to a theory that various parts of the body contain "humors" which bear the hereditary traits and gather in the reproductive organs. *Epigenesis* refers to the theory that organisms are derived from the assembly and reorganization of substances in the egg which eventually lead to the development of the adult. *Preformationism* is a seventeenth-century theory which states that the sex cells (eggs or sperm) contain miniature adults, called homunculi, which grow in size to become the adult. Each postulates a fundamental difference in the manner in which organisms develop from hereditary determiners.

4. Darwin's theory of natural selection proposed that more offspring are produced than can survive and that, in the competition for survival, those with favorable variations survive. Over many generations, this will produce a change in the genetic make-up of populations if the favorable variations are inherited. Darwin did not understand the nature of heredity and variation which led him to lean toward older theories of pangenesis and inheritance of acquired characteristics.

8. Norman Borlaug applied Mendelian principles of hybridization and trait selection to the development of superior varieties of wheat. Such varieties are now grown in many countries, including Mexico, and have helped maintain the world supply of food. This change in worldwide agricultural food production has been called the "Green Revolution."

10. In the last 50 years, human transmission, cytological and molecular genetics have provided an understanding of many aspects of both plant and animal biology including development of pest-resistant crops and identification of hazardous organisms in our food (*E. coli*, for example). In addition, much has been learned about many human diseases. There is promise that a certain amount of human suffering will be minimized by the application of genetics to crop production (disease resistance, protein content, growth conditions) and medicine. Major medical areas of activity include genetic counseling, gene mapping and identification, disease diagnosis, and genetic engineering.

Chapter 2

2. One of the most important concepts to be gained from this chapter is the relationship which exists among chromosomes in a single cell. Chromosomes which are homologous share many properties including: overall length; position of the centromere (*metacentric, submetacentric, acrocentric, telocentric*); *banding patterns*; *type and location of genes*; *autoradiographic pattern*.

 Diploidy is a term often used in conjunction with the symbol *2n*. It means that both members of a homologous pair of chromosomes are present. *Haploidy* specifically refers to the fact that each haploid cell contains *one chromosome of each homologous pair of chromosomes*. *Haploidy* is usually symbolized as *n*. The change from a diploid (*2n*) to haploid (*n*) occurs during *reduction division*

when tetrads become dyads during meiosis I. Referring to the number of human chromosomes, the primary spermatocyte (*2n* = 46) becomes two secondary spermatocytes each with *n* = 23.

6. Refer to the *Essentials* text (Figure 2–3) for an explanation. Notice the different anaphase shapes in the figures. Chromosomes are classified as *metacentric, submetacentric, acrocentric,* or *telocentric* on the basis of centromere location.

8. The mitotic cell cycle is separated into two major portions: interphase (G1, S, G2) and mitosis. Notice that, in contrast to meiosis, there is no pairing of homologous chromosomes in mitosis and the chromosome number does not change. Cell-cycle control is achieved by the interaction of cyclin dependent kinase (cdk) in conjunction with cyclins. There are at least three major checkpoints; G1/S, G2/M, and M (in mitosis).

10. *p53* plays a role in regulating the G1 to S transition. It is a tumor suppressor. Mutations cause a lack of control over the cell cycle and cancer often follows.

12. Compared with mitosis, meiosis provides for a reduction in chromosome number, and an opportunity for exchange of genetic material between homologous chromosomes. In mitosis there is no change in chromosome number or kind in the two daughter cells, whereas in meiosis, numerous potentially different haploid (*n*) cells are produced. During oogenesis, only one of the four meiotic products is functional; however, four of the four meiotic products of spermatogenesis are potentially functional.

14. (a) If there are 16 chromosomes, there should be 8 tetrads.
 (b) After meiosis I and in the second meiotic prophase, there are as many dyads as there are *pairs* of chromosomes. There will be 8 dyads.
 (c) Because the monads migrate to opposite poles during meiosis II (from the separation of dyads), there should be 8 monads migrating to *each* pole.

16. In meiosis, various chromosomal arrangements are possible because of random alignment of homologous chromosomes at metaphase I. In addition, crossing over, which introduces additional variation, is virtually absent in mitotic processes.

18. There would be 16 combinations possible.

20. The cytological origin of the mitotic chromosome (and the meiotic chromosome for that matter) appears during the prophase through a condensation of the dispersed chromatin fibers present during interphase. During interphase, chromatin consists of invisible threads (under light microscopy) of DNA associated with histones. In this form, DNA can function in transcription and replication. Visible mitotic chromosomes appear when each chromosome coils and condenses according to the current *folded fiber* model.

22. The two smaller chromosomes are probably homologous and would be similar for the following characteristics: function, arrangement of genes, time of replication during the S phase, banding, etc.

Chapter 3

2. Start out with the following gene symbols: A = normal (not albino); a = albino. Since albinism is inherited as a recessive trait, genotypes AA and Aa should produce the normal phenotype, while aa will give albinism.

    (a) The parents are both normal and must both be heterozygous (Aa).

    (b) To start out, the normal male could have either the AA or Aa genotype. The female must be aa. Since all the children are normal, one would consider the male to be AA instead of Aa. However, the male could be Aa. Under that circumstance, the likelihood of having six children, all normal, is 1/64.

4. *Pisum sativum* is easy to cultivate. It is naturally self-fertilizing, but it can be crossbred. It has several visible features (e.g., tall or short, red flowers or white flowers) which are consistent under a variety of environmental conditions yet contrast due to genetic circumstances. Seeds could be obtained from local merchants.

6. Symbolism:

w = wrinkled seeds g = green cotyledons

W = round seeds G = yellow cotyledons

P_1: $WWGG \times wwgg$

F_1: $WwGg$

$F_1 \times F_1$: $WwGg \times WwGg$

9/16	$W\_G\_$	round seeds, yellow cotyledons
3/16	W_gg	round seeds, green cotyledons
3/16	$wwG\_$	wrinkled seeds, yellow cotyledons
1/16	$wwgg$	wrinkled seeds, green cotyledons

Seed shape	**Cotyledon color**	**Phenotypes**
3/4 round	3/4 yellow	9/16 round, yellow
	1/4 green	3/16 round, green
1/4 wrinkled	3/4 yellow	3/16 wrinkled, yellow
	1/4 green	1/16 wrinkled, green

8. A test cross involves mating an organism of unknown genotype with a fully homozygous recessive organism. In Problem 7, (c) is a test cross.

16. 1. Factors occur in pairs. 2. Some genes have dominant and recessive alleles. 3. Alleles segregate from each other during gamete formation. When homologous chromosomes separate from each other at anaphase I, alleles will go to opposite poles of the meiotic apparatus. 4. One gene pair separates independently from other gene pairs. Different gene pairs on the same homologous pair of chromosomes (if far apart) or on non-homologous chromosomes will separate independently from each other during meiosis.

12. Homozygosity refers to a condition where both genes of a pair are the same (i.e., AA or GG or hh), whereas heterozygosity refers to the condition where members of a gene pair are different (i.e., Aa or Gg or Bb).

14. The general formula for determining the number of kinds of gametes produced by an organism is 2^n where n = number of *heterozygous* gene pairs.

    (a) 4: AB, Ab, aB, ab

    (b) 2: AB, aB

    (c) 8: ABC, ABc, AbC, Abc, aBC, aBc, abC, abc

    (d) 2: ABc, aBc

    (e) 4: ABc, Abc, aBc, abc

    (f) $2^5 = 32$

ABCDE
ABCDe
ABCdE
ABCde
ABcDE
ABcDe
ABcdE
ABcde
AbCDE
AbCDe
AbCdE
AbCde
AbcDE
AbcDe
AbcdE
Abcde

Notice that there is a pattern that can be used to write these gametes so that fewer errors will occur.

16. Symbols:

Seed shape	*Seed color*
W = round	G = yellow
w = wrinkled	g = green

P_1: $WWgg \times wwGG$

F_1: $WwGg$ cross to $wwgg$

    1/4 $WwGg$ (round, yellow)

    1/4 $Wwgg$ (round, green)

    1/4 $wwGg$ (wrinkled, yellow)

    1/4 $wwgg$ (wrinkled, green)

18. One must think of this problem as a dihybrid F_2 situation with the following expectations:

Expected ratio	**Observed (o)**	**Expected (e)**
9/16	315	312.75
3/16	108	104.25
3/16	101	104.25
1/16	32	34.75

$$\chi^2 = 0.47$$

Looking at the table in *Essentials*, one can see that this χ^2 value is associated with a probability greater than 0.90 for 3 degrees of freedom (because there are now four classes in the χ^2 test). The observed and expected values do not deviate significantly.

    To deal with parts (b) and (c), it is easier to see the observed values for the monohybrid ratios if the phenotypes are listed:

smooth, yellow	315
smooth, green	108
wrinkled, yellow	101
wrinkled, green	32

For the smooth:wrinkled monohybrid component, the smooth types total 423 (315 + 108), while the wrinkled types total 133 (101 + 32).

Expected ratio	**Observed (o)**	**Expected (e)**
3/4	423	417
1/4	133	139

The χ^2 value is 0.35 and in examining the table in *Essentials* for 1 degree of freedom, the p value is greater than 0.50 and less than 0.90. We fail to reject the null hypothesis and are confident that the observed values do not differ significantly from the expected values.

(c) For the yellow:green portion of the problem, see that there are 416 yellow plants (315 + 101) and 140 (108 + 32) green plants.

Expected ratio	Observed (o)	Expected (e)
3/4	416	417
1/4	140	139

The χ^2 value is 0.01 and in examining the table in *Essentials* for 1 degree of freedom, the p value is greater than 0.90. We fail to reject the null hypothesis and are confident that the observed values do not differ significantly from the expected values.

20. Use of the $p = 0.10$ as the "critical" value for rejecting or failing to reject the null hypothesis instead of $p = 0.05$ would allow more null hypotheses to be rejected. As the critical p value is increased, it takes a smaller χ^2 value to cause rejection of the null hypothesis. It would take less difference between the expected and observed values to reject the null hypothesis, therefore the stringency of failing to reject the null hypothesis is increased.

22. Notice that II-4 and II-5 produce a female child (III-4) with the affected phenotype. On these criteria alone, the gene must be viewed as being recessive.

      I-1 (*Aa*), I-2 (*aa*), I-3 (*Aa*), I-4(*Aa*)
      II-1 (*aa*), II-2 (*Aa*), II-3 (*aa*), II-4 (*Aa*), II-5 (*Aa*),
        II-6 (*aa*), II-7 (*AA* or *Aa*), II-8 (*AA* or *Aa*)
      III-1 (*AA* or *Aa*), III-2 (*AA* or *Aa*), III-3 (*AA* or *Aa*),
        III-4 (*aa*), III-5 (probably *AA*), III-6 (*aa*)
      IV-1 through IV-7 all *Aa*.

24. Notice that two normal individuals in II-3 and II-4 have produced a daughter (III-2) with myopia.

      I-1 (*aa*), I-2 (*AA* or *Aa*), I-3 (*Aa*), I-4 (*Aa*)
      II-1 (*Aa*), II-2 (*Aa*), II-3 (*Aa*), II-4 (*Aa*), II-5 (*aa*),
        II-6 (*AA* or *Aa*), II-7 (*AA* or *Aa*)
      III-1 (*AA* or *Aa*), III-2 (*aa*), III-3 (*AA* or *Aa*)

26. (a) First consider that each parent is homozygous (true-breeding in the question) and since in the F₁ only round, axial, violet, and full phenotypes were expressed, they must each be dominant.

    (b) Round, axial, violet, and full would be the most frequent phenotypes: 3/4 × 3/4 × 3/4 × 3/4

    (c) Wrinkled, terminal, white, and constricted would be the least frequent phenotypes: 1/4 × 1/4 × 1/4 × 1/4

    (d) 3/4 × 1/4 × 3/4 × 1/4

    (e) There would be 16 different phenotypes in the test cross offspring just as there are 16 different phenotypes in the F₂ generation.

28. (a) Possible symbols would be (using standard *Drosophila* symbolism):

      straight wings $= w^+$ curled wings $= w$
      short bristles $= b^+$ long bristles $= b$

    (b)
      Cross 1: $w^+/w;b^+/b \times w^+/w;b^+/b$
      Cross 2: $w^+/w;b/b \times w^+/w;b/b$
      Cross 3: $w/w;b/b \times w^+/w;b^+/b$
      Cross 4: $w^+/w^+;b^+/b \times w^+/w^+;b^+/b$ (one parent could be w^+/w)
      Cross 5: $w/w;b^+/b \times w^+/w;b^+/b$

Chapter 4

2. *Incomplete dominance* can be viewed more as a quantitative phenomenon where the heterozygote is intermediate (approximately) between the limits set by the homozygotes. Pink is intermediate between red and white.

Codominance can be viewed in a more qualitative manner where both of the alleles in the heterozygote are expressed. For example in the AB blood group, both the I^A and I^B genes are expressed. There is no intermediate class which is part I^A and I^B.

4. $Pp \times Pp \longrightarrow$
    1/4 *PP* (**lethal**)
    2/4 *Pp* (platinum)
    1/4 *pp* (silver)

Therefore the ratio of surviving foxes is 2/3 platinum, 1/3 silver. The *P* allele behaves as a recessive in terms of lethality (seen only in the homozygote) but as a dominant in terms of coat color (seen in the homozygote).

6. Flower color:
    RR = red; *Rr* = pink; *rr* = white
  Flower shape:
    P = personate; *p* = peloric
  (a) *RRpp* × *rrPP* ⟶ *RrPp*
  (b) *RRPP* × *rrpp* ⟶ *RrPp*
  (c) *RrPp* × *RRpp* ⟶ *RRPp*
            RRpp
            RrPp
            Rrpp
  (d) *RrPp* × *rrpp* ⟶ *rrPp*
            rrpp
            RrPp
            Rrpp

In the cross of the F₁ of (a) to the F₁ of (b), both of which are double heterozygotes, one would expect the following:

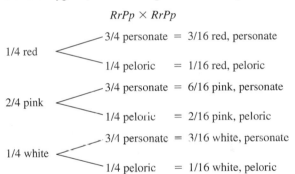

$$RrPp \times RrPp$$

1/4 red
  3/4 personate = 3/16 red, personate
  1/4 peloric = 1/16 red, peloric

2/4 pink
  3/4 personate = 6/16 pink, personate
  1/4 peloric = 2/16 pink, peloric

1/4 white
  3/4 personate = 3/16 white, personate
  1/4 peloric = 1/16 white, peloric

8. (a) *AaBbCc* = gray (*C* allows pigment)
  (b) *A_B_* = gray (*C* allows pigment)
  (c) Use the forked line method for this portion:

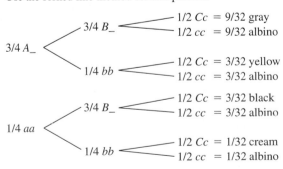

3/4 *A_*
  3/4 *B_*
    1/2 *Cc* = 9/32 gray
    1/2 *cc* = 9/32 albino
  1/4 *bb*
    1/2 *Cc* = 3/32 yellow
    1/2 *cc* = 3/32 albino

1/4 *aa*
  3/4 *B_*
    1/2 *Cc* = 3/32 black
    1/2 *cc* = 3/32 albino
  1/4 *bb*
    1/2 *Cc* = 1/32 cream
    1/2 *cc* = 1/32 albino

Combining the phenotypes gives (always count the proportions to see that they add up to 1.0):
  16/32 albino
  9/32 gray
  3/32 yellow
  3/32 black
  1/32 cream

10. (a) 1/4

  (b) 1/2

  (c) 1/4

  (d) zero

12. P_1: female: RR (red) $\times$ male: rr (mahogany)

  F_1: Rr = females red; males mahogany

    1/2 females (red)

    1/2 males (mahogany)

  F_2: 1/4 RR: 2/4 Rr: 1/4 rr

  Because half of the offspring are males and half are females, one could, for clarity, rewrite the F_2 as:

	1/2 females	*1/2 males*
1/4 RR	1/8 red	1/8 red
2/4 Rr	2/8 red	2/8 mahogany
1/4 rr	1/8 mahogany	1/8 mahogany

14. Symbolism:

  Normal wing margins = sd^+

  scalloped = sd

  (a) P_1: $X^{sd}X^{sd} \times X^+/Y \longrightarrow$

              (assume

              homozygous)

    F_1: 1/2 X^+X^{sd} (female, normal)

      1/2 X^{sd}/Y (male, scalloped)

    F_2: 1/4 X^+X^{sd} (female, normal)

      1/4 $X^{sd}X^{sd}$ (female, scalloped)

      1/4 X^+/Y (male, normal)

      1/4 X^{sd}/Y (male, scalloped)

  (b) P_1: $X^+/X^+ \times X^{sd}/Y \longrightarrow$

    F_1: 1/2 X^+X^{sd} (female, normal)

      1/2 X^+/Y (male, normal)

    F_2: 1/4 X^+X^+ (female, normal)

      1/4 X^+X^{sd} (female, normal)

      1/4 X^+/Y (male, normal)

      1/4 X^{sd}/Y (male, scalloped)

  If the scalloped gene were not X-linked, then all of the F_1 offspring would be wild (phenotypically) and a 3:1 ratio of normal to scalloped would occur in the F_2.

16. It is extremely important that one account for both the mutant genes and each of their wild type alleles.

  (a) P_1: $X^vX^v;+/+ \times X^+/Y;b^r/b^r \longrightarrow$

    F_1: 1/2 $X^+X^v;+/b^r$ (female, normal)

      1/2 $X^v/Y;+/b^r$ (male, vermilion)

    F_2:

Eye color (X)	Eye color (autosomal)
1/4 females, normal	3/4 normal = 3/16
	1/4 brown = 1/16
1/4 females, vermilion	3/4 normal = 3/16
	1/4 brown = 1/16
1/4 males, normal	3/4 normal = 3/16
	1/4 brown = 1/16
1/4 males, vermilion	3/4 normal = 3/16
	1/4 brown = 1/16

    3/16 = females, normal

    1/16 = females, brown eyes

    3/16 = females, vermilion eyes

    1/16 = females, white eyes

    3/16 = males, normal

    1/16 = males, brown eyes

    3/16 = males, vermilion eyes

    1/16 = males, white eyes

  (b) P_1: $X^+X^+;b^r/b^r \times X^v/Y;+/+ \longrightarrow$

    F_1: 1/2 $X^+X^v;+/b^r$ (female, normal)

      1/2 $X^+/Y;+/b^r$ (male, normal)

    6/16 = females, normal

    2/16 = females, brown eyes

    3/16 = males, normal

    1/16 = males, brown eyes

    3/16 = males, vermilion eyes

    1/16 = males, white eyes

  (c) 3/16 = females, normal

    1/16 = females, brown eyes

    3/16 = females, vermilion eyes

    1/16 = females, white eyes

    3/16 = males, normal

    1/16 = males, brown eyes

    3/16 = males, vermilion eyes

    1/16 = males, white eyes

18. (a) Because the denominator in the ratios is 64 one would begin to consider that there are three independently assorting gene pairs. Because there are only two characteristics (eye color and croaking) however, one might consider two gene pairs interacting for one trait.

  (b) Notice that there is a 48:16 (or 3:1) ratio of rib-it to knee-deep and a 36:16:12 (9:3:4) ratio of blue to green to purple eye color. Because of these relationships one would conclude that croaking is due to one gene pair while eye color is due to two gene pairs.

  (c) Croaking: $R\_$ = rib-it; rr = knee-deep

    $A\_B\_$ = blue eyed

    A_bb = purple

    $aaB\_$ and $aabb$ = green

  (d) Parents: $AABBrr \times AAbbRR$

    F_1: $AABbRr$

    F_2: 9/16 $AAB\_R\_$ = blue-eyed, utterer

      3/16 AAB_rr = blue-eyed, mutterer

      3/16 $AAbbR\_$ = purple-eyed, utterer

      1/16 $AAbbrr$ = purple-eyed, mutterer

20. In the first cross let the symbols be as follows:

    $WWAA \times wwaa$

    (white)   (black)

  F_1: $WwAa \times WwAa$ (all solid white)

  F_2: 9/16 $W\_A\_$ = solid white

    3/16 W_aa = solid white

    3/16 $wwA\_$ = spotted

    1/16 $wwaa$ = solid black

  So, W is epistatic to A. It would be possible to isolate a true-breeding strain of black and white spotted cattle; $wwAA$.

22. (a) (i) $BbYy$ = green parents

    (ii, iii) $B\_Y\_$ = green progeny (9/16)

      B_yy = blue progeny (3/16)

      $bbY\_$ = yellow progeny (3/16)

      $bbyy$ = albino progeny (1/16)

  (b) $BByy \times bbYY$ or $BBYY \times bbyy$

24. Cross 1 = complementation (genes are not allelic); Cross 2 = no complementation (genes are allelic), $r2$ must not be allelic to $r3$. Therefore, in cross $r2 \times r3$, complementation should occur and wild-type eye should be the result.

Chapter 5

 2. The *Protenor* form of sex determination involves the XX/XO condition while the *Lygaeus* mode involves the XX/XY condition.

 4. In *primary* nondisjunction half of the gametes contain two X chromosomes while the complementary gametes contain no X chromosomes. Fertilization, by a Y-bearing sperm cell, of those female gametes with two X chromosomes would produce the XXY Klinefelter syndrome. Fertilization of the "no-X" female gamete with a normal X-bearing sperm will produce the Turner syndrome.

6. Because attached-X chromosomes have a mother-to-daughter inheritance and the father's X is transferred to the son, one would see daughters with the white eye phenotype and sons with the miniature wing phenotype.

8. A *Barr body* is a darkly staining chromosome seen in some interphase nuclei of mammals with two X chromosomes. There will be one less Barr body than number of X chromosomes. The Barr body is an X chromosome which is considered to be genetically inactive.

10. The Lyon hypothesis states that the inactivation of the X chromosome occurs at random early in embryonic development. Such X chromosomes are in some way "marked" such that all progeny cells have the same X chromosome inactivated.

12. Phenotypic mosaicism is dependent on the heterozygous condition of genes on the two X chromosomes. Dosage compensation and the formation of Barr bodies occurs only when there are two or more X chromosomes. Males normally have only one X chromosome; therefore, such mosaicism cannot occur. Females normally have two X chromosomes. There are cases of male calico cats which are XXY.

14. In mammals, the scheme of sex determination is dependent on the presence of a piece of the Y chromosome. If present, a male is produced. In *Bonellia viridis*, the female proboscis produces some substance which triggers a morphological, physiological, and behavioral developmental pattern which produces males.

   To elucidate the mechanism, one could attempt to isolate and characterize the active substance by testing different chemical fractions of the proboscis. Secondly, mutant analysis usually provides critical approaches into developmental processes. Depending on characteristics of the organism, one could attempt to isolate mutants which lead to changes in male or female development. Third, by using micro-tissue transplantations, one could attempt to determine which "centers" of the embryo respond to the chemical cues of the female.

16. The presence of the Y chromosome provides a factor (or factors) which leads to the initial specification of maleness, specifically the development of testes. Subsequent expression of secondary sex characteristics must be dependent on the interaction of the normal X-linked *Tfm* allele with testosterone released from testes. Without such interaction, differentiation takes the female path. It is possible that *Tfm* produces an androgen receptor which is distributed in various pre-genital tissues. Mutations in the *Tfm* gene could knock out the supposed receptor. Using tagged antibodies to the receptor, one could establish by *in situ* antigen-antibody labeling, whether the receptor is present in mutant organisms. Interestingly, XXY male mice heterozygous for *Tfm* show variable proportions of androgen-insensitive and androgen-sensitive cells due to random X inactivation.

Chapter 6

4. (a) It is *possible* that two parents of moderate height can produce offspring that are much taller or shorter than either parent because segregation can produce a variety of gametes. Therefore, offspring are as illustrated below:

$$rrSsTtuu \qquad \times \qquad RrSsTtUu$$
$$\text{(moderate)} \qquad\qquad \text{(moderate)}$$

Offspring from this cross can range from very tall *RrSSTTUu* (12 "tall" units) to very short *rrssttuu* (8 "small" units).

(b) If the individual with a minimum height, *rrssttuu*, is married to an individual of intermediate height *RrSsTtUu*, the offspring can be no taller than the height of the tallest parent.

6. (a) Because the extreme phenotypes (6 cm and 30 cm) each represent 1/64 of the total, it is likely that there are three gene pairs in this cross.

   The genotypes of the parents would be combinations of alleles which would produce a 6 cm (*aabbcc*) tail and a 30 cm (*AABBCC*) tail while the 18 cm offspring would have a genotype of *AaBbCc*.

(b) A mating of an *AaBbCc* (for example) pig with the 6 cm *aabbcc* pig would result in the following offspring:

Gametes (18 cm tail)	Gamete (6 cm tail)	Offspring
ABC		AaBbCc (18 cm)
ABc		AaBbcc (14 cm)
AbC		AabbCc (14 cm)
Abc	abc	Aabbcc (10 cm)
aBC		aaBbCc (14 cm)
aBc		aaBbcc (10 cm)
abC		aabbCc (10 cm)
abc		aabbcc (6 cm)

In this example, a 1:3:3:1 ratio is the result. However, had a different 18 cm-tailed pig been selected, a different ratio would occur:

$$AABbcc \times aabbcc$$

Gametes (18 cm tail)	Gamete (6 cm tail)	Offspring
ABc	abc	AaBbcc (14 cm)
Abc		Aabbcc (10 cm)

8. For height, notice that average differences between MZ twins reared together (1.7 cm) and those MZ twins reared apart (1.8 cm) are similar (meaning little environmental influence) and considerably less than differences of DZ twins (4.4 cm) or sibs (4.5 cm) reared together. These data indicate that genetics plays a major role in determining height.

   However, for weight, notice that MZ twins reared together have a much smaller (1.9 kg) difference than MZ twins reared apart, indicating that the environment has a considerable impact on weight. By comparing the weight differences of MZ twins reared apart with DZ twins and sibs reared together, one can conclude that the environment has almost as much an influence on weight as genetics.

10. (a) The mean is computed by adding the measurements of all of the individuals, then dividing by the number of individuals. This gives a value for the mean of 140 cm.

(b) For the variance, use the formula given below (as in the text):

$$s^2 = V = n\Sigma f(x^2) - (\Sigma fx)^2 / n(n - 1)$$
$$= 374.18$$

(c) The *standard deviation* is the square root of the variance or 19.34.

(d) The *standard error of the mean* is the standard deviation divided by the square root of *n*, or about 0.70.

12. The formula for estimating heritability is

$$H^2 = V_G/V_P \text{ where } V_G \text{ and } V_P$$

are the genetic and phenotypic components of variation, respectively.

V_p is the combination of genetic and environmental variance. Because the two parental strains are inbred, they are assumed to be homozygous and the variance of 4.2 and 3.8 considered to be the result of environmental influences. The average of these two values is 4.0. The F_1 is also genetically homogeneous and gives us an additional estimation of the environmental factors.

By averaging with the parents

$$[(4.0 + 5.6)/2 = 4.8]$$

we obtain a relatively good idea of environmental impact on the phenotype. The phenotypic variance in the F_2 is the sum of the genetic (V_G) and environmental (V_E) components. We have estimated the environmental input as 4.8, so 10.3 (V_p) minus 4.8 gives us an estimate of (V_G), which is 5.5. Heritability then becomes 5.5/10.3 or 0.53. This value, when viewed in percentage form, indicates that about 53% of the variation in plant height is due to genetic influences.

14. $h^2 = (7.5 - 8.5/6.0 - 8.5) = 0.4$
    Selection will have little relative influence on olfactory learning in *Drosophila*.

16. Chromosome 2 seems to confer considerable resistance to the insecticide, somewhat in the heterozygous state and more in the homozygous state. Thus, some partial dominance is occurring.

Chapter 7

4. The fact that there is a significant maternal age effect associated with Down syndrome indicates that nondisjunction in older females contributes disproportionately to the number of Down syndrome individuals. In addition, certain genetic and cytogenetic marker data indicate the influence of female nondisjunction.

6. Because an allotetraploid has a possibility of producing bivalents at meiosis I, it would be considered the most fertile of the three. Having an even number of chromosomes to match up at the metaphase I plate, autotetraploids would be considered to be more fertile than autotriploids.

8. American cultivated cotton has 26 pairs of chromosomes; 13 large, 13 small. Old world cotton has 13 pairs of large chromosomes and American wild cotton has 13 pairs of small chromosomes. It is likely that an interspecific hybridization occurred followed by chromosome doubling. These events probably produced a fertile amphidiploid (allotetraploid).

10. While there is the appearance that crossing over is suppressed in inversion "heterozygotes," the phenomenon extends from the fact that the crossover chromatids end up being abnormal in genetic content. As such, they fail to produce viable (or competitive) gametes or lead to zygotic or embryonic death. Notice in the *Essentials* text the crossover chromatids end up genetically unbalanced.

12. In a work entitled *Evolution by Gene Duplication*, Ohno suggests that gene duplication has been essential in the origin of new genes. If gene products serve essential functions, mutation—and therefore evolution—would not be possible unless these gene products could be compensated for by products of duplicated, normal genes. The duplicated genes, or the original genes themselves, would be able to undergo mutational "experimentation" without necessarily threatening the survival of the organism.

14. The primrose, *Primula kewensis*, with its 36 chromosomes, is likely to have formed from the hybridization and subsequent chromosome doubling of a cross between the two other species, each with 18 chromosomes.

16. Let b = bent bristles; b^+ = normal bristles
    (a) $\_/b \times b^+/b^+ \longrightarrow$
    F_1: $\_/b^+$ = normal bristles
        b/b^+ = normal bristles
    F_2: $\_/b^+ \times b/b^+ \longrightarrow$
        $\_/b^+$ = normal bristles
        $\_/b$ = bent bristles
        b^+/b^+ = normal bristles
        b/b^+ = normal bristles
    (b) $\_/b^+ \times b/b \longrightarrow$
    F_1: $\_/b$ = bent bristles
        b/b^+ = normal bristles
    F_2: $\_/b \times b/b^+ \longrightarrow$
        $\_/b^+$ = normal bristles
        $\_/b$ = bent bristles
        b^+/b = normal bristles
        b/b = bent bristles

18. (a) In all probability, crossing over in the inversion loop of an inversion (in the heterozygous state) had produced defective, unbalanced chromatids, thus leading to stillbirths and/or malformed children.

    (b) It is probable that a significant proportion (perhaps 50%) of the children of the man will be similarly influenced by the inversion.

    (c) Since the karyotypic abnormality is observable, it may be possible to detect some of the abnormal chromosomes of the fetus by amniocentesis or CVS. However, depending on the type of inversion and the ability to detect minor changes in banding patterns, all abnormal chromosomes may not be detected.

20. (a) Reciprocal translocation
    (c) If the breakpoints for the translocation occurred within genes, then an abnormal phenotype may be the result. In addition, a gene's function is sometimes influenced by its position; its neighbors in other words. If such "position effects" occur, then a different phenotype may result.

    (d) It is likely that the translocation is the cause of the miscarriages. Segregation of the chromosomal elements will produce approximately half unbalanced gametes. The chance of a normal child is approximately one in two; however, half of the normal children will be translocation carriers. Sometimes, depending on the breakpoints of the translocation, genetically unbalanced children are carried to term. In many such cases, they suffer from developmental abnormalities.

    ...The genetic circumstances result from the reciprocal translocation which then causes unbalanced chromosomal complements to be present in embryos. Such genetically unbalanced embryos fail to thrive.

    ...This question is difficult to answer given the possibility that she may well have some healthy children. It all depends on segregation patterns of the chromosomes. Information is provided by the genetic counselor, but the final decision in these types of cases must rest with the couple involved.

    ...The chance of having a genetically balanced child would be approximately 0.5. However, half of that 0.5 will be translocation carriers.

Chapter 8

2. Within any region, the occurrence of two events is less likely than the occurrence of one event. If the probability of one event is $1/X$, the probability of two events occurring at the same time will be $1/X^2$.

4. dp———cl————————ap
   3 mu.   39 mu.

6. Notice that the most frequent classes are *PZ* and *pz*. These classes represent the parental (non-crossover) groups, which indicates that the original parental arrangement in the test cross was *PZ/pz* × *pz/pz*. Adding the crossover percentages together (6.9 + 7.1) gives 14%, which would be the map distance between the two genes.

8. (a) P₁: *sc s v/sc s v* × + + +/Y
P₁: + + +/*sc s v* × *sc s v*/Y

(b) The map distances are determined by first writing the proper arrangement and sequence of genes, then computing the distances between each set of genes. *sc   v   s*
                                                                                    +   +   +

sc − v = 33%(map units)
v − s = 10%(map units)

(c) The coefficient of coincidence = .727, which indicates that there were fewer double crossovers than expected, therefore positive chromosomal interference is present.

10. The map distance of a gene to the centromere in *Neurospora* is determined by dividing the percentage of second division asci (tetrads) by two. Patterns other than *BBbb* or *bbBB* are second division as discussed in the text and represent a crossover between the gene in question and the centromere. In the data given, the percentage of second division segregation is 20/100 or 20%. Dividing by 2 (because only two of the four chromatids are involved in any single crossover event) gives 10 map units.

12. Since *Stubble* is a dominant mutation (and homozygous lethal), one can determine whether it is heterozygous (*Sb/+*) or homozygous wild type (*+/+*). One would use the typical test cross arrangement with the *curled* gene so the arrangement would be + *cu/+ cu*.

14. The map for parts (a) and (b) is the following:

$$d \ldots b \ldots pr \ldots vg \ldots c \ldots adp$$
$$31 \quad 48 \quad 54 \quad 67 \quad 75 \quad 83$$
Map Units

The expected map units between *d* and *c* would be 44, *d* and *vg* would be 36, and *d* and *adp* 52, however because there is a theoretical maximum of 50 map units possible between two loci in any one cross, that distance would be below the 52 determined by simple subtraction.

16. (a) The *short* gene is on chromosome 2.

(b) The map distance between *black* and *short* is 15.

Chapter 9

2. (a) The requirement for physical contact between bacterial cells during conjugation was established by placing a filter in a U-tube so that the medium can be exchanged but the bacteria cannot come in contact. Under this condition, conjugation does not occur.

(b) By treating cells with streptomycin, an antibiotic, it was shown that recombination would not occur if one of the two bacterial strains was inactivated. However, if the other was similarly treated, recombination would occur. Thus, directionality was suggested, with one strain being a donor strain and the other being the recipient.

(c) An F⁺ bacterium contains a circular, double-stranded, structurally independent, DNA molecule which can direct recombination. In Hfr cells, the F factor is integrated into the bacterial chromosome.

4. Mapping the chromosome in an Hfr × F⁻ cross takes advantage of the oriented transfer of the bacterial chromosome through the conjugation tube. For each F type, the point of insertion and the direction of transfer are fixed, therefore breaking the conjugation tube at different times produces partial diploids with corresponding portions of the donor chromosome being transferred. The length of the chromosome being transferred is contingent on the duration of conjugation, thus mapping of genes is based on time.

6. In an Hfr × F⁻ cross, the F factor is directing the transfer of the donor chromosome. It takes approximately 90 minutes to transfer the entire chromosome. Because the F factor is the last element to be transferred and the conjugation tube is fragile, the likelihood for complete transfer is low.

8. In their experiment, a filter that would not allow contact was placed between the two auxotrophic strains. F-mediated conjugation requires contact and without that contact such conjugation cannot occur. The treatment with DNase showed that the filterable agent was not naked DNA.

10. Starting with a single bacteriophage, one lytic cycle produces 200 progeny phage, three more lytic cycles would produce (200)⁴ or 1,600,000,000 phage.

12. Notice that the incorporation of loci *a⁺* and *b⁺* occurs much more frequently than the incorporation of *b⁺* and *c⁺* together (210 to 1) and the incorporation of all three genes *a⁺b⁺c⁺* occurs relatively infrequently. If *a* and *b* loci are close together and both are far from locus *c*, then fewer crossovers would be required to incorporate the two linked loci compared to all three loci. If all three loci were close together, then the frequency of incorporation of all three would be similar to the frequency of incorporation of any two contiguous loci, which is not the case.

14. (a) When nutrients A and B are added, selection is for *c*. When nutrients B and C are added, selection is for *a*. When nutrients A and C are added, selection is for *b*.

(b) *b a* _____ *c* _____ F

Chapter 10

2. Prior to 1940, most of the interest in genetics centered on the transmission of similarity and variation from parents to offspring (transmission genetics). While some experiments examined the possible nature of the hereditary material, abundant knowledge of the structural and enzymatic properties of proteins generated a bias which worked to favor proteins as the hereditary substance. In addition, proteins were composed of as many as twenty different subunits (amino acids), thereby providing ample structural and functional variation for the multiple tasks which must be accomplished by the genetic material. The tetranucleotide hypothesis (structure) provided insufficient variability to account for the diverse roles of the genetic material.

4. Nucleic acids contain large amounts of phosphorus and no sulfur, whereas proteins contain sulfur and no phosphorus. Therefore, the radioisotopes ³²P and ³⁵S will selectively label nucleic acids and proteins, respectively.

The Hershey and Chase experiment is based on the premise that the substance injected into the bacterium is the substance responsible for producing the progeny phage and therefore must be the hereditary material. The experiment demonstrated that most of the ³²P-labeled material (DNA) was injected while the phage ghosts (protein coats) remained outside the bacterium. Therefore, the nucleic acid must be the genetic material.

6. Some viruses contain a genetic material composed of RNA. The tobacco mosaic virus is composed of an RNA core and a protein coat. "Crosses" can be made in which the protein coat and RNA of TMV are interchanged with another strain (Holmes ribgrass). The source of the RNA determines the type of lesion, thus, RNA is the genetic material in these viruses. Retroviruses contain RNA as the genetic material and use an enzyme known as *reverse transcriptase* to produce DNA, which can be integrated into the host chromosome. See Figure 9.1.

8. Examine the structures of the bases in the *Essentials* text. The other bases would be named as follows:

Guanine: 2-amino-6-oxypurine
Cytosine: 2-oxy-4-aminopyrimidine
Thymine: 2,4-dioxy-5-methylpyrimidine
Uracil: 2,4-dioxypyrimidine

10. In addition to creative "genius" and perseverance, model-building skills, and the conviction that the structure would turn out to be "simple" and have a natural beauty in its simplicity, Watson and Crick employed the X-ray diffraction information of Franklin and Wilkins, and the base ratio information of Chargaff.

12. Three main differences between RNA and DNA are the following: (1) uracil in RNA replaces thymine in DNA; (2) ribose in RNA replaces deoxyribose in DNA; and (3) RNA often occurs as both single- and partially double-stranded forms, whereas DNA most often occurs in a double-stranded form.

14. In order for hydrogen bonds to form between complementary base pairs, complementary strands of nucleic acids must be in proximity. For a given concentration of nucleic acids, the more copies of a given type of nucleic acid that are present, the higher the likelihood that complementary strands will be close to each other. Conversely, unique sequences of nucleic acids have a lower probability of interacting because there are fewer of them; thus, the likelihood of forming hydrogen bonds (hybridizing) is less.

16. Without knowing the exact bonding characteristics of hypoxanthine or xanthine, it may be difficult to predict the likelihood of each pairing type. It is likely that both are of the same class (purine or pyrimidine) because the names of the molecules indicate a similarity. In addition, the diameter of the structure is constant which, under the model to follow, would be expected. In fact, hypoxanthine and xanthine are both purines.

Because there are equal amounts of A, T and H, one could suggest that they are hydrogen bonded to each other; the same may be said for C, G, and X. Given the molar equivalence of erythrose and phosphate, an alternating sugar-phosphate-sugar backbone as in "earth-type" DNA would be acceptable. A model of a triple helix would be acceptable, since the diameter is constant. Given the chemical similarities to "earth-type" DNA, it is probable that the unique creature's DNA follows the same structural plan.

18. Since cytosine pairs with guanine and uracil pairs with adenine, the result would be a base substitution of G:C to A:T after DNA replication.

20. Repetitive sequences renature relatively quickly because the likelihood of complementary strands interacting increases. For curve *A* in the problem, there is evidence for a rapidly renaturing species (repetitive) and a slowly renaturing species (unique). The fraction which reassociates faster than the *E. coli* DNA is highly repetitive and the last fraction (with the highest $C_0t_{1/2}$ value) contains primarily unique sequences. Fraction *B* contains mostly unique, relatively complex DNA.

Chapter 11

2. Under a conservative scheme, the first round of replication in ^{14}N medium produces one dense double helix and one "light" double helix, in contrast to the intermediate density of the DNA in the semiconservative mode. Therefore, after one round of replication in the ^{14}N medium, the conservative scheme can be ruled out. After one round of replication in ^{14}N under a dispersive model, the DNA is of intermediate density, just as it is in the semiconservative model. However, in the next round of replication in ^{14}N medium, the density of the DNA is between the intermediate and "light" densities.

4. The *in vitro* replication requires a DNA template, a divalent cation (Mg^{++}), and all four of the deoxyribonucleoside triphosphates: dATP, dCTP, dTTP, and dGTP.

6. ϕX174 is a well-studied, single-stranded virus (phage) which can be easily isolated. It has a relatively small DNA genome (5500 nucleotides) which, if mutated, usually alters its reproductive cycle.

8. All three enzymes share several common properties. First, none can initiate DNA synthesis on a template but all can elongate an existing DNA strand, assuming there is a template strand. Polymerization of nucleotides occurs in the 5' to 3' direction where each 5' phosphate is added to the 3' end of the growing polynucleotide. All three enzymes are large complex proteins with a molecular weight in excess of 100,000 daltons and each has 3' to 5' exonuclease activity. Refer to *Essentials* for a listing of enzyme properties.

10. Okazaki fragments are relatively short (1000 to 2000 bases in prokaryotes) DNA fragments which are synthesized in a discontinuous fashion on the lagging strand during DNA replication. Such fragments appear to be necessary because template DNA is not available for 5' → 3' synthesis until some degree of continuous DNA synthesis occurs on the leading strand in the direction of the replication fork. The isolation of such fragments provides support for the scheme of replication shown in the *Essentials* text.

DNA ligase is required to form phosphodiester linkages in gaps which are generated when DNA polymerase I removes RNA primer and meets newly synthesized DNA ahead of it.

Notice in the text that the discontinuous DNA strands are ligated together into a single continuous strand. Primer RNA is formed by RNA primase to serve as an initiation point for the production of DNA strands on a DNA template. None of the DNA polymerases are capable of initiating synthesis without a free 3' hydroxyl group. The primer RNA provides that group and thus can be used by DNA polymerase III.

12. Eukaryotic DNA is replicated in a manner which is very similar to that of *E. coli*. Synthesis is bidirectional, continuous on one strand and discontinuous on the other, and the requirements of synthesis (four deoxyribonucleoside triphosphates, divalent cation, template, and primer) are the same. Okazaki fragments of eukaryotes are about one-tenth the size of those in bacteria.

Because there is a much greater amount of DNA to be replicated and DNA replication is slower, there are multiple initiation sites for replication in eukaryotes and increased amouts of DNA polymerase in contrast to the single replication origin in prokaryotes. Replication occurs at different sites during different intervals of the S phase. The proposed functions of four DNA polymerases are described in the text.

14. *Gene conversion* is likely to be a consequence of genetic recombination in which nonreciprocal recombination yields products that appear to be one allele "converted" to another. Gene conversion is now considered a result of heteroduplex formation that is accompanied by mismatched bases. When these mismatches are corrected, the "conversion" occurs.

16. (a) In *E. coli*, 100 kb are added to each growing chain per minute.

Therefore, the chain should be about 4,000,000 bp.

(b) Given $(4 \times 10^6$ bp$) \times 0.34$ nm/bp $= 1.36 \times 10^6$ nm or 1.3 mm

Chapter 12

2. (a) GGG = 3/4 × 3/4 × 3/4 = 27/64
   GGC = 3/4 × 3/4 × 1/4 = 9/64
   GCG = 3/4 × 1/4 × 3/4 = 9/64
   CGG = 1/4 × 3/4 × 3/4 = 9/64

$$CCG = 1/4 \times 1/4 \times 3/4 = 3/64$$
$$CGC = 1/4 \times 3/4 \times 1/4 = 3/64$$
$$GCC = 3/4 \times 1/4 \times 1/4 = 3/64$$
$$CCC = 1/4 \times 1/4 \times 1/4 = 1/64$$

(b) Glycine: GGG and one G_2C (adds up to 36/64)
Alanine: one G_2C and one C_2G (adds up to 12/64)
Arginine: one G_2C and one C_2G (adds up to 12/64)
Proline: one C_2G and CCC (adds up to 4/64)

(c) With the wobble hypothesis, variation can occur in the third position of each codon.
Glycine: GGG, GGC
Alanine: CGG, GCC, CGC, GCG
Arginine: GCG, GCC, CGC, CGG
Proline: CCC, CCG

4. As in the previous problem, the procedure is to find those sequences which are the same for the first two bases but which vary in the third base. Given that AGG = arg, then information from the AG copolymer indicates that AGA also codes for arg and GAG must therefore code for glu. Coupling this information with that of the AAG copolymer, GAA must also code for glu, and AAG must code for lys.

6. Apply the most conservative pathway of change.

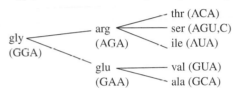

gly (GGA)
arg (AGA) — thr (ACA), ser (AGU,C), ile (AUA)
glu (GAA) — val (GUA), ala (GCA)

8. Because Poly U is complementary to Poly A, double-stranded structures will be formed. In order for an RNA to serve as a messenger RNA it must be single-stranded, thereby exposing the bases for interaction with ribosomal subunits and tRNAs.

10. Given the sequence GGA, by changing each of the bases to the remaining three bases, then checking the code table, one can determine whether amino acid substitutions will occur.

G G A > gly	G G U > gly
U G A > **term**	G G C > gly
C G A > **arg**	G G A > gly
A G A > **arg**	G G G > gly
G U A > **val**	U G U > **cys**
G C A > **ala**	C G U > **arg**
G A A > **glu**	A G U > **ser**
G G U > gly	G U U > **val**
G G C > gly	G C U > **ala**
G G A > gly	G A U > **asp**

12. The central dogma of molecular genetics—and to some, all of biology—states that DNA produces, through transcription, RNA, which is "decoded" (during translation) to produce proteins.

14. RNA polymerase from *E. coli* is a complex, large (almost 500,000 daltons) molecule composed of subunits (α, β, β', σ) in the proportion $\alpha2$, β, β', σ for the holoenzyme. The β subunit provides catalytic function and the sigma (σ) subunit is involved in recognition of specific promoters. The core enzyme is the protein without the sigma subunit.

16. (a) Sequence 1: GAAAAAACGGUA
Sequence 2: UGUAGUUAUUGA
Sequence 3: AUGUUCCCAAGA

(b) Sequence 1: *glu-lys-thr-val*
Sequence 2: *cys-ser-tyr*
Sequence 3: *met-phe-pro-arg*

(c) Because there are no punctuation codons in Sequence 1, it must be in the middle. Sequence 2, having a termination codon (UGA), must be in the terminal portion. Sequence 3, with the AUG starting codon, must be in the initial portion.

Chapter 13

2. Transfer RNAs are "adaptor" molecules in that they provide a way for amino acids to interact with sequences of bases in nucleic acids. Amino acids are specifically and individually attached to the 3' end of tRNAs which possess a three-base sequence (the anticodon) to base-pair with three bases of mRNA. Messenger RNA, on the other hand, contains a copy of the triplet codes which are stored in DNA. The sequences of bases in mRNA interact, three at a time, with the anticodons of tRNAs.

Enzymes involved in transcription include the following: RNA polymerase (*E. coli*), and RNA polymerase I, II, III (eukaryotes). Those involved in translation include the following: aminoacyl tRNA synthetases, peptidyl transferase, and GTP-dependent release factors.

4. The sequence of base triplets in mRNA constitutes the sequence of codons. A three-base portion of the tRNA constitutes the anticodon.

6. An amino acid in the presence of ATP, Mg^{++}, and a specific aminoacyl synthetase produces an amino acid-AMP enzyme complex (+ PP_i). This complex interacts with a specific tRNA to produce the aminoacyl tRNA.

8. Phenylalanine is an amino acid which, like other amino acids, is required for protein synthesis. While too much phenylalanine and its derivatives cause PKU in phenylketonurics, too little will restrict protein synthesis.

10. Tyrosine is a precursor to melanin, skin pigment. Individuals with PKU fail to convert phenylalanine to tyrosine. Although tyrosine is obtained from the diet, at the population level, individuals with PKU have a tendency for less skin pigmentation.

12. (X = precursor to AA)

trp-8 *trp*-2 *trp*-3 *trp*-1
X $\longrightarrow$ AA $\longrightarrow$ IGP $\longrightarrow$ I $\longrightarrow$ TRY

14. Some proteins are made up of subunits, each different type of subunit (polypeptide chain) being under the control of a different gene. Under this circumstance, the *one-gene:one-polypeptide* might be more reasonable.

Many functions of cells and organisms are controlled by stretches of DNA which either produce no protein product (operator and promoter regions, for example) or have more than one function (as in the case of overlapping genes and differential mRNA splicing). A simple statement regarding the relationship of a stretch of DNA to its physical product is difficult to formulate.

16. Sickle-cell anemia is coined a molecular disease because it is well understood at the molecular level; at the level of a base change in DNA which leads to an amino acid change in the β chain of hemoglobin. It is a *genetic* disease in that it is inherited from one generation to the next. It is not contagious as might be the case of a disease caused by a microorganism. Diseases caused by microorganisms may not necessarily follow family blood lines whereas genetic diseases do.

18. It is possible for an amino acid to change without changing the electrophoretic mobility of a protein under standard conditions. If the amino acid is substituted with an amino acid of like charge and similar structure, there is a chance that factors which influence electrophoretic mobility (primarily net charge) will not be altered. Other techniques, such as chromatography of digested peptides, may detect subtle amino acid differences.

22. Enzymes function to regulate catabolic and anabolic activities of cells. They influence (lower) the *energy of activation* thus allowing chemical reactions to occur under conditions which are compatible with living systems. Enzymes possess *active sites* and/or other domains which are sensitive to the environment. The active site is considered to be a crevice, or pit, which binds reactants, thus enhancing their interaction. The other domains mentioned above may influence the conformation and therefore function of the active site.

24. One can conclude that the amino acid is not involved in recognition of the codon.

Chapter 14

2. Each gene and its product function in an environment which has also evolved, or co-evolved. Deviations from the norm, caused by mutation, are likely to be disruptive because of the complex and interactive environment in which each gene product must function. However, on occasion a beneficial variation occurs.

4. Let *II* indicate a *mutagenized* second chromosome

$$\text{Females } Cy\ L/Pm \quad \times \quad \text{males } +/+$$

$$\text{F}_1 \text{ males } Cy\ L/II \quad \times \quad Cy\ L/Pm$$
(individual crosses)
$$\text{F}_2 \text{ females } Cy\ L/II \quad \times \quad Cy\ L/II$$
(individual crosses)

F$_3$ genotypes:
  Cy L/Cy L (lethal)
  Cy L/II (Curly, Lobe)
  II/II (dies if recessive lethal)

6. Frameshift mutations are likely to change more than one amino acid in a protein product because as the reading frame is shifted, new codons are generated. In addition, there is the possibility that a nonsense triplet could be introduced, thus causing premature chain termination. If a single pyrimidine or purine has been substituted, then only one amino acid is influenced.

8. Because mammography involves the use of X-rays and X-rays are known to be mutagenic, it has been suggested that frequent mammograms may do harm. Nevertheless, the risk of cancer from mammograms is more than offset by the benefit of detection of cancer.

10. X-rays are of higher energy and shorter wavelength than UV light. They have greater penetrating ability and can create more disruption of DNA.

12. Your study should include examination of the following short-term aspects: immediate assessment of radiation amounts distributed in a grid within the bomb site as well as a control area not receiving bomb-induced radiation, radiation exposure as measured by radiation sickness and evidence of radiation poisoning from tissue samples, abortion rates, birthing rates, and chromosomal studies.

Long-term assessment should include: sex-ratio distortion (males being more influenced by X-linked recessive lethals than females), chromosomal studies, birth and abortion rates, cancer frequency and type, and genetic disorders. In each case data should be compared to the control site to see if changes are bomb-related.

In addition, to attempt to determine cause-effect, it is often helpful to show a dose response. Thus, by comparing the location of individuals at the time of exposure to the matrix of radiation amounts, one may be able to determine whether those most exposed to radiation suffer the most physiologically and genetically. If a positive correlation is observed, then statistically significant conclusions may be possible.

14. Each involves a ballooning of trinucleotide repeats. Genetic anticipation is the occurrence of an earlier age of onset of a genetic disease in successive generations.

Chapter 15

2. Under *negative* control, the regulatory molecule interferes with transcription, while in *positive* control, the regulatory molecule stimulates transcription. Negative control is seen in the *lactose* and *tryptophan* systems.

4. Refer to the *Essentials* text and to Figure 15.2 to get a good understanding of the lactose system before starting.

$I^+O^+Z^+$ = **Inducible**
$I^-O^+Z^+$ = **Constitutive**
$I^+O^cZ^+$ = **Constitutive**
$I^-O^+Z^+/F'I^+$ = **Inducible**
$I^+O^cZ^+/F'O^+$ = **Constitutive**
$I^sO^+Z^+$ = **Repressed**
$I^sO^+Z^+/F'I^+$ = **Repressed**

6. To determine which gene is the structural gene, look for the IE function and see that it is related to c^-. Therefore, c codes for the **structural gene**. Because when b is mutant, no enzyme is produced, b must be the **promoter**.

Notice that when genes a and d are mutant, constitutive synthesis occurs, therefore one must be the operator and the other gene codes for the repressor protein. To distinguish these functions, one must remember that the repressor operates as a diffusible substance and can be on the host chromosome or the F factor (functioning in *trans*). However, the operator can only operate in *cis*. In addition, in *cis*, the constitutive operator is dominant to its wild-type allele, while the mutant repressor is recessive to its wild-type allele.

Notice that the mutant a gene is dominant to its wild-type allele, whereas the mutant d allele is recessive (behaving as wild type in the first row). Therefore, the a locus is the **operator** and the d locus is the **repressor** gene.

8. (a) Because activated CAP is a component of the cooperative binding of RNA polymerase to the *lac* promoter, absence of a functional *crp* would compromise the positive control exhibited by CAP.

   (b) Without a CAP binding site there would be a reduction in the inducibility of the lac operon.

10. *Organization of DNA*: Changes in DNA/chromosome structure can influence overall gene output: ***transcription, processing and transport, translation***.

12. Transcription factors are modular structures which are not part of the RNA polymerase but are needed for the initiation of transcription. They generally have two functional domains: one that binds to DNA and the other that activates transcription through a variety of protein–protein interactions.

14. Alteration of chromatin structure through interaction of enhancers with a variety of transcription factors creates bending or looping which has regulatory capacity. In altered configurations transcription can be stimulated. By contrast, posttranscriptional forms of regulation include modification of the RNA product: 5' capping, 3' polyadenylation, intron removal, and alternative processing. In addition, a variety of processes influence mRNA's availability for translation; stability, alternative processing, etc.

16. The first two sentences in the problem indicate an inducible system where oil stimulates the production of a protein(?) which turns on (positive control) genes to metabolize oil. The different results in strains 2 and 4 suggest a *cis*-acting system. Because the operon by itself (when mutant as in strain 3) gives constitutive synthesis of the structural genes, *cis*-acting system is also supported. The *cis*-acting element is most likely part of the operon.

Chapter 16

2. Reverse transcriptase is often used to promote the formation of cDNA (complementary DNA) from an mRNA molecule. Eukaryotic mRNAs typically have a 3′ poly-A tail for which a poly-dT segment provides a double-stranded section that serves to prime the production of the complementary strand.

4. The question of protein/DNA recognition and interaction is a difficult one to answer. Much research has been done to attempt to understand the nature of the specificity of such interactions. In general it is believed that the protein interacts with the major groove of the DNA helix. This information comes from the structure of the few proteins which have been sufficiently well studied to suggest that the DNA major groove and "fingers" or extensions of the protein form the basis of interaction.

6. The segment contains the palindromic sequence below which is recognized by the restriction enzyme *Bam*HI.

CCTAGG

GGATCC

8. A filter is used to bind the DNA from the colonies and a labeled probe is used to detect, through hybridization, the DNA of interest. Cells with the desired clone are then picked from the original plate and the plasmid is isolated from the cells.

10. One factor has to do with the possibility that the reverse transcriptase may not completely synthesize the DNA from the RNA template. The other reason may be that the 3′ end of the copied DNA tends to fold back on itself, thus providing a primer for the DNA polymerase. Additional preparation of the cDNA requires some digestion at the folded region. Since this folded region corresponds to the 5′ end of the mRNA, some of the message is often lost.

12. With repeated sequences in the genome, chromosome walking is complicated because the clone hybridizes to multiple regions. One can "chromosome jump" over repeated sequences. One can block repeats by addition of complementary repetitive DNA.

Chapter 17

2. General similarities and differences:

mtDNA	cpDNA
circular	circular
double-stranded	double-stranded
semiconservative replication	semiconservative replication
animal (16 to 18 kb)	>100 kb
plant (>100 kb)	genes (rRNAs, tRNAs, etc.)
genes (rRNAs, tRNAs, etc.)	
diverse (introns in some)	
variations in genetic code	

4. Puffs represent active genes as evidenced by staining and uptake of labeled RNA precursors as assayed by autoradiography.

6. While greater DNA content per cell is associated with eukaryotes, one cannot universally equate genomic size with an increase in organismic complexity. There are numerous examples where DNA content per cell varies considerably among closely related species. Because of the diverse cell types of multicellular eukaryotes, a variety of gene products is required, which may be related to the increase in DNA content per cell. In addition, the advantage of diploidy automatically increases DNA content per cell. However, seeing the question in another way, it is likely that a much higher *percentage* of the genome of a prokaryote is actually involved in phenotype production than in a eukaryote.

Eukaryotes have evolved the capacity to obtain and maintain what appears to be large amounts of "extra" perhaps "junk" DNA. This concept will be examined in subsequent chapters of the text. Prokaryotes on the other hand, with their relatively short life cycle, are extremely efficient in their accumulation and use of their genome.

Given the larger amount of DNA per cell and the requirement that the DNA be partitioned in an orderly fashion to daughter cells during cell division, certain mechanisms and structures (mitosis, nucleosomes, centromeres, etc.) have evolved for *packaging* the DNA. In addition, the genome is divided into separate entities (chromosomes) to perhaps facilitate the partitioning process in mitosis and meiosis.

8. The endosymbiont theory states that mitochondria and chloroplasts evolved from free-living bacteria-like structures that entered a symbiotic relationship with a host cell. In time, both the bacteria-like particle and the host cell became interdependent through the loss and sharing of products. In addition to the size (sedimentation properties) of ribosomes, the ribosomes of mitochondria and chloroplasts show antibiotic sensitivity which is similar to present-day bacteria, thus supporting the endosymbiont theory.

10. (a) Since there are 200 base pairs per nucleosome (as defined in this problem) and 10^9 base pairs, there would be 5×10^6 nucleosomes.

(b) Given that there are 6 nucleosomes per solenoid, there would be $.833 \times 10^5$ solenoids.

(c) Since there are 5×10^6 nucleosomes and nine histones (including H1) per nucleosome, there must be $9(5 \times 10^6)$ histone molecules: 4.5×10^7.

(d) Since there are 10^9 base pairs present and each base pair is 3.4 Å, the overall length of the DNA is 3.4×10^9 Å. Dividing this value by the packing ratio (50) gives 6.8×10^7 Å.

12. One base pair occupies 0.34 nm, therefore the equation would be as follows:

$$52 \ \mu m/(0.34 \ nm/bp) \times 1000 \ nm/\mu m = 152,941 \text{ base pairs}$$

Chapter 18

2. While greater DNA content per cell is associated with eukaryotes, one can not universally equate genomic size with an increase in organismic complexity. There are numerous examples where DNA content per cell varies considerably among closely related species. Because of the diverse cell types of multicellular eukaryotes, a variety of gene products is required, which may be related to the increase in DNA content per cell. In addition, the advantage of diploidy automatically increases DNA content per cell. However, seeing the question in another way, it is likely that a much higher percentage of the genome of a prokaryote is actually involved in phenotype production than in a eukaryote. Eukaryotes have evolved the capacity to obtain and maintain what appears to be large amounts of "extra" perhaps "junk" DNA. Prokaryotes on the other hand, with their relatively short life cycle, are extremely efficient in their accumulation and use of their genome. Given the larger amount of DNA per cell and the requirement that the DNA be partitioned in an orderly fashion to daughter cells during cell division, certain mechanisms and structures (mitosis, nucleosomes, centromeres, etc.) have evolved for *packaging* the DNA. In addition, the genome is divided into separate entities (chromosomes) to perhaps facilitate the partitioning process in mitosis and meiosis.

4. While the β-globin gene family is a relatively large (60 kb) sequence and restriction analyses show that it is composed of six genes, one is a pseudogene and therefore does not produce a product. The five functional genes each contain two similarly-sized introns which when included with non-coding flanking regions (5′ and 3′), and spacer DNA between genes, accounts for the 95% mentioned in the question.

6. The number of combinations is determined by a simple multiplication of the number of genes in each class: V × D × J × C. Thus in this case the answer would be 10 V × 30 D × 50 J × 3 C = 45,000.

Chapter 19

2. Glyphosate (a herbicide) inhibits EPSP, a chloroplast enzyme involved in the synthesis of the amino acids phenylalanine, tyrosine, and tryptophan. To generate glyphosate resistance in crop plants, a fusion gene was created which introduced a viral promoter to control the EPSP synthetase gene. The fusion product was placed into the Ti vector and transferred to *A. tumifaciens*, which was used to infect crop cells. Calluses were selected on the basis of their resistance to glyphosate. Resistant calluses were later developed into transgenic plants.

   There is a remote possibility that such an "accident" can occur. However, in retracing the steps to generate the resistant plant in the first place, it seems more likely that the trait will not "escape" from the plant; rather, the engineered *A. tumifaciens* may escape, infect and transfer glyphosate resistance to pest species.

4. *Drosophila* is a unique experimental organism in that there is a vast knowledge of its genetics, it is easily cultured and genetically manipulated, and it contains unique chromosomes, polytene chromosomes, which allow visual landmarks. Coupled with probe-labeling (sequence tagged sites), the visible landmarks (chromomeres) and ease of manipulation allow one to actually see where important genes are located in chromosomes. *Drosophila* also contains P elements which allow sequence markers to be inserted into the genome. Microdissection of chromosomes is also useful in developing specific clones for sequencing. In addition, techniques have been developed (*in situ* hybridization) to allow scientists to actually determine the distributions of gene activities in all tissues of the organism.

6. Positional cloning relies on segregation (Mendelian) and linkage analysis. Given the numerous limitations associated with such analyses in human populations (family and sample size, etc.), it is unlikely that this technique will be successfully applied to genetically complex traits in the near future.

8. It will hybridize to the normal DNA sequence because it is fully complementary to the bottom normal strand.

10. The answer provided here is based on the condition that individual I-2 is a carrier and the son, II-4, has the disorder. The 3 kb fragment occurs in the normal I-1 father and the normal son II-1. The affected son, II-4, has the 4 kb fragment. One daughter, II-2, is a carrier while the other daughter, II-3, is not a carrier.

12. The two major problems described here are common concerns related to genetic engineering. The first is the localization of the introduced DNA into the target tissue and target location in the genome. Inappropriate targeting may have serious consequences. In addition, it is often difficult to control the output of introduced DNA. Genetic regulation is complicated and subject to a number of factors including upstream and downstream signals as well as various posttranscriptional processing schemes. Artificial control of these factors will prove difficult.

Chapter 20

2. Many of the appendages of the head, including the mouth parts and the antennae are evolutionary derivatives of ancestral leg structures. In *spineless aristapedia*, the distal portion of the antenna is replaced by its ancestral counterpart, the distal portion of the leg (tarsal segments). Because the replacement of the arista (end of the antenna) can occur by a mutation in a single gene, one would consider that one "selector" gene distinguishes aristal from tarsal structures. Notice that a "one-step" change is involved in the interchange of leg and antennal structures.

4. There are several somewhat indirect methods for determining transcriptional activity of a given gene in different cell types. First, if protein products of a given gene are present in different cell types, it can be assumed that the responsible gene is being transcribed. Second, if one is able to actually observe, microscopically, gene activity, as is the case in some specialized chromosomes (polytene chromosomes), gene activity can be inferred by the presence of localized chromosomal puffs.

   A more direct and common practice to assess transcription of particular genes is to use labeled probes. If a labeled probe can be obtained which contains base sequences that are complementary to the transcribed RNA, then such probes will hybridize to that RNA if present in different tissues. This technique is called *in situ* hybridization and is a powerful tool in the study of gene activity during development.

6. Because in *ftz/ftz* embryos, the engrailed product is absent and in *en/en* embryos *ftz* expression is normal, one can conclude that the *ftz* gene product regulates, either directly or indirectly, *en*.

8. The fact that nuclei from almost any source remain transcriptionally and translationally active substantiates the fact that the genetic code and the ancillary processes of transcription and translation are compatible throughout the animal and plant kingdoms. Because the egg represents an isolated, "closed" system which can be mechanically, environmentally, and to some extent biochemically manipulated, various conditions may be developed which allow one to study facets of gene regulation. Combinations of injected nuclei may reveal nuclear–nuclear interactions which could not normally be studied by other methods.

Chapter 21

2. Review Chapter 2 in the text and note that the following stages of the cell cycle are discussed: G1, G0, S, G2. The G1 stage begins after mitosis and is involved in the synthesis of many cytoplasmic elements. In the S phase DNA synthesis occurs. G2 is a period of growth and preparation for mitosis. Most cell cycle time variation is caused by changes in the duration of G1. G0 is the non-dividing state.

4. Kinases regulate other proteins by adding phosphate groups. Cyclins bind to the kinases, switching them on and off. Several cyclins, including D and E, can move cells from G1 to S. At the G2/mitosis border a CDK1 (cyclin dependent kinase) combines with another cyclin (cyclin B). Phosphorylation occurs, bringing about a series of changes in the nuclear membrane, cytoskeleton, and histone 1.

6. A tumor suppressor gene is a gene that normally functions to suppress cell division. Since tumors and cancers represent a significant threat to survival and therefore Darwinian fitness, strong evolutionary forces would favor a variety of co-evolved and perhaps complex conditions in which mutations in these suppressor genes would be recessive. Looking at it in another way, if a tumor suppressor gene makes a product that regulates the cell cycle favorably, cellular conditions have evolved in such a way that sufficient quantities of this gene product are made from just one gene (of the two present in each diploid individual) to provide normal function.

8. A translocation involving exchange of genetic material between chromosomes 9 and 22 is responsible for the generation of the "Philadelphia chromosome." Genetic mapping established that certain oncogenes were combined to form a hybrid gene that encodes a 200 kd protein which has been implicated in the formation of chronic myelocytic leukemia.

10. Any agent which causes damage to DNA is a potential carcinogen since cell cycle control is achieved by gene (DNA) products, known as proteins. Since cigarette smoke is known to contain an agent which changes DNA, in this case transversions, numerous modified gene products (including cell cycle controlling proteins) are likely to be produced. The fact that many cancer patients have such transversions in *p53* strongly suggests that cancer is caused by agents in cigarette smoke.

Chapter 22

2. Calculate *p* and *q*, then apply the equation $p^2 + 2pq + q^2$ to determine genotypic frequencies in the next generation.

$$p = \text{frequency of } A$$
$$= 0.2 + .3$$
$$= 0.5$$
$$q = 1 - p = 0.5$$
Frequency of $AA = p^2$
$$= (.5)^2$$
$$= .25 \text{ or } 25\%$$
Frequency of $Aa = 2pq$
$$= 2(.5)(.5)$$
$$= .5 \text{ or } 50\%$$
Frequency of $aa = q^2$
$$= (.5)^2$$
$$= .25 \text{ or } 25\%$$

The initial population was not in equilibrium, however, after one generation of mating under the Hardy–Weinberg assumptions, the population is in equilibrium and will continue to be so (and not change) until one or more of the Hardy–Weinberg assumptions is not met.

4. In order for the Hardy–Weinberg equations to apply, the population must be in equilibrium.

6. (a) The equilibrium values will be as follows:
Frequency of $1/1$ $= p^2 = (.7755)^2$
$$= .6014 \text{ or } 60.14\%$$
Frequency of $1/\Delta32$ $= 2pq$
$$= 2(.7755)(.2245)$$
$$= .3482 \text{ or } 34.82\%$$
Frequency of $\Delta32/\Delta32 = q^2 = (.2245)^2$
$$= .0504 \text{ or } 5.04\%$$

Comparing these equilibrium values with the observed values strongly suggests that the observed values are drawn from a population in equilibrium.

(b) The equilibrium values will be as follows:
Frequency of AA $= p^2 = (.877)^2$
$$= .7691 \text{ or } 76.91\%$$
Frequency of AS $= 2pq = 2(.877)(.123)$
$$= .2157 \text{ or } 21.57\%$$
Frequency of SS $= q^2 = (.123)^2$
$$= .0151 \text{ or } 1.51\%$$

Comparing these equilibrium values with the observed values suggests that the observed values may be drawn from a population which is not in equilibrium.

$$\chi^2 = \frac{\Sigma(o - e)^2}{e}$$
$$= 1.47$$

There is one degree of freedom even though there are three classes. Since the χ^2 value calculated here is smaller, the null hypothesis (the observed values fluctuate from the equilibrium values by chance and chance alone) should not be rejected. Thus the frequencies of *AA, AS, SS* sampled a population which is in equilibrium.

8. (a) $q_{g+1} = .278; p_{g+1} = .722$
(b) $q_{g+1} = .289; p_{g+1} = .711$
(c) $q_{g+1} = .298; p_{g+1} = .702$
(d) $q_{g+1} = .319; p_{g+1} = .681$

10. (a) $p_1 = 0.6 + 0.2(0.1 - 0.6) = 0.5$
(b) $p_1 = 0.2 + 0.3(0.7 - 0.2) = 0.35$
(c) $p_1 = 0.1 + 0.1(0.2 - 0.1) = 0.11$

12. *Inbreeding depression* refers to the reduction in fitness observed in populations which are inbred. With inbreeding comes an increase in the number of homozygous individuals and a decrease in genetic variability. Genetic variability is necessary for a genetic response to environmental change. As deleterious genes become homozygous, more individuals are less fit in the population.

14. While inbreeding increases the frequency of homozygous individuals in a population, it does not change the *allele* frequencies. There will be fewer heterozygotes in the population to compensate for the additional homozygotes.

16. (a) q is 0.01
(b) $p = 1 - q$ or .99
(c) $2pq = 2(.01)(.99) = 0.0198$ (or about 1/50)
(d) $2pq \times 2pq = 0.0198 \times 0.0198$
$$= 0.000392 \text{ or about } 1/255$$

18. There are 50,000 births, therefore 100,000 genes (gametes) involved. The frequency of mutation is 2×10^{-5}.

Chapter 23

2. During speciation individuals or groups of potentially interbreeding organisms become genetically distinct from other members of the species. Members of different populations with substantial genetic divergence are, at first, not reproductively isolated from each other although gene flow may be restricted. The distinction between such groups is not absolute in that one group may blend with other groups of the species. Any process which favors changes in gene frequencies has the potential of generating substantial genetic differences.

Factors such as selection, migration, genetic drift, or even mutation, may be important in generating genetic change. One would certainly include geographic isolation as a major barrier to gene flow and thus an important process in such formation.

Natural selection occurs when there is non-random elimination of individuals from a population. Since such selection is a strong force in changing gene frequencies, it should also be considered as a significant factor in generating genetic variation among populations.

4. All of the amino acid substitutions (Ala → Gly, Val → Leu, Asp → Asn, Met → Leu) require only one nucleotide change. The last change from Pro (CC−)→ Lys (AAA,G) requires two changes (the minimal mutational distance).

6. Construct a chart similar to the one below which indicates the number of base changes between each pair:

	H	C	G	O
H	—	—	—	—
C	1	—	—	—
G	3	2	—	—
O	7	6	4	—
B	12	11	9	10

Generate the first cluster by selecting the closest related pair (human and chimp). Considering the human–chimp cluster as one entity, "average" its (human and chimp) distance to the next closest primate (gorilla). Continue to treat each new cluster (human–chimp–gorilla) as a single entity, comparing it to the next closest organism until all the organisms are exhausted. Construct the tree on the basis of the average mutational distances as described in the text.

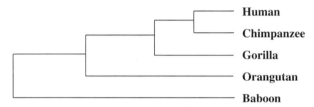

- **Human**
- **Chimpanzee**
- **Gorilla**
- **Orangutan**
- **Baboon**

8. In looking at the figure, notice that the $\Delta T_{50}H$ value of 4.0 on the right could be used as a decision point such that any group which diverged above that line would be considered in the same genus while any group below would be in a different genus. Under this rule, one would have the chimpanzee, pygmy chimpanzee, human, gorilla, and orangutan in the same genus. If one assumed that 3.7 is close enough to be considered above 4.0, given considerable experimental error, one could provide a scheme where the orangutan is not included with the chimpanzee, pygmy chimpanzee, human, and gorilla.

10. The text lists several cornerstones of the *neutral mutation theory*:
    (a) The relatively uniform rate of amino acid substitution in different organisms (under different types of selection).
    (b) There is no particular pattern to the substitutions indicating that selection is not eliminating some variations.
    (c) The rate of mutation is relatively high and has remained relatively constant for millions of years even though environments have fluctuated greatly over that period of time.
    (d) Certain regions of molecules and certain functions of those molecules should logically be less likely to have amino acid substitutions influence the phenotype.
    (e) The rate of amino acid substitution in some proteins is much too high to have been produced by selection. The *selectionists* suggest that even though amino acid substitutions *appear* to be neutral, it is more likely that their influence has just not been determined. In addition, they point out that many polymorphisms are clearly maintained in the population *by* selection.

Chapter 24

2. The frequency of the lethal gene in the captive population ($q^2 = 5/169$ and $q = .172$) is approximately double that in the gene pool as a whole ($q = 0.09$). Applying the formula $q_n = q_0/(1 + nq_0)$, one can estimate that it would take 10 generations to reduce the lethal gene's frequency to .063 in the captive population with no intervention (random mating assumed). Since condors produce very few eggs per year, a more proactive approach seems justified. First, if detailed records are kept of the breeding partners of the captive birds, then knowledge of heterozygotes should be available.

   Breeding programs could be established to restrict matings between those carrying the lethal gene. Such "kinship management" is often used in captive populations. If kinship records are not available, it is often possible to establish kinship using genetic markers such as DNA microsatellite polymorphisms. Using such markers, one can often identify mating partners and link them to their offspring.

   By coupling knowledge of mating partners with the likelihood of producing a lethal genetic combination, selective matings can often be used to minimize the influence of a deleterious gene. In addition, such markers can be used to establish matings which optimize genetic mixing, thus reducing inbreeding depression.

4. Both genetic drift and inbreeding tend to drive populations toward homozygosity. Genetic drift is more common when the effective breeding size of the population is low. When this condition prevails, inbreeding is also much more likely. They are different in that inbreeding can occur when certain population structures or behaviors favor matings between relatives, regardless of the effective size of the population. Inbreeding tends to increase the frequency of both homozygous classes at the expense of the heterozygotes. Genetic drift can lead to fixation of one allele or the other, thus producing a single homozygous class.

6. Inbreeding depression, over time, reduces the level of heterozygosity, usually a selectively advantageous quality of a species. When homozygosity increases (through loss of heterozygosity), deleterious alleles are likely to become more of a load on a population. Outbreeding depression occurs when there is a reduction in fitness of progeny from genetically diverse individuals. It is usually attributed to offspring being less well-adapted to the local environmental conditions of the parents.

   Even though forced outbreeding may be necessary to save a threatened species where population numbers are low, it significantly and permanently changes the genetic make-up of the species.

Glossary

A-DNA An alternative form of the right-handed double-helical structure of DNA in which the helix is more tightly coiled, with 11 base pairs per full turn of the helix. In this form, the bases in the helix are displaced laterally and tilted in relation to the longitudinal axis. It is not yet clear whether this form has biological significance.

abortive transduction An event in which transducing DNA fails to be incorporated into the recipient chromosome. (See *transduction*.)

acentric chromosome Chromosome or chromosome fragment with no centromere.

acquired immunodeficiency syndrome (AIDS) An infectious disease caused by a retrovirus designated as human immuno-deficiency virus (HIV). The disease is characterized by a grad-ual depletion of T lymphocytes, recurring fever, weight loss, multiple opportunistic infections, and rare forms of pneumonia and cancer associated with collapse of the immune system.

acridine dyes A class of organic compounds that bind to DNA and intercalate into the double-stranded structure, producing local disruptions of base pairing. These disruptions result in additions or deletions in the next round of replication.

acrocentric chromosome Chromosome with the centromere located very close to one end. Human chromosomes 13, 14, 15, 21, and 22 are acrocentric.

active immunity Immunity gained by direct exposure to anti-gens followed by antibody production.

active site That portion of a protein, usually an enzyme, whose structural integrity is required for function (e.g., the substrate binding site of an enzyme).

adaptation A heritable component of the phenotype that con-fers an advantage in survival and reproductive success. The process by which organisms adapt to the current environmen-tal conditions.

additive genes See *polygenic inheritance*.

additive variance Genetic variance that is attributed to the sub-stitution of one allele for another at a given locus. This vari-ance can be used to predict the rate of response to phenotypic selection in quantitative traits.

albinism A condition caused by the lack of melanin production in the iris, hair, and skin. In humans, most often inherited as an autosomal recessive.

aleurone layer In seeds, the outer layer of the endosperm.

alkaptonuria An autosomal recessive condition in humans caused by the lack of an enzyme, homogentisic acid oxidase. Urine of homozygous individuals turns dark upon standing due to oxidation of excreted homogentisic acid. The cartilage of homozygous adults blackens from deposition of a pigment derived from homogentisic acid. Such individuals often devel-op arthritic conditions.

allele One of the possible mutational states of a gene, distin-guished from other alleles by phenotypic effects.

allele frequency Measurement of the proportion of individuals in a population carrying a particular allele.

allele-specific nucleotide (ASO) Synthetic nucleotides, usually 15 to 20 bp in length that under carefully controlled condi-tions will hybridize only to a complementary sequence with a perfect match. Under these same conditions, ASOs with a one-nucleotide mismatch will not hybridize.

allelic exclusion In plasma cell heterozygous for an immuno-globulin gene, the selective action of only one allele.

allelism test See *complementation test*.

allolactose A lactose derivative that acts as the inducer for the *lac* operon.

allopatric speciation Process of speciation associated with geographic isolation.

allopolyploid Polyploid condition formed by the union of two or more distinct chromosome sets with a subsequent doubling of chromosome number.

allosteric effect Conformational change in the active site of a protein brought about by interaction with an effector molecule.

allotetraploid Diploid for two genomes derived from different species.

allozyme An allelic form of a protein that can be distinguished from other forms by electrophoresis.

alpha fetoprotein (AFP) A 70-kd glycoprotein synthesized in embryonic development by the yolk sac. High levels of this protein in the amniotic fluid are associated with neural tube defects such as spina bifida. Lower than normal levels may be associated with Down syndrome.

alternative splicing generating different protein molecules from the same pre-mRNA by changing the number and order of exons in the mRNA product.

Alu sequence An interspersed DNA sequence of approximately 300 bp found in the genome of primates that is cleaved by the restriction enzyme *Alu* I. *Alu* sequences are composed of a head-to-tail dimer, with the first monomer approximately 140 bp and the second approximately 170 bp. In humans, they are dispersed throughout the genome and are present in 300,000 to 600,000 copies, constituting some 3 to 6 percent of the genome. See *SINEs*.

amber codon The codon UAG, which does not code for an amino acid but for chain termination.

Ames test An assay developed by Bruce Ames to detect mu-tagenic and carcinogenic compounds using reversion to his-tidine independence in the bacterium *Salmonella typhimurium*.

amino acid Any of the subunit building blocks that are co-valently linked to form proteins.

aminoacyl tRNA   Covalently linked combination of an amino acid and a tRNA molecule.

amniocentesis   A procedure used to test for fetal defects in which fluid and fetal cells are withdrawn from the amniotic layer surrounding the fetus.

amphidiploid   See *allotetraploid.*

anabolism   The metabolic synthesis of complex molecules from less complex precursors.

analog   A chemical compound structurally similar to another, but differing by a single functional group (e.g., 5-bromo-deoxyuridine is an analog of thymidine).

anaphase   Stage of cell division in which chromosomes begin moving to opposite poles of the cell.

anaphase I   The stage in the first meiotic division during which members of homologous pairs of chromosomes separate from one another.

aneuploidy   A condition in which the chromosome number is not an exact multiple of the haploid set.

angstrom   Unit of length equal to 10^{-10} meters. Abbreviated Å.

annotation Analysis of genomic nucleotide sequence data to identify the protein-coding genes, the non-protein coding genes, their regulatory sequences and their function(s).

antibody   Protein (immunoglobulin) produced in response to an antigenic stimulus with the capacity to bind specifically to the antigen.

anticipation   A phenomenon first observed in myotonic dystrophy, where the severity of the symptoms increases from generation to generation and the age of onset decreases from generation to generation. This phenomenon is caused by the expansion of trinucleotide repeats within or near a gene.

anticodon   The nucleotide triplet in a tRNA molecule that is complementary to, and binds to, the codon triplet in an mRNA molecule.

antigen   A molecule, often a cell surface protein, that is capable of eliciting the formation of antibodies.

antiparallel   Describing molecules in parallel alignment, but running in opposite directions. Most commonly used to describe the opposite orientations of the two strands of a DNA molecule.

apoptosis   A genetically controlled program of cell death, activated as part of normal development, or as a result of cell damage.

ascospore   A meiotic spore produced in certain fungi.

ascus   In fungi, the sac enclosing the four or eight ascospores.

asexual reproduction   Production of offspring in the absence of any sexual process.

assortative mating   Nonrandom mating between males and females of a species. Selection of mates with the same genotype is positive; selection of mates with opposite genotypes is negative.

ATP   Adenosine triphosphate.

attached-X chromosome   Two conjoined X chromosomes that share a single centromere.

attenuator   A nucleotide sequence between the promoter and the structural gene of some operons that can act to regulate the transit of RNA polymerase, reducing transcription of the related structural gene.

autogamy   A process of self-fertilization resulting in homozygosis.

autoimmune disease   The production of antibodies that results from an immune response to one's own molecules, cells, or tissues. Such a response results from the inability of the immune system to distinguish self from nonself. Diseases such as arthritis, scleroderma, systemic lupus erythematosus, and juvenile-onset diabetes are examples of autoimmune diseases.

autonomously replicating sequences (ARS)   Origins of replication, about 100 nucleotides in length found in yeast chromosomes. ARS elements are also present in organelle DNA.

autopolyploidy   Polyploid condition resulting from the replication of one diploid set of chromosomes.

autoradiography   Production of a photographic image by radioactive decay. Used to localize radioactively labeled compounds within cells and tissues.

autosomes   Chromosomes other than the sex chromosomes. In humans, there are 22 pairs of autosomes.

autotetraploid   An autopolyploid condition composed of four similar genomes. In this situation, genes with two alleles (*A* and *a*) can have five genotypic classes: *AAAA* (quadruplex), *AAAa* (triplex), *Aaaa* (duplex), *Aaaa* (simplex), and *aaaa* (nulliplex).

auxotroph   A mutant microorganism or cell line which requires a substance for growth that can be synthesized by wild-type strains.

back-cross   A cross involving an F_1 heterozygote and one of the P_1 parents (or an organism with a genotype identical to one of the parents).

bacteriophage   A virus that infects bacteria (synonym is *phage*).

bacteriophage mu   A group of phages whose genetic material behaves as an insertion sequence that can cause inactivation of host genes and rearrangement of host chromosomes.

balanced lethals   Recessive, nonallelic lethal genes, each carried on different homologous chromosomes. When organisms carrying balanced lethal genes are interbred, only organisms with genotypes identical to the parents (heterozygotes) survive.

balanced polymorphism   Genetic polymorphism maintained in a population by natural selection.

Barr body   Densely staining nuclear mass seen in the somatic nuclei of mammalian females. Discovered by Murray Barr, this body is thought to represent an inactivated X chromosome.

base analog   See *analog.*

base substitution   A single base change in a DNA molecule that produces a mutation. There are two types of substitutions: *transitions*, in which a purine is substituted for a purine or a pyrimidine for a pyrimidine; and *transversions*, in which a purine is substituted for a pyrimidine, or vice versa.

β-galactosidase   A bacterial enzyme encoded by the *lacZ* gene that converts lactose into galactose and glucose.

bidirectional replication   A mechanism of DNA replication in which two replication forks move in opposite directions from a common origin of replication.

biodiversity   The genetic diversity present in populations and species of plants and animals.

biometry   The application of statistics and statistical methods to biological problems.

biotechnology   Commercial and/or industrial processes that utilize biological organisms or products.

bivalents   Synapsed homologous chromosomes in the first prophase of meiosis.

Bombay phenotype   A rare variant of the ABO system in which affected individuals do not have A or B antigens, and thus appear as blood type O, even though their genotype may carry unexpressed alleles for the A and/or B antigens.

bottleneck Fluctuation in allele frequency that occurs when a population undergoes a temporary reduction in size.

BrdU (5-bromodeoxyuridine) A mutagenically active analog of thymidine in which the methyl group at the 5′ position in thymine is replaced by bromine.

buoyant density A property of particles (and molecules) that depends upon their actual density, as determined by partial specific volume and degree of hydration. Provides the basis for density gradient separation of molecules or particles.

CAAT box A highly conserved DNA sequence found in the un-translated promoter region of eukaryotic genes. This sequence is recognized by transcription factors.

cAMP Cyclic adenosine monophosphate. An important regulatory molecule in both prokaryotic and eukaryotic organisms.

CAP Catabolite activator protein. A protein that binds cAMP and regulates the activation of inducible operons.

carcinogen A physical or chemical agent that causes cancer.

carrier An individual heterozygous for a recessive trait.

catabolism A metabolic reaction in which complex molecules are broken down into simpler forms, often accompanied by the release of energy.

catabolite activator protein See *CAP*.

catabolite repression The selective inactivation of an operon by a metabolic product of the enzymes encoded by the operon.

cdc mutation A class of *c*ell *d*ivision *c*ycle mutations in yeast that affect the timing and progression through the cell cycle.

cDNA DNA synthesized from an RNA template by the enzyme reverse transcriptase.

cDNA library A collection of cloned cDNA sequences.

cell cycle Sum of the phases of growth of an individual cell type; divided into G1 (gap 1), S (DNA synthesis), G2 (gap 2), and M (mitosis).

cell-free extract A preparation of the soluble fraction of cells, made by lysing cells and removing the particulate matter, such as nuclei, membranes, and organelles. Often used to carry out the synthesis of proteins by the addition of specific, exogenous mRNA molecules.

CEN In yeast, fragments of chromosomal DNA, about 120 bp in length, that when inserted into plasmids confer the ability to segregate during mitosis. These segments contain at least three types of sequence elements associated with centromere function.

centimeter A unit of length equal to 10^{-2} meter. Abbreviated cm.

centimorgan A unit of distance between genes on chromosomes. One centimorgan represents a value of 1 percent crossing over between two genes. Abbreviated cM.

central dogma The concept that information flow progresses from DNA to RNA to proteins. Although exceptions are known, this idea is central to an understanding of gene function.

centric fusion See *Robertsonian translocation*.

centriole A cytoplasmic organelle composed of nine groups of microtubules, generally arranged in triplets. Centrioles function in the generation of cilia and flagella and serve as foci for the spindles in cell division.

centromere Specialized region of a chromosome to which sister chromatids remain attached after replication, and the site to which spindle fibers attach during cell division. Location of the centromere determines the shape of the chromosome during the anaphase portion of cell division. Also known as the primary constriction.

centrosome Region of the cytoplasm containing the centriole.

chaperone A protein that regulates the folding of a polypeptide into a functional conformation.

character An observable phenotypic attribute of an organism.

charon phages A group of genetically modified lambda phages designed to be used as vectors (carriers) for cloning foreign DNA. Named after the ferryman in Greek mythology who carried the souls of the dead across the River Styx.

chemotaxis Negative or positive response to a chemical gradient.

chiasma (pl., chiasmata) The crossed strands of nonsister chromatids seen in diplotene of the first meiotic division. Regarded as the cytological evidence for exchange of chromosomal material, or crossing over.

chi-square (χ^2) analysis Statistical test to determine if an observed set of data fits a theoretical expectation.

chloroplast A cytoplasmic self-replicating organelle containing chlorophyll. The site of photosynthesis.

chorionic villus sampling (CVS) A technique of prenatal diagnosis that intravaginally retrieves fetal cells from the chorion and uses them to detect cytogenetic and biochemical defects in the embryo.

chromatid One of the longitudinal subunits of a replicated chromosome, joined to its sister chromatid at the centromere.

chromatin Term used to describe the complex of DNA, RNA, histones, and nonhistone proteins that make up uncoiled chromosomes characteristic of the eukaryotic interphase nucleus.

chromatography Technique for the separation of a mixture of solubilized molecules by their differential migration over a substrate.

chromocenter An aggregation of centromeres and heterochromatic elements of polytene chromosomes.

chromomere A coiled, beadlike region of a chromosome most easily visible during cell division. The aligned chromomeres of polytene chromosomes are responsible for their distinctive banding pattern.

chromosomal aberration Any change resulting in the duplication, deletion, or rearrangement of chromosomal material.

chromosomal mutation See *chromosomal aberration*.

chromosomal polymorphism Alternative structures or arrangements of a chromosome that are carried by members of a population.

chromosome In prokaryotes, an intact DNA molecule containing the genome; in eukaryotes, a DNA molecule complexed with RNA and proteins to form a threadlike structure containing genetic information arranged in a linear sequence and visible during mitosis and meiosis.

chromosome banding Technique for the differential staining of mitotic or meiotic chromosomes to produce a characteristic banding pattern or selective staining of certain chromosomal regions such as centromeres, the nucleolus organizer regions, and GC- or AT-rich regions. Not to be confused with the banding pattern present in polytene chromosomes, which is produced by the alignment of chromomeres.

chromosome map A diagram showing the location of genes on chromosomes.

chromosome puff A localized uncoiling and swelling in a polytene chromosome, usually regarded as a sign of active transcription.

chromosome theory of inheritance The idea put forward by Walter Sutton and Theodore Boveri that chromosomes are the carriers of genes and the basis for the Mendelian mechanisms of segregation and independent assortment.

chromosome walking   A method for analyzing long stretches of DNA, in which the end of a cloned segment of DNA is subcloned and used as a probe to identify other clones that overlap the first clone.

cis configuration   The arrangement of two genes or two mutant sites within a gene on the same homolog, such as

$$\frac{a^1 \quad a^2}{+ \quad +}$$

Contrasts with a *trans* arrangement, where the mutant alleles are located on opposite homologs.

cis-trans test   A genetic test to determine whether two mutations are located within the same cistron.

cistron   That portion of a DNA molecule that codes for a single polypeptide chain; defined by a genetic test as a region within which two mutations cannot complement each other.

cline   A gradient of genotype or phenotype distributed over a geographic range.

clonal selection   Theory of the immune system that proposes that antibody diversity precedes exposure to the antigen, and that the antigen functions to select the cells containing its specific antibody to undergo proliferation.

clone   Identical molecules, cells or organisms all derived from a single ancestor by asexual or parasexual methods. For example, a DNA segment that has been enzymatically inserted into a plasmid or chromosome of a phage or a bacterium and replicated to form many copies.

cloned library   A collection of cloned DNA molecules representing all or part of an individual's genome.

code   See *genetic code.*

codominance   Condition in which the phenotypic effects of a gene's alleles are fully and simultaneously expressed in the heterozygote.

codon   A triplet of nucleotides that specifies or encodes the information for a single amino acid. Sixty one codons specify the amino acids used in proteins, and three codons signal termination of growth of the polypeptide chain.

coefficient of coincidence   A ratio of the observed number of double-crossovers divided by the expected number of such crossovers.

coefficient of inbreeding   The probability that two alleles present in a zygote are descended from a common ancestor.

coefficient of selection   A measurement of the reproductive disadvantage of a given genotype in a population. If for genotype *aa*, only 99 of 100 individuals reproduce, then the selection coefficient (*s*) is 0.1.

colchicine   An alkaloid compound that inhibits spindle formation during cell division. Used in the preparation of karyotypes to collect a large population of cells inhibited at the metaphase stage of mitosis.

colinearity   The linear relationship between the nucleotide sequence in a gene (or the RNA transcribed from it) and the order of amino acids in the polypeptide chain specified by the gene.

competence   In bacteria, the transient state or condition during which the cell can bind and internalize exogenous DNA molecules, making transformation possible.

complementarity   Chemical affinity between nitrogenous bases as a result of hydrogen bonding. Responsible for the base pairing between the strands of the DNA double helix.

complementation test   A genetic test to determine whether two mutations occur within the same gene. If two mutations are introduced into a cell simultaneously and produce a wild-type phenotype (i.e., they complement each other), they are often nonallelic. If a mutant phenotype is produced, the mutations are noncomplementing and are often allelic.

complete linkage   A condition in which two genes are located so close to each other that no recombination occurs between them.

complexity   The total number of nucleotides or nucleotide pairs in a population of nucleic acid molecules as determined by reassociation kinetics.

complex locus   A gene within which a set of functionally related pseudoalleles can be identified by recombinational analysis (e.g., the *bithorax* locus in *Drosophila*).

concatemer   A chain or linear series of subunits linked together. The process of forming a concatemer is called concatenation (e.g., multiple units of a phage genome produced during replication).

concordance   Pairs or groups of individuals identical in their phenotype. In twin studies, a condition in which both twins exhibit or fail to exhibit a trait under investigation.

conditional mutation   A mutation that expresses a wild-type phenotype under certain (permissive) conditions and a mutant phenotype under other (restrictive) conditions.

conjugation   Temporary fusion of two single-celled organisms for the sexual transfer of genetic material.

consanguineous   Related by a common ancestor within the previous few generations.

consensus sequence   A basically common, although not necessarily identical, sequence of nucleotides in DNA or amino acids in proteins.

conservation genetics   The branch of genetics concerned with the preservation and maintenance of wild species of plants and animals in their natural environments.

continuous variation   Phenotype variation exhibited by quantitative traits distributed from one phenotypic extreme to another in an overlapping or continuous fashion.

cosmid   A vector designed to allow cloning of large segments of foreign DNA. Cosmids are hybrids composed of the *cos* sites of lambda inserted into a plasmid. In cloning, the recombinant DNA molecules are packaged into phage protein coats, and after infection of bacterial cells, the recombinant molecule replicates and can be maintained as a plasmid.

coupling conformation   See *cis configuration.*

covalent bond   A nonionic chemical bond formed by the sharing of electrons.

cri-du-chat syndrome   A clinical syndrome in humans produced by a deletion of a portion of the short arm of chromosome 5. Afflicted infants have a distinctive cry which sounds like that of a cat.

crossing over   The exchange of chromosomal material (parts of chromosomal arms) between homologous chromosomes by breakage and reunion. The exchange of material between nonsister chromatids during meiosis is the basis of genetic recombination.

cross-reacting material (CRM)   Nonfunctional form of an enzyme, produced by a mutant gene, which is recognized by antibodies made against the normal enzyme.

C-terminal amino acid   The terminal amino acid in a peptide chain which carries a free carboxyl group.

C terminus   The end of a polypeptide that carries a free carboxyl group of the last amino acid. By convention, the structural formula of polypeptides is written with the C terminus at the right.

C value   The haploid amount of DNA present in a genome.

C value paradox   The apparent paradox that there is no relationship between the size of the genome and the evolutionary complexity of species. For example, the *C* value (haploid genome size) of amphibians varies by a factor of 100.

cyclins   A class of proteins found in eukaryotic cells that are synthesized and degraded in synchrony with the cell cycle, and regulate passage through stages of the cycle.

cytogenetics   A branch of biology in which the techniques of both cytology and genetics are used to study heredity.

cytokinesis   The division or separation of the cytoplasm during mitosis or meiosis.

cytological map   A diagram showing the location of genes at particular chromosomal sites.

cytoplasmic inheritance   Non-Mendelian form of inheritance involving genetic information transmitted by self-replicating cytoplasmic organelles such as mitochondria, chloroplasts, etc.

cytoskeleton   An internal array of microtubules, microfilaments, and intermediate filaments that confers shape and the ability to move on a eukaryotic cell.

dalton (Da)   A unit of mass equal to that of the hydrogen atom, which is 1.67×10^{-24} gram. A unit used in designating molecular weights.

Darwinian fitness   See *fitness*.

deficiency (deletion)   A chromosomal mutation involving the loss or deletion of chromosomal material.

degenerate code   Term used to describe the genetic code, in which a given amino acid may be represented by more than one codon. Some amino acids (leucine) have six codons, others (isoleucine) have three, etc.

deletion   See *deficiency*.

deme   A local interbreeding population.

denatured DNA   DNA molecules that have been separated into single strands.

de novo   Newly arising; synthesized from less complex precursors rather than having been produced by modification of an existing molecule.

density gradient centrifugation   A method of separating macromolecular mixtures by the use of centrifugal force and solutions of varying density. In buoyant density gradient centrifugation, using cesium chloride, the cesium solution establishes a gradient under the influence of the centrifugal field, and the macromolecules such as DNA sediment in the gradient until the density of the cesium chloride solution equals their own.

deoxyribonuclease   A class of enzymes that breaks down DNA into oligonucleotide fragments by introducing single-stranded breaks into the double helix.

deoxyribonucleic acid (DNA)   A macromolecule usually consisting of antiparallel polynucleotide chains held together by hydrogen bonds, in which the sugar residues are deoxyribose. The primary carrier of genetic information.

deoxyribose   The five-carbon sugar associated with the deoxyribonucleotides found in DNA.

dermatoglyphics   The study of the surface ridges of the skin, especially of the hands and feet.

determination   A regulatory event that establishes a specific pattern of future gene activity and developmental fate for a given cell.

diakinesis   The final stage of meiotic prophase I in which the chromosomes become tightly coiled and compacted and move toward the periphery of the nucleus.

dicentric chromosome   A chromosome having two centromeres.

dideoxynucleotide   A nucleotide containing a deoxyribose sugar lacking a 3′ hydroxyl group. Stops further chain elongation when incorporated into a growing polynucleotide; used in the Sanger method of DNA sequencing.

differentiation   The process of complex changes by which cells and tissues attain their adult structure and functional capacity.

dihybrid cross   A genetic cross involving two characters in which the parents possess different forms of each character (e.g., tall, round × short, wrinkled peas).

diploid   A condition in which each chromosome exists in pairs; having two of each chromosome.

diplotene   A stage of meiotic prophase immediately after pachytene. In diplotene, one pair of sister chromatids begins separating from the other, and chiasmata become visible. These overlaps move laterally toward the ends of the chromatids (terminalization).

directional selection   A selective force that changes the frequency of an allele in a given direction, either toward fixation or toward elimination.

discontinuous replication of DNA   The synthesis of DNA in discontinuous fragments on the lagging strand of the replication fork. The fragments, known as Okazaki fragments, are joined by DNA ligase to form a continuous strand.

discontinuous variation   Phenotypic data that fall into two or more distinct, non-overlapping classes.

discordance   In twin studies, a situation where one twin expresses a trait but the other does not.

disjunction   The separation of chromosomes at the anaphase stage of cell division.

disruptive selection   Simultaneous selection for phenotypic extremes in a population, usually resulting in the production of two phenotypically discontinuous strains.

dizygotic twins   Twins produced from separate fertilization events; two ova fertilized independently. Also known as fraternal twins.

DNA   See *deoxyribonucleic acid*.

DNA footprinting   See *footprinting*.

DNA gyrase   One of the DNA topoisomerases that functions during DNA replication to reduce molecular tension caused by supercoiling. DNA gyrase produces, then seals, double-stranded breaks.

DNA ligase   An enzyme that forms a covalent bond between the 5′-end of one polynucleotide chain and the 3′-end of another polynucleotide chain. Also called polynucleotide-joining enzyme.

DNA polymerase   An enzyme that catalyzes the synthesis of DNA from deoxyribonucleotides and a template DNA molecule.

DNase   Deoxyribonucleosidase, an enzyme that degrades or breaks down DNA into fragments or constitutive nucleotides.

dominance   The expression of a trait in the heterozygous condition.

dominant suppression   A form of epistasis in which a dominant allele at one locus suppresses the effect of a dominant allele at another locus, resulting in a 13:3 phenotypic ratio.

dosage compensation   A genetic mechanism that regulates the levels of gene products at certain loci on the X chromosome in mammals such that males and females have equal amounts of a gene product. In mammals, this is accomplished by random inactivation of one X chromosome.

double-crossover   Two separate events of chromosome breakage and exchange occurring within the same tetrad.

double helix   The model for DNA structure proposed by James Watson and Francis Crick, involving two antiparallel, hydrogen-bonded polynucleotide chains wound into a right-handed helical configuration, with 10 base pairs per full turn of the double helix. Often called B-DNA.

Duchenne muscular dystrophy   An X-linked recessive genetic disorder caused by a mutation in the gene for dystrophin, a protein found in muscle cells.

duplication   A chromosomal aberration in which a segment of the chromosome is repeated.

dyad   The products of tetrad separation or disjunction at the first meiotic prophase. Consists of two sister chromatids joined at the centromere.

dystrophin   See *Duchenne muscular dystrophy*.

effective population size   The number of individuals in a population that have an equal probability of contributing gametes to the next generation.

effector molecule   Small, biologically active molecule that acts to regulate the activity of a protein by binding to a specific receptor site on the protein.

electrophoresis   A technique used to separate a mixture of molecules by their differential migration through a stationary phase (such as a gel) in an electrical field.

endocytosis   The uptake by a cell of fluids, macromolecules, or particles by pinocytosis, phagocytosis, or receptor-mediated endocytosis.

endomitosis   Chromosomal replication that is not accompanied by either nuclear or cytoplasmic division.

endonuclease   An enzyme that hydrolyzes internal phosphodiester bonds in a polynucleotide chain or nucleic acid molecule.

endoplasmic reticulum   A membranous organelle system in the cytoplasm of eukaryotic cells. The outer surface of the membranes may be ribosome-studded (rough ER) or smooth ER.

endopolyploidy   The increase in chromosome sets that results from endomitotic replication within somatic nuclei.

endosymbiont theory   The proposal that self-replicating cellular organelles such as mitochondria and chloroplasts were originally free-living organisms that entered into a symbiotic relationship with nucleated cells.

enhancer   Originally identified as a 72-bp sequence in the genome of the virus, SV40, that increases the transcriptional activity of nearby structural genes. Similar sequences that enhance transcription have been identified in the genomes of eukaryotic cells. Enhancers can act over a distance of thousands of base pairs and can be located 5′, 3′ or internal to the gene they affect, and thus are different from promoters.

environment   The complex of geographic, climatic, and biotic factors within which an organism lives.

enzyme   A protein or complex of proteins that catalyzes a specific biochemical reaction.

epigenesis   The idea that an organism develops by the appearance and growth of new structures. Opposed to preformationism, which holds that development is the growth of structures already present in the egg.

episome   A circular genetic element in bacterial cells that can replicate independently of the bacterial chromosome or integrate and replicate as part of the chromosome.

epistasis   Nonreciprocal interaction between genes such that one gene interferes with or prevents the expression of another gene. For example, in *Drosophila*, the recessive gene *eyeless*, when homozygous, prevents the expression of eye color genes present in the genome.

epitope   That portion of a macromolecule or cell that acts to elicit an antibody response; an antigenic determinant. A complex molecule or cell can contain several such sites.

equational division   A division of each chromosome into longitudinal halves that are distributed into two daughter nuclei. Chromosome division in mitosis is an example of equational division.

equatorial plate   See *metaphase plate*.

euchromatin   Chromatin or chromosomal regions that are lightly staining and are relatively uncoiled during the interphase portion of the cell cycle. Euchromatic regions contain most of the structural genes.

eugenics   The improvement of the human species by selective breeding. Positive eugenics refers to the promotion of breeding of those with favorable genes, and negative eugenics refers to the discouragement of breeding among those with undesirable traits.

eukaryotes   Those organisms having true nuclei and membranous organelles and whose cells demonstrate mitosis and meiosis.

euphenics   Medical or genetic intervention to reduce the impact of defective genotypes.

euploid   Polyploid with a chromosome number that is an exact multiple of a basic chromosome set.

evolution   The origin of plants and animals from preexisting types. Descent with modifications.

excision repair   Removal of damaged DNA segments followed by repair. Excision can include the removal of individual bases (base repair) or a stretch of damaged nucleotides (nucleotide repair). The gap created by excision is filled by polymerase and the ends are ligated to form an intact molecule.

exon (extron)   The DNA segment(s) of a gene that are transcribed and translated into protein.

exonuclease   An enzyme that breaks down nucleic acid molecules by breaking the phosphodiester bonds at the 3′ or 5′ terminal nucleotides.

expressed sequence tags (ESTs)   All or part of the nucleotide sequence of cDNA clones. Used as markers in construction of genetic maps.

expression vector   Plasmids or phages carrying promoter regions designed to cause expression of inserted DNA sequences.

expressivity   The degree or range in which a phenotype for a given trait is expressed.

extranuclear inheritance   Transmission of traits by genetic information contained in cytoplasmic organelles such as mitochondria and chloroplasts.

F^+ cell   A bacterial cell having a fertility (F) factor. Acts as a donor in bacterial conjugation.

F^- cell   A bacterial cell that does not contain a fertility (F) factor. Acts as a recipient in bacterial conjugation.

F factor   An episome in bacterial cells that confers the ability to act as a donor in conjugation.

F′ factor   A fertility (F) factor that contains a portion of the bacterial chromosome.

F_1 generation   First filial generation; the progeny resulting from the first cross in a series.

F_2 generation   Second filial generation; the progeny resulting from a cross of the F_1 generation.

F pilus   See *pilus*.

facultative heterochromatin   A chromosome or chromosome region that becomes heterochromatic only under certain conditions, e.g., the mammalian X chromosome.

familial trait   A trait transmitted through and expressed by members of a family.

fate map   A diagram or "map" of an embryo showing the location of cells whose development fate is known.

fertility (F) factor   See *F factor*.

filial generations   See *F_1, F_2 generations*.

fingerprint   The pattern of ridges and whorls on the tip of a finger. The pattern obtained by enzymatically cleaving a protein or nucleic acid and subjecting the digest to two-dimensional chromatography or electrophoresis.

FISH   See *fluorescence in situ hybridization*.

fitness   A measure of the relative survival and reproductive success of a given individual or genotype.

fixation   In population genetics, a condition in which all members of a population are homozygous for a given allele.

fluctuation test   A statistical test developed by Salvadore Luria and Max Delbrück to determine whether bacterial mutations arise spontaneously or are produced in response to selective agents.

fluorescence *in situ* hybridization (FISH)   A method of *in situ* hybridization that utilizes probes labeled with a fluorescent tag, causing the site of hybridization to fluoresce when viewed in ultraviolet light under a microscope.

flush-crash cycle   A period of rapid population growth followed by a drastic reduction in population size.

fmet   See *formylmethionine*.

folded-fiber model   A model of eukaryotic chromosome organization in which each sister chromatid consists of a single fiber, composed of double-stranded DNA and protein, which is wound up like a tightly coiled skein of yarn.

footprinting   A technique for identifying a DNA sequence that binds to a particular protein, based on the idea that the phosphodiester bonds in the region covered by the protein are protected from digestion by deoxyribonucleases.

formylmethionine (fmet)   A molecule derived from the amino acid methionine by attachment of a formyl group to its terminal amino group. This is the first amino acid inserted in all bacterial polypeptides. Also known as *N*-formyl methionine.

founder effect   A form of genetic drift. The establishment of a population by a small number of individuals whose genotypes carry only a fraction of the different kinds of alleles in the parental population.

fragile site   A heritable gap or nonstaining region of a chromosome that can be induced to generate chromosome breaks.

fragile X syndrome   A genetic disorder caused by the expansion of a CGG trinucleotide repeat and a fragile site at Xq27.3, within the *FMR-1* gene.

frameshift mutation   A mutational event leading to the insertion of one or more base pairs in a gene, shifting the codon reading frame in all codons following the mutational site.

fraternal twins   See *dizygotic twins*.

G1 checkpoint   A point in the G1 phase of the cell cycle when a cell becomes committed to initiate DNA synthesis and continue the cycle or withdraw into the G0 resting stage.

G0   The G-zero stage of the cell cycle is a point in G1 where cells withdraw from the cell cycle and enter a nondividing but metabolically active state.

gamete   A specialized reproductive cell with a haploid number of chromosomes.

gap genes   Genes expressed in contiguous domains along the anterior-posterior axis of the *Drosophila* embryo, which regulate the process of segmentation in each domain.

gene   The fundamental physical unit of heredity whose existence can be confirmed by allelic variants and which occupies a specific chromosomal locus. A DNA sequence coding for a single polypeptide.

gene amplification   The process by which gene sequences are selected and differentially replicated either extrachromosomally or intrachromosomally.

gene conversion   The process of nonreciprocal recombination by which one allele in a heterozygote is converted into the corresponding allele.

gene duplication   An event in replication leading to the production of a tandem repeat of a gene sequence.

gene flow   The gradual exchange of genes between two populations, brought about by the dispersal of gametes or the migration of individuals.

gene frequency   The percentage of alleles of a given type in a population.

gene interaction   Production of novel phenotypes by the interaction of alleles of different genes.

gene mutation   See *point mutation*.

gene pool   The total of all alleles possessed by reproductive members of a population.

generalized transduction   The transduction of any gene in the bacterial genome by a phage.

genetic anticipation   The phenomenon of a progressively earlier age of onset and increasing severity of symptoms for a genetic disorder in successive generations.

genetic background   All genes carried in the genome other than the one being studied.

genetic code   The nucleotide triplets that code for the 20 amino acids or for chain initiation or termination.

genetic counseling   Analysis of risk for genetic defects in a family and the presentation of options available to avoid or ameliorate possible risks.

genetic drift   Random variation in allele frequency from generation to generation. Most often observed in small populations.

genetic engineering   The technique of altering the genetic constitution of cells or individuals by the selective removal, insertion, or modification of individual genes or gene sets.

genetic erosion   The loss of genetic diversity from a population or a species.

genetic equilibrium   Maintenance of allele frequencies at the same value in successive generations. A condition in which allele frequencies are neither increasing nor decreasing.

genetic fine structure   Intragenic recombinational analysis that provides mapping information at the level of individual nucleotides.

genetic load   Average number of recessive lethal genes carried in the heterozygous condition by an individual in a population.

genetic polymorphism   The stable coexistence of two or more discontinuous genotypes in a population. When the frequencies of two alleles are carried to an equilibrium, the condition is called balanced polymorphism.

genetics   The branch of biology that deals with heredity and the expression of inherited traits.

genome   The array of genes carried by an individual.

genomic imprinting   A condition where the expression of a trait depends on whether the trait has been inherited from a male or a female parent.

genomics   The study of genomes, including nucleotide sequence, gene content, organization, and gene number.

genotype   The specific allelic or genetic constitution of an organism; often, the allelic composition of one or a limited number of genes under investigation.

germplasm   Hereditary material transmitted from generation to generation.

Goldberg-Hogness box   A short nucleotide sequence 20 to 30 bp 5′ to the initiation site of eukaryotic genes to which RNA polymerase II binds. The consensus sequence is TATAAAA. Also known as TATA box.

graft versus host disease (GVHD)   In transplants, reaction by immunologically competent cells of the donor against the antigens present on the cells of the host. In human bone marrow transplants, often a fatal condition.

gynandromorph   An individual composed of cells with both male and female genotypes.

gyrase   One of a class of enzymes known as topoisomerases. Gyrase converts closed circular DNA to a negatively supercoiled form prior to replication, transcription, or recombination.

H substance   The carbohydrate group present on the surface of red blood cells. When unmodified, it results in blood type O; when modified by the addition of monosaccharides, it results in type A, B, and AB.

haploid   A cell or organism having a single set of unpaired chromosomes. The gametic chromosome number.

haplotype   The set of alleles from closely linked loci carried by an individual and usually inherited as a unit.

Hardy-Weinberg law   The principle that both gene and genotype frequencies will remain in equilibrium in an infinitely large population in the absence of mutation, migration, selection, and nonrandom mating.

heat shock   A transient response following exposure of cells or organisms to elevated temperatures. The response involves activation of a small number of loci, inactivation of some previously active loci, and selective translation of heat shock mRNA. Appears to be a nearly universal phenomenon observed in organisms ranging from bacteria to humans.

helicase   An enzyme that participates in DNA replication by unwinding the double helix near the replication fork.

helix-turn-helix motif   The structure of a region of DNA-binding proteins in which a turn of four amino acids holds two alpha helices at right angles to each other.

hemizygous   Conditions where a gene is present in a single dose in an otherwise diploid cell. Usually applied to genes on the X chromosome in heterogametic males.

hemoglobin (Hb)   An iron-containing, oxygen-carrying protein occurring chiefly in the red blood cells of vertebrates.

hemophilia   An X-linked trait in humans associated with defective blood-clotting mechanisms.

heredity   Transmission of traits from one generation to another.

heritability   A measure of the degree to which observed phenotypic differences for a trait are genetic.

heterochromatin   The heavily staining, late replicating regions of chromosomes that are condensed in interphase. Thought to be devoid of structural genes.

heteroduplex   A double-stranded nucleic acid molecule in which each polynucleotide chain has a different origin. These structures may be produced as intermediates in a recombinational event, or by the *in vitro* reannealing of single-stranded, complementary molecules.

heterogametic sex   The sex that produces gametes containing unlike sex chromosomes.

heterogeneous nuclear RNA (hnRNA)   The collection of RNA transcripts in the nucleus, representing precursors and processing intermediates to rRNA, mRNA, and tRNA. Also represents RNA transcripts that will not be transported to the cytoplasm, such as snRNA (small nuclear RNA).

heterokaryon   A somatic cell containing nuclei from two different sources.

heterosis   The superiority of a heterozygote over either homozygote for a given trait.

heterozygote   An individual with different alleles at one or more loci. Such individuals will produce unlike gametes and therefore will not breed true.

Hfr   A strain of bacteria exhibiting a high frequency of recombination. These strains have a chromosomally integrated F factor that is able to mobilize and transfer part of the chromosome to a recipient F⁻ cell.

histocompatibility antigens   See *HLA*.

histones   Proteins complexed with DNA in the nucleus. They are rich in the basic amino acids arginine and lysine and function in the coiling of DNA to form nucleosomes.

HLA   Cell surface proteins, produced by histocompatibility loci, involved in the acceptance or rejection of tissue and organ grafts and transplants.

hnRNA   See *heterogeneous nuclear RNA*.

holandric   A trait transmitted from males to males. In humans, genes on the Y chromosome are holandric.

Holliday structure   An intermediate in bidirectional DNA recombination seen in the transmission electron microscope as an X-shaped structure showing four single-stranded DNA regions.

homeobox   A sequence of about 180 nucleotides that encodes a 60-amino-acid sequence called a *homeodomain*, which is part of a DNA-binding protein that acts as a transcription factor.

homeotic mutation   A mutation that causes a tissue normally determined to form a specific organ or body part to alter its differentiation and form another structure. Alternatively spelled: homoeotic.

homogametic sex   The sex that produces gametes that do not differ with respect to sex chromosome content; in mammals, the female is homogametic.

homogeneously staining regions (hsr)   Segments of mammalian chromosomes that stain lightly with Giemsa following exposure of cells to a selective agent. These regions arise in conjunction with gene amplification and are regarded as the structural locus for the amplified gene.

homologous chromosomes   Chromosomes that synapse or pair during meiosis. Chromosomes that are identical with respect to their genetic loci and centromere placement.

homozygote   An individual with identical alleles at one or more loci. Such individuals will produce identical gametes and will therefore breed true.

homunculus   The miniature individual imagined by preformationists to be contained within the sperm or egg.

human immunodeficiency virus (HIV)   A human retrovirus associated with the onset and progression of acquired immunodeficiency syndrome (AIDS).

hybrid   An individual produced by crossing two parents of different genotypes.

hybridoma   A somatic cell hybrid produced by the fusion of an antibody-producing cell and a cancer cell, specifically, a myeloma. The cancer cell contributes the ability to divide indefinitely, and the antibody cell confers the ability to synthesize large amounts of a single antibody.

hybrid vigor   See *heterosis*.

hydrogen bond   An electrostatic attraction between a hydrogen atom bonded to a strongly electronegative atom such as oxygen or nitrogen and another atom that is electronegative or contains an unshared electron pair.

hypervariable regions   The regions of antibody molecules that attach to antigens. These regions have a high degree of diversity in amino acid content.

identical twins   See *monozygotic twins*.

Ig   See *immunoglobulin*.

imaginal disc   Discrete groups of cells set aside during embryogenesis in holometabolous insects which are determined to form the external body parts of the adult.

immunoglobulin   The class of serum proteins having the properties of antibodies.

inborn error of metabolism   A biochemical disorder that is genetically controlled; usually an enzyme defect that produces a clinical syndrome.

inbreeding   Mating between closely related organisms.

inbreeding depression   A decrease in viability, vigor or growth in progeny following several rounds of inbreeding.

incomplete dominance   Expression of heterozygous phenotype which is distinct from, and often intermediate to, that of either parent.

incomplete linkage   Occasional separation of two genes on the same chromosome by a recombinational event.

independent assortment   The independent behavior of each pair of homologous chromosomes during their segregation in meiosis I. The random distribution of maternal and paternal homologs into gametes.

inducer   An effector molecule that activates transcription.

inducible enzyme system   An enzyme system under the control of a regulatory molecule, or inducer, which acts to block a repressor and allow transcription.

initiation codon   The triplet of nucleotides (AUG) in an mRNA molecule that codes for the insertion of the amino acid methionine as the first amino acid in a polypeptide chain.

insertion sequence   See *IS element*.

in situ **hybridization**   A technique for the cytological localization of DNA sequences complementary to a given nucleic acid or polynucleotide.

intercalary deletion   A form of chromosome deletion where material is lost from within the chromosome. Deletions that involve the end of the chromosome are called terminal deletions.

intercalating agent   A compound that inserts between bases in a DNA molecule, disrupting the alignments and pairing of bases in the complementary strands (e.g., acridine dyes).

interference   A measure of the degree to which one crossover affects the incidence of another crossover in an adjacent region of the same chromatid. Negative interference increases the chances of another crossover; positive interference reduces the probability of a second crossover event.

interphase   That portion of the cell cycle between divisions.

intervening sequence   See *intron*.

intron   A portion of DNA between coding regions in a gene which is transcribed, but which does not appear in the mRNA product.

inversion   A chromosomal aberration in which the order of a chromosomal segment has been reversed.

inversion loop   The chromosomal configuration resulting from the synapsis of homologous chromosomes, one of which carries an inversion.

in vitro   Literally, in glass; outside the living organism; occurring in an artificial environment.

in vivo   Literally, in the living; occurring within the living body of an organism.

IS element   A mobile DNA segment that is transposable to any of a number of sites in the genome.

isoagglutinogen   An antigenic factor or substance present on the surface of cells that is capable of inducing the formation of an antibody.

isochromosome   An aberrant chromosome with two identical arms and homologous loci.

isolating mechanism   Any barrier to the exchange of genes between different populations of a group of organisms. In general, isolation can be classified as spatial, environmental, or reproductive.

isotopes   Forms of a chemical element that have the same number of protons and electrons, but differ in the number of neutrons contained in the atomic nucleus.

isozyme   Any of two or more distinct forms of an enzyme that have identical or nearly identical chemical properties, but differ in some property such as net electrical charge, pH optima, number and type of subunits, or substrate concentration.

kappa particles   DNA-containing cytoplasmic particles found in certain strains of *Paramecium aurelia*. When these self-reproducing particles are transferred into the growth medium, they release a toxin, paramecin, which kills other, sensitive strains. A nuclear gene, K, is responsible for maintaining kappa particles in the cytoplasm.

karyokinesis   The process of nuclear division.

karyotype   The chromosome complement of a cell or an individual. Often used to refer to the arrangement of metaphase chromosomes in a sequence according to length and position of the centromere.

kilobase (kb)   A unit of length consisting of 1000 nucleotides. Abbreviated kb.

kinetochore   A fibrous structure with a size of about 400 nm, located within the centromere. It appears to be the site of microtubule attachment during division.

Klenow fragment   A part of bacterial DNA polymerase that lacks exonuclease activity, but retains polymerase activity. It is produced by enzymatic digestion of the intact enzyme.

Klinefelter syndrome   A genetic disorder in human males caused by the presence of an extra X chromosome. Klinefelter males are XXY instead of XY. This syndrome is associated with enlarged breasts, small testes, sterility, and, occasionally, mild mental retardation.

knockout mice   In producing knockout mice, a cloned normal gene is inactivated by the insertion of a marker, such as an antibiotic resistance gene. The altered gene is transferred to embryonic stem cells, where the altered gene will replace the normal gene (in some cells). These cells are injected into a blastomere embryo, producing a mouse that is bred to yield mice that are homozygous for the mutated gene.

lac **repressor protein**   A protein that binds to the operator in the *lac* operon and blocks transcription.

lagging strand   In DNA replication, the strand synthesized in a discontinuous fashion, 5′ to 3′ away from the replication fork. Each short piece of DNA synthesized in this fashion is called an Okazaki fragment.

lampbrush chromosomes   Meiotic chromosomes characterized by extended lateral loops, which reach maximum extension during diplotene. Although most intensively studied in amphibians, these structures occur in meiotic cells of organisms ranging from insects through humans.

lariat structure   A structure formed by an intron via a 5′2′ bond during processing and removal of that intron from an mRNA molecule.

leader sequence   That portion of an mRNA molecule from the 5′-end to the beginning codon; may contain regulatory or ribosome binding sites.

leading strand   During DNA replication, the strand synthesized continuously 5′ to 3′ from the origin of replication toward the replication fork.

leptotene   The initial stage of meiotic prophase I, during which the chromosomes become visible and are often arranged in a bouquet configuration, with one or both ends of the chromosomes gathered at one spot on the inner nuclear membrane.

lethal gene   A gene whose expression results in death.

leucine zipper   A structural motif in a DNA-binding protein that is characterized by a stretch of leucine residues spaced at every seventh amino acid residue, with adjacent regions of positively charged amino acids. Leucine zippers on two polypeptides may interact to form a dimer that binds to DNA.

LINEs   Long interspersed elements are repetitive sequences found in the genomes of higher organisms, such as the 6-kb L1 sequences found in primate genomes.

linkage   Condition in which two or more nonallelic genes tend to be inherited together. Linked genes have their loci along the same chromosome, do not assort independently, but can be separated by crossing over.

linkage group   A group of genes that have their loci on the same chromosome.

linking number   The number of times that two strands of a closed, circular DNA duplex cross over each other.

locus   The site or place on a chromosome where a particular gene is located.

lod score   A statistical method used to determine whether two loci are linked or unlinked. A lod (log of the odds) score of 4 indicates that linkage is 10,000 times more likely than non-linkage. By convention, lod scores of 3–4 are taken as signs of linkage.

long terminal repeat (LTR)   Sequence of several hundred base pairs found at the ends of retroviral DNAs.

Lutheran blood group   One of a number of blood group systems inherited independently of the ABO, MN, and Rh systems. Alleles of this group determine the presence or absence of antigens on the surface of red blood cells. Gene is on human chromosome 19.

Lyon hypothesis   The random inactivation of the maternal or paternal X chromosome in somatic cells of mammalian females early in development. All daughter cells will have the same X chromosome inactivated, producing a mosaic pattern of expression of genes on the X chromosome.

lysis   The disintegration of a cell brought about by the rupture of its membrane.

lysogenic bacterium   A bacterial cell carrying the DNA of a temperate bacteriophage integrated into its chromosome.

lysogeny   The process by which the DNA of an infecting phage becomes repressed and integrated into the chromosome of the bacterial cell it infects.

lytic phase   The condition in which a temperate bacteriophage loses its integrated status in the host chromosome (becomes induced), replicates, and lyses the bacterial cell.

major histocompatibility loci   See *MHC*.

map unit   A measure of the genetic distance between two genes, corresponding to a recombination frequency of 1 percent. See *centimorgan*.

mapping functions   Map distance estimates from recombination when the recombination frequency in a region exceeds 15–20 percent, and double crossovers are undetectable.

maternal effect   Phenotypic effects on the offspring produced by the maternal genome. Factors transmitted through the egg cytoplasm that produce a phenotypic effect in the progeny.

maternal influence   See *maternal effect*.

maternal inheritance   The transmission of traits via cytoplasmic genetic factors such as mitochondria or chloroplasts.

mean   The arithmetic average.

median   The value in a group of numbers below and above which there is an equal number of data points or measurements.

meiosis   The process in gametogenesis or sporogenesis during which one replication of the chromosomes is followed by two nuclear divisions to produce four haploid cells.

melting profile   See T_m.

merozygote   A partially diploid bacterial cell containing, in addition to its own chromosome, a chromosome fragment introduced into the cell by transformation, transduction, or conjugation.

messenger RNA   See *mRNA*.

metabolism   The sum of chemical changes in living organisms by which energy is generated and used.

metacentric chromosome   A chromosome with a centrally located centromere, producing chromosome arms of equal lengths.

metafemale   In *Drosophila*, a poorly developed female of low viability in which the ratio of X chromosomes to sets of autosomes exceeds 1.0. Previously called a superfemale.

metamale   In *Drosophila*, a poorly developed male of low viability in which the ratio of X chromosomes to sets of autosomes is less than 0.5. Previously called a supermale.

metaphase   The stage of cell division in which the condensed chromosomes lie in a central plane between the two poles of the cell, and in which the chromosomes become attached to the spindle fibers.

metaphase plate   The arrangement of mitotic or meiotic chromosomes at the equator of the cell during metaphase.

MHC   Major histocompatibility loci. In humans, the HLA complex; and in mice, the H2 complex.

micrometer   A unit of length equal to 1×10^{-6} meter. Previously called a micron. Abbreviated μm.

micron   See *micrometer*.

migration coefficient   An expression of the proportion of migrant genes entering the population per generation.

millimeter   A unit of length equal to 1×10^{-3} meter. Abbreviated mm.

minimal medium   A medium containing only those nutrients that will support the growth and reproduction of wild-type strains of an organism.

mismatch repair A process of excision repair, during which an unpaired base or bases is excised, followed by the synthesis of a new segment, using the complementary strand as a template.

missense mutation A mutation that alters a codon to that of another amino acid, causing an altered translation product to be made.

mitochondrion Found in the cells of eukaryotes, a cytoplasmic, self-reproducing organelle that is the site of ATP synthesis.

mitogen A substance that stimulates mitosis in nondividing cells (e.g., phytohemagglutinin).

mitosis A form of cell division resulting in the production of two cells, each with the same chromosome and genetic complement as the parent cell.

mode In a set of data, the value occurring in the greatest frequency.

monohybrid cross A genetic cross between two individuals involving only one character (e.g., $AA \times aa$).

monosomic An aneuploid condition in which one member of a chromosome pair is missing; having a chromosome number of $2n - 1$.

monozygotic twins Twins produced from a single fertilization event; the first division of the zygote produces two cells, each of which develops into an embryo. Also known as identical twins.

mRNA An RNA molecule transcribed from DNA and translated into the amino acid sequence of a polypeptide.

mtDNA Mitochondrial DNA.

multigene family A gene set descended from a common ancestor by duplication and subsequent divergence from a common ancestor. The globin genes represent a multigene family.

multiple alleles Three or more alleles of the same gene.

multiple-factor inheritance See *polygenic inheritance*.

multiple infection Simultaneous infection of a bacterial cell by more than one bacteriophage, often of different genotypes.

mu phage A phage group in which the genetic material behaves like an insertion sequence, capable of insertion, excision, transposition, inactivation of host genes, and induction of chromosomal rearrangements.

mutagen Any agent that causes an increase in the rate of mutation.

mutant A cell or organism carrying an altered or mutant gene.

mutation The process that produces an alteration in DNA or chromosome structure; the source of most alleles.

mutation rate The frequency with which mutations take place at a given locus or in a population.

muton The smallest unit of mutation in a gene, corresponding to a single base change.

nanometer A unit of length equal to 1×10^{-9} meter. Abbreviated nm.

natural selection Differential reproduction of some members of a species resulting from variable fitness conferred by genotypic differences.

neutral mutation A mutation with no immediate adaptive significance or phenotypic effect.

nonautonomous transposon A transposable element that lacks a functional transposase gene.

noncrossover gamete A gamete that contains no chromosomes that have undergone genetic recombination.

nondisjunction An error during cell division in which the homologous chromosomes (in meiosis) or the sister chromatids (in mitosis) fail to separate and migrate to opposite poles; responsible for defects such as monosomy and trisomy.

nonsense codon The nucleotide triplet in an mRNA molecule that signals the termination of translation. Three such codons are known: UGA, UAG, and UAA.

nonsense mutation A mutation that changes an amino acid codon into a termination codon; UAG (amber codon), UAA (ochre codon), or UGA (opal codon). Leads to premature termination during translation of mRNA.

NOR See *nucleolar organizer region*.

normal distribution A probability function that approximates the distribution of random variables. The normal curve, also known as a Gaussian or bell-shaped curve, is the graphic display of the normal distribution.

N-terminal amino acid The terminal amino acid in a peptide chain that carries a free amino group.

N terminus The end of a polypeptide that carries a free amino group of the first amino acid. By convention, the structural formula of polypeptides is written with the N terminus at the left.

nu body See *nucleosome*.

nuclease An enzyme that breaks bonds in nucleic acid molecules.

nucleoid The DNA-containing region within the cytoplasm in prokaryotic cells.

nucleolar organizer region (NOR) A chromosomal region containing the genes for rRNA; most often found in physical association with the nucleolus.

nucleolus A nuclear organelle that is the site of ribosome biosynthesis; usually associated with or formed in association with the NOR.

nucleoside A purine or pyrimidine base covalently linked to a ribose or deoxyribose sugar molecule.

nucleosome A complex of four histone molecules, each present in duplicate, wrapped by two turns of a DNA molecule. One of the basic units of eukaryotic chromosome structure. Also known as a *nu* body.

nucleotide A nucleoside covalently linked to a phosphate group. Nucleotides are the basic building blocks of nucleic acids. The nucleotides commonly found in DNA are deoxyadenylic acid, deoxycytidylic acid, deoxyguanylic acid, and deoxythymidylic acid. The nucleotides in RNA are adenylic acid, cytidylic acid, guanylic acid, and uridylic acid.

nucleotide pair The pair of nucleotides (A and T, or G and C) in opposite strands of the DNA molecule that are hydrogen-bonded to each other.

nucleus The membrane-bounded cytoplasmic organelle of eukaryotic cells that contains the chromosomes and nucleolus.

null hypothesis Used in statistical tests, it states that there is no difference between the observed and expected data sets. Statistical methods such as Chi-square analysis are used to test the probability of this hypothesis.

nullisomic Describes an individual with a chromosomal aberration in which both members of a chromosome pair are missing.

ochre codon A codon that does not code for the insertion of an amino acid into a polypeptide chain, but signals chain termination. The ochre codon is UAA.

Okazaki fragment The small, discontinuous strands of DNA produced during DNA synthesis on the lagging strand.

oligonucleotide A linear sequence of nucleotides (about 10–20) connected by 5' to 3' phosphodiester bonds.

oncogene   A gene whose activity promotes uncontrolled proliferation in eukaryotic cells.

open reading frame (ORF)   The nucleotides between the start and stop codons which encode amino acids.

operator region   A region of a DNA molecule that interacts with a specific repressor protein to control the expression of an adjacent gene or gene set.

operon   A genetic unit that consists of one or more structural genes (that code for polypeptides) and an adjacent operator gene that controls the transcriptional activity of the structural gene or genes.

origin of replication (ori)   Sites along the length of the chromosome where DNA replication begins.

outbreeding depression   Reduction in fitness in the offspring produced by mating genetically diverse parents. It is thought to result from a lowered adaptation to local environmental conditions.

overdominance   The phenomenon where heterozygotes have a phenotype that is more extreme than either homozygous genotype.

overlapping code   A genetic code first proposed by George Gamow in which any given nucleotide is shared by three adjacent codons.

pachytene   The stage in meiotic prophase I when the synapsed homologous chromosomes split longitudinally (except at the centromere), producing a group of four chromatids called a tetrad.

pair-rule genes   Genes expressed as stripes around the blastoderm embryo during development of the *Drosophila* embryo.

palindrome   A word, number, verse, or sentence that reads the same backward or forward (e.g., *able was I ere I saw elba*). In nucleic acids, a sequence in which the base pairs read the same on complementary strands in the $5' \rightarrow 3'$ direction. For example:

$$5'\text{GAATTC}3'$$
$$3'\text{CTTAAG}5'$$

These often occur as sites for restriction endonuclease recognition and cutting.

pangenesis   A discarded theory of development that postulated the existence of pangenes, small particles from all parts of the body that concentrated in the gametes, passing traits from generation to generation, blending the traits of the parents in the offspring.

paracentric inversion   A chromosomal inversion that does not include the centromere.

parasexual   Condition describing recombination of genes from different individuals that does not involve meiosis, gamete formation, or zygote production. The formation of somatic cell hybrids is an example.

parental gamete   See *noncrossover gamete*.

parthenogenesis   Development of an egg without fertilization.

partial diploids   See *merozygote*.

partial dominance   See *incomplete dominance*.

patroclinous inheritance   A form of genetic transmission in which the offspring have the phenotype of the father.

pedigree   In human genetics, a diagram showing the ancestral relationships and transmission of genetic traits over several generations in a family.

P element   Transposable DNA elements found in *Drosophila* that are responsible for hybrid dysgenesis.

penetrance   The frequency (expressed as a percentage) with which individuals of a given genotype manifest at least some degree of a specific mutant phenotype associated with a trait.

peptide bond   The covalent bond between the amino group of one amino acid and the carboxyl group of another amino acid.

pericentric inversion   A chromosomal inversion that involves both arms of the chromosome and thus involves the centromere.

phage   See *bacteriophage*.

phenocopy   An environmentally induced phenotype (nonheritable) that closely resembles the phenotype produced by a known gene.

phenotype   The observable properties of an organism that are genetically controlled.

phenylketonuria (PKU)   A hereditary condition in humans associated with the inability to metabolize the amino acid phenylalanine. The most common form is caused by the lack of the enzyme phenylalanine hydroxylase.

Philadelphia chromosome   The product of a reciprocal translocation that contains the short arm of chromosome 9 carrying the *c-abl* oncogene and the long arm of chromosome 22 carrying *bcr*.

phosphodiester bond   In nucleic acids, the covalent bond between a phosphate group and adjacent nucleotides, extending from the $5'$ carbon of one pentose (ribose or deoxyribose) to the $3'$ carbon of the pentose in the neighboring nucleotide. Phosphodiester bonds form the backbone of nucleic acid molecules.

photoreactivation enzyme (PRE)   An exonuclease that catalyzes the light-activated excision of ultraviolet-induced thymine dimers from DNA.

photoreactivation repair   Light-induced repair of damage caused by exposure to ultraviolet light. Associated with an intracellular enzyme system.

phyletic evolution   The gradual transformation of one species into another over time; vertical evolution.

pilus   A filamentlike projection from the surface of a bacterial cell. Often associated with cells possessing F factors.

plaque   A clear area on an otherwise opaque bacterial lawn caused by the growth and reproduction of phages.

plasmid   An extrachromosomal, circular DNA molecule (often carrying genetic information) that replicates independently of the host chromosome.

pleiotropy   Condition in which a single mutation simultaneously affects several characters.

ploidy   Term referring to the basic chromosome set or to multiples of that set.

point mutation   A mutation that can be mapped to a single locus. At the molecular level, a mutation that results in the substitution of one nucleotide for another.

polar body   A cell produced at either the first or second meiotic division in females that contains almost no cytoplasm as a result of an unequal cytokinesis.

polycistronic mRNA   A messenger RNA molecule that encodes the amino acid sequence of two or more polypeptide chains in adjacent structural genes.

polygenic inheritance   The transmission of a phenotypic trait whose expression depends on the additive effect of a number of genes.

polylinker   A segment of DNA that has been engineered to contain multiple sites for restriction enzyme digestion. Polylinkers are usually found in engineered vectors such as plasmids.

polymerase chain reaction (PCR)   A method for amplifying DNA segments that uses cycles of denaturation, annealing to primers, and DNA polymerase-directed DNA synthesis.

polymerases   The enzymes that catalyze the formation of DNA and RNA from deoxynucleotides and ribonucleotides, respectively.

polymorphism   The existence of two or more discontinuous, segregating phenotypes in a population.

polynucleotide   A linear sequence of more than 20 nucleotides, joined by 5′-to-3′ phosphodiester bonds. See *oligonucleotide*.

polypeptide   A molecule made up of amino acids joined by covalent peptide bonds. This term is used to denote the amino acid chain before it assumes its functional three-dimensional configuration.

polyploid   A cell or individual having more than two sets of chromosomes.

polyribosome   See *polysome*.

polysome   A structure composed of two or more ribosomes associated with mRNA, engaged in translation. Formerly called polyribosome.

polytene chromosome   A chromosome that has undergone several rounds of DNA replication without separation of the replicated chromosomes, forming a giant, thick chromosome with aligned chromomeres producing a characteristic banding pattern.

population   A local group of individuals belonging to the same species, which are actually or potentially interbreeding.

position effect   Change in expression of a gene associated with a change in the gene's location within the genome.

postzygotic isolation mechanism   Factors that prevent or reduce inbreeding by acting after fertilization to produce nonviable, sterile hybrids or hybrids of lowered fitness.

preadaptive mutation   A mutational event that later becomes of adaptive significance.

preformationism   The discredited idea that an organism develops by growth of structures already present in the egg or sperm.

prezygotic isolation mechanism   Factors that reduce inbreeding by preventing courtship, mating or fertilization.

Pribnow box   A 6-bp sequence 5′ to the beginning of transcription in prokaryotic genes, to which the sigma subunit of RNA polymerase binds. The consensus sequence for this box is TATAAT.

primary protein structure   Refers to the sequence of amino acids in a polypeptide chain.

primary sex ratio   Ratio of males to females at fertilization.

primer   In nucleic acids, a short length of RNA or single-stranded DNA that is necessary for the functioning of polymerases.

prion   An infectious pathogenic agent devoid of nucleic acid and composed mainly of a protein, PrP, with a molecular weight of 27,000 to 30,000 daltons. Prions are known to cause scrapie, a degenerative neurological disease in sheep, bovine spongiform encephalopathy (BSE or mad cow disease) in cattle, and are thought to cause similar diseases in humans, such as Kuru and Creutzfeldt-Jakob disease.

probability   Ratio of the frequency of a given event to the frequency of all possible events.

proband   An individual in whom a genetically determined trait of interest is first detected. Formerly known as a propositus.

probe   A macromolecule such as DNA or RNA that has been labeled and can be detected by an assay such as autoradiography or fluorescence microscopy. Probes are used to identify target molecules, genes, or gene products.

product law   The law that holds that the probability of two independent events occurring simultaneously is the product of their independent probabilities.

progeny   The offspring produced from a mating.

prokaryotes   Organisms lacking nuclear membranes, meiosis, and mitosis. Bacteria and blue-green algae are examples of prokaryotic organisms.

promoter   Region having a regulatory function and to which RNA polymerase binds prior to the initiation of transcription.

proofreading   A molecular mechanism for correcting errors in replication, transcription, or translation. Also known as editing.

prophage   A phage genome integrated into a bacterial chromosome. Bacterial cells carrying prophages are said to be lysogenic.

propositus (female, **proposita**)   See *proband*.

protein   A molecule composed of one or more polypeptides, each composed of amino acids covalently linked together.

proteomics   The study of the expressed proteins in the cell at a given time.

proto-oncogene   A cellular gene that normally functions to initiate or maintain cell division. Proto-oncogenes can be converted to oncogenes by alterations in structure or expression.

protoplast   A bacterial or plant cell with the cell wall removed. Sometimes called a spheroplast.

prototroph   A strain (usually microorganisms) that is capable of growth on a defined, minimal medium. Wild-type strains are usually regarded as prototrophs.

pseudoalleles   Genes that behave as alleles to one another by complementation, but that can be separated from one another by recombination.

pseudoautosomal inheritance   Inheritance of alleles located within the regions of the Y chromosome that are homologous to the X chromosome. Because these alleles are located on both the X and Y chromosome, their pattern of inheritance is indistinguishable from that of autosomal inheritance.

pseudodominance   The expression of a recessive allele on one homolog caused by the deletion of the dominant allele on the other homolog.

pseudogene   A nonfunctional gene with sequence homology to a known structural gene present elsewhere in the genome. They differ from their functional relatives by insertions or deletions and by the presence of flanking direct repeat sequences of 10 to 20 nucleotides.

punctuated equilibrium   A pattern in the fossil record of brief periods of species divergence punctuated with long periods of species stability.

quantitative inheritance   See *polygenic inheritance*.

quantitative trait loci (QTL)   Two or more genes that act on a single polygenic trait.

quantum speciation   Formation of a new species within a single or a few generations by a combination of selection and drift.

quaternary protein structure   Types and modes of interaction between two or more polypeptide chains within a protein molecule.

R point   The point (also known as the restriction point) during the G1 stage of the cell cycle when either a commitment is made to DNA synthesis and another cell cycle, or the cell withdraws from the cycle and becomes quiescent.

race   A genotypically or geographically distinct subgroup within a species.

rad   A unit of absorbed dose of radiation with an energy equal to 100 ergs per gram of irradiated tissue.

radioactive isotope   One of the forms of an element, differing in atomic weight and possessing an unstable nucleus that emits ionizing radiation during decay.

random amplified polymorphic DNA (RAPD)   A PCR method that uses random primers about 10 nucleotides in length to amplify unknown DNA sequences.

random mating   Mating between individuals without regard to genotype.

reading frame   Linear sequence of codons (groups of three nucleotides) in a nucleic acid.

reannealing   Formation of double-stranded DNA molecules from dissociated single strands.

recessive   Term describing an allele that is not expressed in the heterozygous condition.

reciprocal cross   A paired cross in which the genotype of the female in the first cross is present as the genotype of the male in the second cross, and vice versa.

reciprocal translocation   A chromosomal aberration in which nonhomologous chromosomes exchange parts.

recombinant DNA   A DNA molecule formed by the joining of two heterologous molecules. Usually applied to DNA molecules produced by in vitro ligation of DNA from two different organisms.

recombinant gamete   A gamete containing a new combination of genes produced by crossing over during meiosis.

recombination   The process that leads to the formation of new gene combinations on chromosomes.

recon   A term used by Seymour Benzer to denote the smallest genetic units between which recombination can occur.

reductional division   The chromosome division that halves the diploid chromosome number. The first division of meiosis is a reductional division.

redundant genes   Gene sequences present in more than one copy per haploid genome (e.g., ribosomal genes).

regulatory site   A DNA sequence that is involved in the control of expression of other genes, usually involving an interaction with another molecule.

rem   Radiation equivalent in man; the dosage of radiation that will cause the same biological effect as one roentgen of X-rays.

renaturation   The process by which a denatured protein or nucleic acid returns to its normal three-dimensional structure.

repetitive DNA sequences   DNA sequences present in many copies in the haploid genome.

replicating form (RF)   Double-stranded nucleic acid molecules present as an intermediate during the reproduction of certain viruses.

replication   The process of DNA synthesis.

replication fork   The Y-shaped region of a chromosome associated with the site of replication.

replicon   A chromosomal region or free genetic element containing the DNA sequences necessary for the initiation of DNA replication.

replisome   The term used to describe the complex of proteins, including DNA polymerase, that assembles at the bacterial replication fork to synthesize DNA.

repressible enzyme system   An enzyme or group of enzymes whose synthesis is regulated by the intracellular concentration of certain metabolites.

repressor   A protein that binds to a regulatory sequence adjacent to a gene and blocks transcription of the gene.

reproductive isolation   Absence of interbreeding between populations, subspecies, or species. Reproductive isolation can be brought about by extrinsic factors, such as behavior, and intrinsic barriers, such as hybrid inviability.

resistance transfer factor (RTF)   A component of R plasmids that confers the ability for cell-to-cell transfer of the R plasmid by conjugation.

resolution   In an optical system, the shortest distance between two points or lines at which they can be perceived to be two points or lines.

restriction endonuclease   Nuclease that recognizes specific nucleotide sequences in a DNA molecule, and cleaves or nicks the DNA at that site. Derived from a variety of microorganisms, those enzymes that cleave both strands of the DNA are used in the construction of recombinant DNA molecules.

restriction fragment length polymorphism (RFLP)   Variation in the length of DNA fragments generated by a restriction endonuclease. These variations are caused by mutations that create or abolish cutting sites for restriction enzymes. RFLPs are inherited in a codominant fashion and can be used as genetic markers.

restrictive transduction   See *specialized transduction*.

retrovirus   Viruses with RNA as genetic material that utilize the enzyme reverse transcriptase during their life cycle.

reverse transcriptase   A polymerase that uses RNA as a template to transcribe a single-stranded DNA molecule as a product.

reversion   A mutation that restores the wild-type phenotype.

R factor (R plasmid)   Bacterial plasmids that carry antibiotic resistance genes. Most R plasmids have two components: an r-determinant, which carries the antibiotic resistance genes, and the resistance transfer factor (RTF).

RFLP   See *restriction fragment length polymorphism*.

Rh factor   An antigenic system first described in the rhesus monkey. Recessive *r/r* individuals produce no Rh antigens and are Rh negative, while *R/R* and *R/r* individuals have Rh antigens on the surface of their red blood cells and are classified as Rh positive.

ribonucleic acid   A nucleic acid characterized by the sugar ribose and the pyrimidine uracil, usually a single-stranded polynucleotide. Several forms are recognized, including ribosomal RNA, messenger RNA, transfer RNA, and heterogeneous nuclear RNA.

ribose   The five-carbon sugar associated with the ribonucleotides found in RNA.

ribosomal RNA   See *rRNA*.

ribosome   A ribonucleoprotein organelle consisting of two subunits, each containing RNA and protein. Ribosomes are the site of translation of mRNA codons into the amino acid sequence of a polypeptide chain.

RNA   See *ribonucleic acid*.

RNA editing   Alteration of the nucleotide sequence of an mRNA molecule after transcription and before translation. There are two main types of editing: *substitution editing*, which changes individual nucleotides, and *insertion/deletion editing*, in which individual nucleotides are added or deleted.

RNA polymerase   An enzyme that catalyzes the formation of an RNA polynucleotide strand using the base sequence of a DNA molecule as a template.

RNase   A class of enzymes that hydrolyzes RNA.

Robertsonian translocation   A form of chromosomal aberration that involves the fusion of long arms of acrocentric chromosomes at the centromere.

roentgen   A unit of measure of the amount of radiation corresponding to the generation of 2.083×10^9 ion pairs in one cubic centimeter of air at 0°C at an atmospheric pressure of 760 mm of mercury. Abbreviated R.

rolling circle model   A model of DNA replication in which the growing point or replication fork rolls around a circular template strand; in each pass around the circle, the newly synthesized strand displaces the strand from the previous replication, producing a series of contiguous copies of the template strand.

rRNA   The RNA molecules that are the structural components of the ribosomal subunits. In prokaryotes, these are the 16S, 23S, and 5S molecules; and in eukaryotes, they are the 18S, 28S, and 5S molecules.

RTF   See *resistance transfer factor*.

S$_1$ nuclease   A deoxyribonuclease that cuts and degrades single-stranded molecules of DNA.

satellite DNA   DNA that forms a minor band when genomic DNA is centrifuged in a cesium salt gradient. This DNA usually consists of short sequences repeated many times in the genome.

SCE   See *sister chromatid exchange*.

secondary protein structure   The alpha helical or pleated-sheet form of a protein molecule brought about by the formation of hydrogen bonds between amino acids.

secondary sex ratio   The ratio of males to females at birth

secretor   An individual having soluble forms of the blood group antigens A and/or B present in saliva and other body fluids. This condition is caused by a dominant, autosomal gene unlinked to the *ABO* locus (*I* locus).

sedimentation coefficient   See *Svedberg coefficient unit*.

segment polarity genes   Genes that regulate the spatial pattern of differentiation within each segment of the developing *Drosophila* embryo.

segregation   The separation of homologous chromosomes into different gametes during meiosis.

selection   The force that brings about changes in the frequency of alleles and genotypes in populations through differential reproduction.

selection coefficient (*s*)   A quantitative measure of the relative fitness of one genotype compared with another.

selfing   In plant genetics, the fertilization of ovules of a plant by pollen produced by the same plant. Reproduction by self-fertilization.

semiconservative replication   A model of DNA replication in which a double-stranded molecule replicates in such a way that the daughter molecules are composed of one parental (old) and one newly synthesized strand.

semisterility   A condition in which a proportion of all zygotes are inviable.

sex chromatin body   See *Barr body*.

sex chromosome   A chromosome, such as the X or Y in humans, which is involved in sex determination.

sexduction   Transmission of chromosomal genes from a donor bacterium to a recipient cell by the F factor.

sex-influenced inheritance   Phenotypic expression that is conditioned by the sex of the individual. A heterozygote may express one phenotype in one sex and the alternate phenotype in the other sex.

sex-limited inheritance   A trait that is expressed in only one sex even though the trait may not be X-linked.

sex ratio   See *primary* and *secondary sex ratio*.

sexual reproduction   Reproduction through the fusion of gametes, which are the haploid products of meiosis.

Shine-Dalgarno sequence   The nucleotides AGGAGG present in the leader sequence of prokaryotic genes that serve as a ribosome binding site. The 16S RNA of the small ribosomal subunit contains a complementary sequence to which the mRNA binds.

shotgun experiment   The cloning of random fragments of genomic DNA into a vehicle such as a plasmid or phage, usually to produce a library from which clones of specific interest can be selected.

sibling species   Species that are morphologically almost identical, but which are reproductively isolated from one another.

sickle-cell anemia   A genetic disease in humans caused by an autosomal recessive gene, fatal in the homozygous condition if untreated. Caused by an alteration in the amino acid sequence of the beta chain of globin.

sickle-cell trait   The phenotype exhibited by individuals heterozygous for the sickle-cell gene.

sigma factor   A polypeptide subunit of the RNA polymerase that recognizes the binding site for the initiation of transcription.

SINEs   Short interspersed elements are repetitive sequences found in the genomes of higher organisms, such as the 300-bp *Alu* sequence.

single-stranded binding proteins (SSBs)   In DNA replication, proteins that bind to and stabilize single-stranded regions of DNA that result from the action of unwinding proteins.

sister chromatid exchange (SCE)   A crossing over event that can occur in meiotic and mitotic cells; involves the reciprocal exchange of chromosomal material between sister chromatids (joined by a common centromere). Such exchanges can be detected cytologically after BrdU incorporation into the replicating chromosomes.

site-directed mutagenesis   A process that uses a synthetic oligonucleotide containing a mutant base or sequence as a primer for inducing a mutation at a specific site in a cloned gene.

small nuclear RNA (snRNA)   Species of RNA molecules ranging in size from 90 to 400 nucleotides. The abundant snRNAs are present in 1×10^4 to 1×10^6 copies per cell. snRNAs are associated with proteins and form RNP particles known as snRNPs or *snurps*. Six uridine rich snRNAs known as U1–U6 are located in the nucleoplasm, and the complete nucleotide sequence of these is known. snRNAs have been implicated in the processing of pre-mRNA and may have a range of cleavage and ligation functions.

snurps   See *small nuclear RNA (snRNA)*.

solenoid structure   A level of eukaryotic chromosome structure generated by the supercoiling of nucleosomes.

somatic cell genetics   The use of cultured somatic cells to investigate genetic phenomena by parasexual techniques involving the fusion of cells from different organisms.

somatic cells   All cells other than the germ cells or gametes in an organism.

somatic mutation   A mutational event occurring in a somatic cell. In other words, such mutations are not heritable.

somatic pairing   The pairing of homologous chromosomes in somatic cells.

SOS response   The induction of enzymes to repair damaged DNA in *E. coli*. The response involves activation of an enzyme that cleaves a repressor, activating a series of genes involved in DNA repair.

spacer DNA   DNA sequences found between genes, usually repetitive DNA segments.

specialized transduction   Genetic transfer of only specific host genes by transducing phages.

speciation   The process by which new species of plants and animals arise.

species   A group of actually or potentially interbreeding individuals that is reproductively isolated from other such groups.

spheroplast   See *protoplast*.

spindle fibers   Cytoplasmic fibrils formed during cell division that are involved with the separation of chromatids at anaphase and their movement toward opposite poles in the cell.

spliceosome   The nuclear macromolecule complex within which splicing reactions occur to remove introns from pre-mRNAs.

spontaneous mutation   A mutation that is not induced by a mutagenic agent.

spore   A unicellular body or cell encased in a protective coat that is produced by some bacteria, plants, and invertebrates; is capable of survival in unfavorable environmental conditions; and can give rise to a new individual upon germination. In plants, spores are the haploid products of meiosis.

SRY   A gene (sex-determining region of the Y) found near the pseudoautosomal boundary of the Y chromosome. Accumulated evidence indicates that this gene is the testis-determining factor (TDF).

stabilizing selection   Preferential reproduction of those individuals having genotypes close to the mean for the population. A selective elimination of genotypes at both extremes.

standard deviation   A quantitative measure of the amount of variation in a sample of measurements from a population.

standard error   A quantitative measure of the amount of variation in a sample of measurements from a population.

sterility   The condition of being unable to reproduce; free from contaminating microorganisms.

strain   A group with common ancestry that has physiological or morphological characteristics of interest for genetic study or domestication.

structural gene   A gene that encodes the amino acid sequence of a polypeptide chain.

sublethal gene   A mutation causing lowered viability, with death before maturity in less than 50 percent of the individuals carrying the gene.

submetacentric chromosome   A chromosome with the centromere placed so that one arm of the chromosome is slightly longer than the other.

subspecies   A morphologically or geographically distinct interbreeding population of a species.

sum law   The law that holds that the probability of one or the other of two mutually exclusive events occurring is the sum of their individual probabilities.

supercoiled DNA   A form of DNA structure in which the helix is coiled upon itself. Such structures can exist in stable forms only when the ends of the DNA are not free, as in a covalently closed circular DNA molecule.

superfemale   See *metafemale*.

supermale   See *metamale*.

suppressor mutation   A mutation that acts to restore (completely or partially) the function lost by a previous mutation at another site.

Svedberg coefficient unit   A unit of measure for the rate at which particles (molecules) sediment in a centrifugal field. This unit is a function of several physico-chemical properties, including size and shape. A sedimentation value of 1×10^{-13} sec is defined as one Svedberg coefficient (S) unit.

symbiont   An organism coexisting in a mutually beneficial relationship with another organism.

sympatric speciation   Process of speciation involving populations that inhabit, at least in part, the same geographic range.

synapsis   The pairing of homologous chromosomes at meiosis.

synaptonemal complex (SC)   An organelle consisting of a tripartite nucleoprotein ribbon that forms between the paired homologous chromosomes in the pachytene stage of the first meiotic division.

syndrome   A group of signs or symptoms that occur together and characterize a disease or abnormality.

synkaryon   The nucleus of a zygote that results from the fusion of two gametic nuclei. Also used in somatic cell genetics to describe the product of nuclear fusion.

syntenic test   In somatic cell genetics, a method for determining whether or not two genes are on the same chromosome.

T_m The temperature at which a population of double-stranded nucleic acid molecules is half-dissociated into single strands. This is taken to be the melting temperature for that species of nucleic acid.

TATA box   See *Goldberg–Hogness box*.

tautomeric shift   A reversible isomerization in a molecule brought about by a shift in the localization of a hydrogen atom. In nucleic acids, tautomeric shifts in the bases of nucleotides can cause changes in other bases at replication and are a source of mutations.

TDF (testis-determining factor)   The product of a gene on the Y chromosome that controls the developmental switch point for the development of the indifferent gonad into a testis. See *SRY*.

telocentric chromosome   A chromosome in which the centromere is located at the end of the chromosome.

telomerase   The enzyme that adds short, tandemly repeated DNA sequences to the ends of eukaryotic chromosomes.

telomere   The terminal chromomere of a chromosome.

telophase   The stage of cell division in which the daughter chromosomes reach the opposite poles of the cell and re-form nuclei. Telophase ends with the completion of cytokinesis.

telophase I   In the first meiotic division, when duplicated chromosomes reach the poles of the dividing cell.

temperate phage   A bacteriophage that can become a prophage and confer lysogeny upon the host bacterial cell.

temperature-sensitive mutation   A conditional mutation that produces a mutant phenotype at one temperature range and a wild-type phenotype at another temperature range.

template   The single-stranded DNA or RNA molecule that specifies the nucleotide sequence of a strand synthesized by a polymerase molecule.

terminalization   The movement of chiasmata toward the ends of chromosomes during the diplotene stage of the first meiotic division.

tertiary protein structure   The three-dimensional structure of a polypeptide chain brought about by folding upon itself.

test cross   A cross between an individual whose genotype at one or more loci may be unknown and an individual who is homozygous recessive for the genes in question.

tetrad   The four chromatids that make up paired homologs in the prophase of the first meiotic division. The four haploid cells produced by a single meiotic division.

tetrad analysis   Method for the analysis of gene linkage and recombination using the four haploid cells produced in a single meiotic division.

tetranucleotide hypothesis   An early theory of DNA structure proposing that the molecule was composed of repeating units, each consisting of the four nucleotides containing adenine, thymine, cytosine, and guanine.

theta structure   An intermediate in the bidirectional replication of circular DNA molecules. At about midway through the cycle of replication, the intermediate resembles the Greek letter theta.

thymine dimer   A pair of adjacent thymine bases in a single polynucleotide strand that have chemical bonds formed between carbon atoms 5 and 6. This lesion, usually caused by exposure to ultraviolet light, inhibits DNA replication unless repaired by the appropriate enzymes.

topoisomerase   A class of enzymes that convert DNA from one topological form to another. During DNA replication, these enzymes facilitate the unwinding of the double-helical structure of DNA.

totipotent   The ability of a cell or embryo part to give rise to all adult structures. This capacity is usually progressively restricted during development.

trait   Any detectable phenotypic variation of a particular inherited character.

trans configuration   The arrangement of two mutant sites on opposite homologs, such as

$$\frac{a^1 \quad +}{+ \quad a^2}$$

Contrasts with a *cis* arrangement, where the sites are located on the same homolog.

transcription   Transfer of genetic information from DNA by the synthesis of an RNA molecule copied from a DNA template.

transcriptome   The set of mRNA molecules present in a cell at any given time.

transdetermination   Change in developmental fate of a cell or group of cells.

transduction   Virally mediated genes transfer from one bacterium to another, or the transfer of eukaryotic genes mediated by retrovirus.

transfer RNA   See *tRNA*.

transformation   Heritable change in a cell or an organism brought about by exogenous DNA.

transgenic organism   An organism whose genome has been modified by the introduction of external DNA sequences into the germ line.

transition   A mutational event in which one purine is replaced by another, or one pyrimidine is replaced by another.

translation   The derivation of the amino acid sequence of a polypeptide from the base sequence of an mRNA molecule in association with a ribosome.

translocation   A chromosomal mutation associated with the transfer of a chromosomal segment from one chromosome to another. Also used to denote the movement of mRNA through the ribosome during translation.

transmission genetics   The field of genetics concerned with the mechanisms by which genes are transferred from parent to offspring.

transposable element   A DNA segment that translocates to other sites in the genome, essentially independent of sequence homology. Usually such elements are flanked by short, inverted repeats of 20 to 40 base pairs at each end. Insertion into a structural gene can produce a mutant phenotype. Insertion and excision of transposable elements depends on two enzymes, transposase and resolvase. Such elements have been identified in both prokaryotes and eukaryotes.

transversion   A mutational event in which a purine is replaced by a pyrimidine, or a pyrimidine is replaced by a purine.

trinucleotide repeat   A tandemly repeated cluster of three nucleotides (such as CTG) in or near a gene, that undergo an expansion in copy number, resulting in a disease phenotype.

triploidy   The condition in which a cell or organism possesses three haploid sets of chromosomes.

trisomy   The condition in which a cell or organism possesses two copies of each chromosome, except for one, which is present in three copies. The general form for trisomy is therefore $2n + 1$.

tRNA   Transfer RNA; a small ribonucleic acid molecule that contains a three-base segment (anticodon) that recognizes a codon in mRNA, a binding site for a specific amino acid, and recognition sites for interaction with the ribosomes and the enzyme that links it to its specific amino acid.

tumor suppressor gene   A gene that encodes a gene product that normally functions to suppress cell division. Mutations in tumor suppressor genes result in the activation of cell division and tumor formation.

Turner syndrome   A genetic condition in human females caused by a 45,X genotype (XO). Such individuals are phenotypically female but are sterile because of undeveloped ovaries.

unequal crossing over   A crossover between two improperly aligned homologs, producing one homolog with three copies of a region and the other with one copy of that region.

unique DNA   DNA sequences that are present only once per genome. Single copy DNA.

universal code   The assumption that the genetic code is used by all life forms. In general, this is true; some exceptions are found in mitochondria, ciliates, and mycoplasmas.

unwinding proteins   Nuclear proteins that act during DNA replication to destabilize and unwind the DNA helix ahead of the replicating fork.

variable number tandem repeats (VNTRs)   Short repeated DNA sequences (2–20 nucleotides) present as tandem repeats between two restriction enzyme sites. Variations in the number of repeats creates DNA fragments of differing lengths following restriction enzyme digestion.

variable region   Portion of an immunoglobulin molecule that exhibits many amino acid sequence differences between antibodies of differing specificities.

variance   A statistical measure of the variation of values from a central value, calculated as the square of the standard deviation.

variegation   Patches of differing phenotypes, such as color, in a tissue.

vector   In recombinant DNA, an agent such as a phage or plasmid into which a foreign DNA segment will be inserted.

viability   The measure of the number of individuals in a given phenotypic class that survive, relative to another class (usually wild type).

virulent phage   A bacteriophage that infects and lyses the host bacterial cell.

VNTR   See *variable number tandem repeats*.

W, Z chromosomes   Sex chromosomes in species where the female is the heterogametic sex (WZ).

western blot   A technique in which proteins are separated by gel electrophoresis and transferred by capillary action to a nylon membrane or nitrocellulose sheet. A specific protein can be identified through hybridization to a labeled antibody.

wild type The most commonly observed phenotype or genotype, designated as the norm or standard.

wobble hypothesis An idea proposed by Francis Crick stating that the third base in an anticodon can align in several ways to allow it to recognize more than one base in the codons of mRNA.

writhing number The number of times that the axis of a DNA duplex crosses itself by supercoiling.

X inactivation In mammalian females, the random cessation of transcriptional activity of one X chromosome. This event, which occurs early in development, is a mechanism of dosage compensation. Molecular basis of inactivation is unknown, but involves a region called the X-inactivation center (XIC) on the proximal end of the p arm. Some loci on the tip of the short arm of the X can escape inactivation. See *Barr body, Lyon hypothesis*.

XIST A locus in the X-chromosome inactivation center that may control inactivation of the X chromosome in mammalian females.

X linkage The pattern of inheritance resulting from genes located on the X chromosome.

X-ray crystallography A technique to determine the three-dimensional structure of molecules through diffraction patterns produced by X-ray scattering by crystals of the molecule under study.

YAC A cloning vector in the form of a yeast artificial chromosome, constructed using chromosomal elements including telomeres (from a ciliate), centromeres, origin of replication, and marker genes from yeast. YACs are used to clone long stretches of eukaryotic DNA.

Y chromosome Sex chromosome in species where the male is heterogametic (XY).

Y linkage Mode of inheritance shown by genes located on the Y chromosome.

Z-DNA An alternative structure of DNA in which the two antiparallel polynucleotide chains form a left-handed double helix. Z-DNA has been shown to be present along with B-DNA in chromosomes and may have a role in regulation of gene expression.

zein Principal storage protein of maize endosperm, consisting of two major proteins, with molecular weights of 19,000 and 21,000 daltons.

zinc finger A DNA-binding domain of a protein that has a characteristic pattern of cysteine and histidine residues that complex with zinc ions, throwing intermediate amino acid residues into a series of loops or fingers.

zygote The diploid cell produced by the fusion of haploid gametic nuclei.

zygotene A stage of meiotic prophase I in which the homologous chromosomes synapse and pair along their entire length, forming bivalents. The synaptonemal complex forms at this stage.

Index